HANDBOOK OF INDUSTRIAL MATERIALS

2nd Edition

HANDBOOK OF INDUSTRIAL MATERIALS

2nd Edition

ELSEVIER
ADVANCED
TECHNOLOGY

Published by
ELSEVIER ADVANCED TECHNOLOGY
Mayfield House, 256 Banbury Road, Oxford OX2 7DH, UK

© 1992 ELSEVIER SCIENCE PUBLISHERS LTD.

ISBN 0-946395-83-7

Printed in Great Britain by Galliard (Printers) Ltd, Great Yarmouth

Preface

OVER THE past decade there have been revolutionary changes in the use of materials in almost all industries, due to many factors, not the least of which is cost evaluation. Labour intensive industries must endeavour to stabilise costs by finding less expensive materials to offset the spiralling wages bill. This calls for careful investigation to safeguard the durability of the product.

Manufactuers supply machinery or equipment within the parameters of advanced technology where standards are paramount, face the vying problems of increased cost and increased specifications necessary to meet the demands of new engineering disciplines. It is essential for all industrial manufacturers who are supplying original equipment, capital goods, ancillary equipment, or durables, to be up-to-date with, and keep abreast of developments in every type of material whether ferrous, non-ferrous, natural plastic or composite, if they are to remain competitive in changing markets.

The HANDBOOK OF INDUSTRIAL MATERIALS has been produced for that very purpose and embodies in one volume a wealth of information and data of immense value to designers, production engineers, chief engineers, R. and D. personnel, purchasing officers, metallurgists, technical executives, technicians, and others concerned with the selection and use of materials.

The Publishers

Contents

Materials Newsletters from Elsevier Advanced Technology

SECTION 1

Ferrous Metals

CAST IRONS
CARBON STEELS
BS 970 – REPLACING EN STEELS
ALLOY STEELS
SHEET STEELS
CASTING STEELS

Cast irons

THE COMMON form of cast iron, known as *grey iron* because of the colour of its fracture face, may contain between 1.5 and 4.3% carbon and a silicon content between 0.3 and 5% together with appreciable proportions of manganese, sulphur and phosphorus. Such irons are characterised by a free graphite content in flake form distributed through the iron matrix, yielding what is virtually a brittle structure. The strength and brittleness of the cast iron depend on the quantity and size of the graphite flakes. The lower the carbon content, or the smaller the size of the flakes, the greater the strength of the iron. Conversely, the higher the carbon content, the better the castability. Grey cast irons, however, have the distinct advantage of low shrinkage rate on solidification due to the fact that the flake graphite expands (or has an expanding effect). This enables highly satisfactory castings to be produced (particularly with modern casting methods), with minimum metal wastage.

British grey cast irons are specified in BS 1452, covering seven grades see Table 1.

Spheroidal-graphite (SG) irons

The strength of grey iron can be increased by reducing the size or modifying the form of the graphite flakes. The second method is adopted in the formation

TABLE 1 – LEADING PROPERTIES OF GREY CAST IRONS TO BS 1452

BS 1452 Grade	0·1% Proof Strength (MPa)	Tensile Strength (MPa)	Tensile Elongation (%)	Compressive Strength (MPa)	Impact Strength (J)	Typical Brinell Hardness (HB)	Elastic Modulus (GPa)	Modulus of Rigidity (GPa)	Shear Strength (MPa)
150	98	150	0.6–0.75	600	8–14	145–180	100	40	173
180	117	180	0.5–0.70	672	8–14	155–220	109	44	207
220	143	220	0.4–0.63	768	19–26	165–245	120	48	253
260	169	260	0.6	864	19–26	165–295	128	51	299
300	195	300	0.5	960	16–47	180–300	135	54	345
350	228	350	0.5	1080	16–47	200–310	140	56	403
400	260	400	0.5	1200	16–47	200–330	145	58	460

of spheroidal-graphite or SG irons, also known as *nodular irons* where the free graphite, on solidification of the iron, is rendered in the form of nodules or spheroids. This is usually done by the addition of magnesium or nickel-magnesium to the melt in an electric furnace.

The overall effect is to yield a much more ductile iron (*ie* a less brittle material) and the matrix structure can be further controlled by the process heat treatment. Thus the matrix may be *ferritic* for maximum ductility and good machinability; or *pearlitic* for maximum strength, at the expense of ductility and, to a lesser extent, machinability (see Tables 2 and 3).

The mechanical properties of SG irons approach those of steel more closely than those of grey irons, with no clearly defined yield point (*ie* good ductility over a wide range of the stress-strain curve). They retain, however, good casting characteristics.

British Standards for SG irons are specified in BS 2789, in nine grades (see Table 3).

TABLE 2 – HEAT TREATMENT OF SG IRONS

Temperature	Time	Cooling	Matrix	BS Grades (BS 2798)
900°C	2–4 hours		Fully ferritic	350/22, 400/18
700°C	6–12 hours		Predominantly ferritic	420/12
900°C	2–4 hours	In furnace	Ferritic and pearlitic	450/10, 500/7
900°C	2–4 hours	Air cool	Pearlitic	600/3, 700/2, 800/2
900°C	2–4 hours	Quench and temper	Tempered martensitic	900/2

TABLE 3 – LEADING PROPERTIES OF SG IRONS TO BS 2789

BS 2789 Grade	0.2% Proof Strength (MPa)	Tensile Strength (MPa)	Tensile Elongation (%)	Compressive Strength (MPa)	Impact Strength (J)	Brinell Hardness (HB)	Elastic Modulus (GPa)	Modulus of Rigidity (GPa)	Fatigue Limit (MPa)[c]
350/22L40	220	350	22	600	12[a]	−160	160	62	180
350/22	220	350	22	—	17[a]	−160	—	—	—
400/18L20	250	400	18	700	12[a]	−179	160	62	200
400/18	250	400	18	800	14[a]	−179	165	65	220
420/12	270	420	12	800	80[b]	−212	165	65	220
450/10	320	450	10	—	—	160–221	—	—	—
500/7	320	500	7	900	60[b]	170–241	170	66	240
600/3	370	600	3	1000	40[b]	192–269	175	68	260
700/2	420	700	2	1100	30[b]	229–302	180	70	280
800/2	480	800	2	1200	20[b]	248–352	185	72	300
900/2	600	900	2	1300	—	302–359	185	72	—

[a] Charpy V-notch test pieces tested at 20°C except for 350/22L40 (−40°C) and 400/18L20 (−20°C).
[b] Unnotched 10 mm square test bars.
[c] Wohler, unnotched

Malleable irons

Malleable irons are noted for their excellent machining qualities, with strengths similar to that of grey cast iron and appreciably better ductility. The malleable qualities of these irons are developed by closely controlled heat treatment. Malleable irons fall into three distinct groups:

(i)   Whiteheat
(ii) Blackheat
(iii) Pearlitic

Whiteheat malleable iron has a white crystalline appearance on fracture, the matrix being ferritic in thin sections, and a ferritic surface merging into a ferritic-pearlitic core interspersed with nodular graphite in the case of thicker sections. It generally has a higher carbon content than the other two types of malleable iron, which tends to give it superior casting properties.

Blackheat malleable iron has a black appearance on fracture due to the graphite nodules, the matrix being entirely ferritic, independent of section size. It has generally superior machining properties.

Pearlitic malleable iron is produced in a similar manner to blackheat but the treatment is so controlled as to produce an iron matrix which can vary from lamellar pearlitic to tempered martensitic, depending on property requirements. Its strength is superior to the other two types but it is more difficult to produce (*eg* it can only be produced satisfactorily in an electric furnace).

The relevant British Standard for malleable cast irons is BS 6681; see Table 4 for mechanical properties of grades covered by this standard.

Austenitic cast irons

Austenitic cast irons or *alloy irons* are produced in eight main grades to British Standard specifications. Four of these are flake graphite (*ie* similar in formulation to grey iron) and four nodular graphite (*ie* similar to SG irons).

Table 5 details BS cast irons to BS 3468. Table 7 summarises general characteristics produced by various alloying elements in alloy irons.

Meehanite

Meehanite metal is the trade name for a series of flake and nodular cast irons produced under licence to standards imposed by the International Meehanite Metal Co Ltd. The various types are embraced in four general classifications (see Tables 15, 16, 17 and 18).

(i) General engineering
(ii) Heat resisting
(iii) Abrasion resisting
(iv) Corrosion resisting

Their mechanical and physical properties are detailed in Tables 19 and 20.

TABLE 4 – MECHANICAL PROPERTIES OF MALLEABLE CAST IRONS TO BS 6681

BS 6681 Grade	Type	Test bar Diameter (mm)	0.2% proof Strength (mPa)	Tensile Strength (mPa)	Elongation %	Brinell Hardness (HB)	Elastic Modulus (GPa)	Modulus of rigidity (GPa)	Fatigue Ratio (% UTS)	Impact[a] Strength (J)
W 35–04	Whiteheart	9	—	340	5	230 max.				Notched 3–4
		12	—	350	4					
		15	—	360	3					
W 38–12	Whiteheart	9	170	320	15	200 max.	165	69	40	Unnotched decarberised layer: 41–54
		12	200	380	12		to	to	to	
		15	210	400	8		186	83	60	
W 40–05	Whiteheart	9	200	360	8	220 max.				
		12	220	400	5					
		15	230	420	4					
W 45–07	Whiteheart	9	230	400	10	220 max.				Core: 11–14
		12	260	450	7					
		15	280	480	4					
B 30–06	Blackheart	12–15	—	300	6	150 max.	158	62	40	9.5–19.0
B 32–10	Blackheart	12–15	190	320	10		to	to	to	
B 35–12	Blackheart	12–15	200	350	12		173	69	60	
P 45–06	Pearlitic	12–15	270	450	6	150–200	158	62	40	—
P 50–05	Pearlitic	12–15	300	500	5	160–220	to	to	to	
P 55–04	Pearlitic	12–15	340	550	4	180–230	179	83	60	
P 60–03	Pearlitic	12–15	390	600	3	200–250				
P 65–02	Pearlitic	12–15	430	650	2	210–260				
P 70–02	Pearlitic	12–15	530	700	2	240–290				

[a] Izod test on 10 mm square bars. Notched bars except where stated. Values for Whiteheart malleable iron vary widely due to the extensive decarberised surface

TABLE 5 – CHEMICAL COMPOSITION AND TENSILE STRENGTH OF FLAKE GRAPHITE AUSTENITIC CAST IRONS

Grade	Chemical Composition						Tensile Strength min N/mm^2
	C max %	Si %	Mn %	Ni %	Cr %	Cu %	
L – Ni Mn 13 7	3.0	1.5–3.0	6.0–7.0	12.0–14.0	0.2 max	0.5 max	140
L – Ni Cu Cr 15 6 2	3.0	1.0–2.8	0.5–1.5	12.5–17.5	1.0–2.5	5.5–7.5	170
L – Ni Cu Cr 15 6 3	3.0	1.0–2.8	0.5–1.5	13.5–17.5	2.5–3.5	5.5–7.5	190
L – Ni Cr 20 2	3.0	1.0–2.8	0.5–1.5	18.0–22.0	1.0–2.5	0.5 max	170
L – Ni Cr 20 3	3.0	1.0–2.8	0.5–1.5	18.0–22.0	2.5–3.5	0.5 max	190
L – Ni Si Cr 20 5 3	2.5	4.5–5.5	0.5–1.5	18.0–22.0	1.5–4.5	0.5 max	190
L – Ni Cr 30 3	2.5	1.0–2.0	0.5–1.5	28.0–32.0	2.5–3.5	0.5 max	190
L – Ni Si Cr 30 5 5	2.5	5.0–6.0	0.5–1.5	29.0–32.0	4.5–5.5	0.5 max	170
L – Ni 35	2.4	1.0–2.0	0.5–1.5	34.0–36.0	0.2 max	0.5 max	120

TABLE 6 – CHEMICAL COMPOSITION OF SPHEROIDAL GRAPHITE AUSTENITIC CAST IRONS

Grade	C max %	Si %	Mn %	Ni %	Cr %	P max %	Cu max %
S – Ni Mn 13 7	3.0	2.0–3.0	6.0–7.0	12–14	0.2 max	0.08	0.5
S – Ni Cr 20 2	3.0	1.5–2.0	0.5–1.5	18–22	1.0–2.5	0.08	0.5
S – Ni Cr 20 3	3.0	1.5–3.0	0.5–1.5	18–22	2.5–3.5	0.08	0.5
S – Ni Si Cr 20 5 2	3.0	4.5–5.5	0.5–1.5	18–22	1.0–2.5	0.08	0.5
S – Ni 22	3.0	1.0–3.0	1.5–2.5	21–24	0.5 max	0.08	0.5
S – Ni Mn 23 4	2.6	1.5–2.5	4.0–4.5	22–24	0.2 max	0.08	0.5
S – Ni Cr 30 1	2.6	1.5–3.0	0.5–1.5	28–32	1.0–1.5	0.08	0.5
S – Ni Cr 30 3	2.6	1.5–3.0	0.5–1.5	28–32	2.5–3.5	0.08	0.5
S – Ni Si Cr 30 5 5*	2.6	5.0–6.0	0.5–1.5	28–32	4.5–5.5	0.08	0.5
S – Ni 35	2.4	1.5–3.0	0.5–1.5	34–36	0.2 max	0.08	0.5
S – Ni Cr 35 3	2.4	1.5–3.0	0.5–1.5	34–36	2.0–3.0	0.08	0.5

* The chromium content should be reduced to the range 2.75% to 3.75% for applications in which the best possible scaling and erosion resistance are not essential but for which ductility, machinability and ease of welding are important.

TABLE 7 – MECHANICAL PROPERTIES OF SPHEROIDAL GRAPHITE AUSTENITIC CAST IRONS

Grade	Tensile Strength min N/mm²	0.2% Proof Stress min N/mm²	Elongation min %	Minimum Mean Impact Value on 3 Tests	
				V-notch (Charpy) in accordance with BS 131 : Part 2 J	U-notch (Mesnager) in accordance with BS 131 : Part 3 J
S – Ni Mn 13 7	390	210	15	16	
S – Ni Cr 20 2	370	210	7	13	16
S – Ni Cr 20 3	390	210	7	—	—
S – Ni Si Cr 20 5 2	370	210	10	—	—
S – Ni 22	370	170	20	20	24
S – Ni Mn 23 4	440	210	25	24	28
S – Ni Cr 30 1	370	210	13	—	—
S – Ni Cr 30 3	370	210	7	—	—
S – Ni Si Cr 30 5 5	390	240	—	—	—
S – Ni 35	370	210	20	—	—
S – Ni Cr 35 3	370	210	7	—	—

TABLE 8 – MECHANICAL PROPERTIES OF FLAKE GRAPHITE AUSTENITIC CAST IRONS

Grade	Tensile Strength* N/mm²	Compressive Strength N/mm²	Elongation %	Elastic Modules GN/m²	Brinell Hardness
L – Ni Mn 13 7	140–220	630–840	—	70–90	120–150
L – Ni Cu Cr 15 6 2	170–210	700–840	2	85–105	140–200
L – Ni Cu Cr 15 6 3	190–240	860–1 100	1–2	98–113	150–250
L – Ni Cr 20 2	170–210	700–840	2–3	85–105	120–215
L – Ni Cr 20 3	190–240	860–1 100	1–2	98–113	160–250
L – Ni Si Cr 20 5 3	190–280	860–1 100	2–3	110	140–250
L – Ni Cr 30 3	190–240	700–910	1–3	98–113	120–215
L – Ni Si Cr 30 5 5	170–240	560	—	105	150–210
L – Ni 35	120–180	560–700	1–3	74	120–140

* The minimum tensile strengths indicated are mandatory.

TABLE 9 – PHYSICAL PROPERTIES OF FLAKE GRAPHITE AUSTENITIC CAST IRONS

Grade	Nominal Density Mg/m³	Thermal Coefficient of Expansion (20°C–200°C) m/(m°C) × 10⁻⁶	Thermal Conductivity W/(m°C)	Specific Heat J/(g°C)	Specific Electrical Resistance Ωmm²/m	Relative Permeability (μ)
L – Ni Mn 13 7	7.3	17.7	37.7–41.9	0.46–0.50	1.4	1.02
L – Ni Cu Cr 15 6 2	7.3	18.7	37.7–41.9	0.46–0.50	1.6	1.03
L – Ni Cu Cr 15 6 3	7.3	18.7	37.7–41.9	0.46–0.50	1.1	1.05
L – Ni Cr 20 2	7.3	18.7	37.7–41.9	0.46–0.50	1.4	1.04
L – Ni Cr 20 3	7.3	18.7	37.7–41.9	0.46–0.50	1.2	1.04
L – Ni Si Cr 20 5 3	7.3	18.0	37.7–41.9	0.46–0.50	1.6	1.10
L – Ni Cr 30 3	7.3	12.4	37.7–41.9	0.46–0.50	—	—
L – Ni Si Cr 30 5 5	7.3	14.6	37.7–41.9	0.46–0.50	1.6	>2
L – Ni 35	7.3	5.0	37.7–41.9	0.46–0.50	—	—

TABLE 10 – PROPERTIES AND TYPICAL APPLICATIONS
OF FLAKE GRAPHITE AUSTENITIC CAST IRONS

Grade	Properties*	Typical Applications
L – Ni Mn 13 7	Non-magnetic.	Pressure covers for turbine generator sets, housings for switchgear, insulator flanges, terminals and ducts.
L – Ni Cu Cr 15 6 2	Good resistance to corrosion, particularly in alkalis, dilute acids, sea water and salt solutions. Good heat resistance, good bearing properties, high thermal expansion, non-magnetic at low chromium contents.	Pumps, valves, furnace components, bushings, piston ring carriers for light alloy metal pistons.
L – Ni Cu Cr 15 6 3	Better corrosion and erosion resistance than L – Ni Cu Cr 15 6 2.	
L – Ni Cr 20 2	Similar to L – Ni Cu Cr 15 6 2, but more corrosion resistant to alkalis. High coefficient of thermal expansion.	As for L – Ni Cu Cr 15 6 2, but preferable for pumps handling alkalis, vessels for caustic alkalis, uses in the soap, food, artificial silk and plastics industries. Suitable where copper-free materials are required.
L – Ni Cr 20 3	As L – Ni Cr 20 2, but more resistant to erosion, heat and growth.	As L – Ni Cr 20 2, but preferred also for high temperature applications.
L – Ni Si Cr 20 5 3	Good resistance to corrosion, even to dilute sulphuric acid. More heat resistant than L – Ni Cr 20 2 and L – Ni Cr 20 3. This grade is not suitable for use in the temperature range 500–600°C.	Pump components and flue gas dampers.
L – Ni Cr 30 3	Resistant to heat and thermal shock up to 800°C. Good corrosion resistance at high temperatures; excellent erosion resistance in wet steam and salt slurry; average thermal expansion.	Pumps, pressure vessels, valves, filter parts, exhaust gas manifolds and turbo charger housings.
L – Ni Si Cr 30 5 5	Particularly resistant to corrosion, erosion and heat; average thermal expansion.	Pump components and flue gas dampers.
L – Ni 35	Resistant to thermal shock; low thermal expansion.	Parts with dimensional stability (eg machine tools), scientific instruments and glass moulds.

* The properties given depend on chemical composition.

TABLE 11 – MECHANICAL PROPERTIES OF SPHEROIDAL GRAPHITE AUSTENITIC CAST IRONS

Grade	Tensile Strength* N/mm²	0.2% Proof Stress* N/mm²	Elongation* %	Elastic Modulus GN/m²	Charpy 'V'-notch Impact Strength	Brinell Hardness
S – Ni Mn 13 7	390–460	210–260	15–25	140–150	15.0–27.5	130–170
S – Ni Cr 20 2	370–470	210–250	7–20	112–130	13.5–27.5	140–200
S – Ni Cr 20 3	390–490	210–260	7–15	112–133	12.0	150–255
S – Ni Si Cr 20 5 2	370–430	210–260	10–18	112–133	14.9	180–230
S – Ni 22	370–440	170–250	20–40	85–112	20.0–33.0	130–170
S – Ni Mn 23 4	440–470	210–240	25–45	120–140	24.0	150–180
S – Ni Cr 30 1	370–440	210–270	13–18	112–130	17.0	130–190
S – Ni Cr 30 3	370–470	210–260	7–18	92–105	8.5	140–200
S – Ni Si Cr 30 5 5	390–490	240–310	1–4	91	3.9–5.9	170–250
S – Ni 35	370–410	210–240	20–40	112–140	20.5	130–180
S – Ni Cr 35 3	370–440	210–290	7–10	112–123	7.0	140–190

* The minimum properties indicated are mandatory except the elongation value for S – Ni Si Cr 30 5 5.

TABLE 12 – PHYSICAL PROPERTIES OF SPHEROIDAL GRAPHITE AUSTENITIC CAST IRONS

Grade	Nominal Density Mg/m³	Thermal Coefficient of Expansion (20–200°C) m/(m°C) × 10⁻⁶	Thermal Conductivity W/(m°C)	Specific Electrical Resistance Ωmm²/m	Relative Permeability (μ)
S – Ni Mn 13 7	7.3	18.2	12.6	1	1.02
S – Ni Cr 20 2	7.4	18.7	12.6	1	1.04
S – Ni Cr 20 3	7.4	18.7	12.6	1	1.05
S – Ni Si Cr 20 5 2	7.4	18.0	12.6	—	—
S – Ni 22	7.4	18.4	12.6	1	1.02
S – Ni Mn 23 4	7.4	14.7	12.6	—	—
S – Ni Cr 30 1	7.4	12.6	12.6	—	—
S – Ni Cr 30 3	7.4	12.6	12.6	—	—
S – Ni Si Cr 30 5 5	7.4	14.4	12.6	—	—
S – Ni 35	7.6	5.0	12.6	—	—
S – Ni Cr 35 3	7.6	5.0	12.6	—	—

**TABLE 13 – PROPERTIES AND TYPICAL APPLICATIONS OF
SPHEROIDAL GRAPHITE CAST IRONS**

Grade	Properties*	Typical Applications
S – Ni Mn 13 7	Non-magnetic.	Pressure covers for turbine generator sets, housings for switchgear insulator flanges, terminals and ducts.
S – Ni Cr 20 2	Similar to L – Ni Cr 20 2 in relation to composition, corrosion and heat resistance.	Pumps, valves, compressors, bushings, turbosupercharger housings, exhaust gas manifolds.
S – Ni Cr 20 3	Similar to S – Ni Cr 20 2, but better erosion and heat resistance.	
S – Ni Si Cr 20 5 2	Good resistance to corrosion even to dilute sulphuric acid. Good heat resistance. This grade is not suitable for use in the temperature range 500–600°C.	Pump components, flue gas dampers subject to high mechanical stress.
S – Ni 22	High coefficient of thermal expansion; lower corrosion and heat resistance than S – Ni Cr 20 2. Good impact properties down to –100°C. Non–magnetic.	Pumps, valves, compressors, bushings, turbosupercharger housings, exhaust gas manifolds.
S – Ni Mn 23 4	Good impact properties down to –196°C. Non-magnetic.	Castings for refrigeration engineering for use down to –196°C.
S – Ni Cr 30 1	Similar to S – Ni Cr 30 3; good bearing properties.	Pumps, boilers, filter parts, exhaust gas manifolds, valves turbosupercharger housings.
S – Ni Cr 30 3	Similar to L – Ni Cr 30 3.	Pumps, boilers, valves filter parts, exhaust gas manifolds, turbosupercharger housings.
S – Ni Si Cr 30 5 5	Similar to L – Ni Si Cr 30 5 5.	Pump components, flue gas dampers subject to high mechanical stress.
S – Ni 35	Low thermal expansion similar to L – Ni 35, but more resistant to thermal shock.	Parts with dimensional stability (*eg* machine tools), scientific instruments, glass moulds.
S – Ni Cr 35 3	Similar to S – Ni 35.	Parts of gas turbine housings, glass moulds.

*The properties given depend on the chemical composition.
The creep resistance of all grades except grade S – Ni Mn 13 7 will improve by the addition of 1% molybdenum.

TABLE 14 – CAST IRON ALLOYING ELEMENTS

Element	Favourable Effects	Remarks
Nickel	(i) reduces chilling effect of castings (ii) improves machinability (iii) improves structure, ie smaller graphite flakes with more uniform dispersion	Up to 5% added to grey iron to achieve (i), (ii) and (iii). 5%–8% nickel has a hardening effect.
Chromium	(i) increases hardness (ii) improves resistance to corrosion, heat and wear (iii) reduces graphite content	Increases chilling effect and reduces machinability. Up to 1% usual. High-chrome irons may be used for special heat or corrosion resistant duties.
Molybdenum	(i) strengthens and hardens iron (ii) improves casting properties (iii) improves and refines structure	Machinability greatly reduced
Nickel + Chromium	(i) improves strength and hardness	Improvement obtained without loss of machinability.
Vanadium	(i) improves toughness, hardness and resistance to wear (ii) increases hardness (iii) acts as a deoxidizer	Reduces size and quantity of carbon flakes.
Copper	(i) improves resistance to corrosion (ii) improves toughness and hardness (iii) improves machinability (iv) improves casting characteristics	(i) small quantities (up to 5%) also have the effect of increasing graphite carbon content (ii) larger proportions of copper
Zirconium	(i) improves fluidity for casting (ii) increases density (iii) improves strength of castings	Acts as a deoxidizer and graphitizer
Titanium	(i) improves strength and refines grain structure	Also a very effective deoxidizer
Aluminium	(i) renders cast iron suitable for nitriding	Also acts as a graphitizer and can result in unsound castings

TABLE 15 – GENERAL ENGINEERING TYPES

Type	Tensile Strength tonf/in^2 (min)	0.1% Proof Stress tonf/in^2 (min)	Elongation % (min)	Hardness Brinell	Izod Impact ft lb (min) to BS 131	Service Uses
Flake Graphite Types						
GM*	24			230		Crankshafts, forging, blanking and heading dies, high pressure valve bodies, lathe spindles.
GA*	22			220		High speed gearing and flywheels, press and drawing dies, compressor and diesel engine cylinders, camshafts, rolls.
GB	20			210		Diesel, gas and steam engine components, machine tools, die casting machines, drums and crane parts.
GC	18			195		Machine tool beds, frames, bodies, crankcases, cylinders and cylinder blocks.
GD	16			185		Lighter type of engineering castings, small liners, valves, manifolds, brake drums, clutch plates.
GE	13.5			170		General run of castings where soundness and machinability are important.
GF	10			150		Freely machinable grade for small repetition machining in sections down to ⅛ inch.
Nodular Graphite Types						
SP	38	26	2	200		Highly stressed machine parts, *eg* heavy duty gears, press dies crankshafts.
SPF	33	22	7	180		Good general purpose material for cast-to-form dies, crane wheels, rotors, shafts.
SF	27	19	15	160	8	For applications involving shock, for machine components, levers, housings, brackets.
SFF	25	16	20	140	12	For duties involving heavy shock and sub-zero applications. Freely machinable.
SH/N	47	32	2	270		Normalised for exceptional strength and toughness; crankshafts, camshafts

*Types GM and GA may be heat-treated to tensile strengths up to 35 tonf/in^2 and hardened to values up to 600 Brinell.
These types are particularly amenable to flame hardening, induction hardening and nitriding for applications such as wire drawing equipment and machine tool slides.

Fig. 1

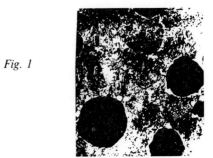

Type GA Meehanite, flake
form graphite Mag. × 500.

Type SP Meehanite, spheroidal
form graphite Mag. × 500.

TABLE 16 – HEAT RESISTING TYPES

Type	Tensile Strength tonf/in^2	Transverse Modulus of Rupture tonf/in^2	BHN	Coefficient of Thermal Expansion per °F × 10^{-6} from 100°F to 1 000°F	Service conditions
Flake Graphite Types					
HE	14	28	<170	6.66	Thermal shock. Used for melting pots, ingot moulds, and hot plates, etc.
HD	16	36	>220	7.09	Furnace parts, roasting and burning retorts, annealing pots etc. Maximum temperature to 620°C
White/Mottled Types					
HR	17	33	<500	7.43	Furnace skid bars, oil refinery tube supports, furnace and burner parts, muffle retorts, etc. Maximum temperature of service up to 800°C
HR1	18	34	<500	7.43	Resistance to scaling and growth for use at temperatures up to 750°C
Nodular Graphite Types					
HS	up to 35		>220		Provides maximum resistance to scaling and growth up to 900°C, possessing good mechanical and heat resisting properties without being brittle. Used for hot gas valves, furnace rails and skids, melting pots, exhaust manifolds and fire grates, etc

TABLE 17 – ABRASION RESISTING TYPES

Type	Tensile Strength tonf/in^2	Brinell Hardness	Service uses
Flake Graphite Types			
WA	22.5	up to 350	Machinable at lower hardness ranges. May be hardened by heat treatment to 600 Brinell where required. Typical uses include cones, sheaves, dies, crane and truck wheels, hoist drums and rolls for sundry purposes.
WB	20	up to 550	Tough, hard wearing iron, possessing good strength and shock resisting properties. Supplied in machinable or non-machinable condition according to service requirements. Typical applications include cement mixer parts, cams, dies, sheaves, pug mill blades, rolling mill guides, gears and rollers, etc.
WEC	14	up to 550 on outer face	Chill cast, combining a hard abrasion-resisting exterior with a tough grey back. Typical uses include crusher rolls, chute and wear plates for coke and coal industries, mine car wheels, stamp mill dies and shoes, etc.
White Mottled Types			
WH	18	up to 600	Its purpose is to resist pure abrasion. Hardness up to 600 according to service conditions. Typical examples are muller tyres, rolls and plates; pug-mill knives, grizzly discs, dredging, pumping and ash disposal parts, etc.
Nodular Graphite Types			
SH/H	50–75	280–450	Heat treated to required hardness level by quench and temper for conditions involving high stresses and impact/abrasive wear. Typical uses include crusher jaws, wear plates and liners, pulverisers, etc.
WSH 1	35–75	600–400 after heat treatment	Pearlitic in the as-cast condition and thus readily machinable. Suitably alloyed so as to be completely martensitic after heat treatment. Offers high strength together with moderate hardness for applications requiring a degree of impact strength such as crushers, dies, sheaves, gears, etc.
WSH 2	17–25	700–650 after heat treatment	Contains hard carbides in a matrix of pearlite in the as-cast condition, thus is more difficult to machine. In the heat treated condition it possesses a matrix of martensite and carbides offering very high hardness values suited to a variety of abrasion resisting applications, *eg* slurry pumps, impeller blades, wear plates, etc, for coke handling, brick manufacture, refractory and ceramics production and shot blasting plant.

TABLE 18 – CORROSION RESISTING TYPES

Type	Tensile Strength tonf/in^2	Elongation %	Brinell	Service uses
Flake Graphite Types				
CB$_3$	up to 20		200	Concentrated sulphuric acid. Typical uses include pumps, valves, fittings, stills and storage vessels. Gives excellent service in sulphuric acid contact plants in 93–105% sulphuric acid in temperatures up to 95°C
CC	up to 20		190	A low-cost general utility material possessing sufficient reagent resistance to reduce to a minimum the necessity for replacement and avoid heavy capital expenditure. Typical uses include acid pans, kettles, pumps, valves evaporators and retorts, etc. Useful also for handling water (mine, sea) and general solutions of less than 2 pH.
KC	up to 20		180	Possesses reasonably good resistance to alkalis. Typical uses include pumps, valves, fittings carbonators, causticizers, evaporators and pans, etc.
Austenitic Nodular Graphite Types				
CRS	<25	up to 20	<250	Resist corrosion by acids such as formic, acetic and oxalic much better than ordinary grey iron, and are not corroded by dilute alkali solutions at any temperature. Recommended for industrial water where the pH is low, or in strongly acidic conditions. Also in applications where components are subjected to high velocity or abrasion. Having the graphite in nodular form, it provides superior mechanical and corrosion resisting properties.

TABLE 19 – MECHANICAL AND PHYSICAL PROPERTIES OF THE ENGINEERING TYPES OF MEEHANITE FLAKE IRON

Property	Units	'GM'	'GA'	'GB'	'GC'	'GD'	'GE'	'GF'
Tensile strength[1] (min)	tonf/in²	24.5	22.5	20.0	18.0	16.0	13.5	10.0
Modulus of elasticity	lbf/in² × 10⁶	22.0	20.0	18.0	17.0	16.0	14.0	13.5
Brinell Hardness[2]	kp/mm²	220/280	210/240	200/230	190/220	180/210	170/200	150/180
Transverse modulus of rupture	tonf/in²	44.0	40.0	35.0	32.0	30.0	28.0	26.0
Compression strength	tonf/in²	90.0	80.0	75.0	70.0	60.0	50.0	40.0
Shear strength (min)	tonf/in²	24.5	22.5	20.0	18.0	16.0	13.5	10.0
Fatigue limit[3]	tonf/in²	±12.0	±10.0	±9.0	±7.5	±6.5	±6.0	±5.5
Impact strength[4]	ft lb	30.0	25.0	20.0	15.0	10.0	7.0	5.0
Impact strength[5]	mkp/cm²	1.0	0.8	0.6	0.5	0.4	0.2	0.1
Damping capacity[6]	%	20.0						32.0
Coefficient of thermal expansion	10⁶/degC	13.0						12.0
Thermal conductivity	cal/cm/sec/degC	0.10						0.12
Specific gravity	gm/cm³	7.48	7.43	7.37	7.25	7.13	7.10	7.05
Patternmakers contraction allowance[7]	%	1.4						0.9
Casting section	inch	0.75	0.50	0.40	0.30	0.25	0.20	0.15

Notes:

(1) Standard 1.2 inch diameter separately cast bar tested according to BS 1452
(2) Depending on section thickness of castings
(3) Wöhler test to BS 3518 using 0.331 inch diameter un-notched test bar
(4) Izod 0.798 inch diameter un-notched test bar
(5) Charpy un-notched DVM test piece
(6) 22 000 lb/in² torsional stress energy dissipated during first cycle
(7) Depending upon design and size of casting
(8) Types 'GM' to 'GC' respond best to hardening operations

TABLE 20 – MECHANICAL AND PHYSICAL PROPERTIES OF THE NODULAR TYPES OF MEEHANITE IRON

Property	Units	'SFF'	'SF'	'SPF'	'SP'	'SH/N'	'SH/H'	'HS'
Tensile strength[1] (min)	tonf/in^2	25.0	27.0	33.0	38.0	47.0	50–75	25.0
0.2% proof stress	tonf/in^2	16.0	19.0	22.0	26.0	32.0	32–55	20.0
Elongation (min)	%	20.0	15.0	7.0	2.0	2.0	2–1	2.0
Brinell Hardness	kp/mm^2	140/180	150/200	170/240	200/280	240/300	280/450	220/250
Impact strength[2] (min)	mkp/cm^2	2.2	2.0					
Impact strength[3] (min)	mkp/cm^2	10.0	8.0	6.0	4.0	2.0		
Impact strength[4] (min)	ft lb	12.0	8.0					
Fatigue limit[5]	tonf/in^2	±13.0	±14.0	±15.0	±16.5	±19.5	±20 –22	
Compression strength	tonf/in^2	40.0	50.0	57.0	63.0	70.0	82.0	
Modulus of elasticity	lbf/in$^2 \times 10^6$	23.0					26.0	
Modulus of rigidity	lbf/in$^2 \times 10^6$	9.3					9.6	
Poisson's ratio		0.28					0.29	
Coefficient of thermal expansion	10^{6}/degC	10.0					13.0	
Thermal conductivity[6]	cal/cm/sec/degC	0.096	0.096	0.094	0.088	0.083		
Specific gravity	gm/cm^3	7.1	7.1	7.1	7.15	7.20		

Notes:
(1) Standard round proportional test pieces taken from separately cast test samples according to BS 2789
(2) Charpy notched DVM test piece. Minimum impact strength value for type 'SFF' at −40°C is 2.0 mkp/cm^2
(3) Charpy un-notched DVM test piece
(4) Izod 10 × 10 mm notched test piece
(5) Wohler test to BS 3518 using 0.417 inch diameter un-notched test bar
(6) Alloying, due to nickel-magnesium nodularising practice, reduces these values by some 8–10%
(7) Type 'SP' responds best to hardening operations.

Carbon steels

THE DESCRIPTION *carbon steel* implies a plain carbon steel or simple alloy or iron and carbon with other minor constituents – manganese, silicon, sulphur, phosphorus and possibly traces of nickel and chromium. The main difference between carbon steel and cast iron is that in the former all the carbon is in chemically combined form. The physical properties of the steel are governed primarily by the carbon content and further modified by heat treatment. The effect of the minor constituents is usually small, although sulphur and phosphorus degrade the mechanical properties and are therefore normally reduced to as small a percentage as possible.

Carbon steels are broadly classified as low carbon (0.05 to 0.30%), medium carbon (0.3 to 0.6%) and high carbon (over 0.6%). The main characteristics of steels within this grouping are summarised in Table 1.

Low carbon steels are generally used where ductility or ease of forming is more important than strength and hardness. They are also more easily welded. The hardness of low carbon steels cannot be improved to any marked extent by heat treatment, but annealing can be used to soften the material or relieve cold working stresses.

Medium carbon steels are heat-treated and their properties depend very largely on whether or not they are so treated. Heat treatment can result in doubling the strength and hardness with reduction in elongation, although good ductility may still be retained.

High carbon steels possess great hardness with maximum mechanical properties being realised by heat treatment. They are used mainly for tools, dies, springs, etc, with lower proportions of carbon for items requiring toughness and high carbon content for those requiring extreme hardness.

Carbon steels specified in BS 970 are basically low to medium carbon and mainly intended for structural purposes – see Table 2. Mild steel plates, bars and sections for welding are specified in BS 4360.

TABLE 1 – GENERAL CLASSIFICATION AND PROPERTIES OF CARBON STEEL

Carbon Content %	Maximum Tensile Strength tons/in^2 (kg/mm^2)	Yield Point tons/in^2 (kg/mm^2)	Hardness Brinell	Characteristics	Typical Applications
LOW ⇑ 0.02–0.07 ⇓	22–35 (35–45)	13–22 (20–35)	100–160	Soft. Fairly resistant to corrosion. Difficult to machine.	Sheets for cold pressing, etc cold-drawn tubes
0.07–0.15				Tough and ductile, easily hot worked. Difficult to machine.	Sheets for deep drawing boiler plates. Rivets, screws, etc.
0.15–0.25				Less ductile	Structural work, deep pressings, case hardening, heavy plates, bolts and nuts.
MEDIUM ⇑ 0.25–0.35 ⇓	Rolled 30–35 (47–87)	Rolled 20–33 (31–52)	140–240	Readily machinable and forgeable. Respond to heat treatment.	Parts requiring medium strength and toughness.
0.35–0.55				Satisfactory machine-ability. Good response to heat treatment.	High tensile machined parts and forgings.
0.55–0.70				More difficult to machine. High strength and stiffness. Heat treatable.	Dies, springs, high strength parts. Hard parts to resist wear and abrasion.
HIGH ⇑ 0.70–0.80 ⇓	Depends on heat treatment			High strength but reducing toughness. Heat treatable.	Band saws, lap saws and tools.
0.80–1.00				Very hard and tough but becoming brittle.	Edge tools, punches, rock drills, springs, etc.
1.00–1.50				Exceptionally hard but brittle. Difficult to machine.	Edge tools, razor blades, cutting and turning tools, dies, etc.

TABLE 2 – CARBON STEEL COMPOSITIONS BS 970

Designation *	Composition % (Maximum)							Tensile strength tons/in^2	Yield Stress tons/in^2
	C	Si	Mn	S	P	Ni	Cr		
040A04	0.20		0.80	0.06	0.06			20	
080A27	0.25	0.35	1.00	0.06	0.06			25–35	
080A25	0.30	0.35	1.00	0.06	0.06			28–38	
060M30	0.34	0.35	1.00	0.06	0.06			32–55	15–16
080A35	0.38	0.35	0.90	0.06	0.06			35	18
080M40	0.44	0.35	1.00	0.06	0.06			40–50	25
150M19	0.23	0.35	1.70	0.06	0.06	0.40	0.25	40–50	28
150M36	0.40	0.35	1.70	0.05	0.05			40–50	26
045M10	0.15	0.35	0.70	0.05	0.05			32	
212M14	0.18	0.35	1.20	0.05	0.05			32	

* See also section on *En Steels*   KEY...   Si = silicon   S = sulphur   Ni = nickel
C = carbon   Mn = manganese   P = phosphorus   Cr = chrome

TABLE 3 – CHEMICAL COMPOSITION OF CARBON AND CARBON
MANGANESE STEELS TO BS 970

Steel	C %	Si %	Mn %	S %	P %
070M20	0.16–0.24	—	0.50–0.90		
070M26	0.22–0.30	—	0.50–0.90		
080M30	0.26–0.34	—	0.60–1.00		
080M36	0.32–0.40	—	0.60–1.00		
080M40	0.36–0.44	—	0.60–1.00		
080M46	0.42–0.50	—	0.60–1.00		
080M50	0.45–0.55	—	0.60–1.00	The sulphur and phosphorous contents of these steels shall be 0.05% maximum unless otherwise agreed.	
070M55	0.50–0.60	—	0.50–0.90		
120M19	0.15–0.23	—	1.00–1.40		
150M19	0.15–0.23	—	1.30–1.70		
120M28	0.24–0.32	—	1.00–1.40		
150M28	0.24–0.32	—	1.30–1.70		
120M36	0.32–0.40	—	1.00–1.40		
150M36	0.32–0.40	—	1.30–1.70		
220M07	0.15 max	—*	0.90–1.30	0.20–0.30	0.07 max†
230M07	0.15 max	—*	0.90–1.30	0.25–0.35	0.07 max†
240M07	0.15 max	—*	1.10–1.50	0.30–0.60	0.07 max†
216M28	0.24–0.32	0.25 max	1.10–1.50	0.12–0.20	0.06 max
212M36	0.32–0.40	0.25 max	1.00–1.40	0.12–0.20	0.06 max
225M36	0.32–0.40	0.25 max	1.00–1.40	0.20–0.30	0.06 max
216M36	0.32–0.40	0.25 max	1.30–1.70	0.12–0.20	0.06 max
212M44	0.40–0.48	0.25 max	1.00–1.40	0.12–0.20	0.06 max
225M44	0.40–0.48	0.25 max	1.30–1.70	0.20–0.30	0.06 max

*A maximum silicon content can be specified by agreement between the purchaser and the supplier.
†Higher phosphorous content can be specified by agreement between the purchaser and the supplier.

TABLE 4 – BS970 CARBON STEELS – MECHANICAL PROPERTIES IN THE NORMALISED CONDITION RELATED TO TENSILE STRENGTH

Tensile strength min. R_m, tonf/in²	26				28				30				32				33				34				35			
Brinell Hardness Range	—				126–179 Not applicable to 070M26				—				143–192 Not applicable to 080M36				—				—				152–207 Not applicable to 080M46			
Steel	LRS*	R_e*	A*	I*	LRS*	R_e*	A*	I*	LRS*	R_e*	A*	I*	LRS*	R_e*	A*	I*	LRS*	R_e*	A*	I*	LRS*	R_e*	A*	I*	LRS*	R_e*	A*	I*
070M20	10	13	21	—	6	14	21	—	—	—	—	—	—	—	—	—	—	—	—	—	—	—	—	—	—	—	—	—
070M26	—	—	—	—	10	14	20	—	—	—	—	—	—	—	—	—	—	—	—	—	—	—	—	—	—	—	—	—
080M30	—	—	—	—	—	—	—	—	10	15	19	—	—	—	—	—	—	—	—	—	—	—	—	—	—	—	—	—
080M36	—	—	—	—	—	—	—	—	—	—	—	—	2½	16	20	—	—	—	—	—	—	—	—	—	2½	18	16	20
080M40	—	—	—	—	—	—	—	—	—	—	—	—	6	16	20	—	—	—	—	—	—	—	—	—	6	18	16	15
080M46	—	—	—	—	—	—	—	—	—	—	—	—	10	16	18	—	10	16	17	—	—	—	—	—	10	18	15	—
080M50	—	—	—	—	—	—	—	—	—	—	—	—	—	—	—	—	—	—	—	—	—	—	—	—	—	—	—	—
070M55	—	—	—	—	—	—	—	—	10	17	19	—	—	—	—	—	—	—	—	—	—	—	—	—	—	—	—	—
120M19	—	—	—	—	—	—	—	—	—	—	—	—	4	19	20	25	10	19	17	—	—	—	—	—	—	—	—	—
150M19	—	—	—	—	—	—	—	—	—	—	—	—	—	—	—	—	—	—	—	—	—	—	—	—	6	21	18	30
120M28	—	—	—	—	—	—	—	—	—	—	—	—	—	—	—	—	—	—	—	—	10	20	17	—	6	21	16	25
150M28	—	—	—	—	—	—	—	—	—	—	—	—	—	—	—	—	—	—	—	—	—	—	—	—	—	—	—	—
120M36	—	—	—	—	—	—	—	—	—	—	—	—	—	—	—	—	—	—	—	—	—	—	—	—	—	—	—	—
150M36	—	—	—	—	—	—	—	—	—	—	—	—	—	—	—	—	—	—	—	—	—	—	—	—	—	—	—	—

(continued)

TABLE 4 – BS 970 CARBON STEELS – MECHANICAL PROPERTIES IN THE NORMALISED CONDITION RELATED TO TENSILE STRENGTH – contd.

Tensile strength min. R_m, tonf/in²	36				37				38				39				40				45			
Brinell Hardness Range	—				—				174–223				—				179–229				201–255			
Steel	LRS*	R_e*	A*	I*	LRS*	R_e*	A*	I*	LRS*	R_e*	A*	I*	LRS*	R_e*	A*	I*	LRS*	R_e*	A*	I*	LRS*	R_e*	A*	I*
070M20	—	—	—	—	—	—	—	—	—	—	—	—	—	—	—	—	—	—	—	—	—	—	—	—
070M26	—	—	—	—	—	—	—	—	—	—	—	—	—	—	—	—	—	—	—	—	—	—	—	—
080M30	—	—	—	—	—	—	—	—	—	—	—	—	—	—	—	—	—	—	—	—	—	—	—	—
080M36	—	—	—	—	—	—	—	—	—	—	—	—	—	—	—	—	—	—	—	—	—	—	—	—
080M40	—	—	—	—	—	—	—	—	—	—	—	—	—	—	—	—	—	—	—	—	—	—	—	—
080M46	—	—	—	—	—	—	—	—	—	—	—	—	—	—	—	—	2½	20	14	—	—	—	—	—
080M50	—	—	—	—	10	18	14	—	—	—	—	—	—	—	—	—	6	20	14	—	—	—	—	—
070M55	—	—	—	—	—	—	—	—	—	—	—	—	10	20	13	—	—	—	—	—	2½	23	12	—
120M19	—	—	—	—	—	—	—	—	—	—	—	—	—	—	—	—	—	—	—	—	—	—	—	—
150M19	—	—	—	—	—	—	—	—	—	—	—	—	—	—	—	—	—	—	—	—	—	—	—	—
120M28	10	21	16	—	—	—	—	—	6	23	16	—	—	—	—	—	—	—	—	—	—	—	—	—
150M28	—	—	—	—	—	—	—	—	6	23	15	25	—	—	—	—	—	—	—	—	—	—	—	—
120M36	—	—	—	—	10	22	16	—	—	—	—	—	—	—	—	—	—	—	—	—	—	—	—	—
150M36	—	—	—	—	—	—	—	—	—	—	—	—	10	23	15	—	6	25	14	—	—	—	—	—

*LRS denotes limiting ruling section, inches.
R_e denotes yield stress, tonf/in², minimum.
A denotes elongation, % min on $5.65\sqrt{S_o}$ gauge length
I denotes Izod impact value, ft lbf, min, only applicable if fine grain controlled material is ordered

TABLE 5 – BS 970 CARBON STEELS – MECHANICAL PROPERTIES IN THE HARDENED AND TEMPERED CONDITION RELATED TO TENSILE RANGE

Heat Treatment Condition Symbol	P					Q					R					S					T				
Tensile Strength Range R_m tonf/in^2	35–45					40–50					45–55					50–60					55–65				
Brinell Hardness Range	152–207					179–229					201–255					223–277					248–302				
Steel	*LRS**	R_e*	*A**	*I**	$R_{p0.2}$	*LRS**	R_e*	*A**	*I**	$R_{p0.2}$	*LRS**	R_e*	*A**	*I**	$R_{p0.2}$	*LRS**	R_e*	*A**	*I**	$R_{p0.2}$	*LRS**	R_e*	*A**	*I**	$R_{p0.2}$
070M20	¾	23	20	30	22	—	—	—	—	—	—	—	—	—	—	—	—	—	—	—	—	—	—	—	—
070M26	1⅛	23	20	30	21	½	27	16	25	26	—	—	—	—	—	—	—	—	—	—	—	—	—	—	—
080M30	2½	22	18	25	20	¾	27	16	25	26	—	—	—	—	—	—	—	—	—	—	—	—	—	—	—
080M36	—	—	—	—	—	1⅛	26	16	25	24	½	30	16	25	29	—	—	—	—	—	—	—	—	—	—
080M40	—	—	—	—	—	2½	25	16	25	23	¾	30	16	25	29	—	—	—	—	—	—	—	—	—	—
080M46	—	—	—	—	—	4	24	16	—	22	1⅛	29	16	—	27	½	34	14	—	33	—	—	—	—	—
080M50	—	—	—	—	—	—	—	—	—	—	2½	28	14	—	26	1⅛	32	14	—	30	½	37	12	—	36
070M55	—	—	—	—	—	—	—	—	—	—	4	27	14	—	25	2½	31	14	—	29	¾	37	12	—	36
120M19	4	23	18	35	21	1⅛	29	16	35	27	¾	33	16	25	32	—	—	—	—	—	—	—	—	—	—
150M19	6	22	18	40	20	2½	28	16	40	26	1⅛	33	16	30	31	—	—	—	—	—	—	—	—	—	—
120M28	—	—	—	—	—	4	27	16	30	25	1⅛	33	16	25	31	—	—	—	—	—	—	—	—	—	—
150M28	—	—	—	—	—	6	26	16	35	24	2½	31	16	30	29	½	37	16	25	36	—	—	—	—	—
120M36	—	—	—	—	—	4	27	18	30	25	1⅛	33	16	25	31	¾	37	14	25	36	—	—	—	—	—
150M36	—	—	—	—	—	6	26	18	35	24	2½	31	16	30	29	1⅛	36	14	30	34	½	41	12	25	40
216M28	2½	23	20	25	21	¾	28	18	25	27	—	—	—	—	—	—	—	—	—	—	—	—	—	—	—
212M36	4	22	20	25	20	2½	26	18	25	24	½	32	16	40	31	—	—	—	—	—	—	—	—	—	—
225M36	—	—	—	—	—	2½	26	18	25	24	1⅛	31	16	25	29	—	—	—	—	—	—	—	—	—	—
216M36	4	22	20	25	20	2½	26	18	25	24	1⅛	31	16	25	29	—	—	—	—	—	—	—	—	—	—
212M44	—	—	—	—	—	4	26	18	25	24	2½	30	16	25	28	½	35	14	20	34	—	—	—	—	—
225M44	—	—	—	—	—	—	—	—	—	—	4	29	16	25	27	1⅛	34	14	20	32	½	39	12	20	38

* *LRS* denotes limiting ruling section, inches.
R_e denotes yield stress, tonf/in^2, min.
A denotes elongation, % min on $5.65\sqrt{S_o}$ gauge length.
I denotes Izod impact value, ft lbf, min, only applicable to carbon and carbon manganese steels if fine grain controlled material is ordered.
$R_{p0.2}$ denotes 0.2% proof stress, tonf/in^2, min, only applicable when specifically ordered.

TABLE 6 – BS 970 CARBON STEELS – MECHANICAL PROPERTIES IN THE NORMALISED CONDITION RELATED TO RULING SECTION

Limiting Ruling Section Steel	Up to and including 2½ in					Up to and including 4 in					Up to and including 6 in					Up to and including 10 in				
	R_m*	R_e*	A*	I*	HB*	R_m*	R_e*	A*	I*	HB*	R_m*	R_e*	A*	I*	HB*	R_m*	R_e*	A*	I*	HB*
070M20	—	—	—	—	—	—	—	—	—	—	28	14	21	—	126–179	26	13	21	—	—
070M26	32	16	20	—	143–192	—	—	—	—	—	—	—	—	—	—	28	14	20	—	—
080M30	35	18	16	20	152–207	—	—	—	—	—	32	16	20	—	143–192	30	15	19	—	—
080M36	—	—	—	—	—	—	—	—	—	—	—	—	—	—	—	32	16	18	—	—
080M40	40	20	14	—	179–229	—	—	—	—	—	35	18	16	15	152–207	33	16	17	—	—
080M46	—	—	—	—	—	—	—	—	—	—	—	—	—	—	—	35	18	15	—	—
080M50	45	23	12	—	201–255	—	—	—	—	—	40	20	14	—	179–229	37	18	14	—	—
070M55	—	—	—	—	—	—	—	—	—	—	—	—	—	—	—	39	20	13	—	—
120M19	—	—	—	—	—	32	19	20	25	143–192	—	—	—	—	—	30	17	19	—	—
150M19	—	—	—	—	—	—	—	—	—	—	35	21	18	30	152–207	33	19	17	—	—
120M28	—	—	—	—	—	—	—	—	—	—	35	21	16	25	152–207	34	20	17	—	—
150M28	—	—	—	—	—	—	—	—	—	—	38	23	16	25	174–223	36	21	16	—	—
120M36	—	—	—	—	—	—	—	—	—	—	38	23	15	—	174–223	37	22	16	—	—
150M36	—	—	—	—	—	—	—	—	—	—	40	25	14	—	179–229	39	23	15	—	—

R_m denotes tensile strength, tonf/in², min.

R_e denotes yield stress, tonf/in², min.

A   denotes elongation, % min on 5.65 $\sqrt{S_o}$ gauge length.

I   denotes Izod impact value, ft lbf, min, only applicable if fine grain controlled material is ordered.

HB   denotes Brinell hardness number – range.

TABLE 7 – BS 970 CARBON STEELS – MECHANICAL PROPERTIES IN THE HARDENED AND TEMPERED CONDITION RELATED TO RULING SECTION

Limiting Ruling Section	Up to and including 1/2 in							Up to and including 3/4 in							Up to and including 1 1/8 in						
Steel	C*	R_m*	R_e*	A*	I*	HB*	$R_{p0.2}$	C*	R_m*	R_e*	A*	I*	HB*	$R_{p0.2}$	C*	R_m*	R_e*	A*	I*	HB*	$R_{p0.2}$
070M20	—	—	—	—	—	—	—	P	35–45	23	20	30	152–207	22	—	—	—	—	—	—	—
070M26	Q	40–50	27	16	25	179–229	26	Q	40–50	27	16	25	179–229	26	P	35–45	23	20	30	152–207	21
080M30	R	45–55	30	16	25	201–255	29	—	—	—	—	—	—	—	Q	40–50	26	16	25	179–229	24
080M36	—	—	—	—	—	—	—	R	45–55	30	16	25	201–255	29	R	45–55	29	16	—	201–255	27
080M40	—	—	—	—	—	—	—	—	—	—	—	—	—	—	—	—	—	—	—	—	—
080M46	S	50–60	34	14	—	223–277	33	—	—	—	—	—	—	—	S	50–60	32	14	—	223–277	30
080M50	T	55–65	37	12	—	248–302	36	—	—	—	—	—	—	—	—	—	—	—	—	—	—
070M55	—	—	—	—	—	—	—	T	55–65	37	12	—	248–302	36	Q	40–50	29	16	35	179–229	27
120M19	—	—	—	—	—	—	—	R	45–55	33	16	25	201–255	32	R	45–55	33	16	30	201–255	31
150M19	—	—	—	—	—	—	—	—	—	—	—	—	—	—	R	45–55	33	16	25	201–255	31
120M28	—	—	—	—	—	—	—	—	—	—	—	—	—	—	—	—	—	—	—	—	—
150M28	S	50–60	37	16	25	223–277	36	S	50–60	37	14	25	223–277	36	R	45–55	33	16	25	201–255	31
120M36	—	—	—	—	—	—	—	—	—	—	—	—	—	—	—	—	—	—	—	—	—
150M36	T	55–65	41	12	25	248–302	40	—	—	—	—	—	—	—	S	50–60	36	14	30	223–277	34
216M28	—	—	—	—	—	—	—	Q	40–50	28	18	25	179–229	27	—	—	—	—	—	—	—
212M36	R	45–55	32	16	40	201–255	31	—	—	—	—	—	—	—	R	45–55	31	16	25	201–255	29
225M36	—	—	—	—	—	—	—	—	—	—	—	—	—	—	R	45–55	31	16	25	201–255	29
216M36	—	—	—	—	—	—	—	—	—	—	—	—	—	—	—	—	—	—	—	—	—
212M44	S	50–60	35	14	20	223–277	34	—	—	—	—	—	—	—	—	—	—	—	—	—	—
225M44	T	55–65	39	12	20	248–302	38	—	—	—	—	—	—	—	S	50–60	34	14	20	223–277	32

(continued)

TABLE 7 – BS 970 CARBON STEELS – MECHANICAL PROPERTIES IN THE HARDENED AND TEMPERED CONDITION RELATED TO RULING SECTION – contd.

Limiting Ruling Section — Steel	Up to and including 2½ in							Up to and including 4 in							Up to and including 6 in						
	C*	Rm*	Re*	A*	I*	HB*	Rp0.2	C*	Rm*	Re*	A*	I*	HB*	Rp0.2	C*	Rm*	Re*	A*	I*	HB*	Rp0.2
070M20	—	—	—	—	—	—	—	—	—	—	—	—	—	—	—	—	—	—	—	—	—
070M26	—	—	—	—	—	—	—	—	—	—	—	—	—	—	—	—	—	—	—	—	—
080M30	P	35–45	22	18	25	152–207	20	—	—	—	—	—	—	—	—	—	—	—	—	—	—
080M36	—	—	—	—	—	—	—	—	—	—	—	—	—	—	—	—	—	—	—	—	—
080M40	Q	40–50	25	16	25	179–229	23	Q	40–50	24	16	—	179–229	22	—	—	—	—	—	—	—
080M46	—	—	—	—	—	—	—	—	—	—	—	—	—	—	—	—	—	—	—	—	—
080M50	R	45–55	28	14	—	201–255	26	R	45–55	27	14	—	201–255	25	—	—	—	—	—	—	—
070M55	S	50–60	31	14	—	223–277	29	—	—	—	—	—	—	—	—	—	—	—	—	—	—
120M19	—	—	—	—	—	—	—	—	—	—	—	—	—	—	—	—	—	—	—	—	—
150M19	Q	40–50	28	16	40	179–229	26	P	35–45	23	18	35	152–207	21	P	35–45	22	18	40	152–207	20
120M28	—	—	—	—	—	—	—	—	—	—	—	—	—	—	—	—	—	—	—	—	—
150M28	R	45–55	31	16	30	201–255	29	Q	40–50	27	16	30	179–229	25	Q	40–50	26	16	35	179–229	24
120M36	—	—	—	—	—	—	—	—	—	—	—	—	—	—	—	—	—	—	—	—	—
150M36	R	45–55	31	16	30	201–255	29	Q	40–50	27	18	30	179–229	25	Q	40–50	26	18	35	179–229	24
216M28	P	35–45	23	20	25	152–207	21	P	35–45	22	20	25	152–207	20	—	—	—	—	—	—	—
212M36	Q	40–50	26	18	25	179–229	24	P	35–45	22	20	25	152–207	20	—	—	—	—	—	—	—
225M36	Q	40–50	26	18	25	179–229	24	—	—	—	—	—	—	—	—	—	—	—	—	—	—
216M36	Q	40–50	26	18	25	179–229	24	Q	40–50	26	18	25	179–229	24	—	—	—	—	—	—	—
212M44	R	45–55	30	16	25	201–255	28	R	45–55	29	16	25	201–255	27	—	—	—	—	—	—	—
225M44	—	—	—	—	—	—	—	—	—	—	—	—	—	—	—	—	—	—	—	—	—

*C denotes heat treatment condition symbol.
R_e denotes yield stress, tonf/in^2, min.
I denotes Izod impact value, ft/lbf, min, only applicable to carbon and carbon manganese steels if fine grain controlled material is ordered.

R_m denotes tensile strength, tonf/in^2, range.
A denotes elongation, % min on 5.65 $\sqrt{S_o}$ gauge length.
HB denotes Brinell hardness number – range.
$R_{p0.2}$ denotes 0.2% proof stress, tonf/in^2, min, only applicable when specifically ordered.

BS 970 – replacing En steels

THE LONG-FAMILIAR 'En' series of steels is now rendered obsolete by the new system of specifying steels in BS 970, Parts 1 to V. This is based on digit series, separated by a letter. The first three digits denote the family of steel:

000–199 – carbon and carbon manganese steels: the actual digit value specifies the manganese content in 1/100ths percent (*eg* 090 – would designate an 0.9% manganese content).

200–240 – free-alloy steels: the second and third digits designating the minimum or mean sulphur content in 1/100ths percent (*eg* 040 would designate 0.4% sulphur)

250 – silicon-manganese spring steels

300–499 – heat-resisting stainless steels and value steels

500–999 – alloy steels: (these are classified in groups of ten, or multiples, according to alloy type).

The three digits are then followed by a capital letter:

A – denoting that the steel will be supplied to close limits of chemical composition.

H – denoting that the steel will be supplied to hardenability requirements.

M – denoting that the steel will be supplied to specified mechanical properties.

S – denoting a stainless steel (alloy).

The letter is then followed by a further two digits denoting the mean carbon content in /100ths percent (*eg* – 30 would designate 0.30% carbon).

The following Tables will be found useful for 'converting' obsolete 'En' numbers to the equivalent new specifications. It should be noted, however, that the new series are not necessarily exact equivalents of the old as there may be small differences incorporated in the new series composites. The 'remarks' apply mainly to usage, etc, associated with the original 'En' numbers.

TABLE 1 – 'CONVERSION' OF 'En' CARBON AND
CARBON-MANGANESE STEELS TO BS 970 SERIES

Old En	Remarks	New BS 970 1983
1A	Sulphurised free-machining mild steel	220M07
1B	Sulphurised free-machining mild steel	240M07
2A/1		040A04
2A		040A10
2B	20-ton mild steels	040A12
2C		040A22 or 050A20
2D		040A22 or 050A20
3	30-ton mild steel. Common mild steel for structural use	080A27
3A	30-ton mild steel	070M20
3C		070M20
4	40-ton mild steel	080A25
5	Harder grade of mild steel suitable for heat treatment in small sections	080M30
5A		080A27
5B		080A30
5C		080A32
8	High tensile carbon steel for heat treatment to 40/45-tons	080M40
8A	35-ton steel (normalised)	080A35
8B		080A37
8C		080A40
8D		080A42
8M	Sulphurised free-cutting steel	212M36 or 212M44
8BM		212A37
8DM		212A42
14A	40/50-ton steel, hardened and tempered	150M19
14B		150M28
15	'Economy' type molybdenum steel for high tensile bolts, etc.	150M36
15AM	Sulphurised free-cutting steel	216M36
15B		120M36
32A	Plain carbon case hardening steel for high surface hardness	145M10
32C		080M15
32M	32-ton steel	210M15
43A		080M50
43B		080A47
43C		080A52
43D		060A42
43E		080A67
44		060A96
44B		060A96
45	Oil-hardening spring steel	250A53
45A		250A58
201	Economy case-hardening steel for small sections	130M15
202		214M15

TABLE 2 – 'CONVERSION' OF 'En' ALLOY STEELS TO BS 970 SERIES

Old En	Remarks	BS 970 1983
11		526M60
12	Deep hardenable : suitable for small drop forgings	503M40
12B		503H37
12C		503A42
13		785M19
16	Cheaper alternative to nickel-chrome steel En23	605M36
16B		605A32
16C		605A37
16D		605M30
16M	Sulphurised free-cutting steel	606M36
17		608H37 or
		608M38
18	'Economy' type chromium steel	530M40
18A		530A30
18B		530A32
18C		530A36
18D		530A40
19	Improved impact quality compared with En18 with better impact strength	709M40
19A		708M40
19B		708A37
19C		708A42
23	Standard nickel-chrome steel for 50/60-ton tensile range. Used for larger heat-treated forgings, etc.	653M31
24	General purpose high strength steel with good hardening range	817M40
25	Higher alloy steel hardenable to 100 tons tensile	826M31
26	High strength alloy steel	826M40
27		830M31
30B	Krupp-type oil-hardening 100-ton steel	835M30
31	Tool steel for dies, etc.	534A99
34	Nickel case hardening steel with strong core	665A17
35		665M23
35A		665A22
35B		665A24
36A		665A12
36C		832M13
39A		659A15
39B	High-alloy case hardening steel for severe duties	835M15
40B	For nitride case hardening of wear components at high temperatures	722M24
40C		897M39
41A		905M31
41B		908M39
47		735A50
48		527A60
100		945M38
100C		945A40
110	'Economy' steel used in American 'triple alloys'	816M40
111	'Economy' steel used in American 'triple alloys'	640M40
111A		640A35
206		523A14
207		527A19
351		635H15
352		637H17
353	'Economy' case-hardening steels based on scrap residuals	815A16
354	available to the steelmaker. Must be carefully heat-treated	870A16
355	to develop their best properties. Use best limited to small sections	822A17
361		805A17
362		805M20
363		805M25

TABLE 3 – 'CONVERSION' OF 'En' STAINLESS
STEELS TO BS 970 SERIES

Old En	Remarks	BS 970
52		401S45
54	Heat resistant steel for valves, etc on i/c engines	331S40
54A		331S42
56A	Martensitic stainless steel, sulphurised free-cutting	410S21
56B	Rust-resistant steel for steam turbine blades at moderate superheat	420S29
56C	Martensitic stainless steel	416S37
56D	Martensitic stainless steel	420S45
56AM	Martensitic stainless steel, sulphurised free-cutting	416S21
56BM	Martensitic stainless steel, sulphurised free-cutting	416S29
57	Superior rust resistant properties to EN56B	431S29
58A		302S25
58B		321S12
58C		321S20
58E		304S15
58F	Austenitic stainless steels	347S17
58G		347S17
58H		315S16
58J		316S16
58M		303S21
59	High alloy exhaust valve steel	443S65
60		430S17

*Tapping an electric furnace at the Darnall Steel
Works of Sanderson Kayser Ltd.*

Alloy steels

ALLOY STEELS are distinguished from carbon steels by having a high proportion of alloying elements other than carbon. For the purposes of classification the element levels which establish a steel as an alloy steel are:

Silicon	2.0%	Copper	0.4%
Sulphur and phosphorus	0.2%	Lead	0.1%
Manganese and silica	2.0%	Molybdenum	0.1%
Aluminium	0.3%	Nickel	0.5%
Chromium	0.5%	Tungsten	0.3%
Cobalt	0.3%	Vanadium	0.1%

(Alloy steels are then sub-classified as low, medium and high alloy steels, mainly based on the increasing proportions of the alloying elements but also on the phase characteristics of the steel during heating and cooling.)

Alloy steels may also be classified according to the number of alloying elements present (other than carbon), *ie*:

ternary steels – containing one alloying element

quaternary steels – containing two alloying elements

complex steels – containing more than two elements

In general the desirable effect required from an alloying element is that it should enter into solid solution to strengthen the ferrite and/or be carbide-forming. Strengthening the ferrite will increase the strength of the steel. Carbide-forming elements reduce the carbon content of the steel matrix, improving its toughness and shock-resistance (*ie* inhibiting brittleness). In the case of pearlitic steels, however, carbide-forming elements tend to increase the proportion of pearlite without necessarily reducing the carbon content, but again strengthen the steel. The chief carbide-forming elements are chromium, molybdenum, niobium, tungsten and vanadium. Lesser carbide-forming elements are manganese, titanium and zirconium.

The general effect of individual alloying elements is summarised briefly, the elements being listed in alphabetical order and not in order of significance as a constituent of alloy steels:

Aluminium – acts as a deoxidiser to increase resistance to oxidation, and in particular resistance to scaling. It may, however, also have a tendency to reduce the strength of the steel unless used only in small proportions.

Chromium – in small proportions (*eg* 0.5 to 2%) improves wear resistance and resistance to oxidation and scaling. Tensile strength, yield strength and hardness are normally also improved, especially with increasing chromium content, although there may be some loss of ductility. With still higher proportions of chromium the steel becomes highly resistant to corrosion and with above 12% may be classified as a stainless steel.

Carbon – the inevitable alloying element in steel, largely determines the room temperature hardness of the steel. The acceptable carbon content is largely governed by the duty or application of the steel.

Cobalt – provides air hardening and resistance to scaling. The addition of 8 to 10% cobalt to tool steels improves their cutting qualities. For exceptional resistance to scaling at elevated temperatures, both cobalt and chromium are added to certain high alloy steels (see section on *High speed steels*).

Copper – has a limited application for improving the corrosion resistance and yield strength of low alloy steels. Unlike the other alloying elements it does not enter into solid solution but is fully dispersed through the ferrite in the form of a copper-rich phase.

Lead – in small proportions (up to 0.25%) improves the machinability of steels without detracting from other desirable mechanical properties. It is mainly used as a 'free cutting' constituent of plain carbon steels (mild steels), rather than in alloy steels; like copper it disperses rather than enters into solid solution (see section on *Free cutting steels*).

Manganese – reduces sulphur brittleness by combining with residual sulphur and can also be considered a hardening agent (up to 1%). In higher proportions (1 to 2%) manganese improves tensile strength and toughness. Above 5% manganese, the steel becomes non-magnetic.

Molybdenum – is a carbide forming element which increases strength, creep resistance and hardenability. It is also a useful addition to nickel-chromium steels to reduce their tendency to temper-brittleness.

Nickel – in proportions up to 5% improves tensile strength and toughness, with no adverse effect on ductility. It also improves hardenability. Higher proportions improve corrosion resistance and strength. It is also widely used in combination with chromium. With a nickel content above 27% the steel becomes non-magnetic.

Silicon – in general has little effect on the mechanical properties of a steel in proportions up to about 3%. Higher proportions of silicon tend to promote strength and hardenability, but reduce ductility.

Sulphur – is a sulphide-forming element, promoting good machinability (see section on *Free cutting steels*).

Tungsten – is a carbide-forming element promoting grain refinement with great hardness and toughness which is maintained at high temperatures – *ie* tungsten steels do not decompose nor form a soft skin when overheated. It is a principal constituent of high speed tool steels (see section on *High speed steels*).

Vanadium – is a carbide-forming element and also a deoxidiser. As a strength-ening agent it is used in conjunction with nickel or chromium (or both). As a de-oxidiser it is a useful constituent of casting steels, decreasing the size of the crystals on solidification and improving the strength and hardness of the casting as well as helping to eliminate blowholes and similar casting faults.

Tables 1–4 show mechanical properties for alloy steels to BS 970 in the hard-ened and tempered condition; Tables 5–8 give mechanical properties for struc-tural steels to BS 4360.

MEDIUM ALLOY STEELS
Into this group come the range of alloy steels with the proportion of alloying elements ranging roughly between 5 and 12%. They do not lend themselves to particular classification except by alloy constitution, type (*eg* as in BS 970) or tensile strength. In general figures the latter comprise:

'40-ton steels' – *eg* nickel steels with 0.6 to 1.0% nickel.

'40/50-ton steels' – *eg* based on 2.75 to 3.25% nickel, 0.9 to 1.2% chromium and 0.45 to 0.70 manganese.

'50-ton steels' – chrome-molybdenum steels.

'60/75-ton steels' – nickel-chrome-molybdenum steels.

Nickel steels
Low nickel steels are suitable for structural applications requiring medium ten-sile strength (although actual tensile strength may be below 40 tons/in^2), and for other stressed components – *eg* axles and shafting, etc. The 40/50-ton steels with higher nickel content and the addition of chromium are more easily hot worked and as such widely used for forgings of stressed components (*eg* crankshafts, connecting rods, etc). Fully tempered they also have good machining qualities and may not need further heat treatment. If they are heat treated, oil quenching is normally used, as water quenching can produce temper brittleness. The addi-tion of manganese reduces this latter possibility.

Nickel-molybdenum steels are suitable for case hardening and are widely used for camshafts, cams, gudgeon pins, roller bearing races, etc. The addition of molybdenum gives a good combination of hard wearing surface, toughness and low distortion in hardening.

Nickel-chromium-molybdenum steels
Medium alloys of this type possess high strength and good resistance to fatigue. They may be of medium carbon (0.25 to 0.35%) or high carbon (0.36 to 0.44%) type with the following additional elements:

Silicon	0.1 to 0.35%
Nickel	2.3 to 2.8%
Molybdenum	0.45 to 0.65%
Manganese	0.45 to 0.7%
Chromium	0.5 to 0.8%

TABLE 1 – ALLOY STEELS TO BS 970 – MECHANICAL PROPERTIES IN THE HARDENED AND TEMPERED CONDITION RELATED TO TENSILE STRENGTH RANGE

Heat Treatment Condition Symbol	Q					R					S					T					U				
Tensile Strength Range R_m tonf/in²	40–50					45–55					50–60					55–65					60–70				
Brinell Hardness Range, HB	179–229					201–255					223–277					248–302					269–331				
Steel	*LRS*	*Re*	*A*	*I*	$R_{p0.2}$	*LRS*	*Re*	*A*	*I*	$R_{p0.2}$	*LRS*	*Re*	*A*	*I*	$R_{p0.2}$	*LRS*	*Re*	*A*	*I*	$R_{p0.2}$	*LRS*	*Re*	*A*	*I*	$R_{p0.2}$
503M40	10	28	15	20	27	—	—	—	—	—	—	—	—	—	—	—	—	—	—	—	—	—	—	—	—
	6	30	17	25	29	2½	34	17	35	33	⅞	38	15	30	37	—	—	—	—	—	—	—	—	—	—
	4	30	17	35	29	—	—	—	—	—	—	—	—	—	—	—	—	—	—	—	—	—	—	—	—
526M60	—	—	—	—	—	—	—	—	—	—	—	—	—	—	—	4	40	11	—	—	—	—	—	—	—
530M40	—	—	—	—	—	4	34	17	40	33	2½	38	15	40	37	1⅛	44	13	40	43	—	—	—	—	—
605M30	—	—	—	—	—	6	34	17	40	33	4	38	15	40	37	2½	44	13	40	43	1⅛	49	12	35	48
605M36	—	—	—	—	—	10	32	15	25	31	4	38	15	40	37	2½	44	13	40	43	1⅛	49	12	35	48
	—	—	—	—	—	6	34	17	40	33	—	—	—	—	—	—	—	—	—	—	—	—	—	—	—
606M36	—	—	—	—	—	4	34	15	40	33	2½	38	13	35	37	1⅛	44	11	30	43	—	—	—	—	—
608M38	—	—	—	—	—	10	32	15	30	31	10	36	13	25	35	—	—	—	—	—	—	—	—	—	—
	—	—	—	—	—	6	34	17	40	33	6	38	15	40	37	4	44	13	40	43	2½	49	12	35	48
640M40	—	—	—	—	—	6	34	17	40	33	4	38	15	40	37	2½	44	13	40	43	1⅛	49	12	35	48
653M31	—	—	—	—	—	—	—	—	—	—	6	38	15	40	37	4	44	13	40	43	2½	49	12	35	48
708M40	—	—	—	—	—	6	34	17	40	33	4	38	15	40	37	2½	44	13	40	43	1⅛	49	12	35	48
709M40	—	—	—	—	—	10	32	15	25	31	10	36	13	20	35	—	—	—	—	—	—	—	—	—	—
	—	—	—	—	—	—	—	—	—	—	6	38	15	40	37	4	44	13	40	43	2½	49	12	35	48
722M24†	—	—	—	—	—	—	—	—	—	—	—	—	—	—	—	10	42	13	30	41	6	49	12	35	48
	—	—	—	—	—	—	—	—	—	—	—	—	—	—	—	6	44	13	40	43	—	—	—	—	—
785M19	10	29	16	30	28	—	—	—	—	—	—	—	—	—	—	—	—	—	—	—	—	—	—	—	—
	6	30	18	40	29	—	—	—	—	—	—	—	—	—	—	—	—	—	—	—	—	—	—	—	—

(continued)

TABLE 1 – ALLOY STEELS TO BS 970 – MECHANICAL PROPERTIES IN THE HARDENED AND TEMPERED CONDITION RELATED TO TENSILE STRENGTH RANGE – contd.

Heat Treatment Condition Symbol	V					W					X					Y					Z				
Tensile Strength Range R_m tonf/in²	65–75					70–80					75–85					80–90					100 (min)				
Brinell Hardness Range, HB	293–352					311–375					341–401					363–429					444 (min)				
Steel	LRS	Re	A	I	$R_{p0.2}$	LRS	Re	A	I	$R_{p0.2}$	LRS	Re	A	I	$R_{p0.2}$	LRS	Re	A	I	$R_{p0.2}$	LRS	Re	A	I	$R_{p0.2}$
503M40	—	—	—	—	—	—	—	—	—	—	—	—	—	—	—	—	—	—	—	—	—	—	—	—	—
	—	—	—	—	—	—	—	—	—	—	—	—	—	—	—	—	—	—	—	—	—	—	—	—	—
526M60	2½	48	8	—	—	—	—	—	—	—	—	—	—	—	—	—	—	—	—	—	—	—	—	—	—
530M40	—	—	—	—	—	—	—	—	—	—	—	—	—	—	—	—	—	—	—	—	—	—	—	—	—
605M30	¾	55	12	35	54	—	—	—	—	—	—	—	—	—	—	—	—	—	—	—	—	—	—	—	—
605M36	¾	55	12	35	54	—	—	—	—	—	—	—	—	—	—	—	—	—	—	—	—	—	—	—	—
	—	—	—	—	—	—	—	—	—	—	—	—	—	—	—	—	—	—	—	—	—	—	—	—	—
606M36	—	—	—	—	—	—	—	—	—	—	—	—	—	—	—	—	—	—	—	—	—	—	—	—	—
608M38	1⅛	55	12	35	54	—	—	—	—	—	—	—	—	—	—	—	—	—	—	—	—	—	—	—	—
	—	—	—	—	—	—	—	—	—	—	—	—	—	—	—	—	—	—	—	—	—	—	—	—	—
640M40	—	—	—	—	—	—	—	—	—	—	—	—	—	—	—	—	—	—	—	—	—	—	—	—	—
653M31	—	—	—	—	—	—	—	—	—	—	—	—	—	—	—	—	—	—	—	—	—	—	—	—	—
708M40	—	—	—	—	—	—	—	—	—	—	—	—	—	—	—	—	—	—	—	—	—	—	—	—	—
709M40	1⅛	55	12	35	54	—	—	—	—	—	—	—	—	—	—	—	—	—	—	—	—	—	—	—	—
	—	—	—	—	—	—	—	—	—	—	—	—	—	—	—	—	—	—	—	—	—	—	—	—	—
722M24†	—	—	—	—	—	—	—	—	—	—	—	—	—	—	—	—	—	—	—	—	—	—	—	—	—
	—	—	—	—	—	—	—	—	—	—	—	—	—	—	—	—	—	—	—	—	—	—	—	—	—
785M19	—	—	—	—	—	—	—	—	—	—	—	—	—	—	—	—	—	—	—	—	—	—	—	—	—

LRS denotes limiting ruling section in inches.
Re denotes yield stress in ton/in² min.
A denotes elongation – % min on 5.65 $\sqrt{S_o}$ gauge length.

I denotes Izod impact value – ft lb min.
$R_{p0.2}$ denotes 0.2% proof stress – ton/in² min – only applicable when specifically ordered.
† Suitable for nitriding.

TABLE 2 – ALLOY STEELS TO BS 970 – MECHANICAL PROPERTIES IN THE HARDENED AND TEMPERED CONDITION RELATED TO TENSILE STRENGTH RANGE

Steel	Q					R					S					T					U				
(Tensile Strength Range R_m tonf/in²)	40–50					45–55					50–60					55–65					60–70				
(Brinell Hardness Range, HB)	179–229					201–255					223–277					248–302					269–331				
	LRS	Re	A	I	$R_{p0.2}$	LRS	Re	A	I	$R_{p0.2}$	LRS	Re	A	I	$R_{p0.2}$	LRS	Re	A	I	$R_{p0.2}$	LRS	Re	A	I	$R_{p0.2}$
816M40	—	—	—	—	—	—	—	—	—	—	10	36	15	25	35	—	—	13	40	43	—	—	—	—	—
817M40	—	—	—	—	—	—	—	—	—	—	6	38	15	40	37	4	44	13	30	41	2½	49	12	35	48
823M30	—	—	—	—	—	—	—	—	—	—	—	—	—	—	—	10	42	13	40	43	4	49	12	35	48
826M31	—	—	—	—	—	—	—	—	—	—	—	—	—	—	—	6	44	13	30	41	10	48	12	25	47
826M40	—	—	—	—	—	—	—	—	—	—	—	—	—	—	—	10	42	13	40	43	6	49	12	35	48
830M31	—	—	—	—	—	—	—	—	—	—	—	—	—	—	—	6	44	13	30	41	10	48	12	25	47
835M30	—	—	—	—	—	—	—	—	—	—	—	—	—	—	—	10	42	13	40	43	6	49	12	35	48
897M39*	—	—	—	—	—	4	34	17	40	33	—	—	—	—	—	6	44	13	30	41	6	48	12	25	47
905M31†	—	—	—	—	—	6	34	17	40	33	2½	38	15	40	37	—	—	—	—	—	—	—	—	—	—
905M39†	—	—	—	—	—	10	32	15	25	31	4	38	15	40	37	2½	42	13	35	43	—	—	—	—	—
945M38	—	—	—	—	—	6	34	17	40	33	4	38	15	40	37	2½	44	13	40	43	1⅛	49	12	35	48

(continued)

TABLE 2 – ALLOY STEELS TO BS 970 – MECHANICAL PROPERTIES IN THE HARDENED AND TEMPERED CONDITION RELATED TO TENSILE STRENGTH RANGE – contd.

Heat Treatment Condition Symbol	V (R_m 65–75, HB 293–352)					W (R_m 70–80, HB 311–375)					X (R_m 75–85, HB 341–401)					Y (R_m 80–90, HB 363–429)					Z (R_m 100 min, HB 444 min)				
Steel	LRS	R_e	A	I	$R_{p0.2}$	LRS	R_e	A	I	$R_{p0.2}$	LRS	R_e	A	I	$R_{p0.2}$	LRS	R_e	A	I	$R_{p0.2}$	LRS	R_e	A	I	$R_{p0.2}$
816M40	1⅛	55	12	35	54	—	—	—	—	—	—	—	—	—	—	—	—	—	—	—	—	—	—	—	—
817M40	2½	55	12	35	54	1⅛	61	11	30	60	1⅛	66	10	25	65	—	—	—	—	—	—	—	—	—	—
823M30	6	55	12	35	54	4	61	11	30	60	2½	66	10	25	65	—	—	—	—	—	1⅛	80	5	8	73
826M31	6	55	12	35	54	4	61	11	30	60	2½	66	10	25	65	—	—	—	—	—	2½	80	7	10	73
826M40	10	54	12	25	53	10	60	11	20	59	—	—	—	—	—	—	—	—	—	—	2½	80	7	10	13
830M31	4	55	12	35	54	6	61	11	30	60	6	66	10	25	65	6	71	10	25	70	4	80	7	10	73
835M30	—	—	—	—	—	2½	61	11	30	60	—	—	—	—	—	—	—	—	—	—	4	80	7	10	73
897M39*	—	—	—	—	—	—	—	—	—	—	—	—	—	—	—	—	—	—	—	—	—	—	—	—	—
905M31†	—	—	—	—	—	—	—	—	—	—	—	—	—	—	—	—	—	—	—	—	6	80	7	15	73
905M39†	—	—	—	—	—	—	—	—	—	—	—	—	—	—	—	—	—	—	—	—	1⅛	80	7	10	78
945M38	1⅛	55	12	35	54	—	—	—	—	—	—	—	—	—	—	—	—	—	—	—	—	—	—	—	—

LRS denotes limiting ruling section in inches.
R_e denotes yield stress in ton/in² min.
A denotes elongation – % min on $5.65\sqrt{S_o}$ gauge length.
I denotes Izod impact value – ft lb min.

$R_{p0.2}$ denotes 0.2% proof stress – ton/in² min – only applicable when specifically ordered.

* Suitable for nitriding in the 85 ton/in² condition: LRS-2½, R_e-75, A-8, I-15, HB-375, min, $R_{p0.2}$-72.

† Nitriding steels.

TABLE 3 – ALLOY STEELS TO BS 970 – MECHANICAL PROPERTIES IN THE HARDENED AND TEMPERED CONDITION RELATED TO RULING SECTION

Limiting Ruling Section Steel	Up to and including ¾ in							Up to and including ⅞ in							Up to and including 1⅛ in							Up to and including 2½ in						
	C	R_m	R_e	A	I	HB	$R_{p0.2}$	C	R_m	R_e	A	I	HB	$R_{p0.2}$	C	R_m	R_e	A	I	HB	$R_{p0.2}$	C	R_m	R_e	A	I	HB	$R_{p0.2}$
503M40	—	—	—	—	—	—	—	S	50–60	38	15	30	223–277	37	—	—	—	—	—	—	—	R	45–55	34	17	35	201–255	33
526M60	—	—	—	—	—	—	—	—	—	—	—	—	—	—	—	—	—	—	—	—	—	V	65–75	48	8	—	293–352	—
530M40	—	—	—	—	—	—	—	—	—	—	—	—	—	—	T	55–65	44	13	40	248–302	43	S	50–60	38	15	40	223–277	37
605M30	V	65–75	55	12	35	293–352	54	—	—	—	—	—	—	—	U	60–70	49	12	35	269–331	48	T	55–65	44	13	40	248–302	43
605M36	V	65–75	55	12	35	293–352	54	—	—	—	—	—	—	—	U	60–70	49	12	35	269–331	48	T	55–65	44	13	40	248–302	43
606M36	—	—	—	—	—	—	—	—	—	—	—	—	—	—	T	55–65	44	11	30	248–302	43	S	50–60	38	13	35	223–277	37
608M38	—	—	—	—	—	—	—	—	—	—	—	—	—	—	V	65–75	55	12	35	293–352	54	U	60–70	49	12	35	269–331	48
640M40	—	—	—	—	—	—	—	—	—	—	—	—	—	—	U	60–70	49	12	35	269–331	48	T	55–65	44	13	40	248–302	43
653M31	—	—	—	—	—	—	—	—	—	—	—	—	—	—	—	—	—	—	—	—	—	U	60–70	49	12	35	269–331	48
708M40	—	—	—	—	—	—	—	—	—	—	—	—	—	—	U	60–70	49	12	35	269–331	48	T	55–65	44	13	40	248–302	43
709M40	—	—	—	—	—	—	—	—	—	—	—	—	—	—	V	65–75	55	12	35	293–352	54	U	60–70	49	12	35	269–331	48
772M24†	—	—	—	—	—	—	—	—	—	—	—	—	—	—	V	65–75	55	12	35	293–352	54	—	—	—	—	—	—	—
785M19	—	—	—	—	—	—	—	—	—	—	—	—	—	—	Z	100 min	80	5	8	444 min	73	—	—	—	—	—	—	—
816M40	—	—	—	—	—	—	—	—	—	—	—	—	—	—	X	75–85	66	10	25	341–401	65	U	60–70	49	12	35	269–331	48
817M40	—	—	—	—	—	—	—	—	—	—	—	—	—	—	W	70–80	61	11	30	311–375	60	V	65–75	55	12	35	293–352	54

(continued)

TABLE 3 – ALLOY STEELS TO BS 970 – MECHANICAL PROPERTIES IN THE HARDENED AND TEMPERED CONDITION RELATED TO THE RULING SECTION – contd.

Limiting Ruling Section	Up to and including 4 in							Up to and including 6 in							Up to and including 10 in						
Steel	C	R_m	R_e	A	I	HB	$R_{p0.2}$	C	R_m	R_e	A	I	HB	$R_{p0.2}$	C	R_m	R_e	A	I	HB	$R_{p0.2}$
503M40	Q	40–50	30	17	35	179–229	29	Q	40–50	30	17	25	179–229	29	Q	40–50	28	15	20	179–229	27
526M60	T	55–65	40	11	—	248–302	33	—	—	—	—	—	—	—	—	—	—	—	—	—	—
530M40	R	45–55	34	17	40	201–255	33	R	—	34	—	—	201–255	33	R	—	32	—	—	201–255	31
605M30	S	50–60	38	15	40	223–277	37	R	45–55	34	17	40	201–255	33	R	45–55	32	13	25	201–255	31
605M36	S	50–60	38	15	40	223–277	37	R	45–55	34	17	40	201–255	33	—	—	—	—	—	—	—
606M36	R	45–55	34	15	40	201–255	33	S	50–60	38	15	40	223–277	37	S	50–60	36	13	25	223–277	35
608M38	T	55–65	44	13	40	248–302	43	S	50–60	38	15	40	223–277	37	S	50–60	36	13	30	223–277	35
								R	45–55	34	17	40	201–255	33	R	45–55	32	15	—	201–255	31
640M40	S	50–60	38	15	40	223–277	37	S	50–60	38	15	40	223–277	37	S	50–60	36	13	20	223–277	35
653M31	T	55–65	44	13	40	248–302	43	R	45–55	34	17	40	201–255	33	R	45–55	32	15	25	201–255	31
708M40	S	50–60	38	15	40	223–277	37	S	50–60	38	15	40	223–277	37	S	50–60	36	13	20	223–277	35
709M40	T	55–65	44	13	40	248–302	43	S	50–60	38	15	40	223–277	37	S	50–60	36	13	20	223–277	35
								R	45–55	34	17	40	201–255	33	R	45–55	32	15	25	201–255	31
722M24†	—	—	—	—	—	—	—	U	60–70	49	12	35	269–331	48	T	55–65	42	13	30	248–302	41
785M19	T	55–65	44	13	40	248–302	43	T	55–65	44	13	40	248–302	43	T	55–65	42	13	30	248–302	41
816M40	—	—	—	—	—	—	—	Q	40–50	30	18	40	179–229	29	Q	40–50	29	16	30	179–229	28
817M40	U	60–70	49	12	35	269–331	48	S	50–60	38	15	40	223–277	37	S	50–60	36	15	25	223–277	35

C   denotes heat treatment condition symbol.
R_m denotes tensile strength – ton/in² range.
R_e denotes yield stress – ton/in², min.
A   denotes elongation – % min, on $5.65\sqrt{S_o}$ gauge length.
I   denotes Izod impact value – ft lb min.
HB   denotes Brinell hardness number – range.
$R_{p0.2}$ denotes 0.2% proof – ton/in², min – only applicable when specifically ordered.
†   Suitable for nitriding.

TABLE 4 – ALLOY STEELS TO BS 970 – MECHANICAL PROPERTIES IN THE HARDENED AND TEMPERED CONDITION RELATED TO RULING SECTION

Limiting Ruling Section / Steel	Up to and including ¾ in							Up to and including ⅞ in							Up to and including 1⅛ in							Up to and including 2½ in						
	C	R_m	R_e	A	I	HB	$R_{p0.2}$	C	R_m	R_e	A	I	HB	$R_{p0.2}$	C	R_m	R_e	A	I	HB	$R_{p0.2}$	C	R_m	R_e	A	I	HB	$R_{p0.2}$
823M30	—	—	—	—	—	—	—	—	—	—	—	—	—	—	—	—	—	—	—	—	—	Z	100 min	80	7	10	444 min	73
	—	—	—	—	—	—	—	—	—	—	—	—	—	—	—	—	—	—	—	—	—	X	75-85	66	10	25	341-401	65
826M31	—	—	—	—	—	—	—	—	—	—	—	—	—	—	—	—	—	—	—	—	—	Z	100 min	80	7	10	444 min	73
826M40	—	—	—	—	—	—	—	—	—	—	—	—	—	—	—	—	—	—	—	—	—	X	75-85	66	10	25	341-401	65
830M31	—	—	—	—	—	—	—	—	—	—	—	—	—	—	—	—	—	—	—	—	—	—	—	—	—	—	—	—
835M30	—	—	—	—	—	—	—	—	—	—	—	—	—	—	—	—	—	—	—	—	—	W	70-80	61	11	30	311-375	60
897M39*	—	—	—	—	—	—	—	—	—	—	—	—	—	—	—	—	—	—	—	—	—	—	85 min	75	8	15	375 min	72
905M31†	—	—	—	—	—	—	—	—	—	—	—	—	—	—	—	—	—	—	—	—	—	S	50-60	38	15	40	223-277	37
905M39†	—	—	—	—	—	—	—	—	—	—	—	—	—	—	V	65-75	55	12	35	293-352	54	T	55-65	44	13	35	248-302	43
945M38	—	—	—	—	—	—	—	—	—	—	—	—	—	—	U	60-70	49	12	35	269-331	48	T	55-65	44	13	40	248-302	43

(continued)

TABLE 4 – ALLOY STEELS TO BS 970 – MECHANICAL PROPERTIES IN THE HARDENED AND TEMPERED CONDITION RELATED TO RULING SECTION – contd.

Limiting Ruling Section / Steel	Up to and including 4 in							Up to and including 6 in							Up to and including 10 in						
	C	R_m	R_e	A	I	HB	$R_{p0.2}$	C	R_m	R_e	A	I	HB	$R_{p0.2}$	C	R_m	R_e	A	I	HB	$R_{p0.2}$
823M30	W	70-80	61	11	30	311-375	60	V	65-75	55	12	35	293-352	54	U	60-70	48	12	25	269-331	47
								U	60-70	49	12	35	269-331	48	T	55-65	42	13	30	248-302	41
								T	55-65	44	13	40	248-302	43							
826M31	W	70-80	61	11	30	311-375	60	V	65-75	55	12	35	293-352	54	U	60-70	48	12	25	269-331	47
								U	60-70	49	12	35	269-331	48	T	55-65	42	13	30	248-302	41
								T	55-65	44	13	40	248-302	43							
826M40	Z	100 min	80	7	10	444 min	73	Y	80-90	71	10	25	363-429	70							
								X	75-85	66	10	25	341-401	65	W	70-80	60	11	20	311-375	59
								W	70-80	61	11	30	311-375	60	V	65-75	54	12	25	293-352	53
								V	65-75	55	12	35	293-352	54	U	60-70	48	12	25	269-331	47
								U	60-70	49	12	35	269-331	48							
830M31	V	65-75	55	12	35	293-352	54	U	60-70	49	12	35	269-331	48	T	55-65	42	13	30	248-301	41
								T	55-65	44	13	40	248-302	43							
835M30								Z	100 min	80	7	15	444 min	73							
897M39*																					
905M31†	R	45-55	34	17	40	201-255	33								R	45-55	32	15	25	201-255	31
905M39	S	50-60	38	15	40	223-277	37	R	45-55	34	17	40	201-255	33							
945M38	S	50-60	38	15	40	223-277	37	R	45-55	34	17	40	201-255	33							

C — denotes heat treatment condition symbol.

R_m — denotes tensile strength – ton/in^2 range.

R_e — denotes yield stress – ton/in^2, min.

A — denotes elongation – % min, on $5.65\sqrt{S_o}$ gauge length.

I — denotes Izod impact value – ft lb min.

HB — denotes Brinell hardness number – range.

$R_{p0.2}$ — denotes 0.2% proof stress – ton/in^2, min – only applicable when specifically ordered.

* — Suitable for nitriding in the 85 ton/in^2 condition.

† — Nitriding steels.

TABLE 5 – MECHANICAL PROPERTIES OF STRUCTURAL STEELS FOR PLATES – BS 4360

Grade	Tensile Strength	Yield Stress Min				Elongation Min on Gauge Length of		Maximum Bend Radius	Charpy V-notch Impact Test		
		Up to and including 16 mm	Over 16 mm up to and including 40 mm	Over 40 mm up to and including 63 mm	Over 63 mm up to and including 100 mm[1]	200 mm	5.65$\sqrt{S_o}$		Temp	Energy Minimum Average	Thickness Minimum
	N/mm²	N/mm²	N/mm²	N/mm²	N/mm²	%	%		°C	J	mm
40A[2]	400/480	—	—	—	—	22[3]	25	1¼T[4]	—	—	—
40B	400/480	230	225	220	210	22[3]	25	1¼T[4]	R.T[5]	27	50
40C	400/480	230	225[6]	220	210	22[3]	25		0	27	50
40D	400/480	260	245	240	225	22[3]	25	1¼T	−10 / −20	41 / 27	75
40E	400/480	260	245	240	225	22[3]	25	1¼T	−20 / −30 / −50	61 / 47 / 27	75
43A1[2]	430/510	—	—	—	—	20[3]	22	1½T	—	—	—
43A	430/510	245	240[7]	230	220	20[3]	22	1½T			
43B	430/510	245	240[7]	230	220	20[3]	22	1½T	R.T.[5]	27	50
43C	430/510	245	240[7]	235	225	20[3]	22	1½T	0	27	50
43D	430/510	280	270	255	240	20[3]	22	1½T	−10 / −20	41 / 27	75
43E	430/510	280	270	255	240	20[3]	22	1½T	−20 / −30 / −50	61 / 47 / 27	75
50A	490/620[8]	—	—	—	—	18[3]	20	1½T	—	—	—
50B	490/620[8]	355	345	340	325	18[3]	20	1T[9]			
50C	490/620[8]	355	345	340	325	18[3]	20	1T[9]	−5 / −15	41 / 27	75
50D	490/620	355	345	340	By agreement	18[3]	20	1½T	−20 / −30	41 / 27	75
50D1	490/620	355	345	—	—	18[3]	20	1½T	−20 / −30	41 / 27	75
		Up to and including 16 mm	Over 16 mm up to and including 25 mm	Over 25 mm up to and including 40 mm	Over 40 mm up to and including 63 mm						
55C	550/700	450	430	415	—	17[3]	19	2T	0	27	19[10]
55E	550/700	450	430	415	400	17[3]	19	1½T	−20 / −30 / −50	61 / 47 / 27	63

1. Minimum yield stress values for material over 100 mm thick to be agreed between the manufacturer and the purchaser. 2. For material under 6 mm thick, only bend tests required. 3. Up to and including 9 mm thick, 16% for Grades 40 and 43 and 15% for Grades 50 and 55. 4. 1T for plates required for cold flanging. 5. Only if specified on the order. 6. Minimum yield stress 230 N/mm² for material up to and including 19 mm thick. 7. Minimum yield stress 245 N/mm² for material up to and including 19 mm thick. 8. Minimum tensile strength 480 N/mm² for material over 63 mm thick. 9. 1½T for material over 25 mm thick. 10. Also applicable to material over 19 mm up to and including 40 mm by agreement between the manufacturer and the purchaser.

TABLE 6 – MECHANICAL PROPERTIES FOR ROUND AND SQUARE BARS OF WELDABLE STRUCTURAL STEELS – BS 4360

Grade	Tensile Strength	Yield Stress Min					Elongation Min on Gauge Length of $5.65\sqrt{S_o}$	Maximum Bend Radius	Charpy V-notch Impact Test		
		Up to and including 25 mm	Over 25 mm up to and including 50 mm	Over 50 mm up to and including 63 mm	Over 63 mm up to and including 100 mm	Over 100 mm up to and including 160 mm①			Temp	Energy Minimum Average	Thickness maximum
	N/mm²	N/mm²	N/mm²	N/mm²	N/mm²	N/mm²	%		°C	J	mm
40A②	400/480	—	—	—	—	—	25	1¼T	—	—	—
40B	400/480	240	230	225	220	210	25	1¼T	R.T③	27	50
40C	400/480	240	230	225	220	210	25		0	27	83
40D	400/480	240	230	225	220	210	25	1¼T	−15	27④	83
40E	400/480	255	245	240	230	225	25	1¼T	{−20 / −30}	{34 / 27}④	83
43A1②	430/510	—	—	—	—	—	22	1½T	—	—	—
43A	430/510	255	245	240	230	—	22	1½T	—	—	—
43B	430/510	255	245	240	230	225	22	1½T	R.T③	27	50
43C	430/510	255	245	240	230	225	22	1½T	0	27	83
43D	430/510	255	245	240	230	225	22	1½T	−15	27④	83
43E	430/510	270	260	255	245	240	22	1½T	{−20 / −30}	{34 / 27}④	83
50A	490/620	—	—	—	—	—	20	1½T	—	—	—
50B	490/620⑤	355	355	345	330	325	20	1T⑥	—	—	—
50C	490/620⑤	355	355	345	330	325	20	1T⑥	0	27④	83
50D	490/620	355	355	345	330	—①	20	1½T	−10	27④	83
		Up to and including 16 mm	Over 16 mm up to and including 25 mm	Over 25 mm up to and including 40 mm	Over 40 mm up to and including 63 mm						
55C	550/700	450	430	415	—		19	2T	0	27⑦	32
55E	550/700	450	430	415	400		19	1½T	{−20 / −30 / −50}	{47 / 41 / 27}	63

1. Minimum yield stress values for round and square bars over 160 mm may be agreed between the manufacturer and the purchaser, as may minimum yield stress values for grade 50D over 100 mm. 2. For round and square bars up to and including 9 mm, only bend tests required. 3. Only if specified by the purchaser. 4. Where normalised round and square bars can be supplied, impact values equivalent to those specified for plates of the same grade and thickness (see Table 5) can be provided. 5. Minimum tensile strength 480 N/mm² for material over 83 mm. 6. 1½T for material over 40 mm thick. 7. Values for round and square bars over 32 mm up to and including 63 mm may be agreed between the manufacturer and the purchaser.

TABLE 7 – MECHANICAL PROPERTIES FOR SECTIONS AND FLAT BARS IN WELDABLE STRUCTURAL STEELS – BS 4360

Grade	Tensile Strength	Yield Stress[1] Min					Elongation min on Gauge Length of		Maximum Bend Radius	Charpy V-notch Impact Test[2]		
		Up to and including 16 mm	Over 16 mm up to and including 25 mm	Over 25 mm up to and including 40 mm	Over 40 mm up to and including 63 mm	Over 63 mm up to and including 100 mm[3]	200 mm L_o	5.65 $\sqrt{S_o}$		Temp	Energy Minimum Average	Thickness Maximum
	N/mm²	N/mm²	N/mm²	N/mm²	N/mm²	N/mm²	%	%		°C	J	mm
40A[4]	400/480	—	—	—	—	—	22[5]	25	1½T	—	—	—
40B	400/480	240	230	225	220	210	22[5]	25	1¼T	R.T.[6]	27	50
40C	400/480	240	230	225	220	210	22[5]	25	1¼T	0	27	50
40D	400/480	240	230	225	220	210	22[5]	25	1¼T	-15	27	25
40E	400/480	255	245	240	230	225	22[5]	25	1¼T	-20 / -30	34 / 27	50
43A1[4]	430/510[7]	255	245	240	230	225	20[5]	22	1½T	—	—	—
43A	430/510[7]	255	245	240[8]	230	225	20[5]	22	1½T	R.T.[6]	27	50
43B	430/510[7]	255	245	240[8]	230	225	20[5]	22	1½T	0	27	50
43C	430/510	255	245	240	230	225	20[5]	22	1½T	-15	27[4]	50
43D	430/510	255	245	240	230	225	20[5]	22	1½T	-20	34	50
43E	430/510	270	260	255	245	240	20[5]	22	1½T	-30	27	50
50A	490/620	355	355	345	340	325	18[5]	20	1½T	—	—	—
50B	490/620[9]	355	355	345	340	325	18[5]	20	1T[10]	—	—	—
50C	490/620[9]	355	355	345	340		18[5]	20	1T[10]	0	27	40
50D	490/620	355	355	345	340	By agreement	18[5]	20	1½T	10	27	40
55C	550/700	450	430	415	—		17[5]	19	2T	0 / 20	27 / 47	19[11]
55E[12]	550/700	450	430	415	400		17[5]	19	1½T	30 / 50	41 / 27	63

1. For universal beams, columns and bearing piles the yield stress obtained on test pieces taken from the web shall be not less than 15 N/mm² greater than the specified minimum value. 2. Where normalised sections and flat bars can be supplied, impact test values equivalent to those specified for plates of the same grade and thickness (see Table 5) can be provided. 3. Minimum yield stress values for flat bars over 100 mm thick to be agreed between the manufacturer and the purchaser. 4. For sections and flat bars under 6 mm thick, only bend tests required. 5. Up to and including 9 mm thick, 18% for Grade 40, 16% for Grade 43, and 15% for Grades 50 and 55. 6. Only if specified by the purchaser. 7. For universal beams, columns and bearing piles the maximum tensile strength may be increased to 540 N/mm². 8. Minimum yield stress 245 N/mm² for universal beams, columns and bearing piles to BS 4 Part 1 with flange thicknesses not exceeding 40 mm. 9. Minimum tensile strength 480 N/mm² for material over 50 mm thick. 10. 1½T for material over 25 mm thick. 11. Values for sections and flat bars over 19 mm up to and including 40 mm may be agreed between the manufacturer and the purchaser. 12. Facilities for normalising sections are limited.

TABLE 8 – MECHANICAL PROPERTIES FOR HOLLOW SECTIONS OF WELDABLE STRUCTURAL STEELS – BS 4360

Grade	Tensile strength	Yield Stress Min up to and including 16 mm thickness[1]	Elongation min. on gauge length of 5.65 $\sqrt{S_o}$	Flattening test	Charpy V-Notch Impact value up to and including 16 mm thickness[1]	
					Temp	Energy minimum Average
	N/mm^2	N/mm^2	%		°C	J
43C	430/540	255	22		0[2]	27
43D	430/540	255	22		−15	27
43E	430/540	270	22		$\begin{cases} -20 \\ -30 \end{cases}$	$\left.\begin{matrix} 34 \\ 27 \end{matrix}\right\}$
					$\begin{cases} -20 \\ -30 \\ -50 \end{cases}$	$\left.\begin{matrix} 61 \\ 47 \\ 27 \end{matrix}\right\}$
50B	490/620	355	20		—	—
50C	490/620	355	20		0	27
50D	490/620	355	20		−10	27
					$\begin{cases} -20 \\ -30 \end{cases}$	$\left.\begin{matrix} 41 \\ 27 \end{matrix}\right\}$
55C	550/700	450	19		0	27
55E	550/700	450	19		$\begin{cases} -20 \\ -30 \\ -50 \end{cases}$	$\left.\begin{matrix} 61 \\ 47 \\ 27 \end{matrix}\right\}$

1. Values for thicknesses over 16 mm are subject to agreement between the manufacturer and the purchaser.
2. Only if specificed by the purchaser.

HIGH ALLOY STEELS
High alloy steels may be described as those containing 12% or more alloying elements. They can be grouped by type (alloy constituents) or by primary performance characteristic, *eg*:

Wear resistant steels.   Corrosion resistant steels.
Heat resistant steels.   High strength steels.

Corrosion resistant steels
These are basically high chrome steels with a normal chromium content of 13% for moderate corrosion resistance properties with the addition of 8% nickel and 17 to 18% chromium for high corrosion resistance properties (see section on *Stainless steels*).

Wear resistant steels
These are austenitic manganese steels with a minimum manganese content of 11.0%. They are extremely hard (up to 500 Brinell) and have high impact strength.

HEAT RESISTANT STEELS
Broadly speaking these fall into three groups:

(i)   Steels resistant to oxidation, scaling and thermal shock.
(ii)   Steels having high resistance to creep at elevated temperatures.
(iii) Steels specifically developed to provide high mechanical properties at elevated temperatures (more properly called high temperature steels).

To some extent (ii) and (iii) represent similar requirements, *ie* (ii) is necessary to maintain (iii). Some types of heat resistant steels may combine all three qualities. Equally, a plain carbon steel may have sufficiently good heat resistance to be classified as a high temperature steel for certain applications (*eg* steam power plant).

As a general rule unalloyed steels are suitable for service temperatures up to 450°C, low alloy steels up to 500°C and medium alloy steels (12% chromium) up to about 700 to 750°C. Austenitic stainless steels are suitable for temperatures up to about 900°C, but special high alloy steels are required for working above this temperature level.

Primary alloying elements in heat resistant steels are nickel (up to 65%) and chromium (12 to 27%), with 1.0% molybdenum (maximum). A high nickel content combines shock resistance and resistance to creep, as well as corrosion resistance (except in a reducing atmosphere or in the presence of sulphurous gases). Nickel, in fact, is the principal heat resisting element; in particular it resists thermal cycling. A high chromium content maximises resistance to oxidation, corrosion and abrasion, but 25 to 30% chromium steels can be comparatively brittle. In general, however, chromium is the principal strength and toughness inducing element.

The inclusion of molybdenum is useful in austenitic steel castings to produce higher creep strength and improve their resistance to acids. For the highest creep strength steels are solution treated and aged.

FREE CUTTING STEELS
The description free cutting steels is given to those grades of steel containing the best proportions of lead and/or sulphur to provide a free machining quality. Because sulphur is the main free cutting constituent such steels are also known as 'sulphurised' or 'sulphurised free cutting' steels.

Sulphur in steel normally consists of inclusions of sulphides (preferably manganese sulphide) which have a low shear stress, reducing or eliminating build-up on the edge of the cutting tool, as well as acting as a lubricant. Lead has the effect of lowering the shear energy required from the tool, assisting chip formation. It also appears to assist in feeding the sulphide (lubricant) to the tool face. The sulphur proportion used may range up to 0.6% maximum. Lead proportion usually ranges from 0.15 to 0.25%, but may be up to 0.35%.

Other elements which may be incorporated in free cutting steels are bismuth, nitrogen, selenium and tellurium. Nitrogen has the effect of promoting chip making by stiffening the ferrite matrix. Selenium is often used to improve the machinability of stainless steels.

HIGH SPEED STEELS

Tool steels comprise *tungsten steels* and steels containing both tungsten and chromium. The latter are known as *high speed steels.* Tungsten steels with a tungsten content in excess of 14% may also be called high speed steels.

With small proportions of tungsten, the tungsten is present as tungsten carbide. With higher proportions the exceptionally hard tungside of iron is formed. This remains hard at elevated temperatures – *eg* with a tungsten tool tip red hot and the chip heated to blue heat. High speed steels containing both tungsten and chromium must be heated to near melting point then cooled at a fairly fast rate in an air blast or oil, and tempered at a fairly high temperature to realise optimum properties.

High speed tool steels usually contain from 14 to 25% tungsten and 3 to 5% chromium together with minor proportions of other elements such as cobalt, manganese, molybdenum, silicon, titanium and vanadium.

Higher proportions of cobalt produce alloys known as *cobalt steels* which can be worked at even higher speeds than tungsten steels although they are weaker and more brittle. These are also known as *super high speed steels,* a typical composition being 20% tungsten, 12% cobalt, 4.5% chromium with manganese, molybdenum, silicon and vanadium as minor consistents.

Stellite is another alloy used for high speed tools, comprising around 60% cobalt, 25% tungsten and 15% chromium, cutting best at red heat.

An alternative form of high speed tool is the *cemented carbide* type where a non-metallic tip or insert of carbides of tungsten, tantalum and titanium is mounted on an ordinary tool steel shank. Such tips are extremely hard and can be operated at higher speeds than metallic tools, although they can suffer from brittleness. A selection of high speed and tool steels are described in Table 9.

MARAGING STEELS

Maraging steels are a comparatively recent development and can best be described as nickel-iron alloys with an exceptionally low carbon content (*circa* 0.01%). They have exceptionally high tensile and yield strength (*eg* up to 200 tons/in^2) with outstanding resistance to fracture.

Typically maraging steels are based on an 18% nickel content with 8 to 9% cobalt, 3 to 5% molybdenum and 0.2 to 0.75% titanium. The last two largely govern the strength level of the steel. Carbon, silicon, sulphur, manganese and phosphorus are held to very low levels.

In an annealed state such an alloy is relatively soft and ductile, with a martensitic structure. The presence of molybdenum and titanium renders the material age hardenable to very high strengths. Full strength can be obtained by simple heat treatment at moderate temperatures, mechanical properties then being retained under service temperatures up to about 410°C. This hardening (maraging)

TABLE 9 – HIGH SPEED AND TOOL STEELS*

Type	Equivalent specifications			Maximum Annealed Brinell Hardness	Forging temp °C	Heat	
	BS	AISI	Werkstoff			Annealing °C	Hardening °C
Molybdemun High Speed Steels							
9/2/1	BM1	M1	3346	248	1 100/900	850	1 210/1 220
6/5/2	BM2	M2	3343	248	1 100/900	850	1 210/1 220
6/5/2 + S	BM2S	M2S		248	1 100/900	850	1 210/1 220
9/5/Mo 8/Co	BM42	M42		285	1 120/980	850	1 180/1 190
Tungsten High Speed Steels							
18/4/1	BT1	T1	3355	255	1 100/900	850	1 270/1 280
5% Co	BT4	T4	3255	285	1 10/900	850	1 280/1 290
10% Co	BT5	T5	3265	321	1 100/900	850	1 290/1 300
High Vanadium	BT15	T15	3215	285	1 100/900	850	1 230/1 240
Hot Work Steels							
3% Co 3% Mo							
3% Cr	BH10A	H10A		255	1 100/900	880	1 100
5% Cr+W	BH12	H12	2606	255	1 100/900	880	1 000
5% Cr	BH13	H13	2344	255	1 100/900	880	1 000/1 040
4% W 4% Cr							
4% Co 2% V	BH19	H19		256	1 100/900	880	1 150/1 200
10% W	BH21	H21	2581	248	1 100/900	850	1 150
Low alloy			2242	248	1 100/850	850	850
2% W	BS1	S1	2542	220	1 100/850	850	870
High Carbon High Chromium Steels							
1% C 5% Cr	BA2	A2	2363	248	1 050/920	850	970
0.8% C							
12% Cr Mo V				255	1 100/900	850/900	980/1 020
High C High Cr	BD2	D2	2601	262	1 050/920	850	1 020
High C High Cr	BD2S	D2S		262	1 050/920	850	1 020
High C High Cr	BD3	D3	2080	262	1 050/920	850	960
5% Cr 5% V		A7		262	1 050/920	870/900	940
Non-Shrinking Tool Steels							
Non distorting	B01	0.1		229	1 000/800	770	800
Non-distorting	B02	0.2		229	1 000/800	770	790
0.9% C 1.75% Mn	B02	0.2		217	1 000/850	770	760/780
Shock Resisting Steels							
2% Si	BS5	S5		220	1 100/870	780/800	880/890
0.43% C 1.80 W							
1.00/ Cr 1.00% Si				207	1 030/850	780/800	850/870
0.35% C 1.80% W							
1.00% Cr 5.00% Si				200	1 030/850	780/800	850/870
3.00% Ni 1.02% Si							
0.65% Cr				269	1 100/800	680	860
Carbon Tool Steels							
Shallow hardening	BW1A/C	W1	Various	201	1 100/700	780	770/800
Medium hardening	BW1A/C	W1	Various	201	1 100/700	780	770/800
Mint die	BW2	W2	Various	201	1 100/700	780	770/800 according to carbon content

(continued)

TABLE 9 – HIGH SPEED AND TOOL STEELS* – *contd.*

Treatment			
Quenching medium	Tempering range °C	Hardness range RC	Applications
Oil or in salt at 550°C	550/580	65/62	Drills, taps and reamers.
,, ,,	560/580	66/63	Most types of cutting tools; also cold forming tools.
,, ,,	550/580	66/63	Unground hobs, milling, shaving and shaping currers, etc.
,, ,,	520/550	68/66	Most types of cutting tools particularly for difficult materials.
,, ,,	550/570	65/62	Most types of cutting tools.
,, ,,	560/580	66/63	Most types of cutting tools for tougher materials and faster machining.
,, ,,	560/580	67/64	Cutting tools for more difficult materials.
,, ,,	550/570	67/65	Most wear resistant of high speed steels. For high speed cutting and difficult materials, *eg* aerospace materials.
Oil	540/680	56/38	Extrusion and hot work tooling for brass and copper alloys.
Oil or air	500/650	53/32	General hot work tool steel with tungsten addition.
,, ,,	500/650	57/38	Popular hot work tool steel for general purposes, including aluminium die-casting. Dies for extruding, stamping and die-casting copper based alloys.
,, ,,	550/680	59/40	
,, ,,	600/680	54/40	10% tungsten general purpose hot work steel.
Oil	200/650	57/34	Low alloy steel for mandrel bars, etc.
,,	200/500	56/47	Shear blades, chisels and higher alloy mandrel bars.
Oil or air	180/220	62/59	General cold work tool steel, combining wear resistance with reasonable degree of toughness.
,, ,,	250/280	57/56	Cold shear blades for thin sheets, trimming dies, cold heading dies, nut forming dies.
,, ,,	180/220	63/60	Cold work tool steel with superior wear resistance to H33 and fair degree of toughness.
,, ,,	180/220	63/60	A free machining version of H42.
,, ,,	180/220	63/60	The popular cold work steel with a high degree of wear resistance.
,, ,,	150/220	63/60	Refractory moulds and other applications where superior wear resistance is necessary.
Oil	180/220	62/58	Gauges, jigs and precision tools; general cold work applications where low distortion is essential.
,,	180/220	62/58	Similar to K4 Spl but with a higher manganese content.
,,	150/250	62/58	Screwing dies, broaches, press and blanking tools, cuts moulds for plastics.
,,	200/250	57/55	Similar to F5 but essentially for cold work applications.
,,	180/260	52/50	Pneumatic chisels and caulking tools, buster chisels, hand chisels, trimming steels, punches, dies, dollies
Water	180/250	53/51	for rock drill forging, hot rivet snaps.
Oil	200/500	57/37	Heavy shear blades, cold punches, pneumatic and hand chisels, coal cutter picks.
Water	100/300	65/58	Shallow hardening steels used to a wide range of dies and cold work tooling.
or	100/300	65/58	Deeper hardening grades used for a wide range of tooling.
8% brine	100/300		Shallow hardening with vanadium addition, used for minting dies, etc.

*Based on data by Jessop Saville Ltd

produces insignificant distortion and oxidation, hence it is readily possible to finish machine maraging steel components in the soft (annealed) state prior to heat treatment.

Although non-critical, the heat treatment for maraging is somewhat inflexible and age hardening is rapid – *eg* realising full strength in about three hours at *circa* 480°C. Annealing temperature is usually around 820°C.

CASE HARDENING STEELS

Case hardening steels are classified in BS 970 according to their tensile strengths when heat treated in ¾ inch section (previously 1⅛ inch section), core strength being dependent on size as well as composition. In general the lower alloy steels are more susceptible to the influence of size and need more careful heat treatment than the higher alloy steels. In the latter category case hardness often tends to be lower because of retained austenite, particularly if the carbon content is also relatively high.

Pack carburising is the oldest and simplest method of case hardening. The parts are packed into metal containers with a carburising mixture consisting typically of powdered charcoal and 10% barium carbonate. A proportion of old compound is used in each charge. When the charge is heated, some time must elapse before a uniform temperature is attained because heat is transmitted only slowly through the pack. Variations in case depth are therefore likely to occur and these become most obvious if the carburising period is short.

With gas carburising greater care in operation is needed, but the process permits accurate control of temperature and carburising atmosphere. The charge can be brought to a uniform temperature in a neutral atmosphere, admitting the carburising atmosphere only for the time required to give the desired depth of case; finally the carbon potential of the gas can be lowered to permit diffusion and thus avoid an excess of carbon at the surface.

Liquid bath salts, in which sodium cyanide is the carburising medium, can also be used. They permit rapid heating and are particularly useful when a shallow case is desired, as in thin sections.

Until the introduction of grain size treated steel the usual procedure after carburising was to quench from 850 to 890°C and then quench again from 760 to 820°C.

This double quenching treatment ensured that good mechanical properties were obtained in case and core but was liable to cause excessive distortion. With grain size treated steel, single quenching treatments can be used and these give less distortion than double quenching. Direct quenching from the carburising furnace gives very little distortion but there may be a considerable amount of retained austenite, particularly in the highly alloyed steels. Less austenite is retained if after carburising the parts are transferred to a furnace standing at about 825°C and then quenched after temperature equalisation.

Alternatively, the parts may be allowed to cool down after carburising and then given a single quench from the region of 825°C. If the parts have been slowly cooled after carburising there will be very little distortion, but the micro-structure may show grain boundary carbide and an incompletely refined core. Rapid cooling after carburising will give a better structure but rather more distortion.

A low temperature treatment at 150 to 200°C after hardening improves toughness and causes only a slight reduction in hardness. It also reduces the susceptibility of the case to grinding cracks.

NITRIDED STEELS
Nitrided steels are superior to un-nitrided steels in their resistance to the corrosive action of fresh water, sea water, steam and moist atmospheres. In general the best resistance to corrosion is shown by the steels with the hardest case. Maximum corrosion resistance is obtained if the 'as nitrided' surface remains intact, but a lightly ground surface still gives good results. It must be emphasised however that at their best nitrided steels have a corrosion resistance which falls far short of that of the stainless steels, particularly when exposed to mineral acids. Stainless steels can be surface hardened by nitriding but the corrosion resistance is very much reduced as a result.

Nitriding increases the fatigue strength of steel and reduces its susceptibility to the stress concentration effect of surface notches. Fatigue cracks usually start at the surface, often because the stress is highest there, but also because stress raisers on the surface can initiate cracks. Nitriding increases the strength of the surface material and also gives it a residual compressive stress. Usually when a nitrided component fails by fatigue the initial crack forms below the surface. Nitrided steel shows a very good resistance to corrosion fatigue.

The main considerations in selecting a suitable nitrided steel for a given purpose are the hardness of the case and the strength of the core. For maximum resistance to wear the usual choice is a steel which gives the hardest case. For conditions where spalling of the case can be a hazard it may be advisable to sacrifice some surface hardness to obtain a tougher case. Apart from its lower hardness, the case of these steels has all the advantages, such as retention of hardness at elevated temperatures and high fatigue strength, which are a consequence of the nitriding process.

SAE STEELS
The American SAE steel compositions are designated by a digit system descriptive of the composition; the first digit indicates the type of steel, ie:

Carbon steels	1	Nickel steels	2
Nickel-chromiumsteels	3	Molybdenum steels	4
Chromium steels	5	Chrome-vanadium steels	6
Silicon-manganese steels	9		

The second digit (usually) indicates the nominal percentage of the main alloying element. In some cases, ie where a specific alloy cannot be properly described otherwise, the second and third digits may be descriptive of the alloying element(s)†.

The last two digits indicate the approximate carbon content in one-hundredths of 1%.

---

†The number may also run to five digits where the carbon content exceeds 1% – ie the carbon percentage can only be expressed in terms of three digits.

TABLE 10 – BS ALLOY STEELS FOR CASE HARDENING – BS 970 : 1971

Steel	En steel replaced	Type
523A14 523M15	206	½% chromium
527A19 527M20	207	¾% chromium
635A14 635H15 635M15	 351	¾% nickel chromium
637A16 637H17 637M17	 352	1% nickel chromium
655A12 655H13 655M13	 36A	3¼% nickel chromium
659A15 659H15 659M15	 39A	4% nickel chromium
665A17 665H17 665M17 665A19 665H20 665M20 665A22 665H23 665M23 665A24	 34 35A 35 35B	1¾% nickel molybdenum
805A15 805A17 805H17 805M17 805A20 805H20 805M20 805A22 805H22 805M22 805A24 805H25 805M25	 361 362 363	½% nickel chromium molybdenum
815A16 815H17 815M17	 353	1½% nickel chromium molybdenum
820A16 820H17 820M17	 354	1¾% nickel chromium molybdenum
822A17 822H17 822M17	 355	2% nickel chromium molybdenum
832H13 832M13	 36C	3½% nickel chromium molybdenum
835A15 835H15 835M15	 39B	4% nickel chromium molybdenum

SAE (Society of Automotive Engineers) numbers agree with AISI numbers (American Iron and Steel Institute), although the latter may be associated with a prefix, viz:

A for alloy steels
B for free cutting steels
C for plain carbon and open hearth steels, etc.

TABLE 11 — ALLOY STEELS FOR CASE HARDENING TO BS 970

Steel	Carbon		Manganese		Nickel		Chromium		Molybdenum	
	% min	% max	% min	% max	% min	% max	% min	% max	% min	% max
½% Cr steel										
523A14	0.12	0.17	0.30	0.50	—	—	0.30	0.50	—	—
¾% Cr steel										
527A19	0.17	0.22	0.70	0.90	—	—	0.70	0.90	—	—
¾% Ni Cr steel										
635A14	0.12	0.17	0.70	0.90	0.70	1.00	0.50	0.75	—	0.10
1% Ni Cr steel										
637A16	0.14	0.19	0.70	0.90	0.90	1.20	0.70	1.00	—	1.10
3¼% Ni Cr steel										
655A12	0.10	0.15	0.40	0.60	3.00	3.50	0.75	1.00	—	—
4% Ni Cr steel										
659A15	0.13	0.18	0.30	0.50	3.90	4.30	1.00	1.30	—	—
1¾% Ni Mo steels										
665A17	0.15	0.20	0.45	0.65	1.60	2.00	—	0.25	0.20	0.30
665A19	0.17	0.22	0.45	0.65	1.60	2.00	—	0.25	0.20	0.30
665A22	0.20	0.25	0.45	0.65	1.60	2.00	—	0.25	0.20	0.30
665A24	0.22	0.27	0.45	0.65	1.60	2.00	—	0.25	0.20	0.30
½% Ni Cr Mo steels										
805A15	0.13	0.18	0.70	0.90	0.40	0.70	0.40	0.60	0.15	0.25
805A17	0.15	0.20	0.70	0.90	0.40	0.70	0.40	0.60	0.15	0.25
805A20	0.18	0.23	0.70	0.90	0.40	0.70	0.40	0.60	0.15	0.25
805A22	0.20	0.25	0.70	0.90	0.40	0.70	0.40	0.60	0.15	0.25
805A24	0.22	0.27	0.70	0.90	0.40	0.70	0.40	0.60	0.15	0.25
1½% Ni Cr Mo steel										
815A16	0.14	0.19	0.70	0.90	1.20	1.60	0.90	1.20	0.10	0.20
1¾% Ni Cr Mo Steel										
820A16	0.14	0.19	0.70	0.90	1.60	2.00	0.90	1.20	0.10	0.20
2% Ni Cr Mo steel										
822A17	0.15	0.20	0.50	0.70	1.80	2.20	1.40	1.70	0.15	0.25
4% Ni Cr Mo steel										
835A15	0.13	0.18	0.30	0.50	3.90	4.30	1.00	1.30	0.15	0.25

TABLE 12 – BS 970 – ALLOY STEELS – MECHANICAL PROPERTIES FOR ¾ INCH DIAMETER TEST BARS IN THE SINGLE QUENCHED CONDITION

Steel	Tensile strength tonf/in², R_m min	Elongation A% min on 5.65 $\sqrt{S_o}$ gauge length	Izod ft lbf, min
523M15	25	13	40
527M20	50	12	15
635M15	50	12	20
637M17	60	10	15
655M13	65	9	30
659M15*	85	8	25
665M17	50	12	30
665M20	55	11	20
665M23	60	10	12
805M17	50	12	20
805M20	55	11	15
805M22	60	10	10
805M25	65	9	—
815M17	70	8	20
820M17	75	8	20
822M17*	85	8	20
832M13	70	8	25
835M15*	85	8	25

*Heat treatment in test piece size

TABLE 13 – MECHANICAL PROPERTIES AND HEAT TREATMENTS PREVIOUSLY SPECFIED FOR ALLOY STEELS FOR DOUBLE QUENCHED 1⅛ INCH DIAMETER TEST BARS – BS 970 : 1955

Steel	En steel	Hardening temperature °C	Tensile strength, R_m tonf/in², min	Elongation A% min on 4 $\sqrt{S_o}$ gauge length	Equivalent elongation, A% min on 5.65 $\sqrt{S_o}$ gauge length	Izod ft lbf, min
635M15	351	780—820	45	18	13	30
637M17	352	780—820	55	15	11	20
655M13	36A	760—780	55	15	11	35
659M15	39A	760—780	85	12	8	25
665M17	34	760—780	45	18	13	40
665M23	35	760—780	55	15	11	22
805M17	361	780—820	45	18	13	25
805M20	362	780—820	55	15	11	15
815M17	353	780—820	65	12	8	20
820M17	354	780—820	75	12	8	20
822M17	355	780—820	85	12	8	25
832M13	36C	760—780	65	13	9	30
835M15	39B	760—780	85	12	8	25

STAINLESS STEELS

The generic description *stainless steel* is applied to a wide range of iron alloys containing more than 10% chromium. They can be divided into various groups: those containing chromium and those containing nickel and chromium; purpose-types, *eg* acid resisting steels, cutlery steels, etc; or metallurgical 'families'. The last method of classification is now normally used and groups stainless steels as austenitic, ferritic and martensitic.

**TABLE 14 – DIRECT HARDENING ALLOY STEELS TO BS 970
INCLUDING NITRIDING STEELS**

Steel	En steel Replaced	Type
503M40	12	
503A37	12B	
503A42	12C	1% nickel
503H37	—	
503H42	—	
526M60	11	¾% chromium
530M40	18	
530A30	18A	
530A32	18B	
530A36	18C	
530A40	18D	1% chromium
530H30	—	
530H32	—	
530H36	—	
530H40	—	
534A99	31	1½% chromium (1% carbon)
535A99	31	
605M30	16D	1½% manganese molybdenum (water hardening)
605M36	16	
605A32	16B	
605A37	16C	1½% manganese molybdenum
605H32	—	
605H37	—	
606M36	16M	1½% manganese molybdenum (free cutting)
608M38	17	1½% manganese molybdenum (higher molybdenum)
608H37	—	
640M40	111	
640A35	111A	1¼% nickel chromium
640H35	—	

(continued)

TABLE 14 – DIRECT HARDENING ALLOY STEELS TO BS 970
INCLUDING NITRIDING STEELS – *contd.*

Steel	En steel Replaced	Type
653M31	23	3% nickel chromium
708M40	19A	
708A37	19B	
708A42	19C	1% chromium molybdenum
708H37	—	
708H42	—	
709M40	19	1% chromium molybdenum (higher molybdenum)
722M24	40B	3% chromium molybdenum (suitable for nitriding)
785M19	13	1½% manganese nickel molybdenum
816M40	110	1½% nickel chromium molybdenum (low molybdenum)
817M40	24	1½% nickel chromium molybdenum
823M30	—	2% nickel chromium molybdenum
826M31	25	2½% nickel chromium molybdenum (medium carbon)
826M40	26	2½% nickel chromium molybdenum (high carbon)
830M31	27	3% nickel chromium molybdenum
835M30	30B	4% nickel chromium molybdenum
897M39	40C	3¼% chromium molybdenum vanadium (suitable for nitriding in the 85 ton condition)
905M31	41A	1½% chromium aluminium molybdenum nitriding (medium carbon)
905M39	41B	1½% chromium aluminium molybdenum [nitriding (high carbon)]
945M38	100	
945A40	100C	1½% manganese nickel chromium molybdenum

TABLE 15 – EXAMPLES OF PROPRIETARY NITRIDING STEELS

Type No*	Typical Composition									Typical Core Strength tonf/in² kgf/mm²		Approx Hardness of case (HV)
	C	Si	Mn	Cr	Ni	Al	Mo	V	En No			
LK1	0.50	0.35	0.50	1.60	—	1.10	0.20	—	—	80	126	1 050
LK3	0.40	0.35	0.50	1.60	—	1.10	0.20	—	En41B	60	94	1 050
LK5	0.30	0.35	0.50	1.60	—	1.10	0.20	—	En41A	50	79	1 050
LK7	0.20	0.35	0.50	1.60	—	1.10	0.20	—	—	40	63	1 050
HCM3	0.40	0.30	0.50	3.00	—	—	1.00	0.25	En40C	90	142	900
HCM5	0.30	0.30	0.45	3.00	—	—	0.40	—	En40B	60	94	800
HCM7	0.20	0.30	0.45	3.00	—	—	0.40	—	En40A	50	79	800
GK3	0.35	0.30	0.50	2.00	—	—	0.25	0.15	—	60	94	750
GK5	0.25	0.30	0.50	2.00	—	—	0.25	0.15	—	50	79	750
GK7	0.18	0.30	0.50	2.00	—	—	0.25	0.15	—	40	63	750
HK5	0.25	0.25	0.65	2.00	—	0.60	1.00	0.50	—	55	87	1 050
LKP	0.24	0.30	0.60	1.30	3.50	1.00	0.30	—	—	80	126	1 050

*Firth Brown Limited

Austenitic stainless steels

These are the most widely used of the stainless steels and possess high resistance to corrosion, good toughness, ductility and weldability. Chrome-nickel steels work harden but cannot be hardened by heat treatment.

A 'standard' composition is 18/8 (18% chromium, 8% nickel) having a moderate to low ultimate strength of 40 to 45 tons/in² or 35 to 40 tons/in² in the softened condition. Corresponding yield strength figures are 35 to 40 tons/in² and 15 to 20 tons/in². Stabilising elements such as tungsten or columbium may be introduced to prevent intergranular corrosion, increase the strength and improve corrosion resistance, but they may reduce ductility and toughness. Molybdenum may also be added to improve resistance to acids and it tends to reduce weld decay. There are numerous other alternative compositions to British, European and American specifications with chromium, cadmium, manganese up to 25% and nickel up to 20% or more (see Tables).

Austenitic stainless steels are normally softened at temperatures above 1010°C to prevent carbon precipitation and inhibit intergranular corrosion. Softening is then followed by quick cooling to prevent transformation of the austenite. Quenching produces softening rather than hardening – ie solution annealing.

Further heat treatment may be applied after annealing or welding to stabilise the steel, although this is normally only required where maximum resistance to corrosion is required in the temperature range 400 to 870°C. An austenitic stainless steel which has not been stabilised can be stress relieved after cold working by heating to 350 to 450°C, depending on the particular requirements. Similar softening treatment may be used to relieve internal stresses after welding, although this is not necessary with all austenitic stainless steels. Higher temperatures (*eg* above 700°C) may be specified for minimising the chance of subsequent stress corrosion cracking in welds.

Ferritic stainless steels

Ferritic stainless steels normally have a chromium content of 17 to 20%. They have good ductility and moderate strength (although subject to brittle failure at sub-zero temperatures). Corrosion resistance is also good – superior to that of martensitic stainless steels but inferior to that of austenitic steels. They have limited suitability for welding, but can be resistance-welded successfully.

The low carbon content of ferritic stainless steels makes them particularly suitable for forming without cracking. Again they can only be hardened by cold working, not by heat treatment. Unlike the two other types of stainless steels they are magnetic and they have a lower coefficient of thermal expansion, virtually the same as that of mild steel.

TABLE 16 – ALLOY STEELS – SAE STEELS – EQUIVALENT SPECIFICATIONS

BS 970	En	AISI and SAE	Steel type
220 M07	1A	1064,1065,1070	Free cutting mild steel
240 M07	1B	1074,1075,78,86	Free cutting higher speed
	2	1080,84,85	G.P. cold forming steel
040 A09	2A/1		Special cold forming steel
040 A10	2A	1045,46,49,50,53	Special cold forming steel
040 A12	2B	1049,50,54,55	Special cold forming steel
040 A20	2C	1095	Special cold forming steel
040 A22	2D	9255,9260	Special cold forming steel
	2E		Special cold forming steel
080 A27	3	6150	'20' carbon steel hot rolled
070 M20	3A,3C	5147	'20' carbon steel hot rolled or normalised
	3B	9254	'20' carbon steel cold drawn
	3D		'20' carbon steel cold drawn (HT)
080 A25	4, 4A		'25' carbon steel (normalised)
080 M30	5,5K,5D		'30' carbon steel
080 A27	5A		'30' carbon steel
080 A30	5B		'30' carbon steel
080 A32	5C		'30' carbon steel
216 M28	7		Semi-free cutting carbon
	7A	403,410,420	Semi-free cutting '15' carbon
080 M40	8,A,B,C,D,E,K	416	'40' carbon steel
212 M36	8M,AM,BM,CM,DM	420F	'40' carbon steel free cutting

(continued)

TABLE 16 – ALLOY STEELS – SAE STEELS – EQUIVALENT SPECIFICATIONS
– contd.

BS 970	En	AISI and SAE	Steel type
	9,9K	431	'55' carbon steel
	10	302,321,304,316 347,348	'55' carbon ¾% nickel
	11		'60' carbon chromium
	12,A,B,C		1% nickel steel
	13	51442	Mn, Ni, Cr
150 M19	14A,a/1,B		Carbon manganese
150 M36	15,A,B		Carbon manganese
216 M36	15AM	1137	Carbon manganese free cutting
605 M36	16,		Manganese molybdenum
606 M36	16M		Manganese molybdenum free cutting
608 H37	17		Manganese molybdenum
530 M40	18	5130,32,35,40	1% chromium
709 M40	19	4137,4140,4142	1% chromium molybdenum
	20A,20B		1% Cr Mo (Higher Mo)
	21,21A		3% Nickel
	22		3½% Nickel
653 M31	23		3% nickel chromium
817 M40	24		1½% Ni-Cr-Mo
826 M31	25		2½% Ni-Cr-Mo
826 M40	26		2½% Ni-Cr-Mo
830 M31	27		3% Ni-Cr-Mo
	28		3½% Ni-Cr-Mo
	29A,B		3% Ni-Cr-Mo
835 M30	30B		4¼% Ni-Cr-Mo
534 A99	31	51100,52100	1% carbon chromium
045 M10	32A,B	1010,1012,1013, 1016	Carbon case hardening
212 M14	32M	1115	Carbon case hardening free cutting
	33		3% Ni case hardening
665 A17	34	4617	2% Ni-Cr-Mo case hardening
665 A22	35,A,B	4620,4621	2% Ni-Mo case hardening
665 A12	36A		3% Ni-Cr and Ni-Cr-Mo
	37		5% Ni case hardening
	38		5% Ni-Mo case hardening
659 A15	39A		4¼% Ni-Cr and Ni-Cr-Mo
835 A15	40A		3% Cr-Mo nitriding
897 M39	40C		3% Cr-Mo-V nitriding
905 M31	41A,41B		1½% Cr-AL-Mo nitriding
	42		Carbon spring steel

Martensitic stainless steels

Martensitic stainless steels are the least corrosion resistant but have the specific advantage that they can be heat treated to enhance their mechanical properties, the actual properties depending on the degree of temper achieved. Thus they offer a wider range of properties than the other two types.

As well as being used for general engineering purposes, the martensitic struc-

TABLE 17 – STAINLESS STEELS : TYPICAL TYPES AND PROPERTIES

BS 970	Code	En No	Remarks	Condition	Tons per sq in		Elongation %
					Yield	Ultimate	
416S21	Stainless iron 1	56A	A martensitic steel easy to manipulate	As drawn Softened	28 16	33 26	12 35
416S29	Stainless iron 2	56B	A similar steel to above, but harder	As drawn Softened Air hardened	30 18 up to 55 max	35 28 up to 70 max	10 — —
	Stainless iron (W)		A weldable martensitic steel	As drawn Softened Air hardened	30 16 up to 45	35 29 up to 60	10 35 —
431S29	20/2	57	A martensitic steel harder to manipulate than stainless iron but offering greater resistance to corrosion, especially sea water	As drawn Softened Hardened	50 38 65	55 48 70	10 25 15
430S15	S.S.17	60	A ferritic steel more corrosion resisting than stainless iron	Softened	20	33	25
430S16	S.S.20	61	Similar to above but a little more resistant to corrosion	Softened	22	35	23
	S.S.27	—	A ferritic steel having excellent resistance to scaling at high temperatures and in sulphurous gases	Softened	25	36	20
302S25	18/8	58A	An austenitic steel suitable for manipulating and welding. Must be softened after welding	As drawn Softened	40 15	45 35	25 50
304S15	18/8 Low	58E	As above but with very low carbon Need not be softened after welding	As drawn Softened	35 15	40 35	25 30
321S31 347S17	18/8/Ti 18/12/Nb	58G	Special welding qualities. Need not be softened after welding	As drawn Softened	45 18	50 38	20 40
316S16	18/8/M	58J	For resistance to certain concentrations of acetic and sulphuric acids	As drawn Softened	45 21	50 43	20 40
	18/8/MT	—	As above, but need not be softened after welding	As drawn Softened	45 21	50 43	20 40
	'316'	—	Similar to 18/8/M	As drawn Softened	45 20	50 40	20 40
—	25/20		An austenitic steel having good heat-resisting properties	As drawn Softened	45 22	50 40	30 45
—	23/16/T	—	Similar to 25/20. Can be welded without subsequent softening	As drawn Softened	45 25	50 43	30 40
—	16/6/H	—	An austenitic/martensitic steel suitable for hardening	Softened Hardened	20 up to 70	55 up to 80	30 15

TABLE 18 – FERRITIC AND MARTENSITIC STAINLESS STEELS TO BS 970 – CHEMICAL COMPOSITION

Grade	C %	Si Max %	Mn Max %	Ni %	Cr %	Mo Max %	S %	P Max %	Se %
403S17	0.08 max	0.08	1.0	0.5 max	12.0/14.0	—	0.03 max	0.04	—
430S15	0.10 max	0.8	1.0	0.5 max	16.0/18.0	—	0.03 max	0.04	—
410S21	0.09/0.15	0.8	1.0	1.0 max	11.5/13.5	—	0.03 max	0.04	—
420S29	0.14/0.20	0.8	1.0	1.0 max	11.5/13.5	—	0.03 max	0.04	—
420S37	0.20/0.28	0.8	1.0	1.0 max	12.0/14.0	—	0.03 max	0.04	—
420S45	0.28/0.36	0.8	1.0	1.0 max	12.0/14.0	—	0.03 max	0.04	—
416S21	0.09/0.15	1.0	1.5	1.0 max	11.5/13.5	0.6	0.15/0.30	0.04	—
416S41	0.09/0.15	1.0	1.5	1.0 max	11.5/13.5	0.6	0.03 max	0.04	0.15/0.30
416S29	0.14/0.20	1.0	1.5	1.0 max	11.5/13.5	0.6	0.15/0.30	0.04	—
416S37	0.20/0.28	1.0	1.5	1.0 max	12.0/14.0	0.6	0.15/0.30	0.04	—
431S29	0.12/0.20	0.8	1.0	2.0/3.0	15.0/18.0	—	0.03 max	0.04	—
441S29	0.12/0.20	1.0	1.5	2.0/3.0	15.0/18.0	0.6	0.15/0.30	0.04	—
441S49	0.12/0.20	1.0	1.5	2.0/3.0	15.0/18.0	0.6	0.03 max	0.04	0.15/0.30

TABLE 19 – MARTENSITIC STAINLESS STEELS TO BS 970 – MECHANICAL PROPERTIES IN THE HARDENED AND TEMPERED CONDITION

Heat Treatment Condition Symbol	P					R					S					T				
Tensile Strength Range R_m ton/in^2	35–45					45–55					50–60					55–65				
Brinell Hardness Number Range, HB	152–207					201–255					223–277					248–302				
Grade	LRS	R_c	A	I	$R_{p0.2}$	LRS	R_c	A	I	$R_{p0.2}$	LRS	R_c	A	I	$R_{p0.2}$	LRS	R	$_eA$	I	$R_{p0.2}$
410S21	2½	24	20	40	22	—	—	—	—	—	—	—	—	—	—	—	—	—	—	—
	6	24	20	25	22	—	—	—	—	—	—	—	—	—	—	—	—	—	—	—
420S29	—	—	—	—	—	2½	34	15	25	32	1⅛	38	13	20	36	—	—	—	—	—
	—	—	—	—	—	6	34	15	20	32	—	—	—	—	—	—	—	—	—	—
420S37	—	—	—	—	—	2½	34	15	25	32	2½	38	13	20	36	—	—	—	—	—
	—	—	—	—	—	6	34	15	20	32	6	38	13	10	36	—	—	—	—	—
420S45†	—	—	—	—	—	2½	34	15	25	32	2½	38	13	20	36	—	—	—	—	—
	—	—	—	—	—	6	34	15	20	32	6	38	13	10	36	—	—	—	—	—
416S21	6	24	15	25	22	—	—	—	—	—	—	—	—	—	—	—	—	—	—	—
416S41	6	24	15	25	22	—	—	—	—	—	—	—	—	—	—	—	—	—	—	—
416S29	—	—	—	—	—	6	34	11	20	32	1⅛	38	10	10	36	—	—	—	—	—
416S37	—	—	—	—	—	6	34	11	20	32	6	38	10	10	36	—	—	—	—	—
431S29	—	—	—	—	—	—	—	—	—	—	—	—	—	—	—	2½	44	11	25	41
	—	—	—	—	—	—	—	—	—	—	—	—	—	—	—	6	44	11	15	41
441S29	—	—	—	—	—	—	—	—	—	—	—	—	—	—	—	2½	44	8	15	40
441S49	—	—	—	—	—	—	—	—	—	—	—	—	—	—	—	2½	44	8	15	40

LRS denotes limiting ruling section in inches.
R_c denotes minimum yield stress in ton/in^2.
A denotes elongation — % min on $5.65\sqrt{S_o}$ gauge length.

I denotes minimum Izod impact value in ft lb.
$R_{p0.2}$ denotes minimum 0.2% proof stress in ton/in^2 only applicable when specifically ordered.

TABLE 20 – AUSTENITIC STAINLESS STEELS TO BS 970 – CHEMICAL COMPOSITION

Grade	C Max %	Si %	Mn %	Ni %	Cr %	Mo %	Ti %	Nb %	S %	P Max %	Se %
304S12	0.03	0.2/1.0	0.5/2.0	9.0/12.0	17.5/19.0	—	—	—	0.03 max	0.045	—
304S15	0.06	0.2/1.0	0.5/2.0	8.0/11.0	17.5/19.0	—	—	—	0.03 max	0.045	—
302S25	0.12	0.2/1.0	0.5/2.0	8.0/11.0	17.0/19.0	—	—	—	0.03 max	0.045	—
321S12	0.08	0.2/1.0	0.5/2.0	9.0/12.0	17.0/19.0	—	5C/0.7	—	0.03 max	0.045	—
321S20	0.12	0.2/1.0	0.5/2.0	8.0/11.0	17.0/19.0	—	5C/0.9	—	0.03 max	0.045	—
347S17	0.08	0.2/1.0	0.5/2.0	9.0/12.0	17.0/19.0	—	—	10C/1.00	0.03 max	0.045	—
315S16	0.07	0.2/1.0	0.5/2.0	9.0/11.0	16.5/18.5	1.25/1.75	—	—	0.03 max	0.045	—
316S12	0.03	0.2/1.0	0.5/2.0	11.0/14.0	16.5/18.5	2.25/3.00	—	—	0.03 max	0.045	—
316S16	0.07	0.2/1.0	0.5/2.0	10.0/13.0	16.5/18.5	2.25/3.00	—	—	0.03 max	0.045	—
320S17	0.08	0.2/1.0	0.5/2.0	11.0/14.0	16.5/18.5	2.25/3.00	4C/0.5	—	0.03 max	0.045	—
317S12	0.03	0.2/1.0	0.5/2.0	14.0/17.0	17.5/19.5	3.00/4.00	—	—	0.03 max	0.045	—
317S16	0.06	0.2/1.0	0.5/2.0	12.0/15.0	17.5/19.5	3.00/4.00	—	—	0.03 max	0.045	—
303S21	0.12	0.2/1.0	1.0/2.0	8.0/11.0	17.0/19.0	—	—	—	0.15/0.30	0.045	—
303S41	0.12	0.2/1.0	1.0/2.0	8.0/11.0	17.0/19.0	—	—	—	0.03 max	0.045	0.15/0.30
325S21	0.12	0.2/1.0	1.0/2.0	8.0/11.0	17.0/19.0	—	5C/0.9	—	0.15/0.30	0.045	—
326S36	0.12	0.2/1.0	1.0/2.0	10.0/13.0	16.5/18.5	2.25/3.00	—	—	0.03 max	0.045	0.15/0.30
310S24	0.15	0.2/1.0	0.5/2.0	19.0/22.0	23.0/26.0	—	—	—	0.03 max	0.045	—

TABLE 21 – AUSTENITIC STAINLESS STEELS TO BS 970 – MECHANICAL PROPERTIES IN THE SOFTENED CONDITION

Grade	R_m	A	HB	$R_{p0.2}$
304S12	30	40	183	11.0
304S15	30	40	183	11.0
302S25	33	40	183	13.5
321S12	32	40	183	12.5
321S20	33	40	183	13.5
347S17	33	40	183	13.5
315S16	30	40	183	11.0
316S12	30	40	183	11.0
316S16	30	40	183	11.0
320S17	32	40	183	12.5
317S12	30	40	183	11.0
317S16	30	40	183	11.0
303S21	33	40	183	13.5
303S41	33	40	183	13.5
325S21	33	40	183	13.5
326S36	33	40	183	13.5
310S24	35	40	207	14.0

Note. The 1.0% proof stress will be found to be not less than 2 ton/in^2 higher than the 0.2% proof stress specified above.

R_m denotes minimum tensile strength in ton/in^2.
A denotes elongation, % min on 5.65 $\sqrt{S_o}$ gauge length.
HB   denotes maximum Brinell harness number.
$R_{p0.2}$ denotes minimum 0.2% proof stress in ton/in^2, only applicable when specifically ordered.

TABLE 22 – AUSTENITIC STAINLESS STEELS TO BS 970 – MECHANICAL PROPERTIES IN THE COLD DRAWN CONDITION

Grade	Section Size – Diameter or Width across Flats (in)		Properties		
	Over	Up to and including	R_m	A	$R_{p0.2}$
304S15 304S25	—	¾	56	12	45
321S12 321S20	¾	1	51	15	36
316S16 303S21	1	1¼	47	20	29
303S41 325S21	1¼	1½	45	28	22
326S36	1½	1¾	42	28	20

R_m denotes minimum tensile strength in ton/in^2.
A denotes elongation, % min on 5.65 $\sqrt{S_o}$ gauge length.
$R_{p0.2}$ denotes minimum 0.2% proof stress in ton/in^2, only applicable when specifically ordered.

TABLE 23 – STAINLESS, HEAT RESISTING OR VALVE STEELS TO BS 970

Grade	En Steel Replaced	Nearest AISI Steel		Type
302S25	58A	302	A	Cr Ni 18/9, C 0.12
303S21	58M	303	A	Cr Ni 18/9, S bearing, free machining
303S41	58M	303 Se	A	Cr Ni 18/9, Se bearing, free machining
304S12		304 L	A	Cr Ni 18/10, C 0.03
304S15	58E	304	A	Cr Ni 18/9, C 0.06
310S24		310	A	Cr Ni 25/20
315S16	58H	—	A	Cr Ni Mo 17/10/1½, C 0.07
316S12		316 L	A	Cr Ni Mo 17/12/2½, C 0.03
316S16	58J	316	A	Cr Ni Mo 17/11/2½, C 0.07
317S12		—	A	Cr Ni Mo 18/15/3½, C 0.03
317S16		317	A	Cr Ni Mo 18/13/3½, C 0.06
320S17	58J	—	A	Cr Ni Mo 17/12/2½ + Ti, C 0.08
321S12	58B & 58C	321	A	Cr Ni 18/9/Ti, C 0.08
321S20	58B & 58C	321	A	Cr Ni 18/9/Ti, C 0.12
325S21	58M	—	A	Cr Ni 18/9/Ti, S bearing, free machining
326S36		—	A	Cr Ni Mo 17/11/2½ Se bearing, free machining
331S40	54	—	V	Ni Cr W 14/14/2½
331S42	54A	—	V	Ni Cr W 14/14/2½ + Mo
347S17	58F & 58G	347	A	Cr Ni 18/9/Nb, C 0.08
349S52		—	V	Cr Mn Ni, 21/4 N
349S54		—	V	Cr Mn Ni, 21/4 N, S bearing
352S52		—	V	Cr Mn Ni, 21/4 N + Nb
352S54		—	V	Cr Mn Ni, 21/4 N + Nb, S bearing
381S34		—	V	Cr Ni, 21/11 + N
401S45	52	—	V	Si Cr, 3/8
403S17		403	F	13 Cr, C 0.08 max
410S21	56A	410	M	13 Cr, C 0.12
416S21	56AM	416	M	13 Cr, C 0.12, S bearing, free machining
416S29	56BM	—	M	13 Cr, C 0.17, S bearing, free machining
416S37	56CM	—	M	13 Cr, C 0.24, S bearing, free machining
416S41	56AM	416 Se	M	13 Cr, C 0.12, Se bearing, free machining
420S29	56B	—	M	13 Cr, C 0.17
420S37	56C	420	M	13 Cr, C 0.24
420S45	56D	—	M	13 Cr, C 0.32
430S15	60	430	F	17 Cr, C 0.10
431S29	57	431	M	17 Cr, 2½ Ni, C 0.15
441S29		—	M	17 Cr, 2½ Ni, C 0.15, S bearing, free machining
441S49		—	M	17 Cr, 2½ Ni, C 0.15, Se bearing, free machining
443S65	59	—	V	Cr Ni Si, 20/1½/2

F denotes ferritic steel; M denotes martensitic steel; A denotes austenitic steel; V denotes valve steel

TABLE 24 – AMERICAN IRON AND STEEL INST STAINLESS STEELS

AISI No	Analyses %							Properties and Applications
	C	Mn Max	Si Max	Cr	Ni	P Max	S Max	
201	0.15	5.5–7.5 Other: N 0.25 max	1.0	16–18	3.5–5.5	0.060	0.03	Low nickel alternative for 301
202	0.15 max	7.5–10.0 Other: N 0.25 max	1.0	17–19	4–6	0.060	0.03	Low nickel alternative for 302.
301	0.15 max	2.0	1.0	16–18	6–8	0.045	0.03	Rapid work hardening. Railroad cars, trailer bodies, aircraft structurals.
302	0.15 max	2.0	1.0	17–19	8–10	0.045	0.03	General purpose chromium-nickel. Trim, food handling equipment, jewellery, aircraft cowling, antennas, springs, architectural, cookware (302 clad).
302B	0.15 max	2.0	2.0–3.0	17–19	8–10	0.045	0.03	Higher scaling resistance at elevated temperatures than 302. Furnace parts, still liners.
303	0.15 max	2.0	1.0	17–19	8–10	0.200	0.15 min	Free machining; lower galling tendency than 302. Screw machine products, shafts, valves.
303Se	0.15 max	2.0 Other: Se 0.15 min	1.0	17–19	8–10	0.200	0.06	Somewhat better transverse mechanical properties and slightly lower machinability than 303.
304	0.08 max	2.0	1.0	18–20	8–12	0.045	0.03	General purpose alloy for welding. Chemical and food processing equipment, recording wire.
304L	0.03 max	2.0	1.0	18–20	8–12	0.045	0.03	For welding in corrosive conditions where inter-granular carbide precipitation must be avoided.
305	0.12 max	2.0	1.0	17–19	10–13	0.045	0.03	Low-work-hardening rate. Maximum plasticity for forming, spinning and cold heading.
308	0.08 max	2.0	1.0	19–21	10–12	0.045	0.03	Welding rod and electrodes.
309	0.20 max	2.0	1.0	22–24	12–15	0.045	0.03	High scale resistance and good strength at high temperatures. Aircraft heaters, heat treating equipment, annealing covers, furnace parts.
309S	0.08 max	2.0	1.0	22–24	12–15	0.045	0.03	For welding with high scale resistance.
310	0.25 max	2.0	1.5	24–26	19–22	0.045	0.03	Like 309 but even higher heat resistance. Excellent for general use. Hydrogenation tubes, heat exchangers, furnace parts, combustion chambers.

(continued)

TABLE 24 – AMERICAN IRON AND STEEL INST STAINLESS STEELS – *contd.*

AISI No	Analyses %							Properties and Applications
	C	Mn Max	Si Max	Cr	Ni	P Max	S Max	
310S	0.08 max	2.0	1.5	24-26	19-22	0.045	0.03	Like 309S but has even higher heat resistance.
314	0.25 max	2.0	1.5-3.0	23-26	19-22	0.045	0.03	Highest heat resistance. Uses similar to 310. Resistant to carburizing.
315	0.08 max	2.0 Other: Mo 2-3	1.0	16-18	10-41	0.045	0.03	Higher resistance to corrosives. High creep resistance High temperature parts, chemical pulp handling, photographic and food equipment.
316L	0.03 max Other: Mo 1.75-2.3	2.0	1.0	16-18	10-14	0.045	0.03	Version of 316 for welding where intergranular carbide precipitation must be avoided. Permits stress relief at 1 550-1 650°F.
317	0.08 max	2.0 Other: Mo 3-4	1.0	18-20	11-15	0.045	0.03	Higher in corrosion resistance than 316.
321	0.08 max	2.0 Other: Ti 5 × C min	1.0	17-19	9-12	0.045	0.03	For welding subject to severely corrosive conditions and service in 800-1 650°F range. Aircraft exhaust manifolds, boiler shells, expansion joints, process equipment.
347	0.08 max	2.0 Other: Cb 10 × C min	1.0	17-19	9-13	0.045	0.03	For uses similar to 321. More widely used for welded process equipment.
348	0.08 max Other: Cb 10 × C min Ta 0.1 max	2.0	1.0	17-19	9-13	0.045	0.03	Equivalent to 347. For atomic energy applications due to low retentivity.
403	0.15 max	1.0	0.5	11.5-13	—	0.040	0.03	Steam turbine blades, jet engine compressor blades, highly stressed parts.
405	0.08 max	1.0 Other: Al 0.1-0.3	1.0	11.5-14.5	—	0.040	0.03	Non-hardenable. For welded assemblies where air-hardening of type 410 or 403 is objectionable.
410	0.15 max	1.0	1.0	11.5-13.5	—	0.040	0.03	General-purpose corrosion and heat resistance. Hardenable by heat treatment. Coal screens, low priced cutlery, machine parts, valve trim.
414	0.15 max	1.0	1.0	11.5-13.5	1.25-2.5	0.400	0.03	High-strength version of 410. Rules and straight edges, mild springs, scraper knives.
416	0.15 max Other: Mo 0.6 max Zr 0.6 max	1.25	1.0	12-14	—	0.060	0.15 min	Free machining. Bolts, nuts and screws, carburettor parts, fishing reels, golf club heads, instrument parts, screw machine parts, valve trim.

(continued)

TABLE 24 – AMERICAN IRON AND STEEL INST STAINLESS STEELS – *contd.*

AISI No	Analyses %							Properties and Applications
	C	Mn Max	Si Max	Cr	Ni	P Max	S Max	
416Se	0.15 max Other: Se 0.15 min	1.25	1.0	12–14	—	0.060	0.06	Slightly higher transverse properties and slightly lower machinability than 416.
420	Over 0.15	1.0	1.0	12–14	—	0.040	0.03	High hardness from heat treatment. General purpose cutlery grade. Surgical instruments.
430	0.12 max	1.0	1.0	14–18	—	0.040	0.03	General-purpose non-hardenable chromium type. Decorative trim, nitric acid tanks, annealing baskets, auto trim, sinks.
430F	0.12 max Other: Mo 0.6 max Zr 0.60 max	1.25	1.0	14–18	—	0.060	0.15 min	Free machining bars. Screw machine parts, screws and other fasteners.
430FSe	0.12 max Other: Se 0.15 min	1.25	1.0	14–18	—	0.060	0.06	Free machining, retaining forgeability. Fine finish quality. Similar to 430F.
431	0.20 max	1.0	1.0	15–17	1.25–2.50	0.040	0.03	Special-purpose hardenable type. Aircraft fittings, bolts, paper machinery, windshield wiper arms, beater, bars, springs.
440A	0.60–0.75 Other: Mo 0.75 max	1.0	1.0	16–18	—	0.040	0.03	High strength and corrosive resistance. Hardenable by heat treatment. Toughest high chrome cutlery grade, bearings, seaming rolls, surgical tools.
440B	0.75–0.95 Other: Mo 0.75 max	1.0	1.0	16–18	—	0.040	0.03	Cutlery grade. Uses similar to 440A.
440C	0.95–1.20 Other: Mo 0.75 max	1.0	1.0	16–18	—	0.040	0.03	Ball bearing grade. Also bushings, valve parts, cutlery.
446	0.20 max Other: N 0.25 max	1.5	1.0	23–27	—	0.040	0.03	Resistance to high-temperature scaling, hot sulphur-bearing gases and oxidising acids.
501	Over 0.10 Other: Mo 0.40–0.65	1.0	1.0	4–6	—	0.040	0.03	Moderate resistance to scaling, but limited corrosion resistance. Oil industry equipment -- stills, heat exchangers, hot lines.
502	0.10 max Other: Mo 0.40–0.65	1.0	1.0	4–6	—	0.040	0.03	Similar uses to 501.

TABLE 25 – BRITISH, AMERICAN AND GERMAN EQUIVALENT STEEL SPECIFICATIONS

BS 970	En	Type of Steel	SAE	AISI	Werkstoff	DIN
670M20	2A	'20' C steel (hot rolled or normalised)	1020	C1020	0402	C22
080M30	5	'30' C steel	1030	C1030	0501	C35
	6A	Bright C steel	1035	C1035	0503	C45
080M40	8	'40' C steel	1040	C1040	0503	C45
	9	'55' C steel	1055	C1055	0601	C60
	11	'60' C-Cr steel	5160	5160	8161	58 Cr-V4
150M19	14B	C-Mn steel	1027	C1027	5066	30 Mn 4
			1330	1330		
530M40	18	1% Cr steel	5140	5140	7035	41 Cr 4
530A22	18B	1% Cr steel	5132	5132	7035	34 Cr 4
530A36	18C	1% Cr steel	5135	5135	7034	37 Cr 4
530A40	18D	1% Cr steel	5140	5140	7035	41 Cr 4
709M40	19	1% Cr-Mo steel	4140	4140	7220	34 Cr-Mo 4
708M40	19A	Cr-Mo steel	4140	4140	7225	42 Cr-Mo 4
708A42	19C	1% Cr-Mo steel	4140	4140	7225	42 Cr-Mo 4
653M31	23	3% Ni-Cr steel			5755	22(31)Ni-Cr 14
817M40	24	1½% Ni-Cr-Mo steel	4340	4340	6582	34 Cr-Ni-Mo 6
410S21	56A	Cr-rust-resisting steel	51410	410	4006	× 10 Cr 13
420S29	56B	Cr-rust resisting steel	51410	410	4021	× 20 Cr 13
416S37	56C	Cr-rust-resisting steel	51420	420	4021	C 20 Cr 13
420S45	56D	Cr-rust-resisting steel	51420	420	4034	× 40 Cr 13
416S21	56AM	Cr-rust-resisting steel	51416	416	4024	× 20 Cr 13
			51416 Se (S1416 Se)	416 Se		
816M40	110	Low Ni-Cr-Mo steel			6582	34 Cr-Ni-Mo 6

ture is associated with cutlery steels which when hardened are (almost) wholly martensitic in microstructure.

Steels of this type are not readily machinable unless hardened and tempered. Hardening is done at high temperatures. Tempering temperatures above 450 to 600°C reduce the corrosion resistance of the steel. Tempering below 250°C may be applied for stress relief. Weldability is generally poor.

More recently precipitation hardening steels of the martensitic type have been developed with precipitation hardening characteristics. These are capable of hardening by heat treatment to a strength of 65 to 80 tons/in^2 with the additional advantages of having superior resistance to corrosion (*ie* comparable to austenitic stainless steels) and good weldability.

These precipitation hardening steels fall into two basic types – martensitic and semi-austenitic. In the latter case the martensitic structure is developed by heat treatment.

HIGH STRENGTH STEELS

As a general definition, high strength steels are those with a minimum tensile strength of 100 tons/in^2. Whilst such figures can be achieved fairly readily by

TABLE 26 – CORROSION RESISTANCE OF FOUR BASIC STAINLESS STEELS

Medium	Type				Medium	Type			
	302	316	430	410		302	316	430	410
Organic Substances					**Acids (cont)**				
Acetone	S	S	S	L	Boric	M	S	M	—
Benzol	S	S	S	S	Butyric	S	S	S	—
Camphor	S	S	S	S	Carbolic	M	M	X	X
Carbon disulphide	S	S	—	—	Chloracetic	X	X	—	—
Carbon tetrachloride	M	M	M	M	Chlorosulphonic (conc)	M	M	—	X
Carbon tetrachloride					Chlorosulphonic (10%)	L	—	—	X
(vapours refluxed)	M	M	M	M	Chromic (50%)	X	X	—	—
Coffee	S	S	S	S	Chromic	X	X	X	X
Copal varnish	S	S	S	S	Citric	S	S	—	L
Ethyl alcohol	S	S	S	S	Cresylic	S	S	—	—
Ethyl chloride	S	S	S	S	Chromic (plus 10%				
Ethyl ether	S	S	S	S	potassium ferricyanide	—	—	—	L
Food pastes	S	S	S	S	Formic	X	M	—	X
Formaldehyde	M	M	M	M	Gallic	S	S	S	—
Fruit juices	S	S	S	S	Hydrobromic	X	X	—	—
Furfural	S	S	—	—	Hydrocyanic	S	S	X	X
Glue	M	M	M	M	Hydrochloric	X	X	X	X
Ink	M	M	M	M	Hydrofluoric	X	X	X	X
Iodoform dressing	X	S	—	—	Lactic	S	S	S	L
Methyl alcohol	S	S	S	S	Lactic plus salt	M	M	—	—
Methyl chloride	S	S	—	—	Malic	S	S	—	L
Milk — fresh or sour	S	S	S	S	Molybdic	S	S	—	—
Mustard	M	M	M	M	Nitric (conc)	S	S	S	S
Naphtha	S	S	S	S	Nitric (conc) plus 2% HCl	S	—	X	X
Oils – mineral and					Nitrous (conc)	S	S	S	L
vegetable	S	S	S	S	Oleic	S	S	S	S
Paraffin (molten)	S	S	S	S	Oxalic	M	M	X	X
Paregoric compound	S	S	—	L	Phosphoric	S	S	X	—
Petrol	S	S	S	S	Phosphoric (10%)	S	S	L	L
Pine tar oil	S	S	—	—	Picric (conc)	S	S	S	S
Quinine bisulphate	L	S	—	X	Pyrogallic (conc)	S	S	S	S
Quinine sulphate	S	S	—	L	Pyroligneous (conc)	S	S	—	—
Rosin (molten)	S	S	S	S	Stearic (conc)	S	S	S	S
Soaps	S	S	S	S	Succinic (molten)	X	—	—	—
Sodium salicylate	S	S	S	S	Sulphuric (conc)	S	S	X	X
Soy bean oil	S	S	—	—	Sulphuric (dil)	M	M	X	X
Tomato juice	M	M	M	M	Sulphuric 15% (plus 2%				
Trichlorethylene	M	M	M	M	potassium dichromate)	S	S	—	—
Tung oil	S	S	—	—	Sulphurous (conc)	M	S	L	L
Vinegar at 21.1°C	M	M	M	M	Tannic (conc)	S	S	S	S
Vinegar (plus 0.5%					Tartaric (conc)	M	M	M	M
salt, 93.3°C)	M	M	M	M	Trichloracetic acid (10%)	S	S	—	—
Acids					Uric (conc)	S	S	S	S
Acetic	M	M	M	M	**Salts**				
Acetic vapour	M	M	X	X	Alum	M	M	—	—
Arsenic 40.6°C	S	S	—	—	Aluminium chloride	X	—	—	X
Arsenic 107.2°C	L	—	—	—	Aluminium flouride	L	—	—	X
Arsenious	S	S	S	S					
Benzoic	S	S	S	S					

TABLE 26 – CORROSION RESISTANCE OF FOUR BASIC STAINLESS STEELS
– *contd.*

Medium	302	316	430	410	Medium	302	316	430	410
Salts (cont)					**Salts (cont)**				
Aluminium sulphate	S	S	—	L	Hydrogen peroxide	M	M	M	—
Aluminium sulphate (sat plus 1% sulphuric acid)	S	S	—	X	Lactic acid salts	S	S	—	—
					Lead acetate	S	S	S	—
Aluminium sulphate (sat plus 1% sodium carbonate)	S	S	—	S	Magnesium carbonate	S	S	S	S
Ammonium alum	S	S	S	—	Magnesium chloride	M	M	X	X
Ammonium alum (sat — slightly ammoniacal) 93.3°C	S	S	—	X	Magnesium sulphate	S	S	S	L
					Magnesium hydroxide	S	S	—	S
					Magnesium nitrate	S	S	S	—
Ammonium bromide	M	S	L	L	Mercurous nitrate	S	S	S	S
Ammonium carbonate	S	S	S	S	Mercuric chloride	M	M	X	X
Ammonium chloride	M	M	L	L	Mercuric cyanide	S	S	—	L
Ammonium hydroxide	S	S	S	S	Nickel nitrate	S	S	S	S
Ammonium mono-phosphate	S	S	—	—	Phosphorus trichloride	S	S	—	—
Ammonium nitrate	S	S	S	S	Potassium bromide	S	S	—	L
Ammonium oxalate	S	S	—	S	Potassium carbonate	S	S	S	S
Ammonium sulphate	S	S	S	L	Potassium chloride	M	M	L	L
Ammonium sulphate (plus 5% sulphuric acid)	S	S	—	X	Potassium chlorate	S	S	S	—
					Potassium cyanide	S	S	S	S
Barium carbonate	S	S	—	—	Potassium dichromate	S	S	S	S
Barium chloride	S	S	—	L	Potassium ferricyanide	S	S	S	S
Barium hydrate	S	S	—	—	Potassium ferricyanide (boiling)	S	S	—	—
Bleaching powder	M	M	X	X	Potassium hypochlorite	X	M	X	X
Bordeaux mixture	S	S	—	—	Potassium iodide	S	S	—	—
Calcium carbonate	S	S	S	S	Potassium iodide (sat plus 0.1% sodium carbonate evaporated to dryness)	S	S	—	L
Calcium chlorate	S	S	—	—					
Calcium chloride	M	M	X	X	Potassium hydrate	S	S	S	S
Calcium hypochlorite	X	M	X	X	Potassium nitrate	S	S	S	S
Calcium hypochlorite made alkaline with NaOH	M	M	—	—	Potassium oxalate	S	S	S	S
					Potassium permanganate	S	S	S	—
					Potassium sulphate	S	S	S	S
Calcium hydroxide or oxide	S	S	S	S	Silver bromide	S	S	S	S
Copper carbonate	S	S	S	S	Silver nitrate	S	S	S	S
Copper chloride	X	—	—	S	Silver cyanide	S	S	S	—
Copper cyanide	S	S	S	S	Sodium acetate	S	S	—	—
Copper nitrate	S	S	S	X	Sodium bicarbonate	S	S	S	—
Copper sulphate (plus 2% sulphuric acid)	S	S	—	L	Sodium bichromate	S	S	S	—
Copper sulphate	S	S	S	—	Sodium bisulphate	S	S	—	—
					Sodium borate	S	S	S	—
Creosote	X	M	—	—	Sodium bromide	S	S	—	L
Creosote (plus 3% salt)	X	—	—	—	Sodium carbonate (10%)	S	S	S	S
Ferric chloride (10%)	X	M	X	X	Sodium carbonate (50%)	S	S	S	—
Ferric nitrate	S	S	S	S	Sodium chlorate (10%)	S	S	S	—
Ferrous sulphate	S	S	S	S	Sodium chlorate (25%)	S	S	S	—
Ferric sulphate	S	S	S	—	Sodium chloride	M	M	—	—
Glauber's salt	S	S	—	—					

(continued)

TABLE 26 – CORROSION RESISTANCE OF FOUR BASIC STAINLESS STEELS
– *contd.*

Medium	Type				Medium	Type			
	302	316	430	410		302	316	430	410
Salts (cont)					**Miscellaneous (cont)**				
Sodium chloride					Glycerine	D			
(2% aerated)	S	S	L	—	Glycerin	S	S	S	S
Sodium citrate	S	S	S	S	Gold cyanide				
Sodium fluoride	L	—	—	—	electroplating solution	S	S	—	—
Sodium hydroxide	S	S	S	S	Hydrogen sulphide				
Sodium hypochlorite					204.4°C	M	M	—	—
(Dakin's solution)	M	M	X	X	Iodine	X	M	X	X
Sodium hypochlorite					Lead (molten)	X	X	—	—
(sat – slightly					Linseed oil	S	S	—	L
alkaline) 93.3°C	S	S	—	L	Lysol	M	M	X	X
Sodium lactate	S	S			Meats	S	S	—	—
Sodium nitrate	S	S	S	S	Mercury	S	S	—	—
Sodium nitrite	S	S	—	—	Mine water	M	M	M	M
Sodium peroxide 100°C	S	S	—	—	Nickel sulphate				
Sodium phosphate	S	S	—	—	electroplating solution	S	S	—	—
Sodium sulphate	S	S	S	S	Sauerkraut brine	X	S	—	—
Sodium sulphide	S	S	S	S	Sea water	M	M	—	X
Sodium sulphite	S	S	S	S	Silver cyanide				
Sodium thiosulphate					electroplating solution	S	S	—	—
(plus 4% potassium					Steam and air (refluxed)	S	S	—	L
meta bisulphate)	S	S	S	S	Steam, CO$_2$	S	S	—	L
Sodium thiosulphate 20%					Steam, SO$_2$, CO$_2$ and air	M	M	L	L
(plus acetic acid 20%)	M	M	M	X	Sulphur dioxide	M	M	—	—
Soda ash (10%) 93.3°C	S	S	S	S	Syrup	S	S	—	—
Soda ash (50%) 93.3°C	S	S	S	S	Vegetable juices	S	S	S	S
Stannic chloride	X	X	—	X	Water	S	S	S	S
Stannous chloride	L	—	X	X	X-ray developing solution	M	M	—	—
Sulphur (molten) (500°F)	S	S	S	—	Zinc (molten)	X	X	X	X
Sulphur chloride	L	—	—	—					
Sulphur oxychloride	S	S	—	—					
Titanium tetrachloride	S	—	—	—					
Zinc chloride	X	L	—	X					
Zinc sulphate	S	S	S	S					
Miscellaneous									
Aluminium (molten)	X	X	X	—					
Ammonia	S	S	S						
Baking oven gases	S	S	S	S					
Beer	S	S	S	S					
Bromine	X	X	X	X					
Bromine water	X	X	X	X					
Cadmium (molten)	X	X	—	—					
Carbonated beverages	S	S	S	S					
Chlorine (wet and dry)	X	X	X	X					
Cider	S	S	S	S					
Copper sulphate									
electroplating solution	S	S	—	—					
Copper cyanide									
electroplating solutions	S	S	—	—					

KEY:
S = suitable
L = limited suitability (slight attack)
M = moderate attack (may or may not be suitable)
X = attacked (not suitable)

When specifying materials for service in aggressive media, the following points should be borne in mind:
1. Allowance should be made for the effects of heat transfer, erosion, galvanic effects, aeration and the effects of minor impurities present in solutions.
2. Materials when in the work-hardened condition may be susceptible to stress corrosion cracking in some environments.
3. It is recommended that 'in plant' trials with sample test coupons are undertaken before a final decision is made.

heat treatment of suitable alloys, it is necessary to ensure that the other mechanical properties are acceptable for producing a practical high strength steel.

The two main types in this category are the maraging steels and the direct hardening steels (*nickel steels* and *chrome/molybdenum steels*). The main difference is that maraging steels age harden by simple heat treatment with minimum distortion, whereas the direct hardening steels require heat treatment after machining and may suffer from relatively high distortion etc, requiring subsequent further machining.

HIGH TEMPERATURE STEELS

High temperature steels are basically alloys that are heat resistant. In this respect there is a distinction between a steel which is resistant to oxidation and one which is formulated to maintain high strength at elevated temperatures, although in fact the two properties may be combined in one material.

As far as oxidation resistance is concerned, plain carbon steels and low alloy steels remain suitable for temperatures up to about 450 and 550°C, respectively. For higher temperature services 12% chrome alloy steels are the usual choice, increasing the maximum service temperature to about 730 to 750°C before oxidation becomes significant.

For stressed applications at elevated temperatures it is important that both the tensile strength and creep resistance remain favourable, together with resistance to oxidation. Ferritic alloy steels in this category may be non-transformable or transformable. The non-transformable steels do not, however, maintain strength at high temperatures, their primary application being for components requiring good oxidation resistance only. Thus most high temperature ferritic steels are of the transformable type (*ie* through hardening and tempering), the alloy constituents being developed for optimum performance in this respect.

The other important group comprises the austenitic steels containing 18% and more chromium, and 10% or more nickel. These, in general, provide the best high temperature characteristics as regards strength and creep resistance as well as excellent resistance to scaling.

Sheet steels

Sheet (or strip) steels are used extensively in the automotive, domestic appliance and general engineering industries. They are available in a wide variety of grades and are obtainable coated for improved corrosion resistance. The sheet steel 'family' may be divided into various classes; they may be obtained hot or cold rolled, coated or non-coated and can have specific requirements based on formability or mechanical strength.

Cold rolled thickness is usually limited to about 3 mm, whereas hot rolled is available up to 19 mm in some grades. The hot rolled steels are used for applications such as construction, pipe and tube manufacture and also form the input for subsequent steel processing operations. Cold rolled steels are used where factors such as surface finish, close thickness tolerance and weight saving are important. Table 1 gives details of steels to BS 1449 Part 1.

Recent developments in sheet steels have included the introduction of the 'high strength steels' which include high strength low alloy steels (HSLA), rephosphorised steels, and bake hardenable steels. These steels have been developed mainly for vehicle manufacture and have improved strength over standard forming grades along with good formability. Bake hardenable steels develop maximum strength on stoving of paint during car body manufacture.

Sheet steels are available with a wide variety of coatings, one of the most commonly used being zinc coated or 'galvanised' steel. The zinc is applied either by hot dipping or electrolytic deposition. The zinc coating produced by hot dipping is much thicker than that formed during electrodeposition. Hot dip coated steels are therefore used where improved corrosion resistance is required. However, surface finish and subsequent paintability is generally not as good as that for electrozinc coated.

Other types of coatings for sheet steels include:

— hot dip zinc/aluminium coated for external painted applications.
— hot dip zinc and electrolytically coated tin for 'tin plate'.
— hot dip tin/lead or 'Terne' coated for fuel tanks and gas meters.
— electrolytically coated zinc/nickel for automotive bodies.
— adhesively bonded plastic for domestic appliances.

TABLE 1 – SUMMARY OF MATERIAL GRADES, CHEMICAL COMPOSITIONS, MECHANICAL PROPERTIES AND TYPES OF CARBON AND CARBON-MANGANESE SHEET AND STRIP STEELS (FROM BS 1449 PT 1)

Section no	Material grade	Rolled condition	Chemical composition							
			C		SI		Mn		S	P
			min.	max.	min.	max.	min.	max.	max.	max.
2	Materials having specific requirements based on formability									
			%	%	%	%	%	%	%	%
	1	HR, HS, —, —	—	0.08	—	—	—	0.45	0.030	0.025
	1	—, —, CR, CS	—	0.08	—	—	—	0.45	0.030	0.025
	2	HR, HS, CR, CS	—	0.08	—	—	—	0.45	0.035	0.030
	3	HR, HS, CR, CS	—	0.10	—	—	—	0.50	0.040	0.040
	4	HR, HS, CR, CS	—	0.12	—	—	—	0.60	0.050	0.050
	14	HR, HS, —, —	—	0.15	—	—	—	0.60	0.050	0.050
	15	HR, HS, —, —	—	0.20	—	—	—	0.90	0.050	0.060
3	Materials having specific requirements based on minimum strengths									
	Carbon–manganese steels									
	34/20	HR, HS, CR, CS	—	0.15	—	—	—	1.20	0.050	0.050
	37/23	HR, HS, CR, CS	—	0.20	—	—	—	1.20	0.050	0.050
	43/25	HR, HS, —, —	—	0.25	—	—	—	1.20	0.050	0.050
	50/35	HR, HS, —, —	—	0.20	—	—	—	1.50	0.050	0.050
	Micro-alloyed steels									
	40/30	HR, HS, —, CS	—	0.15	—	—	—	1.20	0.040	0.040
	43/35	HR, HS, —, CS	—	0.15	—	—	—	1.20	0.040	0.040
	46/40	HR, HS, —, CS	—	0.15	—	—	—	1.20	0.040	0.040
	50/45	HR, HS, —, CS	—	0.20	—	—	—	1.50	0.040	0.040
	60/55	—, HS, —, CS	—	0.20	—	—	—	1.50	0.040	0.040
	40F30	HR, HS, —, CS	—	0.12	—	—	—	1.20	0.035	0.030
	43F35	HR, HS, —, CS	—	0.12	—	—	—	1.20	0.035	0.030
	46F40	HR, HS, —, CS	—	0.12	—	—	—	1.20	0.035	0.030
	50F45	HR, HS, —, CS	—	0.12	—	—	—	1.20	0.035	0.030
	60F55	—, HS, —, CS	—	0.12	—	—	—	1.20	0.035	0.030
	68F62	—, HS, —, —	—	0.12	—	—	—	1.50	0.035	0.030
	75F70	—, HS, —, —	—	0.12	—	—	—	1.50	0.035	0.030

*a = thickness

(continued)

TABLE 1 – SUMMARY OF MATERIAL GRADES, CHEMICAL COMPOSITIONS, MECHANICAL PROPERTIES AND TYPES OF CARBON AND CARBON-MANGANESE SHEET AND STRIP STEELS (FROM BS 1449 PT 1) — *contd.*

Types of steel	Yield strength R_e min.	Tensile strength R_m, min.	Elongation, A, min. Original gauge length L_o			Bend mandrel diameter (180° bend)
			50 mm	80 mm	200 mm	
	N/mm²	N/mm²	%	%	%	
Extra deep drawing aluminium-killed steel Extra deep drawing aluminium-killed stabilized steel Extra deep drawing Deep drawing Drawing or forming Flanging Commercial						
Available as rimmed (R), balanced (B) or kilted (K) steels. Grain-refined balanced (B) or killed (K) steel.	200 230 250 350	340 370 430 500	29 28 25 20	(27) (26) (23) (18)	21 20 16 12	2a 2a 3a 3a
Grain-refined niobium- or titanium-treated fully killed steels having high yield strength and good formability	300 350 400 450 550	400 430 460 500 600	26 23 20 20 17	(24) (21) (18) (18) (15)	18 16 12 12 10	2a 2a 3a 3a 3.5a
The steels including F in their designations in place of the oblique line offer superior formability for the same strength levels	300 350 400 450 550 620 700	400 430 460 500 600 680 750	28 25 22 22 19 18 15	(26) (23) (20) (20) (17) (16) (13)	20 18 14 14 11 10 8	0a 0.5a 1a 1.5a 1.5a 2a 3a

Casting steels

CASTING STEELS are specified in the following British Standards:

BS 592   –   Carbon steel castings for general engineering.

BS 1398   –   Carbon-molybdenum casting steels for higher service temperatures in general engineering.

BS 1456   –   Pearlitic manganese casting steels for general engineering.

BS 1458   –   Low cost, low alloy casting steels responding to hardening.

BS 1459   –   Medium alloy casting steels combining high strength and toughness with good ductility.

BS 1461   –   Casting steels for service temperatures above 400°C.
and
BS 1462

BS 1504   –   Castings for chemical, petroleum and allied application.

BS 1506   –   Castings for chemical, petroleum and allied application.

BS 1617   –   Mild steel castings with high magnetic permeability.

BS 1760   –   Carbon steel castings suitable for surface hardening.

BS 1956   –   Chromium steel castings for high resistance to wear and abrasion.

BS 1462   –   Casting steels for service temperatures above 400°C.

TABLE 1 – SELECTION GUIDE TO BS CASTING STEELS

Properties	British Standards				
Abrasion resistance	1457	1760	1956		
Corrosion resistance	1632	1631	1630		
	1504 846	845	821	801	713
Creep resistance	1398	1631	1463	1462	1461
	1504 629	625	623	622	240
Ductility	1617	592A			
Engineering					
General engineering properties	592				
Hardness	1760	1956			
Heat resistance	1648				
High tensile strength	1458	1459	1504		
High yield stress	1458	1459	1504		
Machinability	592				
	1504–101				
	A and B				
Magnetic permeability	1617				
Notch ductility (Izod)	1456	1458			
Pressure tightness	1504	1506			
Structural strength	1459	1458	1504	1506	
Surface hardness	1760				
Wear resistance	1457	592C	1760	1956	

TABLE 2 – EXAMPLES OF CASTING STEELS

BS 4659 Type	AISI Type	Nearest Werkstoff Number	Constituents									Applications
			C	Si	Mn	Cr	Ni	W	Mo	V	Co	
BH21	H21	2581	0.33	0.30	0.35	3.00	—	9.5	0.40	0.35	—	Hot stamping dies, hot upsetting and gripper dies, hot extrusion dies.
BH21A	—	—	0.25	0.30	0.30	3.00	2.25	9.5	0.40	0.35	—	Pressure die casting dies for high melting point alloys, extrusion dies including port-hole dies, hot brass stamping dies.
BSI	SI	2547	0.50	0.65	0.30	1.50	—	2.25	—	0.20	—	Pressure die casting nozzles, lower temperature range hot punching, piercing and trimming, header and gripper dies, shear blades.
BH23	H23	2625	0.38	0.30	0.35	12.0	—	12.0	—	1.00	—	Master hobs for beryllium copper. Pressure die casting of yellow metals.
BH13	H13	2344	0.38	1.05	0.35	5.25	—	—	1.35	1.00	—	Pressure die casting dies for aluminium and magnesium. Mandrels, bolsters, scavenger pads, adapter rings and aluminium extrusion dies. Hot upsetting and gripper dies, master hobs for hot hobbing beryllium copper. Pressure die casting nozzles.

(continued)

TABLE 2 – EXAMPLES OF CASTING STEELS – *contd.*

BS 4659 Type	AISI Type	Nearest Werkstoff Number	Constituents									Applications
			C	Si	Mn	Cr	Ni	W	Mo	V	Co	
BH19	H19	—	0.40	0.25	0.35	4.25	—	4.25	0.40	2.20	4.25	Pressure die casting dies for yellow metals, extrusion dies and port-hole dies (rear section) bridge dies, hot stamping dies (extended life) hot upsetting dies.
BT1	TI	3355	0.75	—	—	4.10	—	18.00	—	1.10	—	Cores, zinc pressure die casting dies, die inserts.
BM2	M2	3343	0.80	—	—	4.00	—	6.00	5.00	2.00	—	Punches (hot die forging), punches (drop forging).
—	—	2766	0.32	—	—	1.20	4.10	—	0.22	—	—	Zinc pressure die casting dies, racks and pinions. Extrusion die holders and hot upsetting die cases and bolsters.
—	—	6582	0.38	—	—	1.05	1.50	—	0.25	—	—	Large section drop forging dies, die casting die bolsters, nut, bolt and rivet heading (lower temperature).
—	L2	2241	0.50	—	0.65	1.10	—	—	—	0.20	—	Zinc pressure die casting dies, die bolsters and ejector pins, simple hot stamping and pressing dies.

TABLE 3 – SUMMARY OF COMPOSITIONS OF BS CASTING STEELS

BS	Composition (%)							
	C	Mn	Si	S	P	Ni	Cr	Mo
592A	0.25	0.90	0.60	0.06	0.06	0.40	0.25	0.15
592B	0.35	1.00	0.60	0.06	0.06	0.40	0.25	0.15
592C	0.45	1.00	0.60	0.06	0.06	0.40	0.25	0.15
1398A	0.15–0.25	0.50–1.00	0.20–0.50	0.05	0.05	0.40	0.25	0.40–0.70
1456A	0.18–0.25	1.20–1.60	0.50	0.05	0.05	—	—	—
1456B1	0.25–0.33	1.20–1.60	0.50	0.05	0.05	—	—	—
1458A	0.30–0.40	0.50–0.75	0.60	0.05	0.05	0.75–1.00	0.50–0.80	0.15–0.20
1458B	0.30–0.40	0.50–0.75	0.60	0.05	0.05	1.30–1.80	0.90–1.40	0.15–0.40
1461	0.25	0.30–0.70	0.75	0.04	0.04	0.40	2.50–3.50	0.35–0.60
1462	0.20	0.40–0.70	0.75	0.04	0.04	0.40	4.00–6.00	0.45–0.65
1617A	0.15	0.50	0.60	0.05	0.05	0.40	0.25	0.15
1617B	0.25	0.50	0.60	0.05	0.05	0.40	0.25	0.15
1760A	0.40–0.50	1.00	0.60	0.05	0.05	0.40	0.25	0.15
1760B	0.50–0.60	1.00	0.60	0.05	0.05	0.40	0.25	0.15
1956A&B	0.45–0.55	0.50–1.00	0.75	0.05	0.05	0.40	0.80–1.20	0.15

TABLE 4 – PROPERTIES OF CASTING STEELS TO BS 592

Property	Grade A	Grade B	Grade C
Tensile strength, ton/in^2 minimum kg/mm^2 minimum	28 44	32 50.5	35 55
Yield stress or 0.5% proof stress, ton/in^2 minimum kg/mm^2 minimum	15 23.5	17 27	19 30
Elongation, % minimum on 5.65 $\sqrt{S_o}$	22	18	14
Angle of bend	120°	90°	—
Radius of bend	1½t	1½t	
Izod impact value, ft lb minimum	15	15	10

TABLE 5 – PROPERTIES OF LOW ALLOY CASTING STEELS TO BS 1398

Property	Grade A	Grade B	Grade C	Grade D	Grade E
Tensile strength, ton/in^2 minimum kg/mm^2 minimum	30 47	31 49	35 55	33 52	35 55
Yield stress or 0.5% proof stress, ton/in^2 minimum kg/mm^2 minimum	18 28	18 28	21 33	20 31.5	20 31.5
Elongation, % minimum on 5.65 $\sqrt{S_o}$	18	17	17	17	12
Angle of bend	120°	120°	120°	120°	120°
Radius of bend	1½t	1½t	3t	3t	3t
Izod impact value, ft lb minimum	15	25	20	—	—

TABLE 6 – PROPERTIES OF 1½% MANGANESE CASTING STEELS TO BS 1456

Property	Grade A	Grade B1	Grade B2
Tensile strength, ton/in^2 minimum ton/in^2 maximum kg/mm^2 minimum kg/mm^2 maximum	35 45 55 71	40 50 63 78.5	45 55 71 86.5
Yield stress or 0.5% proof stress, ton/in^2 minimum kg/mm^2 minimum	22 34.5	24 38	32 50.5
Elongation, % minimum on 5.65 $\sqrt{S_o}$	16	13	18
Izod impact value, ft lb minimum	25	20	20

TABLE 7 – PROPERTIES OF ALLOY STEEL CASTINGS TO BS 1458

Property	Grade A	Grade B	Grade C
Tensile strength,			
ton/in^2 minimum	45	55	65
ton/in^2 maximum	55	65	75
kg/mm^2 minimum	71	87	102
kg/mm^2 maximum	87	102	118
Yield stress or 0.5% proof stress,			
ton/in^2 minimum	32	38	45
kg/mm^2 minimum	51	60	71
Elongation, % minimum on			
5.65 $\sqrt{S_o}$	11	8	6
Izod impact value, ft lb minimum	25	20	15

TABLE 8 – PROPERTIES OF CHROMIUM-MOLYBDENUM STEEL CASTINGS TO BS 1461

Property	
Tensile strength,	
ton/in^2 minimum	40
ton/in^2 maximum	50
kg/mm^2 minimum	63
kg/mm^2 maximum	79
Yield stress or 0.5% proof stress,	
ton/in^2 minimum	24
kg/mm^2 minimum	38
Elongation, % minimum on	
5.65 $\sqrt{S_o}$	13
Angle of bend	120°
Radius of bend	3t
Izod impact value, ft lb minimum	20

TABLE 9 – PROPERTIES OF 5% CHROMIUM-MOLYBDENUM STEEL CASTINGS TO BS 1462

Property	
Tensile strength,	
ton/in^2 minimum	40
kg/mm^2 minimum	63
Yield stress or 0.5% proof stress,	
ton/in^2 minimum	27
kg/mm^2 minimum	43
Elongation, % minimum on	
5.65 $\sqrt{S_o}$	13
Angle of bend	90°
Radius of bend	3t
Izod impact value, ft lb minimum	20

TABLE 10 – PROPERTIES OF 9% CHROMIUM-MOLYBDENUM STEEL CASTINGS TO BS 1463

Property	
Tensile strength, ton/in^2 minimum kg/mm^2 minimum	40 63
Yield stress or 0.5% proof stress, ton/in^2 minimum kg/mm^2 minimum	27 43
Elongation, % minimum on 5.65 $\sqrt{S_o}$	13

TABLE 11 – PROPERTIES OF HIGH PERMEABILITY CARBON STEEL CASTINGS TO BS 1617

Property	Grade A	Grade B
Tensile strength, ton/in^2 minimum ton/in^2 maximum kg/mm^2 minimum kg/mm^2 maximum	22 28 35 44	26 32 41 50
Yield stress or 0.5% proof stress, ton/in^2 minimum kg/mm^2 minimum	12 19	14 22
Elongation, % minimum on 5.65 $\sqrt{S_o}$	22	22
Angle of bend	120°	120°
Radius of bend	1½t	1½t

TABLE 12 – PROPERTIES OF 13% CHROMIUM STEEL CASTINGS TO BS 1630

Property	Grade A	Grade B
Tensile strength,		
$\quad$ ton/in^2 minimum	35	45
$\quad$ kg/mm^2 minimum	55	71
Yield stress or 0.5% proof stress,		
$\quad$ ton/in^2 minimum	24	30
$\quad$ kg/mm^2 minimum	38	48
Elongation, % minimum on		
$\quad$ 5.65 $\sqrt{S_o}$	15	11
Angle of bend	120°	—
Radius of bend	2t	—

TABLE 13 – PROPERTIES OF AUSTENITIC CHROMIUM-NICKEL STEEL CASTINGS TO BS 1631

Property	Grade A	Grade B	Grade C	Grade D
Tensile strength,				
$\quad$ ton/in^2 minimum	31	31	28	31
$\quad$ kg/mm^2 minimum	49	49	44	49
0.5% proof stress,				
$\quad$ ton/in^2 minimum	13.5	13.5	12	13.5
$\quad$ kg/mm^2 minimum	21	21	19	21
Elongation, % minimum on				
$\quad$ 5.65 $\sqrt{S_o}$	26	22	26	26

TABLE 14 – PROPERTIES OF FERRITIC STEEL CASTINGS TO BS 4242

Property	Grade A	Grade B	Grade C
Tensile strength,			
ton/in^2 minimum	28	30	30
kg/mm^2 minimum	44	47	47
Yield stress,			
ton/in^2 minimum	15	18	18
kg/mm^2 minimum	23.5	28	28
Elongation, % minimum on			
5.65 $\sqrt{S_o}$	22	18	20
Charpy-V-notch impact test			
at temperature of	–40°C	–50°C	–60°C
Minimum average value,			
ft lb	15	15	15
kg m	2.1	2.1	2.1
Minimum individual value,			
ft lb	10	10	10
kg m	1.4	1.4	1.4

TABLE 15 – PROPERTIES OF 1% CHROMIUM STEEL CASTINGS TO BS 1956

Property	Grade A	Grade B	Grade C
Tensile strength,			
ton/in^2 minimum	45	—	—
kg/mm^2 minimum	71	—	—
Elongation, % minimum on			
5.65 $\sqrt{S_o}$	7	—	—

TABLE 16 – STEELS FOR THE CHEMICAL, PETROLEUM AND ALLIED INDUSTRIES

Specification	Grade	Carbon Context (% max)	Chief Alloy Constituent (% max)	UTS (lb/in² min)	Yield (lb/in²)	Elongation (%)
BS 1504–101	A	—	—	58 240	29 120	20
Carbon steel castings	B	—	—	62 720	31 360	20
for structural purposes	C	—	—	78 400	39 200	15
BS 1504–161	A	0.25	—	62 720	31 360	22
Carbon steel castings	B	0.3	—	69 440	35 840	22
for pressure applications						
BS 1504–240	—	0.25	Mo 0.7	67 200	35 840	20
Carbon-molybdenum steel castings						
BS 1504–503	—	0.15	Ni 4.0	64 960	39 200	25
3½% nickel steel castings						
BS 1504–621	—	0.2	Cr 1.5 Mo 0.65	69 440	40 320	20
1¼% chromium-molybdenum steel castings						
BS 1504–622	—	0.18	Cr 2.75 Mo 1.2	69 440	40 320	20
2½% chromium-1% molybdenum steel castings						
BS 1504–623	—	0.25	Cr 3.5 Mo 0.6	89 600	53 760	18
3% chromium-molybdenum steel castings						
BS 1504–625	—	0.2	Cr 6.0 Mo 0.65	89 600	60 480	18
5% chromium-molybdenum steel castings						
BS 1504–629	—	0.2	Cr 10.0 Mo 1.2	89 600	60 480	18
9% chromium-molybdenum steel castings						
BS 1504–713	—	0.2	Cr 13.5	89 600	64 690	18
13% chromium steel castings						
BS 1504–801	—	0.12	(Ni + Cr) 25.0 (min)	67 200	30 240	20
Austenitic chromium-nickel steel castings						
BS 1504–821	Ti	0.12	(Ni + Cr) 25.0 (min)	67 200	30 240	20
Stabilized austenitic chromium-nickel steel castings	Nb	0.12	Ni 8.5 (min) Cr 17–20	67 200	30 240	20
BS 1504–845	B	0.08	Ni 10.0 (min) Cr 18.5 Mo 3.0	67 200	30 240	15
Austenitic chromium-nickel-2½% molybdenum steel castings	Nb	0.12	Ni 10.0 (min) Cr 18.5 Mo 2.75	67 200	30 240	15
BS 1504–846	—	0.08	Ni 14.0 Cr 20.0 Mo 4.0	67 200	30 240	15
Chromium-nickel-3½% molybdenum steel castings						

SECTION 2

Non-Ferrous Metals and Alloys

NON-FERROUS METALS and ALLOYS
METAL MATRIX COMPOSITES
REFRACTORY METALS

Non-ferrous metals and alloys

ALPHABETICAL LISTING OF NON-FERROUS METALS AND ALLOYS IN COMMON USE

Admiralty brass – see *Brass.*
Aich's alloy – copper-zinc alloy for marine services.
Alclad – Dural coated with pure aluminium.
Alcoa – aluminium die casting alloy (American).
Alcomax – see *Magnetic materials* (Section 8).
Aldural – Dural coated with pure aluminium (British).
Allan's alloy – tin-free bronze for diesel engine piston rings.
Alni – see *Magnetic materials* (Section 8).
Alnico – see *Magnetic materials* (Section 8).
Alumel – thermocouple alloy.
Alpax – proprietary aluminium casting alloys (British).

Hot rolling Mill for initial reduction of 200 mm thick block to 9.5 mm thick sheet aluminium for transfer to finishing mills. (Birmetals Ltd).

ALUMINIUM AND ALUMINIUM ALLOYS

Symbol: Al
Specific gravity: 2.7
Density: 2.7 g/cm^3 (0.098 lb/in^3)
Tensile strength, soft: 3.1 to 6.3 kg/cm^2 (2.0 to 4.0 ton/in^2)
Tensile strength (99.99% purity), half hard: 7.9 to 9.5 kg/cm^2
    (5.0 to 6.0 ton/in^2)
Tensile strength (99.99% purity), hard: 10.2 kg/cm^2 (6.5 ton/in^2)
Elongation, soft: 45%
Elongation (99.99% purity), half hard: 12%
Elongation (99.99% purity), hard: 6%
Modulus of elasticity: 10×10^6 lb/in^2
Modulus of elasticity: 7×10^5 kg/cm^2
Modulus of torsion: 3.5×10^6 lb/in^2
Modulus of torsion: 2.5×10^5 kg/cm^2
Melting point: 658°C
Coefficient of expansion: 24×10^{-6} /degC
Thermal conductivity: 0.5 cal/cm/sec/degC
Specific heat: 0.22
Electrical resistivity: 2.7×10^{-6} ohm cm
Electrical conductivity: 62 to 62.9%

Commercial purity aluminium is available in various grades from 99.99% to 99% purity, the chief impurities being silicon and iron. Leading mechanical properties are summarised in Tables 1A and 1B. Pure aluminium is soft and ductile but work hardened. It is difficult to cast since it does not flow readily in

TABLE 1A – CHEMICAL COMPOSITION OF ALUMINIUM PLATE, SHEET AND STRIP (BS 1470)

Material designation			Alumin-ium %	Copper %	Silicon %	Iron %	Mangan-ese %	Zinc %	Notes
Current BS	Former BS	ISO							
1199	SI	Al 99.99	99.99‡						Cu + Si + Fe = 0.01%
1080A	SIA	Al 99.8‡	99.8‡	0.02	0.15	0.15	0.03	0.06	Cu + Si + Fe + Mn + Zn = 0.2%
1050A	SIB	Al 99.5	99.5‡	0.05	0.3	0.40	0.05	0.10	Cu + Si + Fe + Mn + Zn = 0.5
1200	SIC	Al 99.0	99.0‡	0.10	0.5	0.70	0.10	0.10	Cu + Si + Fe + Mn + Zn = 1.0%

‡Grade of aluminium with impurities limited as shown

the molten state, is weak at temperatures just below its solidification point and has high shrinkage on solidification.

Alloying elements may be included to improve both mechanical properties and casting characteristics. These alloys fall into two groups, each of which may be heat treatable to enhance their properties further, or non-heat treatable, *ie*:

(i)   wrought alloys – heat treatable
                      non-heat treatable (work hardening).
(ii)   casting alloys – heat treatable
                      non-heat treatable.

The chief alloying elements, and their effects are:

copper – increases strength and hardness and also makes the alloy heat treatable.

magnesium – increases hardness and corrosion resistance.

manganese – increases strength.

silicon – lowers the melting point and improves castability. Silicon in combination with magnesium yields a heat treatable alloy.

zinc – improves strength and hardness. Zinc in combination with magnesium yields a high strength heat treatable alloy.

TABLE 1B – MECHANICAL PROPERTIES OF ALUMINIUM PLATE, SHEET AND STRIP (BS 1470)

Material designation				Thickness mm		Tensile strength		Elongation					On 5.65 $\sqrt{S_o}$ over 12.5 mm thick min
Current British standard	Former British Standard	ISO	Condition	Over	Up to and including	Min N/mm²	Max N/mm²	on 50 mm material thicker than					
								0.5 mm min %	0.8 mm min %	1.3 mm min %	2.6 mm min %	3.00 mm min %	
	0		0	0.2	6.0	—	65	30	35	40	45	45	—
1199	SI	AI 99.99	H4	0.2	6.0	80	95	7	8	10	12	12	—
			H8	0.2	6.0	100		3	4	5	6	6	—
			M	3.0	25.0	—	—	—	—	—	—	—	—
1080A	SIA	AI 99.8	O	0.2	6.0	—	90	29	29	29	35	35	—
			H4	0.2	12.5	95	120	5	6	7	8	8	—
			H8	0.2	3.0	125	—	3	4	4	5	—	—
			O	0.2	6.0	55	95	22	25	30	32	32	—
1050A	SIB	AI 99.5	H4	0.2	12.5	100	135	4	5	6	6	8	—
			H8	0.2	3.0	135	—	3	3	4	4	—	—
			M	3.0	25.0	—	—	—	—	—	—	—	—
1200	SIC	AI 99.0	O	0.2	6.0	70	105	20	25	30	30	30	—
			H2	0.2	6.0	95	120	4	6	8	9	9	—
			H4	0.2	12.5	110	140	3	4	5	5	6	—
			H6	0.2	6.0	125	150	2	3	4	4	4	—
			H8	0.2	3.0	140	—	2	3	4	4	—	—

bismuth – improves machinability.

lead – improves machinability.

boron – improves electrical conductivity.

nickel – improves strength at elevated temperatures.

titanium – is a strong grain-refining element and improves strength and ductility.

chromium, vanadium and *zirconium* – may also be used as minor alloying elements for special effects.

Wrought alloys

The general engineering series of wrought aluminium and aluminium alloys are specified in:

BS 1470 – plate, sheet and strip.
BS 1471 – drawn tube.
BS 1472 – forging stock and forgings.
BS 1473 – rivet, bolt and screw stock.
BS 1474 – bars, extruded round tube and sections.
BS 1475 – wire.

Leading data are summarised in Tables 2 to 7.

In the former BS specifications the chemical composition is shown by a number with letter prefix H = heat treatable or N = non-heat treatable. The condition (as specified in the tables of mechanical properties) is also indicated by a letter code, viz:

M = as manufactured. Material which acquires some temper from shaping processes in which there is no special control over thermal treatment or amount of strain hardening.

O = annealed. Material which is fully annealed to obtain the lowest-strength condition.

H1, H2, H3, H4, H5, H6, H7, H8 = strain hardened. Material which is cold worked after annealing (or hot forming) or cold worked and partially annealed/stabilised to secure the specified mechanical properties. The designations are in ascending order of tensile strength.

TB – solution heat treated and naturally aged. Material which is not cold worked after solution heat treatment except as may be required to flatten or straighten it. Properties of some alloys in this temper are unstable.

TD – solution heat treated, cold worked and naturally aged.

TE – cooled from an elevated temperature shaping process and precipitation treated.

TH – solution heat treated cold worked and then precipitation treated.

For a more condensed reference, Table 8 summarises the wrought aluminium and aluminium alloys covered by BS 1470–75, whilst Table 9 is a selection grid indicating the suitability of such alloys for cold forming, machining, etc.

TABLE 2 – CHEMICAL COMPOSITION OF ALUMINIUM ALLOY PLATE, SHEET AND STRIP (BS 1470)

Current British Standard	Former British Standard	ISO	Aluminium %	Copper %	Magnesium %	Silicon %	Iron %	Manganese %	Zinc %	Chromium %	Manganese + chromium %	Chromium and titanium and/or other grain refining elements %	Titanium and/or other grain refining elements %	Condition	Thickness over (mm)	Thickness up to and incl. (mm)	0.2% proof stress min. (N/mm²)	Tensile min. (N/mm²)	Tensile max. (N/mm²)	Elong. 0.5 mm min. %	Elong. 0.8 mm min. %	Elong. 1.3 mm min. %	Elong. 2.6 mm min. %	Elong. 3.0 mm min. %	Elong. on $5.65\sqrt{S_o}$ over 12.5 mm thick min. %
3103	NS3	Al Mn 1	Rem.	0.1	0.1	0.6	0.7	0.8/1.5	0.2	—	—	0.2	—	O	0.2	6.0	—	90	130	20	23	24	24	25	—
														H2	0.2	6.0	—	120	145	5	6	7	9	9	—
														H4	0.2	12.5	—	140	175	3	4	5	6	7	—
														H6	0.2	6.0	—	160	195	2	3	4	4	—	—
														H8	0.2	3.0	—	175	—	2	3	4	4	4	—
5251	NS4	Al Mg 2	Rem.	0.10	1.7/2.4	0.5	0.5	0.5	0.2	0.25	0.5	—	0.2	M	3.0	25.0	—	160	—	18	18	18	20	20	—
														O	0.2	6.0	60	160	200	18	18	18	20	20	—
														H3	0.2	6.0	130	200	240	4	5	6	8	8	—
														H6	0.2	12.5	175	225	275	3	4	5	5	5	—
5154A	NS5	Al Mg 3.5	Rem.	0.1	3.1/3.9	0.5	0.5	0.5	0.2	0.25	0.5	—	0.2	O	0.2	6.0	85	215	275	12	14	16	18	18	—
														H2	0.2	6.0	165	245	295	5	6	7	8	8	—
														H4	0.2	6.0	225	275	325	4	4	6	6	6	—
5083	NS8	Al Mg 4.5 Mn	Rem.	0.10	4.0/4.9	0.40	0.40	0.5/1.0	0.2	0.25	—	—	0.2	M	3.0	25.0	125	275	350	—	—	—	—	—	14
														O	0.2	6.0	125	275	350	12	14	16	16	16	—
														H2	0.2	6.0	235	310	375	5	6	8	10	8	—
														H4	0.2	6.0	270	345	405	4	5	6	8	6	—
2014A	HS15	Al Cu 4 Si Mg	Rem.	3.9/5.0	0.2/0.8	0.5/0.9	0.7	0.4/1.2	0.2	0.10	—	—	0.2	TB	0.2	12.5	245	385	—	13	14	14	14	14	—
														TF	12.5	25.0	375	430	—	—	—	—	—	—	10
														TF	0.2	12.5	380	440	—	6	6	7	7	9	—
														TF	25.0	40.0	360	430	—	—	—	—	—	—	6
														TF	40.0	63.0	345	420	—	—	—	—	—	—	6
2014A CLAD	HC15*	Al Cu 4 Si Mg	Rem.	3.9/5.0	0.2/0.8	0.5/0.9	0.7	0.4/1.2	0.2	0.10	—	—	0.2	TB	0.2	12.5	230	375	—	13	14	14	14	14	—
														TB	12.5	25.0	245	385	—	—	—	—	—	—	10
														TF	0.2	3.0	325	400	—	7	7	8	8	8	—
														TF	3.0	12.5	365	425	—	—	—	—	—	—	—
														TF	12.5	25.0	380	440	—	—	—	—	—	—	6
6082	HS30	Al Si 1 Mg Mn	Rem.	0.10	0.5/1.2	0.7/1.3	0.5	0.4/1.0	0.2	0.25	—	—	0.2	O	0.2	3.0	—	—	155	16	16	16	18	—	—
														TB	0.2	3.0	120	200	—	15	15	15	15	15	—
														TB	3.0	25.0	115	200	—	—	—	—	—	—	15
														TF	0.2	12.5	255	295	—	8	8	8	8	8	—
														TF	3.0	25.0	240	295	—	—	—	—	—	—	8

TABLE 3 – CHEMICAL COMPOSITION OF ALUMINIUM ALLOY DRAWN TUBE (BS 1471)

Material designation (Current British Standard)	Material designation (Former British Standard)	ISO	Aluminium %	Copper %	Magnesium %	Silicon %	Iron %	Manganese %	Zinc %	Chromium %	Titanium and/or other grain refining elements %	Notes	Condition	Wall thickness Over mm	Wall thickness Up to and including mm	0.2% proof stress min. N/mm²	Tensile Strength min. N/mm²	Tensile Strength max. N/mm²	Elongation on 50 mm or $5.65\sqrt{S_0}$ min. %
1050A	T1B	Al 99.5	99.5*	0.05	—	0.3	0.4	0.05	0.10	—	—	Cu+Si+Fe+Mn +Zn=0.5%	O	—	12.0	—	—	—	—
													H4	—	12.0	—	100	95	18
													H8	—	12.0	—	135	135	5
1200	T1C	Al 99.0	99.0*	0.10	—	0.5	0.7	0.1	0.1	—	—	Cu+Si+Fe+Mn +Zn=1%	O	—	12.0	—	—	—	—
													H4	—	12.0	—	110	105	16
													H8	—	12.0	—	140	140	4
5251	NT4	Al Mg 2	Rem.	0.10	1.7/2.4	0.5	0.5	0.5	0.2	0.25	0.2	Mn+Cr=0.5%	O	—	10.0	60	160	200	12
													H4	—	10.0	175	225	—	5
5154A	NT5	Al Mg 3.5	Rem.	0.10	3.1/3.9	0.5	0.5	0.5	0.2	0.25	0.2	Mn+Cr=0.5%	O	—	10.0	85	215	260	15
													H4	—	10.0	200	245	—	8
5083	NT8	Al Mg 4.5 Mn	Rem.	0.10	4.0/4.9	0.40	0.40	0.5/1.0	0.2	0.25	0.2		O	—	10.0	125	275	350	15
													H2	—	10.0	235	310	—	8
6063	HT9	Al Mg Si	Rem.	0.10	0.4/0.9	0.3/0.7	0.40	0.1	0.2	0.10	0.2		O	—	10.0	100	155	155	15
													TB	—	10.0	180	200	—	8
													TF	—	10.0				
2014A	HT15	Al Cu 4 Si Mg	Rem.	3.9/5.0	0.2/0.8	0.5/0.9	0.7	0.4/1.2	0.2	0.10	0.2		TB	—	10.0	290	400	—	8
													TF	—	10.0	370	450	—	6
6061	HT20	Al Mg 1 Si Cu	Rem.	0.15/0.40	0.8/1.2	0.4/0.8	0.7	†0.2/0.8	0.2	†0.04/0.35	0.2	†Either Mn or Cr	H4	—	6.0	160	185	—	5
													TB	—	6.0	115	215	—	12
													TB	6.0	10.0	115	215	—	14
													TF	—	6.0	240	295	—	7
													TF	6.0	10.0	225	295	—	9
6082	HT30	Al Si 1 Mg Mn	Rem.	0.10	0.5/1.2	0.7/1.3	0.5	0.40/1.0	0.2	0.25	0.2		TB	—	6.0	115	215	—	12
													TB	6.0	10.0	115	215	—	14
													TF	—	6.0	255	310	—	7
													TF	6.0	10.0	240	310	—	9

TABLE 4 – CHEMICAL COMPOSITION OF ALUMINIUM AND ALUMINIUM ALLOY FORGING STOCK AND FORGINGS (BS 1472)

Current British Standard	Former British Standard	ISO	Aluminium %	Copper %	Magnesium %	Silicon %	Iron %	Manganese %	Nickel %	Zinc %	Chromium %	Titanium and/or other grain refining elements %	Notes	Condition	Condition of test sample bar	Over mm	Up to and including mm	0.2% Proof stress min. N/mm²	Tensile strength min. N/mm²	Elongation on 50 mm or $5.65\sqrt{S_o}$ min. %
1050A	F1B	Al 99.5	99.5	0.05	—	0.3	0.4	0.05	—	0.10	—	—	Cu+Si+Fe+Mn+Zn=0.5%	M	Forged or extruded	—	150	—	60	22
5251	NF4	Al Mg 2	Rem.	0.10	1.7/2.4	0.5	0.5	0.5	—	0.2	0.25	0.2	Mn+Cr=0.5%	M	Forged or extruded	—	150	60	170	16
5154A	NF5	Al Mg 3.5	Rem.	0.10	3.1/3.9	0.5	0.5	0.5	—	0.2	0.25	0.2	Mn+Cr=0.5%	M	Forged or extruded	—	150	100	215	16
5083	NF8	Al Mg 4.5 Mn	Rem.	0.10	4.0/4.9	0.40	0.40	0.5/1.0	—	0.2	0.25	0.2		M	Forged or extruded	—	150	130	280	12
6063	HF9	Al Mg Si	Rem.	0.10	0.4/0.9	0.3/0.7	0.40	0.1	—	0.2	0.10	0.2		TB	Forged or extruded	—	150	85	140	16
														TB	Forged or extruded	150	200	85	125	13
														TF	Forged or extruded	—	150	160	185	10
														TF	Forged or extruded	150	200	130	150	6
2031	HF12	Al Cu 2 Ni 1 Mg Fe Si	Rem.	1.8/2.8	0.6/1.2	0.5/1.3	0.6/1.2	0.5	0.6/1.4	0.2	—	0.2		TB	Forged	—	150	160	310	13
														TB	Extruded	—	200	145	310	13
														TF	Forged	—	150	300	385	6
														TF	Extruded	—	200	285	385	6
2014A	HF15	Al Cu 4 Si Mg	Rem.	3.9/5.0	0.2/0.8	0.5/0.9	0.7	0.4/1.2	—	0.2	0.10	0.2		TB	Forged	—	150	215	370	13
														TB	Extruded	—	20	230	370	11
														TB	Extruded	20	75	250	390	11
														TB	Extruded	75	150	250	390	8
														TB	Extruded	150	200	230	370	8
														TF	Forged	—	150	395	450	6
														TF	Extruded	—	20	370	435	7
														TF	Extruded	20	75	435	480	7
														TF	Extruded	75	150	420	465	7
														TF	Extruded	150	200	390	435	7
2618A	HF16	Al Cu 2 Mg 1.5 Fe 1 Ni 1	Rem.	1.8/2.7	1.2/1.8	0.25	0.9/1.4	0.2	0.8/1.4	0.2	—	0.2		TF	Forged or extruded	—	200	340	430	5
6082	HF30	Al Si 1 Mg Mn	Rem.	0.10	0.5/1.2	0.7/1.3	0.5	0.40/1.0	—	0.2	0.25	0.2		TB	Forged	—	150	120	185	16
														TB	Extruded	—	150	120	190	16
														TB	Extruded	150	200	100	170	13
														TF	Forged	—	150	255	295	8
														TF	Extruded	—	20	255	295	8
														TF	Extruded	20	150	270	310	8
														TF	Extruded	150	200	240	280	5

TABLE 5 – CHEMICAL COMPOSITION AND MECHANICAL PROPERTIES OF ALUMINIUM AND ALUMINIUM ALLOY, RIVET, BOLT AND SCREW STOCK (BS 1473)

Current British Standard	Former British Standard	ISO	Aluminium %	Copper %	Magnesium %	Silicon %	Iron %	Manganese %	Zinc %	Chromium %	Titanium and/or other grain refining elements %	Condition of supply	Condition of test	Diameter Over mm	Diameter Up to and including mm	0.2% Proof stress N/mm²	Tensile strength N/mm²	Tensile strength N/mm²
RIVET STOCK																		
1050A	R1B	Al 99.5	99.5	0.05	—	0.3	0.4	0.05	0.10	—	—	H5	H5	—	12	—	110	—
5154A	NR5	Al Mg 3.5	Rem.	0.10	3.1/3.9	0.5	0.5	0.5	0.2	0.25	0.2	O or M	O or M	—	25	—	215	—
												H2 annealed and drawn 10-20% reduction in area	H2 annealed and drawn 10-20% reduction in area	—	25	—	245	—
5056A	NR6	Al Mg 5	Rem.	0.10	4.5/5.5	0.3	0.5	0.5	0.2	0.25	0.2	O or M	O or M	—	25	—	255	—
												H2 annealed and drawn 10-20% reduction in area	H2 annealed and drawn 10-20% reduction in area	—	25	—	280	—
2014A	HR15	Al Cu 4 Si Mg	Rem.	3.9/5.0	0.2/0.8	0.5/0.9	0.7	0.4/1.2	0.2	0.10	0.2	H2 annealed and drawn 20-40% reduction in area	TB		12	—	385	—
6082	HR30	Al Si Mg Mn	Rem.	0.10	0.5/1.2	0.7/1.3	0.5	0.4/1.0	0.2	0.25	0.2	H2 annealed and drawn 20-40% reduction in area	TB	—	25	—	200	—
BOLT AND SCREW STOCK																		
5056A	NB6	Al Mg 5	Rem.	0.10	4.5/5.5	0.3	0.5	0.5	0.2	0.25	0.2	H4	H4	—	12	240	310	360
2014A	HB15	Al Cu 4 Si Mg	Rem.	3.9/5.0	0.2/0.8	0.5/0.9	0.7	0.4/1.2	0.2	0.10	0.2	H2 annealed and drawn 20-40% reduction in area	TF	—	12	390	430	—
6061	HB20	Al Mg Si Cu	Rem.	0.15/0.40	0.8/1.2	0.4/0.8	0.7	0.2/0.8	0.2	†0.04/0.35	0.2	TH	TH	—	12	245	310	—
6082	HB30	Al Si Mg Mn	Rem.	0.10	0.5/1.2	0.7/1.3	0.5	1.4/1.0	0.2	0.25	0.2	H2 annealed and drawn 20-40% reduction in area	TF	— / 6	6 / 12	255/270	295/310	—

TABLE 6 – CHEMICAL COMPOSITION AND MECHANICAL PROPERTIES OF ALUMINIUM AND ALUMINIUM ALLOY BARS, EXTRUDED ROUND TUBE AND SECTIONS (BS 1474)

Current British Standard	Former British Standard	ISO	Aluminium %	Copper %	Magnesium %	Silicon %	Iron %	Manganese %	Zinc %	Chromium %	Titanium and/ or other grain refining elements %	Condition	Thickness Over mm	Thickness Up to and including mm	0.2% Proof stress min. N/mm²	Tensile strength min. N/mm²	Tensile strength max. N/mm²	Elongation on 5.65√S_o min. %	Elongation on 50 mm max. %
1050A	E1B	Al 99.5	**99.5**	0.05	—	0.3	0.4	0.05	0.10	—	—	M	—	—	—	60	—	25	23
1200	E1C	Al 99.0	**99.0**	0.10	—	0.5	0.7	0.1	0.1	—	—	M	—	—	—	65	—	20	18
5251	NE4	Al Mg 2	Rem.	0.10	**0.7/ 2.4**	0.5	0.5	0.5	0.2	0.25	0.2	M	—	150	60	170	—	16	14
5154A	NE5	Al Mg 3.5	Rem.	0.10	**3.1/ 3.9**	0.5	0.5	0.5	0.2	0.25	0.2	O	—	150	85	215	275	18	16
												M	—	150	100	215	—	16	14
5083	NE8	Al Mg 4.5 Mn	Rem.	0.10	**4.0/ 4.9**	0.40	0.40	**0.5/ 1.0**	0.2	0.25	0.2	O	—	150	125	275	—	14	13
												M	—	150	130	280	—	12	11
6063	HE9	Al Mg Si	Rem.	0.10	**0.4/ 0.9**	**0.3/ 0.7**	0.40	0.1	0.2	0.10	0.2	O	—	200	—	100	140	15	13
												M	—	200	—	130	—	13	12
												TB	—	150	70	120	—	16	14
												TB	150	200	70	150	—	13	—
												TE	—	25	110	185	—	8	7
												TF	—	150	160	150	—	8	7
												TF	150	200	130	150	—	6	—
2014A	HE15	Al Cu 4 Si Mg	Rem.	**3.9/ 5.0**	**0.2/ 0.8**	**0.5/ 0.9**	0.7	**0.4/ 1.2**	0.2	0.10	0.2	TB	—	20	230	370	—	11	10
												TB	20	75	250	390	—	11	—
												TB	75	150	250	390	—	8	—
												TB	150	200	230	370	—	8	—
												TF	—	20	370	435	—	7	—
												TF	20	75	435	480	—	7	—
												TF	75	150	420	465	—	7	—
												TF	150	200	390	435	—	7	—
6061	HE20	Al Mg 1 Si Cu	Rem.	**0.15/ 0.40**	**0.8/ 1.2**	**0.4/ 0.8**	0.7	**‡0.2/ 0.8**	0.2	**‡0.04 0.35**	0.2	TB	—	150	115	190	—	16	14
												TF	—	150	240	280	—	8	7
6082	HE30	Al Si 1 Mg Mn	Rem.	0.10	**0.5/ 1.2**	**0.7/ 1.3**	0.5	**0.40/ 1.0**	0.2	0.25	0.2	O	—	200	—	—	170	16	14
												M†	—	200	—	110	—	13	12
												TB	—	150	120	190	—	16	14
												TB	150	200	100	170	—	13	—
												TF	—	20	255	295	—	8	—
												TF	20	150	270	310	—	8	7
												TF	150	200	240	280	—	5	—

NOTE 2. In the case of extruded round tube and hollow sections the maximum thickness for which properties apply is 75 mm. † Properties in the M condition are given for purposes of information only. ‡ As indicated in the column headed 'Notes', for material designated 6061 (HE20), it is permissible to have either manganese or chromium to the levels indicated in the respective columns.

TABLE 7 – CHEMICAL COMPOSITION AND MECHANICAL PROPERTIES OF ALUMINIUM AND ALUMINIUM ALLOY WIRE (BS 1475)

Current British Standard	Former British Standard	ISO	Aluminium %	Copper %	Magnesium %	Silicon %	Iron %	Manganese %	Zinc %	Chromium %	Titanium and/or other grain refining elements %	Condition	Diameter Over mm	Diameter Up to and including mm	Tensile strength min N/mm²	Tensile strength max N/mm²
1080A	G1A	Al 99.8	99.8	0.02	—	0.15	0.15	0.03	0.06	—	—	O	—	10	—	90
												M	—	10	125	—
												H8	—	10	—	—
1050A	G1B	Al 99.5	99.5	0.05	—	0.3	0.4	0.05	0.10	—	—	O	—	10	—	95
												M	—	10	135	—
												H8	—	10	—	—
4047A	NG2	Al Si 12	Rem.	0.10	0.2	10.0/13.0	0.6	0.5	0.2	—	—	M	—	10	—	—
4043A	NG21	Al Si 5	Rem.	0.10	0.2	4.5/6.0	0.6	0.5	0.2	—	—	M	—	10	—	—
3103	NG3	Al Mn 1	Rem.	0.1	0.1	0.6	0.7	0.8/1.5	0.2	0.2	0.2	O	—	10	—	130
												M	—	10	175	—
												H8	—	10	—	—
5251	NG4	Al Mg 2	Rem.	0.10	1.7/2.4	0.5	0.5	0.5	0.2	0.25	0.2	O	—	10	—	215
												M	—	10	175	—
												H8	—	10	260	—
5154A	NG5	Al Mg 3.5	Rem.	0.10	3.1/3.9	0.5	0.5	0.5	0.2	0.25	0.2	M	—	10	—	—
5056A	NG6	Al Mg 5	Rem.	0.10	4.5/5.5	0.3	0.5	0.5	0.2	0.25	0.2	O	—	10	—	310
												M	—	10	250	—
												H4	—	10	310	360
												H8	—	10	385	—
5556A	NG61	Al Mg 5.2 Mn Cr	Rem.	0.10	5.0/5.5			0.6/1.0	0.2	0.05/0.20	0.5 min. 0.20 max.	M	—	10	—	—
6063	HG9	Al Mg Si	Rem.	0.10	0.4/0.9	0.3/0.7	0.40	0.1	0.2	0.10	0.2	M	—	10	140	—
												TB	—	10	185	—
												TF	—	10	280	—
												TD	6	—	230	—
													—	10		
2014A	HG15	Al Cu 4 Si Mg	Rem.	3.9/5.0	0.2/0.8	0.5/0.9	0.7	0.4/1.2	0.2	0.10	0.2	TB	—	10	385	—
												TF	—	10	430	—
6061	HG20	Al Mg 1 Si Cu	Rem.	0.15/0.40	0.8/1.2	0.4/0.8	0.7	0.2/0.8	0.2	0.01/0.35	0.2	TH	—	6	370	—
													6	10	355	—

TABLE 8 – SUMMARY OF WROUGHT ALUMINIUM ALLOYS TO BS 1470–75

Type	BS designation	Nominal composition
Pure aluminium	1199	99.99% purity
	1080A	99.8% purity
	1050A	99.5% purity
	1200	99.0% purity
Aluminium-manganese	3103	1.25% manganese
Aluminium-magnesium	5251	2.25% magnesium
	5154A	3.5% magnesium
	5056A	5% magnesium
Aluminium-magnesium-manganese	5083	4.5% magnesium, 0.5 to 1% manganese
Aluminium-magnesium-silicon	6063	0.7% magnesium, 0.5% silicon
	6061	1% magnesium, 0.6% silicon
Aluminium-magnesium-silicon-manganese	6082	1% magnesium, 1% silicon, up to 1% manganese
Aluminium-copper-magnesium-silicon-manganese	2014A	4.5% copper plus other elements
Aluminium-copper-nickel-magnesium-iron-silicon	2031	2.25% copper, 1% magnesium, 1% nickel
	2618A	2.25% copper, 1.5% magnesium, 1% nickel

TABLE 9 – SELECTION GUIDE TO WROUGHT ALLOYS

Condition	BS 1470-75	Tensile strength N/mm^2	Cold forming	Machining	Inert gas shielded arc welding	Resistance spot welding	Resistance to atmospheric attack
Non-heat treated (in the annealed condition)	1199	55–100	VG	P	G	F	VG
	1080A	55–120	VG	P	VG	F	VG
	1050A	62–130	F	F	VG	G	VG
	1200	62–140	VG	F	VG	G	VG
	3103	100–170	VG	G	VG	VG	VG
	5251	170–260	G	G	G	VG	VG
	5154A	200–280	G	G	G	VG	VG
	5056A	250–385	G	G	G	VG	VG
	5083	260–295	G	G	G	VG	VG
Solution heat heat treated	6063	125–155	G	G	G	G	G
	6061//6082	170–215	G	G	G	G	G
	2031	310	F	G	NA	NA	F
	2014A	370–400	F	VG	NA	VG	P
	2014A*	370	F	G	NA	VG	G
Fully heat treated	6063	155–215	G	G	G	G	G
	6061/6082	250–310	F	VG	G	G	G
	2031	385	P	VG	NA	NA	F
	2014A	400–480	P	VG	NA	VG	P
	2014A*	400	P	G	NA	VG	G
	2618A	430	U	VG	NA	NA	F

Key: VG–very good   G–good   F–fair   P–poor   NA–not applicable   U–unsuitable   *–clad.

Casting alloys

Casting alloys are specified in BS 1490. A summary of these alloys is given in Table 10, together with a general selection guide in Table 11.

Heat treatments

Heat treatments can consist of one or more processes, viz:

Solution heat treatment – heating to about 500°C for a short period followed by rapid cooling (usually by quenching in water). This leaves the metal in a soft condition for forging, etc, after which hardening develops naturally with age.

Precipitation heat treatment – heating to about 120 to 200°C for a period, which has the effect of artificial ageing.

ANTIMONY

Symbol: Sb
Density: 6.62 g/cm^3 (0.239 lb/in^3)
Tensile strength (approx): 112 kg/cm^2 (1600 lb/in^2)
Modulus of elasticity: 11.3×10^6 lb/in^2
Modulus of torsion: 200 kg/cm^2 (2800 lb/in^2)
Melting point: 630°C (1167°F)

TABLE 10 – SURVEY OF ALUMINIUM CASTING ALLOYS TO BS 1490

Type	BS designation	Composition and/or additions	Remarks
Aluminium	LMO	99.5% purity aluminium	
Aluminium – 5%	LM4	Copper	Heat treatable
silicon	LM16	Copper and magnesium	Heat treatable
	LM18	None	
	LM22	Copper and manganese	Solution treatable
Aluminium – 7 to 9%	LM24	Copper and manganese	For pressure die-casting
silicon	LM25	Magnesium	Heat treatable
	LM26	Copper and magnesium	
	LM27	Copper	
Aluminium – 10 to	LM2	Copper and iron	For pressure die-casting
12% silicon	LM6	None	
	LM13	Copper and magnesium	Heat treatable
	LM20	None	
Aluminium-silicon	LM28	19% silicon plus copper,	Heat treatable
hypereutectic alloys		magnesium and nickel	
	LM29	23% silicon plus copper,	Heat treatable
		magnesium and nickel	
Aluminium-	LM5	5% magnesium	
magnesium	LM10	10% magnesium	Heat treatable
Aluminium-copper	LM12	10% copper plus silicon and magnesium	

Coefficient of linear expansion: 8.5 to 10.8×10^{-6} degC

Thermal conductivity (circa 20°C): 0.045 cal/cm^2/cm/sec/degC

Electrical resistivity at 0°C: 39 m ohm cm

Antimony is a brittle, lustrous white metal with a bluish tinge. Its structure is laminar or granular, depending on the rate of cooling. It is a poor conductor of heat and electricity. A peculiar characteristic of antimony is expansion on solidification, a property which is also present in its alloys with other metals.

Crude antimony is used in pyrotechnics and as the basis of antimony compounds such as stibuite (antimony sulphide) used in the vulcanisation of rubber; its sulphides and oxides are used as pigments.

Metallic antimony is used as an electro-plated coating (particularly on steel) giving high resistance to abrasion and wear (although softer than nickel or chromium); special plating processes are necessary to avoid brittle plating with

TABLE 11 – GENERAL GUIDE TO THE SELECTION OF ALUMINIUM CASTING ALLOYS

| BS 1490 | Casting characteristics | | | | | General properties | | |
| | Suitability for | | | Fluidity | Resistance to hot tearing | Resistance to corrosion | Machinability | Pressure tightness |
	Sand casting	Gravity die casting	Pressure die casting					
LM0	F	F	F	F	P	P	F	F
LM2	G*	G*	E	G	E	E	F	G
LM4	G	G	G	G	G	G	G	G
LM5	F	F	F*	F	F	F	G	P
LM6	E	E	G	E	E	E	F	E
LM9	G	E	G*	G	E	E	F	G
LM10	F	F	F*	F	G	G	G	P
LM12	F	G	U	F	G	G	E	G
LM13	G	G	F*	G	E	E	F	F
LM16	G	G	G*	G	G	G	G	G
LM18	G	G	G*	G	E	E	F	E
LM20	E*	E	G	E	E	E	F	E
LM21	G	G	G*	G	G	G	G	G
LM22	G*	G	G*	G	G	G	G	G
LM24	F*	F*	E	G	G	G	F	G
LM25	G	E	G*	G	G	G	F	G
LM26	G	G	F*	G	G	G	F	F
LM27	G	E	G*	G	G	G	G	G
LM28	P	F	—	F	G	G	P	F
LM29	P	F	—	F	G	G	P	F

Key: *Not usually cast by this method.

E – excellent   G – good   F – fair   P – poor   U – unsuitable

the possibility of the coating flaking off. The use of antimony plating is limited
by the fact that the coating is toxic.

The main application of antimony is as an alloying element with other metals
to produce casting and bearing alloys, *eg*:

antimony-tin – type metal and Britannia metal.

tin-antimony-copper – bearing alloys.

antimony-bronze (7 to 8% antimony, 2% nickel) – as a substitute to copper-
nickel-tin for alloy for gears.

Regulus metal (basically lead with 6 to 12% antimony) – used for battery
plates, sheet and pipes for the chemical process industry, cable sheathing, and
miscellaneous 'lead' castings.

Antimony is also used as a constituent in some solders, giving increased strength.

The antimony normally available commercially is of 99% or 98 to 99% minimum purity.

ARGENTAN

Nickel-silver (see *Nickel*).

ARMCO

see *Magnetic materials* (Section 8).

ARSENIC

Symbol: As
Specific gravity: 5.7
Density: 5.7 g/cm^3 (0.205 lb/in^3)

Arsenic is usually produced as a byproduct from the treatment of copper,
gold, lead, tin and zinc ores. It occurs naturally as sulphides – realgar (As_2S_2)
and mispickel (FeSAs); and as the oxide orpinent (As_2O_3).

The pure material is a brittle, steel-grey crystalline metalloid. There are also
two allotropes yellow arsenic (sg 3.7) and black arsenic (sg 4.7).

The principal uses of arsenic are:

As a hardener for lead – *eg* in lead sheet and some whitemetal bearing alloys.

As a copper 'improver' – a small (0.3 to 0.5%) addition of aresenic improving
the heat-resistance of copper.

In glass manufacture – as an addition for bronzing, decolouring or the production of opal glass.

Arsenic compounds are used as weed killers, fungicides, insecticides and in
wood-reserving solutions, although their general application in these fields is
limited by their poisonous nature.

BABBIT METAL

See *Bearing alloys* (Section 7).

BAHN METAL

See *Bearing alloys* (Section 7).

BARIUM

Symbol: Ba
Specific gravity: 3.66
Density: 3.66 g/cm^3 (0.13 lb/in^3)
Tensile strength: n/a
Modulus of elasticity: n/a
Modulus of torsion: n/a
Melting point: 704°C (1317°F)
Coefficient of linear expansion: n/a
Specific heat at 20°C: 0.068 cal/g/degC
Thermal conductivity: n/a
Electrical resistivity: n/a

Barium is a soft, malleable silvery-white metal (tinged yellow if it contains nitrogen). It oxidises rapidly in air and can be spontaneously combustible in moist air and hydrogen. It readily burns to barium oxide when heated.

Industrially, barium is used only as an alloy, *eg* binary alloys with nickel for automobile ignition equipment. Ternary alloys with lead and calcium as bearing metals. Ternary alloys with aluminium and magnesium for thermionic valves (now largely replaced by 0.15% barium nickel bearing alloy).

Copper or iron-clad barium wire is also used as a 'getter' material.

Barium compounds are used as extenders in paints, paper, soap and rubber, also in the ceramic and glass industries. Barium paints are used as coatings for walls of buildings housing X-ray and gamma ray equipment. Barium cement is used for making salt water resistant concrete. Barium nitrate is used in the manufacture of pyrotechnic flares.

Barium metal is available commercially in pure form (up to 99.55% purity). Barytes (the spar form) is available both unground and ground and separated (from witherite).

BATH METAL

55/45 copper-zinc alloy.

BELL METAL

See *Bronzes.*

BERYLLIUM

Symbol: Be
Specific gravity: 1.848
Density: 1.848 g/cm^3 (0.067 lb/in^3)
Tensile strength: 2200 to 3920 kg/cm^2 (14 to 25 tons/in^2)
Yield strength: 1900 to 2660 kg/cm^2 (27000 to 38000 lb/in^2)
Modulus of elasticity: 2.8 to 3.08×10^6 kg/cm^2 (40 to 44×10^6 lb/in^2)
Elongation: 1 to 3.5% at room temperature
Melting point: 1277°C (2336°F)
Coefficient of linear expansion: 11.6×10^{-6} degC
Poisson's ratio: 0.024 to 0.030
Specific heat at 20°C: 0.45 cal/g/degC
Thermal conductivity (circa 20°C): 0.35 cal/cm^2/m/sec/degC
Electricity resistivity at 20°C: 4 μ ohm cm

Beryllium is a white metal similar in appearance to aluminium, but lighter. It is resistant to corrosion at room temperatures (although attacked by certain acids and alcohols). The metal is not attacked by hydrogen at any temperature. It is brittle at room temperatures but can be formed, etc by hot working (becoming brittle again on cooling). Hot working temperature is 800 to 1100°C (1470 to 2000°F), and casting temperature 1350°C (2460°F) see Table 12. It is most readily fabricated by hot rolling or extrusion, all working being done whilst hot. It can be cast successfully in the absence of oxygen (eg in a vacuum or a dry, inert atmopshere of argon or helium).

Cast beryllium has a tensile strength of about 2170 bar (31 000 lb/in^2) and a hardness of about 60 to 70 Rockwell B. Tensile strength can be increased to about 2800 bar (40 000 lb/in^2) by extrusion. Mechanical properties are dependent on the orientation and grain size of the crystals and extreme anisotropy may be found in the unwrought metal. Ductility can be improved by slow cooling.

TABLE 12 – WORKING TEMPERATURES FOR BERYLLIUM

Process	Working temperature °C	Remarks
Rolling	900–1000	Limited reduction possible.
Extrusion	200– 230	To avoid recrystallation.
	or	
	900–1000	For maximum reduction.
Spinning	400– 800	Limited formability.
Cupping	400– 800	
Forging	410– 900	Limited workability.
Pressure welding‡	950–1150	

‡Brazing (with aluminium or silver) is usually a better jointing method than fusion welding or pressure welding.

The pure metal is used in plate form for transparencies for X-rays, as moderators in atomic piles, as electrodes for neons and targets for cyclotices. It is also used as a hydrogen-resistant coating for steels worked at high temperatures. The material is also used in sintered powdered form, when it can be worked with tungsten carbide tools. Powdered beryllium burns with a brilliant white flame.

One of the main applications of beryllium is an an alloying element for hardening and improving the mechanical characteristics of cast copper and copper nickel alloys and to improve the stability and grain structure of aluminium alloys. Equally important is the addition of beryllium to improve the heat-treatment characteristics of copper, cobalt, iron and nickel – *eg* heat treatment of beryllium-copper yields a strength some six times greater than that of pure copper whilst not detracting from its electrical conductivity properties to any noted extent. Beryllium-nickel alloys (heat treated) can show mechanical properties superior to the best spring steel as well as being suitable for use at much higher temperatures. Beryllium-iron alloys show exceptional corrosion and heat resistance.

The working of beryllium necessitates precautions since the metal and its compounds are toxic if inhaled or ingested the exception being beryllium oxide. This oxide is used as a refractory material with excellent resistance to thermal shock. Melting point of beryllium oxide (beryllia) is 2570°C (4660°F).

Beryllium metal is available commercially in pure form (99.8% purity), the impurity being mainly oxide.

BIRMA BRIGHT

Proprietary aluminium alloys (British).

BIRMAC

Proprietary aluminium die-casting alloys (British).

BIRMETAL

Proprietary aluminium-based alloys (British).

BISMUTH

Symbol: Bi
Specific gravity: 9.80
Density: 9.80 g/cm^3 (0.354 lb/in^3)
Hardness: 7.0 Brinell
Modulus of elasticity: 3.2×10^5 kg/cm^2 (4.6×10^6 lb/in^2)
Melting point: 271°C (520°F)
Coefficient of linear expansion: 13.3×10^{-6} degC
Specific heat at 20°C: 0.0294 cal/g/degC
Thermal conductivity: n/a
Electrical resistivity: 106.8 μ ohm cm

Bismuth is a coarsely crystalline soft greyish-white metal, often tinged with red. It is characterised by a low melting point and expands on solidification. It is thus widely used as a constituent of fusible alloys with lead, tin and cadmium. Both the expansion characteristics and the actual melting point of these alloys vary with the bismuth content – eg Wood's metal melts in hot water. Bismuth and bismuth alloy are used in extruded fine wire form for thermocouples, pyrometers, galvanometer suspensions, optical hairline sights, etc.

Bismuth is also used as an alloying element (0.1 to 0.5%) in austenitic stainless steels to improve machining qualities, and as a free-cutting addition to certain zinc based alloys. It is also used as a deoxidising agent in cast iron, and for composite alloys in powder metallurgy, particularly for the production of sintered electrical contacts and sintered bearings. The largest application of bismuth, however, is for pharmaceutical usage.

Metallic bismuth can be produced by electrical processes yielding a purity of 99.95%. Somewhat lower purity is usually acceptable for metallurgical purposes.

BORON

Symbol: B
Specific gravity: 2.6

The metalloid element boron is an amorphous powder which can be produced with a purity of about 95%, or of up to 99% by further treatment. Crystalline (adamantine) boron is obtained by fusion of the powdered form with aluminium.

Boron is used as a hardening agent in steels, 0.0025 to 0.003% boron producing increasing hardenability, ultimate strength and elastic limit without affecting ductility. In this respect boron is some 400 times more effective than nickel, 200 times more effective than chromium and nearly 100 times more effective than molybdenum.

Residual (or additional) boron in cast iron (0.02 to 0.12%) improves hardness and wear resistance, and also provides grain refinement. Sodium-boron oxide (borax) has wide application as a fluxing agent. Boron carbide is an extremely hard abrasive.

BRASS

The description *brass* covers a whole range of copper-zinc alloys but is generally taken to refer to those alloys with a copper content of 55 to 80%. Copper-zinc alloys with a higher copper content are generally known as *gilding metals,* or by specific names, *eg* red brass, Dutch metal, pinchbeck, etc.

There are two main groups of brasses. The first group is specially suited for cold rolling into sheets, wire drawing, tube manufacture, pressing or other cold working, and in this group are 'basis quality' brass (copper content about 63%) suitable for a wide range of uses, and cartridge brass (70% copper, 30% zinc) which is a material for cartridge case manufacture, but also has many other uses.

The second group, of which yellow metal or Muntz metal (60% copper, 40% zinc) may be taken as typical, is specially suitable for casting, or hot working by rolling, extruding or hot stamping.

Besides these compositions, almost every one of the other copper-zinc percentages obtainable down to 55% copper is used for some purposes, sometimes with small additions of other metals as hardeners or to enhance resistance to certain corrosive conditions.

Alpha brasses

Zinc will dissolve in molten copper in all proportions of the mixture is molten, giving a uniform liquid solution. This condition of uniform solution may be obtainable also in the brass when solidified if the copper content is not less than about 63%.

Thus if a mixture of 70% copper and 30% zinc is melted together to form cartridge brass, the resultant solidified metal consists of crystals of a uniform 'solid solution' of zinc dissolved in copper, as 'alpha' solid solution. Under carefully controlled annealing conditions with very slow cooling, this alpha solid solution may be made to contain up to 39% zinc before a second constituent, richer in zinc, appears in the solid metal at normal temperatures. In practice a composition of about 63% copper, 37% zinc may be considered the minimum copper content to secure a pure alpha brass.

Alpha brasses combine good strength with considerable ductility when cold, and brasses for making into sheet, strip, wire and tube generally come within this class. They are all suitable for cold working.

Alpha-beta brasses

In a brass with increased zinc content a second zinc-rich constituent, 'beta' solid solution, also becomes apparent as distinct crystals of reddish colour, and the resulting duplex structure is known as an 'alpha-beta' brass.

When cold, the beta crystals are hard and serve to increase the tensile strength of the brass while lowering its ductility, and brasses of 60/40 type are widely used for general-purpose castings. When hot, the presence of the beta constituent renders the brass plastic over a wide temperature range, and as a result brasses containing 57 to 61% copper are readily workable by hot rolling, extruding or hot stamping at temperatures from 600 to 800°C.

By contrast the alpha brasses retain comparatively high strength in compression at high temperatures, so that considerable power is needed to hot work them.

Brass seldom contains less than 55% copper to 45% zinc, as slightly below this ratio only the hard and brittle beta or other constituents are present, so that the metal cannot be worked, and the brass is not of engineering interest except for brazing solder, where the lower melting-point is an advantage.

Gilding metals

As the copper content in copper-zinc alloys is reduced, the warm reddish colour of copper is gradually changed, and the gilding metals containing 95% down to

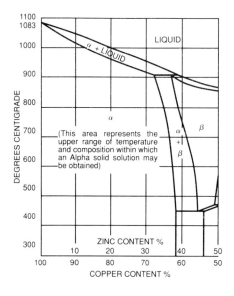

Fig. 1

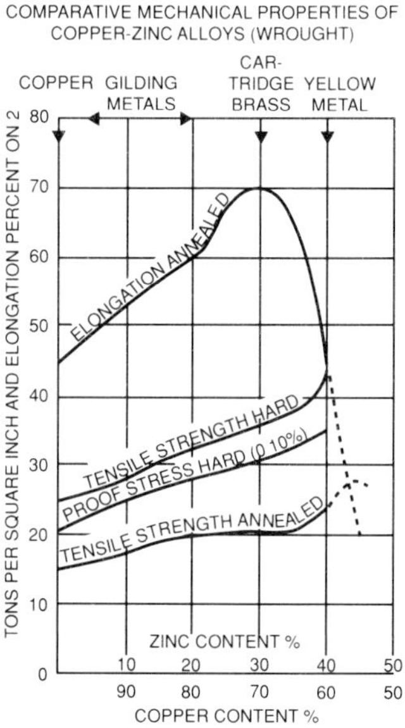

Fig. 2

80% copper have, as their name implies, a range of golden colours, grading to yellow. When the copper content is reduced to about 70% (as in the cartridge brasses), the yellow colour becomes paler, inclining to greenish, but if the copper content is still further reduced, the yellow colour again becomes warmer, and 60/40 brass has a characteristic ochre colour, hence its name 'yellow metal'. When the copper content is reduced to 55%, the metal once more exhibits a golden colour.

Copper-zinc alpha alloys with a copper content of 80 to 95% are called *gilding metals,* the golden colour of which may be further enhanced by polishing or pickling in nitric acid to remove the copper coloured skin formed during heat treatment.

For engineering purposes, the gilding metals are normally supplied as cold rolled sheet or strip, and the mechanical properties are intermediate between those of cartridge brass and copper, see Figure 2.

Mechanical properties

The tensile strength and other mechanical properties of brasses largely depend on the grain and/or crystal structure, which in turn is determined by whether the brass has been cast, forged, extruded, rolled or annealed, and by the amount of subsequent cold work done on it. In general, hot or cold working with subsequent annealing odifies the grain structure of cast brass so that its strength is increased.

The composition, designation and mechanical properties of the main commercial brasses are given in Tables 13 to 18. The following British Standards, some of which have been withdrawn from use, are included for reference.

BS 249   – Leaded brass; rods and sections.
BS 251   – Marine brass; rods, sections, forgings.
BS 251   – Marine brass, rods and sections other than forgings.
BS 265   – Brass sheet; strip, foil.
BS 267   – Brass sheet; strip, foil (70/30).
BS 409   – Marine brass; sheet, strip and plate.
BS 711   – Brass sheet; strip, foil, cold rolled.
BS 944   – Leaded bars.
BS 1001   – Hot stock for forgings.
BS 1949   – 60/40 brass rods, sections, forgings.
BS 2786   – Brass wire for springs.

BRIGHTRAY

Proprietary nickel alloys (British).

BRONZE

Bronzes are basically alloys of copper and tin (tin bronze) but may also contain proportions of other elements such as zinc, lead nickel and phosphorus. Tin content normally lies between 5 and 10% but may range up to 20% in certain alloys. An exception is the traditional 'bell bronze' which contains about 25% tion; also *speculum metal* for decorative plating which contains an even higher tin content.

Tin bronzes containing zinc are known as *gunmetals,* and those containing phosphorus (0.02 to 0.40%) as *phosphor bronze.* In the latter case the percent (%) figure which may be associated with the name refers to the tin content – *eg* 5% phosphor bronze is a 95:5 copper:tin alloy with 0.02 to 0.40% phosphorus. *Bearing bronzes* are usually phosphor bronzes with up to 20% lead content, but also embrace some gunmetals. Some may also incorporate nickel.

Alloys of tin with aluminium, silicon, beryllium, and other elements are, correctly speaking, *special brasses* although they are usually described as bronzes. For example:

Aluminium – bronze approximately 90:10 copper:aluminium; and may also incorporate lead (leaded aluminium bronze) for bearing applications, although this reduces strength and toughness for other applications.

Silicon-bronze – approximately 96:4 copper:silicon, usually with a little manganese or iron. These alloys are distinct from *silicon brasses* which contain a similar proportion of silicon but approximately 10% zinc, the balance being copper.

Beryllium-copper – approximately 97.5:2.5 copper:beryllium, principally known as spring material.

Antimonial-bronze – 80 to 90% copper with nickel and antimony, and possibly also lead and tin. These are mainly associated with bearing metals.

Copper-lead bearing alloys (lead bronze) – approximately 70:30 copper:lead.

TABLE 13 – BRASS SHEET, STRIP AND PLATE**

Description	Composition (%)					Typical mechanical properties								Nearest British or other standard and designation			Remarks
	Cu	Zn	Pb	Sn	Others	0.1% Proof stress (N/mm²) (a)	(h)	Tensile strength (N/mm²) (a)	(h)	Elongation (%) (a)	(h)	Hardness (HV) (a)	(h)	Metric	Imperial	Designation	
Cold rolled sheet and strip																	
Cap copper	95	5	—	—	—	60	355	250	420	50	4	60	130	2870	—	CZ125	Industrial use practically confined to caps for ammunition.
Gilding metals	90	10	—	—	—	77	385	265	450	55	8	60	140	2870	713	CZ101	Used for architectural metal-work, imitation jewellery, etc., on account of golden colours and ability to be brazed and enamelled.
	85	15	—	—	—	90	400	290	465	60	10	65	150	2870	712	CZ102	
	80	20	—	—	—	110	450	310	540	65	12	70	160	2870	711	CZ103	
Cartridge brass	70	30	—	—	—	110	450	325	540	70	15	70	170	2870	267	CZ106	Deep drawing brass having maximum ductility of the Cu-Zn alloys.
2/1 brass	65	35	—	—	—	110	465	325	540	65	10	70	170	2870	266	CZ107	A good cold working alloy.
'Common' or 'Basis' brass	63	37	—	—	—	125	465	355	585	55	7	80	180	2870	—	CZ108	General purpose alloy suitable for simple forming, etc.
60/40 brass	60	40	—	—	—	125	280	355	465	45	25	90	140	2870	—	CZ123	A good hot working alloy which can also be cold worked to a limited extent.
Aluminium	76	22	—	—	2 Al	125	155	340	390	60	50	75	100	2870	—	CZ110	The addition of aluminium and tin, respectively, produces enhanced corrosion resistance when compared with the unalloyed brasses.
Naval brass	62	37	—	1	—	140	340	370	252	45	20	95	160	2870	409	CZ112	
Leaded clock brasses	59	39	2	—	—	170	495	420	590	30	5	110	185	2870	2785	CZ120	In the hard condition these alloys can be accurately punched to shape with a minimum of 'burr', hence their major use as pinions for clocks, watches and instruments. Also suitable for engraving.
	62	36	2	—	—	125	465	340	540	50	7	75	175	2870	2785	CZ119	
	64	35	1	—	—	108	325	325	525	50	7	70	170	2870	2785	CZ118	
Heat treatable alloy brass†	Bal.	20	—	—	6.5 Ni 1.75 Al	400	645	510	695	20	2	175	250	—	—	—	Can be formed in the soft condition and heat treated to develop its full properties. Used for springs, retaining clips, etc.
Hot rolled plate																	
60/40 brass	60	40	0.5	—	—	170		385		40		110		2875	1541	CZ123	Used for tube plates of condensers and similar purposes.
Naval brass	62	37	—	1	—	170		400		35		110		2875	⎰1541	CZ112	These alloys are more resistant to corrosion (especially by sea water) than 60/40 brass. They are also used for the purposes outlined above.
Aluminium brass	76	22	—	—	2 Al	120		340		55		85		2875	409⎱	CZ110	
70/30 arsenical brass	70	30	—	—	0.02–0.06 As	110		325		55		85		2875	1541	CZ105	

*(a) = annealed. (h) = hard.

†figures in column (a) refer to solution heat treated material, whilst those in column (h) refer to solution heat treated, cold rolled and precipitation heat treated material.

TABLE 14 – BRASS RODS AND SECTIONS*

Description	Composition (%)					Typical mechanical properties				Nearest British or other standard and designation			Remarks
	Cu	Zn	Pb	Sn	Others	0.1% Proof stress (N/mm²)	Tensile strength (N/mm²)	Elongation (%)	Hardness (HV)	Metric	Imperial	Designation	
Free cutting brasses	58 62	39 35	3 3	— —	— —	245 245	465 420	20 25	150 130	2874 2874	249 —	CZ121 CZ124	The most suitable materials for high speed machining. The higher copper alloy has improved ductility and impact strength, but both have limited ability to be cold worked.
60/40 brass	60	40	0.30– 0.80	—	—	215	420	30	125	2874	1949	CZ123	Will withstand limited amount of cold working and bending.
Naval brass Leaded naval brass	62 61	37 Rem.	— 0.50– 2.0	1 1	— —	250 250	430 430	30 25	135 135	2874 —	251 —	CZ112 —	The tin addition improves corrosion resistance, especially in sea water. The leaded version has improved machinability.
Free cutting brasses with improved ductility	62 80	36 Rem.	2 0.10– 1.0	— —	— —	215 200	400 350	30 35	125 120	2874 2874	D.T.D. 627 —	CZ119 CZ104	The higher copper and lower lead contents of these alloys improves ductility, while retaining free machining characteristics. Can be used for cold heading, riveting, etc.
Ductile brass	70	30	—	—	—	280	370	40	110	2874	—	CZ106	Can be deformed extensively by cold working.
High tensile brasses (manganese bronzes)	58 66	Rem. Rem.	1 —	0.75 —	0.75 Fe 1.5 Mn 0.75 Fe 4.5 Al 1.5 Mn	240 min. 290 min.	460 min. 540 min.	20 12 min.	— —	2874 2874	250 —	CZ114 CZ116	The alloying additions produce improved mechanical properties compared with unalloyed brasses of similar copper content. The lead in CZ114 improves machinability. Used for fasteners, valve parts, etc.
Brass for-architectural sections	58 58	40 38	2 1	— 0.5	0.4 Al 0.50 Fe 2.0 Mn	200 230	390 460	25 20	120 150	— —	— —	— —	The aluminium containing alloy has a bright yellow colour on the surface of extruded sections, while the manganese containing alloy is chocolate brown. This is due to oxidation effects.

*Copper Development Association

TABLE 15 – BRASS FORGINGS, STAMPINGS AND HOT PRESSINGS*

Description	Composition (%)					Typical mechanical properties				Nearest British or other standard and designation			Remarks
	Cu	Zn	Pb	Sn	Others	0.1% Proof stress (N/mm²)	Tensile strength (N/mm²)	Elon-gation (%)	Hardness (HV)	Metric	Imperial	Desig-nation	
Lead free 60/40 brass	60	40	—	—	—	140	370	40	100	2872	1949	CZ109	These alloys are very plastic at the hot working temperature, therefore very intricate shapes, showing fine surface detail, can be produced. Components made from these alloys have limited cold working ability.
60/40 brass	60	40	0.30–0.80	—	—	140	370	40	100	2872	1949	CZ123	
Naval brass	62	37	—	1	—	150	390	35	110	2872	251	CZ112	The tin addition improves corrosion resistance, particularly in sea water.
High tensile brasses (manganese bronzes)	58	Rem.	1	0.75	0.75 Fe 1.5 Mn	195 min.	460 min.	15 min.	—	2872	—	CZ114	Used in application where high strength is required, such as high pressure gas valves, etc. The lead addition in CZ114 improves machinability.
	66	Rem.		—	0.75 Fe 4.5 Al 1.5 Mn	295 min.	540 min.	12 min.	—	2872	—	CZ116	
High tensile brass (soldering quality)	58	Rem.	1	—	0.75 Fe 1.5 Mn 0.2 max. Al	195 min.	460 min.	15 min.	—	2872	1001	CZ115	Similar to CZ114, but the restriction in aluminium content avoids non-wetting problems during soft soldering operations.
Leaded brass	58	40	2	—	—	170	400	35	100	2872	218	CZ122	The most popular alloy for hot stamping. The lead content ensures free machining characteristics.

*Copper Development Association

TABLE 16 – BRASS TUBES**

Description	Composition (%)					Typical mechanical properties								Nearest British or other standard and designation			Remarks
	Cu	Zn	Pb	Sn	Others	0.1% Proof stress (N/mm²) (a)	(h)	Tensile strength (N/mm²) (a)	(h)	Elong-ation (%) (a)	(h)	Hardness (HV) (a)	(h)*	Metric	Imperial	Desig-nation	
85/15 brass	85	15	—	—	—	80	340	280	430	50	20	65	140	—	—	—	Used for condenser and cooling unit, gauges and instrument tubes. The composition is frequently modified by tin additions.
70/30 arsenical brass	70	30	—	—	0.02–0.06 As	95	385	310	465	60	20	70	165	2871	—	CZ126	Standard compositions for condenser tubes. Admiralty brass has better corrosion resistance. The arsenic is added to inhibit dezincification.
Admiralty brass	70	29	—	1	0.02–0.06 As	110	390	320	460	60	20	75	165	2871	—	CZ111	
Aluminium brass	76	22	—	—	2.0 Al 0.02–0.06 As	140	460	360	560	60	20	75	165	2871	—	CZ110	Possesses excellent corrosion resistance, and is a favoured alloy for condenser tubes.
Free cutting brass	62	36	2	—	—	80	300	340	465	40	10	80	150	2871	—	CZ119	The lead content is responsible for diminished ductility.
Special alloy brass	83	14	—	—	1.0 Ni 1.0 Si 1.0 Al	—	310	—	495	44		—		—	D.T.D. 253A D.T.D. 323B	—	Used for aircraft service pipes.
Extruded and lightly cold drawn only																	
60/40 brass	60	40	—	—	—	110		370		40		75		—	—	—	—
Naval brass	62	37	—	1	—	125		385		40		80		—	—	—	Has better corrosion resistance than 60/40 brass.
High tensile (manganese bronzes)	58	39	—	1.5	1.0 Fe 0.4 Al 0.2 Mn	230		510		25		120		—	—	—	These alloys combine high strength with fair corrosion resistance. Obtainable in heavy gauges only.
	60	37.5	—	1	0.8 Fe 0.7 Mn	215		465		25		100		—	—	—	

*(a)=annealed, (h)=hard.
**Copper Development Association

TABLE 17 – BRASS WIRE**

Description	Composition (%)					Typical mechanical properties								Nearest British or other standard and designation			Remarks
						0.1% Proof stress (N/mm²)		Tensile strength (N/mm²)		Elon-gation (%)		Hardness (HV)					
	Cu	Zn	Pb	Sn	Others	(a)	(h)	(a)	(h)	(a)	(h)	(a)	(h)	Metric	Imperial	Desig-nation	
90/10 brass	90	10	—	—	—	—	—	280	420	50	10	—	—	2873	—	CZ101	Alloys within this range have better corrosion resisting properties than the lower copper content alloys and are used for paper machine plant. They are also used for ornamental purposes, because of their colour and ability to be brazed.
85/15 brass	85	15	—	—	—	—	—	310	510	55	10	—	—	2873	—	CZ102	
80/20 brass	80	20	—	—	—	—	—	310	510	65	15	—	—	2873	—	CZ103	
70/30 brass	70	30	—	—	—	—	—	325	525	70	12	—	—	2873	—	CZ106	Alloys of the higher copper contents, within the range quoted, are the most ductile and suitable for severe cold forming such as heading.
2/1 brass	65	35	—	—	—	—	—	340	570	65	10	—	—	2873	2786	CZ107	
Common brass	63	37	—	—	—	—	—	340	550	60	10	—	—	2873	—	CZ108	
Leaded brass	62	36	2	—	—	—	—	340	400	50	25	—	—	2873	—	CZ119	The lead content is added to impart good machining properties but should be low if the brass is to be cold headed.

*(a)=annealed, (h)=half hard.
**Copper Development Association

TABLE 18 – BRASS AND HIGH TENSILE BRASS (MANGANESE BRONZE) CASTINGS*

Description	Composition (%)					Typical mechanical properties				Nearest British or other standard and designation			Remarks
	Cu	Zn	Pb	Sn	Others	0.1% Proof stress (N/mm²)	Tensile strength (N/mm²)	Elon-gation (%)	Hardness (HB)	Metric	Imperial	Desig-nation	
Brass for general purposes (sand cast)	70–80	Rem.	2.0–5.0	1.0–3.0	0.75 max Fe 1.0 max Ni	80–110	170–200	18–40	45–60	1400	—	SCB1	These alloy groups cover the range of brasses normally employed for the vast majority of general purpose copper-base alloy sand castings, where moderate strength and good corrosion resistance are required at low cost.
	63–70	Rem.	1.0–3.0	1.5 max	0.75 max Fe 1.0 max Ni	70–110	190–220	11–30	45–65	1400	—	SCB3	
Naval brass (sand cast)	60–63	Rem.	0.50 max	1.0–1.5	—	70–110	250–310	18–40	50–75	1400	—	SCB4	The tin addition is made to improve the corrosion resistance in sea water.
Brazable quality brass	83–88	Rem.	0.50 max	—	0.05–0.20 As	80–110	170–190	18–40	45–60	1400	—	SCB6	The high copper content raises the melting point sufficiently to allow the alloy to be jointed by bronze welding.
Brass for gravity die castings	59–63	Rem.	0.25 max	—	0.50 max Al	90–120	280–370	23–50	60–70	1400	—	DCB1	These are the most commonly supplied die cast brasses, naval brass being selected for better corrosion resistance.
	58–63	Rem.	0.5–2.5	1.0 max	1.0 max Ni 0.2–0.8 Al 0.8 max Fe 0.5 max Mn	90–120	300–340	13–40	60–70	1400	—	DCB3	The medium strength high tensile brass listed below is also available as die castings.
Brass for pressure die castings	57–60	Rem.	0.5–2.5	0.5 max	0.3 max Fe 0.5 max Al	90–120	280–370	25–40	60–70	1400	—	PCB1	The low copper content gives this alloy greater plasticity immediately after solidification, thus avoiding hot tearing in the metal mould before ejection.
High tensile brasses	55 min.	Rem.	0.5 max	1.0 max	0.7–2.0 Fe 0.5–2.5 Al 3.0 max Mn 1.0 max Ni	170–280	470–570	18–35	100–150	1400	–	HTB1	High tensile brasses are used where strength and toughness, combined with good corrosion resistance, are of primary importance. Sand castings are employed for most purposes, but die castings (with improved mechanical properties) are also produced. The *beta* brass (HTB3) is particularly susceptible to stress corrosion cracking in sea water.
	55 min.	Rem.	0.20 max	0.20 max	1.5–3.25 Fe 3.0–6.0 Al 4.0 max Mn 1.0 max Ni	400–470	740–810	11–18	150–230	1400	—	HTB3	

*Copper Development Association

TABLE 19 – GUNMETAL AND TIN BRONZE CASTINGS

Description	Composition (%)					Condition	Typical mechanical properties				Nearest British or other standard and designation			Remarks
	Cu	Sn	Zn	Pb	Others		0.2% Proof stress (N/mm²)	Tensile strength (N/mm²)	Elongation (%)	Hardness (HB)	Metric	Imperial	Designation	
88/10/2 (Admiralty) gunmetal	Rem	10	2.00	1.50 max	1.50 max Ni	SC / BC	130-160 / 130-170	270-340 / 250-340	13-25 / 5-16	70-95 / 79-95	1 400	—	G1	Gunmetal is widely employed for pumps, valves and miscellaneous castings, and is also used for statuary. Among the most widely used grades, particularly where pressure tightness is required, is LG2. is, however, recommended when good pressure tightness and optimum mechanical properties are required, particularly for heavy sections.
Nickel gunmetal	Rem	7	2.25	0.30	5.50 Ni	SC / SCHT	140-160 / 280-320	280-340 / 430-480	16-25 / 3-5	70-95 / 160-180	1 400 / 1 400	—	G3 / G3-WP	
83/3/9/5 leaded gunmetal	Rem	3	9.00	5.00	2.00 max Ni	SC	80-130	180-220	11-15	55-65	1 400	—	LG1 / LG4	
85/5/5/5 leaded gunmetal	Rem	5	5.00	5.00	2.00 max Ni	SC / CC	100-130 / 100-140	200-270 / 270-340	13-25 / 13-35	65-75 / 75-90	1 400	—	LG2	
87/7/3/ leaded gunmetal	Rem	7	2.25	3.00	2.00 max Ni	SC / CC	130-140 / 130-160	250-330 / 300-370	16-25 / 13-30	70-85 / 80-95	1 400	—	LG4	
Tin bronze for general purposes	90	10	—	—	0.15 max P	SC	130-160	230-310	9-20	70-90	1 400	—	CT1	A suitable alloy for general sand castings.
Phosphor bronze for bearings	90	10 min	—		0.50 min P	SC / DC / CC / BC	130-160 / 170-230 / 170-280 / 170-230	220-280 / 310-390 / 360-500 / 330-420	3-8 / 2-8 / 6-25 / 4-22	70-100 / 95-150 / 100-150 / 95-150	1 400	—	PB1	Standard phosphor bronzes for bearing applications.
	90	9.5 min	0.50 max	0.75 max	0.40 min P	SC / DC / CC / BC	100-160 / 140-230 / 160-270 / 140-230	190-270 / 270-340 / 330-450 / 280-400	3-12 / 2-10 / 7-30 / 4-20	70-95 / 95-140 / 95-140 / 95-140	1 400	—	PB4	
Phosphor bronze for gears	88	12	0.30 max	0.50 max	0.15 min P	SC / DC / CC / BC	130-170 / 170-200 / 170-250 / 170-200	220-310 / 270-340 / 310-430 / 280-370	5-15 / 3-7 / 5-15 / 5-14	75-110 / 100-150 / 100-150 / 100-150	1 400	—	PB2	This alloy is also utilized for general bearings, where its rigidity is of advantage.
Leaded phosphor bronze	87	7.5	2.00 max	3.50	0.30 min P / 1.00 max Ni	SC / CC	80-130 / 130-200	190-250 / 270-360	3-12 / 5-18	60-90 / 85-110	1 400	—	LPB1	This material is satisfactory for many bearing purposes.
76/9/0/15 leaded bronze	76	9	1.00 max	15	2.00 max Ni	SC / CC	80-110 / 130-190	170-230 / 230-310	4-10 / 8-10	50-70 / 70-90	1 400	—	LB1	This series of leaded bronzes of progressively increasing lead content, and hence increasing plasticity, is of considerable importance for bearings where some measure of plasticity is desired, or where the 'mating' material is a soft steel which would be scored by harder materials.
80/10/0/10 leaded bronze	80	10	1.00 max	10	0.10 max P / 2.00 max Ni	SC / CC	80-130 / 160-220	190-270 / 280-390	5-15 / 5-15	65-85 / 80-90	1 400	—	LB2	
85/5/0/10 leaded bronze	85	5	2.00 max	10	0.10 max P / 2.00 max Ni	SC / CC	60-100 / 130-170	160-190 / 230-310	7-12 / 9-20	55-75 / 60-80	1 400	—	LB4	
75/5/0/20 leaded bronze	75	5	1.00 max	20	0.10 max P / 2.00 max Ni	SC / CC	60-100 / 100-160	160-190 / 190-270	5-10 / 7-16	45-65 / 50-70	1 400	—	LB5	

TABLE 20 – WROUGHT PHOSPHOR BRONZES*

Description	Composition (%)					Typical mechanical properties								Nearest British or other standard and designation			Remarks
	Cu	Sn	Zn	Pb	Others	0.1% Proof stress (N/mm²)		Tensile strength (N/mm²)		Elongation (%)		Hardness (HV)		Metric	Imperial	Designation	
						(a)	(h)	(a)	(h)	(a)	(h)	(a)	(h)†				
Sheet and Strip																	
Bronze coinage alloy	95.5	3.0	—	1.5	—	95	450	310	550	55	8	65	180	—	—	—	Used for British Copper coinage
3% phosphor bronze	97.0	3.0	0.02 0.40	—	—	110	460	320	590	55	8	75	185	2 870	407	PB101	The phosphor bronzes owe the majority of their applications to their good elastic properties, combined with resistance to corrosion and corrosion fatigue. This accounts for their use as springs and instrument components. The usual material is PB102. Alloys with lower and higher tin contents are also available.
5% phosphor bronze	95.0	5.0	0.02 0.40	—	—	120	510	340	630	60	8	80	190	2 870	407	PB102	
7% phosphor bronze	93.0	7.0	0.02 0.40	—	—	140	570	370	650	65	14	85	210	2 870	407	PB103	
Plate																	
3% phosphor bronze	97.0	3.0	0.02 0.40	—	—	90	230	310	400	50	35	75	120	2 875	—	PB101	Condenser tubeplates and vessels, where higher strength than pure copper and improved corrosion resistance are required.
5% phosphor bronze	95.0	5.0	0.02 0.40	—	—	110	260	320	420	55	40	80	130	2 875	—	PB102	
Rod and Bar																	
5% phosphor bronze	95.0	5.0	0.02 0.40	—	—	120	420	340	540	55	20	80	170	2 874	369	PB102	Almost invariably used in a slightly work hardened condition. A much favoured material for engineering components subject to friction.
Free machining phosphor bronze	95.0	5.0	0.15	—	1.0 Pb or 0.5 Te	130	375	340	520	55	17	85	160	—	—	—	The addition of either lead or tellurium considerably improves the machinability compared to PB102.
High tin bronze for bearings	91.5	8.5	0.30	—	—	—	—	390	590	—	—	100	160	—	DTD 265A	—	This is an improvement on the ordinary quality phosphor bronze rod for bearing purposes.
Wire																	
Conductivity bronze	98.5	1.0 1.5	—	—	—	60	465	280	695	55	2	—	—	—	—	—	Used for telephone and trolley wire.
5% phosphor bronze	95.0	5.0	0.02 0.40	—	—	—	—	360	770	50	—	—	—	2 873	384	PB102	Generally employed for light springs.
7% phosphor bronze	93.0	7.0	0.02 0.40	—	—	—	—	390	820	55	—	—	—	2 873	384	PB103	
Fourdinier wire	92.0	8.0	0.02 0.40	—	—	—	—	390	900	60	—	—	—	—	—	—	This quality is used particularly for wire cloth for papermaking machines on account of its good corrosion and abrasion resistance.
Tube																	
5% phosphor bronze	95.0	5.0	0.02 0.40	—	—	—	—	340	620	60	10	100	160	—	—	—	For condenser tubes. Bourdon gauges and piping for fuel systems.
High tin bronze for bearings	91.5	8.5	0.30	—	—	—	—	430	590	—	—	—	—	—	DTD 265A	—	See note above for rod.

*Copper Development Association
†(a) = annealed (h) = hard

TABLE 21 – WROUGHT COPPER-NICKEL ALLOYS

Description	Composition (%)					Typical mechanical properties								Nearest British or other standard and designation			Remarks
	Cu	Ni	Fe	Mn	Others	0.1% Proof stress (N/mm²)		Tensile strength (N/mm²)		Elongation (%)		Hardness (HV)		Metric	Imperial	Designation	
						(a)	(h)	(a)	(h)	(a)	(h)	(a)	(h)				
Sheet and Strip																	
90/10 copper-nickel-iron†	87.5	10	1.5	1.00	—	120	190	320	360	45	40	85	105	2 870	—	CN102	Excellent sea water corrosion resistance. Used for sea water trunking, water boxes, etc, in marine and desalination plant.
80/20 copper-nickel	80.0	20	—	0.05 0.50	—	120	390	340	450	40	15	85	140	2 870	374	CN104	Pressing and deep drawn items requiring strength combined with corrosion resistance.
75/25 copper-nickel	75.0	25	—	0.05 0.40	—	140	390	360	450	40	15	90	145	2 870	374	CN105	The most important commercial use of this grade is for coinage, eg. current British 'silver' coinage.
70/30 copper-nickel†	68.0	30	1.0	1.00	—	150	200	390	430	45	40	100	120	2 870	—	CN107	Excellent sea water corrosion resistance. Withstands erosion better than CN102.
55/45 copper-nickel	55.0	45	—	—	—	190	530	460	650	45	5	120	180	—	—	—	This alloys has a high specific resistance and low (practically zero) temperature coefficient of resistance. Largely used as resistance material.
Plate																	
90/10 copper-nickel-iron†	87.5	10	1.5	1.00	—	120	140	320	340	42	40	85	95	2 875	1 541	CN102	Alloys offering excellent resistance to sea water corrosion and erosion.
70/30 copper-nickel	68.0	30	1.0	1.00	—	150	170	390	400	42	40	100	110	2 875	1 541	CN107	
Tube																	
90/10 copper-nickel-iron	87.5	10	1.5	1.00	—	140	460	320	540	40	13	85	165	2 871	—	CN102	All three alloy have excellent sea water corrosion resistance and are extensively used for marine condensers and in desalination plants. The 30% nickel alloys have better resistance to polluted sea water than the 10% alloy and in addition CN108 is particularly resistant to erosion by entrained silt, etc in cooling waters.
70/30 copper-nickel	68.0	30	1.0	1.00	—	170	570	420	660	42	7	105	190	2 871	—	CN107	
Special 70/30 copper-nickel	66.0	30	2.0	2.00	—	170	570	420	660	42	7	105	190	2 871	—	CN108	
Wire																	
80/20 copper-nickel	80.0	20	—	—	—	—	—	360	—	30	—	—	—	—	—	—	Used for the manufacture of wire-wound resistors, resistance elements and low temperature resistance heating devices.
75/25 copper-nickel	75.0	25	—	—	—	—	—	370	—	30	—	—	—	—	—	—	
55/45 copper-nickel	55.0	45	—	—	—	—	—	480	—	35	—	—	—	—	—	—	

*(a) = annealed   (h) = hard
† These alloys are not normally manufactured in this particular form in the 'hard' condition. The values shown in the (h) column are typical of the 'as manufactured' condition.

TABLE 22 – WROUGHT NICKEL SILVERS

Description	Composition (%)					Typical mechanical properties								Nearest British or other standard and designation			Remarks
	Cu	Ni	Zn	Pb	Others	0.1% Proof stress (N/mm²)		Tensile strength (N/mm²)		Elongation (%)		Hardness (HV)		Metric	Imperial	Designation	
						(a)	(h)	(a)	(h)	(a)	(h)	(a)	(h)*				
Sheet and Strip																	Alloys notable for their good corrosion resistance and attractive colour. The range of alloys becomes progressively whiter with increase of nickel content, although usually there is little advantage in exceeding 18% nickel, except for decorative products in which maximum resistance to stain and tarnish is required. The 12% and 18% alloys, NS104 and 107 respectively, are extensively used for relay contact springs in telecommunication and other equipment, due to their spring properties, solderability and tarnish resistance.
10% nickel silver	63	10	27	—	—	100	600	350	690	65	5	70	210	2870	790	NS103	
12% nickel silver	63	12	25	—	—	110	600	350	710	60	4	75	210	2870	790	NS104	
15% nickel silver	63	15	22	—	—	130	630	360	710	55	4	70	205	2870	790	NS105	
18% nickel silver	63	18	19	—	—	120	630	390	710	52	5	80	210	2870	790	NS106	
18% nickel silver	55	18	27	—	—	160	710	390	790	48	3	95	220	2870	1 824	NS107	
20% nickel silver	63	20	17	—	—	130	630	380	710	50	4	90	205	2870	790	NS108	
25% nickel silver	58	25	17	—	—	130	630	390	710	50	4	90	200	2870	790	NS109	
Rod and bar Leaded 10% nickel brass†	45	10	43	2	—	180	280	460	590	10	30	100	150	2872 2874	—	NS101 NS101	These materials can be hot stamped and extruded into complex shapes and have a white colour.
Leaded 14%	40	14	44	2	—	200	290	510	620	10	25	100	150	2874	—	NS102	

In addition to these two alloys, 10%, 15% and 18% nickel silvers are available in rod and bar form. These alloys, designated NS111, NS112 and NS113 respectively in BS2874, are similar in composition to alloys NS103, NS105 and NS106, except that up to 2% of the zinc is replaced by lead. They are only available in the 'as manufactured' condition, with mechanical properties slightly superior to those of the lead free sheet and strip alloys in the annealed condition.

Wire
All the alloys listed under sheet and strip, namely NS103, NS109, are available as wire in conditions ranging from annealed to hard. The mechanical properties are slightly superior to those obtained in the annealed conditions. They are included in BS2873, with the same designation as in BS2870.

*(a) = annealed   (h) = hard
† These alloys are not normally available in the annealed conditions. The properties quoted in column (a) are typical of the 'as extruded' condition.

TABLE 23 – MISCELLANEOUS ALLOYS

Description	Composition (%)								Typical mechanical properties				Nearest British or other standard and designation			Product Form	Remarks
	Cu	Pb	Mn	Ni	Si	Sn	Ag	Al	Others	0.1% Proof stress (N/mm²)	Tensile strength (N/mm²)	Elongation (%)	Hardness (HV)	Metric	Imperial	Designation		
Copper-lead alloys for bearings																		
80/20 copper-lead	78	20							Up to 2 Sn, Ni and other elements	—	140	—	45	—	—	—	—	These alloys can be bonded to steel shells for bearing purposes. Alloys containing up to 45% lead are in use. Frequently manufactured by powder metallurgy techniques to improve lead distribution.
74/24 copper-lead	74	24				Up to 2			—	—	140	—	40	—	—	—	—	
70/30 copper-lead	69	30					0.6		—	—	140	—	35	—	—	—	—	
60/40 copper-lead	59	40					Up to 1		—	—	120	—	30	—	—	—	—	
Wrought copper-manganese alloys																		
Copper-manganese-nickel ('manganin')	Rem		12	2-4					—	(a) — (h) —	(a) 420 (h) —	(a) 30 (h) —	(a) 85 (h)* —	—	—	—	—	Well known for its high specific resistance and low temperature coefficient of eletrical resistance. Used for instrument work and usually supplied in the annealed condition.
Copper-manganese-aluminium	85		13					2		—	680	10	210	—	—	—	—	A typical composition among the Cu-Mn-Al resistance alloys, which have properties similar to those of the 'manganin' alloys.
Wrought copper-silicon alloys (silicon bronze)																		
Silicon bronze	96		1		3		—		—	(a) 90 (m) 390	(a) 340 (m) 460	(a) 40 (m) 20	(a) 70 (m)† 210	2 870	—	CS101	Sheet, strip and foil	A material with good corrosion resistance and moderate strength. Used mainly in the form of rods and wire for the manufacture of marine fasteners.
														2 872	1 948	CS101	Forging stock and forgings	
														2 873		CS101	Wire	
														2 874	1 948	CS101	Rods and sections	
														2 875	1 541	CS101	Plate	

TABLE 24 – WROUGHT AND CAST ALUMINIUM BRONZES

Description	Composition (%)					Typical mechanical properties								Nearest British or other standard and designation			Remarks
	Cu	Al	Fe	Ni	Others	0.1% Proof stress (N/mm²) (a)	(h)	Tensile strength (N/mm²) (a)	(h)	Elongation (%) (a)	(h)	Hardness (HV) (a)	(h)*	Metric	Imperial	Designation	
Sheet and Strip																	
5% aluminium bronze	95	5	—	—	—	140	540	370	650	65	15	90	190	2 870	—	CA101	May be cold worked to give high strength products of good corrosion and oxidation resistance.
8% aluminium bronze	92	8	—	—	—	170	510	430	660	35	10	100	210	—	—	—	
Tube																	
7% aluminium bronze†	Rem	7	—	—	1.0–2.5 (optional) Fe + Ni +Mn	—	—	420	540	50	10	110	150	2 871	—	CA102	Excellent corrosion resistance. Used for heat exchangers handling corrosive process liquors.
Plate																	
7% aluminium bronze	Rem	7	—	—	1.0–2.5 (optional) Fe + Ni + Mn	—		450 min		31 min		—		2 875	1 541	CA102	Excellent corrosion resistance combined with resistance to impingement and erosion. Used for tube-plates in marine condensers and plant construction when handling corrosive process liquors.
10% aluminium bronze (alloy 'E')	81	9.5	2.5	5.5	1.5 Mn	290		710		20		190		2 875	1 541	CA105	
7% aluminium bronze (alloy 'D')	90.5	7	2.5	—	—	230		540		40		150		2 875	1 541	CA106	
Rod, Bar and Forgings																	
9% aluminium bronze	Rem	9.5	—	—	Up to 4.0 Fe + Ni	280		590		28		160		2 874 / 2 872	2 032 / 2 032	CA103 / CA103	High strength alloys with excellent corrosion resistance. Good oxidation resistance and elevated temperature strength. Used for valve and pump components handling super-heated steam or corrosive process liquors.
10% aluminium bronze	80	10	5	5	—	450		790		15		230		2 874 / 2 872	2 033 / 2 033	CA104 / CA104	
7% aluminium bronze (alloy 'D')	90.5	7	2.5	—	—	250		570		38		160		2 874 / 2 872	—	CA106 / CA106	
Silicon aluminium bronze	Rem	6.2	0.7	—	2.2 Si	250		530		25		—		—	DGS 8 453	—	Ductile alloy having improved machinability and impact strength.
Castings																	
Aluminium bronzes	Rem	9.5	2.5	1.0 max	1.0 max Mn	170–200		500–590		18–40		90–140		1 400	—	AB1	This is the most popular aluminium bronze for die castings and is also frequently sand cast.
	Rem	9.5	5	5	1.5 max Mn	250–300		640–700		13–20		140–180		1 400	—	AB2	The most popular alloy for sand castings, having superior strength and corrosion resistance to AB1.
Copper-manganese-aluminium alloys	73	8	3	3	13 Mn	280–340		650–730		18–35		160–210		1 400	—	CMA1	
	72	9	3	3	13 Mn	380–460		740–820		9–20		200–260		1 400	—	CMA2	Proprietary alloys with good castability and corrosion resistance. Widely used for ships' propellers.

*(a) = annealed   (h) = hard

† This alloy is not normally available in the hard condition. The properties quoted in the (h) column refer to material in the 'as manufactured' condition.

Nickel-silver (white bronze) – copper-nickel-zinc alloys with the possible addition of lead. Copper content is usually of the order of 55 to 63%, the nickel content varying between 10 and 25%, depending on the degree of whiteness required. These are mainly used for ornamental castings and decorative work, and also as spring materials.

There is a wide range of copper-nickel alloys in which copper remains the predominant element (*eg* cupro-nickel) which are not classified as either 'bronzes' or 'special brasses'.

Cast copper-tin alloys containing more than about 5% of tin show the presence of the harder constituent, the amount of which increases with increase of the tin content and is accompanied by improvement of structural strength and resistance to deformation by pounding.

With more than about 12% tin the amount of the delta constituent results in the bronzes being very hard and rather brittle.

Alloys containing higher tin content may be used for more specialized purposes. When lead or other elements are added to the alloy, the tin:copper ratio is often a better criterion than the actual tin content, because the former determines the proportion of the hard constituent.

It is usual for copper-tin bearing bronzes to contain varying percentages of other elements, of which phosphorus, lead, zinc and nickel are the most important.

Phosphorus in small amounts is added to the copper-tin alloys for the purpose of deoxidation, and at the same time to improve the casting properties. Improperly deoxidised bronzes may contain oxides, particularly tin oxide, which are hard and liable to result in scoring in service. A residual phosphorus content in the bronze of about 0.50% or 1% generally indicates the efficient removal of oxygen, with improved crystal structure and wearing qualities.

Further additions of phosphorus in amounts of the order of 0.5% excess after deoxidation results in the formation of a very hard copper phosphide constituent which hardens the matrix but may render it brittle.

Upwards of 1% phosphorus is sometimes added to bronzes as an alternative to high tin content, but for bearings, as opposed to gear-wheels, some authorities consider that a high phosphorus content in bronze causes undue shaft wear. High phosphorus may also introduce casting difficulties and produce low melting point constituents. When lead is also present this may be undesirable, because the lead is not 'held' in diffused particles while solidifying.

Lead in copper-tin bronze bearing alloys

The addition of lead to the copper-tin bronzes has been practised for many years, and for some purposes the leaded bronzes have advantages over the lead-free alloys. The solubility of lead in copper is negligible, so that it is present as a separate microscopic constituent, and should appear in the form of small globules widely distributed.

The principal advantage of lead additions to copper-tin bronzes is the increased plasticity afforded by the lead constituent, which can compensate to some extent for want of fit or alignment of bearings. Leaded bronzes are also an advantage when bearings are operated with unhardened steel shafts. In the event of temporary failure of the oil supply, the lead minimizes the danger of damage to the shaft, and leaded bronzes should therefore be used where lubrication is indifferent.

Additions of 1 or 2% lead are made to the copper-tin bronzes with the object of improving machinability. Lead in amounts of 5% and upwards reduces the dry coefficient of friction of a tin bronze. Lead additions up to 12% further improve plasticity, but conversely reduce the toughness and shock resistance of a bearing, and also its resistance to deformation by pounding.

Zinc acts as a deoxidising agent and bronzes containing zinc do not need phosphorus for this purpose. *Admiralty gunmetal* 88:10:2 copper:tin:zinc is widely used for marine castings. *Phosphor bronzes* are generally superior to zinc containing bronzes for bearing applications.

The presence of nickel in the copper-tin bronzes results in a slight increase of strength and toughness, the optimum improvement being obtained with a nickel addition of about 1 to 2%. The crystal structure of castings is also slightly refined, while in the case of leaded bronzes it has been claimed that the segregation of lead is reduced, thus giving a more even distribution of the lead particles.

High nickel contents are used to produce *white bronze* or copper nickel alloys for corrosion resistant duties.

CADMIUM

Symbol: Cd
Specific gravity: 8.65
Density: 8.65 g/cm^3 (0.313 lb/in^3)
Tensile strength (chill cast): 745 kg/cm^2 (10 650 lb/in^2)
Elongation: 50%
Hardness: 21 to 23 Brinell
Modulus of elasticity: n/a
Melting point: 321°C (610°F)
Coefficient of linear expansion: 29.8×10^{-6} degC
Specific heat at 20°C: n/a
Thermal conductivity: 0.27 cal/cm^2/cm/sec/degC
Electrical resistivity: 6.83 μ ohm cm

Cadmium is a soft white metal similar in appearance to tin, but slightly harder. It is durable enough to draw, extrude or roll. It is relatively inert and finds wide application as a plating or coating metal, particularly on ferrous metals, application being by electro-plating (mainly), spray or hot-dip. The coating is smoother and more uniform than zinc coating (which it resembles). It is particularly attractive in the electrical field (*eg* applied to permanent signals or copper-alloy components) because it takes solder readily. It also has excellent resistance to moist or saline atmospheres.

In metallurgy, cadmium is widely used as an alloying element, *eg*:

Copper with 1% cadmium – for electrical conductors, welding electrodes, etc. Its strength is approximately 50% greater than that of pure copper with similar conductivity.

Copper-lead alloys – which are non-work hardening and can be used as a durable covering for cables, etc, subject to vibration.

Bearing alloys – alloyed with nickel, silver and copper-silver.

Solders – 40:60 cadmium-zinc solder for high strength; silver-copper-cadmium-zinc solder for jointing ferrous to non-ferrous metals; silver-cadmium-lead solders for production line jointing and assembly.

Cadmium is also used as an 'improver' in certain aluminium-copper alloys to delay age-hardening after solution treatment and as a deoxidiser in aluminium, nickel and silver casting alloys. Cadmium amalgam with mercury has an application in dentistry.

Cadmium compounds – *ie* the sulphide and sulpho-selenides – are widely used as pigments.

Cadmium is mainly obtained from zinc ores as a byproduct of smelting, either by precipitation melting and distallation, or by electrolysis. Cadmium produced for the plating industry has a purity of 99.95%, the residual impurities being lead, zinc and thallium.

CAESIUM

Symbol: Cs
Specific gravity: 1.9
Density: 1.903 g/cm^3 (0.688 lb/in^3)
Modulus of elasticity: n/a
Melting point: 28.7°C (83.6°F)
Coefficient of linear expansion: 97×10^{-6} degC
Specific heat at 20°C: 0.04517 cal/g/degC
Electrical resistivity: 20 μ ohm cm

Caesium is a soft silver-white metal with a very low melting point. It is the most strongly lasic (electropositive) metal known. It ignites spontaneously and violently in air and decomposes water at room temperatures. In its metallic form it has to be stored in a vacuum, or dry inert gas, or submerged in an anhydrous liquid hydrocarbon; and it is more usually available in its thoride, hydroxide and nitrate forms.

The principal use of caesium is as a surface coating for photoelectric cells, co-deposited with a resin. The radioactive isotope Caesium 137 has a half-life of 33 years and is used as a source of gamma radiation in radiotherapy.

CALCIUM

Symbol: Ca
Specific gravity: 1.55
Density: 1.55 g/cm^3 (0.056 lb/in^3)

Modulus of elasticity: 2.2 to 2.7×10^5 kg/cm^2 (3.2 to 3.8×10^6 lb/in^2)
Melting point: 838°C (1540°F)
Specific heat: 0.149 cal/g/degC
Coefficient of linear expansion: 22.3×10^{-6} degC
Thermal conductivity (circa 20°C): 0.3 cal/cm^2/cm/sec/degC
Electrical resistivity: 3.91 μ ohm cm

Calcium is a bright, soft, ductile silver-white metal, stable at room temperatures although it oxidises readily and burns when gently heated in air. It can be cut, drawn, machined, extruded or rolled with ease. It can only be cast in a vacuum or inert atmosphere, or under special fluxes.

Calcium is used as a deoxidising and degasifying agent in molten steels, and also to improve the mechanical properties of carbon alloy steels. It cannot be alloyed with iron.

Calcium is also used as a deoxidising agent in copper, aluminium, magnesium, nickel and nickel-chromium alloys and as a lead-hardening agent (*eg* in Bahnmetal). It can be alloyed with aluminium, barium, beryllium, copper, lead and magnesium, but only lead-calcium alloys have any particular applications (*eg* for bearing alloys, cable sheathing and battery plates and grids).

Other applications of calcium are as a reducing agent in the extraction of metals, a dehydrating agent for alcohols, etc, a desulphiting agent for hydrocarbons and for hydration processes.

Commercial metallic calcium is produced with a purity of 98.99%. Crystalline calcium is also produced with a purity of 94 to 97% in a range of particle sizes.

CARPENTER ALLOY

Nickel-iron soft magnetic alloy (American).

CARTRIDGE BRASS

See Brass.

CERIUM

Symbol: Ce
Specific gravity: 6.77
Density: 6.77 g/cm^3 (0.245 lb/in^3)
Tensile strength: 3080 kg/cm^2 (44 000 lb/in^2)
Yield strength: 2275 kg/cm^2 (32500 lb/in^2)
Elongation: 16.5%
Modulus of elasticity: 4.2×10^5 kg/cm^2 (6×10^6 lb/in^2)
Hardness: 31 Vickers (as cast)
Melting point: 804°C (1480°F)
Coefficient of linear expansion: 8×10^{-6} degC
Specific heat at 20°C: 0.045 cal/g/degC
Thermal conductivity at 25°C: 0.026 cal/cm^2/cm/sec/degC
Electrical resistivity: 75 μ ohm cm

Cerium is a rare earth metal and strongly resembles the alkali metals. It is steel-grey in colour, lustrous when freshly cut but tarnishes readily to a lead colour. It burns in air with an intensity greater than that of magnesium. It also has the peculiar characteristic of sparking copiously when rubbed by a rough ferrous metal surface (*eg* a file), a particular application of this property being the use of a cerium alloy (with copper iron and zinc) as lighter flints.

Cerium alloys with base metal are brittle and mainly used for the production of pyrophoric bars. Cerium metal and cerium alloyed with other rare earth metals (excluding thorium) are used as reducing agents and 'getters'. Misch metal (approximately 50% cerium) is particularly well known as a reducing agent and deoxidiser (*eg* in the steel industry).

Cerium is also employed as a minor constituent of certain aluminium and magnesium alloys, having a similar effect on mechanical properties as titanium; 90:10 magnesium-cerium alloy is particularly suitable for castings. Small quantities of cerium (under 0.05%) also improve the hot-working characteristics of many high alloy steels.

Cerium is available in metallic form, but mostly as Misch metal (see previously). It is also produced in powdered form for sintering with iron powder, etc. Auer metal (60% cerium, 35% iron) is a hard alloy normally produced by this process.

CERROBASE

Low melting point casting alloy (American).

CERROBOND

Low melting point alloy (American).

CHROMEL

Nickel-chrome resistance alloy – see *Nickel.*

CHROMIUM

Symbol: Cr
Specific gravity: 7.14
Density: 7.14 g/cm^3 (0.258 lb/in^2)
Melting point: 1830°C (3326°F)
Modulus of elasticity: 2.5×10^6 kg/cm^2 (36×10^6 lb/in^2)
Tensile strength: approx 5.5 to 12 ton/in^2 (8.7 to 19 kg/mm^2)
Hardness (as cast): 125 Brinell
Hardness (electrodeposited): 500 to 1250 Brinell
Specific heat at 20°C: 0.110 cal/g/degC
Coefficient of linear expansion (circa 20°C): 6.2×10^{-6} /degC
Thermal conductivity (circa 20°C): 0.16 cal/cm^2/cm/sec/degC
Electrical resistivity: 12.9 μ ohm cm

Chromium is a steel-grey metal which, in the pure form, is relatively soft (about 100 Brinell) but brittle. The presence of only small traces of carbides, however, results in high to extreme hardness.

The chief uses of chromium are as an alloying element, and for electrodeposition as a decorative, protective or functional coating. The latter includes hard chrome plating and chrome diffusion *eg* chrome-alloy steels and nickel-chromium alloys in particular. Chrome alloys are generally distinguished by excellent hardness, high tensile strength retained at elevated temperatures, absence of scaling at high temperatures, and good resistance to corosion. Nickel-chromium alloys with varying chromium content also have exceptional electrical properties (*eg* for resistance wires, thermocouples, etc). Chromium is a constituent of some magnet steels. Chromium carbide has been developed as a direct competitor to tungsten carbide for tool tips, etc.

Chromium metal is available in both 98 to 99% and 99% purity, each with 0.1% maximum carbon. Chromium is also produced in powder form for sintering in inert atmospheres.

COBALT

Symbol: Co
Specific gravity: 8.85
Density: 8.85 g/cm^3 (0.322 lb/in^3)
Tensile strength (as cast): 2520 kg/cm^2 (36 000 lb/in^2)
Tensile strength (annealed): 2625 kg/cm^2 (37500 lb/in^2)
Tensile strength (zone-refined): up to 9800 kg/cm^2 (140 000 lb/in^2)
Compression yield strength (as cast): 2940 kg/cm^2 (42 000 lb/in^2)
Compression yield strength (annealed): 3900 kg/cm^2 (56 000 lb/in^2)
Modulus of elasticity: 2.1×10^6 kg/cm^2 (30×10^6 lb/in^2)
Melting point: 1495°C (2730°F)
Coefficient of linear expansion: 13.8×10^{-6} degC
Specific heat at 20°C: 0.009 cal/g/degC
Thermal conductivity (circa 20°C): 0.165 cal/cm^2/cm/sec/degC
Electrical resistivity: 6.24 μ ohm cm

Cobalt is a grey-white metal with a slight bluish cast, hard and brittle at room temperature but capable of being rolled, forged, etc when heated. It is magnetic below 1075°C (1960°F) and is strongly resistant to corrosion.

It is particularly useful as an alloying element in the production of tool steels *eg* Stellite (40 to 50% cobalt, 25 to 30% chromium and the balance, steel). Lower cobalt Stellites are used for hard facings, wear resistant parts, etc. Higher cobalt Stellites are used for hard facing valve stems and seats. Certain high temperature nickel steels also have up to 20% cobalt as a constituent. Konel (73% nickel, 17% cobalt, 10% ferro-titanium) is a particularly high-strength alloy at elevated temperatures, generally superior to nickel-chromium steels for such duties.

Vallium, a 60:35:5 cobalt-chromium-molybdenum alloy is a useful modern alloy for high duty die-castings or investment castings, and has good hot working and machining qualities. Cobalt-chromium-aluminium alloy in wire form is an alternative electric heater element to nickel-chromium. Other special alloys exhibit excellent electrical conductivity.

Increasing use is also being made of cobalt as a bright plating material with outstanding corrosion and tarnish resistance. Cobalt salts are widely used as pigments, stains, decolourisers etc in the glass and ceramic industries, and in the paint industry. Radio-active cobalt 60 has certain applications for the treatment of malignant diseases.

The commercial form of the metal normally has a purity of 97 to 99%. Cobalt powder, for powder metallurgy, has a purity of 98 to 99% and is of a size which passes through a 200-mesh screen.

COLUMAX

See *Magnetic materials* (Section 8).

CONSTANTON

Nickel-copper thermoelectric alloy.

COPPER

Symbol: Cu
Specific gravity: 8.82
Density: 8.82 g/cm^3 (0.318 lb/in^3)
Tensile strength (nominal) (as cast): 15.75 kg/mm^2 (10 ton/in^2)
Tensile strength (cold worked and annealed): 22 kg/mm^2 (14 ton/in^2)
Tensile strength (hot worked): 23.6 kg/cm^2 (15 ton/in^2)
Tensile strength (cold worked, 10% reduction): 26.8 kg/mm^2 (17 ton/in^2)
Tensile strength (cold worked, 30% reduction): 31.5 kg/mm^2 (20 ton/in^2)
Tensile strength (cold worked, 60% reduction): 39.4 kg/mm^2 (25 ton/in^2)
Tensile strength (cold worked, 90% reduction [wire]): 45.7 kg/mm^2 (29 ton/in^2)
Melting point: 1083°C (1981°F)
Coefficient of linear expansion: 17×10^{-6} degC
Modulus of elasticity: 1.1×10^6 kg/cm^2 (16×10^6 lb/in^2)
Specific resistance (pure, annealed): 1.7241 μ ohm cm

BS 6017 gives details of 17 grades of copper in various product forms.

Cathode copper

Electrolytically refined cathode copper is of high purity and is the raw material for the production of electrolytic high conductivity coppers, both tough pitch and oxygen-free. It is also used for copper castings of high conductivity and for making alloys, especially those such as cadmium-copper, silver-copper and chromium-copper where low specific resistance is of primary importance.

Electrolytic pitch

Electrolytic tough pitch high conductivity copper is prepared by remelting cathode copper and casting as bars, cakes and billets for fabrication of wire, strip, tube and extruded sections. No special precautions are taken to guard against the introduction of oxygen during melting and casting; on the contrary the cathode is melted and poured under sufficiently oxidising conditions to give the correct oxygen content or 'pitch'. High conductivity tough pitch copper normally contains from 0.02 to 0.04% of oxygen. The chief use of electrolytic tough pitch copper is for wire and various rolled and extruded shapes, such as bar and rods of simple or complicated section for the electrical industry, but it is also fabricated into sheet, strip and tube for purposes where high electrical or thermal conductivity is of importance, such for instance as heat interchangers and some other types of chemical engineering plant.

Fire refined

There is little difference in quality between copper carefully fire refined for high conductivity purposes and the electrolytic grade; both have conductivities of 100% IACS or more in the annealed condition. Certain impurities are, however, difficult to remove entirely by fire refining, and any precious metals which may be present cannot be recovered. Crude or 'blister' copper containing appreciable quantities of gold, silver, selenium, tellurium or bismuth is, therefore, refined electrolytically whenever possible. The traces of such elements remaining in fire refined HC copper are insufficient to affect the conductivity and are not, for most purposes, detrimental from the user's point of view.

Phosphorus deoxidised copper

Phosphorus deoxidised copper, which may be arsenical or non-aresenical, contains between 0.02% and 0.08% of phosphorus, usually about 0.03% to 0.04%. Oxygen is absent, except for traces which may be present as phosphate slag. The absence of cuprous oxide and the inclusion of the strong deoxidising agent, phosphorus, in the metal makes phosphorus deoxidised copper particularly suitable for torch welding and brazing and, as it is easy to cast and sheds its scale after extrusion, it is a favourite material for the manufacture of tubes for plumbing and general purposes. It is much used in the form of plates as well as tubes for chemical plant involving welded joints. The British Standards are not particularly stringent in regard to impurities, but bismuth is limited to a maximum of 0.0030% for normal purposes and to half that amount if the material is to be subjected to heavy working operations at temperatures between 400 and 700°C. Even in small quantities bismuth has an embrittling effect on copper, especially in the temperature range mentioned, and this is much more marked in the case of deoxidised than of tough pitch copper. Deoxidised copper with increased phoshorus content is used as a brazing material for non-ferrous alloys.

TABLE 25 – WROUGHT COPPERS**

Description	Composition (%)				Typical mechanical properties								Nearest British or other standard and designation			Product Form	Remarks
	Cu	P	As	Others	0.1% Proof stress (N/mm²)		Tensile strength (N/mm²)		Elongation (%)		Hardness (HV)		Designation	Imperial	Metric		
					(a)	(h)	(a)	(h)	(a)	(h)	(a)	(h)*					
Electrolytic tough pitch high conductivity copper	99.90 min	—	—	0.03 max impurities	60	325	220	385	55	4	45	115	C101		2870 / 4608	Sheet, strip and foil / Sheet, strip and foil for electrical purposes	Conductors and other fabricated electrical components. C101 and 102 contain approx. 0.05% oxygen, which whilst not affecting ductility or conductivity will cause embrittlement if heated in a reducing atmosphere. For such applications C103 should be used.
Fire refined tough pitch high conductivity copper	99.90 min	—	—	0.04 max impurities	60	325	220	385	55	4	45	115	C102		1432	Strip for electrical purposes	
Oxygen free high conductivity copper	99.95 min	—	—	0.03 max impurities	60	325	220	385	60	4	45	115	C103		2874 / 1433 / 1434† / 2873 / 4109 / 2871 / 1977 / 2875	Rod and bar / Rod and bar for electrical purposes / Commutator bars / Wire / Wire for electrical purposes / Tubes / Tubes for electrical purposes / Plate	
Tough pitch non-arsenical copper	99.85 min	—	—	0.05 max impurities	60	325	220	385	50	4	45	115	C104	2027	2870 / 2875	Plate / Sheet, strip and foil / Plate	General engineering and building applications where high electrical conductivity is not required. The arsenic in C105 slightly improves strength and creep properties at moderately elevated temperatures. Embrittled by heating in reducing atmosphere.
Tough pitch arsenical copper	99.20 min	—	030–0.50	—	60	325	220	385	50	4	50	115	C105	24 / 1541 / 2027	2875 / 2870	Plate / Sheet, strip and foil	

(continued)

TABLE 25 – WROUGHT COPPERS** – contd.

Material	Cu % min	P	As / other	Impurities	60	325	220	385	60	4	45	115	Code	BS No.		Product form	Applications
Phosphorus deoxidised non-arsenical copper	99.85 min	0.013–0.050	—	0.06 max impurities	60	325	220	385	60	4	45	115	C106	2870		Sheet, strip and foil	Widely used in tube and sheet form for general engineering and chemical applications where high electrical conductivity is not required. Due to the absence of oxygen no embrittlement problems arise during brazing. Favoured for applications where joining by welding is involved.
														2871	659 1386 3931 61 1306 2017	Tubes	
														2873		Wire	
														2874		Rods and sections	
														2875	1541 2027	Plate	
Phosphorus deoxidised arsenical copper	99.20 min	0.013–0.050	0.30–0.50	0.07 max impurities	60	325	220	385	60	4	50	115	C107	2870		Sheet, strip and foil	Applications similar to C106, except that the presence of arsenic slightly improves strength and creep properties at moderately elevated temperatures.
														2871	659 1386 3931 61 1306 2017	Tubes	
														2875	24 2027	Plate	
Oxygen free high conductivity copper for electronic valves and semi-conductor devices	99.99 min	—	—	0.008 max impurities	60	325	220	385	60	4	45	115	C110	3839		All wrought forms	A material of high electrical conductivity and very low volatile impurities, making it particularly suitable for electronic vacuum devices.
														4608		Sheet, strip and foil for electrical purposes	
														1433		Rod and bar for electrical purposes	

*(a) = annealed (h) = hard.
† does not cover C103.
**Copper Development Association

TABLE 26 – WROUGHT HIGH COPPER ALLOYS**

Description	Composition (%)					Typical mechanical properties				Nearest British or other standard and designation			Electrical conductivity % IACS	Remarks
	Cu	Cr	Ni	Be	Others	0.1% Proof stress (N/mm²)	Tensile strength (N/mm²)	Elongation (%)	Hardness (HV)	Metric	Imperial	Designation		
Work hardened alloys						(a) (h)	(a) (h)	(a) (h)	(a) (h)*					
Copper-silver	99	—	—	—	0.03–0.1 Ag	60 325	220 385	55 4	45 115	—	—	—	~100 (a) to 97 (h)	Silver increases the softening temperature and has a negligible effect on conductivity. Alloys are normally tough pitch.
Copper-tellurium	99	—	—	—	0.3–0.7 Te	60 265	230 310	50 12	50 100	2874	—	C109	90–98	Free machining properties. Alloys may be tough pitch or deoxidised.
Copper-sulphur	99	—	—	—	0.2–0.5 S	60 265	230 310	40 8	50 100	2874	—	C111	90–98	Free machining properties.
Copper-cadmium	98.5	—	—	—	0.5–1.2 Cd	60 460	280 700	45 4	60 140	2873 2875	— —	C108 C108	90 (a) to 80 (h)	A material of high strength and conductivity for contact wires and overhead conductors.
Heat treatable alloys**						(b) (c) (d)	(b) (c) (d)	(b) (c) (d)	(b) (c) (d)†					
Copper-nickel-silicon	Bal	—	2.0–3.5	—	0.4–0.8 Si	80 480 650	310 635 740	50 15 10	60 170 210	—	D.T.D. 498 & 504	—	40–45	An alloy of moderate conductivity with good resistance to wear.
Copper-nickel-phosphorus	Bal	—	0.8–1.2	—	0.15–0.25 P	60 340 420	230 450 495	45 25 20	60 140 175	—	—	—	55–60	Alloy of moderate strength and conductivity.
Copper-chromium	Bal	0.5–1.0	—	—	—	45 265 430	230 400 510	50 22 20	65 125 160	—	—	—	80–85	The material may be used at temperatures up to 350°C without undue impairment of properties.

(continued)

TABLE 26 – WROUGHT HIGH COPPER ALLOYS** – contd.

Description	Composition (%)					Typical mechanical properties				Nearest British or other standard and designation			Electrical conductivity % IACS	Remarks
	Cu	Cr	Ni	Be	Others	0.1% Proof stress (N/mm²) (a) (h)	Tensile strength (N/mm²) (a) (h)	Elongation (%) (a) (h)	Hardness (HV) (a) (h)*	Metric	Imperial	Designation		
Copper-zirconium	Bal	—	—	—	0.1–0.15 Zr	60 230 420	230 310 495	45 35 10	60 120 150	—	—	—	85–90	High conductivity alloys eminently suitable for elevated temperature service.
Copper-chromium-zirconium	Bal	0.5–1.0	—	—	0.05–0.15 Zr	— — —	— 510 —	— 20 —	— — 155	—	—	—	80	The chromium bearing alloys has slightly improved strength.
Copper-cobalt-beryllium ('Low beryllium copper')	Bal	—	—	0.35–0.7	2–3 Co	— 650 770	310 740 850	32 15 10	80 220 240	—	—	—	45–50	Higher conductivity and lower strength than normal beryllium copper.
Copper-beryllium ('Normal beryllium copper')	Bal	—	—	1.7–1.9	0.05–0.4 Ni + Co	185 930 1080	500 1160 1400	50 5 2	100 370 400	2870 2873	—	CB101 BC101	25–35	A heat treatable alloy of exceptionally high strength and good fatigue properties.

**All the alloys listed are included in BS 4577:1970 — Materials for resistance welding electrodes and ancillary equipment.

*(a) = annealed, (h) = hard.

Tough pitch copper

Ordinary touch pitch copper is available in three British Standard grades and is supplied in all the usual fabricated forms. The conductivity is not specified and may be appreciably below that of high conductivity copper, while oxygen and impurities often reach the somewhat higher proportions allowed by the specifications. These grades are suitable for general purpose use where high electrical conductivity is not required.

High conductivity copper

Oxygen-free high conductivity copper (OFHC) is made by remelting and pouring cathode copper entirely in an atmosphere of carbon monoxide and nitrogen, so that no oxygen can be absorbed. The absence of oxygen in this form of copper renders it immune to 'gassing' or embrittlement when heated in a reducing atmosphere. This makes it suitable for flame welding and brazing and for the preparation of glass-to-metal seals. OFHC copper may also be preferred for the cold impact extrusion of thin-walled radiator tubes, and for other purposes involving heavy degrees of cold work, such as abnormally severe deep drawing or spinning operations. Its recrystallisation temperature is slightly higher than that of tough pitch copper of equal conductivity.

Arsenical copper

Arsenical copper contains between 0.3% and 0.5% of arsenic in sold solution. It may be either tough pitch or deoxidised. The conductivity is relatively low, but the tensile strength is slightly higher than that of the non-arsenical varieties and this increased strength is maintained at temperatures between about 200 and 300°C. Resistance to atmospheric corrosion is enhanced and wastage by oxidation and scaling in fire-box atmospheres is diminished. Arsenical copper is largely used for piping systems and chemical plant operating at moderately elevated temperatures, and similar duties. Much of the copper tube for domestic plumbing is of the phosphorus deoxidised arsenical variety.

Available forms

Copper is readily available in wire, sheet, strip, plate and tube, also in rods and various sections. Apart from these wrought forms the availability and direct application of copper castings are relatively limited. For electrical purposes, however, castings of high conductivity are obtainable, prepared as a rule from cathode copper. Special castings calling for high thermal conductivity are also made. Most copper castings are deoxidised with reagents other than phosphorus, which have a less detrimental effect on the conductivity.

The following British Standards cover wrought copper-based alloys:

BS 2870 Rolled copper and copper alloys – sheet, strip and foil.
BS 2871 Copper and copper alloys – tubes.
            Part 1 Copper tubes for water, gas and sanitation.
            Part 2 Tubes for general purposes.
            Part 3 Tubes for heat exchangers.

TABLE 27 – HIGH COPPER ALLOY CASTINGS**

Description	Composition (%)					Typical mechanical properties				Nearest British or other standard and designation			Electrical conductivity % IACS	Remarks
	Cu	Cr	Ni	Be	Others	0.1% Proof stress (N/mm²)	Tensile strength (N/mm²)	Elongation (%)	Hardness (HV)	Metric	Imperial	Designation		
High conductivity copper	99.9	—	—	—	Deoxidants only	30	155	25	45	1400	—	HCC1	90	Used whenever highest electrical conductivity is required.
Copper with small additions	Bal	—	—	—	1.0 Sn	46	170	25	45	—	—	—	45-50	Used for tuyères, blast furnace coolers and similar applications.
Copper-chromium*	Bal	0.5-1.0	—	—	—	230	340	20	125	—	—	—	80-85	High strength alloy in relation to its conductivity.
Copper-nickel-phosphorus*	Bal	—	0.8-1.2	—	0.15-0.25 P	280	370	15	125	—	—	—	55-60	Alloy of moderate strength and conductivity.
Copper-cobalt beryllium* ('Low beryllium copper')	Bal	—	—	0.35-0.70	2.0-3.0 Co	480	620	10	230	—	—	—	45-50	A heat treatable alloy with conductivity over 45% IACS and having excellent properties at elevated temperatures.
Copper-nickel-silicon*	Bal	—	2.0-3.5	—	0.4-0.8 Si	370	480	10	175	—	—	—	40-45	A heat treatable alloy of moderate conductivity with good resistance to wear.
Copper-beryllium* ('Normal beryllium copper')	Bal	—	—	1.7-2.0	Up to 0.5 Co	540	930	2	380	—	—	—	25-35	The conductivity is higher than that of any other known material with comparable mechanical properties.

*Properties quoted are for the fully heat treated condition.
**Copper Development Association.

TABLE 28 – COMPOSITION AND CODE NUMBERS OF BS WROUGHT COPPER-BASE ALLOYS

BS Desig-nation	Material	Cu	Sn	Pb	Ni	Zn	Others	2870	2871 Pt.1	2871 Pt.2	2871 Pt.3	2872	2873	2874	2875
		Composition (%)						BS No.							
C101	Electrolytic tough pitch high conductivity copper	99.90 min	—	—	—	—	—	•		•			•	•	•
C102	Fire refined tough pitch high conductivity copper	99.90 min	—	—	—	—	—	•		•			•	•	•
C103	Oxygen free high conductivity copper	99.95 min	—	—	—	—	—	•		•			•	•	•
C104	Tough pitch non-arsenical copper	99.85 min	—	—	—	—	—	•	•	•					•
C105	Phosphorus deoxidised non-arsenical copper	99.20 min	—	—	—	—	0.30–0.50 As	•	•	•				•	•
C106	Phosphorus deoxidised non-arsenical copper	99.85 min	—	—	—	—	0.013–0.050 P				•		•	•	•
C107	Phosphorus deoxidised arsenical copper	99.20 min	—	—	—	—	0.30–0.50 As / 0.13–0.050 P					•		•	
C108	Copper-cadmium	Rem	—	—	—	—	0.5–1.2 Cd								
C109	Copper-tellurium	Rem	—	—	—	—	0.30–0.70 Te				•				•
C110†	Oxygen free high conductivity copper for electronic valves and semi-conductor devices	99.99 min	—	—	—	—	—				•				
C111	Copper-sulphur	Rem	—	—	—	—	0.30–0.60 S								
CZ101	90/10 brass	89.0–91.0	—	—	—	Rem	—	•					•		
CZ102	85/15 brass	84.0–86.0	—	—	—	Rem	—	•					•		
CZ103	80/20 brass	79.0–81.0	—	—	—	Rem	—	•					•		
CZ104	Leaded 80/20 brass	79.0–81.0	—	0.1–1.0	—	Rem	—						•		
CZ105	70/30 arsenical brass	70.0–73.0	—	—	—	Rem	0.02–0.06 As						•	•	
CZ106	70/30 brass	68.5–71.5	—	—	—	Rem	—						•	•	
CZ107	2/1 brass	64.0–67.0	—	—	—	Rem	—								
CZ108	Common brass	62.0–65.0	—	—	—	Rem	—	•							
CZ109	Lead free 60/40 brass	59.0–62.0	—	—	—	Rem	—	•					•		
CZ110	Aluminium brass	76.0–78.0	—	—	—	Rem	0.02–0.06 As / 1.80–2.30 Al / 0.02–0.06 As	•				•	•	•	•
CZ111	Admiralty brass	70.0–73.0	0.9–1.3	—	—	Rem							•		•
CZ112	Naval brass	61.0–63.5	1.0–1.4	—	—	Rem							•		
CZ113	Naval brass (special mixture)	57.5–60.5	0.60–1.25	—	—	Rem								•	
CZ114	High tensile brass	56.0–60.0	0.2–1.0	0.5–1.5	—	Rem	0.25–1.2 Fe / 0.3–2.0 Mn				•			•	
CZ115	High tensile brass (soldering quality)	56.0–60.0	0.6–1.1	0.5–1.5	—	Rem	0.25–1.2 Fe / 0.3–2.0 Mn			•	•	•		•	
CZ116	High tensile brass	64.0–68.0	—	—	—	Rem	0.25–1.2 Fe / 4.0–5.0 Al / 0.3–2.0 Mn							•	•
CZ118	Leaded brass 64% copper 1% lead	63.0–66.0	—	0.75–1.5	—	Rem	—							•	
CZ119	Leaded brass 62% copper 2% lead	61.0–64.0	—	1.0–2.5	—	Rem	—								
CZ120	Leaded brass 59% copper 2% lead	58.0–60.0	—	1.5–2.5	—	Rem	—							•	
CZ121	Leaded brass 58% copper 3% lead	56.0–59.0	—	2.0–3.5	—	Rem	—								
CZ122	Leaded brass 58% copper 2% lead	56.5–60.0	—	1.0–2.5	—	Rem	—								
CZ123	Leaded brass 62% copper 3% lead	59.0–62.0	—	0.30–0.80	—	Rem	—						•		
CZ124	60/40 brass	60.0–63.0	—	2.5–3.7	—	Rem	—	•				•		•	
CZ125	Cap Copper	95.0–98.0	—	—	—	Rem	—	•			•			•	•
CZ126	Special 70/30 arsenical brass	69.0–71.0	—	—	—	Rem	0.02–0.06 As		•	•	•	•		•	

(continued)

TABLE 28 – COMPOSITION AND CODE NUMBERS OF BS WROUGHT COPPER-BASE ALLOYS – contd.

Code	Alloy	Cu	Sn/Pb	Ni	Zn	Other elements
CN101	95/5 copper-nickel-iron	Rem		5.0-6.0	—	1.05-1.35 Fe; 0.30-0.80 Mn
CN102	90/10 copper-nickel-iron	Rem		10.0-11.0	—	1.0-2.0 Fe; 0.50-1.0 Mn
CN103	85/15 copper-nickel	84.0-86.0		14.0-16.0	—	0.05-0.50 Mn
CN104	80/20 copper-nickel	79.0-81.0		19.0-21.0	—	0.05-0.50 Mn
CN105	75/25 copper-nickel	Rem		24.0-26.0	—	0.05-0.40 Mn
CN106	70/30 copper-nickel	69.0-71.0		29.0-31.0	—	0.05-0.50 Mn
CN107	70/30 copper-nickel	Rem		30.0-32.0	—	0.40-1.00 Fe; 0.50-1.50 Mn
CN108	66/30/2/2 copper-nickel-iron-manganese	Rem		29.0-32.0	—	1.70-2.30 Fe; 1.50-2.50 Mn
PB101	3% phosphor bronze (copper-tin-phosphorus)	Rem	3.0-4.5	—		0.02-0.40 P
PB102	5% phosphor bronze (copper-tin-phosphorus)	Rem	4.5-6.0	—		0.02-0.40 P
PB103	7% phosphor bronze (copper-tin-phosphorus)	Rem	6.0-7.5	—		0.02-0.40 P
CA101	5% aluminium bronze (copper-aluminium)	Rem				4.5-5.5 Al
CA102	7% aluminium bronze (copper-aluminium)	Rem				6.0-7.5 Al; Ni+Fe+Mn 1.0-2.5 optional but between these limits if present
CA103	9% aluminium bronze (copper-aluminium)	Rem				8.8-10.0 Al
CA104	10% aluminium bronze (copper-aluminium-nickel-iron)	Rem		4.0-6.0	—	8.5-11.0 Al; 4.0-6.0 Fe
CA105	10% aluminium bronze (copper-aluminium-nickel-iron-manganese)	78.0-85.0		4.0-7.0	—	8.5-10.5 Al; 1.5-3.5 Fe; 0.5-2.0 Mn
CA106	7% aluminium bronze (copper-aluminium-iron)	Rem				2.0-3.5 Fe
NS101	Leaded 10% nickel brass	44.0-47.0	1.0-2.5	9.0-11.0	Rem	0.2-0.5 Mn
NS102	Leaded 14% nickel brass	39.0-42.0	1.0-2.25	13.0-15.0	Rem	1.5-3.0 Mn
NS103	10% nickel silver (copper-nickel-zinc)	60.0-65.0		9.0-11.0	Rem	0.05-0.30 Mn
NS104	12% nickel silver (copper-nickel-zinc)	60.0-65.0		11.0-13.0	Rem	0.05-0.30 Mn
NS105	15% nickel silver (copper-nickel-zinc)	60.0-65.0		14.0-16.0	Rem	0.05-0.50 Mn
NS106	18% nickel silver (copper-nickel-zinc)	60.0-65.0		17.0-19.0	Rem	0.05-0.50 Mn
NS107	18% nickel silver (copper-nickel-zinc)	54.0-56.0		17.0-19.0	Rem	0.05-0.35 Mn
NS108	20% nickel silver (copper-nickel-zinc)	60.0-65.0		19.0-21.0	Rem	0.05-0.50 Mn
NS109	25% nickel silver (copper-nickel-zinc)	55.0-60.0		24.0-26.0	Rem	0.05-0.75 Mn
NS111	Leaded 10% nickel silver	58.0-63.0	1.0-2.0	9.0-11.0	Rem	0.1-0.5 Mn
NS112	Leaded 15% nickel silver	60.0-63.0	0.5-1.0	14.0-16.0	Rem	0.1-0.5 Mn
NS113	Leaded 18% nickel silver	60.0-63.0	0.4-0.8	17.0-19.0	Rem	0.1-0.5 Mn
CS101	Copper-silicon (silicon bronze)	Rem				2.75-3.25 Si; 0.75-1.25 Mn
CB101	Copper-beryllium	Rem				1.7-1.9 Be; 0.05-0.40 Co+Ni

*No electrical properties of the high conductivity coppers, namely C101, C102, C103 and C110, are included in BS 2870-5. These standards are not intended to cover material required for electrical applications; for such purposes the following standards should be consulted:

TABLE 29 – REPRESENTATIVE COPPERS AND PROPERTIES*

Nominal Composition	Relevant British Standards	Other National and International Standards	Description
99.99% minimum purity	BS 2873 BS 2874 } C103 BS 1954 BS 1861 BS 3839	DIN1787 OF-Cu CA 102	Oxygen-free H.C. copper used in electronic applications. Immune to embrittlement when heated in reducing atmospheres.
99.90% minimum purity	BS 2873 BS 2874 } C101 and BS 125 C102 BS 4109	DIN1787 } E-Cu } F-Cu ASTM B1 B2 } CA B3 } 110	General-purpose grade of H.C. copper mainly supplied in wire form in coil or on spools. Used in electrical windings, telephone and other cable applications. Finding increasing use as disposable electrode in seam welding of tinned steel plate. Considerable quantities are sold in the tinned condition.
99.90% minimum purity	BS 2873 BS 2874 } C101 and BS 1432 C102 BS 1433	DIN1787 } E-Cu } F-Cu ASTM B124 } CA B133 } 110	General-purpose grade of H.C. copper. Can be supplied in round rod, bar or section for heavy electrical instruction or in wire for cold forming into high conductivity copper rivets and fasteners, etc.
Cu 99.5 S 0.5	BS 2874 C111	DIN17666 CuSP CA 147	Good machinability compared with plain copper. High conductivity retained. Suitable for use on automatic screw machines. Can be plated easily.
99.85% minimum purity 0.013–0.05% phosphorus	BS 2873 BS 2874 } C106 BS 1172	DIN1787 SF-Cu CA 122	Fully de-oxidised with phosphorus, this material is suitable for the more severe cold heading applications
Cu 99.0 Cd 1.0	BS 2873 C108	—	Cadmium copper. High conductivity and higher softening temperature than ordinary copper used mainly as overhead line wire or resistance welding electrodes.
Cu 99.5 Te 0.5	BS 2874 C109	DIN17666 CuTeP	Good machinability compared with plain copper. High conductivity retained. Suitable for use on automatic screw machines.

* IMI (Kynoch)   *(continued)*

TABLE 29 – REPRESENTATIVE COPPERS AND PROPERTIES* – *contd.*

Form, Condition and size (mm)	Tensile Strength N/mm²	t.s.i.	Elongation (Gauge Length)	Density g/cm³	Electrical Conductivity (% IACS at 20°C)	Thermal Conductivity (cal cm/cm²s°C at 20°C)	Coefficient of Expansion (per °C at 20°C)	Modulus of Elasticity (N/mm² at 20°C)
Rod M 6–50	265	17	30(5.65 $\sqrt{A}$)	8.9	100–101.5	0.94	17.7×10^{-6}	118,000
Wire O 0.5–6	230	15	30(250 mm)					
Wire O 0.5–6	230	15	30(250 mm)	8.9	100–101.5	0.94	17.7×10^{-6}	118,000
H 0.5–6	460	30	—					
Rod/Bar M 6–50	265	17	30(5.65$\sqrt{A}$)	8.9	100–101.5	0.94	17.7×10^{-6}	118,000
Wire O 1–6	230	15	30(250 mm)					
½H 1–6	265	17	25(50 mm)					
Rod ½H 2–3	280	18	20(5.65 $\sqrt{A}$)	8.9	80–95	0.89	18×10^{-6}	118,000
M 6–25	300	19.5	16(50 mm)					
Wire O 0.5–6	230	15	30(250 mm)	8.9	70–90	070–0.87	17.7×10^{-6}	118,000
H 0.5–6	460	30	—					
Wire H 0.5–6	620–70	40–45	1–1.5 (250 mm)	8.9	90–97	0.70	17.0×10^{-6}	118,000
½H 2–3	310	20	12(50 mm)	8.9	80–95	0.89	17.7×10^{-6}	118,000
M6–25	280	18	16(5.65 $\sqrt{A}$)					

TABLE 30 – CHEMICAL COMPOSITIONS AND MECHANICAL PROPERTIES OF COPPER AND COPPER ALLOY PLATE (BS 2875)

	Material	Copper (incl. silver) %	Tin %	Lead %	Iron %	Nickel %	Arsenic %	Antimony %	Bismuth %	Phosphorus %	Oxygen %	Selenium %	Tellurium %	Cadmium %	Total impurities %	Condition*	Thickness Over mm	Thickness Up to and including mm	Tensile strength Min hbar	Elongation Min %	Complies with or falls within ISO recommendations
C 101	Electrolytic tough pitch high conductivity copper	99.90 (min)	—	0.005	—	—	—	—	0.0010	—	—	—	—	—	0.03 (excl. O+ Ag)	M or O	10	—	21.0	35	—
																H	10	16	28.0	15	
																	16	25	25.0	20	
C 102	Fire refined tough pitch high conductivity copper	99.90 (min)	—	0.005	—	—	—	—	0.0025	—	—	—	—	—	0.04 (excl. O+ Ag)	M or O	10	—	21.0	35	—
																H	10	16	28.0	15	
																	16	25	25.0	20	
C 103	Oxygen free high conductivity copper	99.95 (min)	—	0.005	—	—	—	—	0.0010	—	—	—	—	—	0.03 (excl. O+ Ag)	M or O	10	—	21.0	35	—
																H	10	16	28.0	15	
																	16	25	25.0	20	
C 104	Tough pitch non-arsenical copper	99.85 (min)	0.01	0.010	0.01	† 0.05	0.02	0.005	0.0030	—	0.10	0.20	0.010	—	0.05 (excl. Ni+ O+ Ag)	M or O	10	—	21.0	35	—
																H	10	16	28.0	15	
																	16	25	25.0	20	
C 105	Tough pitch arsenical copper	99.20 (min)	0.03	0.02	0.02	† 0.15	0.30/0.50	0.01	0.0050	—	0.10	Se + Te 0.030		—	—	M or O	10	—	22.0	35	—
																H	10	16	28.0	15	
																	16	25	25.0	20	
C 106	Phosphorus deoxidised non-arsenical copper	99.85 (min)	0.01	0.010	0.030	† 0.10	0.05	0.005	0.0030	0.013/0.050	—	Se+Te 0.020	0.010	—	0.06 (excl. Ag+ As+ Ni+ P)	M or O	10	—	21.0	35	—
																H	10	16	28.0	15	
																	16	25	25.0	20	
																	25	—	24.0	25	
C 107	Phosphorus deoxidised arsenical copper	99.20 (min)	0.01	0.010	0.030	† 0.15	0.30/0.50	0.01	0.0030	0.13/0.050	—	Se+Te 0.020	0.010	—	0.07 (excl. Ag+ As+ Ni+ P)	M or O	10	—	21.0	35	—
																H	10	16	28.0	15	
																	16	25	25.0	20	
																	25	—	24.0	25	
C 108	Copper cadmium	REM	—	—	—	—	—	—	—	—	—	—	—	0.5/ 1.2	0.05	H	10	16	31.0	13} on $5.65\sqrt{S_o}$	
																	16	25	28.0	18} $\sqrt{S_o}$	

*M = as manufactured   H = hard
† O = annealed

TABLE 31 – CHEMICAL COMPOSITIONS AND MECHANICAL PROPERTIES OF BRASS PLATE (BS 2875)

Designation	Material	Copper %	Tin %	Lead %	Iron %	Zinc %	Aluminium %	Arsenic %	Total impurities %	Condition	Thickness Over mm	Thickness Up to and including mm	Tensile strength Min hbar	Elongation on 5.65 $\sqrt{S_o}$ Min %	Complies with or falls within ISO recommendations
CZ105	70/30 arsenical brass	70.0/73.0	—	0.075	0.06	REM	—	0.02/0.06	0.30	M or O	10	—	28.0	40	—
										H	10	16	36.0	18	
											16	25	34.0	22	
CZ106	70/30 brass	68.5/71.5	—	0.05	0.05	REM	—	—	0.30	M or O	10	—	28.0	40	R426 CuZn30
										H	10	16	36.0	18	
											16	25	34.0	22	
*CZ110	Aluminium brass	76.0/78.0	—	0.07	0.06	REM	1.80/2.30	0.02/0.06	0.30	M	10	—	28.0	36	R426 CuZn 21A12
										O	10	—	28.0	40	
*CZ112	Naval brass	61.0/63.5	1.0/1.4	—	—	REM	—	—	0.75	M or O	10	25	36.0	18	—
											25	125	34.0	18	
											125	—	31.0	18	
										H	10	12.5	40.0	18	
*†CZ123	60/40 brass	59.0/62.0	—	0.3/0.8	—	REM	—	—	0.30	M	10	25	34.0	18	R426 CuZn 40Pb
											25	125	31.0	18	
											125	—	29.0	18	

*M = as manufactured   H = hard
† O = annealed

BS 2872 Copper and copper alloys – forging stock and forgings.
BS 2873 Copper and copper alloys – wire.
BS 2874 Copper and copper alloys – rods and sections (other than forging stock).
BS 2875 Rolled-copper and copper alloys plate.

Copper for electrical purposes is covered in the following British Standards:

BS 1432 Copper for electrical purposes – strip with drawn or rolled edges.
BS 1433 Copper for electrical purposes – rod and bar.
BS 1434 Copper for electrical purposes – commutator bars.
BS 1977 Copper for electrical purposes – tubes (high conductivity).
BS 4109 Copper for electrical purposes – wire for general electrical purposes and for insulated cables and flexible cords.
BS 4608 Copper for electrical purposes rolled sheet, strip and foil.

CROTORITE
Proprietary aluminium bronzes (British).

DELTA ALLOYS
Proprietary alpha-beta brasses (British).

DOW METAL
Proprietary light alloys (American).

DURAL (DURALUMIN)
High strength aluminium alloys.

DUTCH METAL
Alpha-brass.

ELEKTRON
See *Magnesium.*

FANSTERL
Proprietary duplex-melted columbium alloy for high strength at elevated temperatures.

GALLIUM
Symbol: Ga
Specific gravity: 5.9 (6.095 in liquid state)
Density: 5.907 g/cm^3 (0.213 lb/in^3)
Melting point: 29.8°C (85.6°F)
Coefficient of linear expansion: 18×10^{-6} degC

Specific heat at 20°C: 0.079 cal/g/degC
Thermal conductivity: 0.70 to 0.09 cal/cm^2/cm/sec/degC
Electrical resistivity: 17.4 μ ohm cm

Gallium is a low melting point metal with a wide liquid range (boiling point is 2237°C). Metallic gallium is grey in colour tinged with blue-green and has a brilliant lustre when first cut, dulling on exposure to air. In liquid form it looks rather like mercury. It is highly reactive with most other metals and so must be stored in a quartz, graphite or refractory oxide container. Gallium expands on solidification, *ie* the liquid is denser than the solid.

Gallium alloys with tin and zinc or with cadmium have a lower melting point than the metal itself (*ie* down to 17°C). It is thus increasingly used in the production of low melting point alloys. A limitation is that such alloys are normally reactive with other metals.

The wide liquid range of gallium (or gallium alloyed with 5% indium) makes it suitable for use in high temperature direct-reading thermometers (up to 1200°C). Gallium is also used in alloy with cadmium for cadmium arc lamps and to replace mercury in vapour arc lamps. Gallium adheres well to glass and is capable of taking a high polish. It is used both for glass seals and the production of special optical mirrors.

GERMANIUM

Symbol: Ge
Specific gravity: 5.323
Density: 5.323 g/cm^3 (0.192 lb/in^3)
Modulus of elasticity: 1.558×10^{12} dynes/cm^2
Melting point: 937°C (1720°F)
Coefficient of linear expansion: 5.75×10^{-6} degC
Thermal conductivity (circa 20°C): 0.14 cal/cm^2/cm/sec/deg/C
Electrical resistivity: 0.46 μ ohm cm

Germanium is a lustrous grey-white metal of crystalline form which is extremely brittle. It is stable up to moderately high temperatures in air and resistant to attack by common acids (except nitric acid and aqua regia). When heated above the melting point it volatilises and burns.

Germanium crystals are widely used as semi-conductors (diodes, transistors and solid rectifiers). It is also used as a plating material because of its excellent resistance to corrosion and as an additive to aluminium alloys (to improve strength and workability), steels and tin (to improve strength, ductility and hardness). Some use is also made of germanium as a constituent of low melting point alloys which expand on solidification and special solders.

Commercial germanium is available with a minimum purity of 99.95%.

GILDING ALLOYS

See *Brass.*

GOLD

Symbol: Au
Specific gravity (as cast): 19.3
Density: 19.3 g/cm^3 (0.695 lb/in^3)
Tensile strength: 1050 kg/cm^2 (15 000 lb/in^2)
Elongation: 30%
Melting point: 1063°C (1940°F)
Coefficient of linear expansion: 14×10^{-6} degC
Thermal conductivityy: 0.70 cal/cm^2/cm/sec/degC
Specific heat: n/a
Electrical resistivity: 2.4 μ ohm cm

Gold is a 'noble' metal with high resistance to oxidation, attack by acids and alkalis and most chemicals. It is also a good conductor of electricity and finds wide application as a plating material, etc, in the electrical industries. The metal is also extremely malleable and ductile and can be beaten into leaf form of 0.1 μ mm (0.000005 in) thick, or drawn into extremely fine wire. The specific gravity of gold can be raised to above 19.5 by rolling, and 19.65 by hammering. The specific gravity of precipitated gold can be in excess of 20.5.

See also chapter on *Electrical Contact Material* (Section 8).

GUNMETAL

See *Bronze.*

HAFNIUM

Symbol: Hf
Specific gravity: 13.09
Density: 13.09 g/cm^3 (0.473 lb/in^3)
Tensile strength (approx): 4140 kg/cm^2 (60 000 lb/in^2) at normal temperatures
0.2% Yield strength (approx): 2240 kg/cm^2 (32 000 lb/in^2) at normal temperatures
Elongation: approx 25% at room temperature
Modulus of elasticity: n/a
Hardness: 260 Vickers diamond (30 kg load)
Melting point: 2200 to 2250°C (4000 to 4080°F)
Coefficient of linear expansion: 5.19×10^{-6} degC
Specific heat at 20°C: 0.0351 cal/g/degC
Thermal conductivity (circa 20°C): 0.223 watts/cm/degC
Electrical resistivity: 35.1 μ ohm cm

Hafnium metal is usually produced in 'sponge' form rich in oxygen or further refined to crystal bar form. The latter is ductile and can be cold- or hot-worked. The metal work-hardens, and may thus require intermediate annealing when cold working, but can be machined satisfactorily with tungsten carbide tools. Adequate cooling is necessary to prevent the metal overheating and catching fire, and special precautions must be taken with swarf and chips. In finely divided form, hafnium can be spontaneously combustible in air.

Hafnium has a limited application for electric light filaments and also as a getter material. It has little other individual application except as control rods, etc, in the nuclear-energy field.

HASTELLOY
See *Nickel.*

HERCULES ALLOY
Type of aluminium bronze.

HOYT'S ALLOY
Proprietary bearing alloys (British).

ILLIUM
Proprietary corrosion resistant nickel alloys (American).

IMMADIUM
Proprietary high tensile bronzes (British).

INCOLOY
See *Nickel.*

INCONEL
See *Nickel.*

INDIUM
Symbol: In
Specific gravity: 7.31
Density: 7.31 g/cm^3 (0.264 lb/in^3)
Tensile strength: 26.6 kg/cm^2 (380 lb/in^2)
Elongation: 22%
Hardness: 0.9 Brinell
Modulus of elasticity: 1.1×10^5 kg/cm^2 (1.57×10^6 lb/in^2)
Melting point: 156.2°C (313°F)
Specific heat at 20°C: 0.057 cal/g/degC
Coefficient of thermal expansion: 33×10^{-6} degC
Thermal conductivity (circa 20°C): 0.057 cal/cm^2/cm/sec/degC
Electrical resistivity: 8.37 μ ohm cm

Indium is a silver-white metal so soft and weak that it can be shaped and moulded in the fingers. It is stable in dry air but forms a surface (hydroxide) coating in moist air. Its chemical resistance is similar to that of aluminium. Indium is generally regarded as non-toxic but it should not be allowed to come into contact with foodstuffs as it is readily dissolved by many agaric acids. At high temperatures it burns with a brilliant violet coloured flame. It will sublime when heated in a vacuum (or hydrogen).

The chief applications of indium are as a plating material, in nuclear engineering, as a constituent of fusible alloys and in electrical engineering.

Indium plating can be applied to ferrous metals over a non-ferrous undercoating (*eg* copper, cadmium, gold, silver or zinc). The indium coating is then diffused into this undercoat. Heavier thicknesses of undercoating (*eg* 0.001 in) are used for plating bearings; very much reduced thicknesses for decorative/protective coatings. Indium can take a high degree of polishing and plated surfaces can be so treated for reflector surfaces, etc.

Indium is used directly as an alloying element to improve corrosion resistance of certain metals, and/or improve casting characteristics. The addition of indium can also improve the strength and hardness of aluminium, beryllium, copper, gold and lead. A 50:50 tin-indium alloy has the property of 'melting' glass and is widely used for glass-to-glass and metal-to-glass seals.

Metallic indium is available in two grades of purity, 97% and 99%.

INVAR

See *Nickel.*

IRIDIUM

Symbol: Ir
Specific gravity: 22.5
Density: 22.5 g/cm^3 (0.813 lb/in^3)
Tensile strength: 6300 kg/cm^2 (90 000 lb/in^2)
0.2% Yield strength: 2380 kg/cm^2 (34 000 lb/in^2)
Hardness: (annealed) 170 Vickers
Hardness: (hard rolled): 350 Vickers
Modulus of elasticity: 5.3×10^6 kg/cm^2 (76×10^6 lb/in^2)
Melting point: 2455°C (4451°F)
Coefficient of linear expansion: 6.8×10^{-6} degC
Specific heat at 20°C: 0.307 cal/g/degC
Thermal conductivity (circa 20°C): 0.14 ca/cm^2/cm/sec/degC
Electrical resistivity: 5.3 μ ohm cm

Iridium is a hard, dense white metal which is extremely resistant to chemical attack. It is brittle at room temperatures although it can be cold worked to a limited extent if pure. It can be hot-worked successfully at white heat. Machining characteristics are reasonably good.

One of the principal uses for iridium is as a hardener for platinum. Platinum-iridium alloys remain ductile up to 20% iridium content. 75:25 platinum-iridium is a widely used electrical contact material. The tipping of fountain pen nibs with osmoiridium (up to 77% iridium) has now largely been superseded by powder metallurgy techniques producing a tip directly from iridium ruthenium powder.

The radio-active isotope iridium 192 has a half-life of 75 days and finds limited application as a source of low energy rays for the non-destructive testing of welded joints and castings and for other radiographic work.

Iridium is produced entirely from platinum metal refinement and is normally solid in powder form of 99% purity.

K-MONEL

See *Nickel.*

KUNIFER ALLOYS

Proprietary copper-nickel-iron alloys.

KUTHERM

Proprietary electrical alloys (British).

LANGALLOY

See *Nickel.*

LANTHANUM

Symbol: La
Specific gravity: 6.19
Density: 6.19 g/cm^3 (0.224 lb/in^3)
Tensile strength (approx): 2240 kg/cm^2 (32 000 lb/in^2)
Yield strength (approx): 1960 kg/cm^2 (28 000 lb/in^2)
Tensile strength (as cast): 1170 kg/cm^2 (17 000 lb/in^2)
Compression strength (approx): 2100 kg/cm^2 (30 000 lb/in^2)
Hardness (as cast): 51 Vickers
Melting point: 920°C (1688°F)
Coefficient of linear expansion (circa 20°C): 5×10^{-6} degC
Specific heat at 20°C: 0.048 cal/g/degC
Thermal conductivity: 0.033 cal/cm^2/cm/sec/degC
Electrical resistivity: 57 μ ohm cm

Lanthanum is a malleable white metal which, although soft, is not ductile *ie* it can be beaten or hammered, but not drawn or extruded. It is attacked by moist air and decomposes in water (slowly in the case of cold water, violently in the case of hot water). It burns in air. The powder is pyrophoric and a 30:70 lanthanum: aluminium alloy sparks readily when struck by a hard substance.

Lanthanum alloys with iron posseses exceptional hardness, but have not been developed commercially. Limited use of lanthanum has also been made as a minor constituent of aluminium alloys to improve resistance to acids. The main application of lanthanum is in the oxide form in the ceramic and optical glass industries. The radio-active isotope has been used as a strong source of gamma-rays for non-destructive testing of welded joints and castings, etc.

Lanthanum metal is available commercially (in very limited quantities) in the form of rods or bars of 98 to 99% purity.

LEAD

Symbol: Pb
Specific gravity: 11.37
Density: 11.37 g/cm^3 (0.410 lb/in^3)
Melting point: 321°C (620°F)
Coefficient of thermal expansion: 0.000029/degC
Specific heat: 0.0308 cal/g

Lead is a heavy, soft, malleable, easily worked and readily cut metal. It is highly resistant to corrosion. When freshly cut or melted it has a bright metallic lustre, but it oxidises on exposure to the atmosphere and develops a protective skin. The proper-ties of lead at one time made it a first choice material for weatherings, flashings, cladding and roofing.

Lead is a particularly effective sound absorption material, particularly in thin sheets made by the new direct manufacture casting process which produces it in much longer lengths and lighter weights than were previously available. The cost per pound weight, or per decibel sound reduction, is often less with acoustic lead than with many other more rigid panel materials.

Chemical lead

Lead sheet made from chemical lead is produced primarily for the fabrication of chemical plant but may have special applications in building work. The compo-sition can be specified to BS 334 'Lead for Chemical Purposes'. This Standard covers two types of chemical lead: Type A which is lead of 99.99% purity and Type B which covers dilute alloys based on this high purity lead. Such alloys, formed by the addition of quite small quantities of other metals, are for use in fabricating chemical plant where improved strength is an advantage. The general properties of ordinary lead are retained in these alloys.

Lead sheet

Milled lead sheet and flashings for building purposes are specified in BS 1178, thickness now being given in millimetres. Lead flashings are normally available in lengths of 6 m and 12 m throughout the range of thicknesses, and widths from 150 m in increments of 30m. See also Tables 32 and 33.

TABLE 32 – WEIGHTS OF LEAD SHEET

Per square metre (m²)	Standard sheets 12 m x 2.4 m	Half sheets 6 m x 2.4 m (all to nearest 0.5 kg)	Quarter sheets 6 m x 1.2 m or 3 m x 2.4 m
kg	kg	kg	kg
14.18	408.0	204.0	102.0
20.41	588.0	294.0	147.0
25.40	732.0	366.0	183.0
28.36	817.0	408.5	204.0
35.72	1029.0	514.5	257.0
40.26	1159.0	579.5	290.0

TABLE 33 – WEIGHTS OF LEAD FLASHINGS
(to nearest 0.5 kg)*

Width	Code No 3 Thickness 1.25 mm Colour Identification		Code No 4 Thickness 1.80 mm Colour Identification		Code No 5 Thickness 2.24 mm Colour Identification		Code No 6 Thickness 2.50 mm Colour Identification		Code No 7 Thickness 3.15 mm Colour Identification		Code No 8 Thickness 3.55 mm Colour Identification	
	Green		Blue		Red		Black		White		Orange	
mm	12 m	6 m	12 m	6 m	12 m	6 m	12 m	6 m	12 m	6 m	12 m	6 m
	kg	kg	kg	kg	kg	kg	kg	kg	kg	kg	kg	kg
150	25.5	13.0	36.5	18.5	45.5	23.0	51.0	25.5	64.5	32.0	72.5	36.0
180	30.5	15.5	44.0	22.0	55.0	27.5	61.5	30.5	77.0	38.5	87.0	43.5
210	35.5	18.0	51.5	25.5	64.0	32.0	71.5	35.5	90.0	45.0	101.5	50.5
240	41.0	20.5	59.0	29.5	73.0	36.5	81.5	41.0	103.0	51.5	116.0	58.0
270	46.0	23.0	66.0	33.0	82.5	41.0	92.0	46.0	115.5	58.0	130.5	65.0
300	51.0	25.5	73.5	36.5	91.5	45.5	102.0	51.0	128.5	64.5	145.0	72.5
330	56.0	28.0	81.0	40.5	100.5	50.5	112.5	56.0	141.5	70.5	159.5	79.5
360	61.5	30.5	88.0	44.0	109.5	55.0	122.5	61.5	154.5	77.0	174.0	87.0
390	66.5	33.0	95.5	48.0	119.0	59.5	132.5	66.5	167.0	83.5	188.5	94.0
420	71.5	35.5	103.0	51.5	128.0	64.0	143.0	71.5	180.0	90.0	203.0	101.5
450	76.5	38.5	110.0	55.0	137.0	68.5	153.0	76.5	193.0	96.5	217.5	108.5
480	81.5	31.0	117.5	58.5	146.5	73.0	163.5	81.5	205.5	103.0	232.0	116.0
510	87.0	43.5	125.0	62.5	155.5	77.5	173.5	87.0	218.5	109.5	246.5	123.0
540	92.0	46.0	132.5	66.0	164.5	82.5	184.0	92.0	231.5	115.5	261.0	130.5
570	97.0	48.5	139.5	70.0	173.5	87.0	194.0	97.0	244.5	122.0	275.5	137.5
600	102.0	51.0	147.0	73.5	183.0	91.5	204.0	102.0	257.0	128.5	290.0	145.0

* Associated Lead Manufacturers Ltd.

Lead shot

Lead shot sizes up to **BB** are still made by the traditional method of pouring through a sieve in a shot tower. Sizes above **BB** are machine made. Data on lead shot are given in Tables 34 and 35.

British Standards

BS 1178 covers milled lead sheet and strip for building purposes.

LITHIUM

Symbol: Li
Specific gravity: 0.534
Density: 0.534 g/cm^3 (0.019 lb/in^3)
Melting point: 180°C (357°F)
Coefficient of linear expansion at 20°C: 56×10^{-6} /degC
Specific heat at 20°C: 0.79 cal/g/degC
Thermal conductivity (circa 20°C): 0.17 cal/cm^2/cm/sec/degC
Electrical resistivity: 8.55 μ ohm cm

Lithium is the lightest metal which is solid at normal room temperatures. It has a silvery-white lustre when freshly cut, being unstable and tarnishing rapidly on exposure to air. It reacts slowly with water and is readily attacked by acids. It burns at a temperature of 200°C. In the liquid metal state it is highly corrosive and strongly attacks most metals except ferrous metals.

Lithium is used as a strengthening and hardening agent in alloy steels, and also to improve machinability. It is also a hardening agent in lead-based bearing

TABLE 34 – LEAD SHOT SIZES

Name or number	Pellets per ounce	Weight		Diameter	
		grains	grams	m	mm
LG	6	70.00	4.54	0.360	9.14
SG	8	54.70	3.54	0.332	8.43
Special SG	11	39.77	2.58	0.298	7.57
SSG	15	29.17	1.89	0.269	6.83
AAA	35	12.50	0.81	0.203	5.16
BB	70	6.25	0.40	0.161	4.09
1	100	4.38	0.28	0.143	3.63
3	140	3.12	0.20	0.128	3.25
4	170	2.57	0.17	0.120	3.05
5	220	1.99	0.13	0.110	2.79
6	270	1.62	0.10	0.102	2.59
7	340	1.20	0.08	0.095	2.41
8	450	0.97	0.06	0.087	2.21
9	580	0.75	0.05	0.080	2.03

TABLE 35 – COMPARISON OF SHOT SIZES
(approximate only)

English	Cana-dian	American	French (Paris)	Belgian	Swedish	Italian	German	Dutch	Spanish
LG	—	—	—	—	—	—	Posten II	—	—
SG	SSG	OO Buck	—	B8	—	—	Posten III	B8	—
Spec SG	SG	1 Buck	—	B6	—	—	—	B6	—
SSG	AAAA	3 Buck	—	B5	—	—	Posten IV	B5	—
AAA	AAA	4 Buck	5/0	OV 9	—	—	5/0	OV 9	—
BB	Air rifle	Air rifle	1	OV 3	9	00	1	OV 3	1
1	2	2	3	1	7	1 or 2	3	1	3
3	4	4	4	3	5	3	4	3	4
4	5	5	5	4	—	4	5	4	5
5	6	6	6	5	3	5	6	5	6
6	—	—	—	6	2	6	—	6	—
7	7½	7½	7	7	0 or 1	7	7	7	7
7½	8	8	7½	7½	0 or 00	7½	7½	7½	7½
8	—	—	8	8	00	8	8	8	8
9	9	9	9	9	000	9	9	9	9

metals. It may also be used to improve the mechanical properties of aluminium alloys, magnesium alloys and zinc-based alloys. In all cases the lithium content is usually substantially less than 0.5%; although much higher proportions may be incorporated in magnesium-lithium alloys.

Lithium-copper (70:30) alloys are used to deoxidise copper and copper alloys. Lithium-calcium alloys (50:50 or 30:70) are used for the purification of copper and for grain refinement of cast irons. Silver-lithium alloys (up to 0.25% lithium) are used as silver solders for brazing 'difficult' ferrous and non-ferrous metals, and for brazing copper-tungsten contacts.

Lithium metal is used as a catalyst (usually disposed in a suitable fluid medium, such as a mineral oil). Lithium compounds are widely used in the glass, ceramic, chemical, nuclear, refrigeration and air-conditioning industries. Lithium hydride is a source of hydrogen, 1 gram of solid hydride reacting with water to yield approximately 100 ft^3 of hydrogen gas.

Lithium metal of 99.5 to 99.7% purity can be obtained by precipitation of the carbonate which is converted to the chloride and electrolysed using a cathode of stainless steel.

LO-EX

Aluminium piston alloy (American).

MAGNESIUM AND MAGNESIUM ALLOYS

Magnesium-based alloys have high strength and low density, and are also noted for their ease of machining (for a given power normal machining speeds for manganese are about ten times those for steel and twice those for aluminium alloys). Magnesium alloys can be fabricated as castings (sand, gravity die, pressure die), and as forgings, extrusions or sheet. The choice of method depends mainly on the nature of the component, but also on such factors as intricacy of design, strength, dimensional-accuracy, number required and cost.

Magnesium alloys are about one-quarter the weight of steel, and two-thirds that of aluminium. They corrode less in normal atmospheres than iron or steel but for more severe environments may require a protective finish.

The physical properties of pure magnesium are given in Table 36. The principal engineering properties of a range of magnesium-based alloys are given in Table 37. Whilst the older Mg-Al-Zn alloys are only marginally better than the aluminium-based alloys, the zirconium-containing alloy ZRE1, for instance, has approximately the same damping capacity as cast iron with only one quarter of the density.

The new alloy designated ZA, with a damping capacity considerably greater than even that of ZRE1, particularly at the lower stress levels, offers designers new opportunities where high damping capacity is required. Various castings have been made in this alloy, including main housings, instrument mounting plates and gyro unit components.

TABLE 36 – PHYSICAL PROPERTIES OF PURE MAGNESIUM

Atomic weight	24.32
Crystal structure	Close packed hexagonal
	a = 3.2030 Å, c = 5.2002 Å
Specific gravity at 20°C	1.738 g/cm^3
Melting point	650°C (1202°F)
Latent heat of fusion	85.6 cal/g
Latent heat of vaporisation	1315 cal/g
Mean specific heat (0–100°C)	0.25 cal/g/degC
Coefficient of thermal expansion (20–100°C)	26.1×10^{-6}/degC
Thermal conductivity (0–100°C)	0.376 cal/cm^2/sec/cm/degC
Electrical conductivity (20°C)	22.42×10^4 ohm/cm^3
Electrical conductivity (100°C)	17.50×10^4 ohm/cm^3
Electrical resistivity (20°C)	4.4611 μ ohms cm
Modulus of elasticity	6.4×10^6 lb/in^2
Modulus of rigidity	2.5×10^6 lb/in^2
Poisson's ratio	0.35
Vapour pressure (650°C)	2.28 mm Hg
Thermal neutron absorption cross-section	0.059 ± 0.004 barns/atom
Electropositive to most metals	

TABLE 37 – PHYSICAL PROPERTIES OF MAGNESIUM ELEKTRON ALLOYS

Alloy	Specific gravity g/cm³ (20°C)g	Weight lb/in³	Coefficient of thermal expansion 10^{-6}/degC (20 to 200°C)	Thermal conductivity $10^4 \times$ ohm/cm³ (20°C)	Electrical conductivity $10^4 \times$ ohm/cm³ (20°C)	Electrical resistivity μ ohm-cm (20°C)	Specific heat cal/g/degC (20 to 100°C)	Melting range°C
Average all alloys	1.80	0.0645*	27.0	0.25	—	—	—	—
ZA	1.74	0.0620	26.8	0.29	—	5.70	0.25	650
Z5Z	1.81	0.0652	27.3	0.27	15.2	6.60	0.23	550–640
RZ5	1.84	0.0660	27.1	0.26	14.7	6.80	0.23	510–640
TZ6	1.87	0.0675	27.1	0.27	15.2	6.60	0.23	500–630
MSR–A	1.81	0.0652	26.7	0.27	14.6	6.85	0.24	550–640
MSR–B	1.82	0.0658	26.7	0.27	14.6	6.85	0.24	550–640
ZE63A	1.87	0.0675	27.1	0.27	17.9	5.60	0.23	515–630
ZRE1	1.80	0.0647	26.8	0.24	13.7	7.30	0.25	545–640
ZT1	1.83	0.0657	26.7	0.25	13.9	7.20	0.23	550–645
MTZ	1.79	0.0646	27.0	0.22	13.9	7.20	0.25	590–645
A8 A.C.	1.81	0.0652	27.2	0.20	7.5	13.40	0.24	470–600
AZ91 A.C.	1.83	0.0657	27.0	0.20	7.1	14.10	0.24	470–595
C	1.81	0.0652	27.2	0.20	7.5	13.40	0.24	470–595
ZW3	1.80	0.0647	27.1	0.30	18.2	5.50	0.23	600–635
ZW1	1.80	0.0647	27.0	0.32	18.9	5.30	0.23	630–645
ZW6	1.83	0.0657	26.0	0.28	16.7	6.00	0.25	530–635
ZTY	1.76	0.0633	26.4	0.29	15.9	6.30	0.23	610–645
ZM21	1.78	0.0640	27.0	—	—	—	—	—
AM503	1.76	0.0633	26.9	0.34	20.0	5.00	0.25	650–651
AZ31	1.78	0.0640	26.0	0.19	10.0	10.00	0.25	575–630
AZM	1.80	0.0647	27.3	0.19	7.0	14.30	0.24	510–615
AZ855	1.80	0.0647	27.3	0.19	7.0	14.30	0.24	470–600

Key:
*Average weight to all alloys: 1 oz/in³
Modulus of elasticity for all alloys at 20°C: 6.4×10^6 lb/in²
Coefficient of thermal expansion: $26–27 \times 10^6$/degC
Specific heat: 0.23–0.25
Modulus of rigidity: 2.9×10^6 lb/in²

Magnesium alloys exhibit high damping properties – see Figure 3. They are similar to other non-ferrous alloys in that they exhibit no definite yield point but tend to yield slowly when stressed above the limit of proportionality. Based on equal weight (ie specific strength), magnesium alloys compare favourably with other materials, particularly in the case of castings.

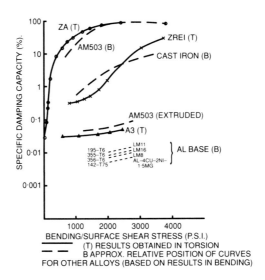

Fig. 3
Approximate damping capacity data
for some magnesium alloys.
(Magnesium Elektron Alloys Ltd).

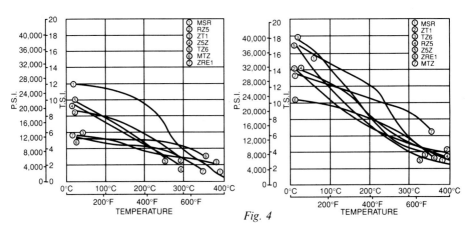

Fig. 4

Effect of temperature on the tensile
0.1% proof stress (offset) of various
magnesium cast alloys.

Effect of temperature on the ultimate
tensile stress of various magnesium
cast alloys.
(Magnesium Elektron Alloys Ltd).

Temporary heating to between 200°C and 350°C is generally required when carrying out forming operations with wrought magnesium alloys, and in most cases some degree of annealing will result during the preheating. As a matter of general good practice, preheating times should be kept as short as possible. The effects of annealing for 30 minutes at various temperatures up to 400°C on the room temperature tensile properties of 18 swg AZ31, ZW1 and ZW3 sheet are shown in Figure 6.

The addition of rare earth and thorium to magnesium alloys has considerably increased the creep strength over that of more conventional alloys and extensive use is now made of magnesium alloys for long time applications at temperatures up to approximately 350°C. In choosing the appropriate alloy both temperature and duration of stress should be considered. Figure 5 shows the short time creep properties of various casting alloys at temperatures ranging from 200 to 315°C.

Magnesium owes its importance in the nuclear energy field to its suitability as a canning material for uranium in gas-cooled, graphite-moderated reactors of the UKAEA Calder Hall and Chapelcross types, nine reactors commissioned by CEGB and also various overseas reactors. The metal has a low thermal neutron absorption cross-section (*ie* its relative affinity for absorbing neutrons is low), good thermal conductivity, does not alloy with uranium, and has satisfactory oxidation resistance to carbon dioxide at operating temperatures.

Canning alloys are required to operate at temperatures from 200 to 480°C, as extensible envelopes deforming without fracture under the forces generated by the growth and distortion of the fuel. High ductility at operating temperatures is, therefore, a prerequisite of canning alloys. Alloys used for ancillary fittings on the other hand need limited ductility but must possess adequate high temperature creep strength to support the stresses due to the weight of the fuel.

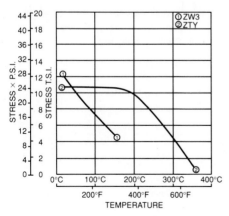

Effect of temperature on the tensile 0.1% proof stress (offset) of rolled ZW3 and ZTY (16 swg).

Fig. 5

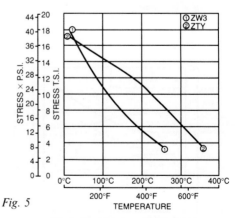

Effect of temperature on the ultimate tensile stress of ZW3 and ZTY (16 swg).
(Magnesium Elektron Alloys Ltd).

The cans are in the form of impact extruded thin wall tube with integral cooling fins which are twisted to improve cooling characteristics, or as a thick wall tube in which the cooling fins are machined in a herringbone pattern to improve both cooling characteristics and reduce flutter in the cooling gas stream.

Alloys of current interest for canning purposes are Magnox AL80, which contains approximately 1% Al and 0.01% Be and particularly in some French reac-

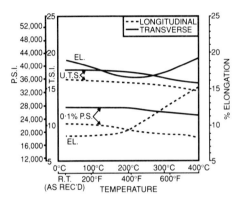

Effect of annealing for 30 minutes at various temperatures up to 400°C on the normal temperature tensile properties of 18 swg ZW1 sheet.

Effect of annealing for 30 minutes at various temperatures up to 400°C on the normal temperature tensile properties of 18 swg ZW3 sheet.

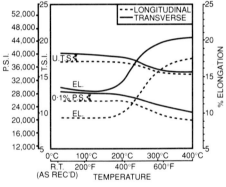

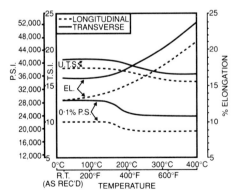

Effect of annealing for 30 minutes at various temperatures up to 400°C on the normal temperature tensile properties of 18 swg AZ31 sheet.

Fig. 6

tors, ZR55, which contains 0.5 to 0.6% Zr. Alloys used for ancillary fittings include AL80, ZR55 (which may be used in the heat treated condition) or a range of magnesium manganese alloys, MN70, MN80, MN125 and MN150 with nominally 0.7, 0.8, 1.3 and 1.5% manganese contents respectively. Other alloys having properties of specific interest for nuclear energy applications are being developed.

All the aforementioned alloys are supplied under strict quality control required by UKAEA specifications.

Forming wrought magnesium

Wrought magnesium products, ie rolled plate and sheet, extruded bar, sections and tubes, and drawn wire lend themselves to all common forming operations. Because of the hexagonal crystal structure of all available magnesium alloys most forming operations have to be carried out hot at temperatures in the range 250 to 350°C (up to 400°C for ZTY). At room temperature only very limited forming is possible, such as plain bending about large radii.

Hot-formed magnesium alloy parts show hardly any springback and are, therefore, dimensionably of good accuracy. Further control is possible by adjusting the temperature of tools and stock.

Magnesium alloy sheet and strip

These can be readily worked hot by most of the well-known manipulations such as bending, deep drawing, press forming, stretch forming, rubber forming, hydraulic pressure forming, spinning and roll forming as well as die drawing into thin sections. Magnesium alloy sheet and strip are normally supplied in the half hard condition in which they lend themselves to all but very severe hot forming for which fully annealed sheet may be required. The good room temperature properties of magnesium alloy sheet as supplied are not very much affected by the heating required for hot forming, but after restoring them to room temperature following hot forming the uts and 0.1% ps are somewhat lower and the elongation higher than before.

If correct forming temperatures are observed and the forming operations are not carried out too rapidly (hydraulic presses having controllable speeds ranging from 4 to 10 ft per minute are preferred), and adequate radii are provided, formed magnesium sheet parts can be designed on the same principles as similar parts in aluminium.

The bending radii depend on the sheet alloy and its condition, sheet thickness, speed of bending and temperature. As a general indication, safe minimum 180g bending radii for straight bending (press-braking) of annealed sheet are 3.5 to 10T cold and about 1T at 300 to 330°C; for hard rolled material the figures for cold bending are roughly double those for annealed sheet. With slow operation (as in the hydraulic press) smaller bend radii are possible. On the other hand with complex bending operations larger radii than those mentioned are required.

In roll-bending of sheet the radii are generally so large that cold bending is normal.

TABLE 38 – MAGNESIUM CASTING ALLOYS*

Typical chemical composition Major alloying elements%	Elektron alloy		Tensile Properties			Compressive Properties		Fatigue Endurance values		Hardness Brinell	Description
			0.2% Proof stress (N/mm²)	Tensile strength (N/mm²)	Elongation %	0.2% Proof stress (N/mm²)	Tensile strength (N/mm²)	Un-notched (N/mm²)	Notched (N/mm²)		
Zn 4.5, Zr 0.7	Z5Z	Precipitation treated Sand cast Chill cast	145 145	230 245	5 7	150-180	355-420	70-85	70-85	65-75	General purpose high proof stress structural alloy with useful properties up to about 150°C.
Zn 4.2, rare earth metals 1.3, Zr 0.7	RZ5	Precipitation treated Sand cast Chill cast	135 135	200 215	3 4	130-150	330-365	85-95	70-80	55-70	A variant of Z5Z, easily cast, weldable, pressure tight, with useful strength at elevated temperatures.
Zn 5.5, Th 1.8, Zr 0.7	TZ6	Precipitation treated Sand cast Chill cast	155 155	255 255	5 5	150-180	325-370	75-80	70-80	65-75	Stronger than Z5Z, as castable as RZ5, weldable, pressure tight.
Ag 2.5, Nd rich earth metals 1.7, Zr 0.6	MSR-A	Solution and precipitation treated Sand cast Chill cast	170 170	240 240	4 4	165-200	310-385	100-110	60-70	70-90	
Ag 2.5, Nd rich earth metals 2.0, Zr 0.6	MSR/ QE22	Solution and precipitated treated Sand cast Chill cast	175 175	240 240	2 2	165-200	310-385	100-110	60-70	70-90	Heat treatable alloys with highest yield strength of any cast magnesium alloys up to 250°C. Pressure tight and weldable.
Ag 2.5, Nd rich earth metals 2.5, Zr 0.6	MSR-B	Solution and precipitation treated Sand cast Chill cast	185 185	240 240	2 2	165-200	310-385	100-110	60-70	70-90	
Zn 5.8, rare earth metals 2.5, Zr 0.7	ZE63	Solution and precipitation treated Sand cast	170	275	5	190-200	430-465	115-125	70-75	60-85	Excellent castability, pressure tight and weldable with high developed properties in thin wall castings.
Rare earth metals 3.0, Zn 2.5, Zr 0.6	ZRE1	Precipitation treated Sand cast Chill cast	95 100	140 155	3 3	85-120	275-340	65-75	50-55	50-60	Creep-resistant up to 250°C. Excellent castability. Pressure tight and weldable.
Th 3.0, Zn 2.2, Zr 0.7	ZT1	Precipitation treated Sand cast Chill cast	(85) (85)	185 185	5 5	85-100	310-325	65-75	55-70	50-60	Creep-resistant up to 350°C. Pressure tight and weldable.

(continued)

TABLE 38 – MAGNESIUM CASTING ALLOYS* – contd.

Composition	Alloy	Condition								Remarks
Al 8.0, Zn 0.5, Mn 0.3	A8	As cast Sand cast	(85)	140	2	280–340	75–85	58–65	50–60	General purpose alloy. Good founding properties. Good ductility, strength and shock resistance.
		Chill cast	(85)	185	4					
		Solution treated Sand cast	80	200	7	325–415	75–90	60–70	50–60	
		Chill cast	80	230	10					
Al 8.0, Zn 0.5, Mn 0.3	A8 High Purity	As cast Sand cast	(85)	140	2	280–340	75–85	58–65	50–60	General purpose alloy. Good founding properties. Good ductility, strength and shock resistance. Good corrosion resistance.
		Chill cast	(85)	185	4					
		Solution treated Sand cast	(80)	200	6	325–420	75–90	60–70	50–60	
		Chill cast	(80)	230	10					
Al 8.0, Zn 0.5, Mn 0.3, Be 0.0015	AZ81	plus be Die cast	85	185	4	—	—	—	—	General purpose pressure die casting alloy.
Al 9.5, Zn 0.5, Mn 0.3	AZ91	As cast Sand cast	(95)	125	—	85–110	77–85	58–65	55–65	General purpose alloy. Good founding properties. Suitable for pressure die castings.
		Chill cast	(100)	170	2					
		Solution treated Sand cast	80	200	4	75–110	77–92	65–77	55–65	
		Chill cast	80	215	5					
		Solution and precipitation treated Sand cast	120	200	—	110–140	70–77	58–62	75–85	
		Chill cast	120	215	2					
Al 7.5–9.5, Zn 0.3–1.5, Mn 0.15 min	C	As cast Sand cast	(85)	125	—	65–90	73–80	58–65	50–60	Cheap general purpose alloy with good average properties.
		Chill cast	(85)	170	2					
		Solution treated Sand cast	(80)	185	4	75–90	77–85	62–73	50–60	
		Chill cast	(80)	215	5					
		Solution and precipitation treated Sand cast	(110)	185	—	90–115	62–73	58–62	70–80	
		Chill cast		215	2					

TABLE 39 – MAGNESIUM WROUGHT ALLOYS*

Typical chemical composition - Major alloying elements %	Elektron alloy	Tensile properties			Compressive properties		Fatigue properties		Impact value		Hardness Vickers	Description
		0.2% Proof stress (N/mm²)	Tensile strength (N/mm²)	Elongation %	0.2% Proof Stress (N/mm²)	Tensile Strength (N/mm²)	Un-notched (N/mm²)	Notched (N/mm²)	Un-notched (J)	Notched (J)		
Zn 3.0. Zr 0.6 ZW3	Sheet & plate 0.5-1.2 mm	160	250	5-6	175						60-70	High strength sheet, extrusion and forging alloy. Weldable under good conditions.
	1.2-6.0 mm	180	265	7-8			80-100					60-70	
	6.0-25 mm	175	260	8-10			80-100					60-70	
	25-50 mm	160	250	8								60-70	
	Extruded bars & Sections 10 mm	200	280	8	200-250	385-465	110-135	85-95	23-31	9.5-12		65-75	
	10-10 mm	225	305	8								65-75	
	Extruded forging stock 10 mm	195	280	8	165-215	370-430			16-27	4.7-9		65-75	
	10-100 mm	205	290	8								65-75	
	Forgings[4]	205	290	7								60-80	
Zn 1.3. Zr 0.6 ZW1	Sheet & plate 0.5-1.2 mm	160	240	5								60-70	High strength sheet and extrusion alloy. Fully weldable.
	1.2-6.0 mm	170	250	6-8								60-70	
	6.0-25 mm	130	230	8								60-70	
	25-50 mm	120	220	8								60-70	
	Extruded bars & Sections 10 mm	170	250	6-8	165-250	295-370	115-125	60-75	16-24	9.5-13		65-75	
	10-50 mm	185	260	6-8								65-75	
	Extruded tube	170	250	5								65-75	
	Forgings[4]	125	200	7								—	
Zn 5.5. Zr 0.6 ZW6	Extruded bars & sections – precipitation treated	230	315	8	200-315							65-80	High strength extrusion and forging alloy. Not weldable.
	Forgings[4] – precipitation treated	180	280	7								—	
Th 0.8. Zn 0.6. Zr 0.6 ZTY	Extruded forging stock 25 mm	130	230	6			75	45				50-60	Creep resistance up to 350°C. Fully weldable.
	25-50 mm	110	200	8								50-60	
	50 mm	95	200	8								50-60	
	Forgings[4]	130	230	6								50-65	

(continued)

TABLE 39 – MAGNESIUM WROUGHT ALLOYS* – *contd.*

Alloy	Form											Remarks
Zn 2.0, ZM21 Mn 1.0	Sheet – soft	(120)	220–265	10–12[5]	–	–	–	–	–	–	–	Medium strength sheet and extrusion alloy, easily formed. Fully weldable by argon arc process.
	– half hard	165	250	5–8[5]	–	–	–	–	–	–	–	
	Plate 6–25 mm	(120)	220	8–10	–	–	–	–	–	–	–	
	Extruded bars, sections & tubes 10 mm	150	230	8	–	–	–	–	–	–	50–65	
	10–75 mm	160	245	10	–	–	–	–	–	–	50–60	
	Forgings⁴	125	200	9	–	–	–	–	–	–	–	
Al 6.0, Zn 1.0, Mn 0.3 AZM	Extruded bars & sections & extruded forging stock 75 mm	180	270	8	130–180	370–420	125–135	90–95	34–43	6.7–9.5	60–70	General purpose alloy. Gas and arc weldable.
	75–150 mm	160	250	7	115–165	340–400	–	–	–	–	55–65	
	Extruded tube	150	260	7	130–180	–	115–125	80–90	16–23	3.4–4	60–70	
	Forgings⁴	160	275	7	130–165	340–400	–	–	–	–	60–70	
Al 3.0, Zn 1.0, Mn 0.3 AZ31	Sheet – soft	(120)	220–265	10–12	–	–	–	–	–	–	50–65	Medium strength sheet, extrusion and tube alloy. Good formability. Weldable.
	Sheet – half hard and plate 0.5–6.0 mm	160	250	5–7	–	–	–	–	–	–	60–70	
	6.0–25 mm	(120)	220	8–10	–	–	–	–	–	–	50–65	
	Extruded bars, sections & tubes 10 mm	150	230	8	–	–	–	–	–	–	50–65	
	10–75 mm	160	245	10	–	–	–	–	–	–	50–60	
	Forgings⁴	–	–	–	–	–	–	–	–	–	–	
Mn 1.5 AM503	Sheet plate 0.5–6 mm	(70)	200	3–5	–	–	–	–	–	–	35–45	Low strength general purpose alloy, gas and argon weldable and good corrosion resistance.
	6–25 mm	(70)	190	5	–	–	–	–	–	–	35–45	
	Extruded bars & sections	130	230	4	–	–	–	–	–	–	35–45	
	Extruded tube	(130)	230	4	–	–	–	–	–	–	45–55	
	Forgings⁴	(105)	200	4	–	–	–	–	–	–	45–55	

TABLE 40 – RELATED SPECIFICATIONS FOR MAGNESIUM CASTING ALLOYS

Alloy	Condition	Min of dep. Procurement Executive (D.T.D. Series)	British specifications		American specifications	
			B.S. Series		ASTM Alloy Designation & temper	ASTM Specification
			Aircraft	General Engineering		
Z5Z	Precipitation treated	—	2 L.127	2970 MAG4-TE	ZK51A-T5	B80-72
RZ5	Precipitation treated	—	2 L.128	2970 MAG5-TE	ZE41A-T5	B80-72
TZ6	Precipitation treated	5015A	—	2970 MAG9-TE	ZH62A-TE	B80-72
MSR A	Solution and precipitation treated	5025A	—	—	—	—
MSR/QE22	Solution and precipitation treated	5055	—	—	QE22A-T6	B80-72
MSR-B	Solution and precipitation treated	5035A	—	—	—	—
ZE6S	Solution and precipitation treated	5045	—	—	ZE63A-T6	—
ZRE1	Precipitation treated	—	2 L.126	2970 MAG6-TE	EZ33A-T5	B80-72
ZT1	Precipitation treated	5005A	—	2970 MAG8-TE	HZ32A-T5	B80-72
AZG	—	—	—	—	AZ63A	B80-72
A8	As cast	—	—	2970 MAG1-M	AZ81 A-F	B80-72
A8	Solution treated	—	3 L.122	2970 MAG1-TB	AZ81A-T4	B80-72
A8 High purity	As cast	684A	—	2970 MAG2-M	—	—
A8 High purity	Solution treated	690A	—	2970 MAG2-TB	—	—
AZ81 plus Be	Die cast	—	—	—	AZ81A	B93-66 (reapproved 1972)
A9V	—	—	—	—	—	—
AZ91	As cast	—	—	2970 MAG3-M	AZ91C-F	B80-72
AZ91	Solution treated	—	3 L.124	2970 MAG3-TB	AZ91C-T4	B80-72
AZ91	Solution and precipitation treated	—	3 L.125	2970 MAG3-TF	AZ91C-T6	B80-72
AZ91 plus Be	Die cast	—	—	—	AZ91B-F	B94-72
AZ92A	—	—	—	—	AZ92A	B80-72
C	As cast	—	—	2970 MAG7-M	—	—
C	Solution treated	—	—	2970 MAG7-TB	—	—
C	Solution and precipitation treated	—	—	2970 MAG7-TF	—	—

(continued)

TABLE 40 – RELATED SPECIFICATIONS FOR MAGNESIUM CASTING ALLOYS
– contd.

American specifications			German specifications		French specifications			
Federal or Military Specification	AMS Specification		Aircraft Number	DIN 1729 Number	Commercial Designation	Air 3380	AFNOR	STANDARD AECMA
QQ-M-56B MIL-M-46062B	4443B		3.5104	—	Z5Z	Z5Z	G-Z5	MG-C-42
—	4439		3.6104	3.5101	RZ5	RZ5	G-Z4TR	MG-C-43
QQ-M-56B MIL-M-46042B	4438B		3.5114	3.5102	TZ6	TZ6	—	MG-C-41
—	—		—	—	—	—	—	—
QQ-M-56B MIL-M-46062A	4418C		3.5164	3.5106	—	—	—	—
QQ-M-56B	—		—	—	MSR-B	—	G-Ag2 5TR	MG-C-51
MIL-M-46062B	4425		—	—	—	—	—	—
QQ-M-56B	4442B		3.6204	3.5103	ZRE1	ZRE1	G-TR3Z2	MG-C-91
QQ-M-56B	4447B		3.6254	3.5105	ZT1	—	G-Th3Z2	MG-C-81
QQ-M-56B	4420J		—	—	—	—	—	—
QQ-M-56B	—		—	3.5812	FT	G-A9	G-A9	MG-C-61
QQ-M-56B	—		—	3.5812	FT	G-A9	G-A9	MG-C-61
—	—		—	—	—	—	—	—
—	—		—	—	—	—	—	—
—	—		—	—	—	—	—	—
—	—		—	—	FT	—	—	MG-C-61
QQ-M-56B	—		—	—	F10	G-A9Z1	G-A9Z1	—
QQ-M-56B	—		—	—	F10	G-A9Z1	G-A9Z1	—
QQ-M-56B MIL-M-46042B	4437A		—	—	F10	G-A9Z1	G-A9Z1	—
QQ-M-38B	4490E		—	—	—	—	—	—
QQ-M-56B MIL-M-46062B	4434G		—	—	—	—	—	—
—	—		3.5194	3.5912	—	—	—	—
—	—		3.5194	3.5912	—	—	—	—
—	—		3.5194	3.5912	—	—	—	—

TABLE 41 – RELATED SPECIFICATIONS FOR WROUGHT MAGNESIUM ALLOYS

| Alloy Condition | | Min of dep. Procurement Executive (D.T.D. Series) | British specifications | | American specifications | |
| | | | B.S. Series | | ASTM Desig-nation | ASTM Specifi-cation |
			Aircraft	General Engineering		
ZW3	Sheet	—	2 L.504	3370 MAG-S-151M	—	—
	Plate	5081A	—	3370 MAG-S-151M	—	—
	Extruded bars & sections	—	2 L.505	3373 MAG-E-151M	—	—
	& forging stock		& L.514			
	Forgings	—	L.514	3372 MAG-F-151M	—	—
ZW1	Sheet	—	L.515	3370 MAG-S-141M	—	—
	Plate	—	—	3370 MAG-S-141M	—	—
	Extruded bars & sections	—	2 L.508	3373 MAG-E-141M	—	—
	Extruded tube	—	2 L.509	3373 MAG-E-141M	—	—
	Forgings	—	—	3372 MAG-F-141M	—	—
ZW6	Extruded bars & sections	5041A	—	3373 MAG-E-161TE	ZK60A-T5	B107-70
	Forgings	—	—	3372 MAG-F-161TE	ZK60A-T5	B91-72
ZTY	Extruded forging stock & forgings	5111	—	—	—	—
ZM21	Sheet - soft	5091A	—	3370 MAG-S-1310	—	—
	- half hard	5101A	—	3370 MAG-S-131M	—	—
	Plate	—	—	3370 MAG-S-131M	—	—
	Extruded bars, sections & tubes	—	—	3373 MAG-E-131M	—	—
	Forgings	—	—	3372 MAG-F-131M	—	—
AZM	Extruded bars & sections & forging stock	—	L.512 & L. 513	3373 MAG-E-121M	AZ61A-F	B107-70
	Extruded tube	—	2 L.503	3373 MAG-E-121M	AZ61A-F	B107-70
	Forgings	—	L. 513	3372 MAG-F-121M	AZ61 A-F	B91-72
AZ31	Sheet - soft			3370 MAG-S-110	AZ31B-O	B90-70
	- half hard & plate	—	—	3370 MAG-S-111M	AZ31B-H24	B90-70
	Extruded bars & sections	—	—	3373 MAG-E-11M	AZ31B-F	B107-70
	Extruded tube	—	—	3373 MAG-E-111M	AZ31B-F	B107-70
AM503	Sheet & plate	118C	—	3370 MAG-S-101M	—	—
	Extruded bars & sections	142B	—	3373 MAG-E-101M	MIA-F	B107-70
	Extruded tube	737A	—	3373 MAG-E-101M	MIA-F	B107-70
	Forgings	—	—	3372 MAG-F-101M	—	—

(continued)

TABLE 41 – RELATED SPECIFICATIONS FOR WROUGHT MAGNESIUM ALLOYS
– *contd.*

American specifications			German specifications		French specifications			
Federal	Military	AMS	Aircraft Number	DIN 9715 Number	Commercial Designation	Air 9052	AFNOR	STANDARD AECMA
—	—	—	—	—	—	—	—	—
—	—	—	—	—	—	—	—	—
—	—	—	—	—	—	—	—	MG-P-43
—	—	—	—	—	—	—	—	MG-P-43
—	—	—	—	—	—	—	—	MG-P-42
—	—	—	—	—	—	—	—	—
—	—	—	—	—	—	—	—	MG-P-42
—	—	—	—	—	—	—	—	MG-P-42
—	—	—	—	—	—	—	—	—
QQ-M-31B	MIL-M-5354A	4352D	—	3.5161	ZK60	—	G-Z5Zr	MG-P-41
QQ-M-40B	MIL-M-5354A	4362B	—	3.5161	ZK60	—	G-Z5Zr	MG-P-41
—	—	—	—	—	—	—	—	—
—	—	—	—	—	—	—	—	—
—	—	—	—	—	—	—	—	—
—	—	—	—	—	—	—	—	—
—	—	—	—	—	—	—	—	—
—	—	—	—	—	—	—	—	—
QQ-M-31B	—	4350H	—	3.5612	M1	G-A6Z1	G-A6Z1	MG-P-63
WW-T-825B	—	—	—	3.5612	M1	G-A6Z1	G-A6Z1	MG-P-63
QQ-M-40B	—	4358A	—	3.5612	M1	G-A6Z1	G-A6Z1	MG-P-63
QQ-M-44B	—	4375F	—	3.5312	F3	G-A3Z1	G-A3Z1	MG-P-62
QQ-M-44B	—	4377D	—	3.5312	F3	G-A3Z1	G-A3Z1	MG-P-62
QQ-M-31B	—	—	—	3.5312	F3	G-A3Z1	G-A3Z1	MG-P-62
WW-T-825B	—	—	—	3.5312	F3	G-A3Z1	G-A3Z1	MG-P-62
QQ-M-54	—	—	—	3.5200	T2	G-M2	G-M2	—
QQ-M-31B	—	—	—	3.5200	T2	G-M2	G-M2	—
WW-T-825B	—	—	—	3.5200	T2	G-M2	G-M2	—
QQ-M-40B	—	—	—	—	—	—	—	—

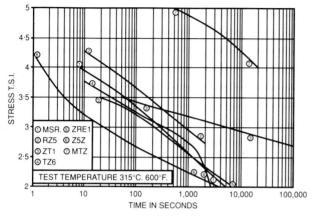

*0.5% total strain short time
creep curves for various
magnesium casting alloys.*

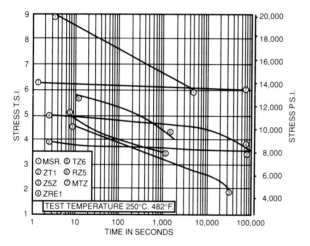

*0.5% total strain short time
creep curves for various
magnesium casting alloys.*

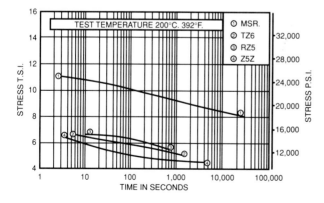

*0.5% total strain short time
creep curves for various
magnesium casting alloys.*

Fig. 7

Extrusions

Extrusions can be formed according to the same principles as apply to sheet. Sections and tubes can be bent in any standard bending equipment. Bending to small radii and the swaging and flaring of tubes must be performed hot. In bending, deep sections flanges may have to be cut and rewelded.

For deep drawing or other operations where graphite lubricants are used, the sheets should be free from chromate films, to assist both the lubricating action of the graphite and its subsequent removal. Sheet is normally supplied chromated unless specially ordered 'unchromated'.

Magnesium can be machined faster than any other metal, the limiting factor usually arising from the geometry of the part rather than the power of the machine or the quality of the tool material. As a result of the remarkable free cutting qualities of the metal, its high thermal conductivity and the low cutting pressure required, magnesium offers the following advantages in machining:

Reduction in machining times, machining speed being limited often by the spindle speed and capacity of the machiner hence economy in equipment and manpower.

Reduction in power consumption, the horsepower required per cubic inch of metal removed varies from 0.15 to 0.3, depending on the operation, and is one-sixth to one-seventh that required for steel and about half that needed for aluminium.

Ability to carry out scurfing and routing operations with ordinary woodworking equipment and with simply jigging.

Excellent surface finish.

Good life of the tool.

Production of well-broken chips which do not clog the machines.

Ease of manipulation owing to the light weight of the articles.

The various magnesium casting and wrought alloys exhibit no marked differences in their machining characteristics.

Chemical etching

Magnesium alloys are very suited to chemical etching and metal may be rapidly removed by immersion in acid of about 5% strength or greater. Sulphuric, nitric or hydrochloric acids are suitable, commercial quality being adequate.

This process may be applied to any form of magnesium product. It is used as a method of making thin walled castings where, for instance, the requirements may not justify the cost of precision casting methods.

Bolting and riveting

Screws and bolts for fastening magnesium alloy assemblies should preferably be

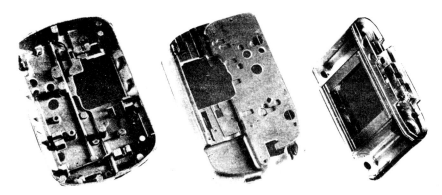

Magnesium alloy pressure die castings.

Trident airframe mounting. Sand casting in
magnesium–zirconium alloy RZ5.
(Sterling Metals).

of zinc- or cadmium-plated steel; or if light alloy bolts are essential they should be of aluminium-base alloy containing 5 or 7% magnesium rather than an aluminium alloy containing copper. Lockwashers are not generally recommended, because they may damage the magnesium parts. Elastic stop nuts should be used instead.

Rivets for magnesium alloys must also be of aluminium alloy with 5 or 7% magnesium: NR6, BS 1473 (BS L58). Experience has shown that these copper-free aluminium-magnesium alloys cause no appreciable bimetallic corrosion, whilst in addition they require no special protection beyond ordinary painting which is, of course, always recommended.

Welding

Magnesium alloys may be welded by the inert gas shielded arc welding (argonarc) process, by the oxyacetylene gas method, and by electric resistance ('spot') welding.

For general use the argonarc method is preferred, this having a number of advantages over gas welding. Gas welding must not be used on alloys containing zirconium.

Tungsten arc (argonarc) welding

Table 42 gives details of the welding compatibility of various alloys together with the recommended filler rod. It should be noted that certain wrought alloys may be satisfactorily welded to a number of the casting alloys.

TABLE 42 – WELDABILITY OF MAGNESIUM ALLOYS

Magnesium wrought alloys							
Alloy	AM503	AZM	AZ31	ZW1	ZW3	ZW6	ZTY
AM503	(AM503) A+	(AM503) C	(AM503) C	(ZW1) B	D	D	D
AZM		(AZM) A	(AZM) A	(ZW1) C	D	D	D
AZ31			(AZ31) A	(ZW1) C	D	D	D
ZW1				(ZW1) A+	(ZW1) B	D	(ZW1) A
ZW3					(Z2Z2C)‡ D	D	(ZW1) C
ZW6						D	D
ZTY							(ZTY) A+

Magnesium casting alloys							
Alloy	A8	ZRE1	MSR	RZ5	TZ6	ZT1	Z5Z
A8	(A8) A	(ZRE1) C	(MSR) C	(ZRE1) C	(ZT1) C	(ZT1) C	D
ZRE1		(ZRE1) A+	(MSR) A	(ZRE1) A	(ZT1) C	(ZT1) C	D
MSR			(MSR) A	(MSR) A	(ZT1) C	(ZT1) C	D
RZ5				(RZ5) A	(ZT1) C	(ZT1) C	D
TZ6					(TZ6) A	(ZT1) A	D
ZT1						(ZT1) A+	D
Z5Z							D

Key:
The recommended filler rod in each instance is shown in brackets.
Weldability code scheme:
A – good (+ sign indicates particularly good weldability)
B – fair
C – possible under certain conditions
D – not recommended
‡For machine welds only; ZW1 rod can be used for manual welds but with reduced weld efficiency.
Note: The nuclear alloys AS80 and ZR55 have good weldability.

(continued)

TABLE 42 – WELDABILITY OF MAGNESIUM ALLOYS – *contd.*

Alloy	AM503	AZM	AZ31	ZW1	ZW3	ZW6	ZTY
			Magnesium cast to wrought alloys				
A8	(AM503) C	(AZM) A	(AZM) A	D	D	D	D
ZRE1	D	(ZRE1) C	(ZRE1) C	(ZW1) A	(ZW1) B	D	(ZTY) C
MSR	D	(MSR) C	(MSR) C	(MSR) A	(MSR) B	D	(MSR) A
RZ5	D	(ZRE1) D	(ZRE1) C	(ZW1) A	(ZW1) B	D	(ZTY) C
TZ6	D	D	D	(ZW1) A	(ZW1) B	D	(ZTY) A
ZT1	D	D	D	(ZW1) A	(ZW1) B	D	(ZTY) A
ZTY	D	D	D	D	D	D	D

Key:
The recommended filler rod in each instance is shown in brackets.
Weldability code scheme:
A – good (+ sign indicates particularly good weldability)
B – fair
C – possible under certain conditions
D – not recommended
‡For machine welds only; ZW1 rod can be used for manual welds but with reduced weld efficiency.
Note: The nuclear alloys AS80 and ZR55 have good weldability.

The tensile strength of most undressed butt welds made by argonarc welding usually exceeds 90% of the parent materials in the annealed state, except in the case of AM503 where efficiencies of about 60% are more usual. The weld efficiency in hard-rolled sheet may be reduced by a few per cent because of annealing effects. In tubular butt welds efficiencies of about 80% can be expected.

MALLORY ALLOYS

Proprietary copper-based alloys.

MANGANESE

Symbol: Mn
Specific gravity: 7.29 to 7.43
Density: 7.29 to 7.43 g/cm^3 (0.26 to 0.27 lb/in^3
Tensile strength: (gamma state) 50.75 kg/mm^2 (32.25 tons/in^2)
Yield strength: (gamma state) 24 kg/mm^2 (15.25 tons/in^2)
Hardness: (gamma state) C.35 Rockwell
Modulus of elasticity: 1.6×10^6 kg/cm^2 (23×10^6 lb/in^2)
Elongation: (gamma state) 40%
Melting point: 1245°C (2273°F)

Coeficient of linear expansion: (alpha state) 22×10^{-6} /degC
Specific heat: 0.115 cal/g/decC
Thermal conductivity: n/a
Electrical resistivity: 185 μ ohm cm

Manganese is a coarse silvery-white metal with a yellow-orange tinge (this colouration may be absent in the electro-deposited metal) which is extremely hard and brittle in the alpha and beta states. It is stable in air at normal temperatures (unless impure), but highly reactive chemically.

Manganese is a 'natural' constituent of most steels and is commonly employed as a deoxidising and hardening agent on account of its additional properties. It is also used to improve the corrosion resistance of magnesium alloys to salt water or salt atmospheres, etc and to improve the casting characteristics of cast iron.

Manganese is a major constituent of a number of important alloys, particularly manganese bronze, noted for its exceptional resistance to salt water corrosion and good mechanical properties; copper-manganese-nickel alloys, used as resistance wires; and copper-manganese-aluminium alloy (Henstler's alloy) which is magnetic. It is also a constituent of certain high nickel alloys (eg Monel). Powdered ferro-manganese is widely used for the production of sintered steels. Other sintered manganese alloys are used for specific applications.

Manganese is toxic and inhalation of dust should be avoided.

The metal is most widely available in the form of ferro-manganese ('crude' manganese metal containing up to 80% manganese), spiegel (10 to 20% manganese) and silico-spiegel (15 to 20% manganese). High purity metal may be produced by electrolysis of the chloride or distillation of the crude metal.

MATRIX ALLOY

See *Printing Metals* (Section 7).

MAZAK

Proprietary zinc-based die casting alloys (British).

MERCURY

Symbol: Hg
Specific gravity: 13.546
Density: 13.546 g/cm^3 (0.4894 lb/in^3)
Melting point: 38.4°C (–37°F)
Boiling point: 357°C (675°F)
Specific heat at 20°C: 0.0323 cal/g/degC
Thermal conductivity (circa 20°C): 0.0196 cal/cm^2/cm/sec/degC
Electrical resistivity: (nominal) 100 μ ohm cm

TABLE 43 – CHEMICAL COMPOSITION OF BS 'MAG' ALLOY INGOTS

| Designation | | Aluminium % | Zinc % | Manganese % | Zirconium % | Rare Earth Metals % | Thorium % | Copper % | Silicon % | Iron % | Nickel % | Magnesium % | Remarks |
BSI	Chemical Symbol												
General Purpose alloys													
MAG 1	Mg-Al8 ZnMn	7.5/8.5	0.3/1.0	0.2/0.4	–	–	–	0.15	0.2	0.03	0.010	Rem	Cu+Si+Fe+Ni 0.35% maximum
MAG 3	Mg-Al10 ZnMn	9.0/10.5	0.3/1.0	0.2/0.4	–	–	–	0.15	0.2	0.03	0.010	Rem	Cu+Si+Fe+Ni 0.35% maximum
MAG 4	Mg-Zn 4.5 Zr	–	3.5/5.5	–	1.0	–	–	0.03	–	–	0.005	Rem	
MAG 7	Mg-Al8.5 Zn 1 Mn	7.5/9.2	0.3/1.5	0.15/0.8	–	–	–	0.30	0.3	0.05	0.020	Rem	Cu+Si+Fe+Ni 0.65% maximum
Special Purpose Alloys													
MAG 2	Mg-Al8 ZnMn	7.5/8.5	0.3/1.0	0.2/0.7	–	–	–	0.005	0.01	0.002	0.001	Rem	
MAG 5	Mg-Zn4 REZr	–	3.5/5.0	–	1.0	1.0/1.75	–	0.030	–	–	0.005	Rem	
MAG 6	Mg-RE3 ZnZr	–	0.8/3.0	–	1.0	2.5/4.0	–	0.030	–	–	0.005	Rem	
MAG 8	Mg-Th3 Zn2Zr	–	1.7/2.5	0.15	1.0	0.10	2.5/4.0	0.030	0.01	0.010	0.005	Rem	
MAG 9	Mg-Zn 5.5 Th2Zr	–	5.3/6.0	0.15	1.0	0.20	1.5/2.3	0.030	0.01	0.010	0.005	Rem	

NON-FERROUS METALS AND ALLOYS

175

TABLE 44 – CHEMICAL COMPOSITION AND MECHANICAL PROPERTIES FOR MAGNESIUM ALLOY CASTINGS
All limits are maxima. Balance Magnesium

BSI	Chemical Symbol	Aluminium %	Zinc %	Manganese %	Zirconium %	Rare Earth Metals %	Thorium %	Copper %	Silicon %	Iron %	Nickel %	Remarks	Condition	0.2% Proof stress minimum Sandcast N/mm²	0.2% Proof stress minimum Chillcast N/mm²	Tensile strength minimum Sandcast N/mm²	Tensile strength minimum Chillcast N/mm²	Elongation minimum Sandcast %	Elongation minimum Chillcast %
General Purpose Alloys																			
MAG 1	Mg-Al8 ZnMn	7.5/9.0	0.3/1.0	0.15/0.4	—	—	—	0.15	0.3	0.05	0.01	Cu+Si+Fe+Ni 0.40% max	M / TB	85 / 80	85 / 80	140 / 200	185 / 230	2 / 6	4 / 10
MAG 3	Mg-Al10 ZnMn	9.0/10.5	0.3/1.0	0.15/0.4	—	—	—	0.15	0.3	0.05	0.01	Cu+Si+Fe+Ni 0.40% max	M / TB / TF	95 / 85 / 130	100 / 85 / 130	125 / 200 / 200	170 / 215 / 215	— / 4 / —	2 / 5 / 2
MAG 4	Mg-Zn 4.5 Zr	—	3.5/5.5	—	0.4/1.0	—	—	0.03	—	—	0.005	—	TE	145	145	230	245	5	7
MAG 7	Mg-Al8.5 Zn 1 Mn	7.5/9.5	0.3/1.5	0.15/0.8	—	—	—	0.35	0.4	0.05	0.02	Cu+Si+Fe+Ni 0.75% max	M / TB / TF	85 / 80 / 110	85 / 80 / 110	125 / 185 / 185	170 / 215 / 215	— / 4 / —	2 / 5 / 2
Special Purpose Alloys																			
MAG 2	Mg-Al8 ZnMn	7.5/9.0	0.3/1.0	0.15/0.7	—	—	—	0.005	0.01	0.003	0.001	—	M / TB	85 / 80	85 / 80	140 / 200	185 / 230	2 / 6	4 / 10
MAG 5	Mg-Zn 4 REZr	—	3.5/5.0	—	0.4/1.0	0.75/1.75	—	0.030	—	—	0.005	—	TE	135	135	200	215	3	4
MAG 6	Mg-RE3 ZnZr	—	0.8/3.0	—	0.4/1.0	2.5/4.0	—	0.030	—	—	0.005	—	TE	95	110	140	155	3	3
MAG 8	Mg-Th3 Zn2Zr	—	1.7/2.5	0.15	0.4/1.0	0.10	2.5/4.0	0.030	0.01	0.010	0.005	—	TE	85	85	185	185	5	5
MAG 9	Mg-Zn 5.5 Th2Zr	—	5.0/6.0	0.15	0.4/1.0	0.20	1.5/2.3	0.030	0.01	0.010	0.005	—	TE	155	155	255	255	5	5

Key: M — as cast; TB — solution heated; TE — precipitation heated only; TF — solution heated and precipitation treated

Mercury is a heavy silvery-white metal which is liquid at normal temperatures. In its solid form (below –39°C) it is soft, ductile and malleable. Because of its wide liquid range and high coefficient of thermal expansion it is widely used in thermometers, but its chief application is in the chemical and pharmaceutical industries. Specific industrial uses of mercury include solders (4 to 8% mercury); as an active element in primary batteries; mercury vapour plant and vacuum plant; rectifiers; fulminates; as amalgams with silver, tin and gold; and prealloys in powder metallurgy.

Mercury vapour is poisonous and the maximum concentration which can be tolerated in a working atmosphere is of the order of 0.1 g/m^3 per eight hour day.

Commercial purity mercury (99.7% mercury) is obtained by distilling the crude metal extracted from cinnabar. It normally contains traces of aluminium, copper, iron, lead, magnesium, chromium, nickel and manganese.

MG

Aluminium-manganesium alloys (British).

MISCH METAL

Rare-earth alloy see *Lanthanum.*

MOLYBDENUM

Symbol: Mo
Specific gravity: 10.22
Density: 10.22 g/cm^3 (0.369 lb/in^3)
Tensile strength (approx) (annealed): 32.3 kg/mm^2 (20.5 $tons/in^2$)
Yield strength (annealed): 5.9 kg/cm^2 (3.75 $tons/in^2$)
Modulus of elasticity: 3.3×10^6 kg/cm^2 (47×10^6 lb/in^2)
Elongation (annealed): 30%
Melting point: 2610°C (4730°F)
Coefficient of linear expansion: 4.99×10^{-6} /degC
Thermal conductivity (circa 20°C): 0.34 cal/g/cm^2/sec/degC
Specific heat at 20°C: 0.066 cal/g/degC
Electrical resistivity: 5.2 μ ohm cm

Molybdenum is a soft silvery-white metal which can be machined and formed readily, and also welded. It work hardens to a brittle state but can be annealed. It cannot be hardened by heat treatment.

The principal use of molybdenum is as an alloying element in alloy steels and tool steels to improve mechanical properties and resistance to creep at elevated temperatures. It is an effective alternative to tungsten in this respect. It also reduces temper brittleness in low alloy steels and case hardening steels. A further application is in austenitic stainless steels to improve corrosion resistance.

Molybdenum metal is used for electrodes, filament supports, resistance elements, etc in the electrical industry; also for glass-to-metal seals, resistance welding electrodes and heavy duty contact breakers. The most ductile form of the metal is produced by powder metallurgy in bar form which can subsequently be reworked into fine wire or thin sheet. More limited use is made of molybdenum alloys, such as molybdenum (60%) cupro nickel (20%) platinum (10%) and tungsten (10%) as an alternative to osmiroidium for tipping pen nibs; and molybdenum-copper-pure iron alloys for corrosion resistant duties in the chemical industry, etc.

MONEL

See *Nickel.*

MUNTZ METAL

60/40 copper/zinc brass

NICKEL AND NICKEL ALLOYS

Symbol: Ni
Specific gravity: 8.88
Density: 8.88 g/cm^3 (0.321 lb/in^3)
Melting point: 1453°C

The latest national specification for pure nickel is DIN 1701: 1974 see Table 45. Composition of typical commercial nickel is given in Table 46, with mechanical properties in Tables 47 and 48. British Standards for the specification of nickel and nickel alloys are:

BS 3071   Nickel-copper alloy castings
BS 3072   Nickel and nickel alloys, sheet and plate
BS 3073   Nickel and nickel alloys, strip
BS 3074   Nickel and nickel alloys, tube
BS 3075   Nickel and nickel alloys, wire
BS 3076   Nickel and nickel alloys, rods

Wrought nickel containing 99% nickel (NA11) has a tensile strength of 3.8×10^8 N/m^2 (25 tons/in^2) in the annealed condition and a hardness of about 125 (Vickers). Low carbon nickel (NA12) has a slightly lower tensile strength and hardness in the annealed condition. The main difference between these two compounds is in the carbon content (NA11 0.15% maximum; NA12 0.02% maximum). Strength can be increased by cold working, although this will reduce ductility. The low carbon metal (NA12) is more ductile and is preferred for deep drawing, etc.

The various nickel alloy types and families are described under separate headings.

TABLE 45 – DIN 1701, GRADE C-Ni 99.95 COMPOSITION, WEIGHT PER CENT

Nickel plus cobalt	99.95000 minimum	Silver	0.00010 maximum
Cobalt	0.00050 maximum	Silicon	0.00050 maximum
Arsenic	0.00050 maximum	Tin	0.00010 maximum
Carbon	0.01500 maximum	Zinc	0.00010 maximum
Copper	0.00100 maximum	Aluminium	0.00010 maximum
Iron	0.02000 maximum	Bismuth	0.00002 maximum
Manganese	0.00010 maximum	Selenium	0.00010 maximum
Phosphorus	0.00050 maximum	Tellurium	0.00005 maximum
Lead	0.00010 maximum	Thallium	0.00010 maximum
Sulphur	0.00150 maximum	Tantalum	0.00010 maximum
Antimony	0.00010 maximum		

TABLE 46 – COMMERCIAL NICKEL

Typical composition

	Wrought %	Cast %
Nickel‡	99.510	Bal
Copper	0.070	0.03 max
Iron	0.140	1.00 max
Manganese	0.130	0.50–1.50
Silicon	0.030	1.00–2.00
Carbon	0.090	0.30 max
Sulphur	0.005 max	0.05 max

‡Includes a small amount of cobalt

Physical properties

Specific gravity	8.880
Weight, lb/in^3	0.321

Thermal properties

Melting range, °C	1435–1445
Specific heat, 27–1000°C	0.110
Coefficient of thermal expansion, millionths per °C,	
25–100°C	13.300
300–600°C	16.500
Thermal conductivity, cgs units	0.145

Electrical properties

Electrical resistance	
20°C μ cm/cm^3	9.0
20°C ohms/sq mil ft	42.5
20°C ohms/cir mil ft	54.0
Temperature coefficient of electrical resistance	
0–100°C	0.004–0.006

Magnetic properties

Saturation induction	6000 gauss (approx)
Magnetic permeability H = 20 oersteds	1000 (approx)
Initial permeability	200 (approx)
Magnetic transformation point, °C	360

Speed of sound

Longitudinal bar, cm/sec	4.7×10^5

Modulus of elasticity

lb/in^2 in tension	$28–30 \times 10^6$
lb/in^2 in torsion	11×10^6

Nickel-copper alloys

These fall into three broad categories:

Low nickel alloys with nickel content less than about 30%. These exhibit excellent resistance to corrosion (particularly sea water attack) and may also incorporate a small percentage of iron (*eg* 2%). Alloys in this range are also used for electrical resistance wires and coinage.

TABLE 47 – MECHANICAL PROPERTIES OF NICKEL

Form	Condition	0.2% proof stress ton/in^2	Maximum stress ton/in^2	Elongation in $^4\sqrt{}$ area %	Hardness DPN
Rounds	Cold-rolled or drawn	28–35	38–45	35–20	180–210
Squares	Ditto annealed	6–8	28–30	50–40	80–110
Rectangles	Hot-rolled	8–12	30–35	50–40	90–120
Hexagons					
Plate	Hot-rolled	9–13	31–45	50–40	100–210
Sheet and strip	Cold-rolled	7–20	25–35	50–35	90–120
	Ditto annealed	4–13	22–36	50–35	90–120
Wire	Cold-drawn, regular temper	40–58	47–63	15– 4	—
	Ditto annealed	4–13	22–35	50–30	—
Tubing	Cold-drawn	22–35	31–42	35–15	150–230
	Ditto annealed	4–13	22–35	50–35	90–120
Castings	As cast	9–13	24–28	15–35	80–120

TABLE 48 – MECHANICAL PROPERTIES NA11 NICKEL (BS 3072)

Condition	0.2% proof stress min	Tensile strength min	Elongation on 50 mm (or 2 in) min	0.2% proof stress min	Tensile strength min	Hardness HV max
	hbar	hbar	%	ton/in^2	ton/in^2	
Cold rolled and annealed	10.5	38.0	40	6.7	24.6	125
Hot rolled and annealed	10.5	38.0	40	6.7	24.6	125

Medium nickel alloys these have a nickel content of between 28 and 34%, corresponding to NA13 nickel-copper alloy for sheet, plate, strip, etc, or a nickel content between 30 and 40% for electrical resistance wires and thermocouples.

High nickel alloys containing about 70% nickel, 30% copper with small additions of iron and manganese, are known as *Monel*. Monel combines good mechanical properties with excellent resistance to corrosion and good high temperature properties. Typical properties of Monel are summarised in Tables 50 and 51. *K-Monel* is a modified form of Monel which is heat treatable and age hardening to enhance its mechanical properties see Table 52. K-Monel is non-magnetic.

TABLE 49 – MECHANICAL PROPERTIES NA12 LOW CARBON NICKEL (BS 3072)

Condition	0.2% proof stress min	Tensile strength min	Elongation on 50 mm (or 2 in) min	0.2% proof stress min	Tensile strength min	Hardness HV max
	hbar	hbar	%	ton/in^2	ton/in^2	
Cold rolled and annealed	8.5	34.5	40	5.4	22.3	110
Hot rolled and annealed	8.5	34.5	40	5.4	22.3	110

TABLE 50 – MONEL – PHYSICAL CONSTANTS

	Monel	'K' Monel
Specific gravity	8.830	8.470
Weight, lb/in^3	0.319	0.306
Melting range °C	1300–1350	1315–1350
Melting range °F	2370–2460	2400–2460
Specific heat	0.127	0.127
Linear coefficient of thermal expansion per degC		
(25 to 100°C)	$10^{-6} \times 14$	$10^{-6} \times 14$
(25 to 300°C)	$10^{-6} \times 15$	$10^{-6} \times 15$
per degF		
(77 to 212°F)	$10^{-6} \times 7.8$	$10^{-6} \times 7.8$
(77 to 572°F)	$10^{-6} \times 8.3$	$10^{-6} \times 8.3$
Thermal conductivity		
gal cal/cm/sec/cm^2 °C at 0 to 100°C	0.062	0.045
Btu/in/hr/ft^2 at 32 to 212°F	180.000	125.000
Modulus in tension, lb/in^2	$10^6 \times 26$	$10^6 \times 26$
Modulus in torsion, lb/in^2	$10^6 \times 9.5$	$10^6 \times 9.5$
Magnetic transformation point		
°C	43–60	below −100
°F	110–140	below −150

Nickel-iron alloys

Nickel-iron alloys with a high percentage of nickel alloyed with pure iron possess unique magnetic and electrical properties (usually after heat treatment), *eg* 'Numetal' and 'Permalloy'. They have been developed specially for the electrical and associated industries.

TABLE 51 – MECHANICAL PROPERTIES OF MONEL

Type	Condition	0.2% Proof Stress ton/in²	Maximum Stress ton/in²	Elongation in $\sqrt[4]{}$ area %	Hardness DPN
Hot-rolled — round, square, rectangle or hexagon; flat	Normal	18–22	35–42	45–30	140–165
Cold-drawn — round	Hard	34–38	42–48	30–15	200–240
Cold-rolled squares or rectangles	Annealed	13–17	33–37	50–40	130–145
Full finish sheet	Normal	20–29	35–38	40–20	130–170
Cold-rolled sheet or strip	Hard	40–49	44–54	15–2	200 min
	Annealed	11–20	31–38	50–30	100–140
Cold-drawn wire	Hard for springs	58–71	62–71	15–2	—
	Annealed	11–18	31–38	50–30	—
Castings:		Yield Point			
Normal Monel	As cast	11–15	23–33	16–40	100–145
'H' Monel (2.75% Si)	As cast	18–23	33–40	8–15	180–225
'S' Monel (3.75% Si)	As cast	40–45	40–45	nil	225–260

TABLE 52 – FORMS AND MECHANICAL PROPERTIES OF 'K' MONEL

Grade code	Form	Condition	0.2% Proof Stress ton/in²	Maximum Stress ton/in²	Elongation in $\sqrt[4]{}$ area %	Hardness DPN	Izod ft lb
A	Rod	Hot-rolled	18–38	40–53	45–30	160–240	120
B	Rod	Hot-rolled and thermally hardened	44–53	60–71	30–20	275–315	40
C	Rod, wire, tube, strip, turbine blading	Cold-worked *not* thermally hardened	31–44	44–55	35–15	190–265	56
D	Rod, wire, tube, strip, turbine blading	Cold-worked and thermally hardened	44–58	62–75	30–156	280–340	26
E	Rod, wire, tube, strip, turbine blading	Cold-worked and softened	18–26	40–49	45–35	155–195	120
F	Rod, wire, tube strip, turbine blading	Cold-worked, softened and thermally hardened	40–49	58–67	30–20	260–280	40

Nickel-iron alloys with a lower nickel content are noted for their very low coefficients of thermal expansion, the actual value being governed by the nickel content. Alloys of this type are known under the proprietary names *Invar, Nilen, Nilvar, Nilgro*, etc.

TABLE 53 – SUMMARY OF COMPOSITIONS OF WROUGHT NICKEL AND NICKEL ALLOYS TO BS 3072

	Nickel (including cobalt) %		Cobalt %		Copper %		Chromium %		Iron %		Molybdenum %		Aluminium %		Carbon %	Silicon %		Manganese %		Magnesium %	Titanium %		Sulphur %	Boron %		Zirconium %
	Min	Max	Min	Max	Min	Max	Min	Max	Min	Max	Min	Max	Min	Max	Max	Min	Max	Min	Max	Max	Min	Max	Max	Max	Min	Max
NA 11	99.0	—	—	—	—	0.25	—	—	—	0.4	—	—	—	—	0.15	—	0.15	—	0.35	0.20	—	0.10	0.010	—	—	—
12	99.0	—	—	—	—	0.25	—	—	—	0.4	—	—	—	—	0.02	—	0.15	—	0.35	0.20	—	0.10	0.010	—	—	—
13	63.0	—	—	—	28.0	34.00	—	—	—	2.5	—	—	—	0.50	0.30	—	0.50	—	2.00	—	—	—	0.020	—	—	—
14	72.0	—	—	—	—	0.50	14.0	17.0	6.0	10.0	—	—	—	—	0.15	—	0.50	—	1.00	—	—	—	0.015	—	—	—
15	30.0	35.0	—	—	—	0.75	19.0	23.0	—	rem	—	—	0.15	0.60	0.10	—	1.00	—	1.50	—	0.15	0.60	0.015	—	—	—
16	38.0	46.0	—	—	1.5	3.00	19.5	23.5	—	rem	2.5	3.5	—	0.20	0.05	—	0.50	—	1.00	—	0.60	1.20	0.030	—	—	—
17	34.5	41.0	—	—	—	0.50	17.0	19.0	—	rem	—	—	—	—	0.10	1.9	2.60	0.8	1.50	—	—	0.20	0.030	—	—	—
18	63.0	—	—	—	27.0	33.0	—	—	—	2.0	—	—	2.00	4.00	0.25	—	1.00	—	1.50	—	0.25	1.00	0.010	—	—	—
19	rem	—	15.0	21.0	—	—	18.0	21.0	—	3.0	—	—	0.80	2.00	0.013	—	1.50	—	1.00	—	1.80	3.00	0.015	0.001	0.03	0.15

TABLE 54 – MECHANICAL PROPERTIES NA13 NICKEL-COPPER ALLOY

Condition	0.2% Proof Stress min hbar	Tensile strength min hbar	Elongation on 50 mm (or 2 in) min %	0.2% proof stress min ton/in²	Tensile strength min ton/in²	Hardness HV max
Cold rolled and annealed	19.5	48.5	35	12.5	31.3	130
Hot rolled and annealed	19.5	48.5	35	12.5	31.3	130

TABLE 55 – NILO ALLOYS

Alloy	Density g/cm³	Melting Range °C	Specific heat at 20°C J/Kg/°C	Thermal Conductivity at 20°C W/m/°C	Thermal Expansion at 20–95°C 10⁻⁶/°C	Electrical Resistivity at 20°C microhm cm	Tensile Strength N/mm²	Hardness HV
NILO alloy 36	8.13	1430 approx	502	10.5	1.5	78	494	135
NILO alloy 42	8.13	1435 approx	502	10.5	5.3	57	494	150
NILO alloy 48	8.20	1450 approx	502	16.7	8.5	44	525	140
NILO alloy 51	8.24	1450 approx	—	16.7	10.0	40	525	140
NILO alloy 475	8.18	1450 approx	—	12.6	8.2	84	525	140
NILO alloy K	8.16	1450 approx	502	16.7	6.0	46	525	160

The addition of small amounts of chromium or cobalt yields alloys with a coefficient of expansion which can be matched to that of glass for glass-to-metal seals, etc – *eg Kovar, Fernico, Nilo K*. See Table 55 for details of 'Nilo' alloys. Other chromium-containing alloys (usually with traces of titanium, tungsten, beryllium, etc) have low temperature coefficients of elastic modulus *eg Elinvar, Ni-Span, Chromovar*.

Nickel-chromium alloys

The 80/20 nickel-chrome alloys have been developed in two distinct groups:

(i)   electrical resistance alloys, typified by 'Brightray', 'Nichrome', 'Pyronica', etc.

(ii) high strength, high temperature structural alloys with excellent creep resistance, typified by the *Nimonic* series of proprietary alloys see Tables 56, 57 and 58.

TABLE 56 – SUMMARY OF NIMONIC ALLOYS

Alloy	(i) Creep tested	(ii) Wrought alloys	(iii) Cast alloys
80 Ni : 20 Cr : Ti : Al	Nimonic 80	Nimonic C	Nimonic CC
80 Ni : 20 Cr : Ti : Al	Nimonic 80A	—	—
62 Ni : 20 Cr : 18 Co : Ti : Al	Nimonic 90	Nimonic B	Nimonic CB
62 Ni : 20 Cr : 18 Co : Ti : Al	Nimonic 95	—	—
80 Ni : 20 Cr : Ti : C	—	Nimonic 75	—
72 Ni : 20 Cr : 8 Fo : Ti : C	—	Nimonic F	Nimonic CF
37 Ni : 18 Cr : 2 Si : Fe	—	Nimonic D3	—

TABLE 57

Alloy	75 (DTD 703) %	F (DTD 714) %	80 (DTD 725) %	90 (DTD 747) %
Carbon	0.08–0.15	0.08–0.15	0.1 max	0.1 max
Titanium	0.20–0.60	0.20–0.60	1.8–2.7	1.8–3.0
Chromium	18–21	18–21	18–21	18–21
Aluminium			0.5–1.8	0.8–1.8
Silicon	1.00 max	1.00 max	1.0 max	1.5 max
Manganese	1.00 max	1.00 max	1.0 max	1.0 max
Iron	5.00 max	5.00–11.00	5.0 max	5.0 max
Cobalt	—	—	2.0 max	15.0–21.0
Copper	0.50 max	0.50 max	—	—
Nickel	Remainder	Remainder	Remainder	Remainder

TABLE 58

Creep criteria	Nimonic 80			Nimonic 80A	Nimonic 90	
Stress, ton/in^2	240.000	19.00	17.00	17.00	19.00	12.50
Stress, kg/mm^2	37.800	29.90	26.80	26.80	29.90	19.70
Temperature, °C	650.000	700.00	750.00	750.00	750.00	815.00
Maximum creep rate, % hour	0.006	0.01	0.01	0.01	0.01	0.01
Time to onset of tertiary creep, hour	75.000	50.00	25.00	50.00	50.00	50.00
Time to rupture, hour	100.000	75.00	35.00	75.00	75.00	75.00

Nickel-chromium-iron alloys

Nickel-chromium-iron alloys with a high nickel content again offer high strength and excellent high temperature performance, but have mainly been developed for their excellent resistance to corrosion under various conditions *eg Inconel* 600 and 625, see Tables 59, 60 and 61 for information relating to *Inconel* 600.

TABLE 59 – INCONEL 600

Typical compositions	Wrought %	Cast %
Nickel*	76.000	Balance
Chromium	15.500	12–15
Iron	8.000	5–10
Manganese	0.250	0.50–1.00
Copper	0.200	0.25 max
Silicon	0.250	0.50–2.00
Carbon	0.080	0.30 max
Sulphur	0.007	0.05 max
Properties		
Specific gravity	8.510	
Weight, lb/in^3	0.307	
Melting range		
°C	1395–1425	
°F	2540–2600	
Specific heat,		
Cgs units		
British thermal units	0.109	
Linear coefficient of thermal expansion,		
per °C (25–100°C)	$10^{-6} \times 11.5$	
Thermal conductivity,		
Cgs units	0.036	
British thermal units	104.500	
Modulus of elasticity,		
in tension or compression, lb/in^2	$10^6 \times 31$	
kg/mm^2	$10^3 \times 22$	

*Includes a small amount of cobalt

TABLE 60 – MECHANICAL PROPERTIES OF INCONEL 600 FOR VARIOUS STANDARD MILL FORMS

Form	0.2% proof stress ton/in^2	Maximum stress ton/in^2	Elongation in $\sqrt[4]{}$ area %	Hardness DPN
Rods:				
hot rolled or forged	20	42	35	160
Cold drawn rod:				
annealed	16	39	40	145
hard	40	53	20	240
Cold rolled sheet and strip:				
annealed	15	40	40	140
hard rolled	45	60	6	260
Cold drawn wire:				
soft annealed	15	40	40	
spring quality	72	78	6	
Tubing:				
annealed	17	39	45	145
cold drawn	40	55	10	250
Castings	15	31	10	175

TABLE 61 – INCONEL 600 COMPARISON OF STRENGTH IN COMPRESSION AND TENSION

Condition	Compression		Tension					
	0.1% proof stress ton/in²	0.2% proof stress ton/in²	0.01% proof stress ton/in²	0.2% proof stress ton/in²	Maxi-mum stress ton/in²	Elonga-tion in ⁴√ area %	Reduc-tion of area %	Hard-ness DPN
Hot rolled	14	19	17	21	41	42	66	170
Cold drawn								
as drawn‡	40	47	44	50	54	19	62	225
annealed	10	12	13	15	41	42	68	142

‡A 275°C low temperature, stress-equalizing anneal followed the cold drawing operation.

TABLE 62 – MECHANICAL PROPERTIES NA14 NICKEL-CHROMIUM-IRON ALLOY (BS 3072)

Condition	0.2% proof stress min hbar	Tensile strength min hbar	Elongation on 50 mm (or 2 in) min %	0.2% proof stress min ton/in²	Tensile strength min ton/in²	Hardness HV max
Cold rolled and annealed	24.0	55.0	30	15.6	35.7	200
Hot rolled and annealed	24.0	55.0	30	15.6	35.7	200

TABLE 63 – MECHANICAL PROPERTIES NA15 NICKEL-CHROMIUM-IRON ALLOY (BS 3072)

Condition	0.2% proof stress min hbar	Tensile strength min hbar	Elongation on 50 mm (or 2 in) min %	0.2% proof stress min ton/in²	Tensile strength min ton/in²	Hardness HV max
Cold rolled and annealed	20.5	51.5	3.0	13.4	33.5	200
Hot rolled and annealed	20.5	51.5	3.0	13.4	33.5	200

Nickel-molybdenum-iron alloys

The addition of molybdenum to nickel, with a trace of iron, produces alloys with outstanding resistance to corrosive attack by acids. Properties may be adjusted to 'tailor' the corrosion resistant properties, or adjust workability. Pro-

TABLE 64 – MECHANICAL PROPERTIES NA16 NICKEL-IRON-CHROMIUM-MOLYBDENUM-COPPER ALLOY (BS 3072)

Condition	0.2% proof stress min hbar	Tensile strength min hbar	Elongation on 50 mm (or 2 in) min %	0.2% proof stress min ton/in^2	Tensile strength min ton/in^2	Hardness HV max
Cold rolled and annealed	24.0	58.5	30	15.6	37.9	200
Hot rolled and annealed	24.0	58.5	30	15.6	37.9	200

TABLE 65 – SUMMARY OF LANGALLOYS

Alloy	Composition*	Tensile strength ton/in^2	Remarks
Langalloy 4R	30% molybdenum, 5% iron	32 – 36	Good resistance to strong mineral acids and corrosive solutions.
Langalloy 5R	17% molybdenum 15% chromium, 5% tungsten, 5% iron	32 – 36	Good resistance to acids, oxidizing agents, inert chlorine and bleaching agents
Langalloy 6R	10% silver, 3% copper	25 – 35	Extremely good resistance to sulphuric acid.
Langalloy 7R	23% chromium, 6% copper, 6% molyb-denum, 5% iron 2% tungsten	25	Excellent resistance to sulphuric, nitric and phosphoric acids and acid mixtures.
Langalloy 8R	15% chromium + iron	—	High strength at elevated temperatures with excellent resistance to creep and thermal shock.
Langalloy 9R	45% iron, 18% chromium	—	Heat resistant alloy with resistance to oxidation.

* – balance nickel.

portions which combine maximum corrosion resistance with good workability are of the order of 66:28:6 nickel:molybdenum:iron, although there are numerous variations to be found in proprietary alloys, *eg* 'Hastelloy' and 'Langalloy'. The composition of Hastelloy B is 65% nickel, 30% molybdenum and 5% iron. Hastelloy C contains 61% nickel, 18% molybdenum, 15% chromium and 6% iron. They have tensile strengths up to 56.5 kg/mm^2 (36 ton/in^2) and are used for the manufacture of chemical plant components where great resistance to corrosion at moderately high temperatures is desired. Another alloy showing great resistance to chemical action (in this case to sulphuric acid) is Hastelloy D, a nickel-silicon alloy containing 90% nickel and 10% silicon, with tensile strength about 26.78 kg/mm^2 (17 ton/in^2). *Langalloys* are summarised in Table 65.

TABLE 66 – MECHANICAL PROPERTIES
NA17 NICKEL-IRON-CHROMIUM-SILICON ALLOY (BS 3072)

Condition	Hardness HV
Cold rolled and annealed	200 maximum
Hot rolled and annealed	200 maximum

TABLE 67 – MECHANICAL PROPERTIES
NA18 NICKEL-COPPER-ALUMINIUM ALLOY (BS 3072)

Condition	0.2% proof stress min hbar	Tensile strength min hbar	Elongation on 50 mm (or 2 in) min %	0.2% proof stress min ton/in²	Tensile strength min ton/in²	Hardness HV min	max
Cold rolled and solution treated	—	—	—	—	—	—	200
Cold rolled, solution treated and precipitation treated	62.0	89.5	15	40.2	58.0	255	—
Hot rolled	—	—	—	—	—	—	270
Hot rolled and solution treated	—	—	—	—	—	—	200
Hot rolled and precipitation treated	69.0	96.5	18	44.6	62.5	275	—
Hot rolled, solution treated and precipitation treated	62.0	89.5	15	40.2	58.0	255	—

TABLE 68 – NOMINAL COMPOSITIONS AND DENSITIES OF HIGH-NICKEL ALLOYS COMMONLY USED WITH SULPHURIC ACID SOLUTIONS

Material	Nickel %	Copper %	Chromium %	Iron %	Silicon %	Manganese %	Carbon %	Other %	Density
Monel	67.0	30.0	—	1.40	0.10	1.00	0.15	—	8.83
'K' Monel	66.0	29.0	—	0.90	0.50	0.85	0.15	A12.75	8.47
'H' Monel	63.0	31.0	—	2.00	3.00	0.75	0.10	—	8.48
'S' Monel	63.0	30.0	—	2.00	4.00	0.75	0.10	—	8.36
Nickel	99.4‡	0.1	—	0.15	0.05	0.20	0.10	—	8.88
Inconel	76.0	0.2	15.50	7.50	0.25	0.25	0.08	—	8.51
Nickel-chromium alloys	80.0	—	20.00	—	—	—	—	—	8.36
Nickel-chromium-iron alloys	65.0	—	15.00	19.00	—	1.60	—	—	8.28
Ni-Resist	13.5–17.5	5.5–7.5	1.75–2.50	Bal	1.0–2.5	1.0–1.5	3.00 max	—	7.50

‡Includes a small amount of cobalt

Superalloys

Superalloy is the name given to the range of alloys specially developed to provide superior high temperature strength to that attainable by alloy steels. Originally these included wrought alloys with an austenitic structure and containing as much as 70% iron, realising high strength at elevated temperatures as a result of strain hardening, but they are now taken to infer high strength alloys where the iron content is replaced by nickel or chromium. They may be in wrought or cast form. The properties of the wrought forms may be capable of further improvement by heat treatment.

Alloys in this field include *Stellite* (cobalt-chromium with small proportions of tungsten and molybdenum), *Hastelloy* (nickel-chromium or nickel-molybdenum with up to 18% iron) – see Table 69; *Nimonic* (nickel-chromium), *Inconel* (nickel-chromium), *Udimet* (nickel-chromium), *IN* (International Nickel) alloys and others.

Many of the high temperature superalloys were then developed to meet the demands of the gas-turbine industry which called for high strength combined with oxidation- and creep-resistance at high temperatures, for turbines and other components. The original 'turbine blade alloy' was 80:20 nickel-chromium and subsequent improvements in high temperature strength have been realised by the addition of small proportions of aluminium and titanium to impart precipitation hardening characteristics, and of other elements such as cobalt, molybdenum, tungsten, niobium and vanadium. The resulting compositions can be quite complex (see Table 69).

Nimonic alloys

The following is a general guide to the application of Nimonic alloys. Properties are given in Table 70, compositions in Table 71 and standards in Table 72.

Nimonic alloy 75 – a high temperature alloy with good mechanical properties and outstanding resistance to oxidation at high temperatures. It is used for sheet metal work in gas-turbines, furnace parts and heat treatment equipment, and in nuclear engineering.

Nimonic alloy 80A – an age hardenable creep resisting alloy for service at temperatures up to about 815°C. Used for gas-turbine blades, rings, discs, etc, die-casting inserts and cores, bolts, nuclear boiler tube supports and reciprocating engine exhaust valves.

Nimonic alloy 81 has similar properties to Nimonic alloy 80A, but with high resistance to corrosion by NaCl, Na_2SO_4 and V_2O_5. Used for gas-turbine components including blades, and for diesel engine valves.

Nimonic alloy 90 – an age hardenable creep resisting alloy for service at temperatures up to about 920°C. Used for gas-turbine blades, discs, etc, and hot-working tools.

Nimonic alloy 105 – an age hardenable creep resisting alloy for service at temperatures up to about 940°C. Used for gas-turbine blades, discs and shafts.

Nimonic alloy 115 – an age hardenable creep resisting alloy for service at temperatures up to 980°C. Used for gas-turbine blades.

TABLE 69 – HASTELLOY ALLOYS

Alloy	Basic Composition %	Tensile Strength		Yield Strength ton/in²	Brinell Hardness	Remarks
		Wrought ton/in²	Cast ton/in²			
Hastelloy B	nickel 62 molybdenum 28 iron 5	58–62	33–37	27–38	210–235	Retains 66% of its strength up to 870°C. Primarily developed for resistance to hydrochloric acid and non-oxidising acids.
Hastelloy C*	nickel 54 molybdenum 17 chromium 15		32–36	20–21	220	Primarily developed for resistance to oxidising agents, acids, salts, hypochlorites and moist chlorine.
Hastelloy C 276	nickel molybdenum chromium					Good strength and oxidation resistance up to 1 050°C
Hastelloy D*	nickel 85 silver 10 copper 3		16–18	16–18	500–550	Resistance to hot acids, oxidising acids and salts.
Hastelloy F*	nickel 47 chromium 22 molybdenum 7	45	33	—	83–86 Rockwell B	Heat treatable alloy designed for resistance to acids, alkalis and chlorine solutions
Hastelloy G*	nickel chromium molybdenum					Corrosion resistant alloy for chemical duties
Hastelloy N	nickel 70 molybdenum 17 chromium 7 iron 5	51	38	—	—	Retains approximately 50% strength up to 870°C
Hastelloy W	nickel 62 molybdenum 25 chromium 5 iron 5	55	24	—	—	Retains approximately 50% strength at 760°C
Hastelloy X	nickel 47 chromium 22 molybdenum 9 iron 18	50	8 at 1 000 °C	23	90 Rockwell B	Maintains high strength and high resistance to oxidation at temperatures up to 1 200°C

* These alloys are primarily high temperature corrosion resistant alloys, but 'superalloys' by definition. They are included for completeness.

Nimonic alloy 263 – an age hardenable creep resisting alloy normally used for gas-turbine rings and sheet metal components for service at temperatures up to about 850°C.

Nimonic alloy 901 – an age hardenable alloy with a maximum service temperature of about 600°C. Used for gas-turbine discs, shafts and other components.

Nimonic alloy PE13 – a high temperature matrix hardened sheet alloy, similar to Nimonic alloy 75, but with improved mechanical properties. Used for sheet metal work for gas-turbines, furnace parts and heat treatment equipment, and in nuclear engineering. Complies with AMS 5536.

TABLE 70 – PHYSICAL AND MECHANICAL PROPERTIES OF NIMONIC ALLOYS*

Alloy	Density g/cm³	Melting Range °C	Specific heat at 20°C J/Kg/°C	Thermal Conductivity at 20°C W/mm/°C	Thermal Expansion at 20°C 10⁻⁶/°C	Electrical Resistivity 10⁻⁶ ohm cm	Tensile Strength N/mm²	Hardness HV
NIMONIC alloy 75	8.37	1340–1380	461	11.7	11.0	109	740	240
NIMONIC alloy 80A	8.19	1320–1365	461	11.2	12.7	124	1214	210
NIMONIC alloy 81	8.06	1305–1375	461	10.9	11.1	127	1050	230
NIMONIC alloy 90	8.18	1310–1370	446	11.5	12.7	118	1175	240
NIMONIC alloy 105	7.99	1290–1345	419	10.9	12.2	131	1140	320
NIMONIC alloy 115	7.85	1260–1315	444	10.6	12.0	139	1231	—
NIMONIC alloy 263	8.36	1300–1355	461	11.7	11.0	115	1000	195
NIMONIC alloy 901	8.16	1280–1345	419	—	12.3	—	1023	160
NIMONIC alloy PE13	8.23	1260–1290	461	—	13.5	118	530	230
NIMONIC alloy PE16	8.02	1310–1355	544	11.7	11.3	110	440	165
NIMONIC alloy PK25	8.02	—	435	11.4	12.1	—	1358	—

*Henry Wiggin & Company Limited

TABLE 71 – NOMINAL COMPOSITIONS (%) OF NIMONIC ALLOYS*

	Ni	C	Mn	Fe	S	Si	Cu	Cr	Co	Mo	Ai	Ti	Others
NIMONIC alloy 75	Bal.	0.13	1.0 max	5.0 max	0.020 max	1.0 max	0.5	19.5	—	—	—	-0.40	—
NIMONIC alloy 80A	Bal.	0.10 max	1.0 max	3.0 max	0.015 max	1.0 max	0.2 max	20.5	2.0 max	—	1.4	2.05	—
NIMONIC alloy 81	Bal	0.05 max	1.0 max	1.0 max	—	1.0 max	0.2 max	30.0	—	—	1.0	1.75	—
NIMONIC alloy 90	Bal	0.09	1.0 max	2.0 max	0.015 max	1.0 max	0.2 max	19.5	16.5	—	1.4	2.35	—
NIMONIC alloy 105	Bal	0.14	1.0 max	2.0 max	—	1.0 max	0.2 max	14.8	20.0	5.0	4.7	1.2	—
NIMONIC alloy 115	Bal	0.16	1.0 max	1.0 max	—	1.0 max	0.2 max	14.25	13.2	3.2	4.8	3.7	—
NIMONIC alloy 263	Bal	0.06	0.4	0.7 max	—	0.25	0.2 max	20.0	20.0	5.58	0.45	2.15	—
NIMONIC alloy 901	42.5	0.1 max	0.5 max	Bal	0.030 max	0.40 max	0.5 max	12.5	1.0 max	5.75	0.35 max	2.9	—
NIMONIC alloy PE13	Bal	0.10	1.0 max	18.5	—	1.0 max	0.5 max	21.75	1.5	9.0	—	—	W 0.6
NIMONIC alloy PE16	43.5	0.06	0.5 max	Bal	—	0.5 max	0.5 max	16.5	—	3.3	1.2	1.2	—
NIMONIC alloy PK25	Bal	0.07	0.5 max	4.0 max	—	0.75 max	0.2 max	18.0	17.5	4.0	2.75	2.9	—
NIMOLOY alloy PK37	Bal	0.12	1.0 max	5.0 max	—	1.5 max	0.5 max	18.5	18.5	—	1.5	2.5	—

*Henry Wiggin & Company Limited

TABLE 72 – NIMONIC ALLOYS – INTERNATIONAL SPECIFICATIONS AND DESIGNATIONS

	BS	AICMA	AMS	Royal Swedish Air Board Spec.	LW No.	DIN Designation	AFNOR No.
NIMONIC alloy 75	HR 5, 203 & 504	Ni-P 91-HT		MH.05	2.4630	NiCR 20 Ti	NC 20T
NIMONIC alloy 80A	HR 1, 201, 401 & 601	Ni-P 95-HT		MH. 07	2.4631	NiCr 20 TiAl	NC 20 TA
NIMONIC alloy 90	HR 2, 202, 501, 502 & 503	Ni-P 96-HT		MH. 45 / MH. 10	2.4632	NiCr 20 Co 18 Ti	NC 20 KTAt
NIMONIC alloy 105	HR 3	Ni-P 61-HT		MH. 14	2.4634	NiCo 20 Cr 15 MoAlTi	NKCD 20 ATv
NIMONIC alloy 115	HR 4	Ni-P 102-HT			2.4636	NiCo 15 Cr 15 MoAlTi	NCK 15 ATD
NIMONIC alloy 263	HR 10	Ni-P 105-HT					NCK 20 D
NIMONIC alloy 901		Fe-PA 99-HT	5660C & 5661A	MH. 16	2.4662	NiCr 15 MoTi	Z8 NC DT42
NIMONIC alloy PE11	5037					X8 NiCrMoTi 38 18	Z8 NC D38
NIMONIC alloy PE13	HR6 & 204	Ni-P 93-HT	5536E, 5754E & 5798	MH. 03	2.4665	NiCr 22 Fe 18 Mo	NC 22 FeD
NIMONIC alloy PE16	5047					X8 NiCrMoTiAl 43 16	NW 11 AC
NIMONIC alloy PK25		Ni-P 94-HT	5751A & 5753		2.4666	NK CD 20ATu	NK CD 20ATw
INCONEL alloy 718		NI-P 100-HT	5383, 5589, 5590, 5596B, 5597A, 5662B, 5663B, 5664A & 5832	MH. 06	2.4668	NiCr 19 NbMo	NC 19 Fe Nb
INCONEL alloy X-750			5542G, 5582, 5598, 5667F, 5668D, 5669, 5671A, 5698B, 5699B & 5778	MH. 04			NC15 FeT(Nb)
NIMOCAST alloy 80	3146						
NIMOCAST alloy 713	HC 203	Ni-C 98-HT	5391 A	MH. 31	2.4670	G·NiCr 13 Al 6 MoNb	NC 13 AD
NIMOCAST alloy PE10	HC 202	Ni-C 103-HT					NC 20 Nb
NIMOCAST alloy PK24	HC 204	Ni-C 104-HT	5397				NK 15 CAT
NIMOCAST alloy 242	3146						

TABLE 73 – NOMINAL COMPOSITIONS OF NIMOCAST CASTING ALLOYS*

	Ni	C	Mn	Fe	S	Si	Cu	Cr	Co	Mo	Ai	Ti	Others
NIMOCAST alloy 80	Bal.	0.07	0.35 max	2.0 max	—	0.4	—	19.5	—	—	1.3	2.45	—
NIMOCAST alloy 242	Bal.	0.35	0.5 max	0.75 max	—	0.4	0.2 max	22.0	10.0	10.5	0.2	0.3 max	—
NIMOCAST alloy PE10	Bal.	0.05 max	0.3	3.0	—	0.25	0.2 max	20.0	—	6.0	—	—	Nb 6.7 W2.5
NIMOCAST alloy 263	Bal.	0.06	0.6 max	0.7 max	—	0.4 max	0.2 max	20.25	20.25	5.85	0.5	2.2	—
NIMOCAST alloy 713	Bal.	0.12	0.2 max	2.0 max	—	0.25 max	—	13.25	—	4.5	6.25	0.9	Nb 2.3
NIMOCAST alloy alloy 713LC*	Bal.	0.05	0.25 max	0.5 max	—	0.5 max	—	11.75	—	4.5	6.0	0.7	Nb 2.0
NIMOCAST alloy PK24	Bal.	0.17	0.2 max	1.0 max	—	0.2 max	—	9.5	15.0	3.0	5.5	4.75	V 1.0

*Henry Wiggin & Company Limited

Nimonic alloy PE16 – an age hardenable creep resisting alloy for service at temperatures up to 750°C. Used for gas-turbine and nuclear applications.

Nimonic alloy PK25 – an age hardenable creep resisting alloy similar to that developed by Special Metals Inc and designated Udimet 500.

Nimocast alloys

Nimocast alloys are a series of nickel-chromium creep resistant casting alloys for such components as cast turbine rotors and stators, centrifugally cast rings and discs. Composition of these alloys is given in Table 73; properties are given in Table 74. Current types are:

TABLE 74 – PHYSICAL AND MECHANICAL PROPERTIES OF NIMOCAST ALLOYS

Alloy	Density g/cm³	Melting Range °C	Specific heat at 20°C J/Kg/°C	Thermal Conductivity at 20°C W/m/°C	Thermal Expansion at 20–95°C 10⁻⁶/°C	Electrical Resistivity at 20°C microhm cm	Tensile Strength N/mm²	Hardness HV
NIMOCAST alloy 80	8.19	1320–1635	461	11.2	12.7	124	734	275
NIMOCAST alloy 242	8.44	1370 approx	—		6.9	—	465	220
NIMOCAST alloy PE10	8.61	1240–1330	380	—	12.8	—	—	—
NIMOCAST alloy 263	8.36	1300–1355	461	11.7	11.0	115	1000	195
NIMOCAST alloy 713	7.88	1290 approx	427	10.5	10.7	144	735	—
NIMOCAST alloy 713LC	8.01	1288–1321	427	—	10.7	—	819	370
NIMOCAST alloy PK24	7.78	1220–1295	—	—	11.0	143	895	—

Nimocast alloy 80 – a casting alloy for service up to about 815°C. Used for diesel engine pre-combustion chambers and other high temperature components.

Nimocast alloy 242 – a casting alloy with outstanding resistance to thermal shock at up to 1050°C. Used for investment cast gas-turbine stator blades.

Nimocast alloy PE10 – an air casting alloy for service at temperatures of up to about 870°C. Used for turbo-charger rotors, gas-turbine components and diesel engine pre-combustion chambers.

Nimocast alloy 263 – a vacuum melted casting alloy for components used in association with the wrought Nimonic alloy 263.

Nimocast alloy 713 – a vacuum melted casting alloy for service at up to 1000°C for gas-turbine stator and rotor blades, turbine and turbo-charger rotors.

Nimocast alloy 713LC – a low carbon modification of Nimocast alloy 713 with good stress rupture properties and excellent room temperature ductility and strength. Used for gas-turbine rotor blades, turbine and turbo-charger rotors.

Nimocast alloy PK24 – a vacuum melted casting alloy for maximum service temperature of about 1040°C. Used for gas-turbine rotor and stator blades.

TABLE 75 – SPECIFICATIONS INDEX

Alloy*	ASTM	ASME	AMS	BS and DTD	DIN designation	Werkstoff No	Royal Swedish Air Board	AFNOR	AICMA
Nickel 200	B160–163	SB160–163	—	3072–76: NA11	17740: Ni 99.2	2.4066	—	—	—
Nickel 201	B160–163	SB160–163	5553	3072–76: NA12	17740: LC-Ni-99	2.4068	—	—	—
Nickel 205	F9	—	5555	—	—	—	—	—	—
Nickel 212	—	—	—	—	17741: NiMn2	2.4110	—	—	—
Nickel 270	F239	—	—	—	—	—	—	—	—
MONEL alloy 400	B127 B163–165	SB127 SB163–165	4544 4574 4675 4730 7233	3072–76: NA13 DTD 10B DTD 200A DTD 204A DTD 477 DTD 192 DTD 196	17743: NiCu30Fe	2.4360		—	—
Cast MONEL alloys	—	—	—	3071: NA1–2–3	—	—	—	—	—
MONEL alloy K-500	—	—	4676	3072–76: NA18	—	—	—	—	—
INCONEL alloy 600	B163 B166–8	SB163 SB166–8	5540 5580 5665 5687 7232	3072–76: NA14 DTD 328A	17742 NiCr15Fe	2.464OLN 2.4816	—	—	—
INCONEL alloy 625	B443–4 B446	SB443–4 SB446	5599 5666 5837	—	—	—	—	—	—
INCOLOY alloy 800	B163 B407–9	SB163 SB407–9	—	3072–76: NA15	X10NiCrAlTi 3220	1.4876	—	—	—
INCOLOY alloy 825	B163 B423–5	SB163 SB423–5	—	3072–76: NA16	NiCr21Mo	2.4858	—	—	—

*Henry Wiggin & Company Limited

Equivalent to the alloy developed by International Nickel Ltd and designated IN100.

Equivalent standards for some of the more important nickel alloys are shown in Table 75.

NICHROME
Proprietary nickel-chrome alloys (American).

NI-HARD
White cast iron alloyed with nickel and chromium.

NILEX (NILVAR)
Low expansion alloy see *Nickel.*

NILO
Proprietary nickel-iron alloys (British).

NIMONIC
Proprietary high-duty nickel alloys (British) see *Nickel.*

NIOBIUM (also known as Columbium)
Symbol: Nb
Specific gravity: 8.57
Density: 8.57 g/cm^3 (0.310 lb/in^3)
Hardness annealed: 80 Vickers
Hardness (wrought): 160 Vickers
Modulus of elasticity: 1.05×10^6 kg/cm^2 (15×10^6 lb/in^2)
Modulus of rigidity: 5.44×10^6 lb/in^2
Tensile strength (annealed): 2800 kg/cm^2 (40 000 lb/in^2)
Tensile strength (wrought): 5950 kg/cm^2 (85 000 lb/in^2)
Yield strength (annealed): 2100 kg/cm^2 (30 000 lb/in^2)
Elongation (annealed): 30%
Elongation (wrought): 5%
Melting point: 2468°C (4474°F)
Coefficient of linear expansion: 7.31×10^{-6} degC
Thermal conductivity: 0.125 $cal/cm^2/cm/sec/degC$
Electrical resistivity: 12 to 17×10^{-6} ohm cm

Unalloyed niobium is exceedingly ductile and easy to fabricate at room temperature. This metal has a moderate density, a high melting point and considerable strength at elevated temperatures. In addition it offers a lower thermal neutron absorption cross section than most other structurally suitable refractory metals. These factors, combined with its excellent corrosion resistance to hot aqueous systems and its capability of handling molten metals, has made niobium an imporant material of construction in the nuclear field.

Niobium is superior to most metals and second only to tantalum in general corrosion resistance. It is highly resistant to most acids and acidic solutions at room temperature. Niobium is completely inert to air, chlorine gas, nitric acid (conc), nitrogen, oxygen, sulphuric acid (20%) and tartaric acid at temperatures up to 100°C (212°F). It is also resistant to many liquid metals at high temperatures.

Other attractive characteristics of columbium are a low ductile-brittle transition temperature and excellent electron emission characteristics at high temperatures.

Niobium is a strong carbide-forming element, the resulting columbium carbide not being soluble in steel. Additions of up to 0.5% columbium can reduce air-hardening and improve heat-resistance, creep characteristics and weldability of wrought chromium steels. It is also a constituent (1 to 5%) of high temperature alloys.

The addition of up to 1% niobium in austenitic stainless steels prevents grain boundary precipitation of carbide and gives improved weldability. It is also added to chrome-aluminium nitriding steels to improve their response to nitriding.

Up to 0.3% niobium may be used to refine the structure of aluminium and aluminium alloys up to 1.5% to harden copper and soft brasses; and in cupro-nickel alloys it increases the annealing temperature. It is also a constituent of some permanent magnet alloys, and nickel-cobalt 'super alloys'.

High columbium alloys have an application in nuclear reactors, both for structural and canning duties.

Niobium metal has only limited availability in the form of sheet and wire. It is more readily available as ferro-niobium.

APPROXIMATE TEMPERATURE LIMITS TO WHICH NIOBIUM IS CORROSION RESISTANT TO LIQUID METALS*

Lithium, magnesium, potassium and sodium	1000°C (1830°F)
Lead	850°C (1560°F)
Mercury	600°C (1110°F)
Bismuth	550°C (1025°F)
Gallium	410°C (750°F)

NORAL
Proprietary aluminium alloys (British).

ORDNANCE ALLOY
Gunmetal see *Bronzes.*

OSMIUM
Symbol: Os
Specific gravity: 22.57

Density: 22.57 g/cm^3 (0.8154 lb/in^3)
Hardness: 800 Vickers diamond
Modulus of elasticity: 5.7×10^6 kg/cm^2 (81×1.0^6 lb/in^2)
Melting point: 2700 ±200°C (4900 ±350°F)
Coefficient of linear expansion: 4.6×10^{-6} /degC
Specific heat at 20°C: 0.031 cal/g/degC
Electrical resistivity: 9.5 μ ohm cm

Osmium is a blue-grey metal in the platinum group, and is the heaviest known metal. It is extremely hard and brittle with exceptional resistance to wear and corrosion, although it can become volatile at temperatures above 100°C, and oxidises readily at a temperature of 950°C. The volatile constituent (the oxide) is toxic.

Because of its hardness the metal is usually worked by powder metallurgy although its applications in this form are strictly limited. Its chief use is in an alloyed form (*eg* with iridium, nickel or cobalt) for electrical contacts or small wear parts.

Osmium of commercial purity is produced in bars and castings with a nominal purity of 99.0%.

See also chapters on *Electrical contacts.*

PALLADIUM

Symbol: Pd
Specific gravity: 12.0
Density: 12 g/cm^3 (0.4343 lb/in^3)
Properties: see Table 76

Palladium is a white, ductile and highly malleable metal and is easily worked hot or cold. It is the most readily fused of the platinum group of metals and is very suitable for cladding other metals and for casting. It is a particular characteristic of the metal that it readily absorbs up to 800 times its own volume of hydrogen at room temperature and will also absorb oxygen when molten. Deoxidiation treatment is necessary to ensure sound castings. Molten palladium is also prone to sulphur contamination which can lead to hot shortness and loss of ductility.

The principal industrial use of palladium is for electrical contacts. Palladium alloys are used for high temperature soldering and brazing, and for jewellery castings (95.5% palladium, 4.5% rutherium).

Commercial palladium is sold as 'sponge' ('black' palladium) or refined metal of 99.5% purity. It is also available as foil, sheets, rods, tubes and wire.

PERMALLOY

Proprietary nickel-iron soft magnetic alloys see *Magnetic materials* (Section 8).

PEWTER

91/7/2 tin/antimony/copper alloy.

TABLE 76 – COMPARATIVE PROPERTIES OF PLATINUM GROUP METALS

	Platinum	Palladium	Rhodium	Iridium	Ruthenium	Osmium
Workability	Soft, ductile; can be fabricated by all normal metalwork-ing methods; can be fuse-welded, brazed.	Ductile; can be worked when cold to thin foil, fine-wire, ribbon and sheet; can be rolled, spun, drawn; fuse-welded, brazed sol-dered.	Can be fabri-cated in form of wire or sheet.	Can be rolled to thin strip or drawn to fine wire after initial hot working.	Difficult to work.	Practically unworkable.
Corrosion resistance	Resistant to most acids, and, under oxidising conditions, to many fused salts, but attacked by strong sulphuric acid at high temperatures, aqua regia and any other acid in which hydrochloric acid is present in oxidising conditions; also attacked by some base elements at high temperatures under reducing conditions.	Less resistant than plat-inum. At-tacked by some oxidis-ing acids and by hot con-centrated sulphuric acid.	Resistant to most acids including aqua regia, and, under oxidising conditions, to a wide variety of fused salts.	Most corro-sion-resistant element known; not attacked by any acid including acid regia; only slowly at-tacked or not attacked by many alkali metals.	Resistant to common acids including aqua regia at room temperatures; resistant to or only slowly attacked by many molten metals, salts and compounds.	Readily soluble in hot, concen-trated nitric acid.
Alloys	Addition of rhodium, iridium or ruthenium greatly increases tensile strength.	Addition of 4.5% ruthenium increases tensile strength.	With plat-inum.	With palla-dium increases resistance to acids, with osmium to form osmi-ridium, a very hard alloy	With palladium and with osmium to increase hard-ness.	

PLATINUM

Symbol: Pt
Specific gravity: 21.45
Density: 21.45 g/cm^3 (0.7763 lb/in^3)
Properties: see Table 76

Platinum is a soft and ductile white metal with exceptional resistance to corro-sion and chemical attack. It is readily worked hot or cold and can be soldered or welded. Cold working increases the hardness. Hardness can also be increased by alloying with small proportions of copper or other metals in the polatinum group, *eg* the annealed hardness figure of 40 for pure platinum can be increased to 110 (5% copper), 105 (5% rutherium), 75 (5% indium) or 57 (5% rhodium). Hardened alloys are used in jewellery.

Platinum and platinum alloys are widely used as electrical contacts, spark plug electrodes, resistance windings, fuse wires, etc; as dies for the production of glass monofilament, anodes and for laboratory ware. Platinum is also a widely used catalyst.

Commercial refined platinum has a purity of 99.99% and is rendered in the form of a grey powder or 'sponge', then cast in ingots or a graphite mould.

PLATINUM GROUP METALS

The platinum group of metals comprises:

(i)   the lighter group rutherium, rhodium and palladium
(ii) the heavier group osmium, indium and platinum

Rutherium/osmium, rhodium/indium and palladium/platinum have essentially similar characteristics, *eg* the first pair are hard and very difficult to work, whilst at the other end of the range palladium and platinum are both soft, ductile metals. All have the characteristics of high resistance to oxidation, corrosion and chemical attack; high melting points; low electrical resistance; high catalyst activity. Platinum, rhodium and palladium have considerable application in the electro-plating field.

Properties of the platinum group metals are compared in Table 76 and specific properties are detailed in Tables 77 and 78.

PLUMBERS SOLDER

See *Solders* and *Low melting point alloys* (Section 9).

PLUTONIUM

Symbol: Pu
Specific gravity: 19.0 to 19.72
Density: 19.7 g/cm^3 (0.709 lb/in^3)
Tensile strength (cast): 45.5 kg/mm^2 (27 tons/in^2)
Yield strength (cast): 28 kg/mm^2 (17.75 tons/in^2)
Compression strength (cast): 84 kg/mm^2 (53.5 tons/in^2)
Modulus of elasticity (cast): 1×10^6 kg/cm^2 (14×10^6 lb/in^2)
Melting point: 640°C (1184°F)
Coefficient of expansion: 55×10^{-6} /degC
Thermal conductivity: 0.020 cal/cm^2/cm/sec/degC
Electrical resistivity: 140×10^{-6} ohm cm

Plutonium is formed from uranium in a nuclear reactor. It is a white metal tarnishing to yellow in air or greenish-black in moist air. The metal is highly radioactive (emitting alpha particles) and is employed in nuclear fuel, nuclear weapons, as a source of neutrons and for transmutation into higher isotopes.

TABLE 77 – SPECIFIC PROPERTIES OF PLATINUM GROUP METALS

	Platinum (Pt)	Iridium (Ir)	Osmium (Os)	Palladium (Pd)	Rhodium (Rh)	Ruthenium (Ru)
Atomic number	78	77	76	46	45	44
Atomic weight	195.09	192.2	190.2	106.4	102.95	101.07
Density, g/cm³ at 20°C	21.45	22.65	22.61	12.02	12.41	12.45
Crystal structure	FCC	FCC	HCP	FCC	FCC	HCP
Melting point, °C	1 769	2 443	3 050	1 552	1 960	2 310
Boiling point, °C	4 530	5 300	5 500	3 200	3 900	3 900
Thermal-neutron capture cross-section barns*	8.8	440	15.3	8.0	156	2.56
Coefficient of linear thermal expansion 0–100°C 10⁻⁶ per degC	9.1	6.8	6.1* (approx)	11.1	8.3	9.1* (approx)
Thermal conductivity at °C, cgs units	0.165	0.353	—	0.160	0.363	—
Electrical resistivity, μ ohm cm³	9.85	4.71	8.12	9.93	4.30	6.71
Temperature co-efficient of electrical resistivity 0–100°C	0.003927	0.0040	0.0042	0.0038	0.0046	0.0046
Hardness (annealed) DPN	40–42	200–240	300–670*	40–42	100–120	200–350*
Ultimate tensile strength (annealed) tons/in²	10	40	—	12	45	24.5
Young's modulus 10⁶ lb/in²	25	75	81	17	46	60
Hardness DPN: annealed electrodeposited	40–42 200–400	200–240 —	300–670 (as cast)	40–42 200–400	100–120 800–900	200–350 900–1 300

*These values depend upon orientation

POLONIUM

Symbol: Po
Atomic number: 89
Atomic weight: 210
Melting point: circa 255°C

Polonium is a radioactive metal with a half life of 136.5 days and is the final stage of the disintegration of radium into lead. It is also known as *radium F.*

POTASSIUM

Symbol: K
Specific gravity: 0.86

TABLE 78 – PROPERTIES OF SOME PLATINUM ALLOYS

Alloy	Density g/cm³ at 20°C	Melting point °C (Solidus)	Resistivity, μ ohm cm at 20°C	Temperature coefficient of resistance (0–100°C) per degC	Ultimate tensile strength tons/in²	Vickers hardness (annealed)
5% Ir-Pt	21.5	1 780	19.6	0.00190	17	90
10% Ir-Pt	21.6	1 800	24.5	0.00130	24	120
20% Ir-Pt	21.7	1 815	32.0	0.00085	45	200
25% Ir-Pt	21.7	1 845	32.5	0.00080	48	240
30% Ir-Pt	21.8	1 885	33.0	0.00070	56	285
5% Ru-Pt	20.8	1 775	30.0	0.00096	26	135
10% Ru-Pt	19.9	1 780	42.0	0.00047	37	200
5% Rh-Pt	20.8	1 820	17.0	0.00210	16	60
10% Rh-Pt	20.0	1 850	18.4	0.00170	21	75
20% Rh-Pt	18.8	1 900	20.0	0.00140	27	90
40% Rh-Pt	16.8	1 950	17.5	0.00140	31	150

Density: 0.896 g/cm³ (0.031 lb/in³)
Melting point: 63°C (146.7°F)
Specific heat at 20°C: 0.177 cal/g/degC
Coefficient of expansion (circa 20°C): 83×10^{-6} /degC
Electrical resistivity: 6.15×10^{-6} ohm cm

Potassium is a silvery-white metal which tarnishes rapidly in air, changing to a wax-like solid. It is highly reactive and has no specific industrial application except as a reagent, and also to a very limited extent as a hardening agent for special leads. Wide use is, of course, made of potassium salts.

RADIUM

Symbol: Ra
Atomic number: 88
Atomic weight: 226.05
Specific gravity: 5.0
Density: 5.0 g/cm³ (0.18 lb/in³)
Melting point: 700°C (1292°F)

Radium is one of the most important radioactive elements with a half life of 1580 years. The metal itself is white, rapidly tarnishing to a black colour in air. Its main application is for therapeutics but it is also used in combination with phosphorescent zinc sulphide in the manufacture of luminous paints for instrument and watch dials, etc and to a limited extent for non-destructive examination of metal castings and assemblies.

RED BRASS

See *Brass*.

REGULUS METAL
Lead-antimony alloy.

RHENIUM
Symbol: R
Specific gravity: 21.04
Density: 21.04 g/cm^3 (0.756 lb/in^3)
Tensile strength (annealed): 118 kg/mm^2 (75 tons/in^2)
Yield strength (annealed): 94.5 kg/mm^2 (60 tons/in^2)
Elongation (annealed): 28%
Modulus of elasticity: 4.7×10^6 kg/cm^2 (66.77×10^6 lb/in^2)
Melting point: 3180°C (5755°F)
Coefficient of expansion: 6.7×10^{-6} /degC
Thermal conductivity (circa 20°C): 0.17 cal/cm^2/cm/sec/degC
Electrical resistivity: 19.3×10^{-6} ohm cm

Rhenium is a silvery-white metal with an extremely high melting point and excellent resistance to corrosion (particularly to salt water and hydrochloric acid). It also has excellent properties as an electrical contact material and can readily be electrodeposited. Its most interesting mechanical property is that the pure metal has a hardness of 180 to 200 (Vickers) which can be increased to about 500 (Vickers) by cold working (*eg* 10 to 15% reduction), and that this hardness is retained at elevated tempeatures (except in an oxidising atmosphere). This characteristic of extreme hardening with cold working also makes it difficult to render rhenium in sheet and wire form, as it requires numerous reduction steps with intermediate annealing treatment. Rhenium can only be rendered ductile by producing it in sintered form. Sintered rhenium will also exhibit a high degree of work-hardening when cold worked.

Rhenium alloys readily with cobalt, nickel, iron, tungsten, tantalum, palladium, indium, rhodium and gold, such alloys having valuable electrical properties.

RHODIUM
Symbol: Rh
Specific gravity: 12.44
Density: 12.44 g/cm^3 (0.447 lb/in^3)
Tensile strength (annealed): 97 kg/mm^2 (61.75 tons/in^2)
Tensile strength (hard): 211 kg/mm^2 (134 tons/in^2)
Modulus of elasticity (hard):^{3}x 10^6 kg/cm^2 (42.5×10^6 lb/in^2)
Melting point: 1966°C (3571°F)
Specific heat at 0°C: 0.059 cal/g/degC
Coefficient of expansion (circa 20°C): 8.3×10^{-6} /degC
Thermal conductivity at 20°C: 0.21 cal/cm^2/cm/sec/degC
Electrical resistivity at 20°C: 4.51×10^{-6} ohm cm

Rhodium is a malleable and ductile silvery-white metal in the pure state, but work-hardens rapidly and is difficult to work either in wrought or cast forms. When electro-deposited, a rhodium coating exhibits exceptionally high hardness (circa 800 Vickers). The metal has exceptional resistance to corrosion (except hot sulphuric acid) and excellent electrical properties. It can be electro-deposited onto most non-ferrous metals (except aluminium, tin, lead and zinc) but not on ferrous metals, except over a primary coating of silver or nickel. Such coatings may be used for tarnish resistance, reflectivity or decorative effects, but the main application is for electrical contacts, etc. Rhodium is also a useful hardening agent when alloyed with platinum.

See also under *Platinum metals group*.

R METAL

Type of Monel.

RR ALLOYS

Aluminium piston alloys (British).

RUBIDIUM

Symbol: Rb
Specific gravity: 1.53
Density: 1.53 g/cm^3 (0.0553 lb/in^3)
Melting point: 38.9°C (102°F)
Specific heat at 20°C: 0.080 cal/g/degC
Coefficient of expansion (circa 20°C): 90×10^{-6} /degC
Electrical resistivity: 12.5×10^{-6} ohm cm

Rubidium is an alkali metal which ignites instantaneously in dry air and also reacts violently with water, generating and igniting hydrogen gas. It can be stored in a vacuum or inert gas, or in a dry hydrocarbon liquid. The metal is soft enough to be cut with a knife.

Rubidium alloys readily with the other alkali metals, also with silver and gold. It forms an amalgam with mercury. Rubidium is used to a limited extent for photo-electric cells, but the principal commercial use is that of rubidium salts as reagents.

RUTHERIUM

Symbol: Ru
Specific gravity: 12.2
Density: 12.2 g/cm^3 (0.441 lb/in^3)
Tensile strength (approx): 55 kg/mm^2 (35 $tons/in^2$)
Hardness (as cast): 220 Vickers
Hardness (forged): 390 Vickers
Modulus of elasticity (approx): 4.2×10^6 kg/cm^2 (60×10^6 lb/in^2)

Melting point: 2500°C (4530°F)
Specific heat (circa 0°C): 0.057 cal/g/degC
Coefficient of expansion: 9.1×10^{-6} /degC
Electrical resistivity: 7.6×10^{-6} ohm cm

Rutherium is a brittle white metal with an extremely high melting point. Its principal applications are alloyed with platinum as a contact material and with platinum or palladium for jewellery. See also section on *Platinum group metals.*

SAMPSON METAL

Proprietary zinc-based bearing alloy (American).

SCANDIUM

Symbol: Sc
Atomic number: 21
Atomic weight: 45.10

Scandium is found in the oxide form (SC_2O_2) in tin-tungsten ore, and is relatively abundant. Little or no commercial development of this metal has taken place although its general characteristics are said to be similar to those of aluminium.

SELENIUM

Symbol: Se
Specific gravity: 4.79
Density: 4.79 g/cm^3 (0.174 lb/in^3)
Melting point: 217°C (423°F)
Modulus of elasticity: 6×10^5 kg/cm^2 (8.4×10^6 lb/in^2)
Specific heat (circa 20°C): 0.084 cal/g/degC
Coefficient of expansion (circa 20°C): 37×10^{-6}/degC
Thermal conductivity (circa 20°C): 7 to 28.3×10^{-4} cal/cm^2/cm/sec/degC
Electrical resistivity: 12×10^{-6} ohm cm

Selenium is classified as a semi-metal, the metallic form being steel grey in colour. Its properties resemble sulphur in many ways (*eg* it sublimes on heating), burns readily in air, and forms selenides (resembling sulphides) with other elements.

Selenium metal is used for rectifiers. Another important use of selenium is for the production of photo-sensitive (photo-voltaire) cells. Ferro-selinium (approximately 10% selenium) is added to steels to improve machinability. Selenium is also added to copper alloys to improve machining properties.

Selenium compounds are used as pigments, glues, deoxidisers, etc.

High purity commercial selenium is available up to 99.99% purity. Normal commercial purity is 99.5%.

SILICON

Symbol: Si
Atomic number: 14
Atomic weight: 28.06

Silicon occurs naturally in the oxide form as sands, quartz, etc and constituents of various rocks. The element silicon can be classified as a semi-metal, but may be crystalline (black, lustrous) or amorphous (dark brown powder).

Silicon is a valuable constituent of aluminium and copper alloys, as well as being a common constituent of other metals and alloys. Silicon is also the basis of silicones.

SILVER

Symbol: Ag
Specific gravity (as cast): 10.5
Density: 10.5 kg/cm^3 (0.378 lb/in^3)
Tensile strength (annealed): 14 kg/mm^2 (9 tons/in^2)
Tensile strength (cold worked): up to 31 to 34 kg/mm^2 (20 to 22 tons/in^2)
Hardness (annealed): 26 Vickers
Hardness (hardened): 95 to 100 Vickers
Melting point: 960.8°C (1761°F)
Modulus of elasticity: 7.9×10^6
Specific heat: 0.057
Coefficient of expansion: 10.5×10^4 /degC (19×10^4 /degF)
Thermal conductivity: 1.0 cal/cm/sec/degC
Electrical conductivity: 62.5×10 4 ohm cm^3 (100%)

Silver is a ductile, malleable metal which can readily be worked, but work-hardens. Cold working increases its density to 10.57 g/cm^3. In the annealed condition the metal has exceptional thermal conductivity and electrical conductivity. It is resistant to attack by alkalis, most organic acids and non-oxidising mineral acids. It does not oxidise at normal or elevated temperatures, but is attacked or tarnished by most sulphur compounds or sulphurous atmospheres.

Commercial purity silver is known as *fine silver* which is 99.9% pure. Standard or *sterling silver* is an alloy containing 7.5% copper.

Silver is widely used as an electrical contact material (see chapter on *Electrical contact materials*); as an electroplated coating for a wide variety of applications; as a catalyst; as anodes, bearing linings, and as an alloying element with other metals.

SILVER SOLDER

See *Solders* (Section 9) and *Brazing materials* (Section 7).

SODIUM

Symbol: Na

Specific gravity: 0.9712
Density: 0.9712 g/cm^3 (0.035 lb/in^3)
Melting point: 97.82°C (208°F)
Specific heat at 20°C: 0.295 cal/g/degC
Coefficient of expansion (circa 20°C): 6.32 cal/cm^2/sec/degC
Electrical resistivity: 4.2×10^{-6} ohm cm

Sodium is an extremely soft alkali metal, silvery-white in colour tarnishing rapidly on exposure to air. It is violently reactive with water and most non-metallic elements, and burns in air when heated. It is normally stored in paraffin. It alloys with potassium to form a liquid alloy at room temperature. This alloy is used for pyrometers.

Metallic sodium is used as a colourant in valve stems, in wire form as a cable conductor, and for other specialised applications *eg* the production of antimony-free lead and solders. It has also been used in molten form as a special duty fluid (*eg* high temperature hydraulic fluid).

SPELTER

High zinc alloys.

STRONTIUM

Symbol: Sr
Specific gravity: 2.60
Density: 2.60 g/cm^3 (0.094 lb/in^3)
Melting point: 768°C (1414°F)
Specific heat at 20°C: 0.176 cal/g/degC
Electrical resistivity: 23×10^{-6} ohm cm

Strontium is a ductile metal which ranges in colour from silver-white to yellow-white. It is used as a fluxing agent in the manufacture of special steels, as a deoxidiser for copper and copper alloys, and as a hardening agent in tin and lead alloys. Strontium salts are more widely used than the metal.

TANTALUM

Symbol: Ta
Specific gravity: 16.6
Density: 16.6 g/cm^3 (0.600 lb/in^3)
Tensile strength (annealed): 35 kg/mm^2 (22.25 tons/in^2)
Tensile strength (hardened): 103 kg/mm^2 (64.5 tons/in^2)
Modulus of elasticity: 1.9×10^6 kg/cm^2 (27×10^6 lb/in^2)
Melting point: 2996°C (5425°F)
Specific heat (circa 20°C): 0.034 cal/g/degC
Coefficient of expansion: 6.5×10^{-6} /degC
Thermal conductivity (circa 20°C): 0.130 cal/mm^2/cm/sec/degC

Tantalum is a heavy metal with a high melting point. It is highly resistant to corrosive attack by virtue of the formation of a tough oxide film (and the most resistant of all metals to attack by acids). Although tough, the pure metal is malleable, ductile and readily machined with high speed tools. It work-hardens only slowly when cold worked. When heated in nitrogen it absorbs gas to form nitrides, becoming very hard (up to 600 Brinell) and brittle.

Tantalum is a refractory elemental metal used in the chemical industry for applications where its stable properties in contact with highly corrosive reagents are important, in high temperature furnace hardware applications, for medical and research purposes, in aircraft and in the electronics industry where the exceptional dielectric properites of tantalum oxide are employed. Tantalum tungsten alloys have been developed to enhance strength and rigidity of tantalum fabrications. Tantalum niobium alloys have been developed to enhance the high strength, temperature characteristics of tantalum.

Tantalum is available both as ferro-tantalum; and as commercial purity (99.95%) tantalum in the form of sheet and wire, as well as various tantalum alloys.

TELLURIUM

Symbol: Te
Specific gravity: 6.24
Density: 6.24 g/cm^3 (0.225 lb/in^3)
Melting point: 450°C (842°F)
Modulus of elasticity: 4.2×10^5 kg/cm^2 (6×10^6 lb/in^2)
Specific heat at 20°C: 0.047 cal/g/degC
Coefficient of expansion: 16.75×10^{-6} /degC
Thermal conductivity (circa 20°C): 0.014 cal/cm^2/cm/sec/degC
Electrical resistivity: 6.436×10^{-6} ohm cm

Tellurium is similar chemically to selenium, exhibiting weak acid and basic properties. The metal is white and lustrous but is crystalline and brittle and readily reduced to a powder. It will burn with a bright blue flame and white 'smoke' (oxide).

Tellerium has useful properties as a carbide stabiliser in the production of cast irons, and also to improve the machinability of copper and alloy steels. It is also a hardening agent for lead and white metal bearing alloys.

THALLIUM

Symbol: Tl
Specific gravity: 11.85
Density: 11.85 g/cm^3 (0.428 lb/in^3)
Melting point: 303°C (577°F)
Specific heat at 20°C: 0.031 cal/g/degC
Coeficient of expansion: 28×10^{-6} /degC
Thermal conductivity (circa 20°C): 0.093 cal/cm^2/cm/sec/degC
Electrical resistivity: 18×10^{-6} ohm cm

Thallium is an extremely soft white metal. The freshly cut surface is lustrous but tarnishes readily. It can be stored in water or oil.

Thallium is used as a constituent of some special fusible alloys, and as alloys with silver, lead, tin and aluminium.

THORIUM

Symbol: Th
Specific gravity: 11.66
Density: 11.66 g/cm^3 (0.421 lb/in^3)
Tensile strength (approx): 22 kg/mm^2 (14 tons/in^2)
Yield strength (approx): 14 kg/mm^2 (9 tons/in^2)
Hardness: 32 to 42 Vickers
Modulus of elasticity: 7 to 20×10^6 lb/in^2
Melting point: 1750°C (3182°F)
Specific heat at 20°C: 0.034 cal/g/degC
Coefficient of expansion: 12.5×10^{-6} /degC
Electrical resistivity: 13×10^{-6} ohm cm

Thorium is a radioactive metal with a half-life of 1.31×10^{10} years. In non-nuclear fields it is used as a deoxidant for molybdenum, a desulphurising agent in steels, an alloying element in magnesium alloys, in welding electrodes and for gas tube electrodes and getters. The pure metal is extremely soft, oxidises in air but the oxide film is self-heating and resists further attacks. The metal burns to its oxide (thoria).

TIN AND TIN ALLOYS

Symbol: Sn
Specific gravity: 7.29
Density (white tin): 7.29 g/cm^3 (0.263 lb/in^3)
Density (grey tin): 5.77 g/cm^3 (0.208 lb/in^3)
Melting point: 231.9°C (450°F)
Specific heat at 25°C (white tin): 0.053 cal/g
Specific heat at 10°C (grey tin): 0.049 cal/g
Specific heat up to 1000°C (liquid tin): 0.0615 cal/g
Thermal conductivity at 0°C: 0.150 cal/cm^2/sec/cm/degC
Coefficient of expansion at 0°C: 19.9×10^{-6} /degC
Modulus of elasticity: 4240 kg/mm^2
Modulus of rigidity: 1640 kg/mm^2
Tensile strength at 15°C: 1.48 kg/cm^2
Elongation: 75%
Shear strength (room temperature): 1.26 kg/cm^2
Hardness at 20°C: 3.9 Brinell
Electrical resistivity: 11.5×10 6 ohm cm

Tin is a soft, lustrous white metal, obtained mainly from cassiterite. Commercial purity or 'standard' tin must contain at least 99.75% tin. The main impurities are lead, antimony, copper, iron and arsenic. High purity tin can be obtained by electrolytic refining. Tin can exist in two forms, *white tin* which is stable above 13.2°C and *grey tin* stable below this temperature. Transformation of white tin to grey tin requires the maintenance of a considerably lower temperature than 13.2°C and is inhibited by the presence of impurities.

Tin has excellent resistance to corrosion and one of its major uses is for protective coatings on ferrous metals and copper and copper alloys. The manufacture of tinplate (tin coated steel) accounts for some 35 to 40% of the world's production of tin. Tin-zinc and tin-nickel alloys are also used as protective coatings for steel. Tin is used as an alloying element with titanium (see *Titanium and Titanium alloys*).

Tin-copper alloys are generally known as bronzes or gunmetals. Tin-lead alloys are used as solders. Tin is also a constituent of white metal bearing alloys, fusible alloys, printing metals and pewter. See also in Section 7: *Bearing metals, Coatings metals, Printing metals.* Section 9: *Solders* and *Low melting point alloys.*

Tin compounds are widely used and extend into a considerable number of fields from ceramic to wood preservatives.

TITANIUM AND TITANIUM ALLOYS

Titanium is a low density high strength metal with a considerable application as a structural material. It has approximately 55% the density of steel with a tensile strength of the order of 26 to 45 tons/in^2 which can be raised to 80 to 90 tons/in^2 by alloying. Strength is also well maintained at elevated temperatures and specific titanium alloys are suitable for use in a stressed condition at temperatures up to or exceeding 500°C. A further attraction is its excellent corrosion resistance and resistance to pitting and stress-corrosion cracking. The stiffness of titanium is lower than that of steel (roughly half that of stainless steel) but appreciably higher than that of aluminium.

Titanium is paramagnetic with very low permeability to magnetic flux. The electrical resistivity approaches that of stainless steel and this is approximately doubled by the addition of alloying elements.

TABLE 79 – PHYSICAL PROPERTIES OF TITANIUM

Specific gravity	4.51	
Density	0.163 lb/in^3	4.51 g/cm^3
Elastic modulus in tension	15.5 × 10^6 il/in^2	10.9 × 10^6 kg/mm^2
Elastic modulus in torsion	6.5 × 10^6 lb/in^2	4.6 × 10^6 kg/mm^2
Poissons ratio	0.321	
Melting point	3 035°F	1 668°C
Thermal conductivity	9 Btu/ft/hr/degF	15 W/mK
Coefficient of thermal expansion	5.0 × 10^{-6}/degF	9.0 × 10^{-6}/degC
Specific heat	0.13 Btu/lb/degF	0.13 cal/g/degC
Electrical resistivity	3% IACS	6 μ ohm–cm
Velocity of sound in titanium	20 100 ft/sec	6 130 m/sec

TABLE 80 – COMPARISON OF TITANIUM AND AEROSPACE ALLOYS

	Mean yield strength lb/in^2	Density lb/in^3	Strength to Weight ratio	Relative to Titanium %
CP Titanium (medium)	47 500	0.1630	292 000	100
Complex titanium alloys	127 500	0.1700	750 000	257
Aluminium alloys	43 500	0.0975	446 000	153
Stainless steel	87 500	0.2850	307 000	108
Magnesium alloys	33 000	0.0614	540 000	185

The largest consumer of titanium continues to be the aerospace industry. The requirements of the industry are met by a wide range of titanium alloys each developed for its specific properties, eg hot strength, fatigue resistance. Some comparisons with other aerospace materials are drawn in Table 80. In contrast, industrial use has tended to focus on the valuable corrosion resistant properties of pure titanium and to exploit these in the widest possikble range of applications for chemical, petrochemical and process engineering industries. Other properties of titanium make it suitable for acoustic and ultrasonic applications and for surgical implants in the human body.

The two crystal forms are alpha-titanium, which is stable up to 882°C (1620°F), above which temperature the body-centred cubic structure known as beta-titanium forms and remains stable up to the melting point of 1668°C (3035°F).

The transformation characteristics of titanium are considerably modified by the addition of alloying elements and this phenomenon can be utlised to produce alloys which have an all-alpha, all-beta, or alpha 8 beta structure at room temperature. As the mechanical properties are influenced by the structure this permits a range of alloys displaying characteristic properties and suited to specific applications.

Alpha alloys are usually of medium to low strength, are reasonably ductile at room temperature, and can be welded but not heat treated. Aluminium is the most favoured substitutional alloy for stabilising the alpha phase (eg 5A1, 2.5 Sn alloy). The interstitial elements, oxygen, nitrogen and carbon are also alpha phase stabilisers.

Beta alloys are mostly of medium to high strength and respond well to heat treatment. Ductility is better than for the alpha-beta alloys. weldability varies with composition. Despite these advantages, beta alloys have not enjoyed a wide acceptance. In some cases this has been due to the presence in relatively high concentrations of heavy beta phase stabilising elements, such as molybdenum. Other beta phase stabilisers are manganese, iron and silicon.

Alpha beta alloys are mostly of medium to high strength and respond well to heat treatment. Ductility is better than for the alpha beta alloys. Weldability varies with composition. Despite these advantages, beta alloys have not enjoyed a wide acceptance. In some cases this has been due to the presence in relatively high concentrations of heavy beta phase stabilising elements, such as molybdenum. Other beta phase stabilisers are manganese, iron and silicon.

Alpha beta alloys are mostly heat treatable, but their hardenability varies considerably. In some cases only shallow hardening is achieved by drastic water quench and in other cases full hardness is developed by air cooling. The former type may be weldable; the latter are generally not weldable. Strength levels are medium to high and this tends to reflect adversely on forming characteristics. The bulk of commercial alloys of titanium are alpha beta types.

Several alloys in all classes, alpha, beta and alpha beta, contain varying amounts of tin and zirconium., These elements are extensively soluble in both the alpha and beta phases and contribute to improved mechanical strength and phase stability.

Composition and mechanical properties

Commercially pure titanium is produced in three grades, soft, medium and hard, the medium grade being the most widely used for general construction work. A considerable number of titanium based alloyus have been developed, aimed principally at improving the desirable properties of.the metal for aerospace applications. Table 81 details various compositions, whilst Table 82 gives the mechanical properties of a selected range of alloys at normal ambient temperatures. Elevated temperature properties will be dependent on the type of alloy. Table 83 outlines elevated temperature properties of two alloys, one a general purpose and the other a high strength type.

Corrosion resistance

In general, titanium displays excellent resistance to metallic chlorides, organic acids and oxidising inorganic acids over a wide range of tempeatures and concentrations see Tables 84 and 85.

The reducing inorganic acids (hydrochloric, sulphuric and phosphoric acid) do not maintain the protective oxide film on the titanium metal surface which is the basis of its passivity, and the corrosion resistance to these acids in their pure condition is not generally good at medium or high concentrations. Very small amounts of oxidising ions (ferric, cupric, nitrate, etc) will serve however to maintain the protective film even when reducing acids form the principal environment. Such ions are frequently present in industrial process liquors and there are many examples of a successful application of titanium under these conditions.

Titanium has the unusual property of remaining corrosion resistant or even increasing its corrosion resistance when anodic potentials of up to 8 volts are applied. The corrosion resistant non-conducting film produced under these conditions makes titanium ideal for use in electro-plating applications. A surface of low resistance may be produced by coating the titanium with platinum, the material then being suitable for use as a non-consumable anode in electro-plating, cathodic protection and electro-chemical manufacturing processes.

TABLE 81 – TITANIUM AND SOME TITANIUM ALLOYS

Grade or Designation	Type	Relevant Specification		Alloying Elements (%)				Tensile strength tons/in²	Remarks
		ASTM	BS	Al	Mo	V	other(s)		
IMI 115	alpha	GR 1	TA 1					20–30	Where exceptional cold forming characteristics required.
IMI 125	alpha	GR 2	TA 2					25–35	Good formability and weldability.
IMI 130	alpha	GR 3						30–40	General construction grade.
IMI 160	alpha	GR 4	TA 6					35–45	Highest strength titanium metal.
IMI 260	alpha	GR 7					0.2 Pd	26	Now defunct.
IMI 230	alpha		TA 24				2.5 Cr	32–35	Welded fabrications, ductile.
IMI 315	alpha	GR 5		5			2.5 Sn	48–52	Welded fabrications.
IMI 318	alpha-beta	GR 6	TA 13	6		4		62–75	Most widely used alloy for structural and engine applications.
IMI 550	alpha-beta			4	4		2 Sn	68–78	High strength creep resistance alloy.
IMI 551	alpha-beta		TA 12	4	4		4 Sn	90 min	High strength structural alloy for airframes.
IMI 679	alpha-beta		Ta 20	2.25	1		11 Sn	67 min	Superior creep resistance at elevated temperatures.
IMI 680	alpha-beta	S213		2.75	4		11 Sn 0.25 Si	80–88	High strength structural alloy for airframes, etc. (now obsolete)
IMI 685	alpha-beta			6	0.5		5 Zr 0.25 Si	64–70	Superior creep strength at elevated temperatures.
6Al-6V 2 Sn	alpha-beta			6		6	2 Sn	71.5 min	High strength at elevated temperature.
8Al-1Mo-1V				8	1	1		68.74	Airframe and engine components.
6Al-2Sn 4 Zr-2Mo	alpha-beta		TA 12	6	2		2 Sn 4 Zr 0.8 Ni	67–74	Superior strength and creep resistance.

TABLE 82 – ROOM TEMPERATURE MECHANICAL PROPERTIES OF TITANIUM AND ESTABLISHED ALLOYS
(Titanium Metal & Alloys Ltd)

Grade	0.1% Proof stress tons/in²	kg/mm²	Ultimate tensile strength tons/in²	kg/mm²	Elongation %	180° Bend radius 16 swg and under) T = sheet thickness	Young's modulus tons/in²	Young's modulus kg/mm²	Density lb/in³	Density g/cm³
Commercially pure, soft (IMI 115)	13 min	20 min	26 max	41 max	30 min	1 T	7 200	11 340	0.163	4.51
Commercially pure, medium (IMI 130)	18 min	28 min	25–35	39–55	20 min	2 T	7 200	11 340	0.163	4.51
Commercially pure, hard (IMI 160)	25 min	39 min	35–45	55–71	18 min	2 T	7 200	11 340	0.163	4.51
Titanium 0.2% palladium alloy (IMI 260)	13 min	20 min	26 max	41 max	30 min	1 T	7 200	11 340	0.163	4.51
6Al–4V alloy (IMI 318)	57 min	90 min	62 min	98 min	10 min		7 100	11 180	0.161	4.45
2Cu alloy (IMI 230*)	25 min	39 min	32–45	50–71	20 min	2T	7 100	11 800	0.163	4.51
11Sn–5Zr complex alloy (IMI 679)*	58 min	91 min	67 min	105 min	10 min		7 100	11 180	0.175	4.85
4Al–4Mo–2Sn complex alloy (IMI 550)†*	56 min	88 min	68 min	106–129	10 min		7 500	11 800	0.166	4.60
11Sn–4Mo–2Al complex alloy (IMI 680)*	69 min	103 min	80 min	126 mn	10 min		6 970	10 950	0.175	4.86
4Al–4Mo–4Sn complex alloy (IMI 551)*†	80.2 min	126.3 min	90.0 min	141.1 min	13 min		7 240	11 400	0.167	4.60
6Al–6V–2Sn complex alloy	62.5 min	98.6 min	71.5 min	112	10 min		7 200	11 340	0.164	4.53

*Proprietary alloy (Imperial Metal Industries (Kynoch) Ltd.)
†Formerly Hylite 30. Proprietary alloy (Jessop-Saville Ltd)

TABLE 83 – SOME ELEVATED TEMPERATURE PROPERTIES OF TITANIUM ALLOYS (Titanium Metal & Alloys Ltd)

Temperature	Time	Stress to produce a total plastic strain of 0.1% tons/in^2		Stress to produce rupture tons/in^2	
°C	hours	(IMI 318)	(IMI 550)†	(IMI 318)	(IMI 550)†
300	100	37.3	46.8	46.5	
	300	36.8	46.5	46.5	
	1 000	31.2	46.0	46.5	
400	100	12.5	33.5	41.5	54.5
	300	9.3	30.0	39.5	54.0
	1 000	6.6	26.5	36.0	52.5
450	100		16.5		49.0
	300		13.0		46.0
	1 000		9.0	33.5	39.0
500	100	2.1	4.0	19.0	30.5
	1 000			11.5	14.5

†Formerly Hylite 50. Proprietary alloy (Jessop-Saville Ltd).

TABLE 84 – CORROSION RESISTANCE OF CP TITANIUM IN METALLIC BRINES (Titanium Metal & Alloys Ltd)

Environment	Concentration %	Temperature (°C)	Corrosion Rate (mpy)*
Aluminium chloride	10	20–105	1.0
	25	20	0.1
	25	100	258.0**
	40	115	4 300.0
Barium chloride	5–20	100	0.1
Calcium chloride	5–80	20–200	0.2
Cupric chloride	1–55	20–105	0.1
Ferric chloride	1–50	20–150	0.7
Mercuric chloride	1–44	20–100	0.4
Manganese chloride	5–20	100	Nil
Magnesium chloride	5–20	20–200	0.4
Nickel chloride	20	95	Nil
Potassium chloride	36 saturation	60–110	0.5
Sea water		20–365	Nil
Sodium chloride	saturation	20–70	Nil
Tin chloride	20	105	Nil
Uranium chloride	saturation	20–90	Nil
Zinc chloride	20–75	105–150	Nil

*Corrosion rates of less than 5 mpy (=0.005 in penetration per year) are considered excellent and suitable for long term applications.
**Corrosion rate of the 0.2% Pd alloy less than 1 mpy

TABLE 85 – CORROSION RESISTANCE OF CP TITANIUM IN ACIDIC MEDIA
(Titanium Metal & Alloy Co Ltd)

Environment	Concentration %	Temperature (°C)	Remarks	Corrosion Rate (mpy)*
H_2SO_4	1	100	Aerated	0.19
	5	100		1 920
	5	190	Chlorine saturated	1.0
	10	20		7.2
	10	100	+1.6% $Fe_2(SO_4)_3$	1.0
	25	35		131
	50	100	Cl_2 present	Nil
	65	38	+1% $CuSo_4$	0.7
HCl	1	60		0.1
	1	100		18
	5	100		1 120
	10	60		351
	10	70	+0.5% $CuSO_4$	1.3
HNO_3	20	205		Nil
	40	100		2.3
	40	200		24
	90	155		6.6

*Fuming nitric acid can, under certain circumstances, cause an explosive pyrophoric reaction and great care should be exercised.

The corrosion of titanium in contact with other metals depends on three factors: the difference in electrode potential, the relative anode/cathode areas and the type of reaction promoted by the coupling. The complicating factor is the effect upon the titanium oxide film. For example, if titanium is coupled to a more noble (cathodic) metal in a corroding environment, its electrode potential is raised and, rather than being increased, the corrosion rate is reduced. On the other hand if titanium is coupled to normally anodic mild sheet and immersed in HCl, the generating of hydrogen by the dissolving steel breaks down the passive oxide film and causes increased corrosion of the titanium.

When in contact with other metals in sea water, the excellent resistance to corrosion of titanium will generally ensure that it is the system cathode and there may be increased corrosion on the other metal.

Titanium comes between 18/8 stainless steel (passive) and silver on the galvanic scale.

Titanium exhibits good resistance to localised attack at gasketed joints, overlapped corners, etc (crevice corrosion) in the majority of environments to which it is normally resistant. Instances of crevice corrosion do occur, however, in some acid solutions, particularly where chlorides are present. Care should be taken in the design of joints, therefore, to avoid crevices, and in the selection of material for gaskets. Materials which give elastically rather than deform plastically (creep) will tend to seal a joint more tightly and thereby prevent the potentially damaging accumulations of the corrosive medium. Natural or synthetic rubbers have given better results as gaskets than PTFE type formulations. Solid PTFE in particular, with its tendency to creep, requires heavy flanges to maintain tight joints.

Traditionally the risk of crevice corrosion in susceptible conditions, has been significantly reduced by the use of titanium 0.2% palladium alloy (T.260). This alloy extended the application of titanium to cover for example chloride containing solutions above their boiling point, where commercially pure titanium proved inadequate.

The rapidly rising cost and scarcity of palladium has created a demand for a low cost titanium alloy with corrosion resistance approximating to that of the Ti-Pd alloy. This demand has now been met by TiCode 12, composition 0.3% Mo, 0.8% Ni. This allow, a development of Timet (Titanium Metals Corporation of America) is currently being produced in the same range of forms as commercially pure titanium and costs relatively little more for the substantial benefits in corrosion resistance.

Ingot production

The great affinity of titanium for oxygen, nitrogen and hydrogen must be countered during the melting of titanium sponge either by providing an inert atmosphere or by carrying out this process under vacuum. As the melting point of titanium is 1668°C it will be appreciated that the technological difficulties to be overcome in the design of these large vacuum melting furnaces are considerable and that the capital cost of such plant is necessarily high. Two methods of melting sponge are commercially available. The first, which is traditional, involves the compaction of the sponge together with any selected scrap into an electrode shape. This electrode is then melted under vacuum into a water or liquid metal cooled copper crucible by striking an arc between the electrode and the crucible. The use of copper for the crucible is necessary because of the extreme reactivity of molten titanium with the majority of conventional refractory materials.

In the second method, the new non-consumable electrode vacuum melting process, has the sponge fed directly into a molten pool produced by an arc struck between the contents of the crucible and a rotating electrode. Unlike the direct consumable electrode melting, temperatures considerably above the melting point of titanium can be achieved in the molten pool, thereby increasing the refining effect of the first. Irrespective of the method used for the first melt, the ingot resulting is then used as an electrode in a second vacuum melting operation to prodeuce a finished ingot. Double vacuum melting has been found necessary to ensure the true homogeneity of the final product, particularly where other elements have been added. Although two melts are satisfactory for most purposes, triple vacuum melting is now regular practice for the production of premium grade aerospace alloys particularly for rotating parts.

Forging and rolling

Forging and rolling of titanium from ingot to semi-finished products are carried out on conventional plant although the differing hot-working characteristics of titanium and the need to protect it against contamination necessitate certain variations in forging and rolling procedures.

The forging of commercially pure titanium is usually commenced at approximately 900°C with the temperature falling to about 700°C during the course of hot-working. While reheating during forging is permissible, it is important that a significant proportion of the deformation is carried out after the final heating. This ensures that a fine grain size and good mechanical properties are obtained the final forging. The forging temperature range for titanium alloys may vary from the foregoing according to the composition and transformation characteristics of the alloy concerned.

In considering the heating technique to be employed prior to the hot-working of titanium, the fact that titanium readily absorbs hydrogen above 150°C, oxygen above 700°C and nitrogen above 800°C must be taken into account. Also all these gases can produce a brittle surface layer on the metal, it is important to minimise their absorption during heating. In practice, heating cycles are kept as short as possible and a dry, slightly oxidising furnace atmosphere is preferred.

TABLE 86 – PRODUCTION SIZES OF TUNGSTEN WIRE (Tungsten Manufacturing Company Limited)

Diameter		Nearest	Weight	
Inches	mm	swg	g/ft	g/m
0.008	0.203	35	0.191	0.627
0.009	0.229	34	0.242	0.794
0.010	2.254	33	0.299	0.981
0.011	0.279	32	0.361	1.184
0.012	0.305	31	0.430	1.411
0.013	0.330	29	0.505	1.657
0.014	0.356	29	0.585	1.919
0.015	0.381	28	0.672	2.205
0.016	0.406	27	0.765	2.507
0.017	0.432	27	0.863	2.831
0.018	4.457	26	0.967	3.173
0.019	0.483	25	1.078	3.537
0.020	0.508	25	1.194	3.917
0.021	0.533	25	1.317	4.321
0.022	0.559	24	1.445	4.741
0.023	0.584	24	1.580	5.184
0.024	0.610	23	1.720	5.643
0.025	0.635	23	1.866	6.122
0.030	0.762	21	2.687	8.816
0.035	0.899	20	3.658	12.001
0.040	1.016	19	4.778	15.676
0.045	1.143	18	6.047	19.839
0.050	1.270	18	7.465	24.491
0.055	1.397	17	9.033	29.636
0.060	1.524	16	10.750	35.269
0.065	1.651	16	12.616	41.391
0.070	1.778	15	14.631	48.002
0.075	1.905	15	16.796	55.105
0.080	2.032	14	19.110	62.697

For small forgings, such as compressor blades, the use of induction or resistance heating is widely adopted.

The hot rolling of rod, plate and sheet is subject to the same general considerations outlined previously. To obtain the optimum mechanical properties and surface finish in titanium sheet, however, the final reduction to finished gauge should be carried out by cold-rolling.

It is customary to supply such cold-rolled sheet in the annealed condition to ensure maximum ductility in the product.

Fire prevention

Storage of coarse titanium turnings and chips is relatively safe. Storage or accumulation of titanium fines constitutes a fire hazard. Clean machines and good workshop practice are usually sufficient to avoid any danger of fire when machining titanium. Titanium chips, turnings or fines should not be allowed to accumulate in machines. If a fire does start its effect can be minimised by isolating the burning material from the bulk. The fire can then be extinguished by covering it gently with a mixture of dry asbestos wool or chalk or with a dry powder extinguisher. No attempt should be made to put out burning titanium with water or an extinguisher of any but the dry powder type.

Forming

Titanium has a low modulus of elasticity. This means it will deform considerably under load and then spring back. This characteristic must be carefully allowed for in forming operations. Hot creep forming overcomes this difficulty and, because the material is more ductile at elevated temperatures, lower forming pressures may be used. Preheating of tools should be employed to obtain maximum results from this type of operation. Cold or warm pressing, stretch forming and spinning can be empoloyed for forming titanium, and handling annealed commercially pure sheet is similar to working 1/4 hard stainless steel.

Bends with radii of 1 to 1½T can be made in ductile grades of titanium without undue difficulty. Deep drawing may only be performed successfully with the softest grades of the metal. Blanks for pressing should be prepared with care and slow cutting speeds are desirable for the shearing, sawing, nibbling or blanking operation. Attention should be given to the preparation of edges of press blanks. Edge cracking will be minimised if the blank cutting blade is kept sharp and close fitting or if the metal is heated before shearing. All burrs must be removed and for more difficult operations filing or polishing of cut edges may be necesssary.

Commercially pure titanium

By precise control of the interstitial elements, carbon, oxygen and nitrogen, it has proved possible to produce commercially pure titanium to specific tensile strength requirements within the overall range 20 to 45 ton/in². The typical room temperature mechanical properties of three grades, soft, medium and hard are detailed in Table 82.

The medium grade is the most widely used for general construction work with the softer grades being utilised where exceptional cold formability is required and the hard grade where higher strength requirements have to be met.

Titanium alloys

A great number of titanium based alloys have been developed over the past years. These are principally aimed at improving the desirable properties of the base metal for aerospace applications. Of those several alloy developments a number have become firmly established and Table 82 summarises the composition and room temperature mechanical properties of these materials.

Whereas the individual titanium melting companies do not generally produce all of these alloys certain companies, who are able to purchase ingot material from any melting source are able to offer this complete range of alloys in their appropriate wrought forms.

Elevated temperature properties

For applications involving temperatures in the range 200 to 550°C it is necessary to assess the hot tensile and creep properties of titanium and its alloys in order to select the most appropriate material. An outline of these properties for two different alloys is presented in Table 83.

TUNGSTEN

Symbol: W
Specific gravity: 19.3
Density: 19.3 g/cm³ (0.697 lb/in³)
Tensile strength (approx): 472 kg/mm² (300 tons/in²)
Modulus of rigidity: 1.4×10^6 kg/cm² (19.6×10^6 lb/in²)
Modulus of elasticity: 3.5×100^6 kg/cm² (50×10^6 lb/in²)
Hardness: 225 (worked) 390 to 480 Vickers (sintered)
Melting point: 3410°C
Coefficient of linear expansion: 4.4×10^{-6}/degC
Specific heat at 20°C: 0.033 cal/g/degC
Thermal conductivity at 0°C: 0.31 cal/cm/degC
Electrical resistivity: 5.65×10^{-6} ohm cm
Electrical conductivity: (% IACS): 31

Tungsten is a refractory metal, steel grey in colour, which can only be produced in rod, sheet, bar or wire form by powder metallurgy because of its exceptionally high melting point. Ince so formed it can be hot worked readily. Some of the main applications of the metal include electrical contact discs; transmitting, receiving and generating valve components; mercury switch electrodes and glass-sealing leads; lamps electrodes; argon-arc and atomic hydrogen welding electrodes; and semiconductor discs.

Tungsten is widely used as an alloying element in high speed steels, tool steels, die steels, 'Stellite', etc; and also in permanent magnet alloys. Tungsten carbide is used as a tip for high speed cutting tools.

Heavy metal is the generic term for a range of tungsten-rich alloys containing copper and nickel. As the name implies, the alloys have high densities and as a result find considerable commercial application. The densities are slightly lower than that of tungsten itself, but the alloys can be fabricated into shapes not possible with pure tungsten.

The alloys are made by the powder metallurgy process of blending, pressing and liquid-phase sintering. The process is versatile and a range of densities from 16.2 to 18.0 g/cm^3 can be offered. Components are easily machined and readily brazed to other metals so detailed shapes and forms can be specified.

Properties will depend on the configuration of the components and to some degree on their size. As a general guide, the ultimate tensile strength is comparable to that of mild steel; hardness ranges from Hv 210 to 290; the coefficient of linear expansion is about 6×10^{-6} in/degC and electrical conductivity is 0.7×10^{-6} ohm cm.

The range of applications of these alloys is very wide and their current uses include high density weights and parts of gyroscopes, gyro-compasses, inertia guidance systems; radiation shields, components and vessels for the handling and transit of radio-active sources, both in medical and industrial applications; counter balances in self-winding watches, aeroengine crankshafts, centrifugal clutch plates, precision instruments; and heavy duty electrical contacts and anti-chatter boring bars.

Commercial tungsten rod has a purity greater than 99.9% with a well-developed, regular, fibrous structure, the grain size being determined by the rod diameter – typically, 35 000 grains/mm at 0.093 in diameter and 8000 grains/mm at 0.187 in diameter. Rod with a finer grain size can be supplied. All rod densities are typically 19.2 g/cm^3. Rods are generally available swaged or centreless ground in densities from 0.5 mm upwards. Tungsten wire is generally available in diameter sizes from 0.2 to 2 mm, black or cleaned, high tensile or annealed. Sheet and foil are available in a wide range of sizes and thicknesses. Tape can be rolled from wire or cut from sheet. Specialised manufacturers can also supply fabricated components in tungsten for the electrical industry, etc.

TUNGUM

Specific gravity: 8.43
Density: 8.43 g/cm^3 (0.253 lb/in^3)
Ultimate tensile stress (annealed): 66 000 lb/in^2
Ultimate tensile stress (precipitation hardened): 67 000 lb/in^2
0.1% proof stress (annealed): 37 200 lb/in^2
0.1% proof stress (precipitation hardened): 45 000 lb/in^2

TABLE 87 – CHEMICAL COMPOSITION OF TUNGUM

Element	%	
	Minimum	Maximum
Copper	81.00	86.00
Aluminium	0.70	1.20
Nickel	0.80	1.40
Silicon	0.80	1.30
Iron		0.25
Lead		0.05
Tin		0.10
Manganese		0.10
Total other impurities		0.50
Zinc	The remainder	

TABLE 88 – PROPERTIES OF TUNGUM

Form	Size and Uses	Properties	
Tungum bar		0.1% minimum proof stress ton/in^2	18
		Minimum uts ton/in^2	35
		Minimum elongation at break, %	20
		0.1% minimum proof stress ton/in^2	15
		UTS ton/in^2	30
		Minimum elongation at break, %	25
Tungum strips	Suitable for springs. (Cold rolled or drawn): Up to and including 0.018 in	Minimum diamond pyramid hardness No	215
	Over 0.018 in up to and including 0.032 in	Minimum uts ton/in^2	50
		Minimum diamond pyramid harness No	215
	Over 0.032 in up to and including 0.104 in	Minimum uts ton/in^2	45
		Minimum diamond pyramid harness No	210
	Over 0.104 in up to and including 0.252 in	Minimum uts ton/in^2	42
		Minimum diamond pyramid hardness No	200
Tungum rods and Wires	Suitable for springs (Hard drawn);	Relevant specification DTD5009	
	Up to and including 0.048 in	Minimum uts ton/in^2	56
	Over 0.048 in up to and including 0.104 in	Minimum uts ton/in^2	53
	Over 0.104 in up to and including 0.252 in	Minimum uts ton/in^2	48
Tungum sheet	(Cold rolled)	UTS ton/in^2	Min 22 Max 30
		Minimum elongation at break, %	30

Tungum is a complex alpha-brass which combines high strength with ductility, good fatigue properties and excellent corrosion resistance. It is widely used in high pressure hydraulics where tensile strength and ease of manipulation are required and when repeated and pulsing pressures are applied. The inherent cleanliness of the bore of the tube also reduces the possibility of system contamination. Due to its excellent corrosion resistance, it has outstanding long term serviceability for intermittent duties and for the same reason requires no external protective treatment.

The aluminium-nickel-silicon-brass alloy tungum has good resistance to corrosion in moving sea water. It is non-magnetic and non-sparking and therefore finds application in marine engineering and for the manufacture of non-sparking tools.

Tungum has high thermal conductivity and good low temperature characteristics. It is not subject to brittle fracture or season cracking. It is a cryogenic material and therefore also suitable for chemical engineering in low temperature processes. In the same field it is used for conveying fluids and gases because of its excellent resistance to corrosion by high concentrations of many chemicals.

Tungum has low friction properties when used in conjunction with steel or cast iron and therefore finds application for control mechanisms such as Bowden cables.

UDIMET

Proprietary nickel-based alloys (American).

URANIUM

Symbol: U
Specific gravity: 19.07
Density: 19.07 g/cm^3 (0.689 lb/in^3)
Tensile strength (as cast): 19.7 kg/mm^2 (12.5 tons/in^2)
Modulus of elasticity: 1.7×10^6 kg/cm^2 (24×10^6 lb/in^2)
Melting point: 1132°C (2072°F)
Specific heat: 0.028 cal/g/degC
Coefficient of expansion: 6.8 to 14.2×10^{-6} /degC
Thermal conductivity at 20°C: 0.071 cal/cm^2/cm/sec/degC
Electrical conductivity: 30×10^{-6} ohm cm

Uranium metal is radioactive with a half life of 4.5×10^9 years. It oxidises slowly and will burn in air if heated to 170°C. At higher temperatures it will combine with nitrogen (to form the nitride) or carbon (to form the carbide). It is corroded by water or steam and dissolved by most acids. It sparks freely if struck with a metallic tool. Pure uranium is quite ductile, but the presence of iron or aluminium induces brittleness.

Non-nuclear uses or uranium are limited but include the use of the oxide in glass manufacture and other uranium compounds as reagents and mordants.

VANADIUM

Symbol: V
Specific gravity: 6.1
Density: 6.1 g/cm^3 (0.220 lb/in^3)
Tensile strength (cold rolled): 83 to 85 kg/mm^2 (53 to 54 tons/in^2)
Tensile strength (hot rolled): 48 kg/mm^2 (30.5 tons/in^2)
Modulus of elasticity: 1.3 to 1.4×10^6 kg/cm^2 (18 to 210×10^6 lb/in^2)
Melting point: 1900°C (3450°F)
Specific heat: 0.119 cal/g/degC
Coefficient of expansion (circa 20°C): 8.3×10^{-6} /degC
Thermal conductivity (circa 100°C): 0.074 cal/cm^2/cm/sec/degC
Electrical resistivity: 24.8 to 26.4×10^{-6} ohm cm

Vanadium (pure) is a ductile, malleable metal, but may become brittle through hydrogen absorption. It can only be produced at high cost, its main application being as an alloying element for cast iron and steels, and to a more limited extent for non-ferrous metals (*eg* aluminium).

WOOD'S METAL

See *Fusible alloys* (Section 9).

Y ALLOYS

Heat treatable aluminium alloys (British).

YTTERBIUM

Symbol: Yb
Specific gravity: 6.96
Density: 6.96 g/cm^3 (0.251 lb/in^3)
Tensile strength: 7.1 kg/mm^2 (4.5 tons/in^2)
Melting point: 824°C (1515°F)
Specific heat at 20°C: 0.035 cal/g/degC
Coefficient of expansion: 25×10^{-6} /degC
Electrical resistivity: 29×10^{-6} ohm cm

Ytterbium is a rare metallic element contained in blomstrandity, gadolinite, polycrase and other rare earths. It is of academic interest only at the present time. Atomic number is 70 and atomic weight 173.04.

YTTRIUM

Symbol: Yt or Y
Specific gravity: 4.47
Density: 4.47 g/cm^3 (0.161 lb/in^3)
Hardness: 30 to 45 Brinell
Tensile strength (annealed): 13.4 kg/mm^2 (8.5 tons/in^2)

Modulus of elasticity: 1.2×10^6 kg/cm^2 (17×10^6 lb/in^2)
Melting point: 1500°C (2748°F)
Specific heat at 20°C: 0.071 cal/g/degC
Coefficient of thermal expansion: 4.52 to 19.22×10^{-6} /degC
Thermal conductivity at 0°C: 0.035 cal/cm^2/cm/sec/degC
Electrical resistivity at 0°C: 57×10^{-6} ohm cm

Yttrium the metal has only so far been prepared in powder of 'scale' form, the principal use being as an addition to improve the oxidation resistance and refine the grain of high temperature steels. In this respect it shows a considerably greater effect than aluminium, but both yttrium and aluminium may be used in combination in chrome-alloy steels.

ZAMAK

Proprietary zinc-based die casting alloys (American).

ZINC

Symbol: Zn
Specific gravity at 25°C: 7.13
Specific gravity at 419.5°C (solid): 6.83
Specific gravity at 419.5°C (liquid): 6.62
Density: 7.00 g/cm^3 (0.253 lb/in^3)
Melting point: 419.5°C (778°F)
Boiling point: 906°C (1660°F)
Thermal conductivity (18°C): 0.27 cal/sec/cm^3/degC
Linear coefficient of thermal expansion (polycrystalline) from 20°C to 250°C: 39.7×10^{-6} /degC
Specific heat (50°C): 0.094
Heat of fusion (419.5°C): 24.09 cal/g
Heat of vaporisation (906°C): 425.6 cal/g
Electrical resistivity (20°C): 5.92×10^{-6} ohm cm
Electrical conductivity: 28.27% IACS
Electrochemical equivalent: 0.3388 mg/coulomb or 2.6886 lb/1000 amp hr
Crystal structure: Close-packed hexagonal

TABLE 89 – PROPERTIES MEASURED AT ROOM TEMPERATURE

Commercial Zinc Minimum purity 98.5%	Strip rolled		Pack rolled	
	Parallel	Perpendicular	Parallel	Perpendicular
Tensile strength (tons/in^2)	9	14	9	12
(kg/mm^2)	14	22	13	19
Elongation %	25	10	23	14
Brinell hardness (kg/mm^2)	45–50		45–50	

The minimum purity of ordinary electrolytic zinc is 99.95%. It is possible to make special high purity metal of 99.99% or greater purity, suitable for the production of die casting alloys, by adjusting the conditions of electrolysis (temperature, current density, etc) or by adding strontium carbonate to the electrolyte, to reduce the amount of lead codeposited on the zinc. Most special high purity zinc is now made in this way, the rest coming from fractional distillation plants.

Main uses of zinc

About 35% of the total world consumption of zinc (including scrap) is used in the form of protective coatings for iron and steelwork. Die casting accounts for another 25%, brassmaking for 20% and sheet zinc for a further 10%. Other important uses of the metal are in the pigments zinc oxide, zinc dust and lithopone.

BS 3436 specifies the minimum zinc content of four grades of ingot zinc as follows:

Zn1: 99.99%
Zn2: 99.95%
Zn3: 99.50%
Zn4: 98.50%

The American Society for Testing and Materials specifies the minimum zinc content of the following grades of slab zinc (ASTM B6-62T):

Special high grade: 99.99
High grade: 99.90%
Intermediate: 99.50
Brass special: 99.00%
Prime Western: 98.00%

BS 1004: 1972 specifies requirements for two zinc die casting alloys A and B, with chemical composition of the ingot metal as shown in Table 90. Specific alloy properties are given in Table 91, and typical mechanical properties of pressure die castings in Table 92.

Shrinkage characteristics and mechanical properties of Alloys A and B at low and high temperatures are given in Tables 93–96.

The mechanical properties of zinc alloy castings depend on the way they are cast and the design of the castings. The mechanical properties of sand or gravity die castings are generally inferior to pressure die castings since the latter have a much finer grain structure. An alternative, *eg* alloy IIzro 12, is preferred for castings in sand or plaster moulds see Table 97. A further alloy (IIzro 14) containing titanium has also been developed with better creep resistance at higher temperatures – see Table 98.

Mazak alloys

Mazak 3 corresponds to Alloy A and Mazak 5 to Alloy B in BS 1004, with typical properties as detailed in Tables 90—96. Some further properties of Mazak 3
and Mazak 5 are given in the following tables.

TABLE 90 – CHEMICAL COMPOSITION OF ZINC ALLOY INGOT METAL

			Alloy A %	Alloy B %
Aluminium			3.90–4.30	3.90–4.30
Copper			—	0.75–1.25
Magnesium			0.04–0.06	0.04–0.06
Impurities				
	Iron	not more than	0.0500	0.0500
	Copper	not more than	0.0300	—
	Lead	not more than	0.0030	0.0030
	Cadmium	not more than	0.0030	0.0030
	Tin	not more than	0.0010	0.0010
	Nickel	not more than	0.0010	0.0010
	Thallium	not more than	0.0010	0.0010
	Indium	not more than	0.0005	0.0005
Zinc (to BS 3436 Zn1)			Remainder	Remainder

TABLE 91 – GENERAL PROPERTIES*

		Alloy A	Alloy B
Specific gravity		6.70	6.70
Density	lb/in^3	0.24	0.24
Melting point	°C	387	388
Solidification point	°C	382	379
Specific heat	cal/g/degC	0.10	1.10
Solidification shrinkage	in/ft	0.14	0.14
	mm/m	11.7	11.7
Thermal expansion coefficient	per degC	27×10^{-6}	27×10^{-6}
Electrical conductivity	ohm cm	15.7×10^{-6}	15.7×10^{-6}
Thermal conductivity	cal/sec/degC at 18°C	0.27	0.26
Compression strength	tonf/in^2	27	39
	kgf/mm^2	42	61
Modulus of rupture	tonf/in^2	43	47
	kgf/mm^2	67	74
Shearing strength	tonf/in^2	14	17
	kgf/mm^2	21	27
Fatigue strength	tonf/in^2	4.7	5.1
(20 × 10^6 cycles)	kgf/mm^2	7	8

*Zinc Development Association.

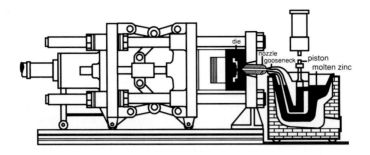

Fig. 8   Diagram of simple hot chamber die casting machine for zinc alloy

TABLE 92 – MECHANICAL PROPERTIES OBTAINABLE FROM PRESSURE DIE CAST ZINC ALLOYS AT 21°C*

Alloy			Tensile strength		Elongation % on 2 in (50 mm)	Impact strength		Brinell Hardness
			tonf/in²	kgf/mm²		ft/lb	kgf/m	
Original value	A	unstabilised	18.5	29.2	15	42	5.80	83
		stabilised	17.7	28.0	17	45	6.20	69
	B	unstabilised	21.7	34.2	9	43	5.90	92
		stabilised	20.2	32.0	10	44	6.10	83
After 12 months ageing at room temperature	A	unstabilised	17.1	27.0	25	43	5.90	67
		stabilised	17.1	27.0	24	41	5.70	54
	B	unstabilised	20.7	32.7	12	42	5.80	74
		stabilised	18.8	29.7	14	45	6.20	72
After 8 years ageing at room temperature	A	unstabilised	16.0	25.3	20	44	6.10	65
		stabilised	15.7	24.8	19	42	5.80	61
	B	unstabilised	18.9	29.9	14	41	5.70	74
		stabilised	—	—	—	—	—	—
After 12 months dry ageing at 95°C	A		15.2	24.0	29	37	5.10	48
	B		16.5	26.0	23	10	1.38	64

*Adapted from Mazak manual.

TABLE 93 – SHRINKAGE DURING NORMAL AGEING*

| | Alloy A | | | | Alloy B | |
| | Air Cooled | | Quenched | | Air Cooled | |
	in/in	mm/m	in/in	mm/m	in/in	mm/m
Shrinkage after 5 weeks	0.00032	0.32	0.0006	0.6	0.00069	0.69
Shrinkage after 6 months	0.00056	0.56	0.0012	1.2	0.00103	1.03
Shrinkage after 5 years	0.00073	0.73	—	—	0.00136	1.36
Shrinkage after 8 years	0.00079	0.79	—	—	0.00141	1.41

*Zinc Development Association.

TABLE 94 – SHRINKAGE AFTER STABILISING*

| | Alloy A Stabilised | | Alloy B Stabilised | |
	in/in	mm/m	in/in	mm/m
Shrinkage after 5 weeks	0.00002	0.2	0.00022	0.22
Shrinkage after 3 months	0.00003	0.3	0.00026	0.26
Shrinkage after 2 years	0.00003	0.4	0.00037	0.37

*Zinc Development Association.

TABLE 95 – MECHANICAL PROPERTIES AT NORMAL AND SUB-NORMAL TEMPERATURES*

| Temp °C | Tensile Strength | | | | Elongation | | Impact Strength | | | |
| | Alloy A | | Alloy B | | Alloy A | Alloy B | Alloy A | | Alloy B | |
	tonf/in²	kg/mm²	tonf/in²	kg/mm²	% on 2 in (50 mm) × ¼ in (6 mm)		ft/lb	m/kg	ft/lb	m/kg
20	18.2	28.8	22.5	35.6	11.0	8	42.0	5.80	44.0	6.10
10	—	—	—	—	—	—	31.0	4.30	41.0	5.70
0	19.2	30.4	24.2	41.4	9.0	8	7.5	1.04	39.0	5.50
−10	—	—	—	—	—	—	3.5	0.48	18.0	2.50
−20	—	—	—	—	—	—	2.5	0.35	3.6	0.53
−40	20.5	32.4	24.2	41.4	4.5	3	2.1	0.29	2.4	0.33

Note: Brinell hardness values vary with temperature in accordance with the changes in tensile properties.
*Zinc Development Association.

TABLE 96 – PROPERTIES OF ZINC ALLOY DIE CASTINGS AT TEMPERATURES UP TO 100°C

Temp °C	Tensile Strength				Elongation		Impact Strength			
	Alloy A		Alloy B		Alloy A	Alloy B	Alloy A		Alloy B	
	tonf/in²	kg/mm²	tonf/in²	kg/mm²	% on 2 in (50 mm) × ¼ in (6 mm)		ft/lb	m/kg	ft/lb	m/kg
20	18.2	28.8	22.5	35.6	11	8	42	5.8	44	6.10
40	16.0	25.2	19.0	30.0	16	13	42	5.8	46	6.35
95	12.7	20.1	15.5	24.5	30	23	43	5.9	—	—

*Zinc Development Association.

TABLE 97 – COMPOSITION AND PROPERTIES OF ILZRO 12 ALLOY

Composition:	Aluminium	11.00–13.00%
	Copper	0.50–01.25%
	Magnesium	0.01–00.03%
	Zinc (to BS 3436 Zn1)	remainder
Properties (sand cast):	Tensile strength	20.7 tonf/in² (32.7 kgf/mm²)
	Elongation	3% over 2 in (50 mm)
	Impact strength	16 ft lb (2.2 m kg)
	Brinell hardness	101
	Specific gravity	6.0
	Melting point	432°C
	Solidification point	380°C
	Solidification shrinkage	0.15 in/ft (12.5 mm/m)
	Thermal expansion coefficient	27×10^{-6}/degC

TABLE 98 – COMPOSITION AND PROPERTIES OF ILZRO 14 ALLOY

Composition:	Copper	1.00–1.50%
	Titanium	0.25–0.30%
	Aluminium	0.01–0.03%
	Zinc	Balance
Physical properties:	Melting range	414–417°C
	Casting temperature	460–470°C
	Soundness	Excellent
	Surface finish	Excellent
	Mechanical stability	(Dry air – 100°C – 35 days) Excellent
	Mechanical stability	(Wet steam – 95°C – 10 days) Excellent
Mechanical properties:	Tensile strength	14.8–15.2 tonf/in² (23.4–24.0 kgf/mm²)
	Elongation, % in 2 in (50 mm)	5–6
	0.2% proof stress	8.90–9.40 tonf/in² (1.23–1.30 kgf/mm²)
	Hardness (Vickers)	79–86

TABLE 99 – FATIGUE LIMIT OF MAZAK
(10^8 REVERSALS AT 1 600 CYCLES/MIN)

Condition	Units	Mazak 3	Mazak 5
After 6 months indoor ageing	lbf/in^2 MN/m^2	6 875.0 47.4	8 175.0 56.4
After 6 hours dry ageing at 95°C	lbf/in^2 MN/m^2	6 500.0 44.8	7 675.0 52.9
After 5 days steam ageing at 95°C	lbf/in^2 MN/m^2	5 950.0 41.0	5 700.0* 39.3*

*Extrapolated value

TABLE 100 – ELONGATION OF MAZAK
COMPARED WITH OTHER DIECAST ALLOYS

	British Standard	Alloy	Elongation % in 2 in
Mazak 3	BS 1004A		15
Mazak 5	BS 1004B		9
Magnesium alloy	BS 2970		2–5
Aluminium alloy	BS 1490	LM2	1–3
Aluminium alloy	BS 1490	LM24	1–3

TABLE 101 – ELECTRICAL CONDUCTIVITY OF MAZAK
COMPARED WITH SOME OTHER MATERIALS

	British Standard	Alloy	Electrical Conductivity % iIACS at 20°C
Mazak 3	BS 1004A		26
Mazak 5	BS 1004B		26
Aluminium alloy	BS 1490	LM2	26
Aluminium alloy	BS 1490	LM6	37
Aluminium alloy	BS 1490	LM24	24
Free-cutting brass	BS 249	60/40	26
Mild steel			10

**TABLE 102 – THERMAL CONDUCTIVITY OF MAZAK
COMPARED WITH SOME OTHER MATERIALS**

Material	British Standard	Alloy	Thermal Conductivity CGS Units
Mazak 3	BS 1004A		0.27 at 18°C
Mazak 5	BS 1004B		0.26 at 18°C
Aluminium alloy	BS 1490	LM2	0.24 at 25°C
Aluminium alloy	BS 1490	LM6	0.34 at 25°C
Aluminium alloy	BS 1490	LM24	0.23 at 25°C
Free-cutting brass	BS 249	60/40	0.26 at 20°C
Mild steel			0.12

**TABLE 103 – COMPATIBILITY OF UNTREATED MAZAK DIE CASTING
ALLOYS WITH VARIOUS MEDIA**

Medium	Medium characteristics	Performance
Aerosol propellants		Excellent
Acid solutions	Weak, sold, quiescent	Fair
	Strong	Not recommended
Alcohols	Anhydrous	Good
	Water mixtures	Not recommended
	Beverages	Not recommended
Alkaline solutions	Up to pH 12.5	Fair
	Strong	Not recommended
Carbon tetrachloride		Excellent
Cleaning solvents	Chlorofluorocarbon	Excellent
Detergents	Inhibited	Good
Diesel oil	Sulphur free	Excellent
Fuel oil	Sulphur free	Excellent
Gas*	Towns, natural, propane, butane	Excellent
Glycerine		Excellent
Inks	Printing	Excellent
	Aqueous writing	Not recommended
Insecticides	Dry	Excellent
	In solution	Not recommended
Lubricants	Mineral, acid free	Excellent
	Organic	Not recommended
Paraffin		Excellent
Perchlorethylene		Excellent
Petrol*		Excellent
Refrigerants	Chlorofluorocarbon	Excellent
Soaps		Good
Trichlorethylene		Excellent

*Chromate passivation treatment recommended because of the possibility of moisture traces being present.

ZIRCONAL

High strength aluminium alloy (German).

ZIRCONIUM

Symbol: Zr
Specific gravity: 6.489
Density: 6.489 g/cm^3 (0.234 lb/in^3)
Modulus of elasticity: 1×10^6 kg/cm^2 (13.7×10^6 lb/in^2)
Melting point: 1852°C (3366°F)
Specific heat at 20°C: 0.067 cal/g/degC
Coefficient of expansion: 5.85×10^{-6} /degC
Thermal conductivity (circa 20°C): 0.211 watts/cm/degC
Electrical resistivity: 40×10^{-6} ohm cm

Zirconium (pure) metal is ductile and malleable but the presence of impurities renders it hard and brittle. The main use of the metal is for nuclear reactor cores and as an alloying element with ferrous metals, aluminium, copper bronzes, and magnesium. It is also used as an alloying element in magnetic alloys and alloyed with lead for lighter flints.

Metallic zirconium is stable at room temperature and is resistant to water and sulphurised steam, alkalis, many acids and other chemicals. It will burn in air if heated to 240 to 270°C, and react violently if powdered metal is heated with copper and lead oxides. Heated zirconium absorbs hydrogen and reacts with chlorine, nitrogen and carbon to form chlorides, nitrides and carbides, respectively. Mechanically, its properties are somewhat similar to low carbon steel, and its chemical properties similar to titanium.

As a refractory metal, zirconium finds uses between tantalum and titanium. Very good corrosion resistance, particularly to sulphuric acid, enables it to be used where tantalum and zirconium are uneconomical. Zirconium is also used in chrome plating applications where considerably increased life is experienced over the traditional titanium heating coils and in the electronics industry as a 'getter' where its reactivity to gases at high temperature is employed.

Commercial purity zirconium is of the order of 99.7 to 99.9% pure and is normally marketed as an amorphous black powder. Zirconium compounds are used as pigments, opacifiers, and catalytic agents.

Metal matrix composites

METAL MATRIX composites (MMC) are a family of materials in which second phases, such as ceramic fibres or particles, are introduced into metals during processing. Thus the microstructure is synthesised from its main constituents, rather than formed naturally from the alloy during solidification or thermo-mechanical processing.

Some MMC materials have been established for decades. Examples include cemented carbides (1920s), thoriated tungsten (1920s), electrical contact composites such as silver-graphite (1930s). These are all examples of MMC produced by powder metallurgy processing. A more recent MMC family is that of the low temperature superconductors comprising fine multi-filaments of intermetallic aligned within a metal matrix. It is noteworthy that all these materials use the composite approach to satisfy functional requirements rather than to provide an enhanced load bearing capability. Since the 1970s there has been major world-wide commitment to the development and use of metal matrix composites for structural applications. This has been inspired by the technical possibility of ex-ploiting the high strength and high modulus of fibres and whiskers as reinforce-ments within metal matrices. Initially the emphasis was on the use of carbon or boron fibres or oxide whiskers as reinforcement, but these have largely been superseded by ceramic based reinforcements more appropriate to reinforcing metals. The most commonly used of these reinforcements are based on alumina or silicon carbide and the properties of some representatives of these classes are included in Table 1.

There are now three main classes of ceramic reinforced MMC in use or under development. The largest volume class is that which is used as feedstock mate-rial for conversion to component shape by conventional metallurgical process-ing. These materials are reinforced with particulate, or less commonly whisker or short fibre ceramic, and they are used in either wrought form, such as sheet, or cast to shape. Representative properties of aluminium based MMC in this class are given in Table 2. These materials are characterised by relatively modest property improvements over the metal baseline properties and by low costs. The

TABLE 1 – TYPICAL PROPERTIES OF REPRESENTATIVE REINFORCING FIBRES

	Diameter (μm)	Young's modulus (GPa)	Tensile strength (GPa)	Density (g/cm³)
Boron (mono-filament)	50–200	400	3.5	1.5
Carbon (high modulus)	5–10	690	2.2	1.9
Carbon (high strength)	5–10	230	3.5	1.75
Silicon carbide (mono-filament)	100–150	420	3.7	3.3
Silicon carbide (multi-filament)	8–15	210	2.8	2.5
α-Alumina (multi-filament)	15–25	385	1.4	3.95
α-Alumina (low density multi-filament)	3–4	200	2.0	2.0

1980s have witnessed significant developments in the low cost processing of such materials. Thus, co-spray and melt stirring routes to feedstock production have been developed to compete with the intrinsically more costly powder metallurgy routes. The co-spray routes involve the injection of the particular reinforcing phase into the stream of atomised matrix alloy so that the deposited ingot, coating or pre-form has the reinforcement pre-dispersed within it. Some degree of secondary processing is usually employed to fully consolidate the material and produce the required product form. The melt stirring routes involve stirring the reinforcing phase into the matrix melt. MMC ingot so produced is then nearly at net shape and is processed to final shape, usually by pressure casting or forging routes. Both these are, however, limited in that they cannot handle volume fractions of reinforcement as high as those that can be processed by conventional powder metallurgy.

At the other extreme is the high performance fibre reinforced class, in which major property improvements are achieved, at least in the direction of fibre

TABLE 2 – PROPERTIES OF PARTICULATE AND WHISKER REINFORCED 6061 ALLOY (T6 CONDITION)

Reinforcement	Process	Modulus (GPa)	Yield strength (MPa)
20% SiC particulate	Powder	103	414
30% SiC particulate	Powder	121	435
15% SiC particulate	Melt stir	96	353
13% SiC particulate	Co-spray	89	317
17% SiC whisker	Melt infiltration	110	421
20% SiC whisker	Powder	115	480
Unreinforced	Ingot	69	274

TABLE 3 – LONGITUDINAL TENSILE PROPERTIES OF REPRESENTATIVE FIRBRE REINFORCED MMC

Composite	Modulus (GPa)	Strength (MPa)	Density (g/cm^3)
35% SiC mono-filament in Ti-6A1-4V	215	1725	3.86
40% SiC multi-filament in 1050 Al	128	900	2.84
48% SiC mono-filament in 6061 Al	207	1725	2.6
43% P100 carbon fibre in 6061 Al	338	995	2.47
20% α-Alumina multi-filament in Pb	86	210	9.9

alignment, but usually at considerable cost. Examples of properties achieved are given in Table 3. The third class is that of local reinforcement of components by arrays of fibre. Here the component is usually produced by placing the reinforcement array in a mould prior to casting under pressure. This approach offers the added advantage of placing the reinforcement only where it is needed in the component. It should be noted that this last class and many examples of the second class of MMC are characterised by the production of the component at the same time as the material. This is in contrast to the majority of metallic materials, where the component is fashioned from ready made stock of the alloy.

The profitable use of MMC materials, in common with other composites, demands a high degree of design awareness. This, together with cost considerations, has led to the relatively simple MMC systems of the first class featuring most prominently in trials aimed at mass market applications. The first major industrial MMC deployment, however, came in the form of a selective reinforcement of the ring groove area of an automotive diesel piston. In this case, pistons produced by Art Metal Mfg Co. Ltd for a range of Toyota cars, the cost of the selectively reinforced pistons was no greater than the component that it replaced and performance was enhanced. Whilst other MMC applications have ranged from horse shoes to space shuttle components, the pattern of use is gravitating to those applications where the special advantages of the MMC concept are most fully exploited. Thus the earliest production uses of ceramic reinforced MMC have been in applications where combinations of physical and mechanical properties are optimised. Varying the proportion of ceramic reinforcement can, for example, control both the thermal expansion coefficient and the strength and stiffness of a component. Similarly, particularly for fibre reinforced MMC, the potential for increasing high temperature strength and stiffness is exploited. MMC materials continue to be introduced and improved. Snapshots of developments world wide can be obtained from reviews such as the DTI-sponsored OSTEM missions (*eg* E. A. Feest et al, Metal Matrix Composites Developments in Japan, Harwell Laboratory 1986; S. E. Booth et al, Metal Matrix Composites Developments in the USA, BNFMTC 1987). Current data can be obtained directly from the MMC suppliers. Suppliers within the United Kingdom include:

Alcan International, Banbury; BP Advanced Materials, Sunbury and Cray Advanced Materials, Yeovil.

The metallurgy of MMC property optimisation is still in its infancy. The MMC systems are generally non-equilibrium and this means that irreversible reactions between matrix and reinforcement can lead to irreproducibility of heat treatment response and degradation of properties in service. The selection of reinforcement type and matrix composition has usually been dictated by ease of availability rather than optimisation, and made in the absence of a truly predictive understanding of structure-property relationships. It is therefore to be expected that the performance of materials exploiting the MMC concept will improve markedly as the technology matures.

Metal matrix composites, whilst not fully optimised, now offer a wide range of processing and property options and there is considerable development effort on the part of users in all major engineering sectors from aerospace and defence through to automotive and electronics, to gain competitive advantage from their deployment.

Refractory metals

THE METALLIC elements commonly defined as refractory are to be found in Groups VA and VIA of the periodic system and are distinguished by their very high melting and boiling points but not by oxidation resistance. The other high melting point metals are in Groups VIIA and VIII and are designated precious metals having resistance to oxidation as well as fairly high melting points.

The basic properties of the four prime refractory metals tungsten, molybdenum, tantalum and niobium are given in Table 1.

Although all four metals have considerable refractoriness their applications are often significantly different and will have to be treated as separate subjects; although the higher the melting point the greater the extent to which they are

TABLE 1

Property	Tungsten	Molybdenum	Tantalum	Niobium
Symbol				
Atomic number	74	42	73	41
Atomic weight	183.86	95.95	180.95	92.91
Melting point °C	3410	2610	2996	2468
Boiling point °C	5930	5660	5425	4927
Density gm cm³	19.3	10.22	16.6	8.6
Coefficient of thermal 25–500°C	4.52	5.19	6.6	7.1
Expansion 3 10-6/°C 25–500°C	5.86	8.00	8.7	*circa* 10.0
				(25 to 1500°C)
Specific heat 20°C	0.032	0.066	0.034	0.065
cal/gm°C 2500°C	0.043	0.128	0.053	0.080
				(1500°C)
Thermal conductivity 20°C	0.31	0.34	0.130	0.125
cal/s cm°C 1500°C	0.253	0.22	0.188	—
Resistivity 20°C	5.65	5.5	12.4	15.1
microhm-cm 2000°C	65	60	87	—
Electrical conductivity Cu = 100% IACS%	31	34	13.9	13.3
Thermal neutron capture cross-section				
barns/atom	19.2	2.7	21	1.1

processed as metallic powders. All are used in the manufacture of carbides for the hard metal industry and tungsten carbide is the majority constituent.

Tungsten was first discovered in 1755 and although it ranks 18th in relative abundance in the earth's crust, its use is limited by the cost of converting the metal into a solid workable mass. The Coolidge process was developed in 1909 to produce ductile tungsten wire for incandescent lamp filaments from tungsten powder by pressing, sintering and mechanically hot working. Subsequently it was discovered that the ductility and hence lamp filament application was derived from the accidental presence of trace impurities, which allowed the ultimate development of a non-sag structure. Research to improve filament quality is still continuing by controlling the known impurities and fixing their ratios, which has resulted in the introduction of 'halogen' lamps and 'long life' bulbs.

Pure tungsten rod and sheet of 99.96% plus purity is now produced by sintering fine tungsten powder without additions into blocks of up to about 20 kilograms weight at about 96% of the metal's boiling point. The consolidated metal bars are hot rolled and then hot rotary swaged into rods. These have a rough swaged finish and need to be centreless ground for most applications, which are now manifold. A large but declining application as is tungsten 'distribution points' for the automobile industry; these are abrasive wheel cut from rod, as tungsten is too tough for sawing. The resultant discs are rumbled, polished and cleaned before copper or silver alloy brazing to rivets, arms and brackets. Pure tungsten is also used for electrodes in starter and igniter tubes because of its coefficient of expansion similarity to glass and quartz.

Tungsten rods for argon or TIG non-consumable welding electrodes are made by the same swaging and centreless grinding route. Pure tungsten electrodes are now a minority usage and have been superseded by a form of MMC, in which refractory metal oxides are added to the tungsten powder prior to the sintering, and end up as elongated particles in the finished rods. These additions of ¼ to 4% of oxides lower the work function of the electrodes, producing a more stable arc. The favoured oxides are of thorium, zirconium, lanthanum, cerium and yttrium, in decreasing order of use. Each refractory oxide has a preferential application according to materials being welded and the electric current form and configuration.

An affiliated production to non-sag tungsten wire is the manufacture of vacuum metallising filaments. Techniques have been developed for the non-sag tungsten structures to be formed at 2 to 0.5 mm diameter as well as the fine wire filament sizes. The wire is hot formed into coils, hairpins and baskets and used to retain metals such as aluminium, silver, gold and electrical alloys for evaporation in vacuum chambers for decorative finishes and electronic and electrical applications. Tungsten sheet is formed into trays to contain fluorides and other salts for coating lenses and reflective glass.

Tungsten sheet is only available as pure tungsten down to a few thousands of an inch thick and is made by cross rolling sintered slabs.

Pure tungsten wire and rods can be used as heating elements for furnaces operating at over 2000°C but it is essential to use a reducing atmosphere or high vacuum. The maximum temperature limitation is normally the refractory brickwork and not the tungsten. Crucibles, pots and dishes for high temperature applications can be made by a variety of techniques, including spot-welding and riveting, to supplement the routes already mentioned as well as chemical vapour deposition. A few minor alloys such as tungsten-molybdenum can be mechanically worked but in these tungsten is a minority constituent. Other tungsten rich metals are generally classified as composites and are made by powder metallurgy routes. The tungsten and silver, copper and additives are dealt with in the *Electrical Contacts* section. The generic group of Heavy Metals also defined as High Density Alloys are made by the powder metallurgy route and contain nickel, copper, iron and molybdenum as additions.

Molybdenum is the most commonly used of the refractory metals but is not so abundant in the earth's crust as tungsten. It was first isolated in 1790 by Hjelm but developed commercially about 1910 by the Coolidge process as used for tungsten. Current practice includes melting of molybdenum alloys as well as the powder metallurgy route. Molybdenum assets are an electrical conductivity of about one-third of that of copper, a good mechanical strength and a low vapour pressure of the same order as carbon.

Excluding use as ferro-molybdenum in steel making, the majority of applications are as pure molybdenum and the metal is much easier to work than tungsten; therefore sheet molybdenum is extensively used and fabricated. But the relative ease with which it is oxidised for a high melting point metal somewhat limits its use. Molybdenum wire and ribbon are used in the incandescent lamp industry because of the metal's matched low coefficient of expansion, and finds outlets in quartz and iodine lamps, as well as eyelets, hooks, etc, for supporting filaments. Sheet punched and also sintered molybdenum discs are used as mounting bases for silicon discs. The wire and ribbon, being more ductile than tungsten, is an excellent material for high temperature furnace windings, provided that a vacuum or protective atmosphere is available. Wire is metal sprayed to give protective and low friction coefficient coatings to other metals, particularly steels.

The sheet can be formed and bent to make crucibles, reflectors, furnace boats and missile components. The ease with which molybdenum sheet can be spun or flow turned has increased its application as nozzle and rocket inserts, and hot forged parts also find applications in this field. Molybdenum vacuum evaporation boats also have specific functions. There is a non-sag molybdenum wire available supplementing the use of tungsten non-sag wire.

Molybdenum alloys containing 10 and 30% tungsten are available for more critical applications and can be made by arc casting as well as powder metallurgy. TZM is used containing 0.5% titanium and 0.07% zirconium, which has a higher hot strength than the pure metal and is used for die casting cores and hot work tools.

Among the manifold uses of molybdenum rods are projection welding electrodes, glass making electrodes and high-rigidity boring bars. Most super alloys contain a significant molybdenum content to enhance their high temperature and wear resistance.

Tantalum as a refractory metal is usually thought of in terms of its great corrosion resistance and high ductility. Once again the main production route is by powder metallurgy, either sodium reduction or fused salt electrolysis. It is available also as fabricated sheet, wire and seamless tubing and as some specialised alloys, but again the greatest use is as pure metal. The corrosion resistance of tantalum has been compared to glass, being inert to all acids except hydrofluoric and suffering attack by hot strong alkalis. The metal is used to fabricate reactors, vapour condensers, bayonet heaters and multi-tube heat exchangers. For large installations, because of the high cost of tantalum, the parts are made of steel and coated with thin tantalum sheets of foil, which is feasible because of its high ductility.

Tantalum capacitors are made from powder and are favoured because of their small volume importance in high tech applications, but there is a high price penalty. It is the ability to form a stable thick electrolytic oxide film whose dielectric constant is even higher than that of aluminium which has advanced the use for capacitors. This oxide feature also relates to the use of tantalum and some alloys as surgical sutures and reinforcements, coupled with the resistance to body fluids and tolerance by animal tissues.

Niobium is often coupled with tantalum in technical literature and is still designated as Columbium in the USA, where they refuse to comply with current nomenclature. Although niobium has useful properties similar to tantalum and also akin to molybdenum, its use is restricted because of its high price as a result of extraction costs. It has been used as a fuel element canning materials in nuclear reactors, but all its main uses are as important minority additions to other materials. Niobium carbide is present in many hard metal blends. It is added to austenitic steels and nickel alloys as a weld decay inhibitor and to alloy steels to enhance their properties.

SECTION 3

Non-Metallic Materials

Carbides

CEMENTED CARBIDES are a special class of materials based on a hard refractory metal carbide held in a binding metal or matrix. The properties of the resulting composite are governed by the type of metal carbide, grain size and grain form, and the proportions of carbide to binding metal. Final forming is done by sintering either directly from a compact, or after presintering and machining to the required shape.

The most used carbides are:

(i)   tungsten carbide (WC)
(ii) titanium carbide (TiC)
(iii) tantalum carbide (TaC)
(iv) niobium carbide (NbC)

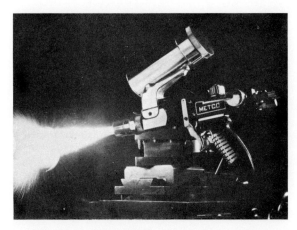

A tungsten carbide, hard-facing alloy being applied to a wire-drawing capstan with a Metco type 5P 'Thermo Spray' gun.

Various binding metals may be used, but the most common is cobalt or cobalt-nickel alloy.

Cemented carbides are well known as tips for metal cutting tools, rock drills, masonry drills, abrasion resistant inserts, etc. Standard grades of tungsten carbide are suitable for cutting almost all metals and all kinds of abrasive non-metallic materials. Special grades are also produced for impact service such as cold heading, piercing, blanking, punching and dies – *eg* components requiring great strength, toughness, shock resistance and resistance to wear.

Carbon

ELEMENTAL CARBON exists naturally in many parts of the world in forms of graphite and diamond. When mined, graphite is not particularly useful as an engineering material, because of its softness, high impurity level and limited size. The manufacture of carbon and graphite has developed for over 80 years with the properties of the resulting products being considerably improved over those of the natural material. These manufactured carbons and graphites are used for a wide and diverse range of engineering applications (see below).

SEM of worm-like exfoliated graphite as used to produce flexible graphite foil. (Insert: typical products made from graphite foil.)

TABLE 1 – PROPERTIES OF CARBON AND GRAPHITE (TYPICAL)

Property	Unit	Carbon	Graphite	Carbon resin impregnated	Carbon metal impregnated
Bulk Density	kg/m^3 × 10^3	1.6	1.7	1.8	2.5
Transverse bend strength	kgF/cm^2	460	300	650	910
	lbF/in^2	6 500	4 200	9 300	13 000
Compressive strength	kgF/cm^2	1 400	600	2 100	2 810
	lbF/in^2	20 000	8 500	30 000	40 000
Shear strength	kgF/cm^2	320	180	460	560
	lbF/in^2	4 600	2 500	6 500	8 000
Dynamic elastic modulus	kgF/cm^2 × 10^3	180	84	230	340
Porosity (apparent)	%	15	15	2	2
Coefficient of thermal expansion	°C^{-1} × 10^{-6}	3.6	4.1	4.0	4.7

By judicious selection of raw materials, and by using a variety of manufacturing processes, carbon and graphite can be modified to give a range of properties suitable for the different applications. The raw materials are selected from naturally occurring graphites, from which some or all of the inorganic impurities have been removed. The original materials include coke derived from coal sources, residues taken from oil distillation (petroleum cokes), and carbon blacks which are obtained from the burning of oil and gases in limited supplies of air.

The raw materials are powdered and bonded together with a pitch or tar, and then moulded under pressure to form a convenient size and shape. The resulting

TABLE 2 – SPECIAL PROPERTIES OF CARBON AND GRAPHITE

Carbon and graphite have a unique combination of special properties which make them the ideal choice for a very wide range of engineering applications. These properties are:

Self-lubricating
Low coefficient of friction
Non-swelling
Chemically inert
Non-contaminating
Cannot weld
Good oxidation resistance
Can be machined to fine tolerances
Good hydrodynamic bearing properties
High resistance to thermal shock
Low wear rate
High strength in compression
Relatively low density

Typical carbon products

product is fired in a protective atmosphere, to approximately 1000°C. During this process, the tar and pitch are converted to a coke thereby cementing the particles of the other raw materials together to form a cohesive mass. Such a product is known as carbon-graphite.

Some grades of carbon are further heat-treated to temperatures of up to 2500°C, at which level the amorphous carbon is transformed into crystalline electrographite. This material has greater purity, improved thermal conductivity, better oxidation resistance, but lower strength.

TABLE 3 – APPLICATIONS OF CARBON AND GRAPHITE

Electrical	Mechanical	Refractory
Brushes Contacts Heating elements	Bearings Compressor components Gland rings Sealing rings Steam joint rings Thrust rings Valves Vanes	Continuous casting dies Furnace furniture Hot pressing dies Semi-conductor components Spark erosion electrodes

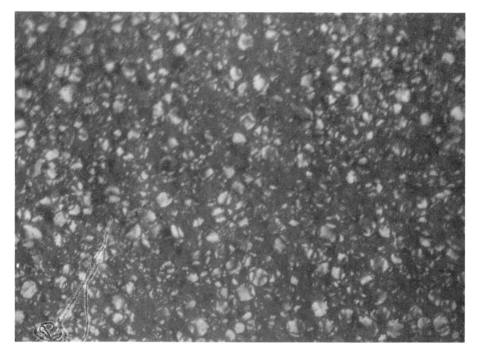

Optical micrograph showing the fine structure of high-performance carbon

Carbon and graphite materials that are produced by this means are porous. The properties, however, can be modified by impregnation with synthetic resins or metals. The impregnants confer increased strength, lower permeability and improved wear resistance characteristics to the basic carbons.

Impregnation with oxidation inhibitors will improve the oxidation resistance and with some base materials both reduce and stabilize friction.

Ceramic fibres

A WIDE variety of ceramic materials are now available in fibre form, and the number is steadily increasing as new production methods are developed. Applications for the fibres can be divided into two main groups, which arise from two of the most attractive properties found in ceramic-based materials.

Being strongly bonded at the atomic level, ceramics have high melting points, good chemical resistance and tend to retain their strength to high temperatures. When produced in fibre form, which reduces thermal mass, this makes ceramic fibre-based products ideal for many refractory applications.

The second property of ceramic fibres which is significantly exploited is their high stiffness. These materials are therefore used as reinforcement for other materials, including metals, polymers and other ceramics and glasses.

Historically, some of the first ceramics to be produced in fibre form are those with compositions in the $Al_2O_3SiO_2$ system. They are often produced by either blowing high velocity gas jets onto a molten stream of the ceramic, or more usually these days, by spinning the molten material. This latter route yields fibres with a longer mean length which imparts extra strength to converted products such as blankets, textiles and felts.

Fibres produced by both techniques are extensively used in refractory applications for insulation and exist in a wide range of compositions and forms. Examples of the latter include: bulk fibre, blankets and felts, moulded shapes, textile products, papers and boards, ropes and braid.

The fibres have the following general properties, which are also common to the made-up forms; high temperature stability, low thermal conductivity, low heat storage, excellent thermal shock resistance, lightweight, excellent corrosion resistance.

The various product forms are used in a wide variety of industries, including: aluminium, appliance, ceramic, chemical process, glass, foundry, heat treatment and forging industries. They are also used in coke ovens, blast furnaces, electric steelmaking, continuous casting, soaking pits, reheat furnaces and annealing. Typical applications include: kiln/furnace insulation, expansion joints, linings for chemical reactors, burner openings, etc.

TABLE 1 – PROPERTIES OF TYPICAL BULK ALUMINA-SILICA FIBRES*

Colour	White	White
Continuous use limit	2300°F	1260°C
Melting point	3260°F	1790°C
Normal packing density	3–12 lb/ft^3	48–192 kg/m^3
Shipping density (approx)	6lb/ft^3	96 kg/m^3
Fibre diameter (mean)	2–3 microns	2–3 microns
Fibre length	to 4 inches	to 100 mm
Relative density	2.53	2.53
Specific heat capacity (2 000°F/1 090°C)	0.27 Btu/lb/°F	1–13 kJ/kg/°C
Fibre tensile strength	4×10^5 lb/in^2	2.76×10^9 N/m^2
Fibre modulus of elasticity	16×10^6 lb/in^2	1.103×10^{11} N/m^2
Fibre surface area	0.5m^2/g	500 m^2/kg

*'Fiberfrax'

As the products are composed of fibres there is no brittle structure to develop stresses during sudden heating or cooling. It is therefore virtually impossible to damage ceramic fibre product forms by thermal shock. The individual fibres themselves are also exceedingly difficult to damage by this phenomenon.

The alumino-silicate fibres generally exhibit excellent resistance to attack from most corrosive agents. Exceptions include hydrofluoric acid, phosphoric acid and strong alkalies. They also resist both oxidising and reducing atmospheres and are generally not wetted by molten aluminium and zinc.

Fibre properties are dependent on composition within the $Al_2O_3SiO_2$. Table 1 indicates typical properties for a range of fibre compositions. Figure 1 shows a plot of thermal conductivity versus mean temperature for Carborundum's Fiberfrax bulk fibre, together with a table of calculated insulation thicknesses for a range of hot and cold face temperatures.

Over the past decade a much wider range of ceramic fibres have been developed for use as reinforcement in ceramic (including glass and glass-ceramic), metal and polymer matrix composites. Composite property improvements include increased strength and stiffness, resistance to thermal fatigue and better high temperature properties. Both short fibres and continuous fibres are produced, in addition to ceramic whiskers (see Ceramics section), and a range of production techniques are used.

Together with the aluminosilicates, aluminoborosilicates and alumina fibres, the most common ceramic fibre compositions available commercially are based on; carbon, silicon carbide, boron, silicon nitride and zirconia with the carbon-based fibres offering perhaps the greatest range of mechanical properties and some of the lowest costs.

Several types of carbon fibre exist, essentially based on their method of manufacture. Two of the most common carbon fibres are those produced from pitch and from precursor polyacrylonitrile (PAN) fibres. The former are generally cheaper, whilst the latter usually have superior mechanical properties. However,

Typical applications

Fiberfrax Bulk Fiber is used as a high temperature fill or packing material in a variety of high temperature applications.

– Expansion joints.
– Furnace base seals.
– Packing around burner tiles.
– Tube seals.
– Glass feeder bowl insulation.

Bulk Fiber is also used for the manufacture of other Fiberfrax product forms.

– Felts
– Board
– Paper
– Vacuum cast shapes
– Cements, castables and moldables
– Laminates

Fiberfrax Bulk Fiber
Thermal Conductivity vs Mean Temperature (per ASTM C-177)**

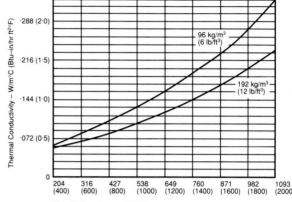

Specifications

Fiberfrax Bulk Fiber conforms to US Coast Guard requirements for 'Incombustible Materials', subpart 164.009. For additional conformations, see list of specification approvals.

FIBERFRAX BULK FIBER – 96 kg/m³ (6 lb/ft³)**

Hot Face °C (°F)	Insulation Thickness – mm (in) Cold Face Temperature – °C(°F)						
	13 (½)	25 (1)	38 (1½)	51 (2)	76 (3)	102 (4)	127 (5)
538 (1000)	176 (348)	123 (254)	99 (211)	86 (186)	70 (158)	61 (142)	55 (131)
649 (1200)	213 (416)	150 (302)	121 (249)	103 (218)	83 (182)	72 (161)	64 (148)
760 (1400)	252 (485)	178 (353)	143 (290)	122 (252)	98 (208)	84 (183)	74 (166)
871 (1600)	290 (554)	207 (405)	167 (333)	143 (289)	114 (237)	97 (207)	86 (187)
982 (1800)	329 (624)	237 (459)	192 (378)	164 (327)	131 (267)	111 (232)	98 (209)
1093 (2000)	367 (692)	267 (512)	217 (422)	186 (366)	148 (298)	126 (258)	111 (231)
1204 (2200)	405 (761)	297 (567)	243 (469)	208 (407)	167 (332)	142 (287)	125 (257)
1260 (2300)	424 (796)	312 (597)	256 (492)	220 (428)	177 (350)	150 (302)	132 (270)

FIBERFRAX BULK FIBER – 192 kg/m³ (12 lb/ft³)**

Hot Face °C (°F)	Insulation Thickness – mm (in) Cold Face Temperature – °C(°F)						
	13 (½)	25 (1)	38 (1½)	51 (2)	76 (3)	102 (4)	127 (5)
538 (1000)	159 (318)	112 (233)	91 (195)	78 (173)	64 (147)	56 (133)	51 (124)
649 (1200)	191 (375)	133 (272)	108 (226)	92 (198)	75 (167)	65 (149)	59 (138)
760 (1400)	223 (433)	157 (314)	126 (259)	108 (226)	87 (188)	74 (166)	67 (152)
871 (1600)	256 (493)	181 (357)	145 (293)	124 (255)	99 (210)	85 (185)	76 (168)
982 (1800)	289 (553)	206 (402)	166 (330)	141 (286)	113 (235)	96 (205)	85 (185)
1093 (2000)	312 (593)	223 (433)	180 (356)	154 (309)	122 (252)	104 (220)	92 (198)
1204 (2200)	344 (651)	248 (478)	201 (393)	172 (341)	137 (279)	117 (242)	103 (217)
1260 (2300)	361 (681)	261 (501)	212 (413)	181 (358)	144 (292)	123 (253)	108 (227)

**All heat flow calculations are based on a surface emissivity factor of .90, an ambient temperature of 27°C (80°F), and zero wind velocity, unless otherwise stated. All thermal conductivity values for Fiberfrax materials have been measured in accordance with ASTM Test Procedure C-177. When comparing similar data, it is advisable to check the validity of all thermal conductivity values and ensure the resulting heat flow calculations are based on the same condition factors. Variations in any of these factors will result in significant differences in the calculated data.

Fig. 1

TABLE 2 – COST AND MECHANICAL PROPERTIES OF VARIOUS FIBRES

Fibre	Dia.(μm)	Cost/kg £	UTS(MNm^{-2})	E(GNm^{-2})	Density
Boron					
Boron	140	£400	3500	438	2.49
Borsic	142	£1430	3100	400	2.71
SiC					
SiC on C	140	£890	4000	430	3.04
SiC on W	100	£1250	3500	420	3.32
Nicalon	12	£400	2500	190	2.55
Tyranno	10	£470	2900	210	2.4
Carbon					
T300	8	£120	2000	330	1.74
P100	10	£1200	2200	690	2.10
P120	10	£1600	2200	820	2.18
Alumina					
Safimax SD	3	£75	2000	300	3.3
Safimax LD	3.5	£75	2000	200	2.0
FP Alumina	20	£308	1800	380	3.96
Sumitomo	20	£423	2200	230	3.25

TABLE 3 – PROPERTIES OF 'SAFFIL' ALUMINA FIBRE – RF AND RG GRADES

	RF Grade†	RG Grade†
Chemical composition: Main components		
Al_2O_3	96–97%	same
SiO_2	3–4%	same
Trace components		
Iron	400 ppm	same
Chromium	60ppm	same
Nickel	140 ppm	same
Sodium	875 ppm	same
Magnesium	130 ppm	same
Calcium	525 ppm	same
Chloride (total)	80 ppm	same
Chloride (leachable)	5 ppm	same
Melting point	>2000°C	same
Maximum use temperature	1600°C	same
Tensile strength	2000 MPa (or 290 × 10^3 psi or 200 kg/mm^2)	1000–2000 MPa (or 145–290 × 10^3 psi or 100–200 kg/mm^2)
Young's modulus	300 GPa (or 43 × 10^6 psi or 30 × 10^3 kg/mm^2)	300–330 GPa (or 43–47 × 10^6 psi or 30–33 × 10^3 kg/mm^2)
Density	3.3 g/cm^3	3.3–3.5 g/cm^3
Median fibre diameter	3 μm	same
Crystal phase	mainly delta alumina	mainly delta to mainly alpha alumina
Hardness (Moh scale)	7	7–9

† All properties quoted are typical of normal production material but should not be taken as a specification.

these differences are gradually being diminished as the properties of the former are improved and the cost of the latter are reduced. The primary advantages of carbon fibres are their high strength and stiffness, combined with low density which makes them ideal for reinforcing other materials, in particular polymers and carbon matrices (the latter to form carbon/carbon composites now used extensively in the aerospace industry). Carbon fibres are available both in single fibre form, in tows and as textiles in a range of different weaves. The primary disadvantage of the carbon fibres is their lack of oxidation resistance at high temperatures.

Examples of ceramic fibre products. (The Carborundum Co Ltd.).

TABLE 3A – TYPICAL PROPERTIES
These values are not intended for use in preparing specifications

	Ceramic Grade	HVR Grade	LVR Grade
Fibre Denier	600, 900, 1200 and 1800	1800	1800
Density, g/cm^3	2.55	2.32	2.45 – 2.55
Tensile Strength, Ksi	430	425	430
Tensile Modulus, Msi	28	27	28
Strain to Failure, average percent	1.5	1.6	1.5
Volume Resistivity, ohm-cm	10^3	>10^6	0.5 – 5.0
Dielectric Constant	9.2	6.4	—
Thermal Conductivity, Kcal/mhr °C, along fibre axis at room temperature	10	—	—
Coefficient of Thermal Expansion, 'C', along fibre axis, 0 to 900 C (32 to 1652 F)	4.0 × 10^6	—	—
Specific Heat, J/g °C	1.14	1.14	—
Thermal Conductivity, Kcal/mhr °C	10	—	—

Specification Writers: Please contact Dow Corning Corporation, Midland, Michigan, before writing specifications on these products.

Silicon carbide fibres were originally developed to overcome the problem of oxidation at elevated temperatures found with carbon fibres. They can be produced by a wide variety of techniques, however, perhaps the most common method involves the pyrolisation of a precursor polymeric (often polycarbosilane) fibre. Fibres are produced in a range of deniers and physical forms, the latter including; continuous fibres, chopped fibres, multifilament tows and as woven cloth. The fibres exhibit high strength and elastic modulus, both of which are retained to high temperatures. Principal applications include high temperature insulation and reinforcement of plastic, ceramic and metal matrix composites. Tables 3A and 3B provide typical properties of two of the most widely used silicon carbide fibres, Nicalon and Tyranno, produced by the Nippon Carbon Co Ltd and Ube Industries Ltd, respectively.

Boron fibres offer high strength and elastic modulus combined with very light weight. They are used in metal matrix composites, however their production route, which involves vapour phase deposition onto fine tungsten wires, means that they are very expensive. The major industrial application at the present time is in the aerospace industry where they are used to reinforce lightweight metal alloy-based structures.

Silicon nitride and zirconia fibres have been developed much more recently than the other ceramic fibres discussed in this section with consequently much lower sales volumes. At present their major applicaitons are in reinforcing com-

TABLE 3B – TYPICAL PROPERTIES OF TYRANNO FIBRE

Filament diameter	8.5 ± 0.5 μm 11.0 ± 0.5 μm
Filaments/yarn	400, 800, 1600
Density	2.3–2.4 g/cm^3
Tensile strength	280 kg/mm^2 2.74 GPa 400 KSI
Tensile modulus	21 000 kg/mm^2 206 GPa 30 MSI
Elongation to break	1.4~1.5%
Specific resistance	10^3 ohm cm
Thermal Expansion coefficient (along fibre axis 0–500°C) Specific heat	3.1×10^{-6}/°C 0.19 cal/deg (60°C) 0.28 cal/deg (400°C)
Heat resistance (more than 95% strength remains after 6 hours exposure)	1300°C (in Nr gas) 1000°C (in air)
Resistance to acueous acid and bace NaOH (30%), HNO$_3$(3N) H$_2$SO$_4$(1BN), HCl(6N)	any change could not be detached after 24 hours dipping at 80°C

posites (both) and insulation (zirconia). Several other ceramic fibres, as well as new versions of existing compositions, are currently being developed in laboratories around the world, largely for specific applications. Examples include; aluminium nitride, boron carbide, boron nitride, magnesia and mullite. Some of these are available in limited quantities now, however it may be some time before the others are available commercially.

Further details on ceramic fibres and their applications may be found in the composites section.

Ceramics

ADVANCED STRUCTURAL ceramics offer some extremely useful properties as engineering materials, these include high hardness, excellent abrasion resistance, chemical inertness and excellent corrosion resistance, retention of high strength and other properties to high temperature and a wide range of useful electrical, magnetic and optical properties. The principal limitation to the use of ceramics is associated with their being brittle materials. This results in low toughness, ie, an extreme sensitivity to the presence of flaws with failure often being catastrophic. The brittleness also imposes severe constraints on processing and makes ceramics considerably more susceptible to thermal and mechanical shock than other classes of materials. Many of these problems can be overcome by suitable design and many companies producing ceramics and ceramic components are now willing to help with generating appropriate design criteria.

Ceramics are generally fabricated by entirely different techniques to either metals or polymers. Their extremely high melting temperatures means that methods based on melting and casting are not usually economically viable, whilst the lack of ductility almost exclusively prohibits the use of secondary forming methods such as forging, rolling etc. Rather, ceramics are typically produced via powder processing techniques, such as die pressing, isostatic pressing, slip casting, tape casting and injection moulding and then sintered to produce a hard, dense body. The characteristics of the final body and the degree of shape complexity capable of being produced vary with the production technique selected. These techniques vary from being fairly well to very well understood and all are used commercially to produce ceramic items.

A wide range of different ceramics are now used in engineering applications, they include: alumina, aluminium nitride, aluminium titanate, beryllia, boron carbide and boron nitride, magnesia, silicates, silicon carbide and silicon nitride, sialon, titanium carbide and titanium diboride, and zirconia. In addition, a range of ceramic-ceramic composites also find applications in modern engineering, the most common composites being: silicon carbide (particle, whisker and fibre) reinforced alumina, silicon carbide (particle, whisker and fibre) reinforced

TABLE 1 – MECHANICAL AND PHYSICAL PROPERTIES

Material		Density (g/cm³)	Conductivity (W/mK)	Expansion (K⁻¹ 10⁻⁴)	Young's mod. (GN/m²)	Hardness (kg/mm²)	Fracture toughness (MPa m$^{1/2}$)	3–PT bend strength (MPa)
Silicon Nitride	HPSN	3.19	25	1.5–3.0	310	1700	5	650–950
	RBSN	2.5	5–13	1.5–3.0	120–250	800	2–3	200–300
	SSN	3.19	5–13	1.5–3.0	300	1700	5	500–800
	SiAlON	3.19	5–13	1.5–3.0	300	2000	7	600–900
Silicon Carbide	SSiC	3.2	90–110	3–4	400	2500	3–4	400–500
	RBSiC	3.1	120–200	3–4	350–380	2000	2–4	250–300
	HPSiC	3.21	90–110	3–4	440	2600	5	600–800
Alumina	85–95%	3.9	15–23	6–9	320	1400–2000	4–6	300–600
Zirconia	PSZ	5.8	2.0	9–10	205	1200	8–10	600–800
	TZP	6.1	2.9	8–10	204	1300	8–19	1200
	ATZ	5.5	5.9	9–10	260	1400	6	2400
Tungsten Carbide		15.0	80–100	5.0	600	1500	8–13	1400–2000

Sources: 1. Morrell
2. Richerson
3. Lucas Cookson Syalon
4. Nilcra
5. Toyo-Soda

silicon nitride and zirconia toughened alumina, although a large number of other composite systems exist. For example, zirconia has been used to increase the toughness of a wide range of ceramic matrices and a numbner of other non-oxide particulates, besides the two based on silicon mentioned above, have been used to improve mechanical properties. Table 1 provides some typical properties for a range of engineering ceramics and ceramic composites.

The advanced ceramics industry is still generally divided into powder produc-ers and component manufacturers. However, there is an increasing trend to-wards the removal of the divisions between the two, largely as a result of powder producers starting to produce finished or semi-finished components. End users interested in utilising ceramics can therefore either purchase their compo-nents directly, with many component manufacturers also willing to produce cus-tom-designed pieces, or end-users can buy the precursor powders and produce their own components if they have the technology. This section of the handbook will concentrate primarily on the properties of sintered ceramic components, however tables illustrating powder properties will be presented in each subsec-tion below, where applicable.

A word of caution must be introduced at this point. One of the greatest diffi-culties associated with the choice of ceramic components is that most of the ceramics described here are produced by a range of companies and in a wide range of grades. Every one of these versions varies in terms of composition and processing conditions etc and, thus, in terms of its properties. Allied to this, there are as yet few standards with respect to the reporting of the properties, with a range of techniques being used to generate the data. Values reported in this handbook, and indeed in much company literature, must therefore only be considered as an approximate indication of the properties and direct compar-isons should be made with extreme care.

Alumina

Alumina, Al_2O_3, is one of the most commonly used engineering ceramic materi-als. This results from its generally low cost and useful 'all round' properties, which include; electrical insulation, high mechanical strength, good abrasion re-sistance, excellent chemical inertness, moderately high thermal conductivity and good nuclear stability. Applications are therefore extensive and diverse, ranging from dies for the wire drawing industry and nozzles for welding and sandblast-ing to substrates for microelectronics and seal faces for chemical plant and other corrosive environments. A list of typical applications for alumina ceramics is given in Table 2, although it should be noted that this list is far from exhaustive.

Alumina ceramics are produced in a variety of grades, usually ranging from about 80% to 99.99% alumina. Specific properties for several grades are shown in Table 3 to provide an indication of how properties change with composition. Table 4 provides details of typical high purity alumina powders and the subse-quent ceramics which can be produced from them.

TABLE 2 – TYPICAL APPLICATIONS OF ALUMINA CERAMICS*

Mechanical engineering	General	Sealing rings and busbar insulator assemblies in electrolytic plating operation. Nozzles in acid recovery plant. Thermocouple sheaths in pickling plants.
	Cigarette making	Abrasion resistant parts for cigarette forming machines.
	Papermaking	Abrasion resistant parts for paper making machinery.
	Cement, chemical, paint and fertiliser production	High density grinding media for milling abrasive materials.
	Machine tools	Tool tips for high speed machining of metal parts.
Wire industry	Fine wire production	Drawing cones for drawing fine copper and alloy wires.
	Insulated conductors	Nozzles for plastic coating fine wires for the electrical and electronic industries.
	Transcontinental telephone cables and power cables	Eyelets, pulleys and guides in the manufacture of the cable.
Electrical and electronics engineering	Radio and telecommunications	Aerial insulators for radio, radar and telephone systems. Sealed terminals in crystal filter units for telephone equipment. Insulators for surge diverters.
	Transformers and power factor correction equipment	Sealed terminals and bushings for oil filled transformers and capacitors.
	Heating elements	Bushes, sleeves, beads and bobbins for insulating industrial and domestic heating elements.
	Power generation — Nuclear	Fuel element thermal and electrical insulation. Thermocouple insulators for instrumentation.
	Coal and oil fired	High tension oil igniters. Rapper bars in electrostatic dust precipitators. Ash ejector nozzles.
Oceanography		Instrument shields for deep water equipment. Rings for lead/zirconite transducers.
Aerospace and aicraft		Fire seals and cable insulators in aircraft engines. Precision bearings in control equipment. Coil formers in hydraulic and pneumatic control equipment. Substrates in electronic engine control equipment.
Textile	Man-made fibre production	Thread guides for textile machinery; eyelets, pins, rollers and specially designed parts.
Chemical engineering		Sealing rings in mechanical seals for pumps. Floats in chemical flow measurement. Plungers in precision metering pumps. Mixer valves for fluid control. Laboratory ware.
Gas industry		Insulators and spark electrode assemblies for ignition of domestics appliances and industrial burners using natural and other gases.

*Royal Worcester Industrial Ceramics Ltd. Source: Original edition.

TABLE 3 – PROPERTIES OF PROPRIETARY ALUMINA CERAMICS OF TABLE 2*

Property	Test particulars	Units	FA	FF	FC	EH	DD	HK
Specific gravity			3.76	3.76	3.69	3.76	3.76	3.83
Porosity	Water absorption	%	nil	nil	nil	nil	nil	nil
Average crystal size	Thermal etched section	microns	4.50	4.50	2.00	4.00	30.0	2.00
Surface finish	CLA after rumbling	micro-inches	15–20	15–20	10–15	–	–	10–15
Stability to acid	CVD specification	mg/cm^2	0.50	0.50	0.10	0.30	0.60	0.00
Al$_2$O$_3$ content		%	95.0	95.0	95.0	96.0	97.0	92.0
Neutron absorption cross section	Gleep test UKAEA	cm^2 gram	–	0.0060	–	0.0029	–	–
Mechanical								
Tensile strength	BS 1598 : 1949	lb/in^2	26 800	27 300	27 000	27 000	–	–
Compressive strength	BS 1598 : 1949	lb/in^2	240 000	240 000	220 000	240 000	–	–
Transverse breaking strength	CVD4 point loading	lb/in^2	45 500	40 000	43 100	44 500	39 500	40 400
Young's modulus	Resonance method BCRA	lb/in$^2 \times 10^6$	46.70	46.70	43.60	47.60	48.80	47.60
Hardness	Rockwell	45N scale	82.00	81.00	85.00	81.00	80.00	83.00
Thermal								
Coefficient of linear expansion	0–200°C	10^{-6}/degC	5.00	5.00	5.00	4.70	4.50	5.00
	0–400°C	10^{-6}/degC	6.50	6.70	7.00	7.00	6.70	6.70
	0–600°C	10^{-6}/degC	7.30	7.50	7.70	7.70	7.80	7.60
	0–800°C	10^{-6}/degC	8.20	8.20	8.40	8.50	8.25	8.20
	0–1 000°C	10^{-6}/degC	8.80	8.90	9.00	9.00	9.20	8.90

(continued)

TABLE 3 – PROPERTIES OF PROPRIETARY ALUMINA CERAMICS OF TABLE 2* – contd.

Property	Condition	Units						
Specific heat	25°C	CGS units	0.175	0.175	0.175	—	—	—
Thermal conductivity	25–40°C	CGS units	0.054	0.058	0.054	0.029	0.068	0.045
Deformation temperature	CVD Spec SP48	°C	1 450	1 450	1 450	1 450	1 650	1 150
Maximum thermal shock	BS1598 : 1964	°C	170	170	160	170	180	190
Electrical								
Power factor	1Mc	%	0.020	0.010	0.045	0.015	0.025	0.050
	70 Mc	%	0.015	0.010	0.040	0.015	0.028	0.030
Dielectric constant	1 Mc		9.20	9.20	8.90	9.10	8.31	9.70
	70 Mc		9.20	9.20	8.90	9.10	8.46	9.70
Temperature coefficient of capacitance	10 200°C (CVD Spec)	10^{-6}/degC	+125	+125	—	—	+116	—
Volume resistivity	50°C	ohm cm	10^{16}	7.0×10^{15}	4.5×10^{14}	1.5×10^{15}	$>10^{16}$	8.4×10^{12}
	100°C		9.5×10^{15}	2.0×10^{15}	3.5×10^{13}	9.5×10^{14}		1.25×10^{12}
	200°C		7.5×10^{14}	5.0×10^{14}	2.5×10^{11}	5.0×10^{14}	1.0×10^{16}	1.0×10^{10}
	300°C		1.0×10^{14}	4.0×10^{13}	3.0×10^{9}	6.5×10^{13}	—	5.0×10^{7}
	400°C		7.0×10^{12}	8.5×10^{11}	$<10^{8}$	9.5×10^{11}	1.0×10^{12}	
	500°C		2.0×10^{11}	5.0×10^{10}	$<10^{8}$	5.5×10^{10}		
	600°C		9.5×10^{9}	4.0×10^{9}	$<10^{8}$	4.0×10^{9}	1.5×10^{10}	
Dielectric strength	20°C (2–3 mm thickness)	KV (peak) mm	20	20	—	—	—	—

* Smiths Industries Ltd (Ceramics Division)

TABLE 4 – PROPERTIES OF ALUMINA CERAMICS BY ROYAL WORCESTER INDUSTRIAL CERAMICS LTD

Property	Test	Units	Regalox	Electrox	Duralox	Fibralox	Nobalox	Crystalox
Nominal alumina content	BCRA method	% by weight	88.000	88.000	95.000	96.000	99.700	99.500
Bulk density	BS 1902 : 1966 Part 1A	g/cm^3	3.450	3.420	3.600	3.700	3.880	3.850
Colour			Deep pink	White	White	White	Peach	Peach
Apparent porosity	BS 1902 : 1966 Part 1A	% by volume	< 0.010	< 0.010	< 0.010	< 0.010	< 0.010	< 0.010
Hardness	ASTM E-18–61	Mohs scale Rockwell 45N	9 74	9 73	9 76	9 80	9 82	9 84
Safe working temperature (no load condition)		°C	1 250	1 250	1 450	1 450	1 700	1 700
Softening temperature		°C	1 400	1 400	1 550	1 550	2 000	2 000
Specific heat (mean value for 20–500°C)	BCRA method	cal/g/degC J/kg/degK	0.253 1 060	0.253 1 060	0.256 1 072	0.255 1 068	0.256 1 072	0.256 1 072
Thermal conductivity 50°C 200°C 400°C	BS 4789 : 1972 App H	$cal/cm^2/sec/$degC/cm (W/m/degK)	0.033(13.8) 0.025(10.3) 0.021(8.6)	0.037(15.5) 0.026(11.0) 0.023(9.0)	0.054(22.5) 0.037(15.5) 0.029(12.2)	0.059(25.0) 0.040(16.9) 0.031(13.0)	0.076(32.0) 0.050(20.9) 0.035(14.8)	0.068(28.5) 0.045(19.0) 0.033(13.6)
Coefficient of linear thermal expansion (mean values) 100–300°C 100–500°C 100–700°C 20–820°C	BS 1598 : 1964 App J		7.3×10^{-6} 7.7×10^{-6} 8.2×10^{-6} 8.1×10^{-6}	6.4×10^{-6} 7.0×10^{-6} 7.5×10^{-6} 7.6×10^{-6}	7.3×10^{-6} 7.8×10^{-6} 8.2×10^{-6} 8.1×10^{-6}	6.9×10^{-6} 7.7×10^{-6} 8.1×10^{-6} 8.1×10^{-6}	7.2×10^{-6} 7.6×10^{-6} 8.0×10^{-6} 8.3×10^{-6}	7.6×10^{-6} 7.8×10^{-6} 8.1×10^{-6} 8.1×10^{-6}
Cross breaking strength	BS 4789 : 1972 App B	lb/in^2 MN/m^2	45 000 310	42 000 290	46 000 320	48 000 330	32 000 220	50 000 345

(continued)

TABLE 4 – PROPERTIES OF ALUMINA CERAMICS BY ROYAL WORCESTER INDUSTRIAL CERAMICS LTD – contd.

Property	Method	Units						
Young's modulus	BCRA method	lb/in² MN/m²	37.8×10^6 26.1×10^4	37.4×10^6 25.8×10^4	46.6×10^6 32.1×10^4	48.5×10^6 33.4×10^4	53.0×10^6 36.6×10^4	52.7×10^6 36.3×10^4
Modulus of rigidity	BCRA method	lb/in² MN/m²	14.4×10^6 9.9×10^4	14.4×10^6 9.9×10^4	18.1×10^6 12.5×10^4	18.7×10^6 12.9×10^4	21.3×10^6 14.6×10^4	20.9×10^6 14.4×10^4
Poisson's ratio	BCRA method		0.31	0.30	0.29	0.30	0.25	0.26
Volume resistivity 200°C 400°C 600°C	BS 1598 : 1964 App F	ohm cm	1.3×10^{11} 1.4×10^8 3.9×10^6	3.5×10^{10} 3.2×10^7 1.1×10^6	1.3×10^{11} 2.4×10^7 1.4×10^6	7.1×10^{13} 1.0×10^{12} 5.0×10^9	1.0×10^{14} 5.6×10^{11} 5.0×10^8	7.9×10^{13} 6.6×10^{12} 3.5×10^{10}
Permittivity 1 MHz 70 MHz 9 368 MHz	BS 1598 : 1964 App B		7.1 7.5 8.5	7.2 7.4 8.5	7.6 7.7 8.8	8.1 8.1 9.4	8.2 8.6 9.7	8.6 8.5 9.6
Loss tangent 1 MHz 70 MHz 9 368 MHz	BS 1598 : 1964 App B		11.5×10^{-4} 9.8×10^{-4} 20.3×10^{-4}	5.1×10^{-4} 6.2×10^{-4} 18.5×10^{-4}	4.7×10^{-4} 5.0×10^{-4} 8.9×10^{-4}	5.5×10^{-4} 5.0×10^{-4} 5.8×10^{-4}	1.9×10^{-4} 1.2×10^{-4} 1.4×10^{-4}	9.6×10^{-4} 8.0×10^{-4} 5.6×10^{-4}
Loss factor 1 MHz 70 MHz 9 368 MHz	BS 1593 : 1964 App B		8.2×10^{-3} 7.4×10^{-3} 17.3×10^{-3}	3.7×10^{-3} 4.6×10^{-3} 15.7×10^{-3}	3.6×10^{-3} 3.9×10^{-3} 7.8×10^{-3}	4.5×10^{-3} 4.1×10^{-3} 5.5×10^{-3}	1.6×10^{-3} 1.0×10^{-3} 1.4×10^{-3}	8.3×10^{-3} 6.8×10^{-3} 5.4×10^{-3}
Dielectric strength (based on 0.1 in thick disc)	BS 1598 : 1964 App D	ac V/mil peak kV/mm	565 22.6	556 22.2	516 20.6	550 22.0	556 22.2	550 22.0
Te value	BS 1598 : 1964 App F	°C	660	600	630	700	630	630

There is actually no hard and fast rule which relates alumina content to any particular industrial field. It is the remainder of the constituents of the material which determine whether it has been designed for use with specific applications in mind, *eg* chromium oxide can be added to improve resistance to abrasion, whilst the use of clays containing sodium silicates improve processability but decrease electrical resistance considerably. The only generalisation that can be used with any degree of accuracy is that as the alumina content of the product is increased, the maximum operational temperature increases.

Aluminium nitride

Aluminium nitride (AIN) is a new ceramic material in terms of commercial exploitation, although it has a history stretching back nearly a century. The primary property of interest currently is its thermal conductivity, which is two to three times higher than that of alumina. Although this is still much lower than the thermal conductivity of beryllia (see below), it does not suffer from the toxicity problems which are associated with this oxide.

Practical applications for aluminium nitride include substrates for integrated circuits, insulation of semiconductor junctions and for vacuum vapourisation chambers and metal melting crucibles. These latter applications arise from the materials' low vapour pressure in a vacuum and excellent thermal resistance in non-oxidising atmospheres. Some of the latest developments are the production of single-crystal aluminium nitride, which, being a III-V family compound, promises exploitable electronic properties, and as a thin film.

Aluminium nitride is a ceramic which is still very much under development, although it is now available commercially from a numer of sources.

Aluminium titanate

The principal property of interest for aluminium titanate (Al_3TiO_5) is its very low thermal expansion coefficient, which is comparable to that of fused silica. However, the thermal properties of the ceramic are very strongly dependent upon the microstructure of the material and it is only in the last few years that major improvements in microstructural control have opened up the possibility of commercial exploitation. Applications now centre on the automotive industry, with major European motor manufacturers considering the material as exhaust port liners in both diesel and petrol engines (to achieve improvements in thermal efficiency), as piston crowns, exhaust manifold inserts and turbocharger liners. Other applications outside of the automotive industry include the non-ferrous metallurgical industries.

Beryllia

Beryllia (beryllium oxide, BeO) has not been exploited as an engineering ceramic to any great degree primarily because of its toxicity. This makes it potentially hazardous to the end-user and also increases the cost of manufacture. However, it should be noted that the toxicity is associated with beryllia dust or powder

TABLE 5 – TYPICAL PROPERTIES OF BERYLLIA

Property	Units		Beryllia 96.00%		99.25%		Comparative data typical alumina	
	1	2	1	2	1	2	1	2
Physical								
Density	g/cm^2	—	2.830	—	2.850	—	3.720	—
Colour			Blue		White		White	
Thermal								
Thermal								
conductivity	W/mK	CGS units						
50°C			239.000	0.57	297.000	0.71	25.000	0.059
100°C			192.000	0.46	264.000	0.63	19.000	0.045
200°C			147.000	0.35	197.000	0.47	—	—
300°C			109.000	0.26	147.000	0.35	—	—
400°C			88.000	0.21	75.000	0.18	10.000	0.024
Thermal expansion	10^{-6}/deg	—						
50–600°C			8.650	—	8.470	—	7.400	—
50–1 000°C			8.890	—	8.850	—	8.200	—
Specific heat	g/cal/g/degC	—	0.300	—→	0.310	—	0.210	—
at 100°C								
Maximum use								
temperature								
(no load)	°C	—	1 700.000	—	1 850.000	—	1 500	—
Electrical								
Dielectric constant	—	—						
1 MHz 20°C			6.580	—	6.450	—	9.000	—
70 MHz 20°C			6.560	—	6.690	—	9.000	—
9 368 MHz 20°C			6.510	—	6.600	—	8.900	—
200°C			6.640	—	6.720	—	9.400	—
400°C			6.840	—	6.940	—	9.900	—
Dielectric loss								
Tangent ($\times 10^4$)	—	—						
1 MHz 20°C			4.800	—	1.700	—	1.000	—
70 MHz 20°C			4.000	—	3.300	—	5.000	—
9 368 MHz 20°C	—	—	10.100	—	2.400	—	10.000	—
200°C			11.700	—	3.100	—	20.000	—
400°C			24.300	—	4.100	—	40.000	—
Volume resistivity	ohm cm	—						
20°C			10^{14}	—	10^{14}	—	10^{14}	—
200°C			4×10^{10}	—	10^{14}	—	10^{11}	—
400°C			4×10^7	—	10^{14}	—	10^9	—
600°C			2×10^6	—	9.2×10^{12}	—	10^8	—
Dielectric strength	KV/mm	—	10–14	—	10–14	—	20	—
Mechanical								
Cross breaking	MN/m^2	lb/in^2	138.000	20 000.0	138.000	20 000.0	358	52 000
strength								
Compressive	MN/m^2	lb/in^2	1 380.000	200 000.0	1 550.000	225 000.0	2 060	300 000
strength								
Young's modulus	MN/m$^2 \times 10^{-6}$	lb/in$^2 \times 10^{-6}$	0.289	42.0	0.358	45–52	0.324	47
Hardness	MOHS	MN/m^2	9.000	628.0	9.000	657.0	9	650

and in a dust-free sintered ceramic form is generally regarded as safe provided that no physical alteration of components is attempted without consulting the manufacturer.

The primary advantage of beryllia is its very high thermal conductivity which is higher than all other ceramics and, indeed, many metals. However, beryllia is also an electric insulator and this combination of properties makes the material suitable for use in the electronics industry, where it is typically used to conduct heat from electronic power devices whilst keeping them electrically isolated, for example, as heat transfer washers, substrates etc.

Other useful properties of beryllia include a low dielectric constant, excellent thermal shock resistance, good chemical stability, a low neutron capture cross-section and lightweight. These properties lead to applications which include; power microwave/radio frequency devices, crucibles, thermocouple tubes, components within the nuclear industry, valve envelopes for portable radio communications, klystron tubes and laser tubes amongst others. Typical property data for beryllia ceramics is provided in Table 5.

Boron carbide

Boron carbide (B_4C) is a material of exceptional hardness which has led to a wide range of applications. In powder form, a major industrial use of boron carbide is as an abrasive, where its performance is substantially better than other powders such as corundum or silicon carbide. The primary disadvantage for this application is the material's cost, though boron carbide grits are still cheaper than diamond.

Ceramic boron carbide is used extensively in wear resistance technology, for example as sandblasting nozzles. Here, the chemical stability of boron carbide also permits the use of nozzles with hard materials slurries. Other advantageous properties of boron carbide include a high elastic stiffness and lightweight. These properties, together with the hardness, has led to the use of boron carbide for lightweight armour systems. Such armour is used for helicopter, other aircraft and personnel protection.

In the nuclear industry boron carbide is used to absorb both thermal and fast neutrons, yielding helium and lithium, with no longlasting radioactive byproducts. By enriching the ^{10}B isotope up to concentrations of above 90 at %, nuetron absorbers of different efficiency can be produced to control the neutron flux of nuclear reactors. Boron carbide is the primary metal for this application at present.

A recent development has led to another new application for boron carbide. By utilising its linear voltage increase with respect to temperature, a B_4C/graphite thermocouple has been developed which can be used to temperatures as high as 2200°C. One immediate application for this device is in hot isostatic pressing, where the temperature cannot be determined pyrometrically.

TABLE 6 – TYPICAL PROPERTIES OF BORON NITRIDE

	Parallel pressing	Perpendicular pressing
Compressive strength, lb/in^2	45 000.000	34 000.000
Transverse strength, lb/in^2, at		
25°C	15 880.000	7 280.000
350°C	14 800.000	6 700.000
700°C	3 840.000	1 900.000
1000°C	2 180.000	1 080.000
Modulus of elasticity, lb/in$^2 \times 10^6$, at		
25°C	12.400	4.900
350°C	8.000	3.200
700°C	1.500	0.500
1000°C	1.600	—
Hardness, Mohs scale	1–2	
Thermal properties		
Sublimation point, at 760 mm Hg	approx 3000°C	
Melting point, under N$_2$ pressure	> 3000°C	
Stability in H$_2$	> 2000°C	
Thermal conductivity, cal/cm/sec/degC, at		
300°C	0.036	0.069
700°C	0.032	0.065
1000°C	0.029	0.064
Resistivity, ohm cm, at		
500°C	5×10^9	
700°C	3×10^7	
1000°C	5×10^5	
1500°C	6×10^2	
2000°C	2×10^3	
Permittivity	4.15–5.00	

Boron nitride

Boron nitride (BN); exists in two crystalline forms; cubic boron nitride which is very hard, second only to diamond, and a hexagonal form which is soft, like graphite. This makes the latter easy to machine to close tolerances by turning, milling, drilling etc, and also leads to one of its major uses, as a filler in a range of other materials to improve lubrication or machinability. For example, hexagonal BN is considered an useful additive to the metal-matrix composite friction materials used in high-performance aircraft brakes. Boron nitride is typically produced as dense components via hot pressing or hot isostatic pressing.

Excellent electrical insulating propeties and the ability to withstand temperatures as high as 2000°C in inert atmospheres (including vacuum) leads to the hexagonal material being used for thermocouple protection tubes and electronic insulating components. Combined with its electrical resistivity is a high thermal conductivity. This, together with a relatively low thermal expansion, results in excellent thermal shock properties.

Physical and mechanical properties of boron nitride are presented in Table 6.

Boron nitride also displays excellent chemical resistance and is not wetted by molten metal, slags and glasses. Combined with the thermal shock resistance, this has led in recent years to the materials use as break rings for the continuous horizontal casting of steel.

The cubic form of boron nitrides has far fewer applications, largely because the cost of machining ceramic forms to close dimensional tolerances is prohibitive.

Magnesia

Magnesia (MgO) is not a widely used ceramic material, being much more expensive than alumina. It displays excellent thermal resistance, a high electrical resistivitry and moderately high thermal conductivity, and is relatively easy to sinter to high density yielding good light transmission. Its principal applications are for crucibles, substrates and some optical devices.

Magnesia powder is also used for special abrasives, as a filler in thermal resistant paint coatings, magnetic recording media and sealants, and also as a sintering aid and stabiliser for a range of other ceramics, including alumina, silicon nitride and zirconia.

Silicates

The silicates comprise a very wide range of compositions, including the clays used in traditional ceramics. However for engineering applications, they can roughly be divided into three groups, the engineering porcelains which are all essentially vitreous refractory alumino-silicates, the magnesia-silicates such as the steatites, and the ternary magnesia-alumino-silicates, such as cordierite.

Engineering porcelains may be classified into various groups, depending on composition or application. Based on the former, we have: high-aluminium, barium, lithium, mullite and zircon porcelains; the two main groups as far as applications are concerned are chemical and hard porcelains. Chemical porcelains may be of high alumina, or alumina with mullite or zirconia replacing the silica content. Iron content is also held to a minimum. These porcelains are strong and hard with good resistance to most chemicals (except alkalis and hydrochloric acid) and to thermal shock. They are used for components in chemical plant, liners and similar duties. Zircon porcelains may have zirconium silicate contents of up to 50% and are denser and stronger, with better chemical resistance. They are strong enough to be used as extrusion dies, nozzles etc although in recent years are increasingly being replaced by advanced ceramics such as zirconia.

Lithium porcelains are noted for their excellent dimensional stability and resistance to thermal shock. They can be formulated with zero (or even negative) coefficients of thermal expansion. They are, however, softer and less refractory than other porcelains.

Hard porcelains are mainly used for electrical insulators of various types. They are typically formulated from 50% kaolin with the remainder equal proportions of feldspar and quartx. They exhibit excellent electrical properties but are brittle and have relatively poor resistance to thermal shock.

Silicon carbide

Silicon carbide (SIC) is a hard, abrasion resistant material which has been commercially available for approximately a century. One of its major applications is in powder or grit form when it is used as an abrasive, being very much cheaper than diamond. More recently, silicon carbide has started to be produced in sintered ceramic form, being fabricated by a range of different processing techniques, such as hot pressing, sintering and reaction bonding. Each of these results in a material with quite substantially different properties, however, as a general rule they exhibit high hardness and strength (retaining the latter to high temperatures), good thermal shock resistance, excellent abrasion resistance and chemical stability and also displays useful electrical properties.

Silicon carbide is used in a wide variety of applications, however, a large proportion of them are tribological in nature. Thus silicon carbide is used extensively for mechanical seals, sliding bearings, wear protection sleeves and as precision ground spheres for regulating valves, or, more generally, where protection against corrosive and abrasive wear, especially in sliding applications, is required.

Other uses for silicon carbide include structural components for high temperature applications owing to good oxidation resistance (for a non-oxide material). Examples of these include components for internal combustion and turbine engines, industrial heat exchanger tubes, kiln or furnace furniture, fixtures and high temperature strength test rigs. Owing to its electrical resistivity the material also finds a major application as a heating element material.

Silicon carbide particulates, whiskers and fibres are being used increasingly to toughen other ceramics. This is discussed further in the Ceramic Composites section.

Silicon nitride

After several years during which silicon nitride ceramics were materials with a 'potential future' but no major applications, due primarily to difficulty in producing dense bodies, improvements in processing have resulted in increasing interest in this material. Just as with silicon carbide there are now a wide variety of processing routes which yield silicon nitride ceramics. These include pressureless sintering, hot pressing, hot isostatic pressing and reaction bonding. Each yields a material with substantially different properties, largely arising from the presence or absence of intergranular glassy phases (which occur due to the use of sintering aids) and porosity.

The first silicon nitrides produced were reaction bonded, a process in which silicon metal powder compacts are nitrided by heating to high temperatures in flowing nitrogen. Whilst this produces a material which has excellent dimensional tolerances (due to only small dimensional changes during reaction bonding), and high purity which results in excellent strength retention on heating to very high temperatures, the actual strength of the material is relatively low due to a necessary high degree of porosity. This limits the application of these silicon nitrides severely.

TABLE 7 – POWDER SPECIFICATION

Grade	SN-E 10	SN-E 05	SN-E 03
Morphology	Equiaxed	Equiaxed	Equiaxed
Particle size	~ 0.2μm	~ 0.4μm	~ 1μm
Specific surface area	12 ~ 10m^2/g	6 ~ 4m^2/g	4 ~ 2m^2/g
Purity	N >38.0% O <2.0% C <0.2% Cl <100 ppm Fe <100 ppm Ca < 50 ppm Al < 50 ppm		
Phase composition	Degree of crystallinity ~ 100% α PHASE > 95% β PHASE < 5%		

Following extensive research, additives were found which permitted first the pressure sintering and subsequently the pressureless sintering of silicon nitride powders. Whilst this has increased the strength of the material substantially over reaction bonded material, strength retention to high temperature is now a problem as the additives tend to result in glassy grain boundary phases which soften. Further work is under way to improve these characteristics and new powders and ceramics are constantly being produced.

Despite the shortcomings described above, silicon nitride is now recognised as a highly useful structural engineering material, having a high strength which is generally retained over a wide temperature range, good thermal shock resistance, and high corrosion and wear resistance. Applications include engine components (silicon nitride turbochargers are now fitted as standard to a number of car model by several different manufacturers), wear resistant parts such as bearings, cutting tools, metallurgical and chemical plant components, amongst others. Typical materials data for several silicon nitride ceramic grades is presented in Table 7.

In addition to silicon carbide, silicon nitride whiskers and fibres are also being used increasingly to toughen other ceramics. Data on whiskers is presented in Table 8, whilst information on fibres and composites can be found in the Ceramic Fibres and Ceramic Composites sections respectively.

Sialon

Sialons are a relatively new class of ceramic materials which are produced by generating a solid solution in the system Si_3N_4–Al_2O_3–SiO_{21}–AlN. The term

Selection of ceramic components: Top: dense silicon nitride and partially stabilised zirconia – cerafine℠ *; Bottom: reaction banded silicon nitride – nitrasil*℠ *(*℠*Trademark T & N plc)*

TABLE 8 – β-SILICON NITRIDE WHISKER

Properties and composition of UBE-SN-WB

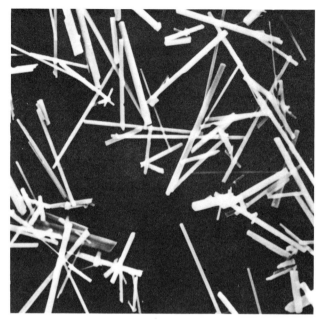

Diameter: 0.1~ 1.5 μm   Aspect ratio: 20~50
Length: 10~ 30 μm   Crystal type: β type

Main applications of UBE-SN-WB

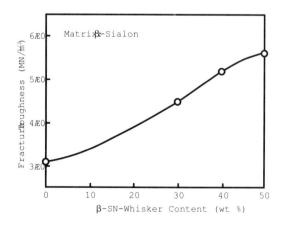

Fracture Toughness of Fibre-Reinforced Ceramic

• Whisker Reinforced Ceramic
• Metal Matrix Composites

Sintering Condition:
 • Hot Press
 • 1750°C × 1 hour,
   250 kg/cm^2

sialon thus arises from the symbols of the compositional elements, Si–Al–O–N. The main advantage of these materials is the large diversity of compositions which can be produced giving rise to the ability to tailor properties for specific applications. These include advanced diesel and gas turbine engines, tooling, extrusion dies, weld shrouds and lathe inserts, amongst others. In general the sialons can be said to exhibit comparatively small thermal expansion coefficients (typically 2–3 x 10^{-6} °C), high strength, high creep and oxidation resistance and high corrosion resistance.

Titanium diboride

Titanium diboride (TiB_2) is a strong, very hard ceramic with an exceptionally high elastic modulus (~550 GPa). This makes it very useful for the production of armour plate, especially for protection against high speed penetrating projectiles. The performance of titanium diboride armour is said to be superior to that of boron carbide, however, its high density restricts its use to the protection of land-based vehicles, such as tanks and armoured limousines rather than the personnel and aero applications of boron carbide armour.

Titanium diboride exhibits metallic character with respect to electrical properties. In addition, it displays excellent chemical stability and possesses a very high melting point (~3200°C). Although the market for the material is currently relatively small, these properties have indicated significant potential for the material for many years now. In the last few years, some of this potential has started to be realised. In addition to the armour plating application described earlier, titanium diboride is now starting to be used in aluminium smelting. It was realised many years ago that if the coventional graphite cathode (which is the crucible containing the aluminium) is replaced by a TiB_2-based crucible, the anode-cathode interpolar distance in the smelting cell could be reduced, since the TiB_2 is wetted by liquid aluminium. The result of these changes is that ohmic losses are reduced and therefore the process becomes more efficient in energy terms – it has been claimed by up to 25%. It is only in the last couple of years, however, that sufficient improvements have been made to the technology to permit industrial application to be seriously considered.

Zirconia

Pure zirconia (ZrO_2) suffers from phase transformations which are accompanied by significant changes in volume. These lead to physical disruption of the body, with a consequent reduction in mechanical properties. The use of alloying additives, such as Y_2O_3, CaO, MgO or CeO, can stabilise the material by suppressing these phase transformations, yielding an industrially useful material. Originally, zirconia was considered only as a refractory material owing to its extremely high melting point of about 2700°C, however, it has received considerable attention over the past decade as a result of its potential to display a much higher toughness than other ceramic materials.

(a)

(b)

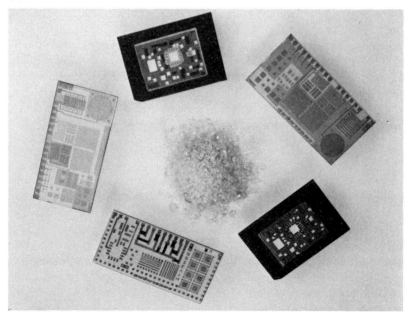

(c)

Some examples of glass–ceramic materials and their uses: (a) armour against ballistics; (b) glass – ceramic metal substrates used in, for example, thin film microwave circuitry; (c) directly bonded glass – ceramic/metal seals.

When the additives mentioned above are added in more limited quantities, then 'partially stabilised zirconia' or PSZ is achieved. These materials rely on the martensitic monoclinic → tetragonal phase transformation occurring in localised regions, usually around propagating cracks. The effect of this is to reduce the effectiveness of the cracks, hence the material is toughened. Table 1 displays toughness values for a range of ceramic materials and illustrates the domination of zirconia.

Other properties for this ceramic include; high strength and hardness, excellent wear resistance, good frictional behaviour, low thermal conductivity, good electrical insulation, a modulus of elasticity similar to that of steel and a thermal expansion coefficient similar to cast iron, and good corrosion resistance. Zirconia powders are also widely available to enable companies to produce components.

Ceramic composites

A very wide range of ceramic-ceramic composites are now produced commercially, and by a number of companies throughout the world. They are usually designed to fill some specific applications, a typical example might be titanium carbide reinforced alumina which is beginning to find significant use as a cutting tool material. With the exception of electrical related applications, most ceramic-composites are intended to provide an enhancement in the mechanical properties over those of the matrix phase on its own.

As stated earlier, the principal limitation to the use of ceramics is associated with their being brittle materials. This results in low toughness, *ie*, an extreme sensitivity to the presence of flaws with failure often being catastrophic. The introduction of a dispersion of another ceramic phase, whether in particle, whisker or fibre form, can inhibit the propagation or cracks, by a variety of mechanisms. This improves the materials fracture toughness (*ie*, more energy is required to cause a crack to propagate through the material).

Levels of usage of ceramic composites are still relatively low compared with several of the monolithic materials, however, considerable time and money is currently being spent on improving compositions, designs, processing and, hence, performance of these materials. Thus, they should not be discounted when selecting potential candidate materials for new applications.

Composites

IN THE general definition composites embrace all reinforced materials, although more specifically the description is applied to materials comprising a matrix including filaments of fibres of another material (or materials) with the object of enhancing the mechanical properties of the composite.

The most common forms of composite are based on a plastic matrix. The fibrous reinforcing material may be in sheet form, as in thermoset plastic laminates; filament form, woven or random, as in glass reinforced plastics; or short fibre form as in 'filled' or reinforced thermoplastics. These materials are well established and widely available.

In the case of thermoset laminate composites, phenolic, melamine and epoxide are the main resin systems used, with paper, cotton fabric, glass fabric and asbestos as the main alternative reinforcing materials (see *Laminated plastics*).

In the case of glass reinforced plastics (GRP) the principal resin systems used are polyester and epoxide with silicons and polyimides for more specialised applications – *ie* again thermoset plastics. Reinforcement takes the form of glass cloth, rovings, chopped strand mat, or a mixture of these disposed through the lay-up. The finished (cured) material is again basically a laminate, but distinguished from those described as *Laminated plastics* by the common employment of glass fibre reinforcement. Even this is not exclusive as carbon fibre is an alternative form of reinforcement (CFRP), and other reinforcement materials may be used – *eg* synthetic fabrics. Such composites may be rendered in flat sheet form like laminated plastics (*eg* for printed circuit boards), although they are more usually applied to specified forms by laying-up, dough moulding, sheet forming (using pre-pregs) or filament winding (see *Reinforced plastics*).

In the case of reinforced thermoplastics used for injection moulding, considerable improvements in mechanical properties can be realised by 'loading' with short fibres of glass, asbestos, or other high strength filaments. Glass fibre is the most common reinforcement, the resulting thermoplastic being referred to as 'glass filled'. Asbestos filled thermoplastics have rather more specialised applications, whilst carbon fibre filling is a more recent, and as yet relatively undevel-

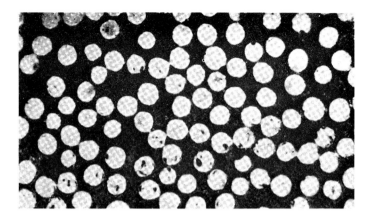

Fibre reinforced thermoplastic.

oped, alternative to glass filling when greater stiffness and rigidity are required (see *Filled plastics*). Filling can be applied to virtually any thermoplastic, but is not necesssarily commercially viable. The main application of glass filling is to acetals and nylons (see *Filled plastics*).

Fibrous reinforcement can be applied to matrices other than plastic – notably ceramics, cement and concrete, and metals. The reinforcement material can be fibrous glass, asbestos, or high modulus filaments such as alumina, carbon boron fibre, silicon carbide or silica nitride 'whiskers' etc, as well as metallic filaments and whiskers. Notably, too, some high strength organic polymer fibres have been developed as reinforcement materials. There is also the possibility of producing aligned composites in a continuous casting process – *eg* a nickel-aluminium eutectic melt yielding directionally oriented aluminium nickelide fibres in a pure aluminium matrix. Equally carbon fibre or similar filament reinforcement can be oriented in a metal composite to produce optimum stress distribution.

Ceramic and metal composites remain relatively undeveloped as general engineering and constructional materials, largely on account of high cost. There are, however, numerous applications of filled and laminated metal forms which qualify as composites under the general description.

Cement and concrete composites are more readily, and cheaply, realised by simple pre-mixing or lay-up techniques. The value of asbestos for strengthening cement has, in fact, been known and explored for a considerable number of years. More recently the addition of small quantities of nylon or polypropylene fibres has been found to reduce the inherent brittleness of cement mortars, with obvious advantages, and this has led to considerable further research in the field of organic fibrous reinforcement of cement itself as an alternative, or supplement, to metallic reinforcement.

TABLE 1 – FIBRE REINFORCED METALS—MANUFACTURING METHODS

Method	Starting point	Process	Remarks
Powder metallurgy	Whiskers or cut fibres mixed with matrix powder.	Pressed and sintered.	Difficult to eliminate porosity and control fibre alignment.
Impaction (pneumatic)	Whiskers or cut fibres mixed with matrix powder.	High pressure impacted.	Limited control on fibre distribution and finished size.
Electroforming	Fibres.	Deposit metal by electrolysis.	Particularly suitable with nickel and aluminium. Difficult to eliminate voids.
Plasma spray	Foil wrapped around a rotary mandrel followed by spiral wound monofilament (eg aluminium fill and continuous boron filament).	Alloy powder fed into plasma torch and sprayed on rotating mandrel.	Single or multiple layer build up. Must be followed by hot pressing.
Vapour deposition	Whiskers and molten metal.	Cast, then hot pressed and diffusion hardened.	Limited scope.
Liquid phase consolidation	Matrix powder and reinforcement mixed.	Heated to above melting point of lowest; consolidated by pressing.	Applicable only to non-eutectic alloys for the matrix.
Vacuum infiltration	Molten matrix metal.	Vacuum infiltration of fibres.	Basically a laboratory method capable of producing high fibre volumes.
Die casting	Continuous filament laid in mould.	Matrix metal gravity or die cast.	Suitable for low fibre volumes but higher volumes restrict flow of metal.
Filament coating	Molten matrix metal.	Continuous filament fed through melt; coated filaments consolidated by diffusion loading.	Largely laboratory or experimental technique.
Diffusion bonding	Lay-up of matrix metal foil and oriented reinforcement.	Heat under pressure in a vacuum.	Considerable scope and control can be varied to suit matrix/filament combinations.
Rolling	Feed of coated continuous fibres and matrix metal strip.	Roll into continuous ribbons or strips.	Strips can be used for sheet fabrication, etc.
Co-extrusion	Continuous wires or filaments and matrix metal.		Difficult in practice and largely experimental.
Controlled solidification	Eutectic matrix alloy and whisker or lamellar reinforcement.	Controlled unidirectional solidification of eutectic alloys.	Largely laboratory or experimental technique.

Basically, the addition of organic polymetric reinforcement to concrete produces a significant increase in impact resistance of the cement matrix with little or no effect on tensile strength. A further virtue is that this improved impact resistance is not affected by ageing.

To improve the other mechanical properties of the cement matrix a high modulus fibre reinforcement can be used, such as asbestos, glass fibre or carbon fibre, as an alternative (or supplement) to conventional metallic reinforcement. Glass fibre composites show improvements in both rupture and impact strength. Asbestos and carbon fibre improve rupture strength, but have little or no effect on impact strength. Glass fibre can be considered a low modulus, high strength reinforcement. Asbestos and carbon fibre are high modulus, high strength reinforcements. This largely explains the difference in the behaviour of composites with these inorganic reinforcements.

The most promising results appear to be around hybrid composites involving both inorganic (*eg* polypropylene) and organic fibrous reinforcement (glass fibre or asbestos). The limit of reinforcement is determined by the amount of fibre which can be added with uniform dispersion without increasing the porosity of the matrix. This will vary with the type and dimensions of the fibre and the method of fabrication. An upper limit of about 2% by weight is indicated with pre-mixes although this proportion can be increased to 5% or more with spray-up methods.

See also *Metal matrix composites* (Section 2).

Cork

CORK IS the outer bark of a species of oak tree characterised by a unique cellular structure with thin thread-like membranes of resinous material linking the cells together. Rather more than 50% of the volume of cork is air trapped within the cells. Density ranges from 9.5 to 13 lb/ft^3, a typical specific gravity being 0.25.

Natural cork is a resilient compressible material and can accommodate up to 30% compression without lateral expansion. Loading simply compresses the air within the cells and the original thickness is recovered when the load is removed. It has a high resistance to penetration by water and other liquids, except strong alkalis, even when boiled in them.

Typical tensile strength of natural cork is 10 to 12 kg/cm^2. Typical compression is 4 to 6% at 7 kg/cm^2 with 3.5% moisture content. Actual compression will vary with the moisture content. Typical coefficients of friction are given in Table 1. Natural cork is used for gaskets, floats, polishing wheels, handles, grips, etc, but the main application of cork is in composition form.

TABLE 1 – COEFFICIENT OF FRICTION OF NATURAL CORK

Cork state	Rubbing on	Coefficient of friction
Dry	Dry natural cork	0.592
Dry	Dry hardwood	0.487
Dry	Dry glass	0.512
Dry	Dry steel	0.452
Dry	Wet steel	0.694
Dry	Oiled steel	0.447
Dry	Hot steel	0.645
Water soaked	Dry steel	0.565
Oil soaked	Oiled steel	0.419

TABLE 2 – TYPICAL CORK COMPOSITIONS

Binder	Particle size	Minimum density lb/ft^3	Minimum tensile strength lb/in^2 (kg/cm^2)	Compression 100 lb/in^2 load %	Typical applications
Gelatin	Coarse	20	150 (10.5)	25–40	Softer gaskets.
Resin	Coarse	20	150 (10.5)	25–35	Gaskets.
Gelatin	Medium	20	150 (10.5)	25–40	Thinner gaskets.
Gelatin	Medium	25	200 (14.0)	15–30	Oil seals and harder gaskets.
Gelatin	Medium	27	225 (16.0)	15–25	Friction discs, friction linings, etc.
Resin	Fine	27	250 (17.0)	10–20	Friction linings, etc.
Resin	Fine	36	350 (24.5)	5–15	Brake pads, clutch inserts, friction linings.
Resin	Fine	40	400 (28.0)	5–10	Semi-hard resilient material for high duty friction linings, valve seats, etc.

Cork compositions are based on cork agglomerate in a large variety of particle sizes in a gelatin or resin binder. A gelatin binder produces a weaker material with higher compressibility. Typical properties are given in Table 2. Cork compositions are widely used for gaskets, friction discs, anti-vibration mounting pads, expansion sheets and many other industrial applications. A particular advantage of cork used as a friction disc material is that the material does not wear smooth or glaze and retains a substantially constant coefficient of friction. Cork composites engage instantly with minimum slip when operating under dry conditions. Operating in an oil bath, engagement is smooth and progressive and can be used at relatively high speeds without excessive wear. Cork composites can be loaded or coated with graphite or other lubricants for minimum friction.

A further category of cork materials is natural cork with various impregnants aimed at increasing the compressibility and load rating of the material (eg for gaskets and anti-vibration mounts). The most useful of these are rubber impregnated or rubber loaded cork where the characteristics can be closely controlled. Compositions of this type can range from highly compressible sponge material to hard, quite rigid solids. The rubber materially improves the strength and compressibility of the compound whilst the cork acts as a reinforcing agent to provide dimensional stability. Maximum service temperature is also increased somewhat compared with that of plain cork compositions, and appreciably increased if a high temperature rubber is used. Cork-rubber compositions, therefore, offer a very attractive range of gasket materials for moderate pressure duties and moderate service temperatures, with a wide choice of properties governed primarily by the choice of basic elastomer.

TABLE 3 – GENERAL PROPERTIES OF CORK GASKET MATERIALS

Material	Specific gravity or density	Tensile strength lb/in²	Compressibility % @ lb/in²	Minimum sealing pressure lb/in²	Maximum‡ internal pressure lb/in²	Maximum service temperature	
						°F	°C
Natural cork.	9.5–13†	140–170	25–40 @ 100			250	120
Cork compositions							
gelatin binder, coarse	20†	150	25–40 @ 100	Low		250	120
gelatin binder, medium	20†	150	25–40 @ 100	Low		250	120
resin binder, fine	23–40†	200–400	10–30 @ 100	Low		250	120
Impregnated cork.	15–20†		30–50 @ 100 / 55–70 @ 300	100	50–100	250	120
	23–30†		20–40 @ 100 / 45–60 @ 300	100–200	50–100	250	120
Cork – rubber compositions							
butyl (typical).	0.70–0.82	200–500	25–40 @ 300	400	50	250–300	120–150
sponged butyl.	0.40–0.55		25–40 @ 40	200	50	250–300	120–150
chloroprene (low density).	0.70–0.80	200–400	25–40 @ 200	400	50	250–300	120–150
chloroprene (medium density).	0.90–1.00	200–500	10–20 @ 300	400	50	250–300	120–150
chloroprene (high density).	1.20–1.30	200–800	5–15 @ 300	400	50	250–300	120–150
sponged chloroprene.	0.50–0.60		25–40 @ 50	200	50	250–300	120–150
nitrile (low density).	0.60–0.75	200–400	30–40 @ 200	400	50	250–300	120–150
nitrile (medium density).	0.95–1.05	200–500	10–20 @ 300	400	50	250–300	120–150
sponged nitrile.	0.45–0.55		30–40 @ 50	200	50	250–300	120–150
sponged styrene.	0.40–0.50		10–25 @ 40	200	50	250–300	120–150
High temperature cork rubbers (typical).			15–25 @ 400	400	50	400–450	200–230
Cork – asbestos (SBR rubber).		350	25		150	250	120

Key:
† Density in lb/ft³
‡ Higher internal pressures may be sealed by suitable flange design.

Photomicrograph of a transverse section of cork by transmitted light showing the cellular structure. Magnification – 47×

Photomicrograph of a single cork cell showing walls, air-filled interior and adjoining cells. Magnification – 950×

Cork board.

Rubber bonded cork can have the same compressibility characteristics as cork, *ie* it does not require clearance to allow for the shape factor necessary for deflection if rubber alone is used. At the same time the rubber bond gives the same compatibility to the composition as rubber and strengthens the relatively fragile characteristics of the cork alone. Rubber bonded cork is, therefore, an extremely useful general purpose material for isolation pads. It is generally rated as suitable for medium or high frequency isolation.

Elastomers

ELASTOMERS MAY be defined as elastic, rubber-like substances, which broadly include natural and synthetic rubbers. More specifically, any synthesised *polymeric* substance can be defined as an elastomer if it can be rendered in a form capable of being stretched at least 100% at normal ambient temperatures, with quick and virtually complete recovery from any extending force (*eg* by ASTM definition, having a permanent set not exceeding 10%). Certain other synthetic substances with more limited stretch, but still semi-elastic in character, may be classified as *plastomers*. The properties of these may be tailored by modification to impart near-elastomeric properties, rendering them suitable for duties normally met by true elastomers.

General properties of elastomers

Tensile stress is a measure of the mechanical strength of the material in tension, and indirectly indicates its likely resistance to deterioration under stress. It does not, however, give any direct indication of fatigue properties, resistance to cracking, or tear strength, etc.

The tensile strength of most elastomers is relatively low (*eg* usually less than 210 bar at normal temperatures) and decreases substantially with increasing temperatures.

Elongation is the percentage increase in length under tension loading, the figure quoted normally being taken as the maximum but not necessarily qualified. This maximum safe elongation for continuous cycling is usually about 80% of the elongation to break.

A more realistic parameter is the *tension modulus* of the material, or the stress produced in the material at a specific elongation (usually 100%). *Shear modulus* and *torsion modulus* can similarly define the stress with specified distortion in shear and torsion respectively.

Resilience is defined as the ability of an elastomer to return to its original shape after loading or compression. Recovery characteristics can, however, vary widely with temperature. Typical elastomers show a marked minimum resilience

TABLE 1 – TYPICAL PROPERTIES OF BASE ELASTOMERS

	Specific gravity	Density lb/in³	Durometer hardness range	Tensile strength lb/in²	Elongation % (max)	Compression set	Resilience	Decrease in tensile strength 212°F	at 350°F (%)	Shrinking point °F	Brittle point °F	Maximum service temp °F	°C	Minimum service temp °F	°C
Natural rubber (polyisoprene)	0.92	0.033	30–90	4 500	650	E	high	-32	-84	-20 – -50	-80	180	82	-60	-50
Nitrile (butadiene acrylonitrile)	1.00	0.036	40–95	4 000	650	G	low–medium	-50	-75	-30 – -70	-40	300	150	-60	-50
Butyl (isobutylene isoprene)	0.92	0.033	40–90	3 000	850	G	low	-57	-87	-2 – -50	-80	300	150	-50	-45
GRS/SBR (styrene butadiene)	0.94	0.034	40–80	3 500	600	G	medium	-44	-72	0 – -50	-80	180	82	-60	-50
Butadiene (BR) (polybutadiene)	0.91	0.033	40–90	3 000	650	G	high	45	-62	-30 – -60	-100	200	93	-150	-100
Chloroprene (neoprene)	1.25	0.045	30–95	4 000	600	G	high	-50	-74	-10 – -50	-85	240	115	-40	-40
Chlorosulphonated polyethylene	1.18	0.043	45–100	4 000	500	F–G	low	-57	-82	-10 – -50	-80	325	163	-40	-40
EP/EPM (ethylene propylene)	0.86	0.031	40–90	2 500	600	G–E	medium	-49	-78	-10 – -40	-70	350	177	-60	-50
Fluorocarbon	1.86	0.067	60–90	3 000	300	G–E	low	-72	-87	-10 – -10	-60	500	260	-10	-23
Isoprene (IR)	0.91	0.033	40–80	4 500	650	E	high	-40	-90	-20 – -50	-80	180	82	-60	-50
Polyacrylic (ACM)	1.09	0.039	40–90	2 500	450	G	medium	-65	-80	-10 – -35	-20	350	177	-20	-30
Polysulphide	1.35	0.049	40–85	1 500	450	P	low	-76	-44	-10 – -45	-70	250	121	-60	-50
Polyurethane	1.25	0.045	35–100	5 000	750	P	high–low	-55	-83	-106 – -30	-60 to -200	240	115	65	-53
Silicon	1.10 1.60	0.040	30–90	1 500	900	G–E	high–low	-17	40	-70 – -180	90 to -180	600	315	178	116
Fluorosilicon	1.40	0.051	40–80	1 500	400	F–G	low	-56	-66	-70	-85	400	204	-90	-68
Epichlorohydrin	1.27 1.36	0.049	40–90	2 500	350	E–F	high–low	-45	67	-15 – -40	10 to 55	300	150	-15 to -80	26 to -62

Key: E – Excellent   G – Good   F – Fair   P - Poor
To convert Fahrenheit to Centigrade, deduct 32, multiply by 5 and divide by 9 (eg 200°F – 32 = 168 × 5 = 840 ÷ 9 = 93.3°C).

within the temperature range –70 to 20°C, but as a general rule resilience increases with increasing temperature.

Permanent set (set) is the permanent change in original dimensions after being subject to tension or compression. It should be noted that the degree of set introduced may also be dependent on the load (*eg* natural rubber may exhibit a permanent set of 10% or more loaded in tension to 80% of its ultimate strength, but substantially less with lower loadings). A subsequent higher loading could then reduce or increase a permanent set.

In most applications it is desirable to limit the amount of permanent set (*eg* to avoid over-stretching) in order to reduce the residual stresses in the elastomer which could reduce its useful life.

Hardness is a property which can vary widely in different compounds and affects the rigidity of the elastomer. It is normally a control factor in compounding an elastomer for a specific duty (*eg* seals). In general, an increase in hardness increases the resistance to wear, abrasion and extrusion and reduces elasticity.

Hardness may be specified in BS degrees, IRHD (International Rubber hardness degrees), or Shore Durometer hardness degrees. The three scales are virtually identical, although BS hardness may be measured to ±1° (but more usually to the nearest 5°), whilst Shore hardness is measured to ±5° and is normally quoted to the nearest 10°.

It is a general characteristic of all elastomers that they become progressively harder at lower temperatures, leading to loss of flexibility and resilience and, ultimately, brittleness. Measurement of the change in hardness with temperature is no reliable guide to performance in this respect as it is the *stiffness* of the elastomer which is the significant parameter. This can be expressed in terms of the *freeze point* and *brittle point.* At elevated temperatures all elastomers lose strength and tend to become softer and more flexible. Normally recovery is complete on reduction of temperature, but if the temperature is high enough permanent changes in the material may be introduced.

The ×2 *freeze point* is determined as the temperature at which the original room temperature stiffness (or 20°C) is doubled. This marks the temperature at which the elastomer suffers from a marked loss of flexibility. The ×10 *freeze point* is similarly determined as the temperature at which the stiffness is 10 times that of the original. This marks the point at which the elastomer has generally lost its elastomeric properties and is on the point of behaving as a brittle material. Stiffness increases very rapidly beyong the ×10 freeze point until the true *brittle point* is reached (*ie* the material is quite brittle and will break if flexed).

The ×2 freeze point represents a safe minimum temperature for using the material as an elastomer (*ie* subject to flexing). The low temperature characteristics of most elastomers can, however, be modified to some extent by compounding.

Ageing is the permanent and progressive deterioration of the properties of the elastomer with age, particularly an increase in hardness and loss of elasticity. This can be widely affected by ambient and working conditions and tempera-

TABLE 2 – GENERAL PROPERTIES OF BASE ELASTOMERS

	Resilience	Impact strength	Abrasion resistance	Tear resistance	Oil resistance	Fuel resistance	Sunlight ageing	Oxidation	Heat ageing	Water swell resistance	Flame resistance	Adhesion to metal	Electrical resistivity
Natural rubber (polyisoprene)	H	E	E	E	P	P	P	G	G	F	P	E	E
Nitrile (butadiene acrylonitrile)	L–M	F	E	G	E	E	P	G	G	E	P	E	F
Butyl (isobutylene isoprene)	L	G	F	G	P	P	E	E	E	E	P	G	E
GRS/SBR (styrene butadiene)	M	E	E	F	P	P	P	F	G	E	P	E	E
Butadiene (BF) (polybutadiene)	H	G	E	G	P	P	P	G	F	E	P	E	E
Chloroprene (neoprene)	H	G	E	G	E	G	G	E	E	G	E	E	F
Chlorosulphonated polyethylene	L	G	E	G	G	F	E	E	E	E	E	E	G
EP/EPM (ethylene propylene)	M	G	G	F	P	P	E	E	E	E	P	G	E
Fluorocarbon	L	G	G	G	E	E	E	E	E	E	E	G	G
Isoprene (IR)	H	E	G	E	P	P	P	G	G	E	P	E	E
Polyacrylic (ACM)	M	P	G	F	E	E	E	E	E	F	P	G	F
Polysulphide	L	P	P	P	E	E	G	G	G	E	P	G	F
Polyurethane	L–H	E	E	E	G	E	E	G	G	E	P	G	G
Silicone	L–H	F	G	G	G	P	E	E	E	E	E	E	E
Fluorosilicone	L	P	P	P	E	E	E	E	E	E	E	G	E
Epichlorohydrin	L–H	G	G	G	E	E	G	G	G	G	F	G	G

Key: E – excellent   F – fair   H – high   L – low
           G – good   P – poor   M – medium

ture, and again can be modified by compounding. Accelerated ageing tests can indicate general ageing characteristics of specific elastomers, but not necessarily under individual applications or use.

Elastomeric materials

Elastomers are most commonly classified under the basic polymer involved. The name may be generic, covering a whole variety of different compounds, all of which will have certain general properties, but specific properties will differ with the extent and degree of polymerisation and copolymerisation in the case of synthetic materials, or compounding in the case of natural rubbers.

Natural rubbers

Natural rubber (polyisoprene) or NR is characterised by high strength, flexibility and resilience, with good flexibility maintained down to about 55°C. It is superior in this respect to all synthetic rubbers, with the sole exception of silicon rubbers. Its main limitation as an industrial material is its lack of resistance to attack by mineral oils and petroleum based fluids.

Natural rubbers can be compounded to meet a wide variety of requirements in the elastomeric field, as well as being rendered as hard, rigid solid materials.

Nitrile rubbers

Nitrile rubbers, also known as Buna N or NBR, form one of the most important groups of synthetic elastomers. Chemically, nitrile is a copolymer of butadiene and acrylonitrile, the acrylo nitrile content typically varying between about 18 and 48%. Nominal designations are low, medium and high nitrile. Resistance to petroleum based oils and hydrocarbons increases with increasing nitrile content, but at the same time low temperature flexibility decreases. In order to obtain good low temperature performance with nitrile rubbers it is usually necessary to sacrifice some high temperature fuel and oil resistance.

Nitrile rubbers have good physical characteristics and are superior to most other rubbers as regards cold flow, tear and abrasion resistance. They are not particularly resistant to ozone, weathering or sunlight, but their properties in this respect can be improved by compounding. Due to their susceptibility to ozone attack, nitrile rubber seals should not be stored near any possible source of ozone (*eg* near an electric motor or electrical equipment, or in direct sunlight).

Nitrile rubbers are particularly suitable for use in contact with petroleum oils and fluids, water, silicon oils and greases and glycol based fire-resistant fluids. They are not normally suitable for use with halogenated hydrocarbons. Trade names include Chemigum, Fr-N, Hycar, Krynac, Nysyn, Paracil, Tylac.

Butyl rubbers

Butyl is a copolymer of isobutylene with isoprene, yielding a group of rubbers which are extremely resistant to water (better than chloroprene and nitrile) and many other fluids, including silicon fluids and greases. A further favourable characteristic is that butyl is very resistant to gas permeation, making it attractive for use in the manufacture of inner tubes and under layers of tubeless tyres, and for seals in vacuum systems. Amongst the Trade names for butyl rubbers are Bucar, Enjay, Hycar 2202 and Polysar.

Buna S

This rubber was originally known as GRS (Government Rubber Styrene) and more recently as SBR (Styrene Butadiene Rubber). It was produced during

TABLE 3 – TYPICAL APPLICATIONS OF NATURAL AND SYNTHETIC RUBBERS

Rubber	Applications
Natural rubber	Vehicle tyres and tubes, seals, anti-vibration mounts, hoses belts.
Nitrile	Oil resistant hoses and liners, gaskets, seals, O-rings, anti-vibration mounts.
Butyl	Inner tubes, anti-vibration mounts, steam and water hose, flexible connectors.
GRS/SBR	Similar to natural rubber.
Butadiene (BR)	Not normally used on its own; only blended with other rubbers.
Chloroprene	Gaskets and seals, weather stripping, tubing, hoses and hose liners.
Chorosulphonated polyethylene	Oil and chemical hose, sealing rings, diaphragms, linings, cable insulation.
RP/RPM	Belting, electrical insulation, weather stripping, dust covers, automobile parts.
Fluorocarbon	O-rings and seals, packings, diaphragms, gaskets.
Isoprene	Similar to natural rubber.
Polyacrylic (ACM)	Hose, seals, gaskets, spark plug covers, wire insulation.
Polysulphide	Inflatables, hose liners, seals, sealants, adhesives, potting compounds.
Polyurethane	Solid tyres, tracks for grease seals, diaphragms.
Silicone	High-temperature O-rings and seals, gaskets, bellows.
Fluorosilicon	O-rings and seals, gaskets.
	Hose for petroleum products, gaskets, low temperature components.
Polyester	Rubber components requiring high hot strength, high flex-fatigue resistance, etc.

World War II as a replacement or substitute for natural rubber, but has generally inferior properties, althouigh its abrasion resistance may be comparable. It can be considered as a direct alternative to natural rubber, and can be used, for example, with castor based hydraulic fluids in automobile brake systems. It is produced under a wide variety of trade names.

Butadiene (BR)

Polybutadiene or butadiene rubber, generally known as BR, is again a direct substitute for natural rubber and has similar, but generally inferior, properties. Low temperature performance is somewhat better than SBR rubber and may be comparable to natural rubbers for many applications. Trade names include Ameripol, Budene, Cisdene, Diene, Duragen, Polysar, Tacktene, E-Br, etc.

Chloroprene

Chloroprene, better known under the Trade name Neoprene, is one of the best general purpose synthetic rubbers. Its main advantage is its excellent resistance to weather ageing. It is also superior to natural rubber in performance at higher temperatures, but tends to harden or stiffen at low temperatures. Resistance to oils is moderate to good. It is excellent for use in contact with refrigerants, may also be used with mild acids, and can be compounded for service at temperatures of –54 to 150°C, although it has a tendency to crystallise when under stress at low temperatures.

Chlorosulphonated polyethylene

This elastomer is also known as CSM (chlorosulphonated monomer). Its mechanical properties tend to be low but the material has good resistance to acids and heat and may be used in such applications. It is also completely resistant to ozone. Its useful temperature range is –54 to 121°C and its trade name is Hypalon.

Ethylene-propylene

Ethylene-propylene is from ethylene and propylene monomers. Ethylene-propylene or EP rubbers (also known as EPM rubbers in the USA) have good temperature characteristics and excellent resistance to many chemicals, such as organic solutions and solvents. EP is not resistant to aromatic solvents, mineral oils, or petroleum products, so that the application of the material to seals is somewhat specialised. Its main use in this respect is for seals used in hydraulic systems based on phosphate ester non-flam fluids and silicon fluids. EP rubbers are also excellent for steam and hot water services (operating within the service temperature limits of the material), dilute acids and alkalis, and ketones. Trade names include: Nordel, Royalene, Enjay EPR, Dutral N, Olemiene, Epsyn, Epcar, Vistalon.

Fluorocarbons

Fluorocarbon elastomers are based on the copolymerisation of vinylidene fluoride and hexafluoropropylene, offering a range of rubbers with outstanding resistance to chemical attack and a maximum continuous service temperature of 200 to 250°C, or up to 320°C for intermittent exposure. They are also particularly suitable for vacuum duties, although low temperature performance is only comparable with that of high nitrile.

Fluorocarbon rubber can be used to advantage where the service conditions demand resistance to heat, acids, petroleum oils, halogenated hydrocarbons, aromatics, etc, or where the cost can be justified. Trade names include: Viton, Fluorel, Kel-F.

Isoprene (IR)

Polyisoprene is a synthetic elastomer, with the same chemical composition as natural rubber. Its properties are therefore generally similar to those of natural rubber. Trade names include: Ameripol SN, Coral, Natsyn, Shell IR.

Neoprene – see Chloroprene.

Polyacrylic

Polyacrylic (ACM) rubbers are polymerised products of acrylic acid esters and form, in effect, a group midway in properties between nitrile and fluorocarbon rubbers. One of the most attractive features from the sealing point of view is their excellent resistance to mineral oils, hypoid oils and greases, up to temperatures of 177°C. Resistance to hot air is slightly superior to nitrile rubber. Excellent resistance to ageing and flex cracking favours their use for rotary shaft seals. Low temperature characteristics are not outstanding and mechanical strength and resistance to water are generally inferior. Trade names include: Cyanacryl, Hycar 1042, Krynac, Polysar, Thiacril.

Polysulphide

Polysulphide elastomer, derived from sodium polysulphide, was one of the earliest synthetic rubbers to be developed. It has excellent resistance to flex cracking and retains good low temperature flexibility, although its other mechanical properties and heat resistance are moderate. It has outstanding resistance to many common solvents and chemicals and high oxygen and ozone resistance, and is also a good adhesive base. Working temperature range is from –54 to 107°C. Its smell makes it objectionable to handle for processing. Trade names include: FA (Polysulphide), ST (Polysulphide), Thiokol, ZR-300.

Polyurethane

This is one of the more recently developed elastomers with exceptional strength, tear and abrasion resistance (better than any of the other rubbers); it retains excellent flexibility at low temperatures. Resistance is good to petroleum products, hydrocarbons, ozone and weathering. Performance is generally unsatisfactory in contact with aqueous solutions of an acid or alkaline nature, chlorinated hydrocarbons, ketones, hot water, steam or glycol, but it may be improved by compounding. Compression and permanent set characteristics also tend to degrade rapidly with increasing temperature, (eg about 110 to 120°C).

Polyurethane rubbers are therefore most attractive for their mechanical strength rather than for their chemical or temperature properties. They may be used to advantage, if compatible, under abrasive conditions. They are now produced under a wide variety of trade names.

Silicon rubbers (SI)

These form a very important group of elastomers, retaining their mechanical performance over a very wide range of temperatures. Flexibility is retained down to temperatures below –73°C while a maximum continuous service temperature of at least 232.2°C can be realised with these compounds. Retention of favourable properties at high temperature is superior to that of any other type of elastomer.

Basically, silicons have poor strength and tear and abrasion resistance, although mechanical performance can be enhanced by special compounding. Resistance is generally good to alkalis, weak acids and ozone, but only moderate as regards oils. Again the chemical properties can be enhanced by compounding (to provide better resistance to oils and fuels, for example). In general, however, silicon rubbers are not recommended for use in contact with hydrocarbons such as petrols and paraffin, the lighter mineral oils or steam above a pressure of about 50 lb/in^2 because considerable swelling of the elastomer can result.

The chief advantage of this type of elastomer is that it retains its flexibility down to very low temperatures, and can also withstand continuous heating at high temperatures without hardening, making it suitable for both high and low temperature seals over a broader range than that covered by the other elastomers. A further application is for rotary seals where the operating temperature may be higher than that permissible with conventional elastomers due to the friction developed. The cost of silicon rubbers is, however, substantially higher than that of most other elastomers.

Fluorosilicon rubbers are a more recent development and even more costly. Working characteristics are generally similar to those of ordinary silicons, but with a more restricted service temperature range. The main advantage offered is that fluorosilicon rubbers can have an oil resistance comparable with or closely approaching that of nitrile rubbers. They can therefore be used for duties beyond the service temperature limits of nitrile rubbers, and where ordinary silicon elastomers do not have the necessary compatibility.

Silicon rubbers are suitable for service in dry air up to temperatures of 230 to 250°C, and may also be compounded to withstand temperatures of up to 380°C for intermittent exposure. Fluorosilicons are suitable for use at temperatures up to 180 to 200°C, even in contact with hydrocarbon fluids.

Trade names include Silastic, SE, Si-O-Flex, K and KW.

TABLE 4

Property	ASTM test	Units	Polyester elastomers*		
Durometer hardness	D–2240	Points	92A	55D	63D
Tensile strength	D–412	lb/in^2	5900.00	6400.00	5800.00
		bar	415.00	450.00	408.00
Elongation at break	D–412	%	800.00	700.00	500.00
Tensile set at 100% strain	D–412	%	18.00	38.00	50.00
Resistance to flex-cut growth					
Ross (Pierced)	D–1052	Cycles with no cut-growth.	>2 × 10^6	>3 × 10^5	n/a**
DeMattia (Pierced)	D–813	Cycles to failure.	>2 × 10^5	>7 × 10^4	n/a
Tear strength					
Split	D–470	lb/in	170.00	200.00	—
		kg/cm	30.00	36.00	—
Die B	D–624	lb/in	631.00	935.00	1055.00
		kg/cm	113.00	167.00	188.00
Die C	D–624	lb/in	700.00	900.00	850.00
		kg/cm	125.00	161.00	152.00
Compression set at: 1 350lb/in^2 (94.9 bar) constant load (after annealing)	D–395A	% at 158°F (70°C)	27.00	4.00	2.00
25% deflection	D–395B	% at 158°F (70°C)	60.00	56.00	n/a
25% deflection (after annealing)†	D–395B	% at 158°F (70°C)	36.00	38.00	n/a
Abrasion resistance Taber, H–18 wheel, 1 000 g load	D–1044	mg/1 000 cyc	100.00	64.00	160.00
NBS	D–1630	% of standard.	800.00	3 540.00	2 800.00
Solenoid brittle point	D–746	°F	<–94.00	< 94.00	<–94.00
		°C	<–70.00	<–70.00	<–70.00
Resilience, Bashore		%	62.00	53.00	43.00
Flexural modulus	D–790	lb/in^2	7000.00	30 000.00	50 000.00
		bar	492.00	2 109.00	3 515.00
Stress at 1% strain [head speed 1 in/min (2.54 cm/min)]	D–638	lb/in^2	100.00	225.00	380.00
		bar	7.00	15.80	26.70
Stress at 3% strain [head speed 1 in/min (2.54 cm/min)]	D–638	lb/in^2	180.00	630.00	950.00
		bar	12.60	44.30	66.80
Stress at 5% strain [head speed 1 in/min (2.54 cm.min)]	D–638	lb/in^2	265.00	1 000.00	1 350.00
		bar	18.60	70.30	94.90
Stress at 10% strain [head speed 1 in/min (2.54 cm/min)]	D–638	lb/in^2	450.00	1 550.00	2 000.00
		bar	31.60	109.00	141.00
Specific gravity	—	—	1.17	1.20	1.22

* Dupont Hytrel.
** Not applicable.
† 16 h at 100°C (212°F) for 92A.
  16 h at 121°C (250°F) for 55D.

Polyester elastomers are readily fabricated by a variety of short cycle techniques, familiar to the plastics industry, into a wide range of products, large and small, that combine the resilience of true elastomers with the strength of high quality plastics.

Polyester elastomer

Polyester elastomers have been described as a product bridging the gap between conventional synthetic rubbers and the more flexible thermoplastics (*eg* fluorocarbons and polypropylenes), whilst retaining basic elastomeric properties. Thus although the harder polymers resemble thermoplastics rather than rubbers, they retain a resilience of 40 to 45%. They also have excellent flex fatigue resistance and good retention of flexibility at low temperatures.

Polyester elastomers show high moduli in tension, compression and flex and at comparable hardness are superior to polyurethane rubbers in load bearing capacity. They also possess excellent hot strength, particularly in the case of the harder polymers, where a tensile strength of 140 to 245 bar is available at 150°C. Brittle point is below –40°C.

Polyester elastomers are fully polymerised materials requiring no post-curing. They have good melt flow properties, melt stability, low mould shrinkage and rapid crystallisation rates. The various hardness grades cover the Durometer range from 92A to 63D (or flexural modulus range from 9000 to 50 000 lb/in^2). They are resistant to oils, fuels and a wide range of chemicals and solvents, but are attacked by phenols, cresols, chlorinated hydrocarbons and strong acids and alkalis. Resistance to fluids generally increases with increasing hardness. Typical specific properties are summarised in Table 4.

Note: see also sections covering Rubber (natural) and Rubber (synthetic).

Felt

THE PRINCIPAL engineering applications of felt are as a vibration insulation material and for felt elements. In the former case it is used in the form of pads cut from the softest felt in thickness from 6 to 25 mm (¼ to 1 in), of relatively small area. Force deflection characteristics are substantially linear up to a compression of 25%, after which stiffness increases substantially. Loading is therefore chosen so that a deflection of 25% is not exceeded on the individual pads.

Anti-vibration mounts

The natural frequency of a felt vibration insulation pad is determined primarily by its thickness with a minimum value of approximately 20 Hz for a 25 mm (1 in) thick pad. The relationship to thickness is not proportional and the actual natural frequency is also influenced by the density of the felt. With denser pads the natural frequency is increased and becomes more influenced by static load.

A figure of 30 Hz is normally adopted as the practical minimum natural frequency for felt vibration insulation pads. In practice this means that density is virtually restricted to frequencies above N2 x 30 or about 40 Hz. Inherent damping is, however, quite high and this is a useful feature for reducing the amplification of vibration at resonance. Typical vibration absorption characteristics are shown in Figure 1.

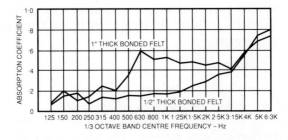

Fig. 1

The main limitation of felt vibration isolation pads is that the material is absorbent and can soak up oil and water. When so loaded its performance will be substantially reduced.

Filter elements

Natural felts are produced by compressing wool or hair or wool/hair mixtures, yielding a wide range of densities and permeabilities. A cross section through such materials would show successive layers of irregular mesh with an extremely tortuous path for filtering in depth. The random size of the pores makes it virtually impossible to establish even a nominal rating or specific resistance except by direct practical tests.

Modern felts are produced from synthetic fibres or mixtures of synthetic and natural fibres, with close control of manufacture to yield consistent densities, pore size, and mesh geometry so that cut-off performance is reasonably predictable. The structure of felts is considerably more open than that of papers, so that whilst filtering in greater depth (thickness), specific resistance is lower and high rates of flow can be achieved with smaller element areas and low pressure drop.

The wide variety of materials from which synthetic felts can be made has also considerably increased their use throughout the chemical and petrochemical industries, breweries, paint industry, food processing, plastics, and dyestuffs industries, etc, since the material can be chosen for compatibility and its mechanical properties controlled during manufacture. Synthetic materials used in the manufacture of felts include nylon, regenerated cellulose (rayon), polyethylene, polypropylene, acrylics and polyesters, some of which have general, and others specific, applications.

None of these is an absorbent material as such, so synthetic felts are, basically, mechanical filters. The same is true of metal felts. Natural felts, on the other hand, are hygroscopic and tend to absorb moisture as well as remove solid particles by screening and mechanical retention.

Felts are most commonly used as filter pads, area and thickness being selected on the basis of specific resistance as against flow rates required and acceptable pressure drop. With filter pads there is usually an optimum thickness for efficient filtering and any greater thickness is usually unnecessary, except for purely physical considerations such as ease of handling. Thinner felts, particularly synthetic felts are similar to cloths and may, in fact, be used as alternatives to cloths or other forms of filter element.

Metal felts

Metal fibres can be produced by drawing, machining and other techniques, and can be processed into felts by mechanical felting or slip casting. This process is generally followed by sintering to increase substantially the mechanical strength of the felt and develop uniform porosity. The sintered felt can then be rolled, pressed or formed into sheets or other shapes, including cylinders and tubes.

Porosity can range from about 10% up to 95% depending on the strength/permeability or strength/weight requirements. Multi-layer laminates can also be produced if necessary with graduated porosity from one side to the other.

Such materials are expensive to produce and have, as yet, relatively restricted application as commercial filter media. Their properties are, however, extremely attractive where high strength elements are required and where sintered metal elements may show limitations.

Fibre

HARD FIBRE is a vulcanised material consisting of cellulose fibres chemically gelatinised and subsequently bonded together into a homogeneous mass. Besides having high strength and rigidity it is practically inert to chemical attack, except by concentrated mineral acids. It is stable at normal temperatures but above 100°C (212°F) there is appreciable evaporation of water from the fibre which can make the material comparatively brittle if this temperature is maintained. Above about 175°C (350°F) fibre begins to char, and at higher temperatures decomposes.

TABLE 1 – PROPERTIES OF TYPICAL HARD FIBRES

Property		Regular, red, black and grey	Super
Tensile strength, with grain	lb/in²	13 000–15 000	16 000–17 500
	kg/cm²	910– 1050	1120– 1225
Tensile strength, across grain	lb/in²	7000– 9000	9500–10 500
	kg/cm²	490– 630	665– 735
Yield point, with grain	lb/in²	7800– 9000	
	kg/cm²	546– 630	
Yield point, across grain	lb/in²	4200– 5400	
	kg/cm²	294– 380	
Ultimate tensile strength with grain	lb/in²	21 400–29 500	
	kg/cm²		
Ultimate tensile strength across grain	lb/in²	16 000–25 500	
	kg/cm²	1500– 1785	
Ultimate shear strength	lb/in²	9700–12 900	
	kg/cm²	680– 840	
Compression strength: perpendicular to grain	lb/in²	27 000–38 500	
	kg/cm²	1890– 2700	
parallel to grain	lb/in²	8200–10 800	
	kg/cm²	575– 755	

Typical dielectric strength (electrical grade) 750 volts per mil thickness (other grades) 100 to 375 volts per mil, depending on thickness.

TABLE 2 – MACHINING FIBRE

Operation	Remarks
Turning	Recommended cutting speed 300 to 500 ft/min. High speed tools: no rake : 15 to 30° clearance. Fairly coarse feed is possible.
Drilling	High speed drills run at highest practical speed. Drills should have generous clearance. The holes produced by drilling fibre will tend to be slightly undersize.
Tapping	Tapping size holes 0.003–0.006 in larger than for tapping steel or brass.
Rough sawing	Bandsaws running at 4000 ft/min. Teeth should be ground straight across.
Smooth sawing	Hollow ground circular saws 120 teeth per inch for fibre up to ¼ inch thick; 80 teeth per inch for fibre over ¼ inch thick. Sawing speed 3000 rev/min (14 in saw).
Shearing	Knife shear for thicknesses up to ¹⁄₁₆ inch; powered square shear or rotary slitter for greater thicknesses.
Rough punching	Metal dies 0.001–0.008 in oversize on the hole size required.
Smooth punching	Smooth shaving dies used on rough punched or sawn stock. Progressive shaving necessary on thicker stock.

Hard fibre is available in various grades, typically coloured red, black or grey. Specific gravity may range from about 1.2 to 1.5, not necessarily dependent on colour (*eg* a 'typical' density for grey fibre is 1.19, but figures of 1.4 to 1.5 may apply with some grey grades). Typical mechanical strength figures are summarised in Table 1.

Fibre is tough and resilient, resistant to mechanical and thermal shock, and can be machined readily. For this high speed tools should be used. The material can also be formed and punched, sheared and sawn (see Table 2). The water content of fibre will vary with atmospheric humidity and for optimum machining properties fibre should be stored in a relatively damp atmosphere (*eg* 50 to

TABLE 3 – TYPICAL WATER ABSORPTION OF FIBRE
(% increase in weight on immersion in water at room temperature)

7 days	Thickness		After 1 hour	After 24 hours	After 7 days
	mm	in			
A	4.76	³⁄₁₆	3.49	14.8	44.4
	9.5	³⁄₈	2.13	12.0	40.5
	12.7	½	1.77	10.1	
B	3.7	⅛	6.84	22.9	44.6
	4.5	³⁄₁₆	3.30	13.7	41.2
	9.5	³⁄₈	2.04	11.4	39.7
	15.87	⅝	1.76	10.7	37.3

Key:
A – Regular grade
B – Superior grade

TABLE 4

Typical original properties† 1.6 mm (1/16 in)	General properties	Fuel and oil resistant	Special oil resistant	Wire reinforced	Oil and petrol resistant	Acid resistant	Moderate service quality	Medium general service quality
Minimum tensile strength	27.5 N/mm^2 (4000 lb/in^2)	27.5 N/mm^2 (4000 lb/in^2)	27.5 N/mm^2 (4000 lb/in^2)	—	27.5 N/mm^2 (4000 lb/in^2)	21 N/mm^2 (3000 lb/in^2)	21 N/mm^2 (3000 lb/in^2)	16.5 N/mm^2 (2400 lb/in^2)
Specific gravity	1.95	1.95	1.95	2.2	1.95	1.90	1.95	1.95
Compressibility BS	8%	9%	8%	8%	9%	9%	9%	7%
Compressibility ASTM	10%	11%	10%	50%	10%	12%	10%	9%
Recovery ASTM	55%	50%	55%	—	50%	45%	50%	60%
Soluble chloride DTD	—		0.02%					
Relaxation stress BS	30 N/mm^2 (305 bar)	29 N/mm^2 (290 bar)	32 N/mm^2 (327 bar)	30.5 N/mm^2 (312 bar)	30 N/mm^2 (305 bar)	—	26 N/mm^2 (260 bar)	23 N/mm^2 (230 bar)
Maximum service temperature	550°C (1000°F)	550°C (1000°F)	550°C (1000°F)	550°C (1000°F)	500°C (950°F)	150°C (300°F)	400°C (750°F)	260°C (500°F)
Typical thickness increase after fluid immersion								
ASTM oil 3 5 hours 150°C	20%	16%	10%	20%	8%		35%	40%
ASTM Fuel A 5 hours 20°C	8%	10%	7%	—	3%		12%	15%
ASTM Fuel B 5 hours 20°C	20%	16%	13%	20%	8%		30%	30%
Fuel DERD 2486 7 days 20°C	—	—	16%	—	—		—	—
Typical thickness increase after acid immersion								
96% sulphuric 18 hours 20°C						6%		—
95% nitric 18 hours 20°C						15%		—
50% nitric 1 hour 65°C						12%		

†'Klinger' and 'Klingerit' materials

60% relative humidity) at normal room temperature. Storage in extremely wet or extremely dry places should be avoided. Typical water absorption characteristics are summarised in Table 3.

Hard fibre stock is generally available in the form of sheets, rods and tubes. Flexible grades of fibre are also available (*eg* for washers, gaskets, joint packings, etc), generally produced by treating vulcanised fibre with an agent which attracts moisture and keeps the fibre soft and relatively pliable. Some typical properties of sheet fibre materials are given in Table 4.

Glass

WHERE A transparent material is required, glass and plastics are the only alternatives. The superior mechanical properties of glass make it virtually the only choice for numerous applications (*eg* where plastic deformation cannot be tolerated).

Glass is an amorphous (non-crystalline) material that is rigid at normal temperatures and softer or almost fluid at elevated temperatures with no definite melting point in between. At normal temperatures it does not plastically deform before failure, the stress-strain curve being a straight line up to the breaking point.

Glass, as a typical ceramic material, has a potential strength in tension comparable with that of steel although such figures are seldom, if ever, realised in practice. This is because the actual ultimate strength is widely influenced by the surface condition. Normal working and handling give rise to stress raisers which normally limit the ultimate tensile stress of commercial glasses to about 700 kg/cm^2 (10 000 lb/in^2), although figures in excess of 98 000 kg/cm^2 (140 0000 lb/in^2) have been achieved in practical tests on glass fibres. This strength figure is largely independent of the composition of the glass, most commercial glasses tending to have similar mechanical properties, although borosilicate glasses are generally better from this point of view since they resist scratching. This means that they are rather less subject to 'notching' and the accidental production of stress raisers than other types.

Types of glass

The principal constituent of glass is sand or silica. Silica itself cannot be melted at reasonable cost so various fluxes are used to lower the melting point for the production of commercial glasses (these fluxes also prevent the re-crystallisation of the silica on cooling). The type of glass is generally known under the name of the flux used.

Soda lime glass (or lime glass) – contains about 25% soda and lime as a fluxing agent, together with smaller proportions of magnesia and alumina. Cost is relatively low and soda lime glass is easy to hot work. It is a general purpose glass used where high resistance to heat or chemical stability is not required.

Soda lead glass (or lead glass) – contains about 68% silica (typically) with 15% lead oxide, the balance being soda, potash and lime. It is a higher cost glass with good hot workability, high electrical resistivity and high refractive indices. The lead content may be varied for specific applications (*eg* X-ray shielding).

Borosilicate glasses – contain about 80% silica with boron oxide as the main flux (14 to 15%) together with soda and alumina. They are characterised by low coefficients of thermal expansion, high heat-shock resistance, high chemical stability and excellent electrical resistivity.

Aluminosilicate glasses – are basically similar to borosilicate glasses with smaller proportions of boron oxide (or phosphorus pentoxide) and up to 20% or more of alumina. They can resist deformation up to much higher temperatures than borosilicate glasses.

Silica glasses – contain a very high proportion of silica (typically 96%) with the balance boron oxide and alumina. Their properties are comparable to those of fused silica (100% silica), but they are less expensive to produce (although considerably more expensive than borosilicate glasses).

Sheet and plate glass

Sheet glass for general purposes – also known as 'plain glass' or 'clear glass' is produced in three grades:

B or OQ – ordinary quality for general glazing duties.
A or SQ – special quality for superior glazing work.
AA or SSQ – specially selected quality for highest grade work where minimum distortion is required.

Plate glass has superior optical properties since the two surfaces are ground and polished to render them truly flat and parallel. Standard grades are:

GG – general glazing quality.
SG – selected quality for better class work (also for mirrors).
SQ – silvering quality for high grade mirrors, etc.

Sheet and plate glass may be specified by weight mg/cm^2 (oz/ft^2) or thickness (Tables 1A and 1B) though plate glass is normally specified by thickness only. Sheet glasses may be produced in fluted or patterned forms, also as coloured or opaque glasses, the latter in a variety of designs. 'One-way' glass is another form

TABLE 1A – WEIGHTS AND THICKNESSES OF SHEET GLASS

Weight	oz/ft^2	10	12	15	18	24	26	32	36
Nominal	in	$1/16$	$1/16$	$5/64$	$3/32$	$1/8$	$1/8$	$5/32$	$3/16$
thickness	mm	1.5	1.6	2	2.5	3	3.5	4	5

TABLE 1B – WEIGHTS AND THICKNESSES OF ROLLED PLATE GLASS

Approx weight oz/ft^2	25–30	40–50	50	60–70	75	100	125	150	175	200
Nominal thickness in	$^1/_8$–$^5/_{32}$	$^3/_{16}$–$^1/_4$	$^1/_4$†	$^5/_{16}$–$^3/_8$	$^3/_8$†	$^1/_2$†	$^5/_8$‡	$^3/_4$‡	$^7/_8$‡	1†
thickness mm	3–4	5–6.5	6.5	8–9.5	9.5	12.5	16	19	22	25

Key:
† Exact   ‡ Exact thickness may be specified.

of opaque glass, generally in the form of neutral tinted plate glass which has the appearance of a black mirror on one side. Other glasses are formulated to reduce the transmission of heat, glare, etc, or to enhance transmission characteristics (*eg* horticultural glass).

Glass can be tempered and toughened. Toughened glass is obtained by carefully controlled heating and cooling to set up differential tensile stresses at the centre of the material cross section and induce residual compressive stresses in the two outer surfaces. This renders the glass in a form much more resistant to tensile stress set up by shock loads. An alternative method of improving the resistance of glass to shock loads is by laminating separate sheets to an elastic core. Laminated construction is also used in the production of bullet-proof glass, when thicknesses may range up to 75 mm (3 inches) or more.

Light transmission

Transparent 6.5 mm (¼ inch) polished plate single glazing typically transmits about 90% of direct light and 85% of diffused light. Light transmission is reduced by surface imperfection, colouring, patterning, etc.

Typical performances of double-glazed windows with 6 mm leaves, covering clear glass and combinations of tinted and clear glass are:

Outer pane	Inner pane	Light transmission %	Solar heat transmission %
Clear	Clear	80	72–73
Green	Clear	67	50
Grey	Clear	40	50
Bronze	Clear	43–44	45

Sound transmission

The sound absorption of glass is only moderate, typical values for single glazing being:

Glass thickness mm	Average sound insulation dB
3	24
4	25
6	28
12	33

As a general rule the sound insulation of single glazing is increased by about 5 dB for each doubling of thickness (*ie* mass per unit area), although there are practical limits to the amount of sound reduction which can be achieved in this manner. Double glazing does not provide much more than 'mass per unit area' performance (*ie* similar to the effect of a single leaf of the same total glass thickness) unless the air space is substantial – *eg* a 200 mm (4 in) air space is considered a practical compromise in this respect (compared with the 3 to 5 mm air space which may be used in conventional double glazing).

Thermal transmission

The thermal transmission or 'U' value of typical single glazing is about 1.0 Btu/ft^2/h/°F or 5.7 W/m^2/h/°C. This can be reduced by as much as 50% by sealed unit double glazing. As far as the 'U' value is concerned there is a reduction of 'U' with increasing air gap up to about 19 mm (¾ inch), after which the 'U' value remains substantially constant (see Table 2).

Mechanical properties of glass

The significant mechanical properties of glasses are virtually limited to tensile strength, hardness and elasticity. Taking the ultimate limit for the former as 700 kg/cm^2 (10 000 lb/in^2), the normal maximum working stress adopted for design purposes is 70 kg/cm^2 (1000 lb/in^2), this allowing an adequate factor of safety for most purposes. For heat treated glasses figures of 140 to 280 kg/cm^2 (2000 to

TABLE 2 – EFFECT OF AIR GAP DOUBLE GLAZED UNITS WITH TYPICAL (4.5 mm) GLASS THICKNESS

Air gap		'U' value	
in	mm	Btu/ft^2 /hr/°F	W/m^2 /hr/°C
⅛	3.0	0.69	3.93
³⁄₁₆	4.5	0.62	3.53
¼	6.5	0.58	3.31
⅜	9.0	0.54	3.08
½	12.5	0.52	2.96
¾	19.0	0.50	2.85
over ¾	over 19.0	0.50	2.85

TABLE 3 – WIND PRESSURES (ON A FREE STANDING WALL)

Velocity km/h	Pressure N/m²	Velocity km/h	Pressure N/m²	Velocity km/h	Pressure N/m²	Velocity km/h	Pressure N/m²
2	0.2	22	26.9	50	139	150	1249
4	0.9	24	32.0	60	200	160	1421
6	2.0	26	37.5	70	272	170	1604
8	3.6	28	43.5	80	355	180	1798
10	5.6	30	50.0	90	450	190	2004
12	8.0	32	56.8	100	555	200	2220
14	10.9	34	64.2	110	672	225	2810
16	14.2	36	71.9	120	799	250	3469
18	18.0	38	80.1	130	938	275	4197
20	22.2	40	88.8	140	1088	300	4995

4000 lb/in²) can be used, provided these are approved by the manufacturers of the glasses concerned. It is normally safe, however, to adopt the lower figures for all commercial glasses, subject to a limit on service temperature.

Safe glazing sizes and suitable glass substance (ie thicknesses) for glazing are normally calculated on wind pressure loads. Typical wind pressure figures adopted are given in Table 3. The aspect ratio of the glass panel is of significance and is normally determined as a *glass factor* defined as the ratio of superficial area to perimeter.

Hardness of glass

The hardness of glasses cannot be measured on the scales common to metals, which are based on indentation tests. Instead such data are based on scratch or abrasion tests and are therefore largely arbitrary. Scratch tests are related to Moh's scale of hardness (a mineral scale which dates from about 1820) and average figures lie between 5 (apatite) and 7 (quartz). Abrasion tests are conducted by sandblasting specimens and comparing the extent of mechanical abrasion on the surface, ie evaluating their respective resistance to abrasion. Whilst such tests are rather more scientific than determining the Moh's hardness number they are essentially relative. Typical practice is to give the abrasion resistance of soda lime plate glass an arbitrary value of unity and evaluate the hardness of other glasses on performance accordingly, as in Table 4.

TABLE 4 – IMPACT ABRASION RESISTANCE (*ie* 'hardness under sandblasting')

Type of glass	Relative hardness
Borosilicate	4.1 to 3.1
96% silica	3.5
Hard lime	2.0
Soda lime	1.2
High lead	0.6

Modulus of elasticity

Since glass under stress shows no yield point or plastic deformation it can be considered to be perfectly elastic up to the point of fracture, the average value for the modulus of elasticity being about 630 000 to 700 000 kg/cm^2 (9 000 000 to 10 000 000 lb/in^2). Unlike tensile strength, however, this can vary with composition, and both higher and lower figures are possible – hard lime glasses generally being higher and high lead glasses lower, for instance. Failing information to the contrary the average figures given may be adopted for most commercial glasses with no great possibility of error. The elastic modulus is significant in mechanical analysis, since it is one of the factors determining the thermal stresses produced in a panel of glass subject to a temperature differential. The other major factor in such cases, apart from the temperature differential itself, is the coefficient of linear thermal expansion. This is largely influenced by the composition of the glass and can vary from 10^{-5} per °C down to as low as 8 x 10^{-7}. Some typical data are summarised in Table 5.

TABLE 5 – COEFFICIENTS OF LINEAR THERMAL EXPANSION (per deg C)

Type of glass	Coefficient of expansion
High lead	91×10^{-7}
Potash soda lead	$90–92 \times 10^{-7}$
Soda lime (lamp bulbs)	92×10^{-7}
Soft green	90×10^{-7}
Soft red	91×10^{-7}
Potash soda lead (tube)	84×10^{-7}
Soda lime (average)	82×10^{-7}
Opal (average)	$70–80 \times 10^{-7}$
Borosilicate (gauge glass)	67×10^{-7}
Borosilicate (general)	$35–45 \times 10^{-7}$
Borosilicate (electrical)	34×10^{-7}
Silica (clear)	8×10^{-7}
Silica (opaque)	8×10^{-7}
Sealing glass	40×10^{-7}

Steady state thermal stresses

Glass subject to different temperatures on both surfaces (*eg* observation windows in furnaces, etc) is under a state of steady thermal stress, the hot surface expanding and pulling the cooler surface into tension. The magnitude of this stress increases in direct proportion to the coefficient of temperature differential. Glass thickness is significant only in that it can affect the temperature differential (*ie* the greater the thickness the greater the temperature differential).

The steady state thermal stress can be expressed mathematically as:

$$S \text{ max} = \frac{\infty E \Delta T}{2(1 - \phi)}$$

where:

S max = maximum stress (tension on the cooler surface,
             compression on the hotter surface)
∞ = coefficient of linear expansion
E   = modulus of elasticity
ΔT = temperature differential between glass surfaces (°C)
ϕ = Poisson's ratio (typically 0.22).

Adopting the nominal safe maximum working stress figure for glass of 70 kg/cm^2 (1000 lb/in^2), this formula can be re-written:

$$\Delta T \max = \frac{2000 (1 - \phi)}{\infty E}$$

where:

   ΔT max = maximum permissible temperature differential
        (*ie* not to exceed a working stress of 1000 lb/in^2).

Steady state thermal stresses are produced when any glass article is heated or cooled, but will eventually disappear when the article attains a uniform higher or lower temperature. Shock stresses are similar temporary stresses, but precipitated by sudden cooling (or heating, although shock stresses are generally related to sudden cooling). Shock stressing is considered separately from transient stressing since it is normally determined on empirical lines, such as by plunging a sample into cold water after oven heating. Thermal shock resistance would then be quoted as a temperature rating for a standard size and each thickness (see also Table 6). This rating refers to the maximum temperature differential

**TABLE 6 – TEMPERATURE DIFFERENTIAL TO STRESS PLATE OF TUBE TO
70 kg/cm^2 (1000 lb/in^2)**

Type	Form	Temperature differential °C
Silica	All forms	200
Borosilicate	Tube, pressed	20
	or blown	40 to 45
Borosilicate	Electrical	70
High lead	All forms	22
Opal	Sheet and formed	20 (average)
Potash soda lead	Tube	19
	Moulded	17
Soda lime	All forms	17 to 20
Hard lime	Moulded	29
Soft green	Sheet	17
Hard green	Tube	39
Soft red	Tube	17
Hard red	Sheet	36

achieved without breaking, *eg* a resistance of 65°C would mean that the specimen could be heated to 80°C and quenched in water at 15°C without breaking. Alternatively thermal shock properties may be determined according to an exact service specification. It might be mentioned as a matter of interest that quenching in a water bath is a more drastic test than chilling with a blast of cold air at the same temperature.

Transient thermal stresses produced by cooling are higher than those produced by heating, the difference becoming more marked as the cooling, or heating process is accelerated. In cooling, the stresses are set up by the chilling of the surface resulting in contraction and the build-up of tensile stresses, resisted by compressive stresses in the hotter centre zone of the glass. On heating, the tensile stresses are concentrated in the centre, resisted by compressive stresses in the surface zones. Since glass fractures only as a result of tensile stresses it is obvious that the former case is the more severe. The degree of stress will also depend on the thickness of the subject which affects the temperature differential, the thermal expansion coefficient and the shape of the subject, which may lead to abrupt changes in stress concentration.

Strength of glass

The normal strength of glass is 70 kg/cm^2 (1000 lb/in^2), which is a normal design load figure. In the case of *annealed* glass, the material will break when the surface tensile stress exceeds 350 kg/cm^2 (5000 lb/in^2). The corresponding figure for surface tensile stress for *tempered* glass is 700 kg/cm^2 (10 000 lb/in^2).

Resistance to corrosion

Hydrochloric acid and hot concentrated phosphoric acid are the only substances which attack glass vigorously. Cold alkaline solutions attack glass very slowly, but the rate of attack increases with increasing temperature. Borosilicate and high silica glasses are the least affected.

Glass is also attacked by superheated water (*ie* above 149°C), this attack being accelerated if the water is alkaline. The rate of attack is seldom serious enough to prevent satisfactory life of boiler gauge glasses, etc.

Viscosity

Glass is sometimes described as a supercooled liquid with an infinitely high viscosity at normal temperatures. At higher temperatures glass becomes increasingly plastic with a real viscosity with four significant viscosity levels.

Viscosity circa 10^{15} *poises* – corresponds to the *strain point* or the highest temperature from which glass can be cooled without becoming permanently strained (unless cooling is very carefully controlled). (Temperatures 380 to 1060°C, depending on the type of glass.)

Viscosity circa 10^{14} *poises* – corresponds to the *annealing point* or the temperature at which internal stresses are quickly relieved. (Temperatures 400 to 1100°C, depending on the type of glass).

TABLE 7 – PHYSICAL PROPERTIES OF VARIOUS TYPES OF GLASS

	Soda lime	Soda lead	Boro-silicate	Alumino-silicate	96% silica	Fused silica
Young's modulus (10^6 lb/in^2)	10.0	9.0	9.5	12.8	9.6	10.5
Poisson's ratio	0.24	0.34	0.20	0.26	0.18	0.17
Specific gravity	2.47	2.85	2.23	2.53	2.18	2.20
Linear coefficient of thermal expansion (per degC)	92×10^{-7}	91×10^{-7}	32.5×10^{-7}	46×10^{-7}	8×10^{-7}	5.6×10^{-7}
Working temperature (°C)						
Annealed glass						
Normal	110	110	230	200	800	900
Maximum	460	380	490	650	1100	1200
Tempered glass						
Normal	220	—	260	400	—	—
Maximum	250	—	290	450	—	—
Volume resistivity (ASTM D257 ohm-cm)						
at 250°C	2.5×10^6	8×10^8	1.3×10^8	3.2×10^{13}	5×10^9	1.6×10^{12}
at 350°C	1.3×10^5	1×10^7	4×10^6	2×10^{11}	1.5×10^8	2.5×10^{10}
Dielectric constant at 1 Mc, 20°C	7.2	6.7	4.6	6.3	3.8	3.8
Loss factor at 1 Mc, 20°C	0.065	0.01	0.026	0.01	0.0019	0.000038

Design and Components in Engineering, May 28, 1964.

Viscosity circa 10^8 *poises* – corresponds to the *softening point,* where glass begins to deform under its own weight. (Temperatures 630 to 1650°C, depending on type of glass.)

Viscosity 10^4 *poises* – corresponds to the *working point* – ie the point at which glass is soft enough for hot working by most of the common methods. (Temperatures 1000 to 1200°, depending on the type of glass.)

Selection of glass

Soda lime glasses – window glass, bottles, lamp envelopes, electrical insulators.

Lead glasses – crystal tableware, electric light bulb stems, TV tubes, neon sign tubing, optical components.

Borosolicate glasses – boiler gauge glasses, glass pipe, electrical insulators, chemical laboratory glassware, baking dishes and ovenware.

High silica glass – high temperature applications.

Forming glass

Glass is produced by melting in pots or tanks. Primary forming is then achieved by rolling, drawing, pressing or blowing. Secondary forming may be by localised heating (lamp-working), re-pressing, sagging, shrinking or sintering. All these involve working the glass in a hot state. Cold working is limited to cutting, grinding and polishing, drilling (the latter normally avoided if possible) and mechanical or chemical surface treatment – *eg* etching.

Blowing is the technique used for producing many types of hollow ware, the basic methods being:

(i)   *Freehand blowing* – for limited numbers of individual articles.
(ii) *Paste mould blowing* – for mass-produced small and medium size articles of circular section requiring a fine surface finish.
(iii) *Hot iron mould blowing* – for non-circular and/or larger articles.
(iv) *Press and blow moulding* – for low cost automated production of all sizes and shapes of articles.

Pressing is an alternative technique for producing both hollow ware and solid forms. The technique is similar to press and blow moulding, a gob of hot glass being pressed in a mould, but without blowing. The three basic types of mould used are:

Block moulds – for low cost production of hollow ware shapes.

Split moulds – for shapes which cannot be withdrawn from a block mould after pressing.

Font moulds – for solid shapes.

With both *blowing* and *pressing* the shapes and sections involved must be designed to provide optimum distribution of glass when formed as this will affect the speed at which the glass takes its shape and the heat transfer necessary before setting.

Glass tubes

Glass tube is produced in a variety of diameter sizes and wall weights (wall thicknesses). Tubing of 50 mm (2 inch) diameter and above is generally classified as *glass cylinder*. Stock sizes in all cases are normally made from standard flint glass.

The internal pressure which glass tubes or cylinders can withstand can be calculated from the Barlow formula for thin-walled tubes, *ie*:

$$P = \frac{2\,St}{d}$$

where: S = tensile strength
         t = tube wall thickness
         d = inside diameter

Working formula:

$$P \text{ (lb/in}^2) = \frac{2000t}{d}$$

where: t and d are in inches

$$P \text{ (bar)} = \frac{70t}{d}$$

where: t and d are in millimetres

Electrical conducting glass

Glass can be rendered electrically conductive by depositing a thin transparent film of metallic oxide on its surface, or on one or more surfaces of a laminated unit.

Electrical conducting glass may be used for high humidity conditions where elimination of condensation is important, the glass being heated by the application of an electric current. For specific applications of this type, laminated glass embodying a metallic foil heating element may be preferred. Another application of electrical conducting glass is for the elimination of static from glass surfaces.

Foamed glass

Foamed glass has a particular application as a heat insulator. It has the advantage of being a rigid cellular material with good compressive strength (approximately 7 to 9 kg/cm^2 (100 to 125 lb/in^2), but depending on density). It can be fabricated in the form of rigid blocks of varying sizes and thicknesses, bevelled lags and shaped segments, and curved segments (*eg* for lagging cylindrical vessels, pipework etc).

Typical properties of a proprietary form (Foamglas) are given in Table 8 and Figure 1.

TABLE 8 – PROPERTIES OF FOAMGLAS

Absorption	0.2% by volume on 2 × 12 × 18 inch block (all on surfaces only).
Acid resistance	Impervious to common acids and acid fumes.
Alkalinity	ph = 7.5
Capillarity	0
Coefficient of expansion	0.0000046/°F
Combustibility	Incombustible.
Composition	A true glass; completely inorganic.
Compressive strength	100 lb/in^2
Density	9 lb/ft^3
Flexural strength	75 lb/in^2
Hygroscopicity	No increase in weight in 246 days in air at 90% relative humidity.
Modulus of elasticity	180 000 lb/in^2
Specific gravity	0.144
Specific heat	0.20 Btu/lb
Thermal conductivity at	
0°F	0.35 Btu/h/ft^2 /°F/in
50°F	0.38 Btu/h/ft^2 /°F/in
300°F	0.55 Btu/h/ft^2 /°F/in
Thermal diffusivity	0.42 ft^2 per day

Note: Values are average for design purposes and where applicable are based on density of 9 lb/ft^3. Density varies from 8 to 10 lb/ft^3.

Glass fibre

Glass fibre is mostly used as a reinforcement material in GRP mouldings, etc,
rendered in the form of chopped strand mat, woven rovings, cloth, tape and tissue. The fibres are formed by drawing or blowing to produce continuous filaments or staple fibres respectively (see Figure 2).

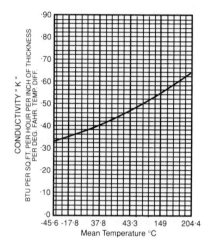

Fig. 1
*Variation of thermal conductivity with
temperature for 'Foamglas'.*

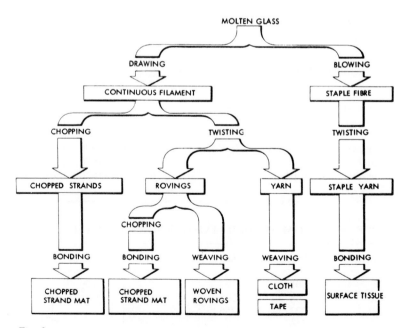

Fig. 2

Filament sizes may range from about 0.0003 to 0.0025 inch, generally classifiable as:

Fine – 0.0127 to 0.01778 mm   (0.0005 to 0.0007 in)
Medium – 0.01778 to 0.0254 mm   (0.0007 to 0.0011 in)
Coarse – 0.0254 to 0.05 mm   (0.0011 to 0.0018 in)

The ultimate strength of a glass filament can be as high as 35000 bar (500 000 lb/in^2) or ten times the strength of steel, although the full strength cannot be realised in practical applications. Another feature of the fine filament form of glass is that it is more flexible (less brittle) than substantial glass forms.

Glass staple fibres are also used to produce glass wool for insulation purposes, although this has largely been replaced for such duties by chopped strand mat.

Glass fibre fabrics may be used in their loom state, or may be chemically treated for the specific application. Chemical treatment is preceded by treatment to remove the protective starch based size (also necessary to avoid incompatibility with thermosetting resins used in GRP fabrications). The most common method of removing size is by batch heat cleaning, heating the fabric to burn off the size at carefully controlled temperatures. Alternatively aqueous de-sizing may be employed, using a solution of various enzymes and wetting agents (see Table 9).

Chemical finishes may then be applied to improve the bonding characteristics with various resins (see Table 10), or for other purposes. The latter include chemical treatment to improve the performance of fabric used at high temperatures (eg high temperature filter cloths) or to stabilise the weave for ease of handling the material. An example of the latter is bitumen or coal tar treatment for roofing membranes and pipe wraps. When treated with silicones or coated with

TABLE 9 – EXAMPLES OF DE-SIZING AND CHEMICAL TREATMENTS

Code	Finish	BSS
165	Heat cleaned	BS 3396 – 1966 – Part 2, Level H
196	Amino silane	
205	Methacrylato chromic chloride Volan type	BS 3396 – 1970 – Part 3, Grade C
208	Vinyl silane	BS 3396 – 1970 – Part 3, Grade S
223	Methacrylato propoxy silane	BS 3396 – 1970 – Part 3, Grade S
256	Epoxy silane	

TABLE 10 – RECOMMENDED FINISHES FOR LAMINATING WORK

Resin system	Polyester	Epoxy	Melamine	Phenolic	Silicon
Code (see Table 8)	205 208 223 256	196 205 256	196	196	165

TABLE 11 – TENSILE PROPERTIES OF GLASS FIBRES COMPARED WITH ALUMINIUM AND STEEL

Material	Specific gravity	Tensile strength		Specific tensile strength		Tensile modulus		Specific tensile modulus	
		lb/in^2 $\times$ 10^3	GN/m^2	lb/in^2 $\times$ 10^3	GN/m^2	lb/in^2 $\times$ 10^3	GN/m^2	lb/in^2 $\times$ 10^3	GN/m^2
Glass fibre 'E' glass	2.5	500	3.450	200	1.380	9	62	3.6	24.1
Aluminium	2.8	67	0.462	24	0.165	10.5	72	3.8	25.7
Steel	7.8	145	0.999	19	0.128	30	207	2.8	26.2

Please note these tables are to be acknowledged to Fothergil & Harvey Ltd (8, 9, 10)

elastomers, glass fabrics can be used up to 290°C, although a maximum service temperature of 200 to 260°C is recommended if the fabric is subject to stressing or flexing. As insulation wraps or mattresses, glass fabrics are used at temperatures up to 650°C.

See also chapter on *Glass Reinforced Plastics* (Section 6).

Types of weave and construction

Fabric constructions describe the weave, yarn count and number of warp and weft yarns per unit length of fabric. Tensile strengths, weight per unit area and fabric thickness are achieved by the correct choice of fabric construction. The choice of cloth construction should be related to the ultimate application, produce performance being directly influenced by balancing or concentrating fabric tensile strengths in either warp or weft directions during manufacture. In the case of laminates, more uniform mouldings have been produced and manufacturing procedures improved where specially developed drape fabrics have been introduced. Once more, the type of weave is the governing factor in achieving the good 'drape' characteristics necessary in double curvature and deep draw mouldings. Where double or compound curvature is involved, a satin weave or twill will follow the contour far more closely than a plain weave. Because of their loose construction, satins and twills tend to wet out more quickly than tight plain weaves.

Satins also tend to produce higher strengths in laminates than do plain weaves due to the lower percentage of yarn crimp in the fabric created by yarn interlacing. Open construction satins or twills would, however, be impractical as they would be far too 'sleazy' for practical handling.

Plain weave

Structurally each warp and weft thread passes over one end or pick and under the next (Figure 3). Some fabrics in plain weave have good resistance to distortion and give reproducible reinforced plastic laminate thickness. Because of

Fig. 3

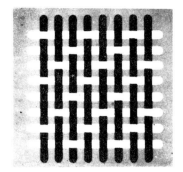

Fig. 4

these attributes, they are widely used as general reinforcements, and in contour or three-dimensional weaving of shapes. Loosely woven scrim fabrics are generally plain weave.

Twill weave

Good drapability and somewhat improved laminate strengths are important characteristics of twill weave 'Tyglas'; the number of warp ends and weft picks which pass over each other can be varied, *eg* 2 x 1 twill; the weft passes under two warp ends and then over one warp end; the interlacing is arranged regularly, forming a distinct diagonal line on the cloth surface (Figure 4).

Satin weave

Excellent drapability, smooth surface, minimum thickness and high laminate tensile and flexural strengths are characteristic of satin weave fabrics. The number of ends and picks which pass over each other before interlacing is greater than with twill, and the interlacing is always with one crossing thread. Consequently, one side of the fabric is comprised mainly of warp yarns and the other of weft yarns (Figure 5).

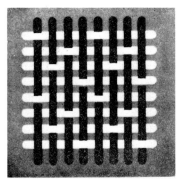

Fig. 5

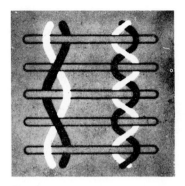

Fig. 6

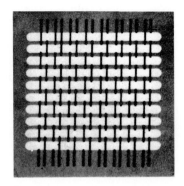

Fig. 7

Leno or gauze

This type of weave is produced by constantly changing the relative position of adjacent warp yarns so that they are twisted between each pick or weft yarn (Figure 6). Gauze weave imparts stability to very open fabrics.

Unidirectional weave

In this fabric tensile strength is directed one way, either warp or weft, with low strength the opposite way (Figure 7). These fabrics can be cross-plied to obtain maximum strength in any direction.

Contour weaving

Contour or 3D (three dimensional) weaving is a process which enables reinforced plastics fabricators to produce components of improved performance.

Conventional lay-up for a complex moulding such as a radome previously consisted of a number of strips or gores cut from flat material which were butted together on a mould. A double curvature shape can be produced by this method but the labour content is very high and identical component reproduction is difficult to achieve. Also, overlapping of the strips or gores affects the microwave performance of the laminate, a condition which is unacceptable in certain aircraft applications.

By contrast, using the 3D weaving technique, shapes with changing configuration can be produced with complete accuracy. Labour costs are considerably reduced and the finished products have a far higher degree of uniformity.

Woven rovings

The term 'woven roving' implies a fabric made from heavy untwisted glass yarns having a resin compatible surface dressing or 'size'. Woven rovings are often used in composite laminates with chopped strand mat, the latter being used to increase the bulk of the laminate. Rovings wet out far more quickly than certain types of mat, pick up a lower percentage of resin, approximately 50% glass, 50% resin, and have far higher physical properties.

A roving laminate, thickness for thickness, is much stronger than all mat laminate and thinner, lighter and less expensive laminate can often be made.

Non-metallic bearing materials

NON-METALLIC materials which are used for bearings include the following:

(i) Carbon and carbon-graphite.
(ii) Rubber.
(iii) Thermoplastics (filled and unfilled).
(iv) Laminated thermoset plastics.
(v) Wood.
(vi) Cermets.
(vii) Jewel bearings.

Carbon bearings

The greatest and best known advantage of carbon is that it needs no lubrication and that it does not seize on to metals. Other properties provide additional advantages for particular bearing applications, or influence the loading factors and methods of fitting which are usually different from ordinary bearing practice – *eg* carbon has a considerable polishing effect on the shaft against which it rubs, resulting in a marked reduction of the rate of wear.

Where normal bearing materials serve the purpose, there are seldom advantages in using carbon. Carbon is used where oil or grease lubricated bearings could be used only with difficulty, if at all. Carbon can run dry or in fluids which are corrosive or unsuitable as lubricants for other bearing materials.

The three principal fields are:

(a) Where contamination is undesirable, as in textile and food machinery, *ie* by avoiding the use of oil or grease.

(b) In furnace and boiler equipment where temperatures are too high for the use of conventional lubricants.

(c) Where the bearings are immersed in liquids such as hot and cold water, sea water, acidic or alkaline solutions or oil solvents such as petrol and benzine.

Carbon bearings run satisfactorily under limited conditions which are, how-ever, so varied that it is impracticable to give maximum load and speed figures which would be universally applicable.

Usually in dry running the maximum permissible load depends on the wear rate that can be allowed. The maximum permissible speed is determined by local heating at the rubbing surface; if surface spot temperature becomes extremely high the carbon may deteriorate too rapidly.

Carbon grades employed as dry bearings are chiefly graphite-carbon and metal-carbon grades. In some special instances electro-graphite may be used.

Carbon and graphite-carbon and electro-graphite

To give a long life with continuous running (ie steady load over long periods) the product PV should not exceed 550 (0.2 metric); P should not exceed about 100 lb/in^2 (7 kg/cm^2) and V about 200 ft/min (1 m/s). Loads above 100 lb/in^2 (7 kg/cm^2) need careful consideration and may not always permit the maximum value of PV. In conditions where short period life (three to six months) is eco-nomically permissible, a PV of 3000 (1.07 metric) can be sustained or even ex-ceeded (see also Table 1).†

TABLE 1 – PV VALUES FOR TYPICAL CARBON BEARINGS

Grade	Continuous dry operation				Continuous lubricated operation		
	P max	V max	PV max		P max	V max	PV max
			Normal	Reduced life			
Plain carbon	100	200	550	3 000	300	200	up to 10^6
Metal-carbons	400*	200	1 500	4 000	400	200–300	up to 10^6
Electro-graphitic	100	200	500	3 000	250	200	up to 10^6

*At low speeds only

Metal-carbon

The strength of metal-carbon grades permits much higher maximum loads but the limit for speed remains unchanged. The load can reach 400 lb/in^2 (28 kg/cm^2) provided the speed is very low, about, 0.1 to 0.2 ft/min (0.0005 to 0.001 m/s). For long period running maximum PV = 1500 (0.54 metric). For short period running maximum PV = 5000 (1.8 metric).

Considerably greater values for PV can be safely used for dry running when the operation is intermittent and especially when it is of short duration with fairly long stationary periods. This applies to all carbon grades.

The maximum operating temperature for carbon bearings is usually about 350°C (but may be lower for some grades), or up to 500°C with reduced loads. In a reducing atmosphere carbon bearings may be run at temperatures from 500 to 1000°C, depending on grade (see also Table 1).†

†PV = P d bearing pressure, lb/in^2
V = rubbing speed, ft/min

TABLE 2 – TEMPERATURE RATINGS FOR TYPICAL CARBON BEARINGS

Grade	Types	Maximum operating temperature
Plain carbons	All	350°C (500–1000°C in reducing atmospheres)
Metal-carbons	Copper Anti-friction metal Lead–bronze	350°C 130°C 350°C
Electro-graphitic	All	450–500°C (500–1000°C in reducing atmospheres).

Dynamic friction

With dry running, the coefficient of dynamic friction depends on the materials and the condition of the surfaces as well as on the operating and loading conditions.

For the general range of carbon grades the coefficient of friction may be 0.25 for heavy loads and diminish to 0.10 for light loads. A further important fact is that the friction tends to diminish with continued and prolonged operation because of the polishing action of the carbon.

With limited lubrication, where the lubricant separating the bearing and shaft exists only as an adsorbed film, boundary lubrication prevails and the coefficient of friction of carbon bearings is variable, although in practice it usually lies within the range 0.01 to 0.10 for heavy loads.

With complete fluid lubrication the hydrostatic pressure is sufficient to support the applied load. The friction varies directly as the viscosity and for thin lubricants, such as petrol, is very small. It also depends on the clearance and ratio of the bearing length to diameter as well as on the load and speed.

The performance of carbon bearings is affected by the surface condition of the carbon which should be impervious.

The shaft surface and finish should be of a high order, and the shaft alignment be as perfect as possible. The bearing should preferably be self-aligning.

Carbon bearings have in several instances been successfully employed at very high speeds with air lubrication. Compressed air at 10 to 100 lb/in^2 is fed to a position half-way along the bearing through a series of small holes. Impervious metal-carbon material is necessary for this application, together with very high surface finishes on the bearing bore and shaft journal.

Suitable materials for the journal are hardened nitrided steel or 'Stellite' puddled on to carbon steel or stainless steel.

Wear

Wear is the most important limiting factor in unlubricated carbon bearings. It is mechanical and thermal in origin.

The shaft surface is of considerable importance and a finely finished and non-corrosive surface is recommended. The finish of the bearing bore is of less importance because the carbon is softer and wears more readily to the same smoothness as the shaft.

The formation or presence of rust on the journal is harmful and should be avoided. Rust is abrasive and can appreciably reduce the bearing life. On the other hand, the fine film of graphite developed on shaft surfaces has a beneficial lubricant effect and lengthens the bearing life.

Broadly speaking, with continuous operation the wear is directly proportional to the load, to the temperature and to the square of the speed. Under various operating conditions the carbon materials generally compare as follows:

Low loads, 0 to 20 lb/in^2 (0 to 1.4 kg/cm^2) – most carbon grades are satisfactory.

Medium loads, 20 to 400 lb/in^2 (1.4 to 28 kg/cm^2), speeds to suit – metal carbon grades wear less than non-metal grades.

Heavy loads up to 800 lb/in^2 (56 kg/cm^2) – suitable for intermittent duty only.

Wet operation

Under proper conditions and fully lubricated, wear is negligible. The construction may sometimes be designed to make fuller use of the working fluid as a lubricant, thereby permitting very high loads and speeds. For example, on some submersible pumps a Mitchell thrust bearing design is highly successful with the water-lubricated Mitchell pads made in a metal-carbon grade.

However, care should be taken to ensure an adequate constant supply of fluid to the bearing. A condition of intermittent fluid supply, possibly involving a cyclic wet to dry running, should be avoided. The transition stage from wet to dry will cause temporary high friction and possible high rate of wear.

Rubber bearings

Soft rubber is used as a bearing surface in a variety of fluid pumps and submerged (water-lubricated) bearings, and possesses certain specific advantages over hard surfaced bearings having a frictional coefficient comparable with that of normal ball bearings. One of the requirements for satisfactory operation is that the rubber surfaces must be flooded with lubricant (usually water), as otherwise friction is high and deterioration of the bearings rapid.

Rubber bearings are manufactured in a variety of sizes, although all follow a similar pattern. The bearing material is vulcanised rubber, either in the form of a lining inside a metal shell or in longitudinal strips consisting of rubber facings bonded to metal backing strips and assembled in dovetailed slots in the bearing housing. In the former case the bearing lining is moulded as a complete unit, this construction being typical of the smaller types. Segmental rubber bearings are common in larger installations where a moulded lining may prove impractical or even unsatisfactory.

Both natural and synthetic rubbers may be used for the bearing material, the synthetic type being almost universal, and the specification is, to a certain extent, adjustable to meet specific service conditions. The normal specification is a grease and oil resistant rubber of soft resilient nature, but which also combines

toughness and wear resistance. Anti-oxidants are incorporated to give good ageing properties. The metal sheet or metal backing strip (usually brass) is bonded
by vulcanisation and this bond is of about the strength as the rubber itself.

Rubber bearings, either moulded or segmental, invariably have longitudinal
grooves running the length of the bearing. The purpose of these is to allow an
adequate flow of lubricant through the bearing and at the same time carry away
any abrasive particles which have worked into the bearings. Under the turning
action of the shaft such particles will be depressed into the soft rubber bearing
surfaces and thence rolled into the trough to be washed away. One of the great
advantages of a rubber bearing over hard surface bearings operating under similar conditions is that shaft scoring and bearing wear are minimised. Abrasive
particles cannot be ground into the shaft since they are backed up only by the
resilient rubber.

The flow of coolant required depends very largely on the type and size of
bearing and its operating conditions. A continuous flow of two gallons (9.1
litres) of water per minute per inch (25.4 mm) of bearing diameter is recommended as a rough working rule. If the bearing is working submerged in water
contaminated with sands, grits, etc, it may be necessary to supply it with cleaner
water under pressure to keep the grooves from clogging and convey the abrasives through the bearing. Immersion in water may not be enough, for good cir-
culation of lubricant is essential for proper cooling and to keep the bearing
surfaces wet. It is also necessary to ensure that the bearing is wet before starting
up.

As far as maximum operating temperature is concerned, a flow of cooling
water of the order specified should restrict temperature rises to about 3°C (maximum). Substantially higher temperature rises can be tolerated and, in any case,
will reach a stable value for any particular installation. A maximum operating
temperature of 70°C is usually quoted. If necessary, coolant flow can be adjusted
to arrive at any particular running temperature.

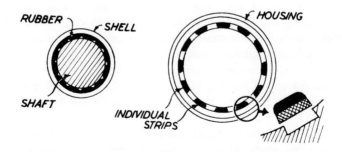

Fig. 1
Typical form of rubber bearing.

No particular limits have been determined for high operating speeds with rubber bearings, speeds in excess of 12 000 r/min or over 4000 ft/min surface speed being readily obtainable. The resilient rubber bearing, too, has one particular advantage for high speed operation, in that it will deflect and allow a shaft to rotate about its true centre of gyration if this is slightly different from its geometric centre, thus tending to compensate for eccentricity and reducing the dynamic stresses in the bearing, and vibration level in the whole machine. On known installations this has proved a particular advantage for shafting operated above 'whirling speed', where the damping action of the bearing as the shaft is speeded up through this critical speed zone is quite notable. The nature of the bearing, in fact, resists shaft oscillation.

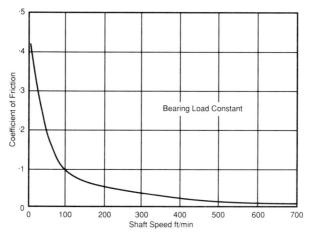

Fig. 2
Typical rubber bearing frictional
coefficient plotted against shaft speed.

For average installations, the maximum bearing load is normally taken as 35 lb/in² of projected bearing area. Higher loadings can be used, particularly for continuous operating with high speed shafting. Lighter loading is recommended for intermittent operation, for shafts running at extremely low speeds or when the bearing is operating under extremely abrasive conditions.

Deflection of the whole rubber bearing under load is small and almost directly proportional to the actual bearing load. Thus deflection at 40 lb/in² loading is roughly twice that at 20 lb/in², and so on. Approximate figures for 35 lb/in² loading are 0.064 in deflection per inch of maximum rubber thickness.

The coefficient of friction is largely independent of bearing loading, but tends to decrease with increasing speed. In a good rubber bearing the average values lie between 0.001 and 0.005, the latter corresponding to low speed running. At extremely low speeds the frictional coefficient tends to rise sharply.

In the moulded shell type of bearings, the shells may be of stainless steel, steel, aluminium, bronze, cast iron, Monel, etc, depending on the service requirements. Design features are capable of numerous variations. The appearance of the bearing is that of a cylindrical shell with a fluted (rubber) lining, the flutes, or water grooves, running from end to end. In some bearings of this type a circumferential groove is formed in the lining, through which water can be fed in (or out) to ensure that there is water at every point of contact between the surfaces of the shaft and bearing.

The larger segmental bearings feature housings in a variety of metals, with the metal backing strips for the rubber inserts normally of brass. In some cases, provision may be made for adjusting the position of the individual inserts to take up wear, etc, whilst each individual strip is, of course, removable for replacement. In advancing a bearing strip, shims are used behind the metal backing according to the amount of take-up required.

The material specified for the shaft which runs in the bearing and also the surface finish of the shafting may be matters of considerable importance. Dealing with corrosion problems first, steels and ferrous metals capable of rusting give unsatisfactory service in water lubricated bearings and are normally to be avoided. Although corrosion affects only the shaft, the roughened surface produced will quickly damage the rubber bearing surfaces. Stainless steels provide a satisfactory solution in such cases and need only be in the form of a sleeve fitted over a plain steel shaft. Similarly, Monel and bronze sleeves have been used for the same purpose, although bronze, Monel or stainless steel shafting are generally better.

Thermoplastic bearings

Virtually any of the so-called engineering thermoset plastics are potential materials for plain bearings in both homogeneous and 'filled' forms. These materials

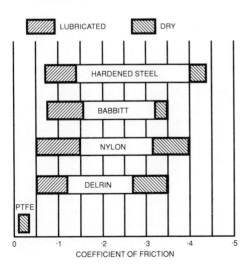

Fig. 3
Coefficients of friction of bearing materials rubbing on a steel shaft.

embrace acetal (copolymers and homopolymers), nylon, polycarbonate, PTFE, PET, styrene-modified PPO and thermoplastic polyester. The usual choice, however, is nylon (or acetal) where maximum strength and resilience are required, or PTFE where minimum friction is the major requirement (see also Figure 3).

All the thermoplastic materials have the advantage that they can run dry or with initial lubrication only, and are also resistant to corrosive media which would attack and pit metal bearing surfaces. PTFE, for example, has good resistance to acids, alkalis and solvents. Nylon, the most popular of the plastic plain bearing materials, is the least satisfactory in respect of chemical inertness and dimensional stability, in fact, although excellent for many applications. Polyacetal has similar mechanical properties to nylon but considerably better dimensional stability and can often be employed where nylon would be unsuitable. The chief limitation of PTFE is its low mechanical strength, although this can be countered by reinforcement (eg glass fibre reinforcement) or the employment of a high strength 'carrier' matrix impregnated with the thermoplastic.

PV values for steel shafts, etc, running on conventional metal bearings may vary from about 800 ft/lb/s up to 2500 to 3000 ft/lb/s, depending on design and materials and the type and method of lubrication. Such ratings can be equalled or bettered by thermoplastic bearings.

In the case of nylon the recommended PV value for continuous operation running dry is usually under 500, although this can safely be increased to 2000 for intermittent operation. The performance of a nylon bearing is materially improved by initial oiling and PV ratings of 2000 to 8000 can be obtained. With continuous oil lubrication, PV values of 50 000 or more can be adopted successfully in design. With water lubrication, on the other hand, the PV rating is usually between 1000 and 4000. Performance is not usually materially improved by

TABLE 3 – TYPICAL VALUES FOR P FOR THERMOPLASTIC BEARINGS
(rated at PV† = 1000)

Bearing diameter mm in	6.3 ¼	12.7 ½	19 ¾	25.4 1	31.7 1¼	38.1 1½	44.4 1¾	30.8 2
r/min								
100	580	300.0	200	160	125.0	110	90.0	80.0
200	290	150.0	190	80	62.5	55	45.0	40.0
300	193	100.0	66	53	42.0	37	30.0	27.0
400	145	75.0	50	40	31.0	28	22.5	20.0
500	116	60.0	40	32	25.0	22	18.0	16.0
600	96	50.0	33	26	21.0	18	15.0	13.0
700	83	43.0	30	23	18.0	16	13.0	11.5
800	72	37.5	25	20	15.0	14	12.0	10.0
900	65	33.0	22	18	14.0	12	10.0	9.0
10 000	58	30.0	20	16	12.5	11	9.0	8.0

†P = bearing pressure, lb/in^2
V = rubbing speed, ft/min

'loading' the nylon with an anti-friction filler, such as graphite or molybdenum disulphide. Although these may result in a slight reduction in frictional coefficient and small improvements to tensile strength and dimensional stability, impact strength and fatigue resistance of the material are usually reduced. In some instances this has led to the failure of loaded nylon bearings at ratings where normal nylon has given a satisfactory performance. It is difficult to give generalised figures for nylon because of the wide variety of different grades available.

PTFE has the lowest coefficient of friction of all the practical bearing materials and remains basically the same whether running dry or lubricated. Lubricant, in other words, has little effect, although it may be desirable in certain cases to act as a coolant. The normal PV rating for PTFE is between 1000 and 3000, although this is uprated to 5000 to 20 000 with reinforced or filled PTFE, depending upon the type, structure and strength of the reinforcement.

The coefficient of friction of polyacetal is slightly lower than nylon and less variable between the dry not-run-in and fully-run-in lubricated state. Typically, for example, nylons show a coefficient of friction of 0.2 to 0.4 on a new surface, rubbing dry against steel; polyacteal has a figure of 0.15 to 0.35 under similar conditions. PV values recommended for polyacetal are 1600 to 2000 running dry, 3000 to 5000 with initial oil only, and 10 000 or above with continuous oil lubrication (see also Table 3).

Limitations of plastic bearings

The chief limitation with thermoplastic bearings is the design limit for permissible compressive stress on the projected area of the bearing surface. For nylon this is usually about 350 to 400 lb/in^2, for polyacetal about twice this figure and for PTFE (unless reinforced) very much lower. Unless a large bearing size can be accommodated, therefore, the load carrying capacity of the thermoplastics is somewhat restricted. They are therefore most suited to the smaller sizes of bearing where loads are not high and shaft sizes are small.

Thermoset plastic bearings

Fabric based thermoset plastic laminates are generally used for bearings, largely on account of their better strength properties, but also because a fabric can be moulded into quite complex shapes without damaging individual fibres, as would be likely with paper sheets. Such bearings may be moulded direct to any special design, machined from tubular stock, block or strips, or be of composite construction, depending on requirements. As a general rule, tubular bearings which embrace the complete journal are best machined from tube, since moulding to a high degree of accuracy is not practical. Open type bearings are usually in the form of liners which are moulded and then machined to finish. Split bearings may be machined from tube or block, segmental bearings moulded and machined, etc. It is an advantage in all moulded bearings to leave an allowance for subsequent machining to size, as this also removes mould skin and gives a sur-

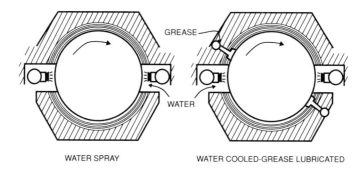

GREASE

WATER

WATER SPRAY   WATER COOLED-GREASE LUBRICATED

Fig. 4
Examples of cooled thermoset plastic bearing design.

face better suited to efficient lubrication and cooling. Provision may have to be made for feeding lubricant to, and dispensing it within, the bearing. Satisfactory performance is largely dependent on satisfactory cooling.

All phenolic laminates are poor conductors of heat and thus local overheating, with subsequent damage to the bearing material, can occur where there is a lack of suitable cooling. The normal maximum recommendation for phenolic laminates used as bearing materials is 80 to 85°C, although in some circumstances operating temperatures up to 110°C may be permitted for short periods.

Water or aqueous fluids are the best coolants, although where water cannot be permitted it is often possible to use oil or graphite. Plastic laminates may be formulated with graphite impregnation to give self-lubricating bearings where no external lubricant can be used. The amount of coolant required depends very much on the type and service of the bearing concerned, and so figures vary enormously. To quote for rolling mill bearings, an application where laminated phenolic bearings have found particular favour, a rough rule is one gallon water per minute per inch of neck diameter for heavy duty work, although under favourable conditions only one-quarter of this amount may actually be required. Smaller bearings require proportionally less coolant. The final criterion is really the working temperature of the bearing. The effect of slight overheating is a darkening of the bearing material surface, which is not necessarily harmful. Excessive bearing temperatures will, however, result in the bearing charring.

Whilst water or aqueous solutions are good coolants, they are not necessarily satisfactory lubricants, and for particularly arduous service conditions grease lubrication may have to be incorporated with additional, and separate, water cooling. Alternatively, a compromise may be reached by employing a fluid which combines the necessary lubricating and coolant properties.

Phenolic laminates are all prone to a certain dimensional instability under intermittent contact with water, and when bearings are designed an allowance is made for the dimensional change due to water absorption in service. The figure normally adopted is a clearance of 0.007 in per inch of shaft diameter, or up to

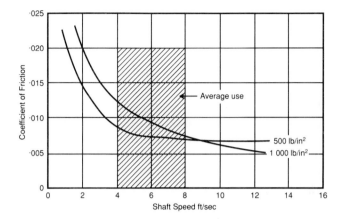

Fig. 5
Coefficient of friction of thermoset plastic bearings.

twice this figure if the bearings are expected to stand idle for long periods immersed in aqueous solutions. On some types of bearing this allowance may be reduced, hence it is important that the designer checks with the manufacturer.

The wall thickness of tubular bearings is usually made about 0.15 times the shaft diameter. Bearing length is generally restricted to less than the shaft diameter, a recommended average for normal design being 0.875 × diameter. Greater wall thicknesses than those specified may be used on open type bearings in which a considerable amount of wear is permissible, but where longer bearing lengths are required it is usual to split the bearing into separate working lengths or short bushes with annular clearance spaces between them.

For minimum friction and wear, bearing surfaces should be finished mirror-smooth, using a slow feed for the final cut. The plastic material must also be fully supported by its metal housing and carefully aligned to avoid excessive loading. Heavy duty bearings are normally improved by careful running-in, when reduced load or speed, or both, and excessive lubrication are generally employed, the bearing and journal being carefully checked against overheating. Light duty bearings seldom require any attention other than a summary check, after an initial period of running, that all is well.

Properly designed and properly installed, a laminated plastic bearing should give a longer life and lower friction (equivalent to a reduction in the driving power required) than plain metal bearings. Plastic bearings can also be used immersed in water or certain chemical solutions which would attack metal bearings. Thus they may offer a solution where metal bearings cannot be used, or alternatively, a longer life and, therefore, cheaper bearings, even though their initial cost may be higher. Like other forms of non-metallic bearing, they have their definite limitations, particularly as regards operating temperatures and maximum permissible loading. Thus their successful application is dependent on use in suitable situations, together with correct design and installation. Best performance is normally realised when they are used with hardened and polished steel shafts and relatively low speed operation, or for light loading only at moderate to high speeds.

See also chapter on *Laminated Plastics,* Section 6.

Wood bearings

Wood continues to have limited application as a bearing material, one advantage being that woods absorb oil or grease and are soft enough to eliminate scoring or seizure of the shaft. The only wood which is still used to any extent for bearings is lignum vitae which is inherently self-lubricating and has particular application for large size, slow speed, low load bearings and submerged bearings (see chapter on *Wood*).

Cermets

Cermets comprising sintered carbides of titanium or tungsten in a cobalt matrix are sometimes used for high temperature bearings and offer high resistance to wear. Such materials are hard and rigid and extremely resistant to corrosion. As bearing materials, however, they have poor conformity and poor impact resistance, as well as being difficult to machine and costly to produce. Their use as bearing materials is thus limited to specialised services where no suitable alternative is available.

Jewel bearings

The use of jewel bearings is confined mainly to the instrument and horological industries. The chief material is synthesised corundum, although spinel is also used to a limited extent where wear resistance is not so important, or where its lower hardness can be of advantage in that it is easier to work. Standard forms and dimensions of bearing jewels are defined in BS 904 : 1948, to which a majority of jewel bearings produced in Britain conform. There is also a small but

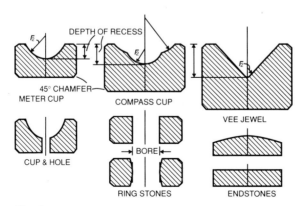

Fig. 6

definite demand for jewel balls for bearing and similar use, ranging in diameter up to ¼ inch. These are made in synthetic sapphire, ruby and agate and in such non-standard materials as glass, tungsten carbide and ceramics.

The forms to which the bearing jewels are cut are, characteristically, cup shapes or ring stones, and end stones, all shapes being circular in plan form; typical shapes are illustrated in Figure 6. The single cup shape is known generally as a meter cup and is normally used in stationary instruments with a vertical pivot, where horizontal thrusts are absent or where there is no likelihood of shocks causing the pivot to jump out of the cup.

The double cup is commonly known as a compass cup and is an elaboration of the meter cup. The centre cup now has a second cup surrounding it, making the total cup deeper and providing a slipway to guide the pivot back to its proper position should it momentarily be jolted out of place. As its name implies, it is the type most commonly used in compasses and instruments of a similar nature.

The V type cup gives even greater depth, and thus support, to the pivot. There is also less tendency for a double ended pivot to jam when run between two Vs with an included angle of 75 to 85g than between two radii (ie between two cup bearings). Since the V bearing can also accommodate appreciable side loads, it is by far the most versatile type and may be used in any position from the vertical to the horizontal. The pivot radius, however, must be nearer the cup radius than in the other two types and friction is higher.

A standard form of cup bearing adapted to take side loads consists of drilling through the cup to engage a further depth of pivot. This has the effect of providing a conventional bearing support with the main bearing surface being still of the cup and cone type. Again friction is higher than with plain cup bearings.

For heavy duty loading (relatively speaking) the bearing type to be preferred approximates to normal engineering standards with the pivot or spindle carried in ring stones. Any end play is then taken up by the provision of end stones, a ring stone and an end stone together forming a complete bearing. The form of these bearing surfaces may be straight or radiused, as shown.

Pivot or spindle materials most commonly employed with jewel bearings include hardened steel, iridium tipped brass and German silver. Steel pivots must be hardened to minimise wear but are generally cheaper and simpler to produce than the alternatives. The main disadvantage with a hardened steel pivot is that it is susceptible to rusting and once rust is formed it will act as an abrasive, promoting a high rate of wear and increased friction. To overcome this, steel pivots must be adequately lubricated. Iridium pivots operate with equal efficiency lubricated or dry. The choice of a non-magnetic material for the pivot is obligatory in compasses and similar instruments.

The frictional coefficients of jewel bearings may range from 0.1 to 0.2, with the mean figure (0.15) representative of good average practice.

Rubber (natural)

RAW RUBBER is sensitive to temperature changes, absorbs water and is readily attacked by organic solvents. If they do not dissolve it, most organic liquids are taken up by the rubber with consequent swelling and softening. Vulcanised rubber is more indifferent to temperature changes, has a higher tensile strength and elongation at break, but tears more easily. Vulcanised rubbers can be classified as either *soft* or *hard,* largely dependent on how far vulcanisation has been allowed to proceed. Soft rubbers normally contain only about 1 to 3% of combined sulphur. Hard rubbers normally contain 25% or more combined sulphur. Both types can be given a variety of different properties (depending on the vulcanisation), and their physical properties can be further modified by the addition of fillers.

Soft rubbers

'Pure' soft rubber has a tensile strength of about 150 to 200 bar (2100 to 2800 lb/in^2), with an elongation at break approximately 900%. This tensile strength figure can be enhanced by compounding, with a reduction in elongation. The effect of various fillers is shown in Table 1. In general, reducing the rubber con-

TABLE 1 – EFFECT OF FILLERS AND PIGMENTS

Filler/pigment (% by volume)		Tensile strength		Elongation at break %	Hardness
		lb/in^2	kg/cm^2		
None (average pure gum compound)		400	28	800	30
Furnace black	(30%)	4200	295	500	70
Channel black	(30%)	4400	310	600	65
Silicon dioxide	(20%)	3200	225	650	65
Whiting	(30%)	2300	160	600	50–55
Calcium silicate	(20%)	2800	200	600	50–55
Clay	(20%)	3400	240	600	50
Wood flour	(27%)	2500	175	600	55

TABLE 2 – QUALITY RANGES OF SOFT RUBBERS

Property	High	Medium	Low
Tensile strength lb/in^2	>2100	800–2100	<1000
Tensile strength kg/cm^2	> 150	60–150	< 70
Elongation at break	>500%	250–500%	< 250%
Hardness	> 65	45–65	< 45
Modulus 300% lb/in^2	>1000	400–1000	< 400
Modulus 300% kg/cm^2	> 70	30–70	< 30

tent and increasing the proportion of filler or pigment will progressively reduce
elongation until the resulting rubber can resemble leather in that it hardly
stretches at all. It will be appreciated that a whole variety of different rubber
proportions can be produced by suitable compounding, although for general use
soft rubbers can be classified as 'high', 'medium' or 'low' with regard to mechanical properties (see Table 2). Typical properties of 'high' quality rubbers are
given in more detail in Table 3, together with one 'low' rubber for comparison. The

TABLE 3 – MECHANICAL PROPERTIES OF TYPICAL VULCANISED RUBBERS

Property	Units	Pure gum compound black	25% carbon black	Heat resistant grade 25% carbon black	80% clay + whiting
Tensile strength	lb/in^2	4400	4400	4400	1160
	kg/cm^2	310	310	310	80
Elongation at break		750%	600%	600%	330%
Modulus 300%	lb/in^2	260	1000	1100	850
	kg/cm^2	18	70	80	60
Tear strength	lb/in^2	1000	2600	2600	430
	kg/cm^2	70	180	180	30
Flex-crack resistance ($\times 10^3$ flexes)	—	700	150	110	10
Resilience (rebound)	—	80%	50%	50%	25%
Permanent set at 200%	—	2%	10%	10%	–41
Hardness	—	40	60	60	80
Modulus of elasticity	lb/in^2	210	500	—	—
	kg/cm^2	15	35	—	—
Energy at break	lb/in^2	5700	10 700	10 700	2100
	kg/cm^2	400	750	750	150
Maximum service temperature	°F	212	212	257	250
	°C	100	100	125	121

Other typical data:
Coefficient of linear expansion *circa* 80–120 $\times$ 10^{-6} in/in/°F
(*Note* fillers reduce the coefficient of expansion)
Specific heat : 8.51
Thermal conductivity : 1.0 Btu/h/ft^2 /in/°F
Electrical conductivity : 13 $\times$ 10 $^{-6}$ ohm/cm

TABLE 4 – MECHANICAL RUBBER SPECIFICATIONS

Com-pound	BS	Hardness range	Tensile strength minimum lb/in²	Elongation at break % minimum	Compression set, % maximum	Max tensile strength change after 7 days at 70°C	Maximum hardness change after 7 days at 70°C
Z11		32–40	3000	700	35		+4
Z12		45–50	3000	600	30		+4
Z13	903	52–60	2500	560	50	±10	+4
Z14		62–70	2000	350	30		+4
Z15		72–80	1500	250	35		+4
Z16		82–89	1250	150	45		+3
NS1		31–40	2500	650	30		+4
NS2		41–50	2500	500	30		+4
NS3	3106	51–60	2000	350	25	±10	+4
NS4		61–70	1500	250	25		+4
NS5		71–80	1250	150	25		+4

properties of mechanical rubbers (*ie* rubbers subject to mechanical loading as in rubber springs, compression blocks, etc) are summarised in Table 4.

Hard rubber

Hard rubbers may be regarded as thermoset materials. They are normally produced by adding 25% or more sulphur to natural or synthetic rubbers and subsequently vulcanising the mixture. Compounding can be varied to provide a wide range of mechanical, chemical and electrical properties. Typical properties of various compounds are given in Table 5.

Hard rubber can be formed by stamping, punching, extrusion, blowing, pressing, compression moulding, transfer moulding (see Table 6). In addition hard rubber can be sawn, drilled, tapped, milled, turned, etc.

Ebonite dust

Ebonite dust is made by reducing a fully-cured ebonite to a fine state of subdivision. The addition of this to a rubber-sulphur mix confers strength and stiffness to the semi-cured moulding, sheet, or extrusions, and enables it to be processed in a steam vulcaniser without distortion. As it is fully cured, it reduces shrinkage in proportion to the amount incorporated.

The better quality ebonite dusts are usually made from pale crepe or smoked sheet, and the principal characteristics are a negligible ash and a complete freedom from grit, which would mar the high quality and high polish obtainable on ebonite articles moulded from stock containing this kind of dust. Such articles are pipe stems and chemical and pharmaceutical apparatus. The better quality dusts pass through 120 mesh screens, and are produced to a specific chemical analysis.

Lower grades of ebonite dust, *ie* those with a higher ash content, are used as stiffening fillers in lower grade ebonite mouldings, and in some cases, subsequent

TABLE 5 – PROPERTIES OF TYPICAL HARD RUBBER COMPOUNDS

Principal characteristics or uses	Tensile strength lb/in²	Colour	Elongation %	SG	Rockwell Pen-Rec	Shore durometer D hardness	Izod impact Strength ft/lb/in notch	Flexural strength MFS/ ib/in²	Heat distortion temp °F	Dielectric strength 60 Hz v/mil	Power factor 1 mega-hertz %	Dielectric constant 1 megahertz	Surface resistance megohms at 74°F 86% RH	Water absorption 48 hrs at room temp
Moulded Articles														
General purpose	8 300	Black	4.0	1.21	108-76	81-87	0.48	12 500	142	435	0.8	3.00	4.8×10^5	0.08
General purpose	7 900	Black	4.0	1.21	107-91	81-87	0.45	11 340	145	496	0.7	2.95	over 10^8	0.06
General purpose	4 500	Black	3.4	1.28	129-79	78-84	0.45	7 125	134	344	1.2	3.80	over 10^8	0.14
General purpose	2 000	Black	2.0	1.54	103-27	76-82	0.27	2 818	Low	377	2.8	4.95	5.08×10^5	0.30
High heat resistance	6 750	Red-brown	2.6	1.65	83-60	85-92	0.35	8 675	283	393	1.2	4.10	5.32×10^3	0.06
High arc resistance, heat and wear resistance	4 000	Red-brown	1.0	1.95	65-41	87-95	0.32	8 400	300	600	2.5	4.80	Arc resistance 248	0.06
High dielectric quality	5 800	Red-brown	2.6	1.71	82-55	86-92	0.38	9 060	246	371	1.2	4.60	4.46×10^4	0.10
Heat resistance, low moisture absorption	5 400	Black	1.8	1.80	82-54	84-92	0.34	10 000	295	420	1.8	4.10	8.02×10^4	0.04
Sheets														
General purpose	9 300	Black	5.0	1.20	112-79	80-86	0.48	16 600	159	487	0.5	2.95	2.23×10^7	0.06
General purpose	8 300	Black	4.1	1.21	108-76	80-86	0.48	12 625	142	435	0.8	3.00	4.81×10^5	0.08
General purpose	6 750	Black	4.2	1.21	126-78	80-86	0.44	12 600	134	474	0.6	3.05	1.24×10^4	0.08
High elongation (semi-hard)	2 550	Black	33.0	1.17		50-60	0.51		Low	437	1.3	3.00	2.66×10^4	0.22
High heat resistance	7 400	Black	1.2	1.43	86-60	86-92	0.53	9 640	221	613	0.6	3.25	over 10^8	0.15
Strength, flow resistance, high dielectric quality	7 450	Yellow-brown	5.3	1.27	112-80	80-86	0.50	14 950	150	415	0.6	3.70	7.98×10^4	0.08
Rods and Tubes														
General purpose	9 700	Black	5.0	1.21	107-70	80-86	0.48	11 375	163	524	0.5	2.90	over 10^8	0.06
General purpose	7 500	Black	3.0	1.21	116-82	80-86	0.47	11 130	146	462	0.9	2.80	over 10^8	0.10
High elongation	3 600	Black	16.0	1.15		65-75	0.48		Low	512	1.4	3.15	over 10^8	0.12
Strength, flow resistance, high dielectric quality	6 500	Yellow-brown	3.8	1.27	112-80	80-86	0.52	9 060	217	374	0.9	3.60	6.73×10^4	0.07

Hard rubber in general has a coefficient of linear expansion of 0.00004/degF.
Note: All values shown are representative average test results.

TABLE 6 – FORMING HARD RUBBERS

Process	Remarks
Pressing	For simple shapes and where a smooth bright finish is required.
Punching/stamping	Production of small parts from flat sheet.
Compression moulding	Suitable for a wide variety of shapes, forms and sizes.
Transfer moulding	Particularly suitable for the production of parts with complex inserts.
Blowing	For the production of hollow parts.

vulcanisation in steam is not required. These grades of ebonite dust may be used for the production of battery boxes or other large mouldings, where the principal advantage is in quickly obtaining rigid mouldings direct from the tool.

The quantity of ebonite dust to use will vary with its application, but it is generally desirable to put in as much as possible, having regard to the flowing properties as a whole, and it can be up to 60% of the total batch.

Cellular rubber

There are three distinct forms of cellular rubber:

(i)   Sponge rubber made from solid rubber chemically blown to produce a sponge-like structure with mainly interconnecting cells.
(ii) Foam rubber made from a liquid form of the rubber, expanded and solidified as a foamed material. The cells may be interconnecting, or discrete, depending on the aeration agent and process used.
(iii) Expanded rubber is made from solid rubber, chemically blown and aerated to produce mainly discrete pockets of entrapped gas. The main difference between expanded rubber and sponge rubber is that the former is stiffer for a similar density of the same cellular rubber.

Applications include gasketing, sealing, insulating, upholstery and cushioning, weatherstripping, dustproofing, shock absorption, sound and vibration damping, flotation and sandwich construction, the rubber base, type and density of the cellular product being chosen accordingly.

The basic mechanical strength of a cellular rubber can be expressed in terms of its deflection under compression. There are several different ways in which this can be done. The British Standard method is to employ a 12 in diameter loading plate or indenter and measure the force or load required to compress the sample to 60% of its original thickness. This load, in kilograms, is read as the Indentation Hardness Index of the material, which can also be expressed as a Hardness Grade. The relationship between Indentation Hardness Index and Hardness Grade is given in Table 9.

TABLE 7 – CHEMICAL RESISTANCE OF HARD AND SOFT RUBBERS*

	Concentration by weight	Maximum temperature		Degree of vulcaniza-tion (soft or hard rubber)	Design of compound (for general chemical service or for specific service)
		°F	°C		
Inorganic Acids:					
Arsenic acid	Any concentration	150	65	Soft	Specific
				Hard	General
Carbonic acid	Up to saturation at atmospheric pressure	150	65	Soft or hard	General
Chlorine water	Up to saturation at	100	38	Soft	Specific
(hypochlorous acid)	atmospheric pressure	150	65	Hard	Specific
Fluoroboric acid	Any concentration	150	63	Soft or hard	General
Fluorosilic acid	Any concentration	150	65	Soft or hard	General
Hydrobromic acid	Any concentration	100	38	Soft	Specific
		150	65	Hard	Specific
Hydrofluoric acid	Up to 50%	150	65	Soft or hard	General
Hydrogen sulphide solution	Up to saturation at atmospheric pressure	150	65	Hard	General
Muriatic acid (hydrochloric)	Any concentration	150	65	Soft or hard	General
Phosphoric acid	Up to 85%	150	65	Soft	Specific
				Hard	General
Sulphuric acid	Up to 50%	150	65	Soft or hard	General
Sulphurous acid (sulphur dioxide water)	Up to saturation at atmospheric pressure	150	65	Hard	Specific
Solutions of Inorganic Salts and Alkalis					
Aluminium chloride	Up to saturation	150	65	Soft or hard	General
Aluminium sulphate	Up to saturation	150	65	Soft or hard	General
Alums	Up to saturation	150	65	Soft or hard	General
Ammonium chloride	Up to saturation	150	65	Soft or hard	General
Ammonium hydroxide	Up to saturation	100	38	Hard	General
Ammonium persulphate	Up to saturation	100	38	Soft	General
Ammonium sulphate	Up to saturation	150	65	Soft or hard	General
Barium sulphide	Up to saturation	150	65	Soft or hard	General
Calcium bi-sulphite	Up to saturation	150	65	Hard	Specific
Calcium chloride	Up to saturation	150	65	Soft or hard	General
Calcium hypochlorite	Up to saturation	150	65	Soft	Specific
				Hard	General
Caustic soda (sodium hydroxide)	Up to saturation	150	65	Soft or hard	General

*Extracted from 'Rubber in Chemical Engineering' (British Rubber Development Board).   (continued)

TABLE 7 – CHEMICAL RESISTANCE OF HARD AND SOFT RUBBERS – *contd.*

	Concentration by weight	Maximum temperature		Degree of vulcaniza-tion (soft or hard rubber)	Design of compound (for general chemical service or for specific service)
		°F	°C		
Caustic potash (potassium hydroxide)	Up to saturation	150	65	Soft or hard	General
Copper chloride (cupric)	Up to saturation	150	65	Hard	General
Copper cyanide (in solution with alkali cyanides)	Up to saturation	150	65	Soft or hard	General
Copper sulphate (cupric)	Up to saturation	150	65	Soft or hard	General
Ferric chloride	Up to saturation	150	65	Soft Hard	Specific General
Ferrous sulphate ('copperas')	Up to saturation	150	65	Soft or hard	General
Nickel acetate	Up to saturation	150	65	Hard	Specific
Plating solutions: Brass Cadmium Copper Gold Lead Nickel Silver Tin Zinc		150	65	Soft or hard	General
Potassium cuprocyanide	Up to saturation	150	65	Soft or hard	General
Potassium dichromate	Up to saturation	150	65	Hard	General
Sodium (or potassium antimonate)	Up to saturation	150	65	Soft or hard	General
Sodium (or potassium bisulphite)	Up to saturation	150	65	Hard	General
Sodium (or potassium acid sulphate)	Up to saturation	150	65	Soft or hard	General
Sodium (or potassium chloride)	Up to saturation	150	65	Soft or hard	General
Sodium (or potassium cyanide)	Up to saturation	150	65	Soft or hard	General
Sodium (or potassium hypochlorite)	Up to saturation	150	65	Soft Hard	Specific General
Sodium (or potassium sulphide)	Up to saturation	150	65	Soft or hard	General
Sodium (or potassium sulphite)	Up to saturation	150	65	Soft or hard	General
Sodium (or potassium thiosulphate)	Up to saturation	150	65	Soft or hard	General
Silver nitrate	Up to saturation	150	65	Soft Hard	Specific if discoloura-tion is to be avoided. General
Tin chloride (either stannous or stannic)	Any aqueous solution	150	65	Soft or hard	General
Zinc chloride	Up to saturation	150	65	Soft or hard	General
Zinc sulphate	Up to saturation	150	65	Soft or hard	General

(continued)

TABLE 7 – CHEMICAL RESISTANCE OF HARD AND SOFT RUBBERS – *contd.*

	Concentration by weight	Maximum temperature		Degree of vulcaniza-tion (soft or hard rubber)	Design of compound (for general chemical service or for specific service)
		°F	°C		
Organic Materials:					
Acetic acid	Any concentration	150	65	Hard	Specific
Acetic anhydride		150	65	Hard	Specific
Acetone	Any concentration	150	65	Soft	Specific
				Hard	General
Amyl alcohol	Any concentration	150	65	Soft	Specific
				Hard	General
Aniline hydrochloride	Any concentration	150	65	Soft or hard	General
Buttermilk		150	65	Hard	Specific
Butyl alcohol	Any concentration	150	65	Soft	Specific
				Hard	General
Casein	Any concentration	150	65	Soft or hard	General
Castor oil		150	65	Hard	Specific
Catsup		150	65	Hard	Specific
Citric acid	Up to saturation	150	65	Soft	Specific
Coconut oil		150	65	Hard	Specific
Cottonseed soil		150	65	Hard	Specific
Dyestuffs		150	65	Hard	Specific
Ethyl alcohol	Any concentration	150	65	Soft	Specific
				Hard	General
Ethylene glycol	Any concentration	150	65	Soft or hard	General
Formaldehyde (formalin)	40% aqueous solution	100	38	Hard	Specific
Formic acid	Any concentration	100	38	Hard	Specific
Furfural		100	38	Hard	Specific
Fruit juices		150	65	Soft or hard	Specific
Gallic acid	Up to saturation	150	65	Soft or hard	General
Glucose	Any concentration	150	65	Soft or hard	General
Glue	Any concentration	150	65	Soft or hard	General
Glycerine	Any concentration	150	65	Soft or hard	General
Lactic acid	Any concentration	150	65	Hard	Specific
Malic acid	Up to saturation	150	65	Soft or hard	Specific
Methyl alcohol	Any concentration	150	65	Soft	Specific
				Hard	General
Mineral oils		100	38	Hard	Specific
Olive oil		150	65	Hard	Specific
Propylalcohol	Any concentration	150	65	Soft	Specific
				Hard	General
Soaps	Any concentration	150	65	Soft or hard	General
Sweet milk		150	65	Hard	Specific
Tannic acid	Up to saturation	150	65	Soft or hard	General
Tartaric acid	Up to saturation	150	65	Soft	Specific
				Hard	General
Triethanolamine	Any concentration	150	65	Soft or hard	General
Vinegar		150	65	Hard	Specific

TABLE 8 – EXAMPLES OF COMMERCIAL GRADES OF EBONITE DUST

Grade†	Ash %	Acetone‡ extract %	Total sulphur %	Free sulphur max %	Specific gravity	Powder density	BS mesh
000	0.5 max	2.0–4.0	26–29	1.0	1.15–1.17	0.27–0.32	120
70/30 S	1.0 max	1.5–3.0	28.5–30.5	2.5	1.18–1.20	0.31–0.35	120
000/E	2.0 max	2.5–3.5	24–27	1.0	1.16–1.17	0.27–0.31	120
231	4.0 max	2.0–4.0	27–31	3.5	1.22–1.24	0.32–0.37	120
Prime B	6.0 max	3.0–5.0	27–30	3.0	1.22–1.25	0.33–0.38	120
Improved C	10.0 max	4.5–7.0	19–21	1.0	1.20–1.24	0.36–0.42	120
Med.A	9.0–12.00	6.0–7.0	19 –24	1.7	1.23–1.25	0.33–0.36	100
Med.B	20–30	10.0–14.0	8 –12	1.0	1.33–1.40	0.32–0.35	100

‡These figures are total acetone extract less the free sulphur.
†James Ferguson & Sons Ltd

TABLE 9 – INDENTATION HARDNESS INDEX AND HARDNESS GRADE

Indentation	8–10½	10½–14	14–17½	17½–22	22–27	27–33	33–41	41–50	50–60 m
Hardness grade	A	B	C	D	E	F	G	H	J

TABLE 10 – ASTM GRADES FOR CELLULAR RUBBERS

Force lb/in2	0.5–2	2–5	5–9	9–13	13–17	17–24
ASTM grade	RN10	RN11	RN12	RN13	RN14	RN15

The (American) ASTM Standard measures the force required to compress a sample of the product one square inch in area to 75% of its original thickness. Corresponding grades are given in Table 10.

Latex

Natural rubber latex can be obtained in a variety of stabilised and concentrated forms, and in combination with other substances, and finds widespread application as adhesives, sealing compounds and coatings (by dip or electrophoretically). Synthetic rubbers can also be rendered in latex form.

Note: see also sections covering *Elastomers* and *Rubber (synthetic)*.

Rubber (synthetic)

TYPES of synthetic rubber and their general properties are given in Table 1. SBR (Buna S) was specifically developed during World War II as a substitute for natural rubber and is one of the most important general purpose synthetic rubbers. It can be compounded with fillers to 'tailor' the mechanical properties to specific requirements, and also oil-extended (*eg* for vehicle tyres). Together SBR and natural rubber account for the bulk of the world's rubber consumption.

Nitrile rubbers are another important general purpose group, particularly where resistance to mineral oils is required. Oil resistance increases with increasing nitrile content, although at the expense of resilience and low temperature

The world's heaviest rubber product – a 20-ton cylindrical fender for mammoth ships, 2.5 m thick and 5.5 m long. It was made for Richard Costain Ltd, London, who placed an order of four fenders with Vredestein.

TABLE 1 – TYPES OF SYNTHETIC RUBBERS

Names	Hardness range	Favourable properties	Unfavourable properties	Temperature range				
				Low		High		
				Continuous °C	Min °C	Continuous °C	Short term °C	
Buna N Nitrile (NBR)	40–90 (low nitrile) 50–95 (high nitrile)	Superior to most synthetic rubbers for compression set cold flow, tear resistance and abrasion resistance. Good to excellent resistance to oils.	Indifferent resistance to weathering and ozone (may be improved by compounding)	-40	-54	121	—	
Buna S (SBR) (GRS)	40–90	Low cost, good average mechanical properties, good resistance to water and alcohols.	Relatively poor resistance to oils, ozone and sunlight.	-40	-50	80–90	100	
Butadiene (BR)	40–90	Similar in properties to natural rubber. Good low temperature flexibility.	Lack of resistance to oils, etc.	-50	-55	75–85	107	
Butyl (IIR)	45–85	Good chemical resistance	Poor resistance to oils, ketones and silicon fluids and grease.	-50	-54	100	—	
Chloroprene (Neoprene)	40–90	Excellent resistance to ozone. Superior resistance to refrigerants, oils, mild acids, etc. Good high temperature properties.	Mechanical properties somewhat inferior to natural rubbers.	0	54	100–120	149	
Chlorosulphonated polyethylene (CSM)	40–70	Good resistance to acids		-45	-54	120	130	
Ethylene propylene (EP) (EPM)	45–85	Excellent resistance to oils, steam, water, dilute acids and alkalis, ketones, etc.	Not resistant to petroleum fluids.	—	54	-149	—	

(continued)

TABLE 1 – TYPES OF SYNTHETIC RUBBERS – *contd.*

Fluorocarbon (FPM)	60–80	Excellent mechanical properties at high temperatures. Excellent resistance to oils, halogenated hydrocarbons, acids, etc. Very low permeability.	Not resistance to ketones or amines. High cost.	-29	-54	200	316
Isoprene rubber (IR)	30–90	Similar to natural rubber.	Less easy to process than natural rubber. Not resistant to mineral oils or petroleum fluids.	—	—	—	—
Polyacrylic (ACM)	—	Outstanding resistance to oils and petroleum fluids, weathering, ozone, etc. High continuous service temperature. Resists flex-cracking.	Poor resistance to moist heat	—	-18	177	—
Polysulphide	—	Excellent low temperature flexibility, solvent resistance and weather and ozone resistance. Resists flex-cracking.	Poor physical properties. Poor resistance to heat.	-50	-54	80–90	107
Polyurethane	50–90	Superior mechanical properties. Superior abrasion resistance.	Softens rapidly above 121°C. Poor resistance to moist heat.	—	—	100–110	121
Silicone (SI)	40–80	Superior high and low temperature performance	Poor tensile strength and tear and abrasion resistance. (May be improved by compounding.) High shrinkage when moulded.	-65	-93	232	375
Fluorosilicone	45–80	Excellent high and low temperature performance. Good oil resistance.	High cost. Moderate physical properties.	—	—	177	250

TABLE 2 – COMPARATIVE PHYSICAL PROPERTIES OF SYNTHETIC RUBBERS

	Buna N (Nitrile)	Buna S (SBR)	Butadiene	Butyl	Chloroprene (Neoprene)	CSM	Ethylene Propylene	Fluorocarbon	Isoprene (Synthetic)	Polyacrylic	Polysulphide	Polyurethane	Silicone	Fluorosilicone
Tensile strength	G	G	E	G	G	F	G	F-G	G	F	F	E	P	P-G
Resilience	F	G	E	F-P	G	G	G	F	E	G	P	F	F-G	G
Tear strength	F	F	E	G	E	F	F	F	E	F	F-P	E	F	F-G
Notch resistance (resistance to cut growth)	G	G	G	E	F	F	F	F	F	G	P	G	F	F
Resistance to compression set	G	G	G	F	G	F	G	G	G	F	P	G-F	F	F
Resistance to cold flow	G	E	E	F	F	F	F	G	G	F-G	P	G-E	G	G
Abrasion resistance	E	E	E	F	E	G	E	G	E	G	P	Ex	F-P	F
Low temperature flexibility	F	G	E	G	F	P-F	G	F	E	P	G	E	Ex	Ex
Permeability to air and gases	G	F	F	F	G	E	F	E	F-P	G	E	G	F-G	G

KEY: Ex – exceptional   E – excellent   G – good   F – fair   P – poor

TABLE 3 – RESISTANT PROPERTIES OF SYNTHETIC RUBBERS

Resistance to:	Buna N (Nitrile)	Buna S (SBR)	Butadiene	Butyl	Chloroprene (Neoprene)	CSM	Ethylene Propylene	Fluorocarbon	Isoprene (Synthetic)	Polyacrylic	Polysulphide	Polyurethane	Silicone	Fluorosilicone
Weathering	F	F	F	G–E	G–E	E	P	E	P	E	F	E	E	E
Ozone	P	P	P	G–E	G–E	E	P	E	P	E	E	E	E	E
Oxidation	G	G	G	E	G	G–E	E	E	G	E	E	E	E	E
Heat	G	F	F	G	G	G	E	E	F	E	P–F	F	E	E
Flame	P	P	P	P	E	G	P	P	P	P	P	P	G–E	E
Water absorption	F–G	F–G	G	F	F	F	E	F–G	F–G	P	F	P	F–G	F–G
Steam	F	F–G	G	F	F	F	G–E	F–G	F–G	P	F	P	F–G	F–G
Oils	E	P	P	P	F–G	F	P	E	P	E	E	G	F–G	G
Solvents	P	P	P	P	P	P	P	G	P	P	G	F	F–G	G
Ketones	P	G	G	G–E	P	G	G	P	G	P	G	G	F–G	G
Dilute acids	G	G	G	E	E	E	F	E	E–G	F	G	F	E	E
Concentrated acids	F	F	F	G–E	G	G	G	E	F	P	P	P	F	F–G
Alkalis	F	G	G	G–E	G	G	P	G	F–G	P	G	F	F	F–G
Alcohols	E	F	F–G	E	F	G	P	E	G	P	G	G	G	G
Vegetable oils	E	P–G	F–G	E	G	G	P	E	P–F	E	E	G	F	F–G

KEY: E – excellent G – good F – fair P – poor

TABLE 4 – COEFFICIENT OF LINEAR EXPANSION OF SYNTHETIC RUBBERS‡

Rubber	Coefficient of expansion $\times$ 10^{-6} in/in/°F	m/m/°C
Butyl	83	46
Chloroprene	76	42
Nitrile	62	34
Fluorocarbon	90	50
Silicone	103	57

‡Approximate figures; actual values will vary with composition.

flexibility. Nitrile rubbers are sometimes sub-classified as low nitrile (18 to 30% nitrile) and high nitrile rubbers (up to 48% nitrile). A major use of nitrile rubbers is for seals.

Butadiene rubber has good low temperature flexibility (comparable with that of natural rubber) and good resistance to abrasion. Its other physical properties are less favourable and it is mainly used in blends with SBR and natural rubber.

Chloroprene (neoprene) is a generic name for a number of different types of chloroprene polymers and copolymers. Rubbers of this type are noted for their exceptional resistance to ozone and their excellent oil, chemical and heat resistance, although their mechanical properties are generally inferior to natural rubber.

The other types listed may be regarded as more specialised in application. Comparative physical properties and resistant properties are summarised in Tables 2 and 3 respectively.

Note: see also sections covering *Elastomers* and *Rubber (natural)*.

Silicones

SILICONES have an organic structure of silicon-oxygen atoms with organic side groups such as methyl, phenol or vinyl. As a consequence they possess the inherent stability of an inorganic substance with the plasticity and 'tailoring' possibilities of an organic substance. By altering the length and complexity of the silica-oxygen chain and varying the organic groups, silicons can be produced in a number of forms from liquids to gums and resins. Their properties in each category can be further varied by the addition of fillers or by compounding.

Silicone fluids

Silicone fluids were the earliest of the commercial silicons developed as lubricants and damping fluids and for specialised applications, *eg* high temperature hydraulic fluids, diffusion pump fluids, clutch fluids, etc. The original silicone lubricants were mainly based on dimethyl silicones with properties 'tailored' by liquid or solid additions to yield a range of lubricants from thin oils to thick

TABLE 1 – SILICONE LUBRICANTS

	Trimethyl silicons	Methyl-alkyl silicons	Fluorosilicons
Forms	Oils – light to medium.	Oils – light to medium.	Oils – light to medium or heavy.
	Greases – light to heavy.	Greases – light to heavy. Fillers.	Greases – light to heavy. Fillers.
Specific applications	General purpose high temperature lubricant.	More severely loaded bearing applications.	Lubricant for machines handling oils, fuels and solvents.
	Roller bearing lubricant		Lubricants for compressors, pumps and valves in chemical service.
Special forms	Lubricants with colloidal graphite. Mist inhibitors. Preservative for plastics.	Wax type lubricants metal sliding on metal.	

TABLE 2 – SILICONE ADHESIVES AND SEALANTS

	Silicone adhesives and sealants	Silicone pressure sensitive adhesives	Silicone gums	Silicone resins
Forms	Liquids to thixotropic pastes.	Liquids.	Pastes and dispersions.	Liquids.
Cure	Air drying or heat curing.	Non-curing (usually).	Heat curing.	Pressure and heat curing.
Applications	Bonding, coating, sealing, caulking, liquid gaskets.	Pressure-sensitive tapes.	Gaskets, dia-phragms, bellows, insulating tapes, insulation and insulation com-ponents, heat shrink tubing.	Laminating
Special forms	Non-adhering liquid gaskets, medical grades, non-curing fuel tank sealants, building construc-tion grades.	Heat curing types for electrical insulation, tapes for bonding PTFE.	Sponges, extra high temperature, extra low temperature, low compression set, hygienic grades, etc.	Low pressure resins for bag moulding.

greases. Dimethyl silicone greases are still the main type chosen for high and low temperature lubricants for rolling bearings. Methyl-alkyl silicons offer superior film performance under higher loads for general applications. Fluorosilicon lubricants are largely preferred for compressors and pumps and valves handling corrosive or high temperature fluids and other applications involving high temperature operation.

The maximum working temperature of fluorosilicon lubricants is about 450°C (730°F) with low temperature limit extending down to as low as –75°C (–110°F). Special non-staining lubricants have also been produced for metal forming operations, as well as solvent and chemical resistant greases.

The cost of silicone lubricants is high, but they may be the only suitable choice for bearings working at very low or very high temperatures. For more general applications the higher cost of silicone lubricants may be offset by the savings gained in longer lubricant life and/or lower feed rates required.

Silicone lubricants are primarily used for applications demanding high temperature stability, superior resistance to moisture, temperature and oxidation, general inertness and long shelf life. They are suitable for the static and dynamic lubrication of metals, plastics, rubbers and leathers.

Adhesives and sealants

Silicone and fluorosilicon adhesives and sealants may be rendered in forms ranging from thin liquids to thixotropic pastes. They can be non-curing although they are more usually either self-curing at room temperature or heat-curing. Pressure sensitive adhesives are more usually non-curing, but may be made heat curable. Gum-like compounds, which are mainly used as sealants, may be air-curing or heat-curing. In the latter case they become, in effect, silicone rubbers.

Resins and varnishes

Heat-curing liquid silicone resins and varnishes are widely used in the manufacture of electrical insulating materials to class H rating, *eg*:

(i)   Impregnation of motor and coil windings.
(ii)   Coating of glass cloth and asbestos paper for flexible, heat resistant insu-
        lation material.
(iii)   Filling with mica for slot liners, armature bar insulation, etc.
(iv)   Manufacture of printed circuit board laminate for high service tempera-
        tures (*eg* glass fibre and asbestos paper laminates).
(v)   As the basic resin for reinforced electrical components or stock blanks,
        etc.
(vi)   Silicon moulding compounds with fillers (*eg* silica, etc).
(vii)   Encapsulating and potting.
(viii) As the basis for cements and coatings for a wide variety of electrical and
        electronic components.

Silicone rubbers

Silicone rubbers may be air-curing (one part), cold-curing by the addition of a
catalyst (two part), or heat-curing. They may be pure silicones or loaded with
fillers to enhance particular properties – *eg* loading with carbon black can produce an electrically conductive rubber which can be used for ignition components, heating elements, microwave shielding, etc.

The original two part cold-curing silicone rubbers were condensation products
with relatively long curing times, curing from the outside inwards. They were dependent on exposure to air, and the curing process was based on moisture.
Modern two part mixtures do not require moisture to cure, cure uniformly over
the whole area and do not produce cure by-products. Curing times are typically
24 hours. They are also suitable for accelerated curing by heat when cure time is
reduced to about 1 hour at 150°C (300°F), or 4 hours at 66°C (150°F). One part
rubbers are primarily used as adhesives and sealants.

The application of silicone rubbers extends from the production of typical
'rubber' compounds to embossing rolls, vinyl fabrics and flexible moulds for
casting alloys with melting points up to about 300°C (600°F) and the casting of
components in rigid polyurethane foam. (See also chapter on *Elastomers.*)

Release agents

Silicone fluids, emulsions, solvent dispersions and greases have a wide application as release agents. They have the advantage of being incompatible (non-ad-
hesive) with most other substances as well as requiring only very thin film
thicknesses to be effective. They are available in brush, spray and aerosol forms.
Silicone film and silicone coated papers are also widely used as release agents for
pressure sensitive adhesives, and for the differential release of unsupported adhesive films, etc.

TABLE 3 – SILICONES FOR ENCAPSULATING AND POTTING

Type	Form	Cure	Applications/remarks
Silicone rubber	One or two part	Air or heat	Various viscosities yielding a wide range of cured densities. Used for environmental protection of components and assemblies.
Resins (solventless)	One or two part.	Air or heat	Used for environmental protection of components and assemblies.
Gels (non-melting)	One or two part.	Air or heat	May be formulated with special dielectric properties. Used for special encapsulations, protection of ignition components, etc.
Greases	Grease.	Non-curing	Protection of terminals, junctions and assemblies.
Varnishes (silicon and organo-silicon)	Liquid.	Air or heat	Air curing for surface coatings, etc. Heat curing for impregnation of coils, etc to class H insulation. Bonding of components.
Rubbers	Pastes and gums or compounded.	Non-vulcanising Vulcanised.	General insulation.
Moulding compounds	Powder.	Compression or transfer moulding.	Environmental protection of integrated circuits, etc.

Moulding resins and compounds

Silicone moulding resins are compounded with inert fillers to yield resilient mouldings noted for their high heat resistance and excellent dielectric properties. They are also used in the manufacture of high temperature laminates. One of the main applications of resin forms is for potting and encapsulation (see Table 3).

Coating resins

Silicone resins can be formulated specifically as long life, high temperature coating materials resistant to heat and chemical attack, with good weathering characteristics and retention of gloss. Equally, and more usually, they can be added to a conventional paint system to improve these qualities.

Surfactants for foamed plastics

Silicones have the ability to control the surface tension of foamed plastics. This property can be used to control both the cell structure of the foamed plastics (ie the response to the foaming agent) and also the development of surface skins. Silicone is mainly used as a surfactant with polyurethane foams to ensure uniformity of the structure when foamed in situ, or in the production of self-skinning polyurethane foams.

Formed-in-place gaskets

Cold curing silicones are finding increasing use as formed-in-place gaskets, the joint surfaces being coated with liquid silicone, which cures to a resilient gasket or seal after the joint has been closed. The gasket can be adhering or non-adhering as required.

Wood

TREES are generally classified under two main grades:

(i) *Hardwoods* – deciduous, broad-leaved trees which shed their leaves in winter (in temperate climates). The living wood is characterised by being porous and moisture-conducting.

(ii) *Softwoods* – coniferous or cone-bearing trees with needle-shaped leaves. The living wood is non-porous and moisture-absorbing.

The terms 'hardwood' and 'softwood' do not relate to the mechanical properties of the timber, although in general hardwoods are denser and stronger than softwoods. On the other hand, the lightest commercial wood – balsa – is a hardwood.

The main individual timbers are described in Table 1. Mechanical properties of a range of structural timbers are given in Table 2.

Complex circular form church roof structure in timber, designed for high strength and stiffness.

TABLE 1 – INDUSTRIAL TIMBERS

Timber	Description	Average weight lb/ft^3	Durability	Working qualities/remarks
Abura (*Mitragyna* spp) Tropical Africa	Pale reddish brown to light brown, fairly fine even texture, slightly inter-locked grain. Medium hardness.	36	Not resistant	Works fairly easily, takes good finish, glues, stains and polishes well. Requires care in nailing and screwing to avoid splitting, especially near edges.
Afara (*Terminalia superba*) Tropical Africa	Pale straw colour to light yellow brown, occasionally with irregular greyish or blackish markings, straight grained, medium textured. Medium hardness.	35	Not resistant	Works easily and finishes clearly. Glues and stains well. Requires filling before polishing. Tendency to split in nailing. Stable.
Alder (*Alnus glutinosa*)		30–40	Not resistant	Used for piles.
Agba *Gossweilerodendron balsamiferum*) Tropical Africa	Yellowish pink and yellowish brown. Grain similar to that of mahogany (*Khaya* ssp). Fairly fine textured. Medium hardness.	30	Resistant	Works, nails, glues and screws well, stains and polishes satisfactorily. Should be selected free from gum, as this is sometimes troublesome. Stable.
Afzelia (*Afzelia* spp) Tropical Africa	Light reddish brown, darkening to reddish brown. Grain irregular, commonly interlocked, texture coarse but even. Hard and strong.	48	Very resistant	Moderately hard to work, takes a good finish. Difficult to nail and screw and recommended to be pre-bored. Glues and polishes satisfactorily. Very stable.
Afrormosia (*Afrormosia* spp)	Yellowish brown with brown markings. Shallowly interlocked grain, and fine even texture. Moderately hard, but strong.	44	Very resistant	Fairly easy to work, finishes smoothly. Glues and polishes well. Needs care in nailing as it has a tendency to split. Stains when in contact with ferrous metals under moist conditions. Very stable.
Ash (*Fraxinus excelsior*) Europe	Whitish with occasional pinkish tinge. Straight grained and rather coarse texture. Moderately hard, excellent bending properties.	44	Not resistant	Works fairly easily and finishes smoothly. Glues, nails, screws and polishes well. Stains satisfactorily. Liable to distort.
Balsa (*Ochroma lagopus*) Ecuador	Almost white	6–16	Resistant	Lightest commercial timber.
Basswood (*Tilia glabra*)	Yellowish white	30		
Beech (*Fagus sylvatica*)	Light yellow	45	Not resistant	Works well and is suitable for steam bending.

(continued)

TABLE 1 – INDUSTRIAL TIMBERS – *contd.*

Birch (*Betula pendula*)	Soft brown	43	Not resistant	Works well.
Boxwood (*Buxus sempervirens*)	Yellow	54-70		Very close and dense grain.
Cedar (*Cedrella mexicana*)	Red	27-31		
Ebony (*Diospyros ebenum*)	Black with shades of brown and grey.	48-77	Resistant	Excellent for turning.
Elm, Rock (*Ulmus thomasi*) Canada, U.S.A.	Pale brown, straight grained with moderately fine texture. Hard, strong and elastic.	50	Moderately resistant	Moderately hard to work, but takes a smooth finish. Takes nails, screws and glues satisfactorily and paints, stains and varnishes well.
Elm, Wych (*Ulmus glabra*) Europe	Medium brown with greenish tinge. Fairly straight grained and medium textured. Moderately hard, strong and elastic.	43	Moderately resistant	Fairly easy to work and takes a good finish. Nails, glues, stains and varnishes well.
Freijo (*Cordia goeldiana*) South America	Yellowish brown, with brown markings. Straight grained, rather coarse texture. Medium hardness.	37	Resistant	Easy to work and finishes well. Takes nails and screws easily. Paints, varnishes well. Very stable.
Cedar (*Cedrela* spp) Tropical America	Pinkish to reddish brown. Usually straight grained, with a moderately coarse and uneven texture. Soft.	30	Resistant	Easy to work and finishes well. Nails, screws and glues excellently. Varnishes, stains and polishes satisfactorily. Stable.
Gabon (*Aucoumea klairena*) Tropical Africa	Pink	28	Not resistant	Mainly used for plywood.
Guarea (*Guarea* spp) West Africa	Light pinkish or orange brown. Straight grained, fine textured. Medium hardness.	37	Moderately resistant	Works easily and takes good finish. Nails, screws and glues well. Gum present which may exude. Not particularly stable.
Hickory (*Carya, hicoria*)	Whitish yellow to yellow brown.	46-47		Particularly used for shafts and handles and bentwork.
Idigbo (*Terminalia ivorensis*) West Africa	Pale yellow to light brown. Grain variable, straight to shallowly inter-locked, texture coarse and uneven. Medium hardness.	36	Resistant	Works fairly easily. Glues, nails and screws moderately well. Requires filling before polishing. Very stable.
Iroko (*Chlorophora excelsa*) Tropical Africa	Light yellow brown when fresh, ageing to medium brown or dark reddish brown. Grain interlocked, texture moderately coarse. Strong and hard.	41	Very resistant	Moderately easy to work, takes a good finish. Nails, screws and glues satisfactorily. Stable timber, though irregular grain may result in distortion.

(continued)

TABLE 1 – INDUSTRIAL TIMBERS – *contd.*

Timber	Description	Average weight lb/ft³‡	Durability	Working qualities/remarks
Jorrah (*Eucalyptus marginata*) Western Australia	Dark red	57	Resistant	Suitable for heavy constructional work.
Lime (*Tilia vulgaris*)	Yellowish white	37–38		Excellent for carving and turning.
Mahogany, African (*Khaya* sp) Tropical Africa	Reddish brown. Grain boardly interlocked, texture coarse, fairly even medium hardness.	35	Moderately resistant	Easy to work, takes a good finish. Nails, screws and glues well. Stable.
Makore (*Mimusops heckelii*) West Africa	Pinkish to dark reddish brown. Grain generally straight, texture fine and even. Hard and fairly strong.	39	Resistant	Works moderately well, but has blunting effect on cutting edges. Nails, screws and glues well. Takes an excellent finish. Stable.
Maple (*Acer*)	Various yellows	37		Excellent for carving.
Obeche (*Triplochiton scleroxylon*) West Africa	Creamy white to pale yellow. Grain often interlocked, texture medium, soft, but has good strength properties for its weight.	24	Not resistant	Works easily and takes a good finish. Nails and screws well, but its holding properties are not great. Has good staining and glueing. Fairly stable.
Oak, White (*Quercus*, sp) Europe, North America	Light brown to medium brown. Grain commonly straight, texture coarse and uneven. Hard and strong with fairly good bending properties.	45	Resistant	Variable in working qualities, but generally medium in working. Nails, screws and glues well. Stains when in contact with ferrous metals under moist conditions.
Plane (*Planatus acerifolia*)	Light yellow to pink	30–40		Excellent for turning.
Poplar (*Populis*)	Whitish or greyish yellow	31–35	Not resistant	Packing case material.
Ramin (*Gonystylus*) Sarawak	Light yellow	42		
Sapele (*Entandrophragma cylindricum*) (Tropical Africa)	Pinkish to reddish brown, grain narrowly interlocked, texture fairly fine. Hard and strong.	40	Moderately resistant	Works fairly well. Nails, screws and glues well, finishes and polishes excellently. Not particularly stable.

(continued)

TABLE 1 – INDUSTRIAL TIMBERS – contd.

Seraya, White (*Parashorea* sp) (North Borneo)	Pale straw coloured. Grain straight, sometimes shallowly interlocked, texture medium to coarse. Medium hardness.	35	Moderately resistant	Easy to work, screws and glues well. Takes a good finish and polish. Stable.
Sycamore (*Acer pseudo platarius*)	White	38–39		
Teak (*Tectona grandis*) Burma, Siam, India, Java	Golden brown to medium brown. Grain generally straight, texture medium, sometimes coarse. Moderately hard, but strong.	41	Very resistant	Moderately easy to work. Nails satisfactorily, tends to split near edges. Very stable.
Willow	Pale yellow	24–25		Very resistant to splitting.
SOFTWOODS				
Cedar, Western, Red (*Thuya plicata*) Canada	Reddish brown, non-resinous wood. Straight grained, medium to coarse textured. Soft	24	Resistant	Easy to work, takes good finish. Nails and screws fairly well, tends to stain ferrous metal under moist conditions. Glues and stains satisfactorily. Stable.
Fir, Douglas (*Pseudotsuga taxifolia*) Canada and U.S.A.	Reddish yellow to light orange red brown. Generally straight grained, sometimes wavy or spiral, texture medium. Medium hardness.	33	Moderately resistant	Fairly easy to work. Nails, screws and glues satisfactorily, requires various finishing treatments as raised grain and resin may be troublesome. Medium stability.
Hemlock (*Tsuga canadensis*)	Pale brown	29–30		
Larch, European (*Larix decidua*) Europe	Light orange red brown. Normally straight grained, medium textured. Medium hardness.	35	Resistant	Fairly easy to work, except when resinous. Nails, screws and glues well. Takes satisfactory finish. Medium stability.
Pine, Parana (*Araucaria augustifolia*) Brazil	Pale straw coloured with brownish and reddish streaks. Straight grained, fine even textured. Variable in density, generally moderately hard.	34	Not resistant	Easy to work, tends to split when nailed and screwed near edges. Takes smooth finish, stains, polishes well. Not particularly stable.
Pine, Pitch (*Pinus* spp) U.S.A. and British Honduras	Pale yellowish brown springwood, with reddish summerwood. Straight grained medium texture. Very resinous. Hard and strong.	41	Very resistant	Moderately hard to work due to its resinous nature. Moderately hard to nail and screw. Needs careful preparation before finishing.

(continued)

TABLE 1 – INDUSTRIAL TIMBERS – *contd.*

Timber	Description	Average weight lb/ft³‡	Durability	Working qualities/remarks
Redwood, European (*Pinus sylvestris*) Europe	Pale yellowish brown to light reddish brown. Straight grained, medium textured, fairly resinous. Moderately hard and strong.	31	Moderately resistant	Fairly easy to work, takes a good finish. Nails, screws and glues well. Takes paint and varnish satisfactorily. Stable.
Sequoia (*Sequoia sempervirens*) American Redwood	Pinkish red	25		
Spruce, Sitka (*Picea sitchensis*) Canada	Light pinkish brown, with high lustre. Straight grained, uniform textured. Commonly free from resin. Moderately soft but strong for its weight.	28	Not resistant	Easy to work and takes fine finish. Nails, screws and glues well. Takes paint, varnish and other treatments readily. Stable.
Yew (*Taxus baccata*)	Pale red	48–50		

‡ At 15% moisture content (unless given as a range).

TABLE 2 – MECHANICAL PROPERTIES OF SOME STRUCTURAL TIMBERS

Wood	Moisture content %	Weight lb/ft³	Fibre stress at elastic limit lb/in²	Modulus of elasticity lb/in²	Modulus of rupture lb/in²	Compressive strength‡ lb/in²	Shear strength
Ash	15	41	8900	1.46×10^6	14 900	7000	1380
Beech	—	46	8900–16 200	1.50×10^6	—	3860–7850	1210–2030
Birch	9–10	44–45	12 300–13 340	2.20×10^6	18 900–19 600	9760–10 680	1800–2680
Elm, English	—	35	5610–7840	1.71×10^6	—	2410–4675	1140–1640
Elm, Dutch	—	35	6100–8540	1.12×10^6	—	2620–4680	1040–1460
Elm, Wych	—	43	9400–14 600	1.14×10^6	—	4250–6860	1060–1660
Oak	15	46	8100–12 600	2.11×10^6	11 600	3850–7210	1170–1760
Mahogany	—	34	8800	1.26×10^6	—	6500	860
Plane	—	40	5900	1.13×10^6	—	3380–5820	1240–1810
Poplar	—	28	5800–6240	1.05×10^6	—	2840	700
Sycamore	—	39	9000–15 400	$1.30–1.95 \times 10^6$	—	3840–6720	1280–2190
Willow	—	28	4800–8280	$0.87–1.6 \times 10^6$	—	2070–3930	620–1070
SOFTWOODS							
Douglas Fir	6–9	30–34	6900–10 600	$1.50–2.20 \times 10^6$	10 300–14 000	7100–10 700	1080–1270
Noble Fir	8.4	26	7700	1.75×10^6	12 000	7240	1090
Silver Fir	—	30	5900–10 900	$1.25–1.47 \times 10^6$	—	3070–5900	740–1240
Hemlock, Western	5.4	28	8000	1.52×10^6	10 000	7910	1170
Larch	—	37	7200–12 600	$1.20–1.45 \times 10^6$	—	3520–6900	890–1400
Pine, Scots	—	33	6000–12 000	$1.24–1.50 \times 10^6$	—	3020–6100	750–1400
Redwood, Baltic	—	—	6500–11 200	$1.24–1.30 \times 10^6$	—	3140–6300	790–1320
Spruce, Norway	—	27	5300–9000	$1.07–1.25 \times 10^6$	—	2610–5670	620–1170
Sitka Spruce	9	26	7200	1.61×10^6	11 200	5770	1210
Yew	—	42	—	—	—	4870–7940	—

‡Parallel to grain.

Mechanical properties of balsa‡

The mechanical properties of balsa wood vary with density. The following formulas can be used to calculate average performance figures for balsa of any density, for pieces with a moisture content of about 12%.

Besides saving time and trouble by obviating the need to refer to empirical data, the use of these formulas enables the performance of balsa of any density to be estimated, *eg* for intermediate densities for which no empirical data is available.

Tensile strength:

Long grain:

Tensile strength (lb/in^2) = 335 × density (lb/ft^3) – 700
Tensile strength (kg/cm^2) = 1.47 × density (kg/m^3) – 49

End grain:

Tensile strength (lb/in^2) = 15.5 × density (lb/ft^3) – Upper‡
Tensile strength (kg/cm^2) = 0.07 × density (kg/m^3) – Upper‡
Tensile strength (lb/in^2) = 10.5 × density (lb/ft^3) – Lower‡
Tensile strength (kg/cm^2) = 0.0445 × density (kg/m^3) – Lower‡

Compression:

End grain crushing strength

(lb/in^2) = 230 × density (lb/ft^3) – 650
(kg/cm^2) = 0.8 density (kg/m^3) – 45

End grain compression to limit of proportionality

(lb/in^2) = 185 × density (lb/ft^3) – 650
(kg/cm^2) = 0.8 × density (kg/m^3) – 45

Long grain compression to limit of proportionality

(lb/in^2) = 13 × density (lb/ft^3) – Upper‡
(kg/cm^2) = 0.057 × density (kg/m^3) – Upper‡
(lb/in^2) = 9 × density (lb/ft^3) – Lower‡
(kg/cm^2) = 0.04 × density (kg/m^3) – Lower‡

Shear strength:

Upper values : Shear

(lb/in^2) = 33 × density (lb/ft^3)
(kg/cm^2) = 0.145 × density (kg/m^3)

Lower values : Shear

(lb/in^2) = 27 × density (lb/ft^3)
(kg/cm^2) = 0.12 × density (kg/m^3)

‡In these configurations actual mechanical values may be expected to lie between the upper and lower values, as calculated, depending on orientation of fibres.

Bending

Stress at limit of proportionality

$$(\text{lb/in}^2) = 157 \times \text{density (lb/ft}^3)$$
$$(\text{kg/cm}^2) = 0.7 \times \text{density (kg/m}^3)$$

Density conversions:

$$\text{Density (kg/m}^3) = \text{density (lb/ft}^3) \times 16$$

$$\text{Density (lb/ft}^3) = \frac{\text{density (kg/m}^3)}{16}$$

‡Solarbo Ltd.

Lignum vitae

Lignum vitae is the hardest and densest of all woods with a normal density range of 72 to 83 lb/ft^3 (1150 to 1330 kg/m^3). Due to the oblique and diagonal arrangement of successive layers of its fibres it cannot be split readily like most other woods and it is also resistant to wear and compression.

TABLE 3 – PHYSICAL PROPERTIES OF LIGNUM VITAE

Density	72 to 84 lb/ft^3 (0.042 to 0.0485 lb/in^3)
Specific gravity	1.17 to 1.32
Modulus of rupture	11 200 lb/in^2
Crushing strength	10 500 lb/in^2
Recommended working pressure	2 000 lb/in^2 maximum
Resin content (average)	28–30%
Expansion in water (typical)	4% by volume

TABLE 4 – RESISTANT PROPERTIES OF LIGNUM VITAE

Resistant to	Affected by
Water	Heat in excess of 66.6°C
Salt water	Alcohol
Weak acids	Ether
Weak alkalis	Many other organic solvents*
Many oils*	Strong acids
Bleaching compounds	Strong alkalis
Liquid phosphorus	
Many chemical solutions*	
(*eg* copper chloride)	
Fruit juices	

*General answers must not be assumed. Spot tests are necessary to confirm results for specific applications.

TABLE 5 – GENERAL FABRICATION
(All machining operations are normally carried out with the material dry)

Operation	Tools	Remarks
Cutting (machining)	High speed steel; carbide tools are not necessary but may be used	7° rake. 15° clearance. Machine dry at high cutting speeds, 1000 ft/min minimum.
Drilling	High speed twist drills	High cutting speeds preferred.
Drilling large holes	Trepanning preferred	Billet removed can be used for smaller parts.
Tapping	Hand taps	Normal technique suitable for internal threads.
Screw cutting (external threads)	Mill cutter	Difficult to cut and normally avoided; a standard die will tear the wood and produce faulty threads.

As cut, the lignum vitae tree produces a log with an outer wood which is yellow in colour and heartwood, in marked contrast, being a very dark greenish-brown. The heartwood accounts for the majority of the cross-section and has a high resin content of between 26 and 30%; the outer wood contains no resin.

The true lignum vitae of commerce is the heartwood of the guaiacum tree. A number of other woods of similar appearance, but inferior characteristics, are marketed under the same name and referred to as bastard lignum vitae. These lack the major properties which make true lignum vitae a highly satisfactory bearing material.

The diameter of the heartwood in typical lumber ranges from 4 to 24 inches maximum, in lengths of 4 to 6 ft. All small components are normally machined from the solid, but segmental construction is preferred for larger parts, eg large bearings up to 3 ft diameter or more. The cost of lignum vitae tends to vary with the source of the supply.

The self-lubricating, non-contaminating and corrosion resistant properties favour its application for bearings, bushings, rolls, sheaves, friction blocks, pulleys and wire guides, where conventional materials may show distinct limitations, eg roll neck bearings; lignum vitae may show a life of up to ten times that of bronze or white metal bearings under similar service conditions. The material is also particularly suitable for submerged bearing applications in aqueous or similar solutions corrosive to metals.

Recommended practice is that bearings should not be machined to finished size until immediately before use. Preferably, the part-machined bearing should be brought into the area in which it is to be installed and left to stand for several days to adjust itself to ambient atmospheric conditions. After this it is finished to size and coated with shellac or paraffin to seal the moisture content and should be installed as soon as possible. If installed at once, this moisture seal is

TABLE 6 – TYPICAL APPLICATIONS OF LIGNUM VITAE BEARINGS

Applications	Working conditions	Reasons for selection	Remarks
Submerged phosphorus pump	1200 r/min 49.9 to 65.6°C continuous	Resistant to phosphorus	Perfectly satisfactory bearing life, five years.
Tanning drums	Room temperature	Low cost	Inferior to bronze bearings.
Rolls in bleach plant		Resistant to bleach solution	Better service than stainless steel.
Food processing plant	Continuous	Non-contaminating, self-lubricating	Perfectly satisfactory long life.
Roll neck bearing steel mill			Longer life than bronze or babbitt.
Bearing in agitator tank aqueous solution	Continuous 75 r/min	Most suitable material available	15 year life.
Foot bearing in acid tank	Continuous 35 r/min	Most suitable material available.	Satisfactory.
Propeller shaft bearings (marine)	Continuous	Self-lubricating. Resistant to sea water.	Highly satisfactory
Pulley bearing	Intermittent loading	Comparison	Longer life than brass or bronze.
Chemical plant	Exposure to copper chloride solution	Most suitable material available	Perfectly satisfactory.
Paint manufacturing plant	Exposure to coal-tar solvents and 15% acid, 65.6°C	Comparison	Not suitable for working temperature involved
Film spool bearings	Constant dry heat 32.2°C		Perfectly satisfactory.
Headboard pulley	Intermittent mechanical loading		Comparable with ball bearing for wear.
Textile machinery	Continuous	Self-lubricating properties	Perfectly satisfactory

not necessary. In the case of a bearing operating under submerged conditions, it should be 'conditioned' by allowing to soak for several days in the appropriate solution, rough finished, re-soaked and then finish machined. If there is any appreciable delay between finishing and installation, a moisture seal coating is again advised.

It is normally machined using conventional metal-cutting machinery and working with a surface speed of at least 1000 surface ft/min. High speed steel cutting tools are used with a recommended rake of 7° and clearance 15°. Chips tend to come off in the form of a heavy yellow dust, which is not harmful. This dust is very oily, however, and will tend to stick and clog on the machine.

The wood is drilled with high speed twist drills. Large holes are best trepanned so that the core removed can be used for other articles instead of becoming waste. Internal threads can be tapped successfully by hand, but external threads should be machined, as a die will tend to tear out pieces of the formed thread.

Leather

LEATHER is a natural fibrous material with good elasticity, strength and durability, although its properties vary with the source of hide and the part of the hide from which it is cut. In general the cheeks, shoulder and belly portions of a hide have a more open textured fibrous structure and lower strength than the butt (Figure 1). The strength of butt leather is also more consistent over the whole of its area.

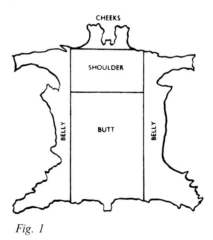

Fig. 1

The main application of leather in engineering is for seals and packings for hydraulic and pneumatic services. Originally engineering leathers were *vegetable tanned* (also known as *oak tanned*) using the heaviest hides available. Vegetable tanned leathers have good strength and substance but their maximum service temperature is limited to 50°C (122°F).

Most modern industrial leathers are of the chrome retan type with a tensile strength in excess of 420 bar (6000 lb/in^2). Maximum service temperature may be as high as 150 to 160°C, depending on the treatment, and also to some extent

TABLE 1 – GENERAL PROPERTIES OF LEATHERS

Type	Tensile strength range		Maximum service temperature	
	kg/cm^2	lb/in^2	°C	°F
Oak tanned	210–420	3000–6000	50	122
Chrome tanned	280–490	4000–7000	75	167
Chrome retan	420+	6000+	150–160	300–320

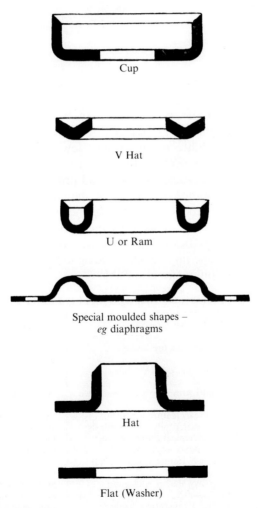

Cup

V Hat

U or Ram

Special moulded shapes –
eg diaphragms

Hat

Flat (Washer)

Fig. 2
Leather seal sections.

Leather cup packing

Leather flange packing

Leather U packing

Leather V packing

Leather washers, gaskets and diaphragms

Typical forms of leather seals and packings.

on the surroundings. Mechanical performance may be further improved or modified by coating or impregnating the material. Silicon rubber impregnation can be used to maintain flexibility down to very low temperatures (–60 to –90°C), although this appreciably increases the cost of the material. Triglyceride and stearine stuffing improves resistance to water and produces a flexible greasy leather for seals and packings, although in this case the impregnant is soluble in oils. Paraffin wax impregnant produces a stiffer leather which is exceptionally water resistant, and is again used for water seals and packings. Acrylic emulsion treatments produce a tough, very flexible grain surface which is oil resistant but permeable to water. Polyurethane impregnant yields good resistance to oils and water, coupled with good flexibility (see also Table 2).

TABLE 2 – PROPERTIES OF IMPREGNATED LEATHERS

Impregnant	Leather properties	Typical applications
Cellulose (surface finish only)	Toughens and seals the leather.	Dry services, *eg* seals for gases.
Acrylic emulsion (surface finish only)	Toughens and increases flexibility of surface.	Oil services – not water resistant.
Paraffin wax	Stiff and hard with very good resistance to water.	Water services.
Polyester (surface finish only)	Stiffens and seals leather for high resistance to water, oils and chemicals.	Rigid leather forms (*ie* too stiff for seals).
Polysulphide	Improves mechanical properties and chemical resistance of leather.	High pressure seals and packings.
Polyurethane	Flexible, greasy leather with excellent resistance to water.	Water services – not oil resistant.
Triglyceride and stearine	Flexible, greasy leather with good resistance to water.	Seals for water services – not oil resistant.
Silicon	Improves low temperature flexibility.	Low temperature seals (*eg* refrigerant seals).

TABLE 3 – COMPATIBILITY OF LEATHERS‡

Fluid	Compatibility		
	Good	Fair	Limited
Acetone	X		
Amyl acetate		X	
Amyl alcohol			X
Aniline			X
Asphalt			X
Beet sugar liquors			X
Benzene			X
Benzol			X
Butyl acetate		X	
Butyl alcohol			X
Cane sugar liquors			X
Castor oil	X		
Chlorinated solvents			X
Cod liver oil	X		
Corn oil	X		
Cotton seed oil	X		
Creosote			X
Diesel fuel	X		
Ethyl acetate		X	
Ethyl alcohol		X	
Ethylene glycol	X		
Fuel oil	X		
Gelatin	X		
Glue	X		
Glycerol	X		
Greases, all types	X		
Hydraulic oils, all types	X		
Kerosene		X	
Ketones	X		
Lacquers		X	
Lactic acid			X
Linseed oil	X		
Lubricating oils	X		
Mercury			X
Methyl alcohol		X	
Mineral oils	X		
Naphtha			X
Nitrobenzene			X
Paraffin		X	
Petrol			X
Petroleum oils, crude	X		
Sewage			X
Soya bean oil	X		
Steam			X
Tar			X
Toluene			X
Trichlorethylene			X
Tung oil	X		
Turpentine		X	

‡ For fluids not mentioned, either leather is not normally suitable or data is lacking.
Consult leather seal or packing manufacturers in such cases for specific information.

The main advantages leather can offer as a seal medium are good conformity, even against rougher surfaces; good flexibility; toughness; resistance to abrasion; and good retention of lubricant. Being a soft material it also has minimal abrading effects on other materials with which it is in rubbing contact. Its main limitations are the extent to which it can be moulded (limiting the form of seal sections which can be reproduced in leather) and lack of rigidity and stretch. Leather seals are therefore limited to cup, flange and hat sections, U-packings and V-rings (chevron seals).

Leather is resistant to a wide range of fluids including oils, but can be affected by many aqueous solutions, acids, alkalis and organic chemicals. A general guide on the subject of the compatibility of leathers is given in Table 3, but as explained previously, the resistance of leathers to specific fluids, etc can be improved by impregnation or stuffing.

SECTION 4

Thermoplastics

Introduction

THERMOPLASTIC MATERIALS are those which soften under the action of heat and harden again to their original characteristics on cooling, *ie* the heating-cooling cycle is fully reversible. Their properties are thus temperature-dependent, but may be modified by plasticisers or other additives, and in some cases by varying degrees of crystallinity.

By definition, thermoplastics are straight and branched linear chain organic polymers with a molecular bond. They are usually isotropic and homogeneous, but sometimes partly crystalline.

'Beetle' nylon 6 mouldings for gear rewind pulley and starter mechanism on a Vickers auxiliary petrol engine.

The following are generally recognised as 'engineering plastics'.

(i)   Nylon
(ii) Acetal – copolymer and homopolymer
(iii) Polycarbonate
(iv) PET
(v)   PPO
(vi) Thermoplastic polyester

This does not, however, exclude other types of plastics from being used in 'engineering' applications.

ABS/SAN/ASA

General

ACRYLONITRILE BUTADIENE Styrene (ABS) products were originally produced as blends of styrene acrylonitrile (SAN) with rubber. They were tough
materials, but in other respects were inadequate and were replaced by terpoly-
mers. Physically ABS is a two phase system in which SAN forms the matrix (*ie*
the rigid phase), and the rubber, usually polybutadiene, is combined as a graft
terpolymer in the disperse phase. In these compounds the rubber provides the
toughness, the acrylonitrile the strength and chemical resistance, and the styrene
the processability. ABS plastics are therefore tough, rigid materials which are
easy to process by most common methods, can be coloured in the mass or
by colour concentrates at the machine, and are also easy to decorate. A major
feature of ABS is its retention of toughness at low temperatures.

The proportions of the components can be varied within limits, giving a wide
range of combinations of properties; toughness is the main one by which ABS is
classified. Generally as toughness (impact, ductility) is increased, other properties, such as tensile strength, are reduced.

Particular properties can be further enhanced by compounding with additives
to produce grades suited to particular applications, in particular for flame
retardant, heat resistant or easy flowing types. As with many plastics, glass fibre
can be added to ABS, but the effect on heat resistance and rigidity is much less
marked than with other materials, and much of the toughness is lost.

ABS blends/alloys

Several manufacturers now produce combinations of ABS with other polymers
(such as PC, nylon or PVC), which combined the properties of the two materials, generally giving a useful improvement in chosen properties, such as heat
resistance. See page 386 for more details.

Properties

ABS materials have an excellent balance of properties well suited to parts needing rigidity, which can be load bearing. High strength and rigidity and low creep

are combined with impact strength which ranges from good to exceptionally high, and low notch sensitivity. The toughness is retained at low temperatures, and unlike many plastics does not show a sudden drop as temperature falls. At the other extreme, ABS can be used at 80 to 85°C with heat resistant grades able to withstand more than 100°C for short periods. Use of ABS as an engineering material covers a large number of applications where these properties are important, but its use when fatigue endurance or outdoor weathering are required is limited, since it is deficient in these properties.

It is worth noting that U-V/weather attack takes the form of chalking and development of micro-cracks in the exposed surface. If the surface is subjected only to compression stresses the performance is much better. As an alternative the surface can be protected, in the case of a sheet based product by laminating onto the sheet a weather resistant film, or for an injection moulding by painting with, for example, a tough and flexible polyurethane coating.

The chemical resistance of ABS is good. It is resistant to alkalis, dilute organic and inorganic acids, aliphatic hydrocarbons and some oils and greases. It is, however, attacked under stress by some oils, causing environmental stress cracking (ESC), and is swollen or partially dissolved by organic solvents such as ketones, ethers and esters, aromatics and chlorinated hydrocarbons, except perchlorethylene.

ABS can be processed by most of the common methods for plastics, the most popular being extrusion (as film or sheet or as pipe), and injection moulding. It is very well suited to this process, being an amorphous material with low specific heat and low shrinkage. It therefore does not suffer from problems of warping on cooling, has high dimensional stability and can be used on very complex moulds with little limitation on gate locations etc; it can also be moulded on fast cycles.

ABS readily gives a high gloss on moulding and gives good reproduction of grained or textured surfaces; sheet can be produced with a high gloss, although for certain applications grades are available which give a semi-matt surface. ABS lends itself readily to decoration by methods such as hot stamping, vacuum metallising, plating, printing and painting, and can readily be assembled by glueing or welding processes such as ultrasonic or friction.

Available as a natural, translucent, straw tinted material and in a great variety of colours, ABS can be self-coloured (preferably with master batch) and one manufacturer now offers a material which can be colourd with lower pigment loadings. Although most physical properties of ABS are little affected by high pigment loadings, impact strength/ductility may be reduced. The material has lower surface resistivity than polystyrene and is therefore less prone to the formation of dust patterns, but for critical applications anti-static versions are available.

ABS can be classified by impact strength, as measured by Izod or Charpy impact tests and can be divided into three broad classes:

Medium impact – 10–15 kJ/m^2 (notched Izod).

High impact – 20–30 kJ/m^2 (notched Izod).

Very high impact – 30–60 kJ/m^2 (notched Izod).

The ductility and toughness of ABS improves with impact level, whilst the highest values of tensile strength, rigidity, hardness and creep resistance are obtained with medium impact grades. Heat resistant grades usually have stiff flow, whilst easy flow grades, useful for thin section moulding, have lower heat resistance.

Certain of these properties have been chosen to define a given type of material in ISO 2580 and BS 4935 for the classification of ABS (see Table 1). Other properties are shown in Table 2, which includes a typical range for those that vary with composition, and values for the properties which are for practical purposes independent of these variations.

Typical properties of some moulding grades are shown in Table 3, and of some heat resistant and other special purpose grades in Table 4. Extrusion materials are usually chosen from very high impact grades with stiff flow, and are sometimes blended with SAN resins on the machine to increase rigidity.

Applications

This combination of properties and processing behaviour has led to the wide use of ABS plastics in applications in motor cars, household appliances, luggage, pipes, and refrigerator liners many of these applications take advantage of its decorative appeal. Some examples are:

Appliances
Housings for vacuum cleaners, kitchen appliances, air conditioning units.
Refrigerator doors' inner liners and other interior parts.
Clocks, hair dryers, electric razors, bathroom cabinets.

Automotive
Radiator grilles, air vents, consoles and shelves, controls, parcel shelves, steering column and instrument housings.

Radio, TV and audio equipment
Housings for TV, radio and video. TV fascias.
Tape and video cassettes, record player chassis, control knobs and small components.

TABLE 1 – PROPERTIES CHOSEN FOR CLASSIFICATION OF ABS IN ISO 2580

	Limits				
	1	2	3	4	Units
AN content in continuous phase	10–30	>30	—	—	%
Vicat softening point (49N load)	<90	90–100	>100–110	>110	°C
Melt flow index (220°C)	<5	5–10	>10–20	>20	g/10 min
Impact strength: (Notched Izod)	30–100	>100–200	>200–300	>300	J/m
,, ,, (Notched Charpy)	3–8	>8–12	>12–18	>18	kJ/m^2
Flexural modulus.	>1.8	1.8—2.3	>2.3—2.8	>2.8	GPa

Business machines
Housings for mini-computers, dictating and photocopying machines. Keyboard housings. Telephones and telecommunications components.

Photography
Cameras, slide and film projectors. Flash units, exposure meters, processing equipment.

Power tools and garden equipment

Leisure and sports equipment

Luggage

TABLE 2 – GENERAL PROPERTIES

Property	Values	Units (SI)
Physical		
Flexural strength	50–85	MN/m^2
Compressive strength	50–90	MN/m^2
Punch shear strength	32–50	MN/m^2
Taber scratch test	10–29	g to initiate scratch
Hardness — (Rockwell R)	90–115	
„ — (Rockwell H)	13–60	
Coefficient of friction:		
ABS/ABS static	0.59	
ABS/ABS dynamic	0.51	
Thermal		
Thermal conductivity	0.16–0.22	W/m^2C
Coefficient of linear		
thermal expansion	6–11	$10^{-5} K^{-1}$
Specific heat (in range 60–140°C)	1.05–1.67	$KJ/(kg.K)$
Electrical properties		
Dielectric constant	2.6–3.0	
Dissipation factors, tan	$6–10x^{-3}$	
Miscellaneous		
Specific Gravity	1.04–1.1*	
Water absorption (24h):	0.3	%
„ „ (equilibrium)	0.7-0.9	%
Mould shrinkage	0.004–0.006	cm/cm^{-1}
	0.4–0.6	%
ASTM flammability (excluding		
Fr grades)	0.4–0.6	mm/s

* Depending on pigment loading or other additives.

TABLE 3 – ABS: TYPICAL PROPERTIES OF SOME MOULDING GRADES

Property	Unit	Medium impact	High impact				Very high impact		Plating	Test standard
Type / Grade		Lustran 240	Ronfalin FX	Terluran 9581	Magnum 3504*	Lustran 648	Cycolac LA	Ronfalin TX	Cycolac EPH 3560	
PHYSICAL										
Tensile yield stress	MPa	49	43	53	38	42	34	37	49	R527
Elongation at yield	%	2.6	6	3	–	2.8	–	6	–	"
Elongation at break	%	18		15	60	25				"
Tensile modulus	GPa	2.70		2.70	2.25	2.30	1.65		2.50	178
Flexural strength	MPa	88	75	82	65	69	54	65	81	"
Flexural elastic modulus	GPa	2.80	2.4		2.25	2.30	1.72	2.00	2.60	2039-2
Hardness — Rockwell R.			100		67		89	91	110	2039-1
— Ball indentation	MPa	120	88	113	87	92	69	75	80	
Impact strength:										
Notched Izod, 23°C	kJ/m^2*	14	22†	22	23	30	40	44†	21.0	180
" " -20°C		6	11†			19	14(-29°)	31†		
" " -40°C		3				8				
Notched Charpy, 23°C	kJ/m^2	8	14	10	15	17	16.7	24	9.0	179
" " -40°C		–		4	10	NB				
Un-notched Charpy, 23°C	kJ/m^2	60		NB	NB	NB				
" " -40°C		20		60	NB	38				
THERMAL										
Thermal conductivity	w/m.K			0.17						ASTM C177
Coefficient of linear thermal expansion		7.5		8–11		9.2	11.0		8.5	ASTM D696
Deflection temperature:										
Under flexural load at 0.5 MPa	°C	–		104		–	96/104‡			75
" " at 1.82 MPa	°C	89	83/92‡	100	-/99	80	86/99‡	84/92‡	92/102‡	
Vicat softening point (9.8N)	°C	105		101	110	101			–	306
" " B50 (49N)	°C	97	90		100	92	90	92	96	
MISCELLANEOUS										
Density	g/cm^3	1.07	1.04	1.05	1.05	1.04	1.02	1.03	1.06	R1183
Flammability		HB	HB	HB	HB	HB	HB	HB	HB	UL subj 94
Water absorption, 24 h	%	0.30		0.4	–	0.30				R62A
Mould shrinkage (flow direction)	%	0.4–0.6	0.4–0.7	0.4–0.7	–	0.4–0.6	0.8	0.4–0.7	0.6	ASTM D955
Melt flow (MFI) 220°/10kg	g/10min	14	50	15	4.5	19	3.5	4.5	15	R1133

* Some data sheets have values quoted in J/m; these are approximately ×10 values shown in kJ/m^2. Values in J/m are shown in kJ/m^2.   ** Gives a matt surface under suitable moulding conditions.

† 6.4 mm bar, otherwise 3.2 mm.   ‡ Un-annealed/annealed.

TABLE 4 – ABS: TYPICAL PROPERTIES OF SOME SPECIAL PURPOSE GRADES

Property / Grade	Type / Unit	Heat resistant	
		High heat	**Heat/impact**
		Cycolac X37	Lustran QE1355
PHYSICAL			
Tensile yield stress	MPa	50	55
Elongation at yield	%		3.2
Elongation at break	%		22
Tensile modulus	GPa	2.25	2.70
Flexural strength.	MPa	85	78
Stress at 3.5% strain	MPa		
Flexural elastic modulus	GPa	2.30	2.90
Hardness — Rockwell R		108	
— Ball indentation	MPa	94	110
Impact strength:			
Notched Izod, 23°C	kJ/m^2*	19.1	35
" " –20°C			14
" " –40°C		7.1 (–29°)	8
Notched Charpy, 23°C	kJ/m^2	8.0	16
" " –40°C			
Un-notched Charpy, 23°C	kJ/m^2		80
" " –20°C			
" " –40°C			35
FDI (Falling dart impact)	J		85
THERMAL			
Thermal conductivity	w/m.K		
Coefficient of linear			
thermal expansion		6.2	7.50
Deflection temperature:			
Under flexural load at 0.5 MPa	°C	115/120†	—
" " " at 1.82 MPa	°C	109/121	92
Vicat softening point 9.8N	°C	—	109
" " " 49N	°C	115	103
" " " B/120	°C		
MISCELLANEOUS			
Density	g/cm^3	1.04	1.06
Flammability		HB	HB
Limiting oxygen index	%		
Water absorption, 24 h	%	—	0.30
Mould shrinkage (flow direction)	%	0.5	0.4–0.6
Melt flow (MFI) 220°/10kg	g/10min	2.7	3
Anti-static rating			6/7

* Some data sheets have values quoted in J/m; these are shown as kJ/m^2.
   Values in J/m are approximately ×10 values shown in kJ/m^2.
† un-annealed/annealed.

(continued)

TABLE 4 – ABS: TYPICAL PROPERTIES OF SOME SPECIAL PURPOSE GRADES – *contd.*

Heat/rigidity	Anti-static	Flame retardant	Glass re'infd	Test standard
Novodur P3T	Lustran QE 1083	Cycolac KJBE	Novodur P2HGV	
54	42	34	62	R527
2.8	2.5		2.0	"
	20			"
3.00	2.60	2.0	6.0	"
	75	55		178
82			105	"
2.80	2.75	2.1	5.0	"
		96		2039–2
115	116	76	135	2039–1
	14	18	—	180
	6	6(−29°)	—	
	3	—		
10	10	6.5	6	179
2		2.7(−29°)	5	
100	65	85 NB	17	
—			—	
70	25		15	
	50			
0.15			0.2	ASTM C177
8.0	7.50	9.5	5.4	ASTM D696
108		—	104	75
101	85	82/93†	101	
—	104	—	—	306
—	95	88		
115			107	
1.04	1.06	1.22	1.18	R1183
HB	HB	VO: 1.47 mm/	HB	UL subj 94
		5V: 3.05 mm		
		30		ASTM D2863
0.3	0.30		0.3	R62A
0.5–0.7	0.4–0.6	0.6	0.1–0.4	ASTM D955
2	28	50	3	R1133
	2/3			Monsanto test

TABLE 5 – ABS: TYPICAL PROPERTIES OF SOME EXTRUSION GRADES

Property	Grade	Unit	Magnum 3105FP	Terluran KR2875	Novodur P3LE	Test standard
PHYSICAL						
Tensile yield stress		MPa	53	53	45	R527
Elongation at yield		%	—	3.0	2.5	"
Elongation at break		%	20	15	—	"
Tensile modulus		GPa	2.70	2.60	2.50	"
Flexural strength		MPa	85	82	—	178
Flexural stress at 3.5% strain		MPa	—	—	67	"
Flexural elastic modulus		GPa	2.75	—	2.20	"
— Ball indentation		MPa	116	100	90	2039-1
Impact strength:						
Notched Izod,	23°C	kJ/m²*	10	30	—	180
Un-notched Charpy,	23°C	kJ/m²	NB	NB	NB	179
" "	−20°C	kJ/m²	—	80–NB	95	
" "	−40°C	kJ/m²	80	65	17	179
Notched Charpy,	23°C	kJ/m²	7	12	10	
" "	−20°C	kJ/m²	6	—	7	
" "	−40°C	kJ/m²	5	4	—	
THERMAL						
Deflection temperature:						
Under flexural load at 0.5 MPa		°C	—	106	100	75
" " at 1.82 MPa		°C	99†	102	97	
Vicat softening point A/120 (9.8N)		°C	112	—	—	306
" " B/50 (49N)		°C	102	104	102	
" " B/120 (49N)		°C	—	—	—	—
MISCELLANEOUS						
Melt flow (MFI)		g/10min	5.5	9	8	R1133

* Some data sheets have values quoted in J/m; these are shown as kJ/m². Values in J/m are approximately ×10 values shown in kJ/m².

† Annealed.

SAN

Styrene acrylonitrile (SAN) as the name suggests, is a copolymer of styrene and acrylonitrile, which keeps the good processing qualities of polystyrene, but has considerably improved strength and chemical resistance, and better heat resistance. It can be regarded as the general purpose material in comparison with its tough relation, ABS.

SAN is a transparent material and all the current grades have good colour; depending on the AN content, this can be almost from water white to very pale straw. Some controlled variation in AN content and in molecular weight is possible in the continuous mass processes used by all manufacturers, so that grades are available to suit most applications. Manufacturers include BASF (Luran), DOW (Tyril), Monsanto (Lustran A) and Montedipe.

Properties

SAN plastics are noted for their excellent rigidity, which is amongst the highest for unfilled thermoplastics. They have high strength and surface hardness, with very low water absorption, and good chemical resistance. Electrical properties are very good and are practically unaffected by frequency. SAN plastics are transparent, with high clarity and gloss; light transmission is over 90% in the visible range, and low for U-V.

Although their impact strength as measured by the Izod or Charpy test is not high, their practical toughness is at least 50% greater than that of PS, since the energy to break depends on both the tensile strength and elongation.

Because they are amorphous, they are easy to mould, with low mould shrinkage, and have excellent dimensional stability; they can also be extruded as sheet or profile.

SAN is available in its natural or crystal tint, or in transparent, translucent or opaque coloured material. It can also be self-coloured, preferably by using a master batch based on the same material. It can also be compounded with glass fibre which considerably increases the already high rigidity.

Outdoor exposure causes a reduction in strength properties, yellowing, and roughening of the surface. These effects can be slowed down by the use of a U-V stabilised version.

Chemical resistance

This is in general very good. SAN resists saturated hydrocarbons, low-aromatic engine fuels and oils, vegetable and animal fats and oils, and aqueous solutions of salts, dilute acids and alkalis. It is attacked by concentrated inorganic acids, aromatic and chlorinated hydrocarbons, esters, ethers and ketones. Grades with higher AN content have much better chemical resistance.

In common with other styrene containing plastics, SAN can readily be bonded with suitable adhesives or welded by hot plate, friction or ultra-sonic methods.

TABLE 6 – TYPICAL PROPERTIES OF SOME GRADES OF SAN

Property	Grade	Unit	Lustran SAN 33	Tyril 790	Lustran SAN 35	Luran 388S	Tyril 602	Luran KR 2556	Luran 35% gl reinforced 378P G7	Test standard
PHYSICAL										
Tensile stress at yield		MPa	75	72	80	84	84	79	110	R 527
Elongation at break		%	2.4	2.6	3.0	3.5	3.5	3.5	2	,,
Tensile modulus		GPa	3.6	3.7	3.7	3.9	3.9	3.8	11.0	,,
Flexural strength		MPa	135	98	135	140	112	135	150	178
Flexural modulus		GPa	3.9	—	4.0	—	—	—	—	,,
Hardness — Rockwell, M scale			—	84	—	86	86	84	93	2039-2
— Ball indentation			140	—	140	175	—	175	240	
Impact strength:										
Notched Izod, 23°C		kJ/m²*	1.2	1.1	2.0	3	1.3	3	4.5	180
Un-notched Charpy, 23°C		kJ/m²	14	16	20	20	20	20	10	179
Notched Charpy 23°C		kJ/m²	—	0.9	—	3.5	1.1	3	4	179
THERMAL										
Deflection temperature:										
Under flexural load at 0.45 MPa		°C	94	101	96	103	103	110	108	75
,, at 1.82 MPa		°C	113	109	111	99	—	104	105	
Vicat softening point A/50 (9.8N)		°C	102	100	104	102	102	117	109	306
,, ,, B/50 (49N)		°C								
,, ,, A/120 (9.8N)		°C	—	—	—	—	111	—	—	1525/306
MISCELLANEOUS										
Melt flow (MFI) 220°C/10kg		g/10min	35	21	15	5.5	6.0	7.0	6	R1133

* Some data sheets have values quoted in J/m; these are shown as kJ/m². Values in J/m are approximately ×10 values shown in kJ/m².
† Unchanged at −40°C.

Typical properties of some of the available grades are given in Table 6, but properties not shown, which are common to all the grades are:

Refractive index   1.56
Flammability   2.5 to 4.5 cm/min
Density   1.07 to 1.08
Water absorption   0.25%
Mould shrinkage   0.3 to 0.6%
Electrical
Volume resistivity 10^{16}
Relative permittivity   2.8
Dissipation factor 0.8×10^{-4}
Tracking index (CTI)   375

Applications include:
  kitchen ware (mixer bowls, beakers, coffee filters, refrigerator fittings);
  radio and hi-fi (record player covers, cassettes, coil formers, tuning scales);
  industrial battery cases;
  sanitary and bathroom fittings;
  toothbrushes and cosmetic packs;
  medical dialysers.

RELATED POLYMERS: ASA/AES/MBS

There are several polymers in which one or more of the constituents of ABS have been replaced by other materials in order to extend the range of applications without losing the favourable properties of ABS.

The main ones commercially are:

  ASA – A plastic with good U-V and weather resistance.
  AES – For good weathering and very high impact.
  MBS – For clarity with impact strength.

ASA

Acrylate styrene acrylonitrile (ASA) polymer was developed by BASF to meet the need for a tough, easily processable material with good U-V and weathering resistance. The weakness of ABS in this respect is due to attack by U-V on the rubbery component (polybutadiene-PB). In ASA this is replaced by an acrylic ester. ASA is an elastomer like PB, but has good U-V resistance. As with ABS, the material is a 2-phase system, with the grafted rubbery component dispersed in the matrix of SAN.

The resultant polymer has most of the good properties of ABS (toughness, rigidity, creep resistance, low shrinkage) and almost as good processing be-

TABLE 7 – TYPICAL PROPERTIES OF SOME ASA GRADES

Property	Type BASF Grade No	Unit	Standard 757 R/RE	Super impact 797S/SE	High heat KR 2853/1 extrusion	Test standard
PHYSICAL						
Tensile yield stress		MPa	56	40	54	R527
Elongation at yield		%	3.1	4.3	3.5	"
Elongation at break		%	15	25	15	"
Tensile modulus		GPa	2.60	2.00	2.40	"
Flexural strength		G	75	60	80	178
— Ball indentation			100	65	100	2039–1
Impact strength:						
Notched Izod	23°C	kJ/m^2*	12	60	30	180
Un-notched Charpy	23°C	kJ/m^2	NB	NB	NB	179
" "	−20°C		50	NB	NB	
" "	−40°C		25	80–NB	40	
Notched Charpy	23°C	kJ/m^2	7	20	9	179
" "	−40°C		1	2	1	
THERMAL						
Thermal conductivity			0.17	0.17	0.17	DIN 52612
Coefficient of linear thermal expansion		$\times 10^{-5}$	8–11	8–11	8–11	ASTM D696
Deflection temperature:						
Under flexural load at 0.45 MPa		°C	101	100	108	75
" " " at 1.82 MPa		°C	97	95	104	
Vicat softening point B/50		°C	98	92	107	1525/306
ELECTRICAL						
Volume resistivity		ohm.cm	10^{14}	10^{14}	10^{14}	IEC 93
Relative permittivity			3.4	3.3	3.6	IEC 250
Dissipation factor		$\times 10^{-3}$	25	26	33	IEC 250
Dielectric strength ($\times$ mm)		kV/mm	90	100	115	IEC 243
MISCELLANEOUS						
Density		g/cm^3	1.07	1.07	1.07	R1183
Flammability			HB	HB	HB	UL subj 94
Water absorption, 24 h		mg	28	28	28	R62A
Mould shrinkage (flow direction)		%	0.4–0.7	0.4–0.7	0.4–0.7	1110
Melt flow (MFI)		g/10min	8	4	5	R1133

haviour; its heat resistance is better. As with ABS the proportions of the components can be varied. allowing the production of grades to suit different applications.

ASA can be processed in a similar way to ABS, but the applications tend to be biased towards outdoor use based on thermo-formed sheet.

Properties

Strength properties and impact resistance are very good and are in the same range as that of ABS. Whilst Vicat and HDT values are in general similar to those of ABS, the retention of strength properties on heating is superior to some types of ABS; there are now development grades with Vicat equal to that of some high heat ABS. Water absorption and dimensional stability are similar to that of ABS, and ASA will also give a high gloss on moulding. Properties of a selection of grades are shown in Table 7.

ASA can be coloured in the mass as with ABS. Mouldability is good, although there are no grades with very soft flow for thin wall moulding. Extrusion behaviour is good. As in the case of ABS, the material must be dry; the finish obtained is egg-shell. ASA can be machined, bonded or decorated similarly to ABS.

Applications

Some examples of uses of ASA are:

Automotive	– Moped engine covers, rear view mirrors and rear lamp mouldings for HGV, caravan window frames.
Garden	– Garden tables and chairs, hose fittings, parts for garden tools, lawn mower covers and housings.
Roads and streets	– Lighting canopies.
Buildings	– Letter boxes, roller brackets and sides for roller blind housings, wash basins.
Sport and recreation	– Boat shells, windsurfer boards.

AES

Acrylonitrile EPDM styrene (AES) materials were originally developed in Japan and the USA and use an ethylene-propylene rubber of the EPDM type to provide the toughness.

A recent similar type of product is Ronfaloy E, produced by DSM, which is claimed to have even better weathering resistance than ASA, and impact strength values higher than those of current ASA grades. Strength properties, dimensional stability and water absorption are similar to those of ABS, and the material can be processed similarly to ABS. Finish obtained can be varied from gloss to matt, depending on the processing conditions.

TABLE 8 – TYPICAL PROPERTIES OF RONFALOY E AND MBS

Property Type	Unit	Ronfaloy E	Cyrolite	Test standard
PHYSICAL				
Tensile yield stress	MPa	35–46	41–45	R527
Elongation at break	%	60	30	"
Flexural strength	GPa	60	66–70	178
Flexural modulus	GPa	2.0–2.2	2.0–2.2	"
Hardness — Rockwell R		85		2039–2
— Ball indentation			105	2039–1
Abrasion — Taber CS17 wheel, 1k. h., 1 kg	mg	30		ASTM D1044
Impact strength:				
Notched Izod, 23°C	kJ/m^2	40–70		180
" " –20°C		6–15		
Un-notched Charpy, 23°C	kJ/m^2		60	179
" " –40°C			34	
Notched Charpy 23°C	kJ/m^2	25–40	6.5	179
" " –20°C		11–18	4.9	
THERMAL				
Coefficient of linear thermal expansion	$\times 10^{-5}$		9	ASTM D696
Deflection temperature:				
Under flexural load at 1.82 MPa	°C	77–83	85	75
" " " at 1.82 MPa	°C	90–95		
Vicat softening point, B/50	°C	81–84	95	306
MISCELLANEOUS				
Density	g/cm^3	1.06	1.11	R1183
Flammability		HB		UL subj 94
Mould shrinkage (flow direction)	%	0.4–0.7	0.3–0.6	1110
Melt flow (MFI) 220°/10kg	g/10min	8–15	1–4	R1133
Transmittance.	%		85	DIN 5036

Chemical resistance is probably slightly better than that of ABS and Table 8 shows a typical range of values for current grades of Ronfaloy E.

Applications

Main applications so far have been on thermo-formed sheet and include:

Camper tops (a severe test of weathering capability).
Coextrusion with ABS to provide a weatherable surface.
Superstructure panels on power boats and cruisers.
Outdoor jacuzzi.
Panels and extruded frame profiles for domestic front doors.

MBS

Methacrylate butadiene styrene (MBS) material has the advantage of being transparent, whereas ABS, ASA and AES are translucent. Until recently this polymer has found its main use as an additive for PVC, to increase toughness, but it is now being marketed by Rohm as a polymer in its own right, called Cyrolite G20, or Cyrolite G20 HiFlo (see Table 8 for typical properties).

Processing is similar to that of ABS, but feed screws need to be carefully cleaned if maximum clarity is to be obtained. Bonding, printing and decorating present no problems.

MBS has good chemical resistance, in particular to oils, fats and fuels. Weather resistance, as in the case of ABS, is not good, due to the presence of polybutadiene rubber.

Applications

The main applications of MBS are for:

Crisper drawers and panels for refrigerators and freezers.
Writing and drawing instruments.
Sheet for blister packaging of medical supplies (MBS can be gamma-ray sterilised without noticeable yellowing).
Disposable medical and surgical instruments and accessories.
Bottles and containers for cosmetics, fine chemicals, medical use.

Acetal

ACETAL RESINS are polymers of formaldehyde. Both homopolymers and copolymers, in which small amounts of comonomers are added during polymerisation, are available. Both have extremely good properties and find wide use as engineering materials. The homopolymers have higher values for physical properties and a slightly higher crystalline melting point; the copolymers have better resistance to alkalis and to hydrolysis at elevated temperature. At one time the copolymer grades generally had better flow, but the latest versions of homopolymer materials have now similar values.

Acetals have a linear structure and high crystallinity. These characteristics provide an excellent combination of properties which has enabled them to replace metals, and other thermoplastics, in many applications. In addition to high rigidity, high strength and impact strength, they have excellent fatigue resistance and low friction behaviour, together with good chemical resistance, including resistance to solvents.

Acetals can be processed by all normal methods for thermoplastics; the most popular conversion methods are injection moulding and extrusion of rod, sheet, etc, for subsequent fabrication. A wide variety of grades is produced with additives, in some cases, where superior properties are needed. Classification is usually by melt flow index as a reflection of the molecular weight.

Properties

Acetals have high strength and stiffness, good toughness properties over a wide temperature range, low creep, and good chemical resistance and electrical properties. They also have low water absorption, which means that they have very good dimensional stability under varying humidity conditions. These properties, combined with the superior fatigue resistance and low friction characteristics, have led to extensive use of acetals in demanding engineering applications.

An important use of acetal is in bearings and gears, where it is competitive with nylon and gives somewhat higher PV values in bearings than nylon. Best results for friction and wear are obtained with acetal against steel. If a non-

TABLE 1 – TYPICAL PROPERTIES OF SOME ACETAL HOMOPOLYMERS

Property	Grade	Unit	General	Hydrolysis resistant 200PL	Low frictional low wear 500AF (Du Pont Delrin)	500CL	Reinforced 570/577	Toughened 100ST	500T	Test standard
PHYSICAL										
Tensile yield stress		MPa	67–72	70	50	70	62	43	57	R527
Elongation at yield		%	5–25	20	5	14	4	70	20	"
Elongation at break		%	15–80	65	15	45	12	200	60	"
Tensile modulus		GPa	3.1	3.1	2.9	3.1	6.2	—	—	"
Flexural strength		MPa	76–84	87	61	78	64	44	78	178
Flexural modulus		GPa	2.8	2.4	2.4	2.8	5.0	1.4	2.4	ASTM D695
Compressive strength (1% deformation)		MPa	22–36	22	31	31	36	8	16	ASTM D621
Deformation under load		%	0.5–0.8	0.7	0.6	0.7	0.4	3.0	0.9	ASTM D732
Shear strength		MPa	66	66	55	66	66	34	48	
Torsional shear modulus		MPa	790–975	730	—	890	1340	370	740	537
Hardness — Rockwell, R			120	120	118	120	118	105	117	2039–2
" " M			93	92	78	90	90	58	79	"
Impact strength										
Notched Izod	23°C	kJ/m²†	7–13	12	2.9	7.8	4.3	91	13.5	180
Notched Izod	-40°C		5–10	9.6	—	—	2.7	—	—	
Un-notched Izod	23°C	kJ/m²	80–NB	NB	45	250	—	NB	NB	179
Notched Charpy	23°C	kJ/m²	8–13	12	3.5	8	3.5	60	14	
Notched Charpy	-40°C		4.0–9.5	9.5	3.0	6	3.0	11	6	
Un-notched Charpy	23°C	kJ/m²	NB	NB	30	NB	30.0	NB	NB	
Flexural fatigue endurance, 50% RH, 1 MHz		MPa	28–31	29	24	28	31	16	25	ASTM D671
Friction coefficient, static			0.20	—	0.08	0.10	—	—	—	Mfr's test
" " dynamic			0.20–0.35	0.20	0.14	0.20	0.35	—	—	"

TABLE 1 – TYPICAL PROPERTIES OF SOME ACETAL HOMOPOLYMERS – contd.

Property	Grade	Unit	General	Hydrolysis resistant 200PL	Low frictional low wear 500AF	Low frictional low wear 500CL	Reinforced 570/577	Toughened 100ST	Toughened 500T	Test standard
THERMAL										
Thermal conductivity		W/m.K	0.3–0.4	0.3	—	—	—	—	—	ASTM D257
Coefficient of linear thermal expansion			11–14	11–14	11–12	11–12	4–8	12	12	ASTM D696
Deflection temperature										
Under flexural load, at 0.5 MPa		°C	170	172	168	170	174	145	169	75
" " at 1.82 MPa			130	130	118	124	158	64	85	
Vicat softening point, B120		°C	162	162	—	162	162	116	139	306
" " A50			174	174	174	174	174	168	171	
ELECTRICAL										
Volume resistivity		Ω.m	1×10^{13}	1×10^{15}	3×10^{14}	5×10^{12}	5×10^{12}	2×10^{14}	2×10^{15}	IEC 93
Dielectric constant			3.7	3.7	3.1	3.5	3.9	4.1	3.6	IEC 250
Dissipation factor $\times10^3$			5–7	5	9	6	5	7	16	IEC 250
Arc resistance‡			200–220	200	180	180	180	120	120	ASTM D495
MISCELLANEOUS										
Density		g/cm³	1.42	1.42	1.54	1.42	1.56	1.34	1.39	R1183
Water absorption, 24 h		%	0.25–0.40	0.32	0.20	0.27	0.25	—	—	62
Mould shrinkage (flow direction)		%	1.9–2.4	2.2–2.4	1.9–2.2	—	1.2–1.4	1.2–1.4	1.7–1.9	1110
Melt flow (MFI)		g/10 min	2–24	—	—	—	—	—	—	R1183

Melt temperature for all these grades is 175°C. All grades are HB (UL94).

†Some data sheets have values quoted in J/m; these are shown as kJ/m².

Values in J/m are approximately ×10 values shown in kJ/m².

‡ All grades show no tracking.

TABLE 2 – TYPICAL PROPERTIES OF ACETAL COPOLYMERS

Property	Unit	General	BASF z2320	Hoe-Cel C2521	Low friction low wear Hoe-Cel C9021TF	Reinforced grades	BASF N2200G5	Toughened	Hoe-Cel S9064	Test standard
PHYSICAL										
Tensile yield stress	MPa	62-71	65	62	49-65	41-125	135(br)	40-53	42	R527
Elongation at yield	%	—	7.0	—	—	2-14	3	60-90	90	¨
Elongation at break	%	15-35	15	35	15-25	3.1-14	9.0	1.7-2.2	1.7	¨
Tensile modulus	GPa	2.7-3.2	2.95	2.75	2.4-3.0	3.2-11		70-150	58	178
Flexural strength	MPa	94-102	—	94	2.4-2.9	70-145		21-60	40	
stress at 3.5% strain	MPa	67-75	—	69	—	—				
Hardness — Rockwell, R		—	—	—	—	—		—	—	2039-2
— Ball indentation		143-150	—	144	120-145	160-215		90-115	86	2039-1
Impact strength										
Notched Izod 23°C	kJ/m²†	4-7	4	—	5-5.5		5.5		—	180
Un-notched Izod 23°C	kJ/m²	80-190	80	—	90-100		32		—	179
Notched Charpy 23°C	kJ/m²	4.0-6.5	3-5	6.5	3.5-5.5	2.5-4.0	—	6-20	10	
Notched Charpy -40°C		3.0-5.5	—	5.5	3.0-4.5	4.5	—	4-5	5	
Un-notched Charpy 23°C	kJ/m²	—	NB	NB	—	30-70	30		NB	
Un-notched Charpy -40°C		—	60-80	—	—	25-60	25			
THERMAL										
Coefficient of linear thermal expansion		11	11	11	11	1-8†	3	15	16	DIN 5232
Deflection temperature:										
Under flexural load at 0.5 MPa	°C	158-164	160	—	150-160	—	—	—	—	
at 1.82 MPa	°C	97-113	110	101	98-110	108-162	162	83-91	83	75
Vicat softening point, B50	°C	150-158	150	151	143-151	151-162	162	124-140	124	306
B50	°C	—	—	—	—	—	—	—	—	
ELECTRICAL										
Volume resistivity	Ωm	$>10^{15}$	10^{15}	$>10^{15}$	$>10^{15}$	$>10^{15}$	10^{14}	10^{14}		IEC 93
Dielectric constant.		3.7-3.9	3.8	3.9	3.6-4.0	3.8-4.8	4	—		IEC 250
Dissipation factor $\times 10^3$		5	2.5	5	5	5	5	—		IEC 250
MISCELLANEOUS										
Density	g/cm³	1.41	1.41	1.41	1.41-1.52	1.47-1.72	1.58	1.38	1.37	R1183
Flammability		HB	HB	HB	HB	HB	HB	HB	HB	UL subj 94
Water absorption, 24h	% or mg	15 mg	0.2%	15	10-15 mg	20-45 mg		—		R62A
23°C/50%RH	% or mg	—	—	0.2%	0.2%		0.15	—	0.25	1110
Mould shrinkage (with flow)	%	1.5-2.5	1.9	2.0-2.2	1.8-2.1	0.3-0.5	0.7	1.7	1.5-1.8	ASTM D955
Melt flow index (MFI) 190/2.16kg	g/10min	1-50	50	2.5	8	4-11	4	7.20	7	R1133

Hoe-Cel is an abbreviation for Hoechst-Celanese.
†Some data sheets have values quoted in J/m; these are shown as kJ/m².
Values in J/m are approximately ×10 values shown in kJ/m².

metallic combination is required, it is better to use acetal/nylon rather than acetal/acetal. Choice between acetal and nylon for gears depends upon the application; acetal gears are superior to those made from nylon in fatigue resistance, dimensional stability and stiffness, whereas nylon gears in conditions of average humidity have better resistance to abrasion and impact fatigue.

Chemical resistance of acetals is good and as mentioned above, the copolymer has better resistance to akalis and to hydrolysis at elevated temperatures. Both types are very resistant to solvents and fuels including those containing methanol. They resist dilute acids but are attacked by strong acids and oxidising agents. Acetals are not subject to problems of ESC (stress cracking in aggressive environments).

Resistance to U-V and weathering is not good, but some protection can be provided by incorporation of suitable additives.

Acetal parts can be assembled by conventional means such as screws or snap fit joints, or by hot tool or ultra-sonic welding. Adhesive bonding is more difficult because of the solvent resistance of acetals, and pressure sensitive adhesives have to be used, preceded by surface preparation. Decoration (including electroplating) is also possible, after surface preparation.

Grades are available to suit most purposes, and are generally divided into:

Basic grades.
Grades with improved sliding properties.
Glass/mineral reinforced products.
Impact modified grades.

Table 1 shows typical values for these types in homopolymer, and Table 2 for copolymers.

Applications

Acetal finds widespread use in automotive engineering because of its excellent fuel resistance in conjunction with its good engineering properties. Examples include fuel tank gauge units, fuel pump parts, window lift mechanisms and cranks, brake servo unit parts, door handles, safety belt mechanisms and clips.

Mechanical engineering applications include rollers, gears, control cams and transmission systems, precision parts for instrumentation, and medical equipment.

Other applications for acetals range from pump impellers and housings, pipe fittings and bath and shower fittings, to ski bindings, jug kettle housings, zip fasteners and gas lighter parts.

Acrylic

ACRYLIC PLASTICS are polymers of methyl methacrylate (PMMA). They are noted for their optical clarity, lustrous appearance, surface hardness and exceptional weathering resistance. They are strong, rigid materials and although the impact resistance of the standard grades is not high, this has not inhibited their use in a wide range of applications.

Where impact resistance is important, impact modified grades can be used. Current types show little or no loss of clarity compared with standard grades, although transmission drops by about 5%. Note that the gain in impact resistance is at the expense of strength values (see Table 1), and there may also be a reduction in weathering resistance.

Acrylic plastics are available as sheet, block, rod or tube (which may be cast or extruded), or moulding powders. In addition, sheet may be in the form of corrugated or double or triple wall. Because of their high light transmission and clarity, acrylics produce brilliant transparent tints and colours.

Properties

Acrylic plastics are amorphous polymers which are rigid and strong, with high creep resistance. They are easy to mould, with low shrinkage, giving parts which are dimensionally stable and have a high gloss. They are also easy to thermo-form.

Their scratch resistance is high for a thermoplastic, and can be improved by a polysiloxane (silicone) coating. Whilst the elongation at fail, and therefore the toughness of the standard grades is low, impact grades are available which are used where resistance to shock is vital. However, for most purposes the standard grades are adequate, as is shown by comparing the impact resistance of acrylic with that of plate glass, measured by a Falling Ball test:

Sheet material	Thickness (mm)	Drop height (mm)
Perspex grade 000†	03	300
Perspex grade 000	5	800
Perspex grade IMO14	3	3200
Plate glass	6.35	250
Toughened glass	6.35	1520

†Standard cast impact modified

TABLE 1 – PROPERTIES OF SOME ACRYLIC MOULDING GRADES

Property	Unit	Standard			Impact modified		Test standard‡
Type / Grade		Rohm 6H	ICI ME135	Rohm HW55	ICI TD525	Rohm zk50	
PHYSICAL							
Tensile strength	MPa	68	72	80	61	20	R527
Elongation at break	%	4	4	3.5	25	50	,,
Tensile (E) modulus	GPa	3.3	3.35†	3.5	2.45†	0.6	,,
Tensile creep resistance (10^4h)	MPa	28		33			DIN 53444
Flexural strength	MPa	110		115?		45	178
Compressive yield stress	MPa	100		110			DIN 53454
Shear modulus G (10Hz)	GPa	1.70	1.70	1.70			DIN 53445
Hardness – Rockwell, M	—	M89	M103	M100	M75		2039–2
– Ball indentation	MPa	180		200		40	2039–1
Martens scratch resistance	N	0.025		0.025			—
Impact strength:							
Un-notched Charpy 23°C	kJ/m²	11	3.6	10	7	NB	179
Notched Charpy 23°C		2		2		7	179
OPTICAL							
Refractive Index		1.491	1.491	1.509	1.491	1.492	DIN 53491
Transmission (Mean value 380–780 n.m)	%	92	92	92	90	85	DIN 5036

(continued)

TABLE 1 – PROPERTIES OF SOME ACRYLIC MOULDING GRADES – *contd.*

	Units						Standard
THERMAL							
Thermal conductivity	W/m.K	0.19		0.19		0.19	DIN 52612
Coefficient of linear thermal expansion (0–50°C)	$10^{-5}K^{-1}$	7	7.1	7	10.3	9	VDE 0304
Deflection temperature:							
Under flexural load at 0.46 MPa	°C	87	—	114	—	—	75
" " at 1.82 MPa	°C	82	98	109	92	68	75
Vicat softening point, B120	°C	92	111	119	106	76	DIN 53460
Max service temperature without load	°C	82		109		68	—
ELECTRICAL							
Volume resistivity	ohm.cm	$>10^{15}$		$>10^{15}$		2×10^{14}	IEC93
Dielectric constant at 50 Hz		3.7	3.7	3.5	3.1	—	IEC250
Dissipation factor at 50 Hz		0.06	0.08	0.06	0.05	0.06	
Tracking resistance	grade	KC>600		KC>600		—	DIN 53480/A
MISCELLANEOUS							
Density	g/cm³	1.18	1.18	1.18	1.16	1.12	R1183
Ignition temperature	°C	430		430		—	DIN 51794
Water absorption, 24h	mg or %	30	0.3%	35	0.3%	42	53495/R62A
Melt flow (MFI) – 230°C/3.8kg	g/10min	3	0.9	1.0	2.6	0.1	DIN 53735

Mould shrinkage: All grades of acrylic plastics have values 0.4–0.7%, depending on conditions.
†Flexural modulus.
‡ISO standards unless otherwise indicated.

TABLE 2 – PROPERTIES OF ACRYLIC SHEET (TYPICAL VALUES), INCLUDING CAST COMPARED WITH EXTRUDED

Property	Unit	Perspex Cast	Perspex Extruded TX	Plexiglas Cast GS	Plexiglas Extruded XT	Altuglas Cast	Perspex Impact modf. IM014	Test standard
PHYSICAL								
Tensile yield stress	MPa	80	70	80	72	65–75	62	R527
Elongation at break	%	—	—	5.5	4.5	3.0	—	:
Tensile modulus	GPa	3.20	3.21	3.3	3.3	3.0	3.37	:
Flexural strength	MPa	116	107	115	105	110	113	178
Flexural fatigue strength, 1MHz:								
un-notched	MPa	—	—	40	30	—	—	
notched	MPa	—	—	20	10	—	—	
Flexural elastic modulus	GPa	3.21	3.16	—	—	—	2.96	178
Compressive yield stress	MPa	—	—	110	103	110	—	ASTM D695
Dynamic shear modulus (10 Hz)	GPa	—	—	1.7	1.7	1.7	—	DIN 53445
Hardness – Rockwell M	GPa	102	101	115	—	95	98.5	DIN 53456
– Indentation	MPa	—	—	200	190	—	—	2039/1
Impact strength:								
Un-notched Charpy	kJ/m^2	12	10	12	12	—	21.7	179
Notched Charpy	kJ/m^2	—	—	2	2	—	1.2	
Friction coefficient:								
Plastic/plastic		—	—	0.8	0.8	—	—	
Steel on plastic		—	—	0.45	0.45	—	—	
OPTICAL								
Refractive Index		1.489	1.490	1.491	1.491	1.493	—	DIN 53491
Transmission (at 3 mm)	%	92	92	92	92	92	92	DIN 5036
THERMAL								
Thermal conductivity	W/m.K	0.17	0.22	0.19	0.19	—	—	DIN 52612
Coefficient of linear thermal expansion	10^5.K^{-1}	7.7	7.8	7	7	—	—	53752-A
Forming temperature	°C	—	—	160–175	150–160	—	—	
Deflection temperature:								
Under flexural load at 0.46 MPa	°C	—	—	113	95	90–100	—	75
at 1.82 MPa	°C	—	—	105	90	—	—	75
Vicat softening point, 5 kg	°C	111	106.5	115	102	129	119	53460/
" " 1kg(0.1mm)	°C	114	106	—	—	—	112	BS 2782:120C
ELECTRICAL				- - As for moulding grades - -				
MISCELLANEOUS								
Density	g/cm^3	1.189	1.186	1.19	1.19	1.18	1.175	R1183
Ignition temperature	°C	—	—	425	430	—	—	51704
Water absorption, 24h	%	0.2	0.21	—	—	0.25–0.3	0.3	R62A
" "	mg	—	—	—	—	30	30	DIN 53495

TABLE 3 – TYPICAL APPLICATIONS OF ACRYLICS

Required property	Applications
Resistance to weather and light.	Glazing on buildings, dome roof lights, conservatories. Transport (reflectors, rear light clusters, wind deflectors). Illuminated outdoor advertising signs.
Appearance, strength and light resistance.	Offices & shops (false ceilings, decorative partitions, lighting fittings). Wash basins, bath tubs.
Clarity and water white colour.	Optical lenses and prisms, drawing instruments (rulers etc.), transparent covers and containers.
Scratch resistance.	Radio, hi-fi scales. Lenses for sun-glasses.

Acrylic resins also have wide application in lacquers, paint finishes and surface coatings.

The copolymer with acrylonitrile (AMMA) is a tough, weather resistant material which is very resistant to organic solvents but has limited resistance to alkalis; it is transparent but yellowish in colour.

The most significant property of acrylic is its visual impact. The light transmission is high, at 92%, the remaining 8% being mostly reflection from the surface. Absorption is low, so the transparency is not affected by thickness in the normal range. The base material is water white and colourless (crystal clear) with a natural brilliance. When coloured, pure colours are obtained and translucent tints have high light transmission with a good dispersion effect.

Acrylics can be used over a wide temperature range. Maximum service temperature with no load is 80 to 95°C, but there are special grades available which can be used up to 105°C. Chemical resistance is good, such as resistance to petrol, salt water and car wash fluids; hence their widespread use for external motor car parts. They have good resistance to alkalis, but limited resistance to organic solvents and they are not resistant to strong acids or concentrated oxidising agents. Water absorption is low and electrical properties are good.

Cast or moulded acrylic parts are easy to work (eg cut, drill or machine) provided that the cutting tools used are designed for plastics, and are good for vacuum metallising, bonding, welding or polishing.

Properties of some of the moulding grades are shown in Table 1, where typical standard, heat resistant and impact grades are shown. There are differences in properties between cast and extruded products, with the cast material generally having a flatter surface and superior properties. The properties of sheet made by the two processes are compared in Table 2.

Applications

Acrylics are used in a wide range of applications where their strength and hardness combine with their optical properties to make them exemplary materials; some examples are shown in Table 3.

Acrylic resins also have wide application in lacquers, paint finishes and surface coatings.

Cellulose plastics

CELLULOSE PLASTICS are a group of polymers based on wood pulp or cotton linters (the short fibres from cotton seed) which are treated with suitable acids in a process called esterification. For example, to produce cellulose nitrate a mixture of sulphuric and nitric acid is used, the level of esterification being controlled according to the application.

The family consists of cellulose nitrate, cellulose acetate (CA) and triacetate, other cellulose asters such as acetobutyrate (CAB) or propionate (CAP), and ethyl cellulose.

The major commercial product today is *cellulose acetate*, which is valued for applications where transparency, toughness and a natural feel have to be combined. It is extensively used in film form for packaging, as a laminated surface on packages, and for moulding.

In Europe, CAB and CAP retain only a certain specialised market, although in the USA they are more widely used; ethyl cellulose is virtually unknown in Europe.

Properties

The cellulose esters have a high transparency (up to 90%), good strength and toughness and good electrical properties. They are produced in a range of grades (to suit different process and application requirements) by varying the esterification level, molecular weight and especially plasticiser content.

Increasing the plasticiser increases flow and toughness, but also creep. The effect of creep is, however, used to advantage, for example in the manufacture of hand tools, where the metal shank can be shot in without problems, because the consequent stresses relax quickly.

Although when compared with CA, CAB and CAP are slightly softer, of lower density and have lower heat distortion temperatures with easier flow, there is considerable overlap in the properties of the many formulations available (see Table 2). However, CAB and CAP do have a wide operating range than CA, which is limited to 0 to 60°C.

TABLE 1 – CELLULOSE ACETATE, TYPICAL PROPERTIES OF SOME GRADES.

Property	Type	Unit	Dexel S100–30	Tenite 007 H3	Dexel S200–22	Tenite 007 MS	Dexel E10MS	Test standard
PHYSICAL								
Tensile yield stress.		MPa	34	40	50	23	29	R527
Elongation at yield.		%	3.1	—	3.6	—	3.2	"
Tensile modulus.		GPa	1.8	—	2.7	—	1.5	"
Flexural yield stress.		MPa	42	63	65	32	31	178
Stress at 3.5% strain		MPa		—			—	"
Flexural modulus.		GPa		2.31		1.59		
Compressive stress at 1% defm.		MPa	31	100	42	55	16	DIN 53454
Hardness — Rockwell R.		MPa	90		110		44	2039–2
— Ball indentation.		MPa	50		78		36	2039–1
Impact strength:								
Notched Izod,	23°C	kJ/m^2*	25	12.8	9	24.6	26.6	180
"	-40°C		—	2.7	—	4.8		180
Un-notched Charpy.	23°C	kJ/m^2	NB		100–NB	—	NB	179
"	-40°C		50		45		34	
Notched Charpy	23°C	kJ/m^2	15		5		23	179
THERMAL								
Thermal conductivity.		W/K.m	0.22		0.21		0.22	DIN 52612
Coefficient of linear thermal expansion.		$\times 10^{5}$	10.2		10.5		12.5	ASTM D696
Deflection temperature:								
Under flexural load	0.45 MPa.	°C	67	86**	87	62**	52	75
"	1.82 MPa.	°C	52	76**	73	54**	45	
Vicat softening point	B/120.	°C	81†		100†		56†	1525/306
Weight loss on heating (72h at 82°C)		%		0.7		4.2		ASTM D706
ELECTRICAL								
Volume resistivity (24h water immersion)		ohm.cm	10^{11}		10^{11}		10^{11}	IEC 93
Dielectric constant	(1 MHz)		4.2		4.1		4.4	IEC 250
Dissipation factor	(1 MHz)	$\times 10^{3}$	64		56		70	IEC 250
Electric strength (24h water immersion)		kV/mm	27		28		24.5	IEC 243
MISCELLANEOUS								
Density.		g/cm^2	1.27	1.29	1.28	1.27	1.25	R1183
Water absorption, 24h.		%	3.2‡	2.3	4.2‡	2.3	2.5‡	R62A
Mould shrinkage (flow direction).		%	↓	↓	0.4–0.7		↑	1110
Melt flow (MFI)		g/10min	9		2.1		15	R1133
Refractive index.					1.49–1.51			DIN 53491

* Some data sheets have values quoted in J/m; these are shown as kJ/m^2. Values in J/m are approximately $\times 10$ values shown in kJ/m^2.
† conditioned 16 hr/80°C.   ‡ from dry.   ** compression moulded.

TABLE 2 – CELLULOSICS, COMPARISON OF CA, CAB, CAP
(Typical values for Tenite products)

Property	Formula	Unit	Cellulose acetate		Acetobutyrate		Acetopropionate		Test standard
			007	042	203	264	311	353	
PHYSICAL									
Tensile yield stress		MPa	47–17	50–23	43–15	43–10	46–25	48–10	R527
Flexural yield stress		MPa	73–24	79–35	64–21	64–12	62–33	64–12	178
Flexural modulus		GPa	2.55–1.28	2.79–1.52	2.07–0.83	2.07–0.69	2.35–1.24	2.42–0.66	"
Hardness — Rockwell, R			112–41	>115–74	112–48	112–35	113–60	114–37	2039–2
Deformation under load (6.9MPa, 24h, 50°C)		%	1–28	1–5	1–35	1–35	1–4	1–30	ASTM D621
Impact strength:									
Notched Izod, 23°C		kJ/m²*	10.7–32	5.3–21	6.4–44	6.4–58	8.5–49	6.4–NB	180
" −40°C		kJ/m²	2.7–5.9	3.2–4.8	3.3–6.4	3.2–14.4	4.3–9.6	2.7–16	
Deflection temperature:									
Under flexural load at 0.45 MPa		°C	94–55	98–57	108–59	108–54	113–75	114–41	75
" " at 1.82 MPa			85–48	91–51	94–51	94–45	101–57	100–38	
Weight Loss on heating (72h at 82°C)		%	0.2–6.9	0.1–2.4	0.1–3.2	0.1–1.5	0.1–0.8	0.1–5.0	ASTM D706
Density		g/cm³	1.29–1.26	1.30	1.22–1.18	1.22–1.15	1.23–1.19	1.23–1.16	R1183
Water absorption, 24h		%	2.3–2.6	2.2	2.2–1.2	2.2–1.2	2.4–1.5	2.6–2.0	R62A

* Some data sheets have values quoted in J/m; these are shown as kJ/m². Values in J/m are approximately ×10 values shown in kJ/m².

The cellulose esters are good electrical insulators, but with their somewhat lower surface resistivity, do not suffer from build-up of static and the consequent unsightly dust patterns of other plastics. They have an inherent gloss and a self-polishing effect which allows surface scratches to disappear with use.

The cellulosics have excellent flow properties and are therefore very well suited to injection moulding, provided they are completely dry – in fact the process was developed with CA. Stiffer flow grades are produced for other processes, such as extrusion, which is used to produce film for a variety of applications, and also to produce sheet. Cellulosics are available in a very wide range of colours, including pearlescent, opaque or of metallic appearance, and can be formulated for outdoor use.

Besides being easy to mould, the cellulose esters are some of the best plastics for fabricating processes such as sawing, turning and milling. They can also be decorated or printed by most standard methods.

Typical properties of a selection of the many grades of CA available are shown in Table 1, and some examples of CAB and CAP in comparison with CA are shown in Table 2. Whilst it is not possible to give examples of all types or formulas that are available, Table 2 does show the range of variations possible within one type, and also the overlap between the properties of CA, CAB and CAP. At the same time, it should be remembered that no table of standard property tests can give a complete picture of the behaviour of a material.

Chemical resistance

CA has the best chemical properties, being resistant to petrol, oils and greases. It is attacked by acids and alkalis and is attacked or dissolved by some solvents, including alcohols and acetone. CAB and CAP will, like CA, resist 10% sulphuric acid, but are not resistant to petrol, aromatic hydrocarbons or turpentine.

Because stresses dissipate rapidly, there is no problem of ESC with the cellulosics.

Applications

Sheet	For sunglasses, safety glasses and spectacle frames
Film	For packaging, over-wrap, advertising and PR material such as laminated printed brochures, graphic arts films and cells
Extruded tube	For transport presentation packaging.
Rod or mouldings	For tool handles, hammer heads and electrical screwdrivers.

Polystyrene (PS)

POLYSTYRENE (PS) was one of the first of the new thermoplastics introduced in the 1940's and still maintains its position as one of the major plastics. This is due to periodical improvement and development of the material and its applications, and because it is still the easiest of all the thermoplastics to process.

Polystyrene is produced by the polymerisation of styrene monomer, using heat and a catalyst. Another name for the nomomer is vinyl benzene:

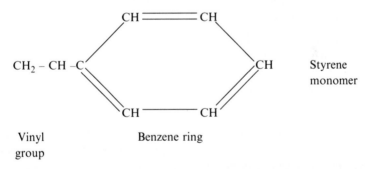

Styrene
monomer

Vinyl
group

Benzene ring

PS is amorphous and transparent, with low water absorption and low mould shrinkage. The bulky benzene ring makes it difficult for the chains to move under stress, so that the polymer is therefore strong and rigid – but brittle. This can be overcome by the incorporation of a synthetic rubber (usually polybutadiene) during polymerisation.

Polystyrene is available in two basic forms:

The homopolymer; general purpose polystyrene (transparent).

Poplymer containing rubber. This is a 'graft' copolymer, where the PS chains are grafted onto a backbone of rubber; it is a two phase system, where the graft copolymer is the disperse phase in a matrix (continuous phase) of polystyrene. This is called variously highly impact polystyrene (HIPS), impact polystyrene (IPS), toughened PS, or the German abbreviation SB or S/B; it is translucent, although in thin sections it will have contact transparency.

Properties
GPPS

This is one of the most rigid thermoplastics, highly transparent, with a light transmission of 89% and a high refractive index of 1.59. It is strong and hard and moulds to a high gloss, but has only reasonably good scratch resistance. Although it is brittle, what impact strength it has is retained to quite low temperatures. Molecular weight can be varied between limits, the highest being used for heat resistant grades with Vicat softening points around 100°C. The lower molecular weight types are used where very easy flow is required for very thin wall mouldings; these usually contain 3 to 5% of a flow-aid to increase melt flow.

Typical properties of some of the grades available are given in Table 1. Properties not shown, which are common to all grades, are:

Flammability	HB
Density	1.05
Water absorption	< 0.1%
Mould shrinkage	0.3–0.6%
Electrical	
Volume resistivity	$>10^{16}$
Rel permittivity	2.5
Dissipation factor	$0.5–0.9 \times 10^{4}$
Tracking index (CTI)	375–425

An important use of GPPS is for biaxially oriented films, made by stretching the film as it comes out of the die of an extruder. As the polymer is cooling whilst this is going on, the chains are stretched and oriented closer together than normal which significantly increases the strength. See page 417.

HIPS

Whilst the incorporation of rubber eliminates the problem of brittleness, this is achieved at the cost of a reduction in other properties. The relation between rubber level and properties can be depicted as:

Impact strength.		Strength.
Toughness.		Rigidity.
Elongation at break.	RUBBER LEVEL	Hardness.
		Ease of moulding to a high gloss.
		Transparency.

In selecting a grade for a particular application, it is usual to choose the one with the lowest impact that will be acceptable, so that all the other properties will be as good as possible. Whilst at one time HIPS grades were neatly divided into medium, high and super-high impact categories, the proliferation of grades designed for different applications has blurred this division. In additions to grades having variations in the rubber level, there are, as with GPPS, high heat and easy flow grades.

TABLE 1 – TYPICAL PROPERTIES OF PS (SOME GP GRADES)

Property	Unit	Easy flow Dow 638	Easy flow Huntsman 308	Standard Atochem 1540	High heat BASF 158K	High heat BP Chemicals HH101	U-V stablsd BASF 165 UV	Test standard
PHYSICAL								
Tensile strength	MPa	35	42	45	55	48	56	R527
Elongation at break	%	1	1	3	3	1.8	2	,,
Tensile modulus	GPa	3.2	—	—	3.2	3.1	3.15	,,
Flexural strength	MPa	—	73	75	103	96	86	178
Flexural modulus	GPa	—	3.4	3.0				
Hardness — Rockwell, L		—	95					2039–2
,, M								
,, — Ball indentation H132/30				74	150		150	2039–1
Impact strength:								
Un-notched Izod	23°C kJ/m²†				10		11	180
,,	−30°C				10		11	
Notched ,, (Izod)	23°C kJ/m²†	1.0	1.2	1.8	2	1.5	2	180
Un-notched Charpy	23°C kJ/m²	11			14	25	14	179
Notched Charpy	23°C kJ/m²	1.7						179
THERMAL								
Deflection temperature:								
Under flexural load at 0.45 MPa	°C				98		84	75
,, at 1.82 MPa	°C	80	78	73	86	93	76	
Vicat softening point A/50 9.8N	°C	82	89	91	106	101	92	306
,, B/50 49N		89		84	101		89	
,, A/120 9.8N								1525/306
Melt flow (MFI)	g/10min	25	17	11	3‡	2.2	2.5‡	R1133

†Some data sheets have values quoted in J/m, these are approximately ×10 values shown in kJ/m². Values in J/m are approximately ×10 values shown in kJ/m².

‡MVI (Melt volume index).

TABLE 2 – TYPICAL PROPERTIES OF PS (SOME HIPS GRADES)

Property	Unit	BASF 427D	BP Chemicals 2220	Huntsman 545	Dow 485	Huntsman 464	Atochem 7240	BP Chemicals 4230	Test standard
PHYSICAL									
Tensile stress at yield	MPa	44	26	18	18	26	21	21	R527
Elongation at break	%	10	35	40	50	40	50	46	¨
Tensile modulus	GPa	2.8	2.7		2.00		2.59	2.15	¨
Flexural strength	MPa	73					40		178
Flexural modulus	GPa		2.6	1.90		2.40	1.75	2.1	¨
Hardness — Rockwell				L75	M25	L85	L67		2039–2
Impact strength:									
Un-notched Izod 23°C	kJ/m²†	22							179
¨ −30°C		20							179
Notched Izod 23°C	kJ/m²†	4	8.7	10	9.5	7.0	15	15	179
Un-notched Charpy 23°C	kJ/m²	55	50		NB			NB	179
Notched Charpy 23°C	kJ/m²	3	4.5		6.5			8.5	179
THERMAL									
Deflection temperature:									
Under flexural load at 0.45 MPa	°C	96							75
¨ ¨ at 1.82 MPa		86							
Vicat softening point A/50 9.8N	°C	102	70	68	84	72	80	70	306
¨ ¨ B/50 49N		96	79	82	82	86	97	79	
¨ ¨ A/120 9.8N					93		90		1525/306
Melt flow (MFI)	g/10min	8‡	14	30	12	18	4.5	9	R1133

†Some data sheets have values quoted in J/m, these are shown as kJ/m². Values in J/m are approximately ×10 values shown in kJ/m².
‡MVI (Melt volume index).

PS has low notch sensitivity, and this is reflected in its low temperature behaviour; impact retention is good, and there is no sudden transition to a much
lower impact strength. HIPS grades loose their impact gradually, until at –60 to
–70°C it is the same as that of GP.

A minor effect on unnotched impact or falling ball impact is flow/heat resistance balance; easy flow materials are likely to have higher values by these tests
than heat resistant grades.

HIPS grades all share the excellent moulding characteristics of GPPS. To
achieve a high gloss at the higher rubber levels it is necessary to have mould
temperatures around 60°C, and this lengthens the cooling time.

The slight difficulty of producing a high gloss on higher impact grades led to the
loss of some applications for ABS, but this has been countered by the development
of high gloss grades where the polymerisation process has been modified to change
the structure of the grafted particles. These grades have an inherent gloss and are
therefore not dependent on mould temperature for the level of gloss. Such grades
were developed by BASF, and are now also marketed by other manufacturers.

Table 2 shows typical properties of a selection of HIPS grades, and Table 3
gives examples of FR and high grades. Properties not shown, such as flammability and water absorption, are close to those for GPPS given above.

Both GP and HIPS types are now made by most manufacturers by a continuous mass process, which allows close control and consistency of molecular
weight, structure and quality to be exercised at the lowest conversion cost. The
exception is expandable PS for production of expanded PS, which is still produced by a suspension process.

Although HIPS is made with different levels of rubber to suit individual applications, GPPS and HIPS can be mixed to provide intermediate levels of rubber,
so that several products for different applications can be produced by a con-
verter from only two stock materials.

All types of polystyrene have excellent processing behaviour and are
converted on a large scale by injection moulding, extrusion as profile, or as sheet
for thermo-forming. The latter process is now the main method of conversion
for packaging applications, frequently with the extruder and thermo-former
in-line with high rates of production, also used on a smaller scale is blow moulding (usually injection blow). A special type of extrusion process is the direct
gassing system which is used for the production of expanded thin sheet for
packaging.

All types of PS are easy to colour with a wide range of colours. Whilst GPPS
can be produced in many transparent, translucent or opaque shades, for obvious
reasons only translucent or opaque colours can be made in HIPS. Colouring can
easily be carried out on the machine, and is sometimes called self-colouring,
using proportioning dosers/blenders between the hopper and the throat of the
extruder or moulding machine. A large proportion of coloured articles is made
using this technique.

Dry colouring using powder pigment was used in the past, but for reasons of

TABLE 3 – TYPICAL PROPERTIES OF PS (SOME SPECIAL HIPS GRADES)

Property	Unit	Flame retdnt (VO- 1.7mm) Atochem ST 108FR (VO- 1.7mm)	Impr. chem. resistance Dow 469	Impr. ESC resistance BASF 2710	BASF 525K	BASF 587N	High gloss Atochem ST 168	High gloss Dow XZ 95164	Test standard
PHYSICAL									
Tensile stress at yield	MPa	28	21	21	35	35	29	26	R527
Elongation at break	%	20	50	>50	24	40	45	30	..
Tensile modulus	GPa	2.63	1.90	1.50	2.10	1.90	—	2.2	..
Flexural strength	MPa	50		32	50	52	62		178
Flexural modulus	GPa	2.99					2.30		..
Hardness — Rockwell		L70					L71		2039—2
— Ball indentation				65	86				
Impact strength:									
Un-notched Izod 23°C	kJ/m²‡			100	20	90	17		179
,, ,, -30°C				60	10	60			179
Notched Izod 23°C	kJ/m²	11.5	8	9	6	11		6	179
Un-notched Charpy 23°C	kJ/m²		NB	NB	60	NB			179
Notched Charpy 23°C	kJ/m²		9	5.5	4.5	7.5			179
THERMAL									
Deflection temperature:									
Under flexural load at 0.45 MPa	°C	80	93	87	87	90	77		75
,, ,, at 1.82 MPa				78	77	80	95		
Vicat softening point A/50 9.8N	°C	95	92	89	97	102	87	93	306
,, ,, B/50 49N		86	101	85	89	95	102	102	
,, ,, A/120 9.8N									
Melt flow (MFI)	g/10min	7.5	2.5	4‡	4‡	3‡	7	8.5	1525/306
Gloss					79%*	74%*		>90%**	R1133

†Some data sheets have values quoted in J/m; these are approximately ×10 values shown in kJ/m². Values in J/m are approximately ×10 values shown in kJ/m².

‡MVI (Melt Volume Index).

*BASF method, cf 20–35 for normal HIPS.

**Dow method, cf 40–50 for normal HIPS.

hygiene and quality this has largely been replaced by master batch, either liquid or solid.

PS has very good electrical properties and this can lead to unsightly dust patterns on mouldings, antistatic versions are available to overcome this problem. Flame retardant grades offering UL94 V0 or V1 ratings for electrical appliances are also available, as are U-V stabilised grades for lighting fittings. PS is not recommended for outdoor use.

Chemical resistance

Polystyrene has good chemical resistance and resists dilute acids, alkalis and most salts. Aliphatic hydrocarbons and the lower alcohols do not visibly attack PS, but are likely to cause ESC (environmental stress cracking) if used in such environments when under stress. This is in fact used as a test for locked-in thermal or moulding stress for GPPS mouldings, using white spirit. PS is attacked or dissolved by many solvents such as aromatic or chlorinated hydrocarbons, ketones and esters.

Applications

Uses for PS range from toys and premium articles, through a very wide range of packaging (food and non-food), to stressed applications like refrigerator liners. Some examples are:

Packaging
Cosmetic and personal care containers.
Pesentation boxes, DIY packs for screws etc.

Food packaging
Dairy products (yoghurt, cottage cheese, butter, eggs).
Fast food take-away packs, food trays, airline lunch trays.
Vending cups for drink machines.

Household
Containers for food storage.

Medical
Petri dishes, blood sample tubes, laboratory containers.

Housings
Domestic appliances, TV cabinets (with FR for the back), computers, audio equipment.

Refrigerators
Doors and inner liners.

Polystyrene films (OPS)

These films can be streched in one direction (uniaxially oriented), or in two (biaxially oriented), and can be based on GPPS (highly transparent), or HIPS (translucent) where increased toughness is required. Biaxially oriented is the preferred type and the increase in properties obtained by this process can be judged from these typical figures comparing the values for the type of GP used with those for the film:

	GPPS	Biax film[†]	
Tensile strength	50– 55	75	MPa
Elongation at fail	2–3	25–50	%

†Dow values

They can be printed and/or bonded by the usual methods, provided that care is taken to keep the heat or solvent to the minimum, to avoid release of the orientation!

Applications

Packaging
Window boxes and cartons, overwraps, food trays and packs, and as meat interleaf.

Publishing and printing
Printed overlays and interleaf for brochures, promotion material, greetings cards. Protective interleaf for slides for overhead projectors, stencils or lithoplates.

Laminating
To HIPS sheet for refrigerator door liners, and to PS foam sheet or other materials, to provide high gloss.

Fluorocarbons

General

FLUORINATED PLASTICS have outstanding chemical, physical, thermal and electrical properties, due to the great stability of the fluorine-carbon bond. They are produced in several variations, obtained by polymerising combinations of monomers with differing fluorine contents and different structures. For each type of polymer there are usually several grades varying in molecular weight and additives.

The polymers obtained all have outstanding properties. PTFE has the best properties but is difficult to process; the other types have easier processing characteristics but at the sacrifice of some heat resistance and/of chemical resistance.

Manufacturers

PTFE	Du Pont	'Teflon'
(polytetrafluoroethylene)	ICI	'Fluon'
	Hoechst-Celanese	'Hostaflon TF'
		'Hostaflon TFM'†
FEP	Du Pont	'Teflon FEP'
(fluoroethylenepropylene)	Hoechst-Celanese	'Hostaflon FEP'
ETFE	Du Pont	'Tefzel ETFE'
(ethylene tetrafluoroethylene)	Hoechst-Celanese	'Hostaflon ET/ETFE'
PFA	Du Pont	'Teflon PFA'
(perfluoroalkoxy)	Hoechst-Celanese	'Hostaflon PFA'
PVDF	Atochem	'Foraflon'
(polyvinylidene fluoride)	Hoechts-Celanese	'Hostaflon PFA'
PCTFE	Atochem	'Voltalef'
(polychlorotrifluoroethylene)		

†modified TF powders for compression moulding

PTFE

These consist of long unbranched chains of polytetrafluoroethylene; these chains contain from 10 to 100 000 molecules of the monomer, compared with an average of 5000 for most other plastics. PTFE thus combines the carbon-fluorine bond with very high molecular weight.

Polymerisation under heat and pressure with suitable initiators is usually carried out in water, either in suspension or emulsion. Depending on the conditions used the polymer can be made available in different forms:

Fine powder for processing by paste-extrusion.
Granular for compression forming into shape followed by sintering.
Filled materials; granular polymer compounded with various fillers – glass fibre, graphite, bronze etc.
Aqueous dispersion: for film casting, coating and impregnation.

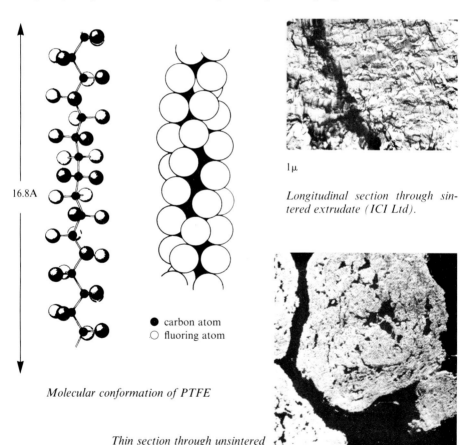

16.8A

● carbon atom
○ fluoring atom

Molecular conformation of PTFE

1 μ

Longitudinal section through sintered extrudate (ICI Ltd).

Thin section through unsintered granular polymer (ICI Ltd)

500μ

Properties

The chains, while essentially straight, are in effect twisted around themselves, so that fluorine atoms form a sheath around the carbon atoms. The compact inter-locking of the fluorine atoms leads to a molecule of great stiffness, which accounts for the high crystalline melting point of the polymer, as well as its remarkable properties.

PTFE resins are characterised by:

Chemical inertness – they are resistant to all known chemicals and solvents other than molten alkali metals, gaseous fluorine and certain complex halo-genated compounds. This gives them exceptional non-adhesiveness.

Heat and cold resistance – products made from PTFE are fully resistant to temperatures ranging from –270°C to 260°C.

Very low friction – with no stick-slip tendency or 'squeaking'.

excellent electrical properties – which hardly vary with frequency or environ-mental conditions.

Low creep, retention of toughness – over a wide temperature range. Can be reinforced with glass fibres to increase rigidity and wear resistance and give re-duced cold flow.

Long term weather resistance, non-flammability, non-toxic and very low water absorption.

Processing

PTFE requires special moulding techniques since it does not melt when heated. It is crystalline in structure up to about 327°C, and then takes on the appear-ance of an amorphous translucid gel.

Techniques commonly used to mould PTFE are similar to those used in pow-der metallurgy. PTFE granules are cold compacted in a mould under high pres-sure. The part thus preformed is then sintered at 370 to 400°C and cooled under controlled conditions.

Paste-extrusion is used for continuous shapes. The paste is made by mixing the PTFE powder with a lubricant, and sintering the coating of PTFE after re-moving the lubricant.

Impregnation or coating is done by dip or spray methods using PTFE disper-sion. A compact layer is first deposited, dried and then sintered at very high temperatures.

Other techniques such as ram extrusion and isotactic moulding of granules are used to obtain basic shapes such as rodstock, tube, film sheet and miscellaneous mouldings.

Generally, to obtain a good finish and/or close tolerances, machining is neces-sary. Machining can also be used for prototypes or when product shape is com-plex. All standard operations can be carried out provided that the normal approach for plastics is followed – lubrication, control of cutting angles depth and speed to minimise heat generation etc.

TABLE 1 – PROPERTIES OF PTFE AND FEP

Property	Unit	Test Method	PTFE	FEP
Specific gravity			2.17	
Tensile strength	lb/in^2	ASTM 638	2 500–4 000	2 700–3 100
	kg/cm^2		175–280	190– 217
Tensile modulus	lb/in^2	ASTM D747	$0.5–0.9 \times 10^5$	0.95×10^5
	kg/cm^2		$35–63 \times 10^3$	66.5×10^3
Compression strength	lb/in^2	ASTM 638	600	
	bar		42	
Impact strength	ft lb/in	ASTM D256	3.0	unbroken
Elongation	%	ASTM D638	200–600	200–300
Hardness	Rockwell D	ASTM D676	50–65	57– 60
Coefficient of friction			0.04	0.08
Water absorption	%	ASTM D570	less than 0.01	less than 0.01
Deflection temperature	°C	ASTM D648	250	
Maximum service temperature	°C	continuous	260	205
Coefficient of expansion	10 ^{5}/°F	ASTM D696	9.9	9.9
	10 ^{5}/°C		5.5	5.5
Thermal conductivity	Btu/hr/ft/ft^2/°F	ASTM C177	1.7	1.4
Dielectric strength	V/mil	ASTM D149	600	600
Volume resistivity	ohm cm	ASTM D257	over 10^{18}	over 10^{18}
Arc resistance	sec	ASTM D495	over 300	over 300

FEP

Was introduced by Du Pont in 1960 to overcome the processing difficulties of PTFE at the cost of slightly reduced thermal properties. It is a copolymer of tetra-fluoroethylene and hexafluoropropylene. It has a big advantage over PTFE in that it can be processed on conventional injection moulding machines and extruders.

The carbon to fluorine ratio is the same for both PTFE and FEP, which leads to their close resemblance in chemical and electrical properties. The main difference is that where PTFE is a straight chain polymer, FEP has some side chains or 'branches', which at the same time reduces its heat resistance and makes it easier to process.

FEP is available as melt-cut granules for melt-extrusion, compression and injection moulding, or as a stabilised aqueous dispersion for coating and impregnation. There is also a foam concentrate mainly used to produce foamed dielectrics for high-speed signal transmission cables, and colour master batches for colouring materials being moulded or extruded.

Properties

FEP has a melting point of 290° compared with the transition of PTFE at 327°C. Where, as mentioned above, PTFE never really melts, FEP can be handled by conventional melt flow processes A result of this difference in behaviour at high temperature is that the continuous service temperature of FEP is lower, at 205g compared with 260°C for PTFE.

In all other properties FEP is very similar to PTFE – see Table 1.

TABLE 2 – PHYSICAL PROPERTIES OF UNFILLED PTFE*

Property	Unit	Value
Mechanical		
Relative density	—	2.1–2.2
Ultimate tensile strength	MN/m^2	20.5–34
	kgf/cm^2	210–350
	lbf/in^2	3000–5000
Elongation at break	%	250–400
Compressive stress at 1% deformation,	MN/m^2	6
25°C(77°F), 100 seconds	kgf/cm^2	63
	lbf/in^2	900
Impact strength (Charpy, –20°C, notched)	$cm\ kg/cm^2$	6.8–9.6
	$ft\ lb/in^2$	3.2–4.5
Hardness, Shore D scale	—	50–55
Electrical		
Permittivity	—	2.05
Loss angle	—	<100 μ radians
Volume resistivity	ohm m	$>10^{16}$
Surface resistivity	ohm	10^{17}
Thermal		
Melting point	°C	327
	°F	620
Coefficient of linear expansion	—	
Thermal conductivity	10^{-4} cal/cm s deg C	6
	Btu in/ft^2 h deg F	1.7
Specific heat	kcal/kg deg C	0.23
	Btu/lb deg F	0.23
Surface		
Coefficient of friction	—	0.02–0.1
Angle of contact with water	degrees(°)	108

*ICI 'Fluon'

Processing

FEP is converted by various melt processes – extrusion, transfer, compression or injection moulding and blow moulding. Compared to other thermoplastics FEP has a high melt viscosity, whilst melt fracture occurs at a low shear rate. Extruders should thus be designed to operate at low shear rates, and large runners and gates should be used in injeciton moulds.

As with PTFE, FEP can be machined to tight tolerances, using any of the standard engineering operations. Because of the high shrinkage on moulding, machining will be necessary if tight tolerances are needed on injection mouldings.

ETFE

ETFE is a modified copolymer of tetrafluoroethylene and ethylene, and has a good balance between the properties of the fluoropolymers and the thermoplastic behaviour of polyethylene.

ETFE is available as a clear, melt-cut granule or as a powder suitable for rotational moulding or coating of corrosion resistance vessels and parts.

Properties

ETFE combines mechanical toughness with chemical, electrical and thermal properties approaching those of PTFE whilst having good processing properties in extrusion or injection moulding and lower density than PTFE.

Use temperature range is from the brittleness temperature of –100°C to a continuous service temperature of 155°C for wire applications, and of 180° for injection moulding grades. This figure is increased to 200° with glass reinforced grades.

These properties make ETFE particularly suitable for wire and cable insulation. Wire coated with the fluoropolymer can take a remarkable amount of physical abuse since it resists cut-through, abrasion and notch propagation.

PFA

A more recent introduction, PFA has properties similar to those of PTFE but has excellent melt processability. Like FEP, PFA has branch chains, which in this case are connected to the main carbon 'backbone' by oxygen linkages which are longer and more flexible than the links holding the side chains of FEP; this gives greater thermal stability. As a result, PFA can stand continuous service temperatures up to 260°C.

PFA is available as clear, melt-cut granules, and foam and colour masterbatches are available as for FEP.

PVDF

This is a homopolymer of 1,1-difluoroethylene which is similar in properties to the other fluorine containing plastics although because of its structure, it is somewhat polar. As a result, it is soluble in organic esters and amines, which is an advantage in the preparation of coatings. It also has low permeability to gases and liquids.

As with the products above, PVDF can be easily processed on conventional plastics machinery. It finds significant use as a film for demanding electrical, pharmaceutical and food applications.

PCTFE

PCTFE is very similar to PTFE in its chemical resistance, being practically inert to most chemical reagents. It has excellent fire resistance. It is almost impervious to gases and liquids. Its operating range is –250°C to +160 to 200°C. It therefore finds use in components for use at near absolute zero.

It is available as powder, or in granule form for compression moulding or extrusion.

Applications for fluorocarbons

Fluorocarbons are used in a wide variety of applications in the chemical industry, mechanical and electrical engineering. Some examples:

Wire and cable coating, especially for high frequency, high temperature and high voltage use. Coaxial spacers, inserts, tube sockets. High quality wrapping tape for motor insulation.

Corrosion resistance linings of pipes, fittings, valves and pumps. Mechanical seals and gaskets. Piston rings and valve seats.

Materials conveying, roll covers.

Valve bodies, extruded standard or corrugated hose, pumps and components.

Transparent roofing film for insulated domes and as weather protection laminating film for use with *eg* ABS sheet.

Automobile transmission rings, power steering seals, dynamic and static bearings.

Coated fabrics for electrical insulation, architectural, belting and other applications.

Nylons (polyamides)

THE NAME 'nylon' is a generic name for a whole range of resins which can be produced by condensing diamines with dibasic acids, or by direct polymerisation of amino acids to form long polyamide chains. These chains consist of 4 to 11 methylene groups separated by an amine group, the number designations of nylons showing the number of carbon atoms in the materials from which they were produced (see Table 1).

Nylons were first used as fibres and this still accounts for a major part of the production. Since their introduction to the plastics processing field, however, they have been the most widely used of all the engineering plastics and this has led to increasing specialisation of types and grades to meet processing and user requirements. Whilst there are considerable differences between the types, they have common basic properties:

TABLE 1

Designation	Chemical structure	Melting point (°C)
Nylon 6	Polycaprolactam	220
Nylon 66	Polyhexamethylene adipamide	255
Nylon 610	Polyhexamethylene sebacamide	215
Nylon 11	Polyundecanoamide	185
Nylon 12	Polydodecanoamide (or polylauryllactam)	212
Trogamid T (PA 6–(3)–T)	Polytrimethylhexamethylene-teraphthalamide	*
Nylon 46	Polytetramethylene adipamide	295
TR55	Complex copolymer containing aliphatic, cycloaliphatic and aromatic components	*

*Amorphous – no sudden melting point.

Toughness, strength and rigidity.
Low friction characteristics and good abrasion resistance.
Good fatigue resistance.
High heat resistance and retention of strength properties at elevated temperatures (particularly in reinforced grades).

In addition to some grades made from their basic type, most suppliers produce impact modified grades and products reinforced with glass or fillers. Although many grades are inherently V2, flame retardant grades to VO are produced, as are super-lubricated grades containing molybdenum disulphide or grapite. Many grades are available in pigmented versions, or in heat or light stabilised form.

General properties

Nylon is one of the tougher thermoplastics even in its basic form, but many manufacturers now make it in impact modified form, producing very tough to super tough materials. It has high strength and low creep characteristics which are retained at elevated temperatures. The most important properties, however, are its low friction and comparatively high PV rating in bearings and sliding parts, even when unlubricated. There is hardly any wear of unfilled products subjected to sliding friction at temperatures up to 80°C. Unlubricated nylon/ nylon or nylon/steel systems do not seize. Sand, dust and other foreign bodies become embedded in nylon bearings and are thereby rendered harmless.

Although nylon does not age in normal use, for parts exposed to outdoor and U-V conditions stabilised grades should be used, preferably pigmented in dark shades.

As a rule, standard grades of nylon can withstand long term exposure at up to 100°C to air, neutral oils, greases, liquid fuels and many solvents. It is preferable, however, to use heat stabilised versions which can be used up to 120°C. If no mechanical stress is involved in the application, unfilled nylon components can withstand short term exposure to temperatures around 150 to 170°C, depending on the type.

A direct competitor to nylon is acetal. Whilst acetal has better fatigue endurance and creep, under average conditions of humidity the nylons are superior in toughness and abrasion resistance; they are also much lower in density and have better flammability properties.

Reinforced materials

Addition of glass fibre and/or certain fillers has the effect of reinforcing the nylon and this reinforcement has enabled nylon based materials to become a major part of the market for reinforced products in Europe.

Reinforced grades retain their strength properties up to quote elevated temperatures, even under load, as is demonstrated by comparing the distortion temperature under flexural load (ISO75), measured at the two standard stress values, for standard and reinforced materials. Typical values for Nylon 66 are:

Stress	Standard °C	Reinforced °C
0.46 MPa	220	230
1.82 MPa	95	230

Besides retaining strength properties, reinforced grades show greatly improved dimensional stability at elevated temperatures, and lower mould shrinkage and coefficient of expansion. Heat stability is improved, and at zero stress, reinforced materials can withstand short time exposure up to 230°C. Typical electrical properties for nylons are shown in Table 2.

Chemical resistance

Nylon displays outstanding resistance to fuels, oils, greases, alcohols, esters, ketones, and aliphatic and aromatic hydrocarbons. It is attacked by acids, particularly inorganic acids, even in very low concentrations, Oils containing acids or acidic solutions of salts may also damage nylon articles; on the other hand nylon is not attacked by alkalis. It is damaged by hot solutions of oxidising agents, but has good resistance to fuels and lubricants even at elevated temperatures.

Nylon is resistant to boiling and can be sterilised. No damage is likely unless it is immersed for several months in hot water or hot aqueous solutions, when hydrolysis will occur. The likelihood of this happening is increased if the solutions contain large amounts of oxygen or oxidising atgents. The stabilised glass reinforced grades of nylon have even higher long term resistance to hot water.

Individual nylons

Nylons 6 and 66

In addition to the normal injection moulding grades, often with internal lubricants for better processing, grades are available from most producers with differing viscosity to suit extrusion. Nylons 6 and 66 are semi-crystalline materials and there are also nucleated grades available which have a finer crystalline structure achieved with faster cooling times. Impact modified grades are available where the already high toughness of the basic material may be insufficient.

TABLE 2 – TYPICAL ELECTRICAL PROPERTIES OF NYLONS†

Property	Type Unit	6	66	610	11	12	46	T
Dielectric strength Short time, 3mm	MV/m	10	12	15?	16	16	12–15	16
Volume resistivity	ohm.cm	10^{13}	10^{13}	10^{14}	10^{14}	10^{13}	10^{14}	$>10^{14}$
Relative permittivity at 1MHz.	—		4.0	4.3	3.4?	3.7	4.0	4.5	3.3
Loss tangent at 1MHz.	—		0.08	0.08	0.1	0.05	0.1	0.1–0.16	0.027

All values after conditioning.
NOTE: Additives (lubricants, impact modifiers, reinforcements) will change these values somewhat. Consult the supplier for actual values on grade chosen for use.

Under normal conditions Nylons 6 and 66 absorb 2 to 3% moisture, which has a minor plasticising effect on the material, giving an increase in toughness (impact, elongation) and a small decrease in strength properties (rigidity, hardness, tensile strength). Moisture absorption is accompanied by a linear swelling of about 0.25% for each 1% of moisture absorbed. These changes obviously occur after moulding, since the as moulded material has a zero moisture content. Consequently the data sheets for these types always quote figures for 'as moulded' and 'conditioned' (usually at 50% RH).

Once having reached equilibrium, it is also possible for the moisture content to reduce (*eg* after long exposure to dry conditions, or after heating) and this will reverse the changes mentioned above.

These effects are considerably reduced in reinforced grades, which is another reason for their widespread use in engineering applications. Glass fibre content starts at 15%, with grades available with contents as high as 50%, the change in properties is more or less proportional to the glass content. In certain cases glass beads or mineral reinforcement are used, with somewhat different effects on the properties. Some manufacturers also include impact modified versions of glass reinforced grades in their range.

With the large number of possible permutations it is difficult to give a full picture of the properties of the materials available, but Tables 3, 4 and 5 give a typical selection of grades from different manufacturers to give an indication of the range available.

Nylons 612, 11 and 12

This group of nylons differs considerably from the previous group in having very much lower water absorption. Their properties are therefore far less affected by the level of humidity and they have much better dimensional stability than 6 or 66; in general they have lower tensile strength and rigidity and lower heat resistance. On the other hand they have better flexibility and low temperature impact retention and their electrical properties are less affected by changes in moisture content. From the moulding point of view they have advantages in lower melt temperatures and little need for conditioning.

As with Nylons 6 and 66, they are available in various modified versions, although not in such a large range. A selection of the products available is shown in Table 6.

Nylon 46

Based on polytetramethylene adipamide this has recently been introduced by DSM, under the name 'Stanyl'. It has a symmetrical chain structure which leads to a high crystallisation rate and gives a high melting point. Even unreinforced, it has a distortion temperature under flexural load of 160°C, whilst when reinforced this goes up to 285°C. Continuous use in temperatures between 130 and 160°C are claimed, which allows the use of the material for high temperature electrical/electronic applications such as those involving high soldering temperatures.

TABLE 3 – TYPICAL PROPERTIES OF SELECTED NYLON 66 MOULDING COMPOUNDS

Property	Unit	General purpose — Standard ICI A125 D	C	General purpose — Fast cycling BIP A25 D	C	Impact modified — Tough Celanese 7020 D	C	Impact modified — Super tough Du Pont S1801 D	C	Flame retardant — ICI A427 D	C	Test standard
PHYSICAL												
Tensile yield stress	MPa	85	60	80	65	61	48	52	41	80	60	R527
Elongation at yield	%			4	15							"
Elongation at break	%	40–50	>100	–	–	65	240	60	210	5	100	"
Flexural elastic modulus	GPa	2.85	1.05	3.1	1.1	2.1	1.0	1.9	0.9	2.8	0.9	178
Hardness — Rockwell R						114	93	112	89			2039 2
Impact strength												
Notched Izod, 23°C	kJ/m²*	5.5	22	13	DNB	26	32	91	107	5	7	180
-40°C	kJ/m²*	4.0						16	14			"
Notched Charpy, 23°C	kJ/m²*							65	70+			179
-40°C	kJ/m²*							25	20			"
Tensile impact strength	kJ/m²							588	1155			ASTM D1822
Friction												
coefficient, plastic/steel dynamic									0.29			Mfr's test
THERMAL												
Coefficient of linear thermal expansion	x10⁵							44				ASTM D696
Deflection temperature:												
Under flexural load at 0.5 MPa	°C	100		220		227		219		105		75
at 1.82 MPa	°C			90		114		83				"
Melt temperature	°C		255		265		257		255		255	1218
MISCELLANEOUS												
Density	g/cm³	1.14		1.14		1.09		1.08		1.18		R1183
Flammability		V2		V2				HB		V0‡		UL subj 94
Water absorption, 24 h	% / mg	1.2		77		1.2		1.2		1.3		R62A / 1110
Mould shrinkage (flow direction)	%	1.5–2.0		1.3–1.8		1.3–1.7		1.3–1.7		1.2–1.3		ASTM D955

* Some data sheets have values quoted in J/m; these are shown as kJ/m². Values in J/m are approximately ×10 values shown in kJ/m².
Conversion used is 10 J/m = 1 kJ/m².
† Sample did not break, but bend.
‡ Down to 0.56 mm thickness

TABLE 4 – TYPICAL PROPERTIES OF SELECTED REINFORCED NYLON 66 MOULDING COMPOUNDS

Property	Type / Grade	Unit	Glass reinforced, standard												Test standard
			ICI A180 15% glass		Du Pont 70 G20 20% glass		Bayer AKV 30G 30%‡		BIP AF307 33% glass		1600 43% glass		ICI A690 50% glass		
			D	C	D	C	D	C	D	C	D	C	D	C	
PHYSICAL															
Tensile yield stress		MPa	130	85	156	103	190	120	170	120	234	172	200	155	R 527
Elongation at break		%	3	6	3	7.3	—		4	6	3	4	2	6	‥
Flexural elastic modulus		GPa	5.4	3.2	6.7	4.7	9.7	7.2	9.7	5.8	11	6.9	12	9.6	178
Hardness — Rockwell R					122	115					100	85			2039-2
″ M					102	85									
— Ball indentation		MPa			250	155									2039-1
Impact strength:															
Notched Izod, 23°C		kJ/m²	5.5	8	7	10	15	13			14	17	11	13	180
″ -40°C					6	6	12	11							
Notched Charpy, 23°C		kJ/m²			8	9			15	38					‥ 179
″ -30°C					6.7	6.4									
THERMAL															
Coefficient of linear thermal expansion		×10⁵			2-3		2				1.4				ASIM D696
Deflection temperature: Under flexural load at 0.5 MPa		°C	215		254		250		>240	>240	257	257	250		75
″ at 1.82 MPa (B/50)		°C						245	240	235					‥
Vicat softening point		°C													306
MISCELLANEOUS															
Density		g/cm³	1.26		1.29		1.36		1.40		1.47		1.57		R1183
Flammability			HB		HB		HB		HB				HB		UL subj 94
Water absorption, 24 h.		%	0.8		1.0		0.7				0.9		0.4		R62A
″ 23°C/50% RH		mg							58						1110
Mould shrinkage (flow direction)		%	0.7-1.4		0.48-1.16†		0.5-1.0		0.7-1.3				0.3-0.8		

(continued)

* Some data sheets have values quoted in J/m; these are shown as kJ/m². Values in J/m are approximately ×10 values shown in kJ/m².
† Across flow.
‡ Very good finish.

TABLE 4 – TYPICAL PROPERTIES OF SELECTED REINFORCED NYLON 66 MOULDING COMPOUNDS – *contd.*

Property / Type Grade	Unit	BIP AS 408 Standard 40% gl bead (D)	(C)	ICI A175 (30%G) Hydrolysis resistant (D)	(C)	ICI T A510 Toughened 28% glass (D)	(C)	Du Pont FR51 Flame retard 25% glass (D)	(C)	Du Pont Minlon 11C140 mineral (D)	(C)	Test standard
PHYSICAL												
Tensile yield stress	MPa	90	45	185	135	46	41	115	70	85	65	R 527
Elongation at break	%	2	3	3	5	30-60	100	2	4	10	30	..
Flexural elastic modulus	GPa	5.8	2.3	8.0	5.3	1.4	0.8	10.4	6.2	5.1	2.1	178
Hardness — Rockwell R								122		120		2039 2
M										86		
— Ball indentation.	MPa							290	150	245	110	2039 1
Impact strength: Notched Izod. 23°C	kJ/m²	5	8	9.5	15	65	80	6.5	8.5	6.4	10.5	180
-40°C						18						..
Notched Charpy. 23°C	kJ/m²							8.0	9.9	6	13	179
-30°C												..
THERMAL												
Coefficient of linear thermal expansion	×10⁵											ASTM D696
Deflection temperature: under flexural load at 0.5 MPa	°C	>240	235					240		220		75
at 1.82 MPa	°C	220	180				65	226	213	147		..
Vicat softening point. (B/50).	°C			242				236				306
MISCELLANEOUS												
Density.	g/cm³	1.44		1.37		1.09		1.44		1.48		R 1183
Flammability		HB		HB		HB		VO(3.2 mm)		HB		UL subj 94
Water absorption, 24 h	mg	54										R62A
23°C/50% RH	%			0.7		0.9		0.42		0.8		..
Mould shrinkage (flow direction)	%	0.9-1.5		0.5-1.0		2.4-2.5		0.2-0.4		1.3-1.5		1110

Special versions: ICI A175 (30%G), ICI T A510, Du Pont FR51, Du Pont Minlon 11C140

* Some data sheets have values quoted in J/m: these are shown as kJ/m²: Values in J/m are approximately ×10 values shown in kJ/m².

† Across flow.

‡ Very good finish.

TABLE 5 – TYPICAL PROPERTIES OF SOME NYLON 6 GRADES

General purpose (columns: Med visc Eni 333HV, Fast cyc BIP L03, No conditioning EN30)

Property	Unit	Standard Injection EMS A23FC (D)	(C)	Transparent if thin Bayer B25T (D)	(C)	Plasticised ATO RMN P40CD (D)	(C)	Med visc, impr crystallinity Eni 333HV (D)	(C)	Fast cyc, low visc, nucleated BIP L03 (D)	(C)	No conditioning EN30 (D)	(C)	Impact modified ICI TB560 (D)	(C)	Impact-glass Bayer BKV 120 (D)	(C)	Test standard
PHYSICAL																		
Tensile yield stress	MPa	85	45	70	35	28		80	40	80	45	60	38	38	32	—	—	R527
Elongation at yield	%	10	30	4	20	40		—		5	25	5	13	—		5	—	''
Elongation at break	%	20	300	>50	>50	360		60	250	—	—	—	—	40–80	100	—		''
Tensile modulus	GPa	3.3	1.5	2.60	0.8			3.0	1.5	2.8	1.0	2.5	1.4	1.5	0.7	6.8	4.2	178
Flexural elastic modulus	GPa	—	—					2.7	0.8			2.1	0.8					2039-2
Hardness — Rockwell R								120	102									
Impact strength: Notched Izod, 23°C	kJ/m²*	2	10	10	NB									65	80	15	24	180
40°C		2	2	11	9									18	—	10	10	''
Notched Charpy, 23°C	kJ/m²					14		4	70	13	>500	12	40					179
40°C						2												''
THERMAL																		
Coefficient of linear thermal expansion	×10⁵	7–8		10		9		7								0.3		ASTM D696
Deflection temperature: Under flexural load at 0.5 MPa	°C	190		185	185	172		180		190		54	54	50		200	200	75
at 1.8 MPa	°C	68				60		65		95								''
Vicat softening point A 9.8N	°C	218				190												306
B 49N	°C	203																''
MISCELLANEOUS																		
Density	g/cm³	1.14		1.13		1.13		1.13		1.13		1.107		1.07		1.28		R1183
Flammability				V2		V2		HB		HB		HB		HB				UL subj 94
Water absorption, 24 h	%	2.0–2.5‡						1.1–1.3				1.3		1.3				R62A
Mould shrinkage (flow direction)	%	1–2.5						0.9–1.1		1.2–2.0		2.1–2.2		2.1–2.2				ASTM D955

Values in J/m are approximately ×10 values shown in kJ/m².

* Some data sheets have values quoted in J/m; these are shown as kJ/m²; Values in J/m are approximately ×10 values shown in kJ/m².
† Annealed 5h at 150°C.
‡ 23°C, 50%RH – DIN 53714.

TABLE 6 – TYPICAL PROPERTIES OF NYLONS, EXCEPT 6 & 66 (Basic grades only)

Property	Type Grade	Unit	610	612 Zytel 151L Du Pont		11 Rilsan BMN Atochem		12 Grilamid L20GM EMS-Chemie		46 Stanyl TW300 DSM		Test standard
				D	C	D	C	D	C	D	C	
PHYSICAL												
Tensile yield stress		MPa	70	61	52	48	34		40	100	40	R527
Elongation at yield		%					20		15		280	..
Elongation at break		%	50	100	250		300–350		300	30		..
Tensile modulus		GPa	1.7	2.0	1.2		1.0		1.40	3.0	1.0	..
Flexural elastic modulus		GPa	1.9							3.5	1.2	178
Hardness — Rockwell R			110						—	123	107	2039-2
— Ball indentation		MPa		114	103		108		—	200	170	2039-1
Impact strength:												
Notched Izod.	23°C	kJ/m²*	6.4	4.3	6.9					10	40	180
..	-40°C	kJ/m²*								5	—	..
Notched Charpy,	23°C	kJ/m²		3.2	2.1		4.0		10	6	33	179
..	-40°C	kJ/m²					2.7		6	5	—	..
Friction coefficient dynamic, mild steel, 200Hz.							0.18					mfr's test
THERMAL												
Coefficient of linear thermal expansion.		°C	12	9		9		12		8		ASTM D696
Deflection temperature: Under flexural load at 0.5 MPa		°C	150	180		150		155				..
.. .. at 1.82 MPa		°C	57	90		55		50		160		75
Vicat softening point .. 9.8N		°C				181		172				..
.. .. 49N		°C				160		140				306
Melt temperature ..		°C	220	212		185		178		288 (B/120)	295	1218
MISCELLANEOUS												
Density		g/cm³	1.08	1.06		1.05		1.01		1.18		R1183
Flammability			V2	V2		V2		V2		V2		UL subj 94
Water absorption, 24 h		%	0.4	0.25		0.3?		0.7		2.3		R62A
.. 23°C/50% RH		%										DIN 53714
Mould shrinkage (flow direction)		%		1.3				0.8–2.0		1.5–2.4		ASTM D955

TABLE 7 – TYPICAL PROPERTIES OF SPECIAL NYLONS

Property	Type	Unit	Trogamid T Dynamit Nobel		Grilamid TR55 EMS-Chemie		Durethan T40 Bayer	Grilon C CR9 EMS-Chemie		Cast nylon	Test standard
			D	C*	D	C		D	C		
PHYSICAL											
Tensile yield stress		MPa	85	75	75		110	50	35	80	R527
Elongation at yield		%	9.5		8		6	10	20	30	"
Elongation at break		%	70	150	50–150		>50	250	300		"
Tensile modulus		GPa	3.0	3.0	2.30		3.00	2.40	0.85	2.4	"
Flexural strength		MPa	125								178
" stress at 3.5% strain		MPa									
Hardness — Rockwell R		MPa	93		70			70	25	118	2039-2
" M											
" — Ball indentation		MPa	140	125							2039-1
Impact strength: Notched Izod, 23°C		kJ/m²*	7				10				180
" -40°C		kJ/m²*					8				
" Notched Charpy, 23°C		kJ/m²	10–15	40	5			5	NB		179
" -40°C		kJ/m²	3–5†		3			2	2		
Tensile impact strength		kJ/m²	(171)							(80)	ASIM D1822
THERMAL											
Coefficient of linear thermal expansion		°C	6		6–9		6.5	12–14		10	ASIM D696
Deflection temperature: Under flexural load at 0.5 MPa		°C	140		143‡		110	103‡		218	75
" at 1.82 MPa		°C	124		134‡		118	54‡		204	
Vicat softening point A (9.8N)		°C	140		155		125	193			306
" " B (49N)		°C	130		155			160			
MISCELLANEOUS											
Density		g/cm³	1.12		1.06		1.18	1.10		1.16	R1183
Flammability			V2		V2			V2		V2	UL subj 94
Water absorption, 24 h		mg or %	40		1.3%					1.4	
" 23°C/50% RH		%			0.5–1.0		2.0%	1.5–2.0%			
Mould shrinkage (flow direction)		%								—	
Melt temperature		°C									

* 4 months at 23°C/50% RH   † –50°C.   ‡ Annealed 5 h at 150°C.

Although it has similar water absorption and shrinkage properties to Nylons 6 and 66, the improved heat resistance results in better strength retention at elevated temperatures. Nylon 46 also has higher impact strength. Typical properties are included in Table 6.

Special nylons

In addition to the types already mentioned there are nylon products having special properties. These materials are generally based on copolymers or sometimes on mixed polymers. (See Table 7 for typical properties.)

Trogamid T (Hüls) and Grilamid TR55 (EMS-Grilon) have somewhat irregular structures; they therefore behave as amorphous materials and are transparent. They are characterised by low moisture absorption and as with Nylons 11 and 12 their strength and electrical properties which are good are largely unaffected by humidity level. Similarly, they have lower melt temperature and distortion temperature under flexural load than 6 or 66. A transparent material called Durethan T40 is produced by Bayer.

Copolyamides such as Grilon C (EMS-Grilon) are highly transparent and virtually impervious to gases and are used in packaging films. There are also block polyamides, which are elastomers used for hydraulic and pneumatic tubing.

Cast nylon is made from caprolactam and is used for large parts, where their higher strength, hardness and resistance to creep compared with moulding grades, is an advantage, and for extruded sections for machining. Although the shrinkage is high, the mouldings have a regular structure and are relatively stress free.

Applications

As well as its large scale use in fibres, nylon is used as a monofilament for brush tufting, wigs, surgical sutures, sports equipment and angling lines. Extrusion applications include cable sheathing (to resist abrasion and chemical attack), flexible tubing for petrol and diesel fuel lines and blow moulded containers where strength, chemical resistance and low gas permeability are required.

The main uses for nylon in plastics processing are in injection moulding of engineering parts and the applications therefore cover a very wide field; some examples are:

Insulators, gears, coil formers, wheels, screws, bolts, and washers.

Automotive steering column switches, items such as ignition distributor parts, fan blades, and many others.

Industrial and domestic plugs and sockets, fixing and fastening devices. Switchgear and components for the electrical and electronic industries, appliance housings.

Bearings and wear parts, seals and valve parts.

Pump housings, typewriter parts, filter system parts, gas lighters and curtain accessories.

Modified PPO/PPE

POLYPROPYLENE ETHER (abbreviated PPO or PPE) has not been used on its own because of the difficulties in processing. Once combined with other thermoplastics these problems disappear and the modified materials become successful engineering plastics. They have high rigidity, strength and hardness with high to very high toughness combined with high heat resistance, excellent mouldability and dimensional stability.

This combination of properties renders them well suited for making large parts which are stressed and/or at high temperatures. In competition with materials such as ABS, most of the grades have higher heat distortion temperatures than high heat ABS, and compared with PBT they have much lower shrinkage.

Manufacturers

Modified PPO has been produced for some years by GE Plastics. Now there are also BASF and Hüls.

Properties

There are two basic ranges:

PPE combined with impact polystyrene (HIPS), such as Noryl (GE Plastics) and Luranyl (BASF).

PPE combined with nylon, such as Noryl GTX (GE Plastics) and Ultranyl (BASF).

Whilst HIPS compounds have many of the characteristics of the styrene based plastics (PS, SAN, ABS), such as high strength and impact, dimensional stability, ease of moulding (to a high gloss), low shrinkage and low water absorption, the PPE brings improved strength characteristics and better heat resistance, with Vicat Softening Points in most cases between 105 and 150°C.

Being amorphous they have low shrinkage and can be moulded on fast cycles. Unlike the semi-crystalline plastics, their resistance to stress does not diminish considerably at elevated temperatures, which means that they can be used at higher temperatures without needing reinforcement, they also have good low temperature impact retention. A great advantage is that flame retardant grades can easily be made without using halogen containing additives and with little change in the low density. Because of the content of PS, the weathering and U-V resistance of modified PPO is poor, so it is not used in outdoor applications where the surface is unprotected.

Electrical properties of modified PPO are similar to those of polystyrene:

Volume resistivity	10^{16}
Dielectric constant	2.6–3.2
Dissipation factor ($\times 10^{-4}$)	
Standard grades	4 at 50 Hz–9 at 1 MHz
To FR reinforced grades	16–20 at 1 MHz
Dielectric strength	20–24 kV/mm

Chemical resistance

Modified PPO products have excellent resistance to alkalis, inorganic salts, detergents and dilute acids. They are highly resistant to hydrolysis (unlike some of the other engineering plastics), but are subject to ESC problems in contact with strong acids or alcohols at higher temperatures. Ketones, esters and ethers will attack, and modified PPO is dissolved by many hydrocarbons. Certain oils/greases in contact will give acceptable results, but this depends on the composition.

Processing

These materials may be extruded, injection moulded or blow moulded without difficulty. By far the bulk of modified PPO is injection moulded, taking advantage of their similarity to polystyrene in their ease of moulding. Pre-drying is only necessary in extreme situations (*eg* where the highest gloss is required). Because of the amorphous nature and low water absorption of these materials, there are virtually no problems such as warping during cooling.

In addition to the standard and flame retardant grades, glass reinforced and FR glass reinforced grades are available, together with some special grades, including structural foam types. Tables 1 and 2 show typical properties of some of these grades.

PPE combined with nylon produces materials which possess a good combination of the properties of standard PPE compounds with those of the nylons, having good strength properties with higher resilience and impact strength than either standard Noryl or Luranyl, and a considerable improvement in Vicat to 190 to 200°C. There is invariably a sacrifice in water absorption which is higher but still good, and in mould shrinkage, which is close to that of the nylons. Typical properties of some of the available grades are shown in Table 3. Trade names include Noryl GTX (GE) and Ultranyl (BASF).

TABLE 1 – TYPICAL PROPERTIES OF SOME MODIFIED PPO UN-REINFORCED GRADES

Property	Unit	Standard				Flame retardant			Test standard
		GE Plastics high gloss 110HG	GE Plastics higher heat 731	BASF genl. purpose KR 2404	BASF higher impact KR 2421	GE Plastics SE1	GE Plastics Vo-150B	BASF 2453	
PHYSICAL									
Tensile yield stress	MPa	45	55	55	48	55	65	60	R527
Elongation at yield	%	3	7	4	5	7	7	6	"
Elongation at break	%	50	50	35	50	50	30	55	"
Tensile modulus	GPa	2.40	2.50	2.50	2.20	2.50	2.40	2.50	178
Flexural strength	MPa	90	98	95	80	98	105	102	
Flexural elastic modulus	GPa	2.50	2.50	—	—	2.50	2.50	—	2039-2
Hardness — Rockwell, R	MPa	115	115	113	115	115	115	110	2039-1
— Ball indentation, 30s value.		87	100	—	—	100	113	—	ASTM D1044
Taber abrasion CS17, 1kg	mg/kHz	20	20	—	—	20	20	—	
Impact strength:									
Notched Izod 23°C†	kJ/m²†	25	20	37	48	20	25	35	180
" 40°C		14	14	NB	NB	14	10	NB	
Un-notched Charpy, 23°C	kJ/m²	—	—	NB	NB	—	—	NB	179
" 40°C		—	—	—	—	—	—	—	
Notched Charpy, 23°C	kJ/m²	15	15	11	25	15	20	11	179
" 40°C.		—	—	7	15	—	—	6	
THERMAL									
Thermal conductivity	W/m.K	0.16	0.22	0.18	0.17	0.22	0.16	0.18	DIN 52612
Coefficient of linear thermal expansion	10^{-5}.OK	7	6	6-7	6-7	6	7	6-7	ASTM D696
Deflection temperature:									
Under flexural load, at 0.45 MPa	°C	110	130	120	123	125	150	112	75
" at 1.82 MPa		125	148	105	106	142	165	95	
Vicat softening point, rate B B/120	°C	120	135	—	—	130	155	—	1525/306
" " B/50		—	—	125	125	—	—	116	53460/306
MISCELLANEOUS									
Density	g/cm³	1.06	1.06	1.06	1.06	1.06	1.06	1.08	R1183
Flammability		HB	HB	HB	HB	V1(1.47mm)	VO(1.55mm)	V1(1.6mm)	UL subj 94
Water absorption, 24h. 23°C/50% RH	% or mg	0.07	0.07	<0.1%	<0.1%	0.08	0.08	<0.15	R62
Mould shrinkage (flow direction)	%	0.5-0.7	0.5-0.7	0.5-0.7	0.5-0.7	0.5-0.7	0.5-0.7	0.5-0.7	1110
Melt flow (MFI) 250°/21.6	g/10min	—	—	50	30	—	—	75	R1133

†Some data sheets have values quoted in J/m these are shown as kJ/m². Values in J/m are approximately ×10 values shown in kJ/m².

TABLE 2 – MODIFIED PPO – TYPICAL PROPERTIES OF SOME GLASS REINFORCED COMPOUNDS AND SPECIALS

Property	Unit	Standard BASF 2403 G2 10% glass	Standard GE plastics GFN2 20% glass	Flame retdnt GE plastics VO3505 35% short gl	Specials (GE Plastics) Auto* PX 1760	Specials — Structural foam Fn 215D Flame retdnt	Specials — Structural foam FN 5007 7% glass	Test standard
PHYSICAL								
Tensile strength (at break)	MPa	70	90	77	35(yield)	22	45	R527
Elongation at yield	%	—	—	2	6	—	—	::
Elongation at break	%	2	3	—	60	8	8	::
Tensile modulus	GPa	4	—	—	—	—	—	
Flexural strength	MPa	4.5	6.50	7.30	1.80	8	2.05	178
Flexural elastic modulus	GPa	112	140	120	65	1.60	65(Br)	
Hardness — Rockwell, R		—	6.0	5.0	1.90	46	2.40	2039-2
— Ball indentation		—	L106	—	L63	1.80		2039-1
Taber abrasion CS17, 1kg	mg/kHz	135	122	147	75			ASTM D1044
Impact strength:			35					
Notched Izod 23°C	kJ/m²	9	8.0	6.0	40			180
-40°C		—	7.0	5.0				
Un-notched Charpy, 23°C	kJ/m²	23	—	—	—	20	16	179
-40°C		21						
Notched Charpy, '' 23°C	kJ/m²	7	9	5	20			179
'' '' -40°C		5						
THERMAL								
Thermal conductivity		0.21	0.24	—	—	0.12	0.12	DIN 52612
Coefficient of linear thermal expansion		0.5–0.6	4.0	2.4–5.1‡	6	7–8	5–6	ASTM D696
Deflection temperature:								
Under flexural load at 0.45 MPa	°C	134	132	116	—	96	103	75
'' '' at 1.82 MPa	°C	123	145	107	130	83	92	
Vicat softening point, rate B B/120	°C			136	115			1525/306
'' '' '' B/50		137	140	123				53460/306
								:: ::
MISCELLANEOUS								
Density	g/cm³	1.14	1.21	1.45	1.06	0.88	0.88	R1183
Flammability		HB	HB	VO(1.6mm) 5V(3.2mm)	HB	VO(6.29mm) 5V ''	HB	UL subj 94
Water absorption, 24h	%	<0.1%	0.06	0.05	0.07			R62
23°C/50%RH	% or mg	0.4–0.5						
Mould shrinkage (flow direction)		0.4–0.5	0.2–0.4	0.1–0.3	0.5–0.7	0.6–0.9	0.4–0.7	1110
Melt flow (MFI)	g/10min	11						R1133

†Some data sheets have values quoted in J/m; these are shown as kJ/m². Values in J/m are approximately ×10 values shown in kJ/m².

‡with/across flow.

*For EEC impact tests on dashboards

TABLE 3 – TYPICAL PROPERTIES OF SOME PPO (PPE)/PA ALLOYS

Property		Unit	STD GE plastics GTX 900	Reinforced GE plastics GTX 830 30% glass	BASF KR 4540 30% glass	Test standard
Type Grade Reinforcement						
PHYSICAL						
Tensile yield stress		MPa	55	160(Br)	140(Br)	R527
Elongation at yield		%	5	—	—	,,
Elongation at break		%	100	3	4	,,
Tensile modulus		GPa	2.00	9.0	8.6	,,
Flexural stress at yield		MPa	85	220(Br)	—	178
Flexural modulus		GPa	2.10	7.0	—	,,
Hardness — Ball indentation			115	141	—	2039–1
Taber abrasion CS17, 1kg		mg/kHz	14	—	—	ASTM D1044
Impact strength:						
Notched Izod	23°C	kJ/m²†	24	8.5	15	180
,, ,,	–20°C		19	7.5		
,, ,,	–30°C				12	
Notched Charpy	23°C	kJ/m²	15	7	—	179
THERMAL						
Thermal conductivity		W/m.K	0.23	0.26	—	DIN 52612
Coefficient of linear thermal expansion			9	2–3	—	ASTM D696
Deflection temperature:						
Under flexural load at 0.45 MPa		°C			250	75
,, ,, ,, at 1.82 MPa		°C	—	225–	200	
Vicat softening point, rate B		°C	220	240		1525/306
,, ,, ,, B/120		°C	190	240		53460/306
,, ,, ,, B/50					220	,, ,,
MISCELLANEOUS						
Density		g/cm³	1.10	1.31	1.35	R1183
Flammability			HB	HB	HB	UL subj 94
Water absorption, 24h		%	0.40	0.50	0.9(satn)	R62
23°C/50%RH		% or mg				
Mould shrinkage (flow direction)		%	1.2–1.6	0.3–0.5	—	1110
Melt flow (MVI)	275°C	15g/10min	—	—	10	R1133

†Some data sheets have values quoted in J/m; these are shown as kJ/m². Values in J/m are approximately ×10 values shown in kJ/m².

Noryl X-tra is a recent introduction by GE, which consists of a range of special products based on high PPE content formulations. So far GE have introduced two types are available:

HH195, which takes full advantage of the heat resistance of PPE.
LSE types which are low smoke resins for electrical applications.

Preliminary data for these grades is given in Table 4.

TABLE 4 – NORYL XTRA, PRELIMINARY DATA

Property	Unit	HH 195	LSE PO 1948	LSE PO 1961	Test standard
PHYSICAL					
Tensile yield stress	MPa	65	70	65	R527
Elongation at yield	%	10			,,
Elongation at break	%	50	20	20	,,
Tensile modulus	GPa	1.90			,,
Flexural strength	MPa	100	100	90	178
Flexural modulus	GPa	2.21	2.20	2.00	,,
Impact strength:					
Notched Izod 23°C.	kJ/m²†	30	25	35	180
–20°C.		7.5			
–30°C.			6	13	
Notched Charpy 23°C.	kJ/m²	8	10	20	179
THERMAL					
Coefficient of linear thermal expansion		7.5	7.5	7.7	ASTM D696
Deflection temperature:					
,, ,, at 1.82 MPa	°C	195	135	135	
Vicat softening point B/120	°C	205	150	150	53460/306
MISCELLANEOUS					
Density	g/cm³	1.04	1.11	1.09	R1183
Flammability		V1 (3.2mm)	Vo	Vo	UL subj 94
Oxygen index (LOI)	%	25			ASTM D2863
Water absorption, 24h	%	0.13			R62
Mould shrinkage (flow direction)	%	0.4–0.6			1110

†Some data sheets have values quoted in J/m; these are shown as kJ/m². Values in J/m are approximately ×10 values shown in kJ/m².

Applications

Modified PPO plastics are used in a wide range of applications:

Automotive

Instrument panels, consoles, speaker housings, ventilator grilles and nozzles, expansion tanks and air inlet grilles.

On-line painted parts (Noryl GTX) such as fenders, valance panels and exterior body parts.

Electrical

Switch cabinets, fuse boxes, small motor housings, radio and television fascias, tube deflection yokes and coil formers.

Business machines and computer housings.

Fluid engineering

Pumps, impellers and water meter parts.

Polyolefins: LDPE/LLDPE/MDPE/HDPE/ polypropylene/EVA

POLYETHYLENES are a family of polymers based on the simplest of chemical structures – a carbon-carbon backbone with hydrogen atoms attached, using ethylene as the starting material:

$$\left[\begin{array}{cccc} H & H & H & H \\ -C & -C & -C & -C- \\ H & H & H & H \end{array} \right]_n$$

The polymer structure is not, however, as simple as this. Some of the polymer chains have branches, and the structure may be complicated in some cases by the presence of higher olefin comonomers; the spread of chain length can also vary. There are different processes for producing the polymer, which accentuate one or another of these characteristics. As a result of these variations polyethylenes are available in a wide range.

Some of these variations change the way the chains pack together and therefore change the density of the material; in addition control of each process allows variation of the molecular weight and this controls the melt flow of the material.

To categorise the grades available, polyethylenes are arbitrarily divided into groups by density, although there is some overlap at the borders of the groups:

		Density range
Low density	(LDPE)	0.910–0.925
Linear low density	(LLDPE)	0.890–0.955
Medium density	(MDPE)	0.926–0.945
High density	(HDPE)	0.946–0.965

In addition there is Ultra-high Molecular Weight (UHMWPE). Note that LLD products usually have a narrower distribution of molecular weight than LD, so that properties at the same density will be different. (Medium density is not normally made as such, but is a blend of LD and HD.)

Polyethylene has limited use as an engineering plastic because its strength properties and heat resistance are low, compared with other thermoplastics. Nevertheless certain grades are used for heavy duty industrial purposes such as gas pipe, cable insulation and semi-bulk materials handling containers.

Properties

The properties of the material depend on the structure of the polymer and the molecular weight. Since as mentioned above this is reflected in the density and MFI, these attributes are used to classify grades.

For a given density, as molecular weight increases, the impact strength and ESC (environmental stress cracking) resistance improve, but the melt flow (MFI) decreases (*ie* the material becomes less easy flowing). On the other hand, for a given MFI an increase in density increases tensile strength, rigidity, heat and chemical resistance, whilst impact and permeability are decreased. The choice of density and MFI is therefore a compromise between ease of processing and the needs of the application.

Polyethylenes are easily fabricated by the usual processes for thermoplastics, although being crystalline materials they have a high specific heat in the temperature range where crystallisation occurs; this causes them to have high shrinkage on moulding, with the attendant risk of warpage. They also have a high coefficient of expansion. As with other thermoplastics, injection moulding grades usually have high MFI (typically 6–20, but some grades are as high as 80) whilst extrusion grades are relatively stiff flowing (0.1–3).

Film may be produced as blown (tubular) film, or as cast film (particularly with LLDPE, making use of its excellent draw down without causing web breaks, for lower-cost products with high clarity). Bottles are produced by extrusion blow moulding, and rotational moulding is extensively used for large mouldings.

Uses of polyethylenes have become closely defined, particularly in extrusion, and grades are often formulated with appropriate additives, such as slip of antistatic, or to allow then to be cross-linked (*eg* for cable insulation). Ageing of PE products by U-V on outdoor exposure can be considerably reduced by the incorporation of finely divided carbon black, up to 2.5%; alternatively, U-V absorption systems can be used. Pigmentation may be helpful, but certain combinations of pigment can in fact accelerate the degradation.

Typical property ranges for polyethylenes are shown in Table 1, and some comparisons of typical grades for specific applications in Table 2.

Ultra high molecular weight PE (UHMWPE)

UHMWPE was, in the past, very difficult to process in injection moulding machines. There is now a granular grade available from Hoechst which they claim can be moulded into complex parts. UHMWPE provides wear resistance, low friction, chemical resistance, physiological inertness and high notched impact strength even at low temperatures.

TABLE 1 – TYPICAL RANGE OF PROPERTIES FOR POLYETHYLENES

Property		Unit	Values	Test standard
PHYSICAL				
Tensile yield stress		MPa	7–10(LD)–10–12(LLD)–19–28(HD)	R5277
Tensile strength at break		MPa	7–18 –20–26 –18–32	,,
Elongation at break		%	100–700 –800–1000 –200–800	,,
Tensile modulus		MPa	160–240 – 800–1300	,,
Hardness — Shore D			42–53 – 60–70	DIN 53505
Impact strength:				
Notched Charpy	23°C	kJ/m^2	NB 2.5–NB	179
	−40°C		NB 2.5–NB	
Brittleness temperature		°C	−10 – −80 —	ASTM D746
THERMAL				
Thermal conductivity		×10^{-5}	0.35 – 0.43	DIN 52612
Coefficient of linear thermal expansion		°C	10 – 25	ASTM D696
Vicat softening point	1kg		70–100 – 120	1525/306
,,	5kg		50–65 – 65–85	
ELECTRICAL				
Volume resistivity		Ohm.cm	>10^{16}	IEC 93
Relative permittivity			2.3–2.5	IEC 250
Dissipation factor		×10^{-4}	<1–7	IEC 250
Electric strength			500–950	IEC 243
Arc resistance			L4	ASTM D495

Note: In some cases figures for ESC resistance and for gas permeabilities are given in data sheets. They are not shown here because test methods vary and the results are therefore not comparable between manufacturers.

TABLE 2a – PE – COMPARISON OF LD AND HD FOR MOULDING

	LD	HD	
MFI	7–70	4–25	g/10 min
Density	0.918–0.926	0.950–0.961	g/cm^3
Tensile strength	10	18–32	MPa
Elongation at break	120–500	400–800	%
Tensile modulus	160–240	950–1250	MPa
Vicat softening Pt 1 kg	94–108	124–128	°C

TABLE 2b – PE – COMPARISON OF LD AND HD FOR BOTTLES

	LD	HD	
MFI	0.25–2.4	0.3–2	g/10 min
Density	0.922–0.930	0.951–0.963	g/cm^3
Tensile strength	10–18	30	MPa
Elongation at break	500–650	800	%
Tensile modulus	205–350	1100–1250	MPa
Vicat 1kg	94–108	128	°C
ESC resistance (ASTM D1093)		44–600h	

TABLE 2c – PE – COMPARISON OF LD AND LLD FOR FILMS

	LD	LLD	
MFI	0.25–3.8	0.9–2.5	g/10 min
Density	0.921–0.928	0.917–0.919	g/cm^3
Tensile strength	10–18	20–26	MPa
Elongation at break	450–700	900–1000	%
Vicat 1 kg	92–100	90–100	°C

Chemical resistance

Because of the polymer structure and its high crystallinity, polyethylene is resistant to a wide variety of chemicals; its water absorption is very low. At temperatures below 60°C it is insoluble in practically all solvents. At room temperature PE is unaffected by alkalis, salt solutions and inorganic acids with exception of strongly oxidising types. Some polar liquids cause swelling and loss in properties, but this is reversed on evaporation of the solvent. PEs are however subject to ESC (environmental stress cracking – ie cracking when exposed to the environment when under stress), to some extent from caustic soda, washing soda or more particularly from silicone and essential oils, aqueous solutions of surfactants, alcohols, organic acids, ESC resistance is better with grades of low MFI.

Assembly and decoration

Polyethylenes have an inert surface and good solvent resistance. They are therefore difficult to bond using adhesives, although there is now a primer available which allows a strong bond to be made with a superglue type of adhesive. PE can be assembled using heat or ultrasonics. Preparation of the surface is necessary before decoration, in order to get adhesion, and this is usually affected by flame treatment, corona discharge or special primers; all common decorating techniques can then be used.

Applications

With development of new processes and grades there is a constant flex of applications between types of PE. In particular, LLDPE with its superior properties and ability to down gauge is taking over many applications which were previously in LD or even HD. This makes it difficult to give other than general indications on the uses of the different types of PE:

- LD is mainly used for films and coatings, wire and cable insulation and sheathing, housewares and some closures
- LLD has taken a large part of the film market and is spreading into other applications, such as closures, where its superior ESC resistance is valuable, especially in pharmaceutical.
- HD is used in many applications for its excellent low temperature impact and/or its lower gas permeability. It is used for bottles, sheet, pipe, crates, materials handling (boxes, drums, semi-bulk containers), petrol tanks. Paper-like film for wrapping, bags and carrier bags.

POLYPROPYLENE (PP)

POLYPROPYLENE is made by polymerisation of propylene monomer, using what are called stereospecific catalysts. It has the repeating unit:

$$
\begin{array}{ccc}
\text{H} & & \text{H} \\
-\ \text{C} & - & \text{C}\ - \\
\text{H} & & | \\
& & \text{CH}_3 \\
& & _n
\end{array}
$$

The CH_3 is called a methyl group and can be arranged around the chain in different ways:

All on one side	isotactic
Alternating between sides	syndiotactic
Randomly on either side	atactic

The purpose of the special catalyst system is to produce as high a percentage of the isotactic as possible, since it is this version that gives a product with interesting properties. Most commercial grades contain around 98% isotactic. Syndiotactic is not produced as such but atactic, which is a largely amorphous, semi-tacky material, is produced in small quantities for particular applications. Polypropylene is a versatile plastic with uses ranging from fibre to film, and from housewares to applications at elevated temperatures as an engineering plastic. This is due to its excellent combination of low density, good hardness and abrasion resistance, rigidity, toughness and flex fatigue resistance, excellent chemical resistance and, above all, heat resistance. This is combined with very good electrical properties and versatile processability.

Properties

PP is produced as:

Homopolymer (*ie* using polypropylene alone).

Copolymer, which may be either;

Random or block, in which small amounts of ethylene are incorporated in the polymer chains.
Heterophase, where ethylene-propylene rubbers reinforce the PP matrix.

These three types differ in their properties.

The versatility of PP has given it a continuing high growth rate, which has encourage progressive development and improvement in the process. As a result there is now a tendency for users to define a desired product as much by the type of manufacturing process as by the producer.

As with PE, within each type the molecular weight can be varied, and between processes there are differences in molecular weight distribution. There is therefore a very large number of grades available; fortunately many of them are designed for particular applications, which somewhat simplifies the choice. It should be noted that as with PE, increase in the molecular weight increases toughness, but in this case it decreases the tensile properties.

Homopolymer PP is in many respects similar to HDPE, which is made by a similar process. It is superior in many properties, having lower density, greater rigidity, considerably higher maximum operating temperature, and it is free from ESC problems. It is however more sensitive than HDPE to oxidation at elevated temperatures, although this deficiency can be largely overcome through suitable stabilisation. Impact strength is rather low and homopolymer PP suffers from notch sensitivity and a dramatic drop in impact at around 0°C.

Copolymers overcome this defect and have greatly improved toughness, at some expense of rigidity and strength properties. The types differ considerably in their property profile:

Random copolymers have a much reduced possibility of crystallisation. They are less rigid than homopolymer or block copolymer, have lower softening temperatures, but have greatly improved transparency, especially what is called contact clarity. Block copolymers have improved low temperature toughness, with little loss in other properties.

Heterophase copolymers show much greater toughness than homopolymer, particularly at low temperatures. The improvement depends on the content of elastomer and can be considerable, but it is achieved at the expense of strength properties.

A particular type of PP is called controlled rheology (CR). In these grades the polymer chains are modified to enhance the flow properties, particularly for thin-walled mouldings for packaging. These grades also have less odour, better impact, and reduced shrinkage and distortion.

Typical range values for the different types of PP are shown in Table 3 and typical values for homopolymer, block and random copolymers from one manufacturer are compared in Table 4. These values are given for injection moulding grades; products intended for extrusion are within the range shown in Table 3. Grades for blow moulding, sheet or BOPP are chosen from low MFI products, whilst those for fibre are usually from high MFI grades. They differ from injection grades in other ways – for example additives for film grades to provide slip. Typical values for extrusion grades are shown in Table 5.

As a crystalline material PP has high specific heat (although lower than that of PE) and it can therefore be moulded on shorter cycles. It has high mould shrinkage, but with lower anisotropy than PE there are fewer problems of warpage, although care is still needed. Orientation of the molecular chains (by stretching or drawing) has a very favourable effect on properties and improves clarity; one of the fastest growing applications is as biaxially oriented film (OPP or BOPP).

PP may be further modified by the incorporation of filler/glass-fibre. Minerals such as talc, calcium carbonate, or milled glass-fibre act as fillers, reducing both shrinkage and the coefficient of expansion and improving the dimensional stability. They also provide a reinforcing effect, improving rigidity and heat resistance, although with some minerals there is a reduction in tensile strength.

To obtain the full effect of adding glass-fibre it is necessary for the fibre surface to be treated with a suitable coupling agent, in which case the reinforcing effect may cause for example, the tensile strength and rigidity to be 3–4 times greater than that of the un-reinforced material.

Mineral/glass-fibre reinforced grades based on homopolymer are available from several producers and some also make reinforced grades based on copolymers. Values for some typical grades are shown in Table 6.

As with most thermoplastics, compounds of PP are available to provide flame retardancy, U-V resistance for outdoor use, and an anti-static surface. PP is also used for larger mouldings, in the form of structural foam.

TABLE 3 – PP: TYPICAL VALUES OF INJECTION MOULDING GRADES

Property	Type	Unit	Homopolymer	Copolymer	'Heavy Duty' ICI	Elastomer blends Hoechst	CR grades ICI	Test standard
MISCELLANEOUS								
Density		kg/m³	902–915	900–912	905	896–900	905	R1183
Melt flow (MFI)	230/2.16	g/10min	0.3–45	0.25–40	2–9	0.7–8	8.5–22	R1133
PHYSICAL								
Tensile yield stress		MPa	30–38	20–30	23	10–23	21–27.5	R527
Elongation at yield		%	15–20	10–21	—	9–17	—	"
Elongation at break		%	>600	>600	—	—	—	
Flexural modulus		GPa	1.1–1.7	1.0–1.5	0.95–1.10	0.3–0.9	0.93–1.20	2039–2
Hardness — Rockwell R			95	66	75	—	75–80	DIN 53505
— Shore D			70–79	50–90		46–64	—	2039–1
— Ball indentation			63–90			28–48	—	DIN 53754
Abrasion — Taber CS17 wheel, 1k.h.1kg		mg	13–16	13–15				
Impact strength:								
Notched Izod	23°C	kJ/m²†	2.0–6.5	6.0–8.0	10–NB(30)‡	60–80	5.0–NB	180
	0°C		2.5–3.5	3.0–8.5	6–10	10–75	4.0–9.0	
	−20°C		2.0	2.0–5.5	4–7.5	6–50	3.0–7.0	
	−40°C					4–6	5–25	
Notched Charpy	23°C	kJ/m²	4–12	6–44		35–NB		179
	0°C		2–5	4–25		22–NB		
	−20°C		2–4	3–10		11–NB		
THERMAL								
Vicat softening point: 9.8N		°C	148–155	137–147	143	100–140	143	306
" " 49N		°C	87–102	72–78		54–58	—	
Deflection temperature:								
Under flexural load at	0.45 MPa	°C	95–130	85–115	95	60–105	90–97	75
" " at	1.82 MPa	°C	60–75	50–70	50	40–55	48–55	

TABLE 4 – PP: COMPARISON OF HOMOPOLYMER WITH BLOCK AND RANDOM COPOLYMER. (Typical values for one producer – Appryl)

Property		Unit	Homo-polymer	Block	Random	Test standard
MISCELLANEOUS						
Density		kg/m^3	905	902	900	R1183
Melt flow (MFI) 230/2.16		g/10min	3–40	5–30	10	R1133
PHYSICAL						
Tensile yield stress		MPa	30–35	28	29	R527
Elongation at break		%	>600	>600	>600	,,
Flexural modulus		GPa	1.2–1.7	0.95–1.3	1.05–1.20	,,
Hardness — Shore D			74	66–70	61	DIN 53505
Impact strength:						
Notched Izod	23°C	kJ/m^2†	4.5–6.5	8.0–50	7.5	180
	0°C		1.5–3.5	4.0–11.0	2.3–3.0	
THERMAL						
Vicat softening point:	9.8N	°C	151	140–151	125–130	306
,, ,, ,,	49N	°C	10	79–85	—	
Deflection temperature:						
Under flexural load at 0.45 MPa		°C	94–98	80–87	73–80	75
,, ,, ,, at 1.82 MPa		°C	64	52–56	—	

†Shown as J/m on data sheets. Values in J/m are approximately ×10 values shown in kJ/m^2.

Whilst PP can be used at high temperature (mouldings can be sterilised at 130°C), this is restricted, as with other crystalline polymers, to low or zero stress situations. At higher stresses operating temperatures are much reduced. Reinforcement with glass-fibre significantly reduces this effect, as can be seen by comparing the values for distortion temperature under flexural load (Table 3) with those in Table 6. The differences in value between the two stress levels in Table 3 are seen to be very much reduced in Table 6 (columns 3 and 4).

A special property of PP associated with its high flexural fatigue resistance is its ability to act as a kind of living hinge in thin sections. This means for example, that both halves of a box can be moulded together so that the top is hinged to the base, and the top can have a snap closure. A PP hinge can withstand millions of flexings and advantage has been taken of this in producing the accelerator pedals for cars.

Chemical resistance

PP has very low water absorption and excellent chemical resistance. It is resistant to aqueous solutions of salts and to non-oxidising acids and alkalis. Strong acids may, however cause shade changes with certain colourants, and at higher temperatures under stress will cause crazing.

Up to 60°C PP is resistant to many solvent, fats and waxes, but is slightly swollen by hydrocarbons. At higher temperatures fats, oils and waxes cause swelling. Certain grades have excellent resistance to hot detergent solutions. PP

TABLE 5 – PP: TYPICAL RANGE OF EXTRUSION GRADES

Property	Unit	Film	Bottles BOPP Sheet	Tapes Rope Fibre	Pipe	Test standard
MISCELLANEOUS						
Density	kg/m^3	905	905	905	912–935	R1183
Melt flow (MFI) 230/2.16	g/10min	2–10	0.4–3.4	4.5–35	0.4–0.8	R1133
PHYSICAL						
Tensile yield stress	MPa	35–37	28–35	25–37	25–33	R527
Elongation at yield	%				15–20	,,
Elongation at break	%		>600	600	800–1000	,,
Flexural modulus	GPa	1.36–1.55	1.1–1.4	1.30–1.55		178
Hardness — Shore D		74	74	74–79	67–72	DIN 53505
— Ball indentation					50–68	2039-1
Impact strength:						
Notched Izod 23°C	kJ/m^2*	2–4.5	4–5.5	2–3		180
Notched Charpy 23°C	kJ/m^2				10–40	179
0°C					3–17	
–20°C					3–7	
THERMAL						
Vicat softening point: 9.8N	°C	153	150–155	150–155		306
,, ,, ,, 49N	°C		93–100	95–105		
Deflection temperature						
Under flexural load at 0.45 MPa	°C	108–114	92–108	90–115	110	75
,, ,, ,, at 1.82 MPa	°C		60	60–63	60–65	56–60

MFI at 190°C/5kg.
*Some data sheets have values quoted in J/m; these are shown as kJ/m^2. Values in J/m are approximately ×10 values shown in kJ/m^2.

is not resistant to strong oxidising agents such as nitric acid or fuming sulphuric acid, nor to halogenated hydrocarbons.

Sterilisation by higher energy radiation can reduce toughness, with copolymers or CR grades performing better in this situation, but trials are recommended.

Electrical properties do not differ significantly between PP types. Typical values are given below; note that the dissipation factor shows only small variation with temperature and frequency:

	Unit	Value	Test standard IEC
Volume resistivity	Ohm.cm	>10^{16}	93
Relative permittivity		2.25	250
Dissipation factor at 50 Hz	10^{-4}	2.0–2.8	250
,, ,, at 10^5 Hz		3.0–4.5	
Tracking resistance	Rating	CTI>600	112
Arc resistance	Rating	L4	DIN 53484

TABLE 6 – PP: TYPICAL PROPERTIES OF SOME FILLED/REINFORCED GRADES

Property	Grade	Unit	Atochem X1240 40% mineral	Hoechst PPN 7780 GV20 20% mild gl	Hoechst PPN 7790 GV2/30 30% coupled gl	ICI GC40S402 40% cpld gl + elastomer	Test standard
MISCELLANEOUS							
Density		kg/m^3	1.23	1.05	1.14	1.22	R1183
Melt flow (MFI)	230/2.16	g/10min	12	1.4	0.8	4.0	R1133
Mould shrinkage						0.2–1.0	
PHYSICAL							
Tensile yield stress		MPa	34	32	—	77	R527
Tensile stress at break		MPa	—	20	80		''
Elongation at break		%	—	50	3	—	''
Flexural modulus		GPa	3.4	[2.4]*	[5.5]*	6.9	178
Hardness — Rockwell R			78	80	110	95	2039–2
— Shore D							DIN 53505
— Ball indentation				16	16		2039–1
Abrasion–Taber CS17 wheel, lk.h,1kg		mg					ASTM D1044
Impact strength:							
Notched Izod 23°C		kJ/m^2	3			16.0	180
0°C			1.5				
−20°C							
−40°C							
Notched Charpy 23°C		kJ/m^2	2.5	5	6	11.0	179
0°C			1.7	—	—		
−20°C			0.5				
THERMAL							
Thermal conductivity		W/m.K		0.25	0.30		DIN 52612
Coefficient of linear thermal expansion		x10^{-5}		6–17	6–7	2.5	ASTM D696
Vicat softening point: 9.8N		°C	155	150	160	158	306
49N		°C	110	90	130		
Deflection temperature: Under flexural load at 0.45 MPa		°C	130	118	155	159	75
'' '' at 1.82 MPa		°C	82	72	130	146	

*Flexural creep modulus — 1 min value.

Assembly and decorating

PP has an inert surface and good solvent resistance. It is therefore difficult to bond using adhesives; but there is now a primer available which allows a strong bond to be made with the superglue type of adhesive. It can be assembled using heat (*eg* friction welding), or by ultrasonics.

Preparation of the surface is necessary before decoration, in order to obtain adhesion; this is usually achieved by flame treatment, corona discharge or the use of special primers, following which the common decorating techniques can be used.

Applications

Weaving tapes for carpet backing, sacks, ropes, industrial fabrics.
Spun fibres of carpets, upholstery. Fibrillated film for concrete reinforcement.
Extruded strapping, sheet, profiles, blow-moulded bottles and containers.
Moulded housewares, domestic appliances (especially washing machine drums and dishwasher parts), closures, toys.
Packaging – BOPP and laminated or co-extruded film for food packs, moulded tubs and pots (*eg* margarine packs).
Automotive battery cases, spoilers, wheel arch liners, brake fluid reservoirs, radiator expansion tanks, ventilator blades and frames.
Tote boxes, luggage, outdoor furniture and stadium seating, artificial grass.
Paint pails, creates, chair shells.
Chemical plant (tanks, filter press plates) and pipe (domestic drain and waste, pressure pipe).
Blow moulded hot water taken, under floor heating pipes.
Sterilisable medical items.
Boxes with integrally hinged lids.

EVA (ETHYLENE VINYL ACETATE)

EVA is a copolymer of ethylene and vinyl acetate in which the proportion of vinyl acetate (VA) is varied according to the application, as well as the variation in density and MFI seen with the polyethylenes. It is a tough, flexible material which retains its properties over a wide temperature range.

Properties

EVA has excellent elasticity and resilience down to low temperatures; its brittle point is at about –60°C. It has excellent flex-crack and ESC resistance. Clarity is good to excellent and chemical resistance is generally good. The rigidity and hardness of EVA can be increased by the incorporation of fillers and high levels can be used without loss of properties.

EVA can be processed by most normal methods for thermoplastics. It is mainly converted into film (especially co-extrusion), by blow moulding and injection moulding. It serves as the base for the production of many master batches for PE, as it blends well with polyethylenes. It is also used for blending with waxes and for adhesives. EVA has an inherent tackiness, which may have to be taken into account when moulding, by reducing moulding temperatures.

A typical range for EVA grades is:

MFI (190/2.16)	0.5–7	g/10 min
Density	0.925–0.950	g/cm^3
VA content	5–28	%
Tensile strength	16–33	MPa
Tensile modulus	20–190	MPa
Elongation at break	750–850	%
Vicat (1 kg)	50–94	°C
Hardness (Shore A)	80–95	

Applications

EVA finds extensive use in packaging, particularly for bags, shrink film, deep freeze bags, and for co-extruded and laminated film.

Other uses include closures, ice trays, gaskets, medical parts, tubing and gloves, and inflatable articles.

Polycarbonate (PC)

POLYCARBONATES ARE tough, rigid, transparent materials made from bisphenol A and carbonyl chloride. The repeating unit in the polymer is therefore:

Bisphenol group   Carbonate group

The bisphenol group with its two benzene rings contributes the rigidity and heat resistance, and the carbonate group the toughness and ductility. Polycarbonates are amorphous materials and have low moulding shrinkage and low water absorption. Heat resistance is very good, as is retention of toughness at low temperatures. The material is inherently self-extinguishing, has a UL 94 classification of V2, and can easily be made VO or VO/5V.

Properties

PC is an excellent engineering plastic and its high strength and rigidity are retained to high temperatures, with only a slow fall off. Impact resistance is very high at ambient temperatures, and although there is a step-wise drop in impact strength between –5 and –20°C, the material still retains good impact resistance, and is tough and ductile down to –150°C. Because of the low shrinkage and low water absorption, dimensional stability is excellent.

The good flammability behaviour of PC is due to its high LOI (Limited Oxygen Index) of 26% or more, and this can easily be increased to 35–40%. This is achieved in some grades without the use of additives containing bromine.

Electrical properties are very good and little affected by additives. Optical properties are also very good, and PC has the best scratch resistance of any of the transparent plastics.

PC is made in a wide range of grades, the basic change being in *molecular weight*, with various versions which include additives for improved mould release, increased U-V resistance and special applications; a series of structural foam grades are also produced. Typical values for the properties of PC are compared between three of the major manufacturers in Table 1 and it is evident that most of the properties are independent of these variations.

The changes in molecular weight have little effect on strength properties. The properties which are affected are MFI (Melt Flow Index) and impact strength. As molecular weight increases, the MFI decreases and the impact strength increases. This can be illustrated by comparing MFI and impact strength (see Table 2). Certain grades for special applications do show differences in properties compared with standard grades, but the only effect of modifying flammability is, in the case of Calibre resins, to marginally reduce the strength properties with an equally marginal increase in rigidity; for Makrolon there appears to be a small drop in the impact value (Table 3).

As with most thermoplastics, glass reinforced grades are available which significantly increase strength properties at the expense of toughness. Typical values of the relevant properties of such grades are shown in Table 4.

Polycarbonates can be coloured and are available in transparent, translucent and opaque colours. They are readily machined or finished by conventional means, and can be printed or decorated without difficultly.

Chemical resistance

Polycarbonates are resistant to dilute mineral and organic acids, animal and vegetable oils and fats, and aliphatic and cyclic hydrocarbons. Resistance to concentrated acids is good to fair, but alkalis will attack the material unless in dilute solution. Aromatic and halogenated hydrocarbons are excellent solvents for PC, which is swollen by other solvents such as benzene, acetone or carbon tetrachloride.

Although PC tableware, for example, can be washed many times in hot water, the material cannot stand continuous exposure to water above 60°C, because hydrolysis will cause slow breakdown of the polymer and therefore degradation of its properties. For intermittent use in such conditions a grade should be chosen which has with high molecular weight (*ie* low MFI); preferably a glass fibre reinforced grade.

Whilst the U-V resistance of PC is good, for outdoor applications it is advisable to use a grade formulated with additives to give improved resistance.

Note that PC is used in Alloys with some other thermo-plastics (see chapter on this subject (p. 469).

TABLE 1 – PC: TYPICAL PROPERTIES RANGE FOR NON-REINFORCED PRODUCTS

Property	Unit	Bayer	Dow	GE plastics	Test standard
PHYSICAL					
Tensile yield stress	MPa	>55	60–62	60	R527
Tensile strength at break	MPa	>65	66–72	70	,,
Elongation at yield	%	6	7	7	,,
Elongation at break	%	90–110	100–120	120	,,
Tensile modulus	GPa	2.3	2.3	2.3	,,
Flexural strength	GPa		97–100	100	178
,, stress at 3.5% strain	GPa	>70			,,
Flexural modulus	GPa		2.4	2.5	,,
Compressive stress at yield	MPa	>80			DIN 53454
Shear strength (yield/break)	MPa		40/60		DIN 53444
Hardness — Rockwell R			118–123		2039–2
,, ,, M			72–59	70–74	
— Ball indentation		110	110	95–110	2039–1
Abrasion					
– Taber CS17 wheel, 1kh, 1kg	mg	15		10	ASTM D1044
Impact strength:					
Un-notched Izod, 23°C	kJ/m²*		NB		180
Notched Izod, 23°C	kJ/m²*	>70->85	75–95	15–70	180
,, ,, –40°C	kJ/m²*			8–13	,,
Un-notched Charpy, 23°C	kJ/m²	NB	NB		179
,, ,, –40°C	kJ/m²	NB	NB		,,
Notched Charpy, 23°C	kJ/m²	>25->35	20–55	20–40	179
THERMAL					
Thermal conductivity		0.21	0.20	0.20	DIN 52612
Coefficient of linear					
thermal expansion	×10⁻³	6.5	6.8	7	ASIM D696
Deflection temperature:					
Under flexural load at 0.45 MPa	°C	135–145	142–146†	125–143	75
,, ,, ,, at 1.82 MPa	°C	126–131	139–143†	110–138	,,
Vicat softening point B/120	°C	145–153		135–148	306
,, ,, ,, B/50	°C		144–148		,,
ELECTRICAL					
Volume resistivity	ohm. cm.	>10¹⁶	>10¹⁶	>10¹⁵	IEC 93
Rel permittivity 1 MHz		2.9	3.0‡	2.9	IEC 250
Dissipation factor 1 MHz	x 10⁻³	11	10‡	8	IEC 250
Dielectric strength (× mm)	kV/mm	>30 (1 mm)	70 (2 mm)	15–20 (3.2 mm)	IEC 243
MISCELLANEOUS					
Density	g/cm³	1.20–1.25	1.20–1.22	1.20–1.24	R1183
Flammability		V2–V0(0.8)	V2–V0(1)	V2–V0(1.04)	UL subj 94
			–V0/5V(3.2)	–V0/5V(3.05)	(mm)
Limited Oxygen Index (LOI)		26–36	26–41	25–40	ASTM D2863
Water absorption (equibm)	%	0.36	0.35	0.35	R62A
Water absorption, 23°C/50% RH	%	0.15	0.15	0.10**	R62
Mould shrinkage (flow direction)	%		0.5–0.7	0.5–0.7	1110
Melt flow (MFI)	g/10 min	15–19 to 2	22–4	—	R1133
Refractive index		1.584	1.586	1.586	DIN 53491
Light transmission	%	—	87–91	88–89	ASTM D1003
Haze	%		0.7–1.5		,, ,,

* Some data sheets have values quoted in J/m, these are shown as kJ/m². Values in J/m are approximately ×10 values shown in kJ/m².
† Annealed.   ‡ 300 kHz.   ** 24h.

Processing of polycarbonates

PC can be processed by all conventional means for thermoplastics, although the most commonly used in injection moulding. The material must be dried just before use, using a de-humidifying drier, to avoid risk of hydrolysis.

Joining of parts can be effected by using adhesive, or friction or ultra-sonic welding.

TABLE 2 – PC: COMPARISON OF MFI AND IMPACT PROPERTIES
(Calibre 200/300 series)

Property Type	Unit					
Grade 200/300.		22	15	10	6	4
MFI.		22	15	10	6	4
Izod impact	kJ/m^2	75	85	90	90	95
Charpy impact	kJ/m^2	20	25	35	50	55
Tensile stress at yield	MPa	62	62	62	62	62
Tensile strength		66	71	71	72	72
Makrolon general purpose grades						
Grade		2400		2600/2800	3100/3200	3118/3119
MFI		15–19		9–13/7–10	5–7/3.5–5.0	2
Impact – Izod	kJ/m^2	>70		>80	>85	>85
„ – Charpy	kJ/m^2	>25		>30	>35	>35

TABLE 3 – PC: EFFECT ON PROPERTIES OF FR ADDITIVES

Calibre grades	Unit	200/300 GP	700 Ignition resistance	800 Superior ignition resistance
Tensile stress at yield	MPa	62	61	60
Tensile strength		66–72	66	59
Hardness – Rockwell R		118	123	122
„ M		72	59	59
Makrolon grades				
		General purpose	Flame retardant	
Impact– Charpy	kJ/m^2	>25 to >35	>20 to >30	
„ Izod	kJ/m^2	>70 to >85	>75	

TABLE 4 – PC: SOME TYPICAL GLASS REINFORCED GRADES
(Properties significantly affected only)

Property	Grade	Unit	Calibre 550	Macrolon 8325	Lexan 3414R	Test standard
PHYSICAL						
Tensile yield stress		MPa	59–66	90	—	R527
Tensile strength at break		MPa	48–63	100	120	,,
Elongation at break		%	4–7	3.3	2	,,
Tensile modulus		GPa	3.1–3.7	6.0	10.0	,,
Hardness — Rockwell M			62		93	2039-2
,, — Ball indentation		MPa		145	145	2039-1
Impact strength:						
Notched Izod,	23°C	kJ/m²*	11	12.5	10	180
Un-notched Charpy,	23°C	kJ/m²	NB	50		179
,, ,,	–40°C	kJ/m²	NB	50		,,
Notched Charpy,	23°C	kJ/m²	12	12	9	179
THERMAL						
Coefficient of linear						
thermal expansion		×10⁻⁵	3.8	2.7	2.0	ASTM D696
Deflection temperature:						
,, ,,	at 1.82 MPa	°C	147	138	148	,,
Vicat softening point	B/120	°C		150	163	1525†/306
,, ,, ,,	B/50	°C	150			,,
MISCELLANEOUS						
Density		g/cm³	1.27	1.35	1.52	R1183
Flammability			V0(1.6)	V1(1.6)	V1(1.47)	UL subj 94
			V0/5V(3.2)		V0(3.05)	(mm)
Limited Oxygen Index	(LOI)		38	35	34	ASTM D2863
Water absorption	(equibm)	%	0.32	0.29	0.23	R62A
Water absorption,	23°C/50% RH	%	0.15	0.13		R62
Mould shrinkage	(flow direction)	%	0.2–0.5		0.1–0.3	1110
Melt flow (MFI)		g/10 min	—	3–6		R1133

* Some data sheets have values quoted in J/m, these are shown as kJ/m². Values in J/m are approximately ×10 values shown in kJ/m².
† Annealed.

Applications

Polycarbonates are used for a large number of applications. These range from household products to engineering applications involving high temperatures (except) those such as for bearings or pumps, where friction is important). Other examples include:

Building – Glazing panels (single and double wall) for roofs and partitions. Safety and security screens.

Lighting – Lamp housing, illuminated traffic bollards and signs.

Electrical engineering – Distributor cabinets, relays, computer parts, holders for fluorescent tubes, door intercoms.

Transport – Crash helmets, headlamp parts, heater and radiator grilles, tail lamps, windshields.

Photographic and Audio/TV equipment – Compact discs, TV parts, video bases, camera bodies and parts.

Safety and health care – Goggles, spectacles, visors.

Blood plasma bottles, breathing masks, centrifuge tubes, dialysers.

Housewares and appliances – Coffee grinders, filters, housings for coffee makers, hair dryer and shaver housings, dishes, baby bottles.

Office equipment – Housings for photocopiers, fax machines, computers, cash registers.

Sport and leisure – Sailboards, ski cases, slalom poles, compass housings, electrical tool housings and parts.

Thermoplastic polyesters: PET/PETP/PETG/PBT

THERMOPLASTIC POLYESTERS are polymers based on polyterephthalates. The most important groups of this type are produced by condensation of terephthalic acid with various diols (glycols), such as polyethylene terephthalate (PET or PETP):

| Terephthalic acid. | Ethylene glycol. | PETP. |

PET and PBT are commercially the most important members of the family; they are semi-crystalline materials with high strength and hardness. They are also tough and have low friction coefficients, good abrasion resistance and good electrical properties. They have inherently good weathering resistance and their U-V resistance is particularly good. Above all, they have excellent behaviour under long term static and dynamic loading, and have high heat resistance.

Highly resistant to a wide range of chemicals, PET and PBT have virtually no susceptibility to environmental stress cracking. Exposure to water vapour at high temperature does, however, present a risk of hydrolysis.

Processing of PET at low temperatures (as in extrusion or blow moulding), leaves the material in the amorphous state when it is transparent. In injection moulding, however, it is difficult to develop the properties of the material, and PET is rarely used unfilled or unreinforced in this process. PBT on the other hand is easier to produce as a grade which will mould well.

Addition of glass fibre enhances many of their properties particularly that of heat resistance under stress and the thermoplastic polyesters thereby become one

TABLE 1 – TYPICAL PROPERTIES OF PETG (KODAR 6763)

Specific gravity.	1.27
Light transmission.	90% (Measured on 250 μm film)
Refractive index.	1.567
Distortion temperature under flexural load (1.82 M.Pa).	63°C
Elongation at break.	180%
Modulus in flexure.	2.0 G.Pa
Rockwell Hardness.	R108
Izod impact: Un-notched.	No break
Notched.	9 kJ/m^2

TABLE 2 – PROPERTIES OF APEC COMPOUNDS

Tensile stress at yield	60–65 MPa
Elongation at yield	7–9%
Young's modulus	2.3 GPa
Notched Izod impact strength	47–NB KJ/m^2
Distortion temperature:	
Under flexural load (1.8 MPa)	136–160°C
,, ,, ,, (0.45 MPa)	161–174°C

of the leading engineering plastics. They compare favourably with the other re-inforced materials in heat resistance, impact strength and rigidity.

Variants include PETG, which is a copolymer made from a mixture of isophthalic and terephthalic acids. As a result the polymer structure is irregular and the product is therefore amorphous. Typical properties of this material are shown in Table 1. A tough clear film is produced by the manufacturer which is called Kodapak.

Another variation is PCT (polycyclohexane terephthalate). A related material is PAR, which is an aromatic polyester carbonate (Table 2).

Properties

Even unreinforced, the thermoplastic polyesters are some of the most rigid thermoplastics. They are strong, with good elongation at break, and have tough-ness comparable with that of medium impact ABS, which is acceptable for many applications. This is combined with high heat stability and high heat resistance when unstressed. With low water absorption and low mould shrinkage, they have very good dimensional stability. Typical properties of the basic, unfilled polymers are shown in Table 3.

They are highly resistant to a wide range of chemicals, including dilute acids and alkalis, oils fats and solutions of mineral salts, and organic solvents such as alcohols, esters, ethers and hydrocarbons.

Both types of thermoplastic polyesters are available as filled or unfilled materials, and in flame retardant grades to UL94. The armophous version of Pet is almost transparent, whilst other types and PBT are semi-crystalline and there-fore translucent.

TABLE 3 – PROPERTIES OF BASIC THERMOPLASTIC POLYESTERS

Property Type	Unit	Un-filled PET	Un-filled PBT
PHYSICAL			
Tensile yield stress	MPa	50	50
Elongation at yield	%		
Elongation at break	%	300	250–300
Tensile modulus	GPa	2.5	2.8
Flexural strength	MPa	80	85
Flexural modulus	GPa	2.5	2.6
Impact strength:			
Notched Charpy, 23°C	kJ/m^2	8	12
„ Izod 23°C	kJ/m^2	—	—
„ „ 23°C	J/m	40	—
Hardness — Rockwell M		106	
Friction coefficient:			
Plastic/plastic			0.12
Plastic/metal			0.10–0.13
THERMAL			
Thermal conductivity			0.21
Coefficient of linear			
thermal expansion	$\times 10^{-5}$		8
Deflection temperature:			
Under flexural load at 0.5 MPa	°C		160
„ „ „ at 1.82 MPa	°C	65	55
MISCELLANEOUS			
Density	g/cm^3	1.34	1.30
Water absorption, 24h	%		
Mould shrinkage (flow direction)	%	0.2–0.5	2.2

Addition of glass and/or mineral fillers has a reinforcing effect, and considerably improves the strength and impact resistance, although the elongation at break is very much reduced. The improvement in heat resistance under stress is very noticeable, being between 50 and 100°C, depending on the grade. These materials are therefore extensively used under stress at elevated temperatures, with PET offering about 10°C higher performance than PBT. The good creep behaviour of the base polymers is reflected in the reinforced grades, which show low creep even at elevated temperatures. Note, however, that grades to be exposed to air at high temperatures need to have stabilisers incorporated to avoid yellowing.

Typical properties of a selection of grades, including a very high impact material from Du Pont, illustrate the range of products available from different manufacturers (see Table 4 for PET and Table 5 for PBT. Typical electrical properties are shown in Table 6).

TABLE 4 – TYPICAL PROPERTIES OF PET COMPOUNDS

Property	Grade Unit	Mineral 40% 3400 ICI	Glass/minl 10/20% PET102F BIP	Glass 30% 4300LT ICI	Glass/FR 30% 9530(FR) GE Plastics	Glass 45% 545 Du Pont	Glass 35% *modf SST 35 Du Pont	Test Method
PHYSICAL								
Tensile yield stress	MPa	53	80	174	145(br)	193	103	R527
Elongation at break	%	—	2.5		2	2	6	,,
Flexural strength	MPa	95	150	250	195	283	145	
Flexural modulus	GPa	10	8	9.6	9.0	13.8	6.9	178
Compressive strength	MPa	—	—	—	—	179	81	ASTM D695
Shear strength.	MPa	—	—	—	—	86	38	ASTM D732
Hardness — Rockwell, R		—	—	—	—	120	—	2039–2
Impact strength:								
Charpy notched	23°C kJ/m²	—	7	—	8	11	—	179
Charpy un-notched	23°C kJ/m²	—	22	—	—	—	—	
Izod un-notched	23°C kJ/m²	12	—	56	—	—	133.5	180
Notched	23°C kJ/m²	1.9	—	11.6	7.5	11.7	23.5	
,,	−40°C J/m	—	—	—	—	12.3	16.0	
Fatigue								
Flexural fatigue endurance 50% RH, 1 MHz	M.Pa					51		mfr's test
Friction coefficient static/dynamic						0.18/0.17†		mfr's test
THERMAL								
Coefficient of linear thermal expansion	×10⁻⁵				4‡	2.4	2.4	ASTM D696
Deflection temperature under flexural load, at 1.82 M.Pa	°C	175	210	230	225	226	220	75
Vicat softening point, 49N	°C	220		241				306
,, ,, B/120								DIN 53460
MISCELLANEOUS								
Density	g/cm³	1.73	1.60	1.60	1.72	1.69	1.51	R1183
Flammability		HB	HB	HB	VO	HB	HB	UL subj 94
Water absorption, 24h	%				0.08	0.04	—	R62A
Mould shrinkage (flow dir'n)	%		0.4–0.8		0.2–0.3	0.2	0.2	ASTM D955
,, (across flow)	%						0.8	0.9

*Some data sheets have values quoted in J/m; these are shown as kJ/m². Values in J/m are approximately ×10 values shown in kJ/m².

†Measured on grade 530

‡In flow direction

TABLE 5 – TYPICAL PROPERTIES OF SOME PBT COMPOUNDS

Property	Unit	Standard 20% glass TZM20 Atochem	Flame retdt 30% glass 3112 Hoe-Cel	STD 30% 'short' gl SK 605 Ciba-Geigy	STD 30% 'long' gl SG 625 Ciba-Geigy	STD 30% gl sph TUM 30 Atochem	STD 45% glass/minl 736 GE Plastics	Impact modfd S1517 Bayer	Test standard
PHYSICAL									
Tensile yield stress	MPa	110	97	133	133	57	110	45	R527
Elongation at yield	%		—	—	7	—		4	"
Elongation at break	%	4	3.5	2–3	2–3	3		>50	"
Tensile modulus	GPa		5.5	8.5	8.5			2.1	¨178
Flexural stress	MPa	165	155	202	214	92	185		
Flexural elastic modulus	GPa	6.44	5.5	—	—	4.1	8.9		
Hardness — Rockwell, R		118			—	118	116		2039-2
" M					—	79			
" — Ball indentation	M.Pa	216	88		228	—			2039-1
Impact strength:									
Un-notched Charpy 23°C	kJ/m²	25		40	40	5			179
" -40°C	kJ/m²	30		32	32	5			
Notched Charpy 23°C	kJ/m²	8		10	19	5			¨179
" -40°C	kJ/m²	8		8	17	5			
Notched Izod 23°C	kJ/m²†		4.8				7.5	57	¨179
THERMAL									
Thermal: conductivity	W/m.K	0.2–0.3	—	0.24	0.24	—			ASTM C177
Coefficient of linear thermal expansion	10⁻⁵	3–4	4.1	3	3	8	2.3	13	ASTM D696
Deflection temperature:									
Under flexural load at 0.5 MPa	°C	218	214	220	225	198	215	120	75
" " " at 1.82 MPa	°C	200	180	205	220	88	205	65	
Vicat softening point, 9.8N	°C	215				215		170	306
" " " 49N	°C	206				192		140	
MISCELLANEOUS									
Density	g/cm³	1.45	1.52(0.7mm)	1.53	1.53	1.54	1.69	1.26	R1183
Flammability		HB	VO	HB	HB	HB	HB	HB	UL subj 94
Water absorption, equibm, 20°C, 50%RH	%	0.20	0.05	0.13	0.20	0.2	0.07		R62A
Mould shrinkage (flow direction)	%	0.6–0.8	0.5–0.7		0.3–0.5		0.3–0.6		ASTM D955

† Some data sheets have values quoted in J/m; these are shown as kJ/m². Values in J/m are approximately ×10 values shown in kJ/m².

TABLE 6 – TYPICAL ELECTRICAL PROPERTIES OF PET, PBT

	Unit	PET	PBT	Test standard
Volume resistivity	ohm.cm	$>10^{15}$	$>10^{15}$	IEC 93
Surface resistivity	ohm	$10^{14}-10^{16}$	$>10^{13}$	IEC 93
Dielectric strength				
(short time, 3.2 mm)	kV/mm	16–24	15–17	D149
Electric strength	MV/m	10–13		IEC 243
Dielectric constant 50 Hz		3.2–4.0	3.8–4.6	IEC 250
,, ,, 1 kHz		—	2.23	
,, ,, 1 MHz		3.4–4.0	3.2–3.9	
Dissipation factor 50 Hz		0.001–0.003	0.002–0.003	IEC 250
,, ,, 1 kHz		0.005–0.008	0.0013	
,, ,, 1 MHz		0.01–0.03	0.016–0.020	
Arc resistance.	sec	63–146		D495
Tracking resistance.	V	210–310		IEC 112

TABLE 7

Type of product	Typical applications
Un-reinforced PBT	— Precision parts for mechanical, sliding, electrical and electronic applications (*eg* housings for electric irons, lamp components, electrical insulation covers, gears).
30% glass reinforced PET	— Ignition components, coil caps, relay bases, gears, sprockets, carburettor components, various pump housings, motor housings, business machine housings and components. Vacuum cleaner parts, furniture parts.
35% glass PET, impact modified	— Automotive components, wheels, housings, sports equipment, yard and shop tool housings, furniture and luggage components.
30% glass PBT halogenated flame retardant grade	— Applications requiring optimum thermomechanical and good electrical performance coupled with flame retardancy (*eg* edge connectors, fuse holders, switch components, electric motor housings, lamp holders).
PETG (un-reinforced)	— Where advantage can be taken of the transparency and toughness of the material. In film or sheet form, for packaging such as thermoformed containers for refrigerated/frozen foods and in protective covers and displays. As blow moulded bottles for shampoo, soaps and oils, and for injection moulding of parts for medical equipment, brush backs and containers.

Processing

Thermoplastic polyesters can be extruded, blow moulded or injection moulded. Apart from the large and rapidly growing market for soft drink bottles and food packaging films, their main applications are in injection moulded parts. Current moulding grades of PET and PBT have good mouldability, even in relatively thin sections, giving good surface finish, and are generally formulated for fast crystallisation, allowing shorter cycles. They are generally used in reinforced form. Care does need to be taken with mould design (as with all fibre loaded materials) and the material must be dry, to avoid hydrolysis during processing. Whilst for PBT mould temperature can be between 30 and 130°C, for PET it has to be in the region 135 to 160°C to obtain good crystallisation; this can affect moulding cycles.

Applications

The applications for these materials are many and varied. The wide range can be indicated from some typical applications for the different types of product (see Table 7).

Polyarylates

THESE MATERIALS are defined as polyesters from bis-phenols and dicarboxylic acids. Bis-phenols have two aromatic rings, compared with none for the glycols used for PET/PBT, and the polymer made with terephthalic acid therefore would have higher heat resistance, but as it would also be a semi-crystalline material it would be difficult to process.

The problem can be overcome by making a copolymer using a mixture of, for example, terephthalic and isophthalic acids with bis-phenol A. This gives a polymer with an irregular chain, thereby inhibiting crystallisation. The polymer is therefore transparent and can be processed, and also self-extinguishing. Other combinations may not have the same properties.

Polyarylates are tough materials with high heat resistance, excellent U-V resistance combined with transparency.

Manufacturers

Amoco	—	'Ardel'
Solvay	—	'Arylef'
Hoechst	—	'Durel'

Properties

As mentioned above, the basic polyarylate is transparent and self-extinguishing, with a high LOI (Limiting Oxygen Index). Some of the more recent types sacrifice some of this quality in favour of improvements in other properties.

Strength properties of polyarylate are good, with high impact strength, good tensile and creep behaviour and an exceptionally high level of recovery after deformation. It has excellent abrasion resistance (better than polycarbonate).

U-V resistance is excellent and is claimed to be better than that of polycarbonate. Most grades have good transparency, which is retained after U-V exposure.

In addition to good retention of impact strength, haze level only increases slowly with U-V exposure.

Chemical resistance

ESC resistance and hydrolytic stability are similar to those of polycarbonate.

TABLE 1 – TYPICAL PROPERTIES OF POLYARYLATES
Ardel grades from Amoco

Property	Unit	D–100	D–170	D–240	Test standard
PHYSICAL					
Tensile yield stress	MPa	69	69	69	R527
Tensile strength at break	MPa			64	,,
Elongation at yield	%	8.4		6	,,
Elongation at break	%	50	73	100	,,
Tensile modulus	GPa	2.07	2.1	2.28	,,
Flexural strength		76			178
Flexural modulus	GPa	2.14		2.34	,,
Hardness — Rockwell, R		126			2039–2
Impact strength:					
Notched Izod −20°C	kJ/m^2*	21.8	21.8	6.9	180
−40°C		15.5			180
Tensile impact strength	kJ/m^2		200	295	ASTM D1822
THERMAL					
Thermal conductivity	W/m. K	0.18			ASTM C117
Coefficient of linear thermal					
expansion	10^{-5}	6–8			ASTM D696
Deflection temperature under					
flexural load at 1.82 MPa	°C	174	170	110	75
ELECTRICAL					
Dielectric strength	kV/mm	16.8		17	IEC 243
Volume resistivity	Ohm. cm	3×10^{16}		8.9×10^{17}	IEC 93
Dielectric constant, 10^6 Hz		3.30		3.37	IEC 250
Dissipation factor, 10^6 Hz		0.02		0.023	IEC 250
OPTICAL					
Light transmission		85		85.9	ASTM D1003
Haze		6		2.9	,,
Refractive index		1.61			
MISCELLANEOUS					
Density	g/cm^3	1.21	1.22	1.26	R1183
Flammability:		V0		V2	UL subj 94
Auto-ignition	°C	545		480	ASTM D1929
LOI	%	36			
Water absorption, 24 h	%	0.27			R62A
Mould shrinkage (flow direction)	%	0.9			1110
Melt flow (MFI) 375°C	g/10 min	4.5	15	9	R1133

* Data sheet values are quoted in J/m. These are shown as kJ/m^2. Values in J/m are approximately ×10 values shown in kJ/m^2.

Most aggressive in this respect are aromatic hydrocarbons, esters, ketones and chlorinated solvents.

Processing

Polyarylates are suitable for injection moulding. Thorough pre-drying is necessary, and care is required in choosing moulding conditions; to minimise risk of thermal degradation it is necessary to use low-compression-ratio screws and little or no back pressure. Melt should be kept between 355–390°C. In extrusion, slow cooling is advised to limit residual stresses and a potential for warpage.

PVC (polyvinyl chloride)

PVC is produced by all the major processes for thermoplastics – emulsion, mass and suspension. The resulting polymer is a powder and is sold as such, or as a compounded pellet. Because of an inherent lack of stability at processing temperatures it is always blended and/or compounded with stabilisers; at the same time the material is often modified with lubricants, fillers, pigments and plasticisers.

PVC is therefore available with a wide range of properties from soft, flexible materials (flexible PVC) to hard, rigid plastics (un-plasticised PVC and uPVC). Modified or toughened grades extend the range still further. There is therefore wide scope for tailoring of properties to suit application requirements. Common characteristics are good impact resistance and strength properties and very good chemical resistance. uPVC is flame retardant, and when suitably stabilised has excellent weathering properties.

PVC may be used as clear transparent, or in a wide variety of translucent or opaque colours.

Properties

The base PVC polymers ('resins') are rigid, strong and tough, although there is a sharp drop in impact strength around 0°C, which can be overcome by modifying the polymer. PVC resins are amorphous and have low water absorption and low mould shrinkage. Heat resistance is limited, with a softening point around 70–80°C, although this can be raised by suitable compounding. Certain modified products and copolymers also have higher softening points.

PVC resins are classified by ISO 1060/1, DIN 7746/7 and ASTM 1755 and 2474. Attributes covered are:

Viscosity (viscosity number or K value).
Apparent density.
Particle size.
Plasticiser absorption (for paste resins, the percentage necessary to produce a material with a defined viscosity).

TABLE 1 – TYPICAL RANGES OF PVC RESINS

Producer Unit	Process	Viscosity Number	K Value g/kg	Plasticiser Absbtn phr	Copolymer	
					Content %	Comonomer
Atochem		80–125	57–70	—	—	—
		50–88	45–60	—	9–14	VA
	C–PVC	72–94	—	—	—	—
EVC ('Corvic')	S	62–125	50–70	130–320	—	—
	Copoly	52–89	46–60	40–80	9–14	VA
('Ravinil')	S	80–170	57–80	160–380	—	—
	M	61–90	50–60	140–220	—	—
(Corvic)	paste	113–160	67–78	—	—	—
(VIPLA)	paste	117–170	68–80	—	—	—
Hoechst	S	88–123	60–70		—	—
	Copoly	85	59		10	VA
					5	cyclohexyl maleinimide
	M	82–112	58–67		—	—
	E	59–78	85–119		—	—
		59	85		7	Butadiene-AN
	toughened	80–125	57–70			CPE
Hydro polymers	S	80–130	57–71	140–420	—	—
	Paste		68–82		—	—

Process:
S   Suspension.
M   Mass.
E   Emulsion.
Copoly Copolymer.

Typical ranges for the products available from some of the producers are shown in Table 1. Apparent density and particle size, being attributes relevant to processing, are not shown in the Table.

PVC is often bought as the resin by the larger processors, and is blended/compound in-house. Alternatively it can be purchased as the ready-to-use blend or compound. For the latter, ISO/DIN designations for uPVC and PVC-P materials include the following properties:

uPVC – Vicat softening point, Charpy notched impact, tensile modulus and density (ISO 1163, DIN 7748).
PVC-P – Shore hardness, Density, tensile stress at 100% elongation and torsional stiffness temperature at 309 MPa (ISO 2898/1, DIN 7749/1).

TABLE 2 – GENERAL PROPERTIES OF MOULDING AND EXTRUSION GRADES

Property	Unit	uPVC	Modified PVC — Toughened	Modified PVC — High impact	C-PVC	Plasticised PVC (Increasing level)		Test standard
PHYSICAL								
Tensile strength	MPa	50–65	45–50	35–50	75	25–30	15–20	;;
Elongation at break	%	20–50	25–70	30–100	10–15	150–200	370–400	;;
Tensile modulus	GPa	2.8–3.0	2.3–2.5	2.0–2.2	3.50	–	–	
Flexural stress at yield	GPa	70–110	70–80	60–80	125	20–23	10–15	178
Hardness — Shore		D83	D81	D75–80	155	D60/A95	D30/A80	2039-2
" — Ball indentation		110–130	95–105	75–100				2039-1
Impact strength:								
Notched Izod, 23°C	kJ/m²	3.5–8.5	10–16	53–130	–	3–4	NB	180
Notched Charpy, 20°C	kJ/m²	2–5	5–10	30–50	2	2–3	NB	179
" -20°C		2–3	3–7	4–10	–			;;
THERMAL								
Thermal conductivity	W/m.K	0.16	0.14	0.14	0.14	0.13	0.13	DIN 52612
Coefficient of linear thermal expansion	x10⁻⁵	7–8	8	8	6	15	21	ASTM D696
Vicat softening point A, B/50	°C	70–90	75–83	70–77	110	37–43	–	1525/306
ELECTRICAL								
Volume resistivity	Ohm-cm	10^{15} to $>10^{16}$	10^{15}	10^{15}	10^{15}	10^{14}–10^{15}	10^{11}	IEC 93
Dielectric strength (1 mm)	kV/mm	20–50				10^{12}	10^{11}	IEC 93
Dielectric constant: 50 Hz		3.2–3.7				32–34	24–26	IEC 250
1 MHz		2.9–3.2				4.2–3.2	8.0–4.0	IEC 250
Dissipation factor	×10⁻³	11–15				0.06–0.03	0.08–0.12	IEC 250
Tracking resistance		KC300–>600						IEC 112
MISCELLANEOUS								
Density	g/mm³	1.39	1.38	1.37	1.55	1.28	1.19	R1183

As well as the basic uPVC, PVC is sold in various modified forms including toughened, chlorinated and plasticised versions. Typical general properties of moulding and extrusion grades of the different types can be found in Table 2, whilst in Table 3 are shown the typical properties of one range of compounds.

Processing

Processing of PVC may be by extrusion (pipe, profile, sheet), calendering (film whether rigid or flexible, flooring), or injection or blow moulding. Injection moulding is not widely practised because the operating temperatures necessary are close to those at which degradation starts; careful control is therefore needed and moulds are usually made from a high grade steel, to avoid problems of corrosion. Paste resins are used for paper and fabric coating and for dip moulding and coating.

As mentioned above, when heated above 150°C, PVS tends to yellow and decompose, giving off hydrogen chloride. Since processing temperatures are usually close to or above this figure it is essential to add stabilisers which inhibit this reaction. Different stabiliser systems are used depending on the application; for example grades for outdoor use will usually contain barium-cadmium systems with a U-V absorber, which would not be suitable for food packaging work.

The addition of lubricants improves the flow properties of PVC and facilitates processing, particularly it there is no plasticiser in the compound. This reduces the harmful effect of high shear forces and enhances stability. Typical lubricants are metal soaps, fatty acid esters and monoglycerides. A small amount of a lubricant of lower compatibility may be used to reduce friction of the blend at the early stages of compounding or processing, when the polymer is still solid.

Fillers are often added to PVC in order to improve properties such as rigidity or hardness, and to reduce costs. PVC can be coloured without difficulty, but where a plasticised compound is being used, only colourants which are insoluble in the plasticiser should be used, to avid the risk of bleeding.

PVC is modified in various ways to improve its properties:

1. Copolymerisation. The most popular is with vinyl acetate, which improves processability (eg for office and drawing equipment, or records).
2. Addition of elastomers such as CPE (chlorinated PE) or polyacrylate, to improve impact strength, particularly at low temperatures. These are usually grafted polymers, which are two-phase and therefore translucent/opaque. Alloys with nitrile rubber are also used for gaskets and other profiles exposed to outdoor weathering.
3. Production as chlorinated PVC (C-PVC), to improve heat resistance and strength properties.
4. Formulation to allow conversion to a cross-linked structure by ionising radiation after normal processing, into such items as, jumper wire insulation. This considerably improves the properties, particularly heat resistance, abrasion resistance and toughness.

TABLE 3A – TYPICAL RANGES OF PVC RIGID COMPOUNDS†

Type	Vicat Softening °C	Density kg/m³	Tensile strength MPa	Tensile modulus GPa	Drop impact	Charpy impact (Unn'chd)	Charpy impact (N'ched)	Weathering	Cold flex
Extrusion/moulding	60–85	1350–1490	44–57						
Bottles — standard	70–75	1325–1380							
— biax orientated	73	1320–1380			91/130‡ 220/150**				
High-impact weathering	81	1.4	45–33	2.5†		60†	70†	70%†	4–5

TABLE 3B – TYPICAL RANGES OF PVC FLEXIBLE COMPOUNDS†

Type	Density kg/m³	Tensile strength MPa	Hardness Shore A GPa	Elongation %	Cold flex
Extrusion	1180–1510	10–24	53–96	200–420	+4 to –38
" — cable grades	1240–1560	14–24	71–94	210–375	+10 to –27
Moulding	1200–1480	11–21	55–100	200–400	+9 to –42
Medical uses	1195–1345	13–24	56–96	270–425	+8 to –41

† Hydro Polymers data†
‡ On 8 oz Boston Round bottles.
** NHPL test on 1.01 ribbed container.

Chemical resistance

uPVC has very good chemical resistance and is frequently used in chemical plant construction for this reason. It is resistant to acids organic and inorganic, alkalis and salts, fats and oils and most alcohols, but is attacked by many solvents. Plasticised PVC will not have the same chemical resistance as uPVC.

Applications

Consumption of PVC in its main applications is roughly divided as follows:

Pipe	25%
Film and sheet	19%
Profiles, tubing	17%
Wire & Cable	10%
Bottles	8%

With the remainder miscellaneous.

Typical of the many applications of PVC are:

Building	– Pipe – particularly large diameter for water distribution and drainage systems. Window frames. Cladding & fencing. Soffits, skirting, coving, handrail capping. Cable insulation & sheathing. Trunking and conduit. Industrial & garden hose.
Furniture and home	– Upholstery covers. Shower curtains. Flooring. Coated fabrics for clothing.
Medical	– Bags for storing intravenous solutions and blood. Gastric tubing, catheters, taps. Components for blood transfusion sets.
Packaging	– Bottles for mineral water, shampoos and toiletry preparations, cosmetics, pharmaceutical, edible oils, fruit squash concentrates.
Miscellaneous	– Records, toys, junctions boxes, industrial mouldings, refrigerator gaskets, shower hoses.

Blends and alloys

OWING TO the wide range of types of plastics offered for engineering applications it is very difficult and expensive for polymer producers to develop a completely new material. To maintain their position in this growing market their research effort is often therefore concentrated on extending the range of applications of existing polymers; the production of blends or alloys with other polymers is proving to be a successful way of achieving this. At the same time the user benefits by having the performance he needs achieved more economically than with a single plastic, since blends/alloys offer either lower material costs, lower conversion costs, or both.

The objective is usually to find a way of mixing two polymers to produce a material with the best properties of each constituent (for example mixing A, which has a high heat distortion, with B, which has good mouldability, to produce a blend with a combination of the two). For such a blend to be successful it must be homogeneous and must remain so during and after processing. This is not easy to achieve with two polymers which are miscible; when they are not miscible it is even more difficult to achieve, and usually requires a compatibiliser to link the two materials – hence the term alloy. Note however that the line where a mix is no longer a blend and becomes an alloy is not well defined, and no attempt will be made here to define in which category the different mixes under discussion belong.

There are a number of products on the market and the number is frequently being increased; most of the major polymer producers either have products or are developing them, so that it is not possible to present a comprehensive survey. Instead, listed in Table 1 are the mixes which are known at present, although not all of them are currently available in Europe. Some examples are then covered in more detail to illustrate the results that are obtained from this technique.

Certain compounds which are combinations of two polymers have already been covered in the sections dealing with the parent polymer and these will not be mentioned again here. The products concerned are:

TABLE 1 – BLENDS AND ALLOYS CURRENTLY KNOWN

Mix	Manufacturer	Trade Name
ABS/PC	Bayer	Bayblend
	BASF	
	Dow	Pulse
	GE Plastics	Cycoloy
ASA/PC	Bayer	Bayblend
	BASF	
ASA/PBT	BASF	Ultrablend
ABS/PA	Monsanto*	Triax
	GE Plastics (development)	
PET/PC	GE Plastics	Xenoy
PBT/PC	GE Plastics	Xenoy
	Bayer	Macroblend
PP/PA	Atochem	Orgalloy
PS/PE	BASF	
Acrylic/PBT	Hoechst (development USA)	Vandar
PC/APE		
PC/PEI		
SMA/PBT	Arco (USA)	
Styrenic/PC	Polysar (USA)	
SB block/PC	Shell Chemical (development)	
and ranges from		
	Du Pont	Bexloy
	Amoco	Mindel

*Monsanto has announced new products under the Triax name, based on
PC/ABS, PA/PE etc, PBT.

Product	*Refer to*
PPE/HIPS	Modified PPO/PPE
PPE/PA	,, ,,
Modified PA	PA
,, PP	PP
,, acrylic	PMMA
,, PVC	PVC

and blends/copolymers of PP/PE/EVA.

PC/ABS

This combination is the most well established of the types listed Table 1. It provides the easy processability and low temperature ductility of ABS, with improved heat resistance and strength properties from the PC. Flame-retardant grades can be made, with good properties.

With a wide choice of both ABS and PC grades, a wide range of PC/ABS products is theoretically possible and each producer has several grades available.

Properties of a selection of grades are shown in Table 2, with typical ABS for

TABLE 2 – PC/ABS COMPOSITIONS COMPARED WITH ABS
(Typical values)

Property	Producer grade	Unit	Dow Pulse A30—105	Dow Pulse A20—95	TYPICAL ABS	ABS HIGH HEAT	GE Plastics Cycoloy C1110	Bayer Bayblend FR 1441	Test standard
PHYSICAL									
Tensile yield stress		MPa	50	44	35-50	50-55	55	59	R527
Elongation at break		%	>80	>80	15-25	20	160	>50	''
Tensile modulus		GPa	2.00	2.05	1.8-2.5	2.25-3.0	2.2	2.60	''
Flexural strength			80	70	55-80	75-85	82		''
Flexural modulus		GPa	2.20	2.25	2.0-2.8	2.3-2.9	2.35		2039-1
Ball indentation			105	97	70-115	95-115			
Impact strength:									
Notched Izod	23°C	kJ/m²*	70	50	14-40	19-35	64	NB	180
	-40°C		50	30	3-10	8		23 (-30°)	
Notched Charpy	23°C	kJ/m²	35	22	8-24	8-16	54 (-30°)		179
	-40°C		20	15	4-10	2-4			
THERMAL									
Deflection temperature under flexural load at 0.45 MPa		°C	126	110	96-104	105-115	118	100	75
'' at 1.82 MPa		°C	104	96	80-90	95-105	107		
Vicat softening point	A/120	°C	144	127				108	1525/306
''	B/50	°C	125	109	90-101	105-115			
MISCELLANEOUS									
Density		g/cm³	1.12	1.12	1.04	1.05	1.14	1.18	R1183
Flammability at thickness, mm			HB	HB	HB	HB	HB	VO	UL subj 94
Water absorption, 24 h		%	0.2-0.6	0.2-0.6	0.4	0.3		1.6	R62A
Mould shrinkage (flow direction)		%	0.3-0.7	0.3-0.7	0.4-0.7	0.4-0.7	0.5-0.7		1110
Melt flow (MFI)		g/10 min	6.9	9.4	4-50	2-3		9	R1133

* Some data sheets have values quoted in J/m. These are shown as kJ/m². Values in J/m are approximately ×10 values shown in kJ/m².
MFI at 250/5.   MFI at 220/10.   MFV at 240/5.

comparison. Note the improvement in impact strength and heat resistance, whilst tensile strength is retained.

ABS/PA

These compositions combine the best properties of the two components. A variety of grades is available, offering very high impact strength and improved chemical and abrasion resistance, with processing behaviour intermediate between those of ABS and PA. Density at 1·06 g/ml remains a considerable advantage compared with many other engineering plastics.

The behaviour of these materials under stress at high temperatures is similar to that of PA, in that at low stress the heat resistance is excellent, for example allowing parts to be baked in paint ovens at up to 200°C without significant distortion. Although the maximum use temperature under high stress is very much reduced (as measured by Vicat softening point under 50N load) it is still higher than that of most ABS. Typical properties of two grades of ABS/PA are shown in Table 3.

TABLE 3 – TYPICAL PROPERTIES OF TWO ABS/PA GRADES
(Monsanto data)

Property	Unit	TRIAX 1120		TRIAX 1180		Test standard
		Dry	Cond'nd	Dry	Cond'nd	
PHYSICAL						
Tensile yield stress	MPa	46	40	51	42	R527
Elongation at break	%	270	290	330	320	,,
Tensile modulus	GPa	1.90	1.18	1.93	1.31	,,
Flexural modulus	GPa	2.07	1.14	1.93	0.83	,,
Hardness — Rockwell R		95	80	101	85	2039–2
— Shore D		75	68	75	65	2039–1
Impact strength:						
Notched Izod 23°C	kJ/m^2*	85	100	103	120	180
Notched Charpy 23°C	kJ/m^2	48	54	54	54	179
Driven Dart						
(32 mm ring, 13 mm dart) 23°C	J	51	52	—	56	Monsanto
—20°C		49	50	—	65	
—40°C		14	46	—	—	
THERMAL						
Thermal conductivity		0.29		0.30		DIN 52612
Coefficient of linear						
thermal expansion	×10^{-5}	17		17		ASTM D696
Deflection temperature						
under flexural load — 0.45 MPa	°C	92	91	93	91	75
Vicat softening point 9.8 N	°C	197	170	210	177	1525/306
,, ,, 49 N	°C	110	—	130	—	
MISCELLANEOUS						
Density	g/cm^3	1.06	1.06	1.06	1.06	R1183
Flammability (1.6 mm)		HB				UL subj 94
Limiting Oxygen Index	%	20		20		
Water absorption, 24 h	%	1.1		1.5		R62A
Mould shrinkage (flow direction)	%	1.0–1.5		1.0–1.5		1110
Melt flow (MFI) 230/3.8	g/10 min	1.5		4.7		R1133

*Some data sheets have values quoted in J/m. These are shown as kJ/m^2. Values in J/m are approximately ×10 values shown in kJ/m^2.

PP/PA

This is a new development, combining two apparently incompatible polymers. The resulting material combines the water resistance and processability of PP with the mechanical and thermal properties of PA.

Initial property values given by Atochem are shown in Table 4 and particular points to note about these materials are:

Moisture absorption at equilibrium is close to that of PP. It is therefore much less than that of PA, and physical properties are correspondingly less affected. For example, flexural modulus:

	Dry %	Equilibrium %	
Orgalloy R6000	2.2	1.5	GPa
PA 6	2.3	0.5	

Impact strength is high; the Charpy impact of R6000 at 25 kJ/m^2 is higher than that of PA6 and very much higher than that of PP at 6.

Distortion temperature under load is between those for PP and PA. Glass reinforced material has a value similar to that for PA/glass, and has a flexural modulus of 7·5.

Mould filling behaviour is close to that for PP and is much more satisfactory than that of nylon.

TABLE 4 – INITIAL DATA FOR PP/PA GRADES (Atochem data)

Property		Unit	Orgalloy R 6000	Orgalloy R 6600	Test standard
PHYSICAL†					
Tensile yield stress		MPa	43	50	R527
Elongation at break		%	296	—	,,
Flexural stress at yield			70	70	,,
Flexural modulus		GPa	1.98	2.20	,,
Impact strength:					
Notched Izod	23°C	kJ/m^2*	12	7	180
Notched Charpy	23°C	kJ/m^2	25	20	179
Unnotched Charpy	23°C	kJ/m^2	NB	NB	179
THERMAL					
Deflection temperature					
under flexural load at 0.45 MPa		°C	140	210	75
,, ,, at 1.82 MPa		°C	60	63	
MISCELLANEOUS					
Density		g/cm^3	1.03	1.03	R1183
Flammability			HB	HB	UL subj 94
Water absorption, 24 h		%	0.40	0.50	R62A
Water absorption, 23°C/50%, equibm		%	0.42	—	R62A
Mould shrinkage (flow direction)		%	0.79	—	1110
Melt flow (MFI)		g/10 min			R1133

*Some data sheets have values quoted in J/m. These are shown as kJ/m^2. Values in J/m are approximately ×10 values shown in kJ/m^2.   † Conditioned

High performance plastics

THESE MATERIALS are thermoplastics with exceptional properties, which are in increasing demand for electronic, automotive and engineering applications. They are used where resistance to long term exposure to high temperatures, combined with good physical properties and processing behaviour, are essential. With an annual growth rate in this part of the market of around 8%, most of the major producers are competing for a share, mainly on a price/performance basis; consequently new grades appear frequently.

Manufacturers

Table 1 lists the product types available (or reported to be available) at the time of writing, together with the producers and trade names.

Properties

The range of properties offered by these materials is quite varied; some are transparent, some are very tough, and many are self-extinguishing in the UL94 sense. All offer high Heat Distortion or Vicat Softening temperatures, with some at extremely high values. This is combined with good to excellent retention of strength and/or rigidity at high temperatures. In many applications these materials can and are replacing metals, with considerable weight saving.

Most types are available with glass or other reinforcement and some of them are not marketed without it. In reinforced form they provide exceptional dimensional stability at elevated temperatures.

All of these plastics are at the higher end of the price range. The choice of material for a given application must therefore be made carefully, and the part must be designed for the most economical use of the material and process time. In most cases there is multi-point data available on the variation of properties with exposure conditions, and many manufacturers can provide assistance on design using CAD (computer aided design) systems.

The main attributes of the major products in this field are summarised below and typical properties of the range of grades currently available are shown in

TABLE 1 – TYPES OF HIGH PERFORMANCE PLASTICS

Type	Name	Producer	Trade name
PAA	Polyaryl amide	Solvay	Ixef
PSU	Polysulphone	BASF	Ultrason
		Amoco	Udel
PEI	Polyetherimide	GE Plastics	Ultem
PAS	Polyaryl sulphone	Amoco	Radel
PES	Polyether sulphone	BASF	Ultrason E
		ICI	Victrex PES
PPS	Polyphenylene sulphide	Phillips petroleum	Ryton
		Bayer	Tedur
		Hoechst	Fortron
		Solvay	Primef
		GE Plastics	Supec
PEEK	Polyether ether ketone	ICI	Victrex PEEK
PEK	Polyether ketone	ICI	Victrex PEK
		Hoechst	Hostatec
		BASF	Ultrapek
PAI	Polyamide imide	Amoco	Torlon
Polyimide		Rhone Poulenc	Kinel-Kerimide
		Du Pont	Vespel
Polyarylate		Amoco	Ardel
HTA	Biphenyl modified PES	ICI	HTA
LCP	Liquid crystal polymers		
		Amoco	Xydar
		BASF	Ultrax
		Bayer	
		Du Pont	LCP
		Hoechst	Vectra
		Rhone Poulenc	Rhodester CL
		ICI	SRP
PEBA	Polyether block amide	Atochem	Pebax
(on border between thermoplastic and thermoplastic elastomer)			

Note Polyamide 4.6 from DSM also has high heat performance (see Blends and Alloys; p. 469).

Tables 2 and 3. Since many of the actual grades on offer are development products, contact should be made with the producers to confirm performance data, particularly if this is required for design purposes.

Processing

Whilst many of the products on offer have good processing characteristics, their high heat resistance implies equally high processing temperatures, with melt temperatures above 300°C, and for injection moulding, a mould temperature above 100°C. Most of these products need pre-drying. This demands more sophisticated equipment and more skilled staff than for ordinary thermoplastics.

TABLE 2 – TYPICAL DATA FOR PPS, PSU, PES, PAS, PEK, PEEK, PAI and PEI

Property	Unit	PPS	PSU	PES
PHYSICAL				
Tensile strength at break	MPa	50 120–180	70 78–120	85 (y) 140
Elongation at break	%	0.5– 1.6	>50 3	35 4
Tensile modulus	GPa	13.0– 19.0	2.5 3.7–7.4	2.9 8
Flexural modulus	GPa	13.0– 17.0	2.7 3.8–7.6	2.6 8.4
Hardness — Rockwell, R		123		
„ „ M				88 98
Abrasion — Taber CS17 wheel, 1 k, h, 1 kg	mg	51		6–8
Impact strength:				
Un-notched Izod 23°C	kJ/m^2*	9– 26.5		NB 54
Notched Izod 23°C	kJ/m^2*	4– 7	7 7	7.5 8.5
Tensile impact strength	kJ/m^2		420 100	
THERMAL				
Coefficient of linear thermal expansion	×10^{-5}	2.0– 3.1	5.6 2.7	5.5 2.4
Deflection temperature				
under flexural load at 1.82 MPa	°C	195 260	174 180	203 210–216
Vicat softening point, A/50	°C		178– 196	226 233
„ „ B/50	°C	260	184– 188	222 230
Maximu use temperature		200–220	160	180 190
ELECTRICAL				
Volume resistivity	Ohm. cm	±10^{16}		>10^{16}
Dielectric constant		3.8– 4.6	3.1 3.5	3.6 3.9
Dissipation factor	×10^{-3}	1.3– 9	5	4.1 4.8
MISCELLANEOUS				
Density	g/m^3	1.9– 1.7	1.25– 1.50	1.37 1.51
Flammability		V0 or V0/V5	V2 V0	V0 (0.43) V0
LOI		46– 53		34 38
Water absorption, 24 h	%	0.07– 0.03	0.3– 0.6	0.43– 0.29
Mould shrinkage (flow direction)	%	0.25– 0.1	0.7 0.2	0.6 0.2
MOULDING				
Mould temperature	°C	140	100–160	150
Melt temperature	°C	315–335	310–390	340 370

(continued)

Note: Where figures are joined by a dash, they show the range for the grades available, the first figure usually being that for the grade with the lowest loading of reinforcement/filler; un-reinforced material may be shown separately or as the first figure. If there is no dash, the first figure is that for the base polymer, the second that for reinforced material.

* Some data sheets have values quoted in J/m. These are shown as kJ/m^2. Values in J/m are approximately ×10 values shown in kJ/m^2.

† Carbon reinforcement.

‡ Except wear resistant grade, which has value of 2.

TABLE 2 – TYPICAL DATA FOR PPS, PSU, PES, PAS, PEK, PEEK, PAI and PEI – *contd.*

PAS 'Radel'	PEK 'Victrex'	PEEK 'Victrex'	PAI 'Torlon'	PEI 'Ultem'	Test standard
83 (y) 86–126 (y)	105 155–225	92– 208	150– 205	90– 160	R527
40 2–6	5– 3	50– 1.3	15– 7	60– 1	,,
2.65 3.8–8.6	4.0 10.5	3.6– 13		3– 11.7	,,
2.75 4.0–8.0	3.7 17.9†	3.7– 13.0	5– 11.7	3.3– 11.7	178
	126 126	125		109– 125	2039–2
	105 109	99– 107			
	3–4			10	ASTM D1044
		NB 67–75	100– 25		180
8.5 5–7.5	7 8	8.3 9.6	14, 8	3 10.5	180
355 59–72					ASTM D1822
4.9 3.1	5.7 1.7	4.7 2.2	3.0– 1.6	5.6 5.1–1.4	ASTM D696
204 213	186 360	160– 315	280	197 195–223	75
					1525‡/306
				219 223–243	
170	260 260	250			
>10¹⁶	10¹⁷	>10¹⁶		>10¹⁵	IEC 93
3.5– 4.2	3.4– 3.9	3.2– 3.3	4.2– 7.3		IEC 250
5.6– 9.4	5 4–5	—	3.1– 6.3	2.5 5.0	IEC 250
1.37– 1.58	1.30– 1.53	1.32– 1.49	1.42– 1.61	1.27 1.29–1.70	R1183
V0(0.58) V0(0.79)	V0 V0	V0 (1.45) V0	V0(0.2) V0(1.2)	V0(0.40) V0(1.6)	UL subj 94
	40– 46	24– 35		44– 54	
0.4 0.2	0.11 0.08	0.5 0.06	0.33– 0.17	0.25 0.28–0.11	R62
0.6– 0.3	0.6– 0.1	1.1– 0.1	0.6– 0.2	0.7 0.6–0.2	1110
150–165	180–190	150–180		400	
345–390	400	360–390		140	

Note: Where figures are joined by a dash, they show the range for the grades available, the first figure usually being that for the grade with the lowest loading of reinforcement/filler; un-reinforced material may be shown separately or as the first figure. If there is no dash, the first figure is that for the base polymer, the second that for reinforced material.

*Some data sheets have values quoted in J/m. These are shown as kJ/m². Values in J/m are approximately ×10 values shown in kJ/m².

†Carbon reinforcement.

‡Except wear resistant grade, which has value of 2.

TABLE 3 – TYPICAL DATA FOR HTA, PAA, POLYIMIDE & PIBA

Property	Unit	HTA	PAA	Polyimide 'Kinel'	PEBA 'Pebax'	Test standard
PHYSICAL						
Tensile strength at break	MPa	86–134	160–260	20 80	27–35	R527
Elongation at break	%	19–4	1–2	0.35–1.0	360–>550	„
Tensile modulus	GPa	2.3	1.2–2.03	2.4–10.0		„
Flexural modulus	GPa	2.5–8.5			0.50–0.370	178
Hardness — Shore D					30–65	R868
„ — Rockwell M		101	86–112	94–119		2039–2
Impact strength:						
Un-notched Izod 23°C	kJ/m²†					180
Notched Izod 23°C	kJ/m²†	12.3–5.8		1.6–5.5		180
Notched Izod −40°C	kJ/m²†					
Tensile impact strength	kJ/m²		6.5–16.1		NB 3.5	ASTM D1822
THERMAL						
Coefficient of linear thermal expansion	×10⁻³	5.4 2.3	2.1–1.1	3–5	25–14	ASTM D696
Deflection temperature at 1.82 MPa	°C	234–252	217–231	>300		
Vicat softening point, A/50	°C	265			60①–190①	75
„ „ B/50	°C	257				306
Maximum use temperature	°C	200	130–140	>200		

(continued)

TABLE 3 – TYPICAL DATA FOR HTA, PAA, POLYIMIDE & PIBA – *contd.*

ELECTRICAL						
Volume resistivity	Ohm.cm	10^{17}	$>10^{16}$	10^{15}	10^{11}–10^{13}[2]	IEC 93
Dielectric constant		3.7	3.70–4.3	3.63	7–12(1 kHz)	IEC 250
Dissipation factor	$\times 10^{-3}$	5		11	20–170 (,, ,,)	IEC 250
MISCELLANEOUS						
Density	g/cm³	1.36 1.58	1.42–1.63	1.40–1.65	1.06–1.14	R1183
Flammability		V0(3,2)V0(1,6)		HB[3]		VL subj 94
LOI		38				
Water absorption, 24h	%	0.9	0.20–0.14	0.15–1.25	1.2–6.5	R62A
Mould shrinkage (flow direction)	%	0.8 0.2	0.4–0.6	0.3 to 0.6–1.0		1110
MOULDING						
Mould temperature	°C	150–180	125	220–260	20–40	
Melt temperature	°C	360–390	280	100–120 (plus post-cure)	160–190 to 240–280 (dep on grade)	

†Some data sheets have values quoted in/J/m.
These are shown as kJ/m². Values in J.m are
approximately ×10 values shown in kJ/m².

[1] ASTM 1525, 9.8N.
[2] Grade 4011 has lower values and can be made semi-conducting.
[3] Grades for comprn/transfer moulding are V1 or V0.

MAIN ATTRIBUTES OF THE MAJOR PRODUCTS

PPS

PPS is a semi-crystalline material which has high strength and rigidity which reach exceptionally high levels in reinforced grades. Un-notched impact strength is good, but owing to its notch sensitivity, impact strength notched is only medium. Heat resistance is very good with low creep, and PPS can be used at temperatures up to 200–220°C. It has low flammability characteristics, being V0 or V0/V5 reinforced, with a high LOI at around 50. Water absorption and mould shrinkage are both low, as is thermal expansion. Chemical resistance is excellent particularly to solvents, even at temperatures approaching 200°C (see Table 2 for typical data).

Processing

PPS is mainly used in injection moulding, with mould temperature at 140°C to fully develop crystallinity, which gives optimum dimensional stability. It has very good flow and is therefore good for thin sections.

Applications

Electrical	– Coil formers, relay bases, low profile electronic components (chip carriers, sockets, contactors).
Automotive	– Under-the-bonnet parts such as fuel injection systems, air filter inlets.
Chemical engineering	– Pumps, meters, heat exchangers operating at 200°+ in hostile environments.
Miscellaneous	– Cases, supports, handles for kitchen equipment. Sockets and reflectors for lighting fittings. Parts for computer terminals and video recorders.

PSU

Polysulphone is an amorphous polymer which (un-reinforced) is transparent, with a refractive index of 1.65. It is strong and rigid, with medium impact strength, Reinforced with glass it has very high strength. Both forms can be used at temperatures up to 160°C. The polymer can be considered to be V2, with the reinforced material at V0. Water absorption of the polymer is low, although this increases when reinforced, whilst thermal expansion (which is low) and mould shrinkage decrease (see Table 2 for typical data).

Processing   – By injection moulding, extrusion and blow moulding.

Applications

Electrical	– Fuse and switch boxes, lamp housings.
Medical	– Applications needing sterilisation (flow meters, filter plates).
Appliances	– View ports for chip fryers, kettles.

PES

PES is very similar to PSu in its properties, apart from heat resistance, PES having higher values of Vicat than PSu. The natural material is transparent. PES can be used at 200°C, or at 160–180°C under stress, without deterioration. Unlike PSu the natural material, as well as the reinforced grades, have a UL rating of VO, with low levels of smoke and toxic gases (see Table 2 for typical data).

Processing

PES can be processed by injection moulding, extrusion or blow moulding. It can also be used in solution coating.

Applications

Electrical	– Industrial control relays, injection moulded printed circuits. High performance terminal blocks.
Automotive	– Car heater fans, bearing cages.
Plumbing	– Hot water pumps, valves.
Lighting	– Reflectors.
Medical	– Lids of laboratory centrifuges, handles for dentists' drills, Surgeons' lamps.

PAS

This material has very similar properties to PES and is therefore suitable for similar applications (see Table 2 for typical data).

PEK and PEEK

These two materials are very similar. They are both semi-crystalline plastics, with very high strength and rigidity and exceptional heat resistance. They therefore retain engineering properties to very high temperatures and can be used up to 300°C (short time) with continuous ratings of 250°C for PEEK and 260°C for PEK. They both offer high hardness and abrasion resistance, and excellent fatigue properties with good creep resistance. PEEK in particular has good wear resistance and low friction. Flammability is low, giving a V0 rating on the UL test, with very low smoke emission during combustion. With an LOI of 40–46%, PEK is very difficult to ignite. PEK and PEEK both have very good chemical resistance (see Table 2 for typical data).

Processing

In spite of its exceptional heat resistance, PEK has low viscosity at processing temperatures and does not therefore present problems in mould filling for injection moulding. Care is necessary however, as melt temperature has to be around 400°C. PEEK is somewhat easier to handle, with melt temperature around 375°C.

Both materials can also be processed by extrusion (rod, profile, film, wire insulation), compression moulding or powder coating.

Applications

Electrical connectors, particularly high density connectors.
Hot water meters.
Motorcar engine components.
Valve and bearing components.
Wire and cable coatings, particularly in fire hazard situations such as oil/gas
well data logging; nuclear installations.
Film and filament for specialised applications.

PAI

PAI is another material with very high strength and rigidity and exceptional
heat resistance. It maintains high strength even at 260°C, and has high retention
of properties after heat ageing. Chemical resistance is excellent, PAI being virtually unaffected by solvents, oils or hydraulic fluids, and by most acids at moderate temperatures.

With very low water absorption, low mould shrinkage and low coefficient of
thermal expansion (especially with glass reinforcement) PAI offers exceptional
dimensional stability (see Table 2 for typical data).

PEI

Polyetherimide is an amorphous plastic which in the un-reinforced state is transparent. Even un-reinforced it has very high strength and rigidity, and has good
heat resistance with a continuous use temperature of 170°C.

PEI has inherent flame resistance, with low smoke evolution. Thermal expansion, water absorption and mould shrinkage are all low, offering very good dimensional stability.

In addition to glass reinforced and internal lubricated grades, PEI is available
in structural foam formulations, which are useful for parts with thick sections
(see Table 2 for typical data).

Applications

Range from medical devices, metallised halogen light reflectors and pump impellers to high performance connectors.

It is used for parts of the air conditioning system of the Airbus, and for the
ashtrays, and for microwave dishes.

HTA

HTA (High Temperature Amorphous) is a relatively new material and as the
name suggests, it is amorphous. It is a biphenyl modified polyethersulphone and
is therefore related to PES, but shows higher load bearing properties at high
temperatures.

Even un-reinforced it has high strength and rigidity and these values are enhanced by glass reinforcement. HTA is a very tough material, with impact
strength approaching that of polycarbonate.

Heat resistance is extremely good, with retention of properties to over 200°C. Having an LOI of 38, HTA has inherent flame resistance, with low smoke. Its chemical resistance is very good, with virtually no stress cracking. Unreinforced, HTA is transparent (see Table 3 for typical data).

Applications

Applications for which HTA has been suggested include connectors, bobbins, relay components, circuit boards, and compressor and jet engine components.

PAA

PAA has very high strength and good impact resistance. Operating temperature limit is currently 130–140°C but it is expected that is will be increased to 150°C.

PAA has good mouldability, with low mould shrinkage at 0·4–0·6%. Together with low thermal expansion and low water absorption this gives it very good dimensional stability (see Table 3 for typical data).

Applications

PAA is being used for anti-vandal seats for buses and railways in France and trials are in hand for automotive under-the-bonnet applications. It is also finding application in electronics, household appliances, hi-fi and sports gear.

POLYIMIDE

This group of products is borderline between thermoplastics and thermosets. Only certain products, such as the 4000 series of Kinel, are proposed for injection moulding and even these need extensive post-cure at high temperatures to fully develop their properties.

The most favourable property of polyimide is its heat resistance, which is excellent, with little loss of properties at 200°C and 70% retention of tensile strength at 250°C. Thermal expansion is low, approximating to that of metals (see Table 3 for typical properties).

Applications

This excellent combination of properties has led to the use of polyimides in many critical applications (see also Chapter on Polyimides in Section 4B – Thermosets). The 4000 series has been for such varied applications as generator bobbins, jet engine acoustic panel spacers, drive components for microwave ovens and rotisseries, and cigar lighter sockets.

PEBA

This material is on the border between normal thermoplastics and thermoplastic elastomers. It is available in a variety of formulations which bridge the gap, as evidenced by the range of values of tensile strength and Shore hardness which are available (see Table 3).

Applications

Principal application areas for PEBA are automobile, sport, medical, agricultural, and mechanical engineering.

Polyarylates

These materials are included in the Chapter on *Thermoplastic Polyesters* (see p. 454).

LCP

These materials (Liquid Crystal Polymers) are a special class of plastics. The molecular chains take the form of discrete rods, which on processing align in the flow direction, giving a reinforcing effect – hence the name of the ICI product (Self-Reinforcing Polymer).

Whilst there are several polymers of this general class available, they differ widely from one another in structure and properties. Typical properties of three of these materials are shown in Table 4. It can be seen that LCP have in common:

High to very high strength and modulus.
High toughness over a wide range of temperatures.
Low thermal expansion.
High heat resistance.
Low mould shrinkage and very low water absorption.
Inherent low flammability.

Of the products available the most well established are the Vectra grades, which are approved for aeronautical applications by the USA FAA.

Processing with LCP presents no problems, as their viscosity at melt temperature is particularly low. The Hoechst company claims that the viscosity of Vectra is half or less than that of polyamide.

Applications

Principal applications are to be found in the electronics field, for connectors, relays, bobbins, surface mounted devices, and other parts needing excellent dimensional stability and high precision. LCP are also used in fibre optics and is in distillation columns.

TABLE 4 – TYPICAL DATA FOR LIQUID CRYSTAL POLYMERS
(Measured in flow direction)

Property	Unit	'Vectra' Hoechst	'SRP' ICI	'Xydar' Amoco	Test standard
PHYSICAL					
Tensile strength at break	MPa	165–207	230 170	98–126	R527
Elongation at break	%	6.6–1.1	8.0 3.2	2.5	,,
Tensile modulus	GPa	9.7–31.0			,,
Flexural modulus	GPa		12 15	11.6–14.4	178
Hardness — Rockwell		M58–M76	M69–M90	R77–R97	2039-2
Impact strength:					
Un-notched Izod 23°C	kJ/m^2†		NB–32	96–51	180
Notched Izod 23°C	kJ/m^2†	52 6.4	27,93 13,21	17.6–10	180
THERMAL					
Coefficient of linear					
thermal expansion	×10^{-3}		1.0 1.4	0.5–1.25	ASTM D696
Deflection temperature					
Under flexural load 1.82 MPa	°C	180 230	170–230	252 300	75
Maximum use temperature		200	160–220		
ELECTRICAL					
Volume resistivity	Ohm.cm	10^{15}–10^{16}	10^{17}	>10^{15}	IEC 93
Dielectric constant		2.6–4.2	3.0–3.5	3.4–3.9	IEC 250
Dissipation factor	×10^{-3}		26–18	29 33	IEC 250
MISCELLANEOUS					
Density	g/cm^3	1.4 1.9	1.38 1.56	1.62–1.75	R1183
Flammability			V0 V0	V0(0.79)	UL subj 94
LOI			36–39		
Water absorption, 24h	%	0.01 0.03	0.08–0.14	0.02–<0.01	R62
Mould shrinkage (flow dir'n)	%	0.05	0.11 0.01		1110
MOULDING					
Mould temperature	°C		40–100		
Melt temperature	°C		310–340	230	

†Some data sheets have values quoted in J/m. These are shown as kJ/m^2. Values in J/m are approximately ×10 values shown in kJ/m^2.

Self-intumescent

Methylpentene (TPX)

METHYLPENTENE POLYMERS are high clarity resins with excellent electrical properties and chemical resistance. They have better resistance to distortion at high temperatures than most other transparent thermoplastics, and have the lowest density of all the thermoplastics (0.834).

Mechanical properties are moderate. TPX has limited creep resistance under load, and although it fails in a ductile manner, this may be preceded by surface cracking or crazing, and whitening of the body of the material. Impact strength is also relatively moderate, but superior to that of general purpose polystyrene. As an optical material it can be considered mid-way between polystyrene and acrylic, having light transmission of 90% and a refractive index of 1.425. Its resistance to ageing and discoloration is less favourable than acrylic, but it can be stabilised. Glass reinforced grades are available. Typical properties are given in Table 1.

TABLE 1 – TYPICAL PROPERTIES OF TPX

Property	Unit	Value	Test standard
Tensile strength.	MPa	28	R527
Elongation at break	%		,,
Tensile modulus	GPa	1.45	,,
Hardness — Rockwell, L		67–74	2039–2
Impact strength, Notched Izod, 23°C	J/m	27–42	180
Coefficient of thermal expansion		11.7×10^{-5}	ASTM D696
Volume resistivity	Ohm.cm	10^{16}	IEC 93
Permittivity		2.12	IEC 250
Dielectric strength	kV/mm	700	IEC 243
Density	g/cm^3	0.83	R1183
Water absorption, 24h	%	0.01	R62A

PBA

POLYBENZOXAZOLE (PBA) represents something of a borderline case between a thermoplastic and a thermoset plastic. Strictly speaking it is a thermoplastic material with a high softening point of 270 to 300°C. It does, however, require 'curing' or cyclising at a temperature of about 300°C. The pre-polymer can be used for coatings and impregnation. The cyclised polymer is used for producing mouldings by compression moulding in a heated press. Both plain and filled mouldings can be produced.

PBA has good mechanical strength and a maximum service temperature of 250 to 270°C.

Typical properties of PBA

Specific gravity	1.39
Tensile strength	90–204 MPa
Flexural strength (ASTM D790)	69–83 MPa
Elongation	approx. 10%
Coefficient of thermal expansion	5×10^{-5}
Water absorption (24 hours – ASTM D570)	0.5%

Ionomers

IONOMERS ARE copolymers of ethylene and methacrylic acid. They are thermoplastic polymers but with ionic cross-links, so that in the operating range of 40 to –40°C they behave almost as though they were thermoset materials. They are tough, transparent and grease resistant materials. At high temperatures (290–330°C) the ionic bonds relax and the polymer can be injection moulded, blow moulded, extruded and thermoformed. It is also an excellent material for heat sealing and for heat stretching and forming.

Ionomers are very good materials for the production of film and this is their major use.

Properties

Ionomer grades differ in their MFI and in whether they contain sodium or zinc ions; the sodium ion types have better optical properties, oil resistance and hot tack, whilst zinc ion types have better adhesion in co-extrusion and foil coating. Ionomers have low temperature heat sealability, outstanding hot tack (up to 10 times that of LDPE) and excellent adhesion to nylon and other packaging films. They also have good low temperature impact strength and good puncture and abrasion resistance.

Chemical resistance

Surlyn A is resistance to alkalis, alcohols and ketones; other organic solvents plasticised or cause swelling. It is also resistant to fats and soils. Resistance to acids is not so good, and oxidising agents such as fuming nitric acid or bromine (liquid) cause degradation.

Weathering behaviour is similar to that of PE, and like PE it can be stabilised by the addition of carbon black. Permeability is similar to that of PE, apart from permeability to carbon dioxide, which is lower. There is little tendency to ESC.

TABLE 1 – TYPICAL PROPERTY OF IONOMERS
(Du Pont data)

Property		Unit	Value	Test standard
MISCELLANEOUS				
Density		g/cm^3	0.94	R1183
Melt flow (MFI)	190/2.16	g/10 min	0.7–14	R1133
PHYSICAL				
Tensile stress at yield		MPa	9–29	R527
Tensile strength at break		MPa	21–35	R527
Elongation at break		%	280–520	,,
Tensile modulus		GPa		,,
Flexural modulus		MPa	70–380	178
Hardness, Shore D			56–68	DIN 53505
Toughness:				
Tensile impact	23°C	10^2/kJ/m^2	7.6–12.8	180
,, ,,	–40°C		5.6–11.8	
Brittleness temperature		°C	–50 to –112	ASTM D746
THERMAL				
Thermal conductivity			0.24	DIN 52612
Coefficient of linear				
thermal expansion		10^{-5}	11–13	ASTM D696
Deflection temperature				75
under flexural load	0.45 MPa	°C	40–47	
Vicat softening point, B		°C	40–47	1525/306

Applications

DuPont Surlyn, a typical ionomers, has been used as the cover material for golf balls and for other sporting applications, and for such automotive applications as bumper guards and rubbing strips. However its major use is for films and coatings, and for composites, where the ionomer serves as a heat seal layer and at the same time improves package durability. It is also used in snack food package structures, in skin and blister packs and in bottles.

Polyvinylcarbazole

POLYVINYLCARBAZOLE (POLY-N-VINYL carbazole) is a thermoplastic
with an excellent resistance to heat and good dielectric properties. It is mostly
used for the manufacture of high quality insulation for high frequency electronics, including radio and television components and data processing machines. It
is resistant to water, alkalis and salt solutions at temperatures up to about
180°C, and is resistant to reducing agent oxidation up to 100°C. It is insoluble
in esters, alcohols, ketons, carbon tetrachloride, aliphatic hydrocarbons and low

TABLE 1 – PROPERTIES OF POLYVINYLCARBAZOLE (BASF 'LUVICAN')

Property	Unit	Test standard	Value[†]
Density	g/ml	1183	1.19
Flexural strength[‡]	MPa	178	700
Impact strength[‡] (notched Charpy)	kj/m^2	179	5
Ball indentation hardness	MPa	53 456	200
Heat distortion temperature			
Martens	°C	53 458	150–170
Vicat	°C	306	200
		Method B	
Volume resistivity≠	Ohm/cm	IEC 93	10^{16}
Dielectric constant, ϵ (10^6 c/s)		IEC 25	3.0–3.1
Dissipation factor≠,* tan δ_r (10^6 c/s)		IEC 250	0.0006—0.0010
Dielectric strength	kV/mm	IEC 243 53 481	50
Refractive index n$_D$^{20}		489	1.696
Water absorption after four days	%	53 476	<0.1*; 0.1–0.2

†   Determined at 20°C unless otherwise stated.
‡   Injection-moulded 50 × 6 × 4 mm rod.
≠   Almost constant between 20 and 100°C.
*   Measured on dried specimens.
** Lower value measured on sintered articles.

Poly-N-vinyl carbazole molecule.

aromatic mineral oils, but is attacked and/or dissolved by aromatic hydrocarbons, chlorinated hydrocarbons, chlorobenzene and methylene chloride. It is slightly attacked (swollen) by petrol and cyclohexanone.

The monomeric vinylcarbazole is used for impregnating porous or absorptive articles such as capacitors and coils in electrical engineering and is particularly suitable for impregnating papers. The crystalline substance melts at 65 to 67°C and polymerises slowly in air. Polymerisation is accelerated by the use of catalysts, such as di-t-butyl peroxide.

Polyvinylcarbazole is toxic in the unpolymerised state, and vinylcarbazole is even more toxic, so skin contact should be avoided.

Polyisobutylene (PIB)

POLYISOBUTYLENES ARE viscous substances at room temperature, their liquid characteristics and cold flow properties varying with molecular weight. High molecular weight polyisobutylenes display the melt elasticity common to all molten polymers and have the feel of cross linked natural rubber.

For practical applications they are blended with other substances, or incorporate fillers to modify the properties for particular applications. Blending is usually with polyolefins, or with waxes for laminating and coating. Filled polyisobutylenes are used as adhesives and sealing compounds. Unfilled ones may be used as viscosity index improvers in oils and also as electrical insulating oils.

TABLE 1 – PROPERTIES OF A RANGE OF POLYISOBUTYLENES (BASF OPPANOL)

Grade	Viscosity, Poise				Molecular weight number average
	At 20°C		At 100°C		
B3	250 ± 30		2.0 ± 0.2		820
	Solution Viscosity				Viscosity average
	Solvent	Concentration g/100 ml	Ubbelohde capillary No	Flow time seconds	
B10	Tetralin	2.69	II	45–49	50 000
B15	Tetralin	2.69	II	65–73	95 000
B50	Isooctane	0.2	I	93–98	380 000
B100	Isooctane	0.2	I	116–127	1 300 000
B150	Isooctane	0.1	I	108–118	2 700 000
B200	Isooctane	0.1	I	123–141	4 700 000

SECTION 5

Thermoset plastics

Introduction

THERMOSET PLASTICS are generally defined as synthetic resins which undergo a permanent, irreversible change on curing — *ie* they solidify into an infusible state so that if subsequently heated they do not soften and become plastic again. This change is reflected in their structure, which is that of a permanently set cross-linked polymer. Curing is generally accomplished by heat and pressure but certain resin groups, *eg* polyesters and epoxides, are capable of cold curing at room temperatures without pressure.

The chief types of thermoset plastics are:–

(i) Alkyds
(ii) Phenolics
(iii) Polyesters
(iv) Epoxides
(v) Aminos — embracing both urea-formaldehyde (UF) and melamine-formaldehyde (MF)
(vi) Polyimides

Silicons are also classifiable as thermoplastic resins.

Alkyds

ALKYD RESINS are reaction products of an alcohol and an organic acid and as such can be considered as modified polyesters. They fall into two main groups — diallyl phthalate (DAP) and diallyl isophthalate (DIAP). They are particularly noted for good dimensional stability, good resistance to heat (maximum service temperature 170°C for continuous exposure, and up to 260°C or more for intermittent exposure) and excellent electrical properties.

Their mechanical properties can be enhanced with mineral or fibre fillers, glass fibre reinforcement giving the highest mechanical and electrical properties. Alkyds are not attacked by fats, oils, alcohols or common solvents, but their resistance to strong acids and alkalis is restricted.

The chief application of alkyd moulding powders, doughs and encapsulating grades is for electrical components. They can be produced in a wide range of

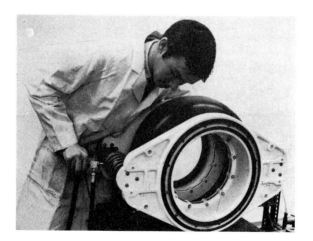

Non-shrink grade of 'Beetle' polyester DMC (dough moulding compound) is used for this moulding. Polyester is closely related to DAP, also used for moulding compounds.

fade-resistant colours and fast curing grades are also available. Typical properties of four representative types of alkyds are given in Table 1.

TABLE 1 — CHARACTERISTIC PROPERTIES OF ALKYDS

Property	Mineral Filled (Dough)	Mineral Filled (Granular)	Glass Filled	High Impact Grades
Specific gravity	2·05–2·15	1·96–2·24	1·9–2·3	1·9–2·1
Density, lb/in³	0·074–0·077	0·067–0·083	0·069–0·083	0·067–0·077
Tensile strength, lb/in²	4 000–5 000	3 000–5 000	3 000–7 000	5 000–10 000
Tensile modulus, lb/in²	—	—	—	—
Flexural strength, lb/in²	8 000–11 000	6 000–10 000	14 000–25 000	12 000–26 000
Flexural modulus, lb/in²	$2·0–2·7 \times 10^6$	$1·6–2·9 \times 10^6$	$1·5–2·5 \times 10^6$	$1·9–2·5 \times 10^6$
Compressive strength, lb/in²	20 000–25 000	16 000–24 000	—	—
Shear yield strength	—	—	—	—
Elongation, % (ultimate) at 2 in/min	—	—	—	—
Impact strength, Izod 73°F (notched), ft lb/in of notch	0·25–0·35	0·30–0·60	1–5	6–20
Hardness (Barcol)	60–70	55–70	60–80	65–80
Thermal conductivity Btu/hr/ft²/degF/ft	0·35–0·60	0·25–0·60	0·2–0·3	0·2–0·3
Coefficient of thermal expansion $\times 10^{-6}$ in/in/ degC, –30°C to 30°C	—	10–15	10–30	10–15
Heat distortion, temp °F at 264 lb/in² stress	350–400	350–400	300–400	400–450
Maximum recommended intermittent service temperature, °F	325	300–375	400	400–450
Maximum recommended continuous service temperature, continuous loading, °F	250	250–300	300	300–400
Mould shrinkage, in/in	0·004–0·008	0·004–0·008	0·001–0·006	0·001–0·004
Water absorption, 24 hr, 1/8 inch thickness. 73°F				
% increase in weight	0·10–0·15	0·08–0·50	0·04–0·25	0·07–0·20
Dielectric constant				
60 Hz	5·4–5·9	5·7–7·0	5·2–7·7	5·2–7·1
10^6 Hz	4·5–4·7	4·7–5·7	5·2–7·1	4·5–6·2
Power factor				
60 Hz	0·03–0·045	0·03–0·05	0·008–0·025	0·02–0·03
10^6 Hz	0·016–0·022	0·016–0·032	0·011–0·025	0·015–0·022
Resistance to acids	Poor	Poor	Poor	Poor
Resistance to bases	Poor	Poor	Poor	Poor
Resistance to solvents	Good	Good	Good	Good
Flammability	Non-burning	Self-extinguishing to nonburning	Self-extinguishing to nonburning	Self-extinguishing to nonburning

Amino resins

AMINO RESINS are condensation products of formaldehyde with urea (urea-formaldehyde, or UF) or melamine (melamine-formaldehyde, or MF). Unfilled

TABLE 1 — TYPICAL PROPERTIES OF UREA MOULDING COMPOUNDS WITH CELLULOSE FILLERS (AMERICAN CYANAMID 'BEETLE')

Filler			Alpha Cellulose	
PROPERTIES				
Processing Methods			Compr/Inject/Trans	
Processing Temp			310°F[a]	(154°C)
Linear Mould Shrinkage	in/in		9.00×10^{-3}	
Density	lb/ft^3	(g/cm^3)	93.6	(1.50)
Tensile Str, Break	lb/in^2	(kg/cm^2)	6.30×10^3	(442)
Elongation, Break				
Tensile Modulus	lb/in^2	(kg/cm^2)	1.35×10^6	(9.49×10^4)
Flexural Str, Yield	lb/in^2	(kg/cm^2)	1.45×10^4	(1.01×10^3)
Flexural Modulus	lb/in^2	(kg/cm^2)	1.45×10^6	(1.01×10^5)
Compressive Str	lb/in^2	(kg/cm^2)	3.40×10^4	(2.39×10^3)
Izod, Notched, R.T.	ft lb/in	(kg cm/cm)	0.30	(1.63)
Hardness (Test)			E96 (Rockwell)	
Therm Cond	BTU in/hr ft^2 °F(cal cm/sec cm^2 °C)		2.92	(1.01×10^{-3})
Linear Therm Expan	in/in°F	(cm/cm °C)	1.61×10^{-5}	(2.90×10^{-5})
Continuous Svc Temp			170°F	(76°C)
Defl Temp, 264 lb/in^2			266°F	(130°C)
U.L. Temp Index	°C/mm		100	
Volume Resistivity	Ohm cm		3.00×10^{11}	
Surface Resistivity	Ohm			
Dielectric Strength	V/10^{-3} in (V/mm)		350	(1.37×10^4)
Dielectric Constant, 10^6 Hz			6.80	
Dissipation Factor, 10^6 Hz			0.030	
Water Absorp 24 hr			0.60%	

TABLE 2 — PROPERTIES OF MELAMINE MOULDING COMPOUNDS WITH CELLULOSE FILLER (PERSTORP GRADE 751)

Filler			Cellulose	
PROPERTIES Processing Methods			Moulding	
Processing Temp			323°F	(161°C)
Linear Mould Shrinkage	in/in		7.00×10^{-3}	
Density	lb/ft^3	(g/cm^3)	93·6	(1·50)
Tensile Str, Break	lb/in^2	(kg/cm^2)		
Elongation, Break				
Tensile Modulus	lb/in^2	(kg/cm^2)		
Flexural Str, Yield	lb/in^2	(kg/cm^2)	1.00×10^4	(703)
Flexural Modulus	lb/in^2	(kg/cm^2)	1.30×10^6	(9.13×10^4)
Compressive Str	lb/in^2	(kg/cm^2)		
Izod, Notched, R.T.	ft lb/in	(kg cm/cm)	0·25	(1·36)
Dielectric Strength	V/10^{-3} in	(V/mm)	300	(1.18×10^4)
Dielectric Constant,10^6 Hz				
Dissipation Factor, 10^6 Hz			0·033	
Water Absorp, 24 hr			0·40%	

resins can be formulated as coatings, adhesives for laminating, and impregnants for paper and textiles, etc. Moulding powders are produced by compounding with fillers, pigments, extenders and other additives.

Amino resin mouldings are noted for their high gloss, hardness, freedom from taste or odour, good chemical resistance, good electrical properties, good heat resistance and self-extinguishing properties if exposed to flame. The principal use of UF mouldings is for domestic and electrical appliances. MF resins are more suitable for domestic tableware, etc as they have better resistance to hot water, weak acids and alkalis, and superior resistance to staining. Both UF and MF resins are produced in cellulose-filled grades; there are also melamine-filled UF resins which offer some of the superior properties of MF at lower cost. Cellulose-filled grades can normally be expected to discolour at moderate temperatures — eg about 80°C in the case of cellulose-filled UF, and 120°C in the case of cellulose-filled MF.

Amino resins are not normally used for stressed components as, although their mechanical strength is quite high, they are essentially brittle materials.

Thermosetting acrylic resins

AMINO RESINS are valuable components in thermosetting acrylic enamels, in which they are used not only to improve the performance of the film but often to lower the cost as well. The acrylic resins used are made by the addition copolymerization of a number of ethylenically unsaturated monomers, which are predominantly esters of acrylic or methacrylic acid. The relative proportions of these are dictated by the hardness and flexibility required, methacrylic ester monomers giving harder, more brittle polymers than acrylic esters. Into this polymer backbone are copolymerized other monomers such as acrylic acid, hydroxyethyl acrylate and acrylamide, giving polymers with pendant carboxyl, hydroxyl and amide groups respectively. In practice, polymers containing both hydroxyl and carboxyl pendant groups are generally used in conjunction with alkylated melamine-formaldehyde resins. The hydroxyl groups form sites for reaction and cross-linking with the amino resin in much the same way as with alkyd resins, and the residual acidity due to the carboxyl groups assists the curing process.

Cross-linking occurs on stoving at about 120°C and the cured films have excellent gloss, weather resistance, and polishing characteristics. For these reasons, they have proved particularly valuable as automobile finishes.

Epoxides

EPOXIDE THERMOSETTING resins are capable of cold curing at normal ambient temperatures without pressure by admixture with a curing agent or hardener. They can also readily be compounded with inert fillers, plasticizers, diluents and flexibilizers to yield a wide range of properties in the cured state, ranging from soft, flexible materials to tough, rigid solids. Epoxides also bond tenaciously to most substrates, including all metals, glass, ceramics, most plastics, wood, paper, etc. Cast resins have good long-term dimensional stability with low mould shrinkage (which can be less than 1%). The cure is also volatile free. Mechanical properties are good, with reasonable impact strength. Electrical properties are excellent to outstanding, as is chemical resistance. They are thus very versatile materials capable of meeting even the most exacting demands of both structural and coating applications, particularly in electronic, electrical, mechanical, nuclear, civil, marine, car and aerospace engineering and in the building industry.

The range of properties offered by typical casting and moulding compounds is indicated in Table 1.

The principal applications of epoxide resin are:-

Casting large high-grade electrical insulators and the insulation around equipment such as transformers, switchgear components, etc.

Impregnating, potting and sealing of electrical windings and components.

Construction of glass-fibre- and carbon-fibre-reinforced structures in electrical, nuclear, mechanical and aerospace engineering.

Compression, transfer and injection moulding of high-accuracy components with excellent electrical properties.

Casting and moulding of plant components for the chemical industry.

Bonding of metals, ceramics, glass, rubber, plastics, wood, etc.

Construction of tools for forming plastics or sheet-metal, patterns for moulding foundry sand, jigs and fixtures for checking accuracy of assembly, etc.

Chemical-resistant flooring and floor surfacing.

Anti-corrosion protection of metals, concrete, wood, etc.

TABLE 1 — TYPICAL PROPERTIES OF EPOXIDES

Generic Type			Epoxy	Epoxy	Epoxy	Epoxy	Epoxy	Epoxy
Processing Method			Casting	Casting	Casting	Casting	Moulding	Moulding
Filler, Blend, Grade, etc.			Unfilled	Silica	Aluminium	Flexible grade	Glass fibre	Mineral
Property	Unit	ASTM Test						
Processing Temp	°F	—	—	—	—	—	300-330	250-325
Linear Mould Shrinkage	in/in	D955	·001-·010	·005-·008	·001-·005	·001-·010	·001-·005	·002-·007
Density	g/cm³	D792	1·10-1·40	1·60-2·00	1·40-1·80	1·05-1·30	1·60-2·00	1·60-2·00
Specific Volume	in³/lb	D792	21-20	15-13	19-15	20-21	15-13	15-13
Tensile Str, yield	10³lb/in²	D638	5·0-12·0	7·0-13·0	7·0-11·5	2·5-10·0	10·0-20·0	5·0-10·0
Elongation %, break	—	D638	3-6	1·3	0·5-3·0	25·0-70·0	1·0-2·0	—
Tensile Modulus	10⁵lb/in²	D638	3·5-4·0	—	—	—	20·0-22·0	16·0-17·0
Flexural Str, yield	10³lb/in²	D790	13-21	8-14	9-23	2-12	8-14	5-7
Flexural Modulus	10⁵lb/in²	D790	2-5	14-30	—	—	25-55	14-22
Compressive Str	10³lb/in²	D695	15-25	15-34	15-33	2-13	18-35	10-15
Izod, notched R.T.	ft lb/in	D256	0·2-1·0	0·3-0·5	0·4-1·5	3·5-5·0	0·3-0·4	0·1-0·2
Hardness (test)			M80-110	M85-120	M55-85	—	M100-110	M100-110
Thermal conductivity	BTU in/hr ft²°F	C177	1·2-1·6	2·9-5·1	4·4-7·3	—	1·1-2·9	1·2-8·7
Specific Heat	BTU/lb°F	C351	0·2-0·3	0·2-0·3	—	—	0·2	—
Linear Therm Expan	10⁻⁵in/in°F	D696	4·5-6·2	2·0-4·0	4·0	2·0-8·0	1·1-3·5	2·0-4·5
Continuous Svc Temp	°F	—	175-190	—	—	100-125	—	—
Volume Resistivity	Ohm cm	D257	1·0 × 10¹²-10¹⁷	1·0 × 10¹³-10¹⁶	—	1·0 × 10¹⁴	1·0 × 10¹⁴	>1·0 × 10¹⁴
Dielectric Strength	V/10⁻³in	D149	300-500	300-550	10-200	250-400	300-400	300-400
Dielectric Constant	50-100 Hz	D150	3·5-5·0	3·2-4·5	—	3·0-5·0	3·5-5·0	3·5-5·0
Dissipation Factor	50-100 Hz	D150	·002-·010	·008-·030	—	·010-·040	·010	·010
Water Absorp %, 24 hr		D570	0·1	0·1	—	—	—	—

The number of epoxide formulations continues to grow, but the main types are those based on bisphenol-A and epichlorobydrin. Amongst the newer materials to appear are the cresol novolacs with enhanced thermal properties and improved resistance to solvents and chemicals, the phenol novolacs with superior electrical properties and the cycloaliphatics with enhanced weather resistance and durability.

Phenolics

PHENOLICS were one of the original thermosetting resins, developed in 1909 as *Bakelite*, since when they have continued to have a wide application as a low cost material. Phenolic resins are reaction products of phenol and formaldehyde, or chemically phenol-formaldehyde (PF). They are variously used in filled grades for moulding not-stressed or lightly-stressed components, the filler normally used being wood flour in proportions of 50–70%. Other fillers which may be used for specific purposes include mica (for enhanced electrical properties); asbestos (for enhanced heat resistance); glass fibre (for enhanced strength and elec-

TABLE 1 — TYPICAL PROPERTIES OF PHENOLIC PLASTICS*

Property	Unit	Unfilled	Wood Flour Filler	Mineral Filler
Specific gravity		1·280	1·250–1·520	1·590–2·090
Density	g/cm³	1·280	1·250–1·520	1·590–2·090
	lb/in³	0·046	0·045–0·055	0·057–0·075
Tensile strength	tons/in²	2·700–4·000	1·800–4·900	1·800–4·450
	kg/mm²	4·250–6·300	2·800–7·700	2·800–8·600
Modulus of elasticity	lb/m²	$7–10 \times 10^6$	$10–15 \times 10^6$	$10–15 \times 10^6$
	kg/m²			
Compressive strength	tons/in²	—	7·000–16·000	8·000–16·000
	kg/m²	—	11·000–25·000	12·500–25·000
Flexural strength	tons/in²	5·350–7·600	3·600–6·700	3·600–9·000
	kg/mm²	8·400–11·900	5·700–10·600	5·700–14·100
Maximum service temperature	°C	120·000	175·000	230·000
	°F	250·000	350·000	450·000
Water absorption (24 hour)	%	0·100–0·200	0·200–0·600	0·100–0·300
Dielectric constant		5·000–6·000	5·000–12·000	5·000–20·000
Coefficient of thermal expansion	degC.	—	$3·7–75 \times 10^{-6}$	$25–40 \times 10^{-6}$

*See also chapter on *Laminated Plastics.*

trical properties and shock-resistant grades); nylon (for enhanced water resistance); and graphite.

Limitations with phenolic resins have been limited colour range (dark or 'solid' colours only) and difficulty of moulding other than by con.pression or transfer methods. Colour limitations have largely been removed by compounding with melamine, whilst the development of thermoset injection moulding techniques and the appearance of easy-flow PF moulding powders has increased both the scope and speed of production of phenolic mouldings.

Like UF resins, phenolic mouldings have good mechanical strength and gloss, but are brittle. They are superior to most other thermoset resins as regards resistance to creep, however.

Phenolic resins with wood flour filler are generally suitable for continuous exposure to temperatures up to about 150°C, or up to 220°C for intermittent exposure. The material is self-extinguishing, but tends to support combustion. At high temperatures phenolic resins char and decompose.

Closed cell phenolic foams are also finding an increasing application for thermal insulation where their fireproof characteristics are particularly favourable. They are resistant to ignition and under high heating have negligible emission of smoke and fumes. K-values and U-values obtainable with phenolic foams also compare favourably with those of other thermal insulation materials — see also chapter on *Insulation, Thermal.*

Parts for a power drill moulded in phenolic (Hooker Chemical Corp).

Polyester

THERMOSETTING POLYESTERS are unsaturated liquid polyester resins normally containing an organic peroxide catalyst which can be activated by an accelerator or 'hardener' to produce curing at normal ambient temperatures. In the
absence of an accelerator, curing can be produced at temperatures of 70–150°C.

The unsaturated resins are obtained by the condensation of dihydric alcohols
with divalent saturated and unsaturated carboxylic acids and are dissolved in a
monomer that is capable of cross-linking with the polyester to form a copolymer. Styrene is the monomer most used for this purpose.

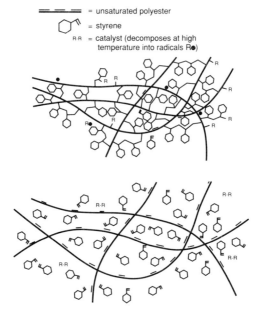

*Resin plus catalyst (top) and resultant polymer with
crosslinking formed by curving (bottom).*

The cure is exothermic and can generate high internal temperatures in a large mass of resin. Temperatures should normally be kept to below 140°C to prevent internal cracking. Control in this respect is possible by formulation and restricting the amount of resin cured at any one time. During curing shrinkage of up to about 8% may occur.

Unfilled polyester resins are normally used as surface coatings, for potting and encapsulation, linings, thread locking compounds and castings. They tend to be brittle although formulations for potting and casting can be made tougher and more elastic. Filled resins are used for filler work (especially in automobile body repairs), castings and industrial mouldings, etc. The most widespread use of polyester resins, however, is in the glass fibre reinforced form for GRP mouldings — see *Glass Reinforced Plastics*.

Polyimides

POLYIMIDES are noted for their high resistance to oxidation and good retention of properties over a wide temperature range. Maximum continuous service
temperature is of the order of 250°C or up to 350–400°C for intermittent exposure. Tensile strength, rigidity and other mechanical properties are also well
maintained even after long exposure to high temperatures.

TABLE 1 — TYPICAL PROPERTIES OF POLYMIDES

Property	Unit	Test method	Base Polymer	15% Graphite Loaded
Specific gravity	—	—	1·43	1·51
Tensile strength	lb/in^2	ASTM D-1708	10 500	6 500
	kg/cm^2		735	455
Elongation	%		5–6	3–4
Compressive strength	lb/in^2	ASTM D-695	24 000	18 000
	kg/cm^2		1 680	1 260
Flexural strength	lb/in^2	ASTM D-790	14 000	17 000
	kg/cm^2		980	1 190
Impact strength	ft lb/in	ASTM D-756	9	—
Tensile modulus		ASTM D-790	460 000	560 000
Hardness	Rockwell	ASTM D-785	E 45–52	E 32–44
Coefficient of expansion	degF	ASTM D-696	26–30 × 10^{-6}	21–30 × 10^{-6}
Maximum service temperature	°C	—	260/480*	260/480*
Heat deflection temperature	°C	ASTM D-648	480*	480*
Thermal conductivity	Btu/hr/ft^2/degF/in	—	245	26–33
Dielectric strength	V/mil	ASTM D-149	560	250
Volume resistivity	ohm cm	ASTM D-257	10^{16}–10^{17}	15 × 10^{15}
Arc resistance	seconds	ASTM D-498	230	—
Water absorption	24 hrs %	ASTM D-570	0·32	—

*intermittent

TABLE 2 — WHICH PLASTIC? TEST BY BURNING A SMALL PIECE

Colour of Flame	Smell	Other Observations	Plastic
No flame	Fishy	Chars with further heating	UF
No flame	Carbolic	Darkens, chars with excess heat	PF
Brilliant white	Camphor	Burns readily	Celluloid (cellulose nitrate)
White	—	Burns like paper	Cellophane
Pale yellow	Acrid (vinegar and burning paper)	Melts into brown drips.	Cellulose acetate
Yellow	Marigold	Smoky flame	Polystyrene
Weak blue	Acrid at first, turning sweeter	—	PVC
Medium blue	Very sweet — peardrops	Remains clear, does not char or change colour	Methacrylic
Blue	Celery	Melts readily and drops droplets when burning	Nylon
Blue	Candle wax	Burns readily and drips	Polyethylene
Very pale green	Sweet	Extensive black ash	Polyvinylidene chloride (Saven)
Green	Burning rubber	Smokes	Rubber hypochloride

Polymides are attractive as non-lubricated bearing materials offering a PV rating in excess of 10^6 lb/in^2 ft/min with a coefficient of friction of 0·05–0·20 (friction decreases with increasing load). In addition the material has excellent resistance to abrasion. Graphite-filled and molybdenum-disulphide-filled polyimides can also be used, showing even lower friction and wear. A further application of polyimides is for high density electrical insulation tapes (class H temperature rating).

Typical tensile strength is of the order of (700 kg/cm^2), 10 000 lb/in^2 with compression strength about 2·5 times this figure. Impact strength is comparable or superior to all unfilled thermoset plastics. When overheated the material chars but does not melt.

Electrical properties are generally good. Water absorption is low, but polyimides will not withstand prolonged immersion in boiling water or steam. They are not attacked by most solvents or acids, alcohols, esters, etc, but are attacked by strong alkalis and aqueous solutions of ammonia.

Silicons

THE PRINCIPAL forms and applications of silicons are described in the separate chapter on *Silicones* in Section 3. Silicon rubbers, which represent the majority form of solid silicons, are also described in the chapter on *Elastomers.*

Silicon thermosetting resins may be cold curing or heat curing, their principal applications being for the production of high-temperature laminates and electrical insulation, etc. They may also be used in both unfilled and filled forms for moulding special duty components. Moulding is usually accomplished by high pressure press cure with ovenpost cure. Low pressure resins are also produced for bag moulding of complex shapes, followed by post curing. Typical properties of a silicon moulding compound are summarized in Table 1.

Organo-silicons are a separate category of inorganic/organic substances with many of the general properties of silicons. Their molecules comprise silicon groups (O-Si-O linked to organic side groups) joined to reactive organic groups.

TABLE 1 — TYPICAL PROPERTIES OF MOULDED SILICON RESINS

Property	Unit	Test method	Value
Specific gravity	—	—	1·88
Tensile strength	lb/in²	ASTM D 638	6 000
	kg/cm²		420
Compressive strength	lb/in²	ASTM D 695	14 500
	kg/cm²		
Impact strength	ft lb/in	ASTM D 256	0·3
Hardness	Rockwell	ASTM D 785	U 80
Coefficient of expansion	degC	—	$2·4$–$2·9 \times 10^{-5}$
Heat distortion temperature	°C	ASTM D 648	450
Thermal conductivity	cal/cm/sec/cm²/degC		$1·2 \times 10^{-3}$
Electrical strength	V/mil	ASTM D 149	> 280 in air
			> 370 in vacuum
Power factor	at 1 MHz	ASTM D 150	–0·0022
Volume resistivity	ohm cm	ASTM D 257	1–2×10^{14}
Water absorption	%	ASTM D 570	0·09

As a result they are particularly effective as adhesives between organic and inorganic materials as well as having special characteristics such as excellent resistance to moisture, water, steam and ultra-violet light. This makes them particularly attractive as surface coatings.

SECTION 6

Plastics (processed)

LAMINATED PLASTICS (INDUSTRIAL LAMINATES)
SANDWICH MOULDINGS
CELLULAR PLASTICS
GLASS REINFORCED PLASTICS (GRP)
FILLED THERMOPLASTICS
CARBON FIBRE REINFORCED PLASTICS (CFRP)

Laminated plastics (industrial laminates)

LAMINATED PLASTICS comprise layers of fibrous material impregnated with and bonded together by a thermosetting resin to form a rigid solid material. They are most commonly rendered in sheet form but are also produced in bars, rods, tubes and standard and special sections. The properties of the resulting material depend on the type and proportions of resin used and the type of fibrous reinforcement employed.

They can be classified generally as decorative and industrial laminates. Industrial laminates may be sub-divided into mechanical and electrical grades.

Types of fibrous reinforcement

In general the reinforcement materials are used in the form of papers or thin fabrics, although there is also a more limited application of felted materials. In manufacture these are preimpregnated with a predetermined quantity of resin and part precured. They are then stacked in layers after cutting to suitable shapes and fully cured under heat and pressure in high pressure moulds. Certain sections, *eg* wrapped tubes, may be oven cured on a mandrel without pressure, although moulded tubes are generally preferred.

General characteristics of reinforcement materials are:

Paper reinforcement – the cheapest type for any resin system, yielding a light-weight material with good mechanical strength, easily cut, drilled and machined. It also has excellent electrical properties.

Cotton fabric reinforcement – this offers superior mechanical but inferior electrical properties. It is therefore used primarily in mechanical grades, although electrical grades may be used where high wear resistance is required (*eg* for contact breaker rotor arms) and the electrical voltages involved are moderate or low.

Weaves used range from coarse (30 threads/inch) to very fine (120 threads/inch); the finer the weave the better the machined finish and the dimensional stability and the closer the tolerance possible. As far as electrical grades are concerned, the finer the weave the lower the water absorption and the better the electrical properties.

Asbestos reinforcement – is used where increased resistance to heat is required, particularly for mechanical applications. Other properties improved are rigidity, dimensional stability and, to a lesser degree, chemical resistance. The asbestos may be in the form of paper, woven fabric or felt. Paper reinforcement has the disadvantage of being prone to edge chipping but gives a finer finish than fabric or felt. In all cases machining is rather difficult because of the abrasive nature of the reinforcement.

Electrical grades of asbestos reinforced laminates are less attractive because of low insulation resistance and electrical strength but they may be used for low voltage applications where their mechanical properties are advantageous.

The resin used with asbestos reinforcement is normally phenolic.

Glass fabric reinforcement – gives greatly increased strength, rigidity and dimensional stability, with high resistance to heat. Water absorption is low and electrical and dielectric properties are good under both dry and humid conditions. Its cost is relatively high, particularly as it is normally used with epoxide or polyamide resins for maximum benefit. Glass fabric reinforcement may, however, be used with any of the resin systems to establish particular requirements.

Nylon fabric reinforcement – offers certain advantages for electrical grades because of its very low water absorption, even under long exposure to extreme humidity. Its mechanical applications are limited by a low service temperature and a tendency to cold flow under stress. Mechanical properties are only moderate and flexural strength is low.

Synthetic fibre reinforcement – other synthetic fibre fabrics may be used for specific duty laminates although they have limited practical use at present. None appears to offer advantages over existing reinforcement materials.

TABLE 1 – RESIN-REINFORCEMENT COMBINATIONS

Resin	Paper	Cotton Fabric	Asbestos Paper or Fabric	Asbestos Felt	Glass Fabric	Synthetic Fabric	Wood
Phenolic	○	○	○	○	○	○	○
Epoxide	○	○			○		
Modified epoxide		○					
Melamine	○	○	○		○		○
Polyimide					○		
Polyester					○		
Silicon			○		○		

Wood veneer reinforcement – is very attractive for decorative laminates although it may also have some structural applications. Its general properties can be considered similar to cotton fabric reinforcement.

Type of Resins

Phenolic resin is the 'traditional' and still the most widely used type for general purpose, mechanical and electrical laminates, with all types of reinforcement. Maximum operating temperature varies quite widely with the type of fibrous reinforcement used. Phenolic laminates have the particular advantage of good machining properties.

Epoxide resins are now very widely used for high performance grades of laminates for both mechanical and electrical duties. Cost and performance are directly related to the epoxide resin system used, as well as to the reinforcement. Thus, modified epoxide laminates with paper reinforcement have exceptionally high resistance to tracking as well as high service temperatures. Flame retardant resin systems are also available.

TABLE 2 – THERMAL RESISTANCE OF INDUSTRIAL LAMINATES

Resin	Reinforcement	Maximum Service Temperature	
		Continuous °C	Intermittent °C
Phenolic	Paper	85	100–105
	Cotton fabric	85	100–105
	Nylon fabric	70	90
	Asbestos	165	200
	Glass fabric	130	150
Epoxide	Glass fabric	130	150
Modified epoxide	Paper	110	130
	Cotton fabric	110	130
	Asbestos	200	230
	Glass fibre	130	150
Melamine	Paper	85	105
	Cotton fabric	85	105
	Glass fabric	140	180
Silicon	Glass fabric	250	300

TABLE 3 – TRACKING RESISTANCE OF ELECTRICAL GRADE LAMINATES

Resin	Reinforcement	Comparative Tracking Index (BS 3781)
Phenolic	Paper or cotton fabric (press finished surface)	80–100
	Paper or cotton fabric (machine finished surface)	100–120
Epoxide	Glass fabric	250–350
Silicon	Glass fabric	250–600
Modified epoxide	Paper	up to 800
Melamine	Glass fabric	600–900

TABLE 4 – SPECIFIC GRAVITY AND DENSITY OF TYPICAL LAMINATES

Resin	Reinforcement	Specific Gravity	Density lb/in^2
Phenolic	Paper	1.36	0.049
	Coarse weave cotton fabric	1.36	0.049
	Medium weave cotton fabric	1.36	0.049
	Fine weave cotton fabric	1.35	0.047
	Asbestos paper	1.82	0.066
	Asbestos fabric	1.64	0.059
	Cotton fabric/PTFE	1.50	0.054
Epoxide	Paper	1.36	0.049
	Fine weave cotton fabric	1.36	0.049
	Glass fibre	1.90	0.068
Modified epoxide	Paper	1.36	0.049
	Fine weave cotton fabric	1.36	0.049
Melamine	Glass fabric	1.86	0.067
Polyimide	Glass fabric	2.00	0.072

In general modified epoxide cotton fabric laminates offer the superior electrical properties of paper or glass fabric laminates together with the mechanical strength and good machining qualities of phenolic cotton fabric laminates.

Polyester resins are normally associated with glass fabric reinforcement and are the cheapest type of glass laminate. Again flame retardant resin systems are available. Mechanical and electrical properties are inferior to epoxide glass fabric laminates, however. Their particular advantage is that they can readily be rendered in self-colour.

Melamine resins have the advantage of providing a hard, scratch-resistant surface and superior electrical properties (particularly in arc-resistance and tracking resistance). They are thus particularly useful for their insulating properties in duties involving high humidity and/or a dirty environment. Again the resins can readily be rendered in self-colour and their hard surface makes them attractive as decorative/structural laminates, mechanical strength being comparable with polyester laminates.

Silicon resins have a particular application for the production of laminates with the highest resistance to heat (see Table 2). Electrical properties are also excellent and maintained over a wide temperature range. Their high cost otherwise limits their application and their mechanical properties are generally inferior to other resin system laminates.

Polyamide resins have superior mechanical, thermal and electrical properties under severe environmental conditions and with glass fabric reinforcement have a maximum service temperature of the same order as a silicon glass laminate with superior mechanical properties (*eg* two to three times the flexural strength at both room and elevated temperatures). This is a relatively new laminate material with considerable potential for high temperature services.

PTFE is used only a 'filler' or impregnant to impart self-lubricating low-friction characteristics to a mechanical grade of laminate for use as a dry bearing material. To realise these properties fully the PTFE must be uniformly disposed through the laminate so that regardless of the subsequently machined shape the PTFE is present at the bearing surface.

Graphite filling is also used as an alternative to PTFE for self-lubricating low-friction laminate materials.

TABLE 5 – GENERAL GUIDE TO TUFNOL INDUSTRIAL LAMINATES

Base material	Resin	TUFNOL Grade	Available forms							
			Sheet	Round tube	Square and rectangular tube	Round rod	Square and rectangular bar	Hexagonal bar	Angle	Channel
Paper	Phenolic	Heron Brand	●							
Paper	Phenolic	Kite Brand	●	●	●				●	●
Paper	Phenolic	Swan Brand				●	●	●		
Paper	Phenolic	Swan CP Brand	●							
Paper	Phenolic	Ketch Brand	●							
Paper	Phenolic	Grade 3P/46	●							
Asbestos paper	Phenolic	Asp Brand	●	●	●	●	●	●		●
Asbestos fabric	Phenolic	Adder Brand	●	●	●	●	●	●	●	●
Fine weave fabric	Phenolic	Carp Brand	●	●	●	●	●	●	●	●
Fine weave fabric	Phenolic	Vole Brand	●	●	●	●	●	●	●	●
Medium weave fabric	Phenolic	Grade 2F/14	●							
Medium weave fabric	Phenolic	Whale Brand	●	●	●	●	●	●	●	●
Medium weave fabric	Phenolic	Bear Brand	●	●	●	●	●	●	●	●
Medium weave fabric	Phenolic/PTFE	Grade 2F/3/PTFE	●	●						
Coarse weave fabric	Phenolic	Crow Brand	●	●	●	●	●	●	●	●
Fine weave fabric	Epoxide	Grade 6F/45	●	●	●	●	●	●		
Glass fabric	Epoxide	Grade 10G/40	●							
Glass fabric	Epoxide	Grade 10G/41	●							
Glass fabric	Epoxide	Grade 10G/42	●							

(continued)

TABLE 5 – GENERAL GUIDE TO TUFNOL INDUSTRIAL LAMINATES – *contd.*

							Guide to properties								
Tensile strength	Cross-breaking strength	Compressive strength (flatwise)	Impact strength	Electric strength (flatwise)	Electric strength (edgewise)	Insulation resistance	Dielectric properties	Water absorption	Dimensional stability	Heat resistance	Thermal insulation	Acid resistance	Alkali resistance	Machining properties	Punching properties
G	G	G	A	E	E	E	E	E	G	A	G	A	A	G	A
G	G	G	A	G	G	G	G	G	G	A	G	A	A	G	A
—	G	G	—	G	G	A	G	G	G	A	G	A	A	A	—
A	A	A	A	G	G	G	G	G	G	A	G	A	A	G	E
G	G	G	A	A	A	A	G	A	G	A	G	A	A	G	G
G	G	G	A	E	E	E	E	G	E	G	G	G	E	G	E
A	G	G	A	F	F	F	F	G	A	E	F	E	E	F	A
F	A	G	G	F	F	F	F	F	E	E	F	E	E	A	A
G	G	G	G	A	A	A	F	G	G	A	G	A	A	G	G
A	G	G	G	F	F	F	F	A	A	A	G	A	A	G	G
F	A	G	A	E	E	E	A	G	A	A	G	A	A	G	A
A	A	G	G	F	F	F	F	A	F	A	G	A	A	G	G
F	A	G	G	A	A	A	A	G	G	A	G	A	A	G	A
F	F	A	G	F	F	F	F	A	F	A	A	A	A	G	G
A	A	G	G	F	F	F	F	F	F	A	G	A	A	G	A
G	G	G	A	E	E	E	E	G	E	G	G	G	E	E	G
E	E	E	E	E	E	E	E	E	E	E	E	E	E	F	F
E	E	E	E	E	E	E	E	E	E	E	E	E	E	F	F
E	E	E	E	E	E	E	E	E	E	E	E	E	E	F	F

TABLE 6A – PHYSICAL PROPERTIES OF TUFNOL SHEET

Base material		Paper				
Resin		Phenolic				Epoxide
TUFNOL Grade		Heron Brand	Kite Brand	Swan CP Brand	Ketch Brand	Grade 3P/46
Ultimate tensile strength		17,000 *1,245*	20,200 *1,420*	13,150 *925*	21,500 *1,510*	20,000 *1,405*
Ultimate cross-breaking strength		22,850 *1,605*	25,100 *1,765*	18,400 *1,295*	20,550 *2,010*	27,000 *1,898*
Impact strength (notched)		0.375 *0.052*	0.40 *0.055*	0.37 *0.051*	0.55 *0.076*	0.37 *0.051*
Ultimate compressive strength (flatwise)		50,600 *3,555*	50,500 *3,550*	34,000 *2,390*	50,000 *3,515*	47,000 *3,305*
Ultimate compressive strength (edgewise)		31,500 *2,215*	29,300 *2,060*	19,000 *1,335*	29,000 *2,040*	30,000 *2,110*
Resistance to compression		0.97	1.05	2.25	1.60	1.2
Ultimate shear strength (flatwise)		15,850 *1,115*	16,200 *1,140*	12,300 *865*	15,950 *1,120*	16,000 *1,125*
Deformation (bending) at 90°C	½ in.	0.084	0.093	—	—	0.11
Water absorption	$^1/_{16}$ in.	8.9	14	19	32	3.0
	⅛ in.	9.7	18	23	42	3.5
	¼ in.	12.9	23	—	50	5.0
	½ in.	19.0	32	—	—	7.5
Electric strength in oil at 90°C (flatwise) Breakdown voltage following 1 minute at proof voltage shown in brackets	$^1/_{16}$ in	53 (19)	43 (19)	31 (19)	29 (19)	—
	⅛ in	80 (25)	56 (25)	42 (25)	40 (25)	—
	¼ in	113 (34)	78 (34)	—	50 (34)	—
Electric strength in oil at 90°C (edgewise) Breakdown voltage following 1 minute at proof voltage shown in brackets		74 (25)	62 (25)	65 (25)	52 (20)	—
Electric strength in oil at 90°C (flatwise) Step-by-step, as BS 2782, Method 201C	$^1/_{16}$ in.	— —	— —	— —	— —	550 *220*
	⅛ in.	— —	— —	— —	— —	425 *170*
	¼ in.	— —	— —	— —	— —	350 *140*
	½ in.	— —	— —	— —	— —	280 *112*
Electric strength in oil at 90°C (edgewise) Step-by-step, as BS 2782, Method 201E		—	—	—	—	70
Insulation resistance after 24 hours' immersion		5×10^{12}	7.5×10^{10}	2.4×10^{10}	8.0×10^{9}	5.0×10^{12}
Power factor at 1MHz		0.035	0.038	0.042	0.046	0.034
Permittivity at 1 MHz		4.75	5.07	4.44	5.45	4.3
Comparative tracking index		—	—	—	—	800
Specific gravity		1.36	1.36	1.36	1.36	1.36

(continued)

TABLE 6A – PHYSICAL PROPERTIES OF TUFNEL SHEET – *contd.*

Asbestos		Fabric						
Paper	Fabric							
Phenolic		Phenolic						Epoxide
Asp Brand	Adder Brand	Carp Brand	Vole Brand	Grade 2F/14	Whale Brand	Bear Brand	Crow Brand	Grade 6F/45
16,400	9,500	16, 150	12,850	9,500	12,050	8,700	10,400	18,000
1,155	*670*	*1,135*	*975*	*670*	*845*	*610*	*730*	*1,270*
23,300	18,500	27, 200	22,550	16,000	20,800	16,700	19,200	26,000
1,640	*1,300*	*1,910*	*1,585*	*1,125*	*1,465*	*1,175*	*1,350*	*1,820*
0.42	2.0	1.05	0.96	0.68	1.25	0.95	1.26	0.42
0.058	*0.275*	*0.145*	*0.133*	*0.094*	*0.173*	*0.131*	*0.174*	*0.058*
51,200	44,400	50, 800	47,400	42,000	45,500	42,800	45,400	42,000
3,600	*3,120*	*3,570*	*3,335*	*2,950*	*3,200*	*3,010*	*3,190*	*2,950*
23,900	23,900	28,400	28,500	31,000	28,800	30,500	26,900	28,000
1,680	*1,680*	*1,995*	*2,005*	*2,180*	*2,025*	*2,145*	*1,890*	*1,970*
0.805	1.40	1.42	1.50	—	1.51	1.44	1.56	1.5
12,300	16,250	17,300	16,200	—	16,000	15,100	15,850	15,000
865	*1,140*	*1,215*	*1,140*	—	*1,125*	*1,060*	*1,115*	*1,050*
—	—	0.137	—	—	—	—	—	0.095
—	—	20	30	21	38	—	38	10
34	40	23	33	28	40	16	41	12
45	60	31	41	38	46	22	49	14
65	75	47	53	52	63	35	65	17
—	—	17 (2)	7 (1)	25 (11.5)	5 (1)	—	5 (1)	—
—	—	23 (6)	9 (2)	35 (17)	9 (2)	21 (nil)	8 (2)	—
—	—	37 (6)	13 (2)	48 (nil)	11 (2)	32 (nil)	9 (2)	—
—	—	36 (15)	16 (1)	50 (20)	16 (1)	26 (nil)	14 (1)	—
—	—	—	—	—	—	—	—	500
—	—	—	—	—	—	—	—	*200*
—	—	—	—	—	—	—	—	350
—	—	—	—	—	—	—	—	*140*
—	—	—	—	—	—	—	—	250
—	—	—	—	—	—	—	—	*100*
—	—	—	—	—	—	—	—	180
—	—	—	—	—	—	—	—	*72*
		—	—	—	—	—	—	70
6.0×10^6	1.5×10^8	8.2×10^9	6.5×10^7	5.0×10^{10}	7.7×10^7	5.0×10^{10}	6.0×10^7	5.0×10^{12}
—	—	—	—	—	—	—	—	0.040
—	—	—	—	—	—	—	-	4.4
—	—	—	—	—	—	—	-	800
1.82	1.64	1.36	1.36	1.33	1.36	1.32	1.36	1.36

(continued)

TABLE 6A – PHYSICAL PROPERTIES OF TUFNOL SHEET – *contd.*

Base material		Fabric	Glass fabric			Units
Resin		PTFE/Phenolic	Epoxide			
TUFNOL Grade		Grade 2F/3/PTFE	Grade 10G/40	Grade 10G/41	Grade 10G/42	
Ultimate tensile strength		8,000 *560*				lb/in² bar
Ultimate cross-breaking strength		15,000 *1,050*				lb/in² bar
Impact strength (notched)		1.0 *0.138*				ft-lb *m-kg*
Ultimate compressive strength (flatwise)		30,600 *2,100*				lb/in² bar
Ultimate compressive strength (edgewise)		15,000 *1,050*				lb/sq.in. *kg/cm²*
Resistance to compression		2.8				%
Ultimate shear strength (flatwise)		7,500 *525*				lb/sq.in. *kg/cm²*
Deformation (bending) at 90°C	½ in.	—				inch
Water absorption	¹⁄₁₆ in.	—				mg
	⅛ in.	30				mg
	¼ in.	—				mg
	½ in.	50				mg
Electric strength in oil at 90°C (flatwise) Breakdown voltage following 1 minute at proof voltage shown in brackets	¹⁄₁₆ in	—				kV
	⅛ in	10 (2)				kV
	¼ in	—				kV
Electric strength in oil at 90°C (edgewise) Breakdown voltage following 1 minute at proof voltage shown in brackets		10 (1)				kV
Electric strength in oil at 90°C (flatwise) Step-by-step, as BS 2782, Method 201C	¹⁄₁₆ in.	— —				volt/mil *kV/cm*
	⅛ in.	— —				volt/mil *kV/cm*
	¼ in.	— —				volt/mil *kV/cm*
	½ in.	— —				volt/mil *kV/cm*
Electric strength in oil at 90°C (edgewise) Step-by-step, as BS 2782, Method 201E		—				kV
Insulation resistance after 24 hours' immersion		5.0×10^7				ohms
Power factor at 1MHz		—				—
Permittivity at 1 MHz		—				—
Comparative tracking index		—				—
Specific gravity		1.50				—

These laminates have been specially developed to meet the requirements of NEMA specifications LI 1-1971. Grades G10, FR4 and G11. Because of the different test methods required by these specifications, the test results relating to these materials are not included in this table. Comprehensive results of tests to NEMA specifications are available in the publication 'TUFNOL Epoxide Laminates'.

TABLE 6B – PHYSICAL PROPERTIES OF TUFNOL SHEET – SI UNITS

Base material	Paper				
Resin	Phenolic				Epoxide
TUFNOL Grade	Heron Brand	Kite Brand	Swan CP Brand	Ketch Brand	Grade 3P/46
Ultimate tensile strength	122	139	91	148	138
Ultimate cross-breaking strength	158	173	127	197	186
Impact strength (notched)	0.51	0.54	0.50	0.75	0.50
Ultimate compressive strength (flatwise)	349	349	234	345	324
Ultimate compressive strength (edgewise)	217	202	131	200	207
Resistance to compression	0.97	1.05	2.25	1.60	1.2
Ultimate shear strength (flatwise)	109	112	85	110	110
Deformation (bending) at 90°C 12.5 mm	2.13	2.36	—	—	2.79
Water absorption 1.6 mm	8.9	14	19	32	3.0
3.0 mm	9.7	18	23	42	3.5
6.0 mm	12.9	23	—	50	5.0
12.5 mm	19.0	32	—	—	7.5
Electric strength in oil at 90°C (flatwise) Breakdown voltage following 1 minute at proof voltage shown in brackets 1.6 mm	53 (19)	43 (19)	31 (19)	29 (19)	—
3.0 mm	80 (25)	56 (25)	42 (25)	40 (25)	—
6.0 mm	113 (34)	78 (34)	—	50 (34)	—
Electric strength in oil at 90°C (edgewise) Breakdown voltage following 1 minute at proof voltage shown in brackets	74 (25)	62 (25)	65 (25)	52 (20)	—
Electric strength in oil at 90°C (flatwise) Step-by-step, as BS 2782, Method 201C 1.6 mm	—	—	—	—	22
3.0 mm	—	—	—	—	17
6.0 mm	—	—	—	—	14
12.5 mm	—	—	—	—	11
Electric strength in oil at 90°C (edgewise) Step-by-step, as BS 2782, Method 201E	—	—	—	—	70
Insulation resistance after 24 hours' immersion	5×10^{12}	7.5×10^{10}	2.4×10^{10}	8.0×10^{9}	5.0×10^{12}
Power factor at 1MHz	0.035	0.038	0.042	0.046	0.034
Permittivity at 1 MHz	4.75	5.07	4.44	5.45	4.3
Comparative tracking index	—	—	—	—	800
Specific gravity	1.36	1.36	1.36	1.36	1.36

(continued)

TABLE 6B – PHYSICAL PROPERTIES OF TUFNOL SHEET – SI UNITS – *contd.*

Base material		Asbestos		Fabric		
		Paper	Fabric			
Resin		Phenolic		Phenolic		
TUFNOL Grade		Asp Brand	Adder Brand	Carp Brand	Vole Brand	Grade Brand
Ultimate tensile strength		113	66	111	95	66
Ultimate cross-breaking strength		161	128	188	155	110
Impact strength (notched)		0.57	2.71	1.42	1.30	0.92
Ultimate compressive strength (flatwise)		353	306	351	327	290
Ultimate compressive strength (edgewise)		165	165	196	197	214
Resistance to compression		0.805	1.40	1.42	1.50	—
Ultimate shear strength (flatwise)		85	112	119	112	—
Deformation (bending) at 90°C	12.5 mm	—	—	3.68	—	—
Water absorption	1.6 mm	—	—	20	30	21
	3.0 mm	34	40	23	33	28
	6.0 mm	45	60	31	41	38
	12.5 mm	65	75	47	53	52
Electric strength in oil at 90°C (flatwise) Breakdown voltage following 1 minute at proof voltage shown in brackets	1.6 mm	—	—	17 (2)	7 (1)	25 (11.5)
	3.0 mm	—	—	23 (6)	9 (2)	35 (17)
	6.0 mm	—	—	37 (6)	13 (2)	48 (nil)
Electric strength in oil at 90°C (edgewise) Breakdown voltage following 1 minute at proof voltage shown in brackets		—	—	36 (15)	16 (1)	50 (20)
Electric strength in oil at 90°C (flatwise) Step-by-step, as BS 2782, Method 201C	1.6 mm	—	—	—	—	—
	3.0 mm	—	—	—	—	—
	6.0 mm	—	—	—	—	—
	12.5 mm	—	—	—	—	—
Electric strength in oil at 90°C (edgewise) Step-by-step, as BS 2782, Method 201E		—	—	—	—	—
Insulation resistance after 24 hours' immersion		6.0×10^6	1.5×10^8	8.2×10^9	6.5×10^7	5.0×10^{10}
Power factor at 1MHz		—	—	—	—	—
Permittivity at 1 MHz		—	—	—	—	—
Comparative tracking index		—	—	—	—	—
Specific gravity		1.82	1.64	1.36	1.36	1.33

TABLE 6B – PHYSICAL PROPERTIES OF TUFNOL SHEET – S! UNITS – *contd.*

Fabric					Glass fabric			Units
Phenolic			Epoxide	PTFE/Phenolic	Epoxide			
Whale Brand	Bear Brand	Crow Brand	Grade 6F/45	Grade 2F/3/PTFE	Grade 10G/40	Grade 10G/41	Grade 10G/42	
83	60	72	124	55				MPa
143	115	132	179	103				MPa
1.69	1.29	1.71	0.57	1.35				J
314	295	313	290	207				MPa
199	210	185	193	103				MPa
!.51	1.44	1.56	1.5	2.8				%
110	104	109	103	52				MPa
—	—	—	2.41	—				mm
38	—	38	10	—				mg
40	16	41	12	30				mg
46	22	49	14	—				mg
63	35	65	17	50				mg
5 (1)	—	5 (1)	—	—				kV
9 (2)	21 (nil)	8 (2)	—	10 (2)				kV
11 (2)	32 (nil)	9 (2)	—	—				kV
16 (1)	26 (nil)	14 (1)	—	10 (1)				kV
—	—	—	20	—				MV/m
—	—	—	14	—				MV/m
—	—	—	10	—				MV/m
—	—	—	7	—				MV/m
—	—	—	70	—				kV
7.7×10^7	5.0×10^{10}	6.0×10^7	5.0×10^{12}	5.0×10^7				ohms
—	—	—	0.040	—				—
—	—	—	4.4	—				—
—	—	—	800	—				—
1.36	1.32	1.36	1.36	1.50				—

These laminates have been specially developed to meet the requirements of NEMA specifications LI 1-1971, Grades G10, FR4 and G11. Because of the different test methods required by these specifications, the test results relating to these materials are not included in this table. Comprehensive results of tests to NEMA specifications are available in the publication 'TUFNOL Epoxide Laminates'.

TABLE 7 – TUFNOL GRADES AND SOME TYPICAL APPLICATIONS

TUFNOL Grade		Kite Brand	Swan Brand	Carp Brand	Vole Brand	Whale Brand	Crow Brand	Heron Brand
Resin		Phenolic	Phenolic	Phenolic	Phenolic	Phenolic	Phenolic	Phenolic
Reinforcement		Paper	Paper	Fine Fabric	Fine Fabric	Medium Fabric	Coarse Fabric	Paper
Available forms								
Sheet		●		●	●	●	●	●
Round tube		●		●	●	●	●	
Square and rectangular tube		●		●	●	●	●	
Round rod			●	●	●	●	●	
Square and rectangular bar			●	●	●	●	●	
Hexagonal bar			●	●	●	●	●	
Angle		●		●	●	●	●	
Channel		●		●	●	●	●	
Typical Applications	**Properties**							
Meter unions	M					●		
Heavy duty rollers	M W					●	●	
Guide plates and star-wheels	M W					●	●	
Conveyor wheels	M W			●		●	●	
Pulleys and sheaves	M W			●		●	●	
Thrust washers	M W					●	●	
Rolling mill bearings	M W					●	●	
Cams and cam followers	M W			●		●	●	
Wear resistant components	M W			●	●	●	●	
Dry-bearings	M W							
Gears – coarse tooth	M W					●	●	
Gears – Medium tooth	M W				●	●		
Gears – Fine tooth	M W S			●				
Water lubricated bearings	M W S							

M   Mechanical strength and toughness   S   High dimensional stability
W   Good wear resistance and low friction   C   Chemical resistance   *(continued)*

In applications where several grades are shown it is usually the
environmental conditions that determine the ultimate choice.

TABLE 7 – TUFNOL GRADES AND SOME TYPICAL APPLICATIONS – *contd.*

Asp Brand	Adder Brand	Grade 2F/14	Bear Brand	Grade 2F/3/PTFE	Grade 2P/45	Grade 2P/46	Grade 6F/45	Grade 10G/40	Grade 10G/41	Grade 10G/42	Grade 6G/91
Phenolic	Phenolic	Phenolic	Phenolic	PTFE/ Phenolic	Epoxide	Epoxide	Epoxide	Epoxide	Epoxide	Epoxide	Poly-imide
Asbestos Paper	Asbestos Fabric	Medium Fabric	Medium Fabric	Medium Fabric	Paper	Paper	Fine Fabric	Glass Fabric	Glass Fabric	Glass Fabric	Glass Fabric
●	●	●	●	●	●	●	●	●	●	●	●
●	●		●	●			●				
●	●		●				●				
●	●		●				●				
●	●		●				●				
●	●		●				●				
★	●		●	★			★				
●	●		●				★				
				●							
			●								
	●										
	●		●	●			●				
				●							
	●										
							●				
			●								

★   Subject to special enquiry

(continued)

TABLE 7 – TUFNOL GRADES AND SOME TYPICAL APPLICATIONS – *contd.*

TUFNOL Grade		Kite Brand	Swan Brand	Carp Brand	Vole Brand	Whale Brand	Crow Brand	Heron Brand
Resin		Phenolic	Phenolic	Phenolic	Phenolic	Phenolic	Phenolic	Phenolic
Reinforcement		Paper	Paper	Fine Fabric	Fine Fabric	Medium Fabric	Coarse Fabric	Paper
Typical Applications	**Properties**							
Piston rings	M W S			●	●	●	●	
Slideways	M W S					●		
Jigs and fixtures	M W S				●	●	●	
Ball race cages	M W S			●	●			
Rotor blades (compressor)	M W S			●	●	●		
High precision machined parts	M W S			●				
Scraper blades	M W S C			●				
Seal rings	M W S C			●				
Pump sleeve bearings	M W S C							
Chemical resistance components	M S C							
Foodstuff applications	M S C E							
Electro chemical machining jigs	M E							
Pipeline insulation	M E	●				●	●	
Bolts, screws and washers	M E	●	●	●		●		
Welding jigs	M E	●				●	●	
Insulating handles	M E	●	●	●		●	●	
Low voltage insulation	E				●	●	●	
Terminal board and tag strips	E	●						●
Coil formers	E	●	●					
High voltage insulation	E	●	●					●
Track resistance	E							
Self extinguishing insulation	E							
High temperature components								

M   Mechanical strength and toughness 　　　 S   High dimensional stability
W   Good wear resistance and low friction 　　 C   Chemical resistance
E   Good electrical properties

(continued)

TABLE 7 – TUFNOL GRADES AND SOME TYPICAL APPLICATIONS – *contd.*

Asp Brand	Adder Brand	Grade 2F/14	Bear Brand	Grade 2F/3/PTFE	Grade 2P/45	Grade 2P/46	Grade 6F/45	Grade 10G/40	Grade 10G/41	Grade 10G/42	Grade 6G/91
Phenolic	Phenolic	Phenolic	Phenolic	PTFE/ Phenolic	Epoxide	Epoxide	Epoxide	Epoxide	Epoxide	Epoxide	Poly-imide
Asbestos Paper	Asbestos Fabric	Medium Fabric	Medium Fabric	Medium Fabric	Paper	Paper	Fine Fabric	Glass Fabric	Glass Fabric	Glass Fabric	Glass Fabric
	•		•	•			•				
			•	•							
			•								•
				•							
							•				
					•		•	•	○	•	•
	•		•				•				
			•	•			•				
	•		○	•							
	•		•		•	•	•	•	•	•	
							•				
							•	•			
											•
•	•										
					•	•		•	•	•	•
					○						
		•			•	•	•	•	•	•	•
					•	•	•				
										•	•
•	•				•	•	•	•	•	•	•

Sandwich mouldings*

THE VERSATILITY of the sandwich moulding process allows three different types of sandwich structure to be produced. A skin of polymer A may be used with a foamed core of polymer A. Alternatively a skin of polymer A may encapsulate a core of polymer B, and B can be either solid or foamed.

Cost benefits can be obtained both by the use of a foam core, which gives a stiffer structure for the same weight of polymer, and by combining different polymers with different relationships of stiffness to cost and/or strength to cost.

Tensile and comprehensive behaviour

Loads in the plane of a sandwich moulding will be carried by skin and core in proportion to their relative strengths and cross sectional areas. Where the core is foamed the tensile or compressive strength of the moulding is likely to be somewhat less than that of a moulding of the same cross sectional area with a solid core.

Compressive loads normal to the plane of a sandwich moulding with a foam core will cause greater indentation than will occur with a solid moulding. However, at the foam densities which occur in sandwich mouldings, puncture of the skin layer is unlikely to occur unless the load is very localised.

Behaviour under bending loads

It has been found possible to represent in a simple manner the behaviour of sandwich mouldings whose *skin and foam core are of the same basic material*. The formulas for solid plastics (*eg* for beams or flat plates) are used but with a reduced flexural modulus $\dfrac{E}{e}$

where E is the modulus of the skin material appropriate to the service conditions, and e is the expansion ratio of the sandwich moulding, defined as

$$\frac{\text{final thickness of moulding}}{\text{unexpanded thickness of moulding}}$$

The accurate calculation of flexural rigidity for a sandwich moulding with *dissimilar materials in skin and core* is complicated because the properties of each

*ICI Limited

material may have a different dependence on time and temperature and also because the skin and core layers will have different states of stress and strain.

Theoretical analyses have been attempted and the results have been published in various technical literature. Comparisons of experimental data for flexural rigidity relating to foamed and unfoamed sandwich structures of dissimilar materials have shown that added theoretical complexity has not necessarily resulted in added accuracy. Indeed it has been observed that the experimental data fit very well the simple empirical relationship:

$$E_{SM} = \frac{E_{skin} + E_{core}}{2e}$$

That is, the effective modulus of the sandwich structure is given by the average of the moduli of the skin and core materials – modified, as previously discussed, by the expansion ratio, e. The values of the moduli should be related to the appropriate conditions of loading time and temperature.

Table 1 lists typical densities and moduli for sandwich moulded materials hav-

TABLE 1

Material		Elastic			
Skin	Core	Specific Gravity	Modulus (GN/m²)	Thickness (mm)	Weight (kg)
Sandwich Moulded					
Polypropylene copolymer	Polypropylene copolymer	0.50	0.72	13.5	3.4
25% coupled glass filled polypropylene	Polypropylene copolymer	0.53	1.58	10.4	2.8
EVA	Polypropylene copolymer	0.51	0.38	16.7	4.3
Plasticised PVC	SAN	0.62	0.97	12.2	3.8
Acrylic	SAN	0.62	1.82	9.9	3.1
Nylon 66 (65% RH)	Nylon 66	0.63	0.72	13.5	4.3
ABS	ABS	0.59	1.33	11.0	3.2
ABS	SAN	0.59	1.62	10.3	3.0
High impact polystyrene	High impact polystyrene	0.59	1.22	11.3	3.3
'Noryl'	High impact polystyrene	0.59	1.31	11.1	3.3
'Noryl'	'Noryl'	0.59	1.39	10.8	3.2
Polycarbonate	SAN	0.62	1.62	10.3	3.2
Solid Thermoplastic					
Polypropylene copolymer		0.90	1.30	11.1	5.0
Acrylic		1.18	3.10	8.3	4.9
Nylon 66 (65% RH)		1.14	1.30	11.1	6.3
High impact polystyrene		1.07	2.20	9.3	5.0
SAN		1.08	3.45	8.0	4.3
ABS		1.05	2.40	9.0	4.7
'Noryl'		1.06	2.50	8.9	4.7
Polycarbonate		1.20	2.40	9.0	5.4
Other materials					
Steel		7.80	207·00	2.0	8.0
Aluminium		2.78	69.00	3.0	4.1
Beech		0.66	13.70	5.1	1.7
Chipboard		0.60	3.20	8.2	2.5
Plywood		0.63	4.20	7.5	2.4

ing an expansion ratio of 1:8:1. Comparative data for solid thermoplastics and other materials are also given and the tables shows how panels having equivalent flexural rigidity compare in weight and thickness.

The table also compares simple panels of area 0.5 m^2 assuming equal flexural rigidity for short loading times at 20°C. The approximate mid-point deflection for such a panel if square and supported along its edges is 0.1 mm per kg uniformly distributed load.

Impact loading

As with all articles made from thermoplastics, the performance of a sandwich moulded article under impact loading cannot be predicted from results on test specimens. Only tests on the final product can determine whether the article will be satisfactory in service. However, a useful guide to the relative performance of different materials and/or combinations of materials can be obtained from simple tests.

ICI have studied the impact strength of sandwich moulded specimens using a falling weight test (BS 2782 Method 306B with minor modifications). This method has been adopted because it most closely approximates what is liable to happen in service. Also, samples of different thickness can be tested easily. Sandwich mouldings with skin and core of the same and different materials have been studied, and the following general comments can be made.

(a) Thickness has a very significant effect on impact strength – see Figs 1 and 2.
(b) With foam core, fractures will occur at a lower energy level than with a solid specimen of the same thickness. The lowering of impact strength is proportionally much greater for tough polymers than for brittle ones.
(c) When two polymers are combined in a sandwich moulding the resistance to crack initiation and fracture is dominated by the behaviour of the skin polymer. A tough skin will improve the behaviour of a brittle core, but a tough core does not compensate for a brittle skin. This effect is important in practice because the corner or edge of a part will frequently take the impact, and this area of the moulding will be wholly or mainly composed of skin polymer.

The behaviour of sandwich moulded samples in falling weight tests is shown in Figs 1 and 2. Fig 1 shows the effects of a foam core where skin and core are the same polymer; in this instance a polypropylene copolymer. The same effect is shown with other tough polymers, such as ABS, high density polyethylene, and high impact polystyrene. In Fig 2 the manner in which a tough skin can improve the performance of a brittle core is shown.

Thermal properties

Thermal conductivity decreases in step with decrease of foam density, and foam cored sandwich mouldings give better thermal insulation than solid mouldings of the same weight per unit area.

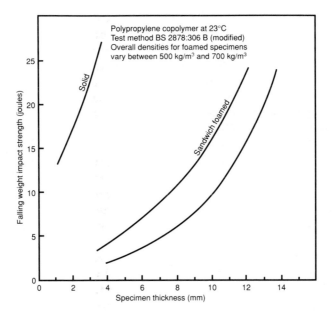

Fig. 1
Impact behaviour of polypropylene sandwich mouldings.

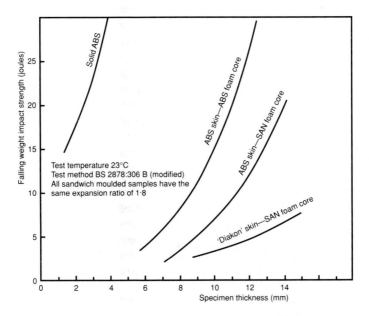

Fig. 2
*Impact behaviour of sandwich mouldings – effect of combining tough
and brittle polymers together.*

As a particular example, a polypropylene composite with an average density of 500 kg/m^3 (compared with 905 kg/m^3) was found to have a thermal conductivity of 0.14 W/m°K compared to 0.21 W/m°K for a solid moulding. This is from 3 to 5 times that of commercially available slabstock of synthetic foams such as polyurethane and expanded polystyrene.

The thermal expansion behaviour of a composite is strongly influenced by the ratio of elastic moduli of its components, as well as by the foam density and the volume ratio of the two components. When the same polymer is used throughout, the overall equilibrium behaviour would be identical with that of solid polymer.

Cellular plastics

THE FOLLOWING abbreviations are used;

ABS	Acrylonitrile butadiene styrene
EVA	Ethylene vinyl acetate
HDPE	High density polyethylene
LDPE	Low density polyethylene
PBT	Polybutylene terephthalate
PC	Polycarbonate
PES	Polyether sulphone
PH	Phenolic
PI	Polyimide
PIR	Polyisocyanurate
PMI	Polyacrylamide
PP	Polypropylene
PPO	Polyphenylene oxide
PS	Polystyrene
PUR	Polyurethane
PZ	Polyphosphazene

Introduction

Cellular plastics are formed by the controlled expansion and solidification of a liquid reacting mixture or polymer melt through the action of a blowing gas. Physical and chemical blowing agents and mechanical frothing are used to entrain the gas in the liquid system. Chemical blowing agents evolve the expansion gas through a chemical reaction (eg the reaction between water and isocyanate to give $CO_2(g)$ in polyurethane foams) or by breakdown at elevated temperatures (eg azodicarbonimide incorporated into thermoplastics which evolves $N_2(g)$). $N_2(g)$ can be incorporated physically into a melt at high pressure which subsequently expands at reduced pressure when the melt is shot into the mould. Once bubbles of gas have been nucleated they expand by a process of diffusion. The resulting cellular structure depends on the amount of blowing gas present and, in the case of closed mould techniques, the volume of the cavity. Large bubbles expand preferentially to small bubbles and unless constrained by the starting conditions they grow until they coalesce to form an approximately closed packed array of spheres. The subsequent behaviour depends in a complex way on the processing conditions, gravitational forces, viscosity and surface energy of the semi-liquid and the dynamics of the solidification process. The resulting cellular structure can range from open to closed cell.

Plastic foams exhibit a wide range of physical, mechanical and chemical properties – the most important factors being the starting polymer, formulation, density and cellular structure.

The modulus and strength of a homogeneous foam are given approximately by

$$E_f = E_p \, (\rho_f/\rho_p)^2$$

$$\sigma_f = \sigma_p \, (\rho_f/\rho_p)^2$$

where E_p and σ_p are the stiffness and strength of the matrix polymer, and ρ_p and ρ_f are the densities of the matrix polymer and the foam.

The classification of foams becomes more complex as new cellular materials become available. Differentiation between foams is based mainly on

(i)   Generic starting polymer, *eg* PUR, PP, PPO, PC
(ii) Cellular structure — open or closed cell and cell size
(iii) Physical properties — high or low density
                          — rigid or flexible
                          — high or low hysteresis
                          — combustion modified or non-flam
(iv) — Application — comfort cushioning
                    — energy management
                    — thermal insulation

Markets for cellular plastics have increased considerably over the last decade; main growth sectors being domestic, automotive, aerospace and industrial. Application include comfort cushioning, packaging, noise control, energy management, thermal insulation and engineering structural components. The growth has been driven by enhanced functional performance, integrated design, part reduction, and legislation (*eg* automobile weight reduction in the USA (CAFE) and drive past noise reduction in Europe (84/424/EEC)). Manufacturers benefit through on-line cost advantages, reduced inventories and minimum defects.

Cushion foams

This sector can be broadly divided into two parts: (i) furniture, and (ii) transportation (automotive, mass transport, aerospace). The main requirements of a cushion foam are comfort, functional life and safety, *eg* flammability. These are governed by the chemical, physical and mechanical characteristics of the foam.

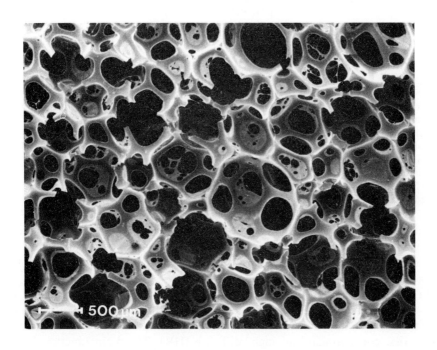

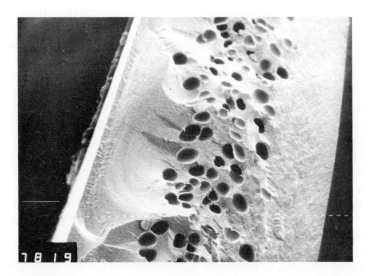

Electron micrographs of a flexible polyurethane cushion foam, density = 40 kg/m⁻³ and a moulded polypropylene integral skin engineering foam, density = 700 kg/m⁻³.

TABLE 1 – COMPARISON OF TYPICAL POLYURETHANE FURNITURE AND EUROPEAN AUTOMOTIVE CUSHION FOAM SYSTEMS (SOURCE: ICI POLYURETHANES)

| | Furniture | | Automotive MDI System | |
	MDI system	MDI/TDI system	Front	Rear
Core density (kg/m^3)	39	37	45	45
Compression set (%)	4	4	8	7
Tensile strength (kN/m^2)	80	65	120	110
Elongation at break (%)	110	125	130	140
Tear strength (N/m)	180	130	280	230
Indentation hardness (N) (Method B ISO 2439)	200	140	300	200
Jig dwell time (min)	3	5	3	3

Static comfort, as measured in terms of the SAG factor, is controlled by the gradient of the indentation force deflection (IFD) response post-yield. Foams with SAG factor greater than about 2.8 are generally rated as having good comfort. Typical values for furniture foams are 3.0.

In terms of volume, the domestic furniture market is the largest user of flexible cellular plastic. In this application polyurethane slabstock foams manufactured from TDI are the dominant material. The foams are divided into two categories: conventional and high resilience (HR). The main functional difference is comfort, HR foams having superior SAG factor. Recent changes in legislation concerning flammability have resulted in the introduction of combustion modified foams, such as CMHR, into the domestic UK market. Hardness values range from hypersoft, with density about 18 kg/m^3, to very firm with density about 80 kg/m^3. Recent changes in technology have resulted in water blown foams which are totally CFC free.

In the manufacture of moulded comfort cushioning MDI is continuing to replace TDI as the isocyanate of choice because it helps to enhance the work environment as well as offering physical and processing benefits. MDI provides rapid hardness build up and shorter demould times by the control of cellular structure is more difficult than with TDI-based foams.

The requirements of an automotive cushion foam are different to those of furniture foams in that they are intended to provide dynamic comfort (*eg* vibration isolation) as well as static comfort. The main design parameters are the resonance frequency of the person seated on the cushion and the amplification at resonance. These are governed by the stiffness and hysteresis of the cushion foam under the quiescent static compression. Moulded polyurethane foam is also dominant in this sector and there has been a trend on the part of vehicle manufacturers to move to full depth cushion. Recent development have centred on dual hardness, density reduction, and cell opening technology. Foam-in-fabric technology, where the liquid in a creaming state is dispensed onto a preformed trilaminate fabric cover in a closed mould, is being introduced for the manufacture of polyurethane automotive and office furniture seating.

Flammability is of concern in all applications but particularly so in public transportation and aerospace. Alternatives to polyurethane foams in this sector include melamine-based foam and silicone room temperature vulcanizate (RTV) foam. Although superior in terms of flammability these foams do not provide the some comfort qualities as polyurethane foams.

Energy management foams

Energy management foams (EMF) are employed in packaging, safety and damage mitigation situations. They are also referred to as energy absorbing (EA) foams. This description is misleading in that in most situations the purpose of the foam component is to ensure that during impact the force experienced by the packaged component does not exceed a specified value. Only in a few situations is the rebound energy (and hence energy absorption) of importance. As in the case of cushion foams, functional performance is related to IFD behaviour post yield. However, in this case what is wanted to minimize the reaction force (the ideal shock mitigator) is a foam that exhibits a zero gradient in the plateau region up to the highest possible deflection. Foams used for this purpose can be divided into two categories, low density rigid and semi-rigid.

The IFD behaviour of low density rigid foams demonstrates that they are better shock mitigators than semi-rigid foams. They also exhibit a lower level of rate and temperature dependence. A typical UK motorcyclist helmet employs a 25-mm-thick 60 kg/m^3 polystyrene shock mitigating liner. However, semi-rigid foams, unlike low density rigid foams, recover after impact with little permanent damage. Consequently semi-rigids are more suitable in applications demanding multiple impact performance, *eg* automotive.

EMF pads have been incorporated into car bumpers for many years. Initially the interest was strong but subsequently it declined in favour of mechanical shock mitigating systems. However, the development of low weight foam back-filled facias has renewed interest in foam technology. This includes pour-in-place semi-rigid MDI-based polyurethanes (typical density 100 kg/m^3) and expandable bead polypropylene back-moulded into polypropylene facias. The polypropylene based foam provides good shock mitigation performance over a wide range of temperatures at reduced density (55 kg/m^3).

**TABLE 2 – EXAMPLES OF FOAMS USED IN ENERGY
MANAGEMENT APPLICATIONS**

Low density rigid	Semi-rigid
Polyurethane	Polyurethane
Polisocyanurate	Polypropylene
Phenolic	Polyethylene
Polystyrene	Polyvinylchloride

Innovations have also taken place in the industrial products packaging sector. They include pour-in-place and moulded all-MDI flexible polyurethane foam systems for electrical products. Moulded low density rigid polyurethane and SR polyethylene units are being used for heavy industrial components, such as gear boxes. It is claimed that the systems give better protection with lower secondary packaging cost.

Acoustic foams

Acoustic foams are used to absorb, attenuate and insulate against noise. The physical processes involved are different for airborne noise and structure-borne noise so that foams selected for these applications have different properties. All foams absorb noise to some degree but they have to be engineered to achieve optimum performance. Acoustic energy is dissipated primarily through the viscous interaction between the air wave and the foam matrix and this is controlled by the air flow resistance R (Rayls) of the absorber. R is related to the thickness l of the absorber by $R = R1 \times l$ where $R1$ is the specific flow resistivity. It is generally accepted that optimum acoustic absorption is obtained when R is about 1.3kRayls, *ie* three times the acoustic impedance of free air. However, this is only a rough guideline since the absorption coefficient is strongly dependent on frequency and the frequency response is influenced by thickness. For a particular foam there is a thickness above which little benefit is obtained by increasing the thickness further. Absorption is also influenced to a small degree by the density, stiffness and damping of the foam. Consequently, material selection and absorber thickness can be designed to meet specific situations.

Plastic foams are replacing traditional materials such as felt and mineral fibre because they are easier to handle, reduce assembly costs and offer enhanced acoustic performance. Many types and grades of foam are employed. They include flexible polyurethane, melamine and crosslinked low density polyethylene. They are used for noise absorption in a wide range of applications such as appliances, transportation, noise shields, noise hoods, duct linings and acoustic components (*eg* loud speakers). Polyurethanes are the most widely used and are available with controlled flow resistance, with or without post-treatment (*eg* carbon impregnation, felted) and different levels of non-flammability. For hostile environments, such as the engine compartments of trucks, cars and small marine craft, plastic film and metal foil coverings prevent the ingress of contaminants. The surface covering affects the acoustic absorption characteristics.

Low density rigid polystyrene and polyurethane and crosslinked low density polyethylene boards are used for under floor applications in buildings to reduce the transmission of impact noise. Thermoformable grades are used to manufacture noise-absorbing automotive headliners and interior panels. Acoustic performance depends on whether the head liner is bonded or non-bonded to the vehicle roof. Bonded systems reduce structural vibration, and hance vibration generated noise, but non-bonded systems, with an air gap, have superior interior compartment acoustic absorption qualities.

TABLE 3 – PHYSICAL PROPERTIES AND NORMAL INCIDENCE ACOUSTIC ABSORPTION COEFFICIENTS OF ACOUSTIC FOAMS COMPARED WITH TRADITIONAL MATERIAL

		Foam		
Property		**PUR[a]**	**PUR[b]**	**Felt**
Core density (kg/m³)		30	80	155
Flow resistivity (kRayl/m)		14	39	0.53
Dynamic modulus (10^5Pa)		4.8	1.3	0.27
Loss factor		0.21	0.42	0.39
Acoustic absorption coefficient				
Thickness (mm)		26	20	10
Frequency (Hz):	250	0.05	0.08	0.05
	500	0.20	0.13	0.05
	1000	0.70	0.40	0.15
	2000	0.80	0.90	0.40
	4000	0.60	0.98	0.70

[a]PUR slabstock acoustic absorption foam with controlled level of reticulation.
[b]RIM PU automotive undercarpet foam.

In some applications, such as automotive carpet systems, the acoustic processes are noise insulation and attenuation. This involves the reduction of airborne noise transmission and the damping of panel motion of reduce vibration transmitted noise. Reaction injection moulded polyurethane and thermoformable polyurethane and thermoplastic foams are used to manufacture large area integrated systems. These consist of a foam isolating layer, heavy layer and carpet. The material requirements for this application are different to those of acoustic absorbers, but like absorbers the system can be optimised to meet the needs of a particular noise control situation.

Thermal insulation foams

By volume, the construction and refrigeration sectors are the largest users of low density rigid cellular plastics. For example, they account for about 50% of the MDI polyurethane market. The major application is thermal insulation but, as described above, low density rigid foams are employed in other areas, such as noise control and shock mitigation. The thermal conductivity of a foam Kf is given approximately by $K_f = K_g + K_s + K_r$ where K_g is the conductivity due to conduction through the gas, K_s by conduction through the cell elements and K_r by radiation. Heat transfer by convection is negligible with cell size less than about 3 mm. K_g is the main contributor to K_f and depends on the composition of the blowing gas and temperature. In the light of the Montreal Protocol concerning the industrial use of CFCs, which have very low thermal conductivity, considerable attention has been given to the development of alternative blowing agents for low density urethane rigid foams, such as HFA in Europe (HCFC–123 or

TABLE 4 – THERMAL INSULATION PROPERTIES OF RIGID LOW DENSITY THERMOSET AND THERMOPLASTIC FOAMS.

Property	Foam				
	PIR[a]	PH[a]	PS/PPO[b]	PUR[a]	PS[c]
Density (kg/m^3)	32	35	60	30	30
Maximum service temperature (°C)	140	120	120	110	100
Thermal conductivity (W/mK)	0.023	0.020	0.030	0.023	0.031
Compressive strength (kPa)	172	172	260	172	180
Tensile strength (kN/m^2)	207	210		240	
Thermal expansion (10^6)	40–80	40		40–80	60

Sources: [a]Kooltherm Insulation Products Ltd.
            [b]Shell Chemicals.
            [c]BASF plc.

–141b in North America). The increased conductivity of the gas K_g has been compensated by the reduction of K_s through improved foam cell structure control.

All closed cell foams offer some degree of thermal resistance and flexible foams such as polyethylene and polyurethane are used for pipe lagging. However, the majority of thermal insulation materials are low density rigid foams. They include polystyrene, polyurethane, polyisocyanurate and phenolic. A recent introduction is an expandable polystyrene (EPS)/polyphenylene oxide blend which has improved maximum operating temperature compared with EPS. Polyurethane, polyisocyanurate and phenolic foams are manufactured from liquid starting materials into buns which are then precision cut into boards or profiles. High density phenolic foam profiles, with enhanced compression strength, are available for high temperature pipe support systems. Face sandwich composites for building applications are manufactured by bonding cut slabstock to plaster board or by foaming on-line. EPS and EPS/PPO blends are available in the form of beads and are expanded using incorporated pentane. They are moulded into boards, profiles and containers, using steam heating. In-situ foaming of low density rigid polyurethane inside an inner and outer casing is used in the manufacture of cold boxes, freezers, refrigerators and thermally controlled vehicles. High density (200 kg/m^3), high compression strength (4200 kpa) phenolic foams (K=0.05 W/mK) are produced for thermal insulation applications and a new phenolic foam filled Aramid honeycomb has been developed for interior use in aircraft, marine and industrial transport.

Microcellular foams

Microcellular foams are used for shoe soles, shock and vibration control, seals and medical applications. They have small cell size, integral skin and densities range form about 250 kg/m^3 to 1000 kg/m^3. In the shoe sector expanded EVA,

PVC, PUR and rubber are competitive materials. Comfort performance is governed by damping, elasticity, and weight. Density reduction is a current development and in terms of lightness microcellular EVA is superior to PUR but has inferior mechanical properties. With outer soles, safety (*ie* non-slip) performance is a requirement. This is controlled by the coefficients of static and dynamic friction. For normal walking, values greater than about 0.3 are indicated for both starting and stopping. In sports shoe applications, higher values may be required but for normal walking too high values degrade comfort. Microcellular PUR sheets for gaskets, engine mounts, etc are produced by continuously casting a mechanical frothed reactive urethane mixture to the desired thickness. It is claimed that the control of density and thickness is superior to that of chemically blown foams.

Structural foams

Engineering structural foams (ESF) offer moulded components with a high strength/weight ratio, freedom from sink marks and the potential for integrated design, part reduction and simplified assembly. Several processing technologies exist. Low pressure moulding is used to manufacture parts with thickness greater than 5 mm from thermoplastics such as PP, HDPE, PS, ABS, PA, PPO and PC. High pressure machines are used for parts with wall thickness less than 5 mm. Low pressure mouldings normally show a surface swirl pattern. Superior quality surface finish can be achieved using multi-component (sandwich) moulding in which two polymers, one for the skin and one for the core are injected through a two channel nozzle. The skin completely covers the core material. Either the skin or the core can be foamed. Multi-component moulding is not feasible with all combinations of thermoplastics because of poor bonding between skin and core.

Integral skin polyurethane ESF, with density in the range 400–800 kg/m^3, is used for the manufacture of housing, furniture and fittings. It is formed by a liquid reaction injection moulding process. The low viscosity of the mixture allows swirl-free surfaces to be achieved and the low cavity pressures means that tooling and conversion costs are low. Stiffness and impact behaviour depend on the chemical system employed and part thickness and density.

Integral rigid PVC foams are extruded in the form of sheets, boards and profiles. The advantage of this material is that it feels, works and looks like wood.

Foam cored sandwich composites are used in structural applications. The tensile and impact properties are provided by the skin and the flexural stiffness is controlled by thickness of skin and core. The function of the foam core is to ensure that in flexure the separation of the skins remains constant, the skins do not slip and remain essentially planar. Recent developments in process technology include resin transfer moulding (RTM) in which oriented fibres, integrated foam box sections and moulded inserts are tailored to a particular application. RTM produces large area parts with class A finish in a shorter time than traditional sandwich composite technology.

TABLE 5 – PROPERTIES OF THERMOPLASTIC ESF WITH 20–30% DENSITY REDUCTION (Source: General Electric Plastics)

Property	Foam			
	PPO	PC	PBT[a]	PEI
Mechanical				
Density (kg/m^3)	850–900	950	1200	900
Tensile strength (MN/m^2)	22	40	91	40
Elongation to break (%)	5–10	5	1.3	4.5
Flexural strength (MPa)	45	65	120	100
Flexural modulus (GN/m^2)	1.8	2.0	5.8	2.9
Un-notched impact strength (Company test) (Nm)	25–35	35		
Thermal				
Heat deflection temperature (1.82MN/m^2) (°C)	83	132	145	175
Flammability (UL94)	V–O	V–O	HB	V–O
Thermal expansion (10^5/°C)	7–8	3–4		4
Electrical				
Dielectric strength (kV/mm)	7.5	10		10
Dielectric constant at 1 MHz	2.2	2.3	2.8	2.6
Dissipation factor at 1 MHz	0.003 9	0.008 1	0.011	0.000 8

[a] 33% Glass filled.

High heat foams

High performance foams are non-burning or highly non-flam foam used in critical applications where high operating temperatures or non-burning qualities are essential, *eg* aerospace, public transport systems, military vehicles and engine components. Different additives can be incorporated into foams to suppress ignition, smoke generation and combustion but the most effective way of producing a non-burning foam is to use a polymer base which is essentially non-burning. Examples of high performance foams are given in Table 6.

Polyphosphazene is a non-halogenated low density closed-cell elastomeric foam with a high level of heat resistance and low smoke generation. One of its main uses have been as extruded boards and sections for thermal and acoustic insulation, *eg* in the pressure hulls of submarines and pipe lagging in surface craft. The closed cell, nitrogen filled, structure impacts low thermal conductivity with a modest level of acoustic absorption. Fluoroelastomer is a sponge-like closed-cell foam designed for gaskets to seal against high temperature corrosive conditions. Its resilience allows it to conform to surface irregularities and provide a gas tight seal after many opening and closing cycles.

Open-cell melamine-based foam is manufactured by incorporating a low boiling agent into a melamine formaldehyde precondensate. Foams can be produced

**TABLE 6 – THERMAL PROPERTIES OF HIGH HEAT ELASTOMERIC,
FLEXIBLE AND RIGID FOAMS.**

Property	Foam			
	SI[a]	FE[b]	PES[c]	PI[d]
Nature	Flex	Elast	Rig	Flex
Density (kg/m^3)	50		50	9.6
Thermal conductivity (W/mK)	0.041		0.05	0.042
Maximum continuous operating temperature (°C)	200	204	226	260

Sources: [a]Dow Corning Corporation.
　　　　[b]Posiflex Ltd.
　　　　[c]BASF plc.
　　　　[d]IMI - Tech Corporation.

with properties ranging from flexible to rigid by controlling the degree of expansion. Thermomechanical post-treatment allows further refinement of the physical properties, such as density, hardness, airflow resistivity. Applications include thermal and acoustic insulation and automotive and aircraft seat cushions. Closed-cell and open-cell flexible silicone foams have also been under development for these applications. They are produced by the platinum catalysed room temperature vulcanisation (RTV) process in which hydrogen generated in the reaction expands the polymer. Density reduction can be achieved by balancing the rates of crosslinking and hydrogen generation during cure. Two component pour-in-place silicone foam sealing systems are available which prevent the movement of fire, toxic gases, smoke, etc. between rooms and floors in buildings.

Polymethacrylimide is a closed-cell rigid foam intended for lightweight structural components. It was developed to offer the aircraft industry a core material for sandwich composites which can be autoclave cured at temperatures and pressures up to 180°C and 0.7N/mm^2. The foam is manufactured in the form of blocks which are then precision cut and machined. A closed-cell rigid thermoplastic polyether sulphone is under development which is resistant to heat deformation and thermal ageing. It offers reliable fire behaviour, low smoke emission and resistance to solvents. The foam can be compacted, moulded and laminated in a single operation. Potential applications are as the core material in sandwich composites for aerospace, marine craft and railway rolling stock. Thermosetting polyimide foam is classified as a flexible foam. It is manufactured in the form of blocks by the microwave heating of a mixture of powdered polyimide prepolymer and blowing agent. The foam is easily worked and requires no special handling. It offers good thermal insulation and acoustic properties and has found extensive use in aerospace and marine applications.

Glass reinforced plastics (GRP)

THE GLASS reinforcement most commonly used in GRP is chopped strand mat, with woven fabric or rovings for stressed areas or special components. The resin type most commonly used is polyester, or for more specialised applications epoxide or silicone. Virtually any type of thermosetting resin can, of course, be used in the formulation of GRP mouldings, depending on compatibility requirements. It is the resin content which determines the thermal properties, chemical resistance, water absorption, and (largely) the electrical properties of the laminate. The glass reinforcement type and glass content largely determine the mechanical properties of the laminate.

Choice of reinforcement material for GRP laminates depends on the particular requirements of the job, cost, ease of working, and other factors. The following is a general guide.

1. *Glass fibre mat.* This is the usual material employed for GRP laminates since it is easy to shape (necessarily easiest to handle), costs considerably less than glass cloth, and with modern soluble binders is readily wetted out. Glass: resin ratio of a similar order to those achieved with cloth are readily possible.

2. *Woven glass cloths.* These generally produce the strongest laminates, but are not always easy to wet out thoroughly. The greatest tensile strength is obtained by using the thinnest cloths with the closest weave, which can aggravate the problem of wetting out. As a consequence, interlaminar adhesion can be poor locally. Inter-layer bonding is usually improved by using thicker cloths and fewer layers.

3. *Woven roving.* These cloths are obtainable in various thicknesses – *eg* in weights from about 1 oz/ft^2 upwards. They do not have the same strength as woven cloths but drape well and wet out fairly easily. Since they cost less than woven cloths they are often used a reinforcement layer or layers in conjunction with glass fibre mat, or as a simple means of adding bulk to increase stiffness.

4. *Rovings.* These again are normally used for reinforcing glass mat layers,

and also to add to bulk and improve stiffness. They are widely used for strength purposes in boat hull construction.

5. *Glass fibre tissue.* This was originally developed to hide the pattern of glass cloths when laid up in matched metal moulds, and also to be used as a 'cushioning' layer between the gel coat and the main reinforcement layers in any moulding or lay-up produced in a female mould. For general use it has two main applications:

(a) As a finishing layer on a chopped strand mat lay-up to cover up the coarse glass pattern of the mat before the gel coat.

(b) As a finish coat applied over a set gel coat. Again this will prevent any glass pattern appearing in the finished surface, and it also provides some reinforcement for the gel coat.

6. *Glass fibre tape.* This is simply woven cloth in tape form, and is the most convenient material to use where narrow strips of reinforcement are required for spiral bindings. It is a strong material, readily handled, and easy to use for small repair jobs, etc. Tape is correctly specified by width and thickness.

Preimpregnated glass fibre

Glass fibre cloth and mat are also produced uniformly impregnated with a resin-catalyst mix which can be activated by heat. In other words, the material is complete in itself for lay-up work which can be cured by heat and pressure. In its original state the material is slightly tacky, which can help in laying up. It is, however, only used for hot press mouldings and is generally unsuitable for amateur work because both heat and pressure are required to cure. Preimpregnated glass fibre is known variously as 'pregreg', sheet moulding material, etc.

Types of glass

Ordinary glass, or soda glass, can vary considerably in composition – mainly as regards its alkali content, which affects its resistance to moisture and chemical attack. A low alkali content implies a high resistance to moisture, whilst a high alkali content implies good resistance to attack by acids. Soda glass may have an alkali content of anything from 8% to 15%, so it does not have particularly good resistance to water, but is strongly resistant to attack by most acidic solutions and chemicals. It is thus well suited for general applications and is specifically known as 'A' grade glass.

Where extreme resistance to moisture is required, a boro-silicate glass is to be preferred. This has an alkali content of less than 0.8% and for all practical purposes can be regarded as completely unaffected by water. It is more expensive to produce than soda glass, and is generally known as 'E' or 'Electrical' grade glass. Glass fibre mat (or cloth) made from 'E' grade glass or boro-silicate glass would, therefore, normally be chosen for mouldings that are continually exposed to water, such as boat hulls, garden pool liners, etc, although this is not necessarily always followed in practice as 'E' grade glass cloth or mat, in fact, may be more difficult to obtain than 'A' grade. On the other hand 'E' grade glass would not be chosen for a job requiring maximum resistance to chemical attack.

Normal supplies of glass-cloths and mats are made from soda glass fibres or 'A' glass. These are suitable for all general applications as the alkali contents is usually reasonably controlled, offering a compromise between good water resistance and good resistance to acis and chemicals. Glass, as is well known, is a fairly inert substance, and in a GRP laminate it will normally be the resin which finally determines the resistance of the finished laminate to chemical attack.

Resin formulations may vary widely, according to the application involved, *eg*

Gel coat resins – formulated to give a tough, resilient film and commonly pigmented.
Thixotropic resins – to eliminate draining off vertical surfaces.
Flame-resistant resins
Light stabilised resins – for the production of translucent sheeting.
Chemical resistant resins – with enhanced chemical resistance, etc.

Fillers

Filler are not normally used in structural mouldings although they may be used when 'through colour' is desirable, or to enhance the properties of the moulding. For general use the usual choice is a mineral filler, such as precipitated chalk, China clay, talc or whitening. These materials are satisfactory provided they are pure and very finely powdered. They will then mix uniformly with the resin, with the effect of increasing its viscosity, *ie* tending to make it become more paste-like. The higher the proportion of filler the more stiff the resin becomes for working. This effect is even more marked with the modern surface-treated calcium carbonate fillers, particularly crystaline types, which have become the preferred choice. They improve the impact strength of the moulding, and produce a harder, more scratch-resistant surface. This type of filler is also used where mouldings in light or bright colours are required, without going to the expense of titanium dioxide or other similar pigments possessing a high refractive index.

Other types of filler which may be used include:

Slate powder – for general 'bulking' of the resin, with the addition of a grey colour.
Mica powder – for casting and potting applications.
Silica flour – to improve abrasion resistance.
Aluminium powder – for 'metallic' mouldings.
Metallic flakes – for opalescent metallic mouldings and colouring.
Iron powder or powdered silver – for moulding magnetic cores, etc.
Graphite – for moulding semi-conductors.
Asbestos powder – for self-extinguishing or flameproof mouldings.
Antimony oxide – for self-extinguishing mouldings.
Sawdust – for producing resin/filler mixes that set with the appearance of solid wood.
Vinyl flakes – for 'pearlescent' decorative effects.

TABLE 1 – COMPARISON OF GRP WITH TRADITIONAL MATERIALS*

Metal	Timber	Concrete
Compared with metals GRP has:	Compared with timber GRP has:	Compared with concrete GRP has:
1. Higher strength/weight ratio. 2. Easier and cheaper manufacture of complex shapes. 3. Good corrosion resistance. 4. Ability to incorporate self colours.	1. Much high strength. 2. Greatly increased strength/ weight ratio. 3. Improved dimensional stability 4. Better weathering properties. 5. Higher water resistance. 6. Ease of fabricating complete structures.	1. Considerable weight saving. 2. Excellent design potential. 3. Better chemical resistance. 4. Superior weathering properties. 5. Higher strength/weight ratio.

*Freeman Chemicals Ltd

Pigments

Pigments may be used for through colouring (*ie* added to the main resin mix), or more usually, for colouring the gel coat only. As a general rule, no more pigment should be added than is necessary to achieve the required degree of colour or opacity as most pigments tend to detract from, rather than enhance, the mechanical properties of the resin. Some types of conventional pigment may also have an inhibiting action on the resin and must be avoided. These include black pigments based on carbon.

Pigments are usually produced in the form of solids which must be ground into a fine paste with a suitable vehicle – in this case the resin. Ready-prepared pigments can be obtained in the form of resin-pastes, and this is the best way of using them, although it is not always possible to know whether the paste is

TABLE 2 – COMPARISON OF GRP WITH OTHER STRUCTURAL MATERIALS*

Materials	Specific gravity	Tensile strength lb/in² × 1000	Flexural strength lb/in² × 1000	Compressive strength lb/in² × 1000	Young's modulus lb/in² × 10⁻⁶	Maximum working temperature °C	Maximum working temperature °F	Maximum glass content possible (% by weight)
Random glass fibre/polyester resin	1.3–1.8	10–25	20–40	15–30	0.8–1.5	175	350	40
Bi-directional glass fibre/polyester resin	1.5–2.1	30–65	40–90	25–45	2.6–3.5	175	350	70
Uni-directional glass fibre/polyester resin	1.6–2.0	60–150	95–170	30–65	4.0–6.0	205	400	80
Uni-directional glass fibre/epoxide resin	1.6–2.0	80–200	100–250	45–70	4.0–8.5	260	500	85
Bi-directional glass fibre/epoxide resin	1.5–2.1	35–65	50–90	45–55	2.3–3.6	230	445	70
Mild steel	7.7–7.8	52–60	22.0†	34–38	30.0	400	750	n/a
Aluminium alloy	2.7–2.8	40–85	—	40–85	9.9–10.5	200	390	n/a
European oak	0.69	—	14.0	7.5	1.5	—		n/a
Basic redwood	0.48	—	12.1	6.5	1.5	—		n/a

*British Plastics Federation   † Yield stress

TABLE 3 – COMPARATIVE MATERIAL PERFORMANCE

Material	Glass content		Specific gravity	Density lb/in³	Thermal Coefficient of expansion in/°F × 10⁻⁶	Tensile Strength lb/in² × 10⁻³	Tensile Modulus lb/in² × 10⁻⁶	Compressive strength lb/in² × 10⁻³	Flexural strength lb/in² × 10⁻³
	% by weight	% by volume							
Fibreglass reinforced									
Uni-directional roving									
wound epoxide	60–90	40–80	1.7–2.2	0.061–0.079	2–6	80–250	4.00–9.00	45.0–70.0	100–270
extrusion polyester	50–75	32–59	1.6–2.0	0.057–0.072	3–8	60–170	3.00–6.00	30.0–70.0	100–180
Bi-directional fabric									
satin weave polyester	50–70	32–52	1.6–1.9	0.057–0.068	5–6	35–58	2.00–3.50	30.0–40.0	50–75
woven roving polyester	45–60	28–41	1.5–1.8	0.054–0.065	6–9	32–50	1.80–2.40	14.0–20.0	30–43
Random mat/spray up									
preform polyester									
hand lay up polyester	25–50	13–32	1.4–1.6	0.052–0.058	10–18	10–24	0.80–1.80	18.0–30.0	20–45
spray up polyester	25–40	14–24	1.4–1.5	0.052–0.054	12–20	9–20	0.80–1.60	18.0–25.0	20–40
Moulding compound									
dough moulding									
compound polyester	6–26	10–24	1.8–2.0	0.065–0.073	13–19	5–10	1.60–2.00	20.0–26.0	6–26
glass/nylon	20–40		1.3–1.5	0.048–0.055	7–18	17–28	0.80–2.00	15.0–24.0	21–40
Thermoplastics									
Nylon			1.14	0.042	55–63	11.5	0.20–0.40	5.0–13.0	8–15
Polyethylene (high density)			0.96	0.035	61–72	4.4	0.08–0.15	2.4	2–3
Polypropylene			0.90	0.033	55–111	5.7	0.15–0.25	8.5–10.0	4–5
Polystyrene (high impact)			1.08	0.040	22–56	6.5	0.50	16.0	3–5

TABLE 4 – APPROXIMATE WEIGHTS AND THICKNESS OF LAMINATES

Material	No of layers	Glass weight oz/ft²	Resin weight oz/ft²	Glass : Resin %	Laminate weight oz/ft²	Laminate weight oz/yd²	Approximate thickness in
1 oz glass mat	1	1.000	2.5	30	3.5	2.0	0.031
	2	2.000	4.0	30	6.0	3.9	0.063
	3	3.000	6.0	30	9.0	5.1	0.094
1.5 oz glass mat	1	1.500	3.5	30	5.0	2.8	0.047
	2	3.000	6.0	30	9.0	5.1	0.094
	3	4.500	9.0	30	13.5	7.6	0.141
2 oz glass mat	1	2.000	4.0	30	6.0	3.4	0.063
	2	4.000	8.0	30	12.0	6.8	0.125
	3	6.000	12.0	30	18.0	10.2	0.188
0.015 in scrim	1	1.000	2.0	30	3.0	1.7	0.035
	2	2.000	3.0	30	5.0	2.8	0.050
	4	4.000	5.5	40	9.5	5.4	0.085
Cloth (0.009 in close weave)	1	0.800	0.6	60	1.4	0.8	0.015
	2	1.600	1.2	60	2.8	1.6	0.030
	4	3.200	2.4	60	5.6	3.2	0.060
Cloth (0.010 in open weave)	1	0.800	1.2	40	2.0	1.1	0.020
	2	1.600	2.4	40	4.0	2.3	0.035
	4	3.200	3.6	45	6.8	3.8	0.055
Surface tissue	1	0.075	1.0	—	—	—	—

TABLE 5 – PROCESSING TECHNIQUES FOR GRP*

Method	Mould	Equipment	Reinforcement
Hand lay-up	Usually one-piece, open	Lambskin roller, brush corrugated roller	Chopped strand mats, woven glass fabric (roving)
Fibre spray-up	Usually one-piece, open	As for hand lay-up and in addition three-component spray equipment	Roving
Cold press	Two-piece, closed plastics die	Press	Chopped strand mat, glass fabric, roving prepregs
Matched metal die moulding	Two-piece, closed chromium-plated steel die	Press	Pressing in wet state: as for cold press moulding: flowable prepregs (resin mats) chopped strand mats. Nonflowable prepregs (chopped strand mats): glass fabric
Filament winding	Core consisting of one or more parts	Winding equipment with impregnation bath (similar to a lathe)	Roving, glass fabric (strips)
Centrifugal casting	Centrifuge		Chopped strand mats, glass fabric (roving)
	Hollow mould consisting of one or more parts	Drive unit for mould, special lance	
Die drawing	Heated dies	Impregnation bath haul-off and cutting device	Roving (glass fabric)
Continuous manufacture of sheet			Chopped strand mats, roving

* BASF

TABLE 6*

Property	Glass weight (sb = soluble binder) (ib = insoluble binder)		— Non-reinforced cured resin	30–35% Chopped strand mat (sb)
	Unit	Test method		
Density at 20°C	g/cm^3	DIN 53479	1.219	1.50
Tensile strength	kg/cm^2	DIN 53455	600	1200
Elongation at break	%	DIN 53455	2.0	3.5
Young's modulus in tension	kg/cm^2	DIN 53457	48 000	110 000
Flexural strength	kg/cm^2	DIN 53452	900	2000
Young's modulus in flexure	kg/cm^2		40 000	100 000
Compressive strength of 3–4 mm specimens	kg/cm^2	Fed Spec LP-406 b/1021.1	1600	1500
Impact strength/notched impact strength	cmkg/cm^2	DIN 53453	9/1.2	280/200
Glass transition temperature according to DIN 7724	°C	DIN 53445	125	
Thermal conductivity	kcal/mh · °C	DIN 52612, B11	0.16	0.19
Specific heat	k/cal/kg · °C	DIN 0335/7.56	0.30	0.27
Thermal coefficient of linear expansion	1/°C	VDE 0304 T.1/§ 4	125·10^{-6}	28·10^{-6}
Dielectric constant				
Dielectric constant ϵ_r dry		DIN 53483	3.4/3.4	3.8/3.8
at 50 Hz/a Hz wet			3.8/3.7	6.6/5.2
Dissipation factor, tan § dry		DIN 53483	0.005/0.006	0.007/0.006
at 50 Hz/1 Hz wet			0.009/0.009	0.140/0.070
Volume resistivity dry	Ω cm	DIN 53482	10^{16}	10^{16}
wet			10^{15}	10^{12}
Surface resistivity	Ω	DIN 53482	10^{14}	10^{13}
Dielectric strength	kV/mm	DIN 53481	35	37
Resistance to tracking		DIN 53480	KA3c	KA3c
Moisture absorption after 30 days at 23°C	%	analogous to MIL-F-9118A	0.60	0.80
Moisture absorption after 2 hours at 100°C	%	MIL-F-9118A	0.40	0.50

*BASF   (continued)

TABLE 6 – *contd.*

40–45% Chopped strand mat (ib)	50–55% Chopped strand mat (ib)	50–55% Chopped strand mat (sb) alternating with roving fabric	55–60% Roving fabric	60–70% Woven glass fabric	70–75% uni-directional roving fabric
1.58	1.70	1.70	1.76	1.88	1.98
1600	1800	3000	3200	3400	6300
3.3	2.4	3.5	3.7	3.4	2.7
124 000	160 000	170 000	200 000	270 000	330 000
2500	2700	3600	3200	4200	4200
120 000	150 000	160 000	190 000	250 000	300 000
1600	1700	1700	2000	2900	2300
300/240	350	360	370	380	390
	150		157	155	
0.21	0.23	0.23	0.24	0.25	0.37
0.26	0.25	0.25	0.24	0.23	0.22
$25 \cdot 10^{-6}$	$20 \cdot 10^{-6}$	$20 \cdot 10^{-6}$	$15 \cdot 10^{-6}$	$12 \cdot 10^{-6}$	$8 \cdot 10^{-6}$
4.2/4.2	3.9/3.9	4.7/4.6	5.1/5.0	4.6/4.6	5.1/5.0
8.0/7.0	5.0/4.7	8.0/7.0		15.0/11.0	
0.006/0.005	0.006/0.005	0.006/0.005	0.006/0.005	0.004/0.004	0.005/0.005
0.200/0.100	0.050/0.030	0.200/0.100		0.300/0.200	
10^{16}	10^{15}	10^{16}	10^{16}	10^{16}	10^{16}
10^{12}	10^{13}	10^{11}	10^{7}	10^{11}	10^{16}
10^{13}	10^{13}	10^{13}	10^{13}	10^{13}	10^{13}
	25			28	
KA3c	KA3c	KA3c	KA3c	KA3c	KA3c
1.20	1.40	0.80	0.40	0.45	0.50
0.50	0.60	0.40	0.30	0.40	0.40

highly concentrated or dilute. A maximum of 5% pigment should not be exceeded, which in the case of a highly concentrated paste usually means a maximum of 10% paste. With a diluted pigment paste, a 10% addition to the resin may not give the required depth of colour or opacity.

Laminates

GRP laminates can be considered competitive with metal, timber and concrete. Table 1 summarises the advantages GRP can offer in this respect. Comparative performance figures are given in Table 2. The figures for GRP cover a representative range, actual performances realised depending on the resin/glass ratio and reinforcing technique. Some comparisons with thermoplastics are drawn in Table 3.

As a general guide for hand lay-up moulding with polyester resins, Table 4 gives approximate weights and thicknesses yielded by typical average skilled labour. Hand lay-up is only one of the many processing techniques available (see Table 5), but it is still the most commonly employed.

Specific mechanical, thermal and electrical properties of GRP depend on the type of resin used, the type of reinforcement and the glass weight. The glass weight which can be accommodated also depends on the type of reinforcement – eg it will be lower for chopped strand mat than fabric or rovings.

Table 6 shows typical results which good workmanship can achieve using various types of glass reinforcement and different glass weights, with polyester resin. The performance of unreinforced resin is also included for comparison.

Filled thermoplastics

THE REINFORCEMENT of thermosets and thermoplastics by ceramic, metallic or polymeric fibres with large length to diameter ratios is a most successful method of improving the mechanical properties of the matrix polymer. Often, a marginal improvement in polymers is obtained by filling the polymer with particulate matter, as in talc reinforced polypropylene, but fibre reinforcement is the best way of improving stiffness, tensile strength, creep resistance and upper service temperature. The most commonly used fibres are of E-glass, but better properties are obtained with carbon fibres, which make the composite very expensive.

The dispersed fibres may be continuous or more usually short (0.125 to 0.5 mm) or long (~12 mm). Short fibre reinforced thermoplastics have been the main replacement for metals in applications combining lightness with strength, particularly when the product is made by injection moulding.

Reinforcement of polymer matrices by fibres marries the strength and stiffness of the fibres with the impact resistance of the plastics matrix. Alone, the fibres are very brittle and their strength and stiffness cannot be fully realised. The polymer matrix protects the fibres as well as transferring stress to them in loaded situations. This transfer of stress is improved if the matrix wets the fibre. Often this is not the case, and good interfacial adhesion is only obtained by coating the fibre with a coupling agent (often a silane coating is used).

Continuous fibre reinforcement gives the highest moduli because the fibres can be almost perfectly aligned in the direction along which the stress acts and their length is sufficient to maximise the transfer of stress from the matrix to the fibre. These composites cannot be injection moulded and do not lend themselves to high production rates.

For a continuous fibre reinforced polymer matrix the composite modules E_c is given under ideal conditions by the rule of mixtures equation;

$$E_c = \Phi_m \, E_m + \Phi_f \, E_f$$

where E_m and E_f are the matric and fibre moduli and Φ_m and Φ_f are the volume fractions of the matrix and fibre materials respectively. In practice, composite moduli are almost as high as E_c for continuous fibre reinforcement. Carbon fibres are generally used for these high performance composites.

Short and long fibre reinforced composites may be processed by injection moulding, but some fibre degradation does occur during the process. The mechanical properties due to long fibre reinforcement are better because after processing the overall fibre length is greater than in the short fibre reinforced composites.

In both cases the maximum tensile modulus is not realised because, in addition to an imperfect stress transfer from the matrix to the fibres as a result of their relative shortness, the alignment of the fibres in one direction induced by processing is not perfect.

Both amorphous and semi-crystalline polymers may be fibre reinforced, but the semi-crystalline matrices show a greater enhancement of heat deflection temperatures because the fibre reinforcement can bridge across the softening amorphous regions to join crystallites that have not yet melted at the high temperatures (see Table 1).

The main advantage of short and long fibre reinforced thermoplastics over continuous fibre composites is that products may be injection moulded. As the composite viscosity is higher than for matrix materials, a loading of over 40–50% fibre (w/w) is not generally used. Barrels of injection moulding machines and extruders must have hardened lines because the fibres are abrasive.

Non-fibrous reinforcements, because of their small length to diameter ratios, enhance matrix properties less. Such fibres include powders, microballoons and glass spheres. Powders may be used to impart hardness, rigidity and wear resistance and may have an effect on other properties. other solids such as molybdenum disulphide, graphite or polytetrafluoroethylene (PTFE) may be added as

TABLE 1 – INCREASES IN HEAT DEFLECTION TEMPERATURE AS A RESULT OF SHORT GLASS FIBRE REINFORCEMENT (30% W/W) (1)

Polymer	Heat deflection temperature at $1.8/MNm^{-2}/°C$	
	Actual	Enhancement
Crystalline		
Polyetherketone	358	+172
Polyetheretherketone	300	+145
Polyamide 66	248	+153
Polybutyleneterephthalate	200	+115
Amorphous		
Polyethersulphone	216	+15
Modified polyphenyleneoxide	145	+15
Polycarbonate	140	+10
ABS	100	+10

lubricants. Microballoons may be used to produce bulk without any marked increase in density, and glass spheres to improve the compressive strength of the resin. The latter may be coated in a conductive material to provide EMI shielding. Property values are given in Table 2 for short fibre glass reinforcement, except where stated. The tables should be used with care and design of products should be carried out using the manufacturers' literature.

Filled acrylonitrile-butadiene-styrene (ABS)

Glass filled ABS for injection moulding contains between 20 and 40% short glass fibres, giving a variation in density of 1.23–1.36 g/cc. The impact strength is greatly improved, with enhancement of flexural modulus and strength. The thermal expansion is halved, giving low mould shrinkage and some electrical properties are improved.

Filled fluorinated ethylene propylene (FEP)

The addition of 25% short glass reduces the impact strength but other mechanical properties are improved.

Filled polytetrafluoroethylene (PTFE)

Fillers are used to improve the mechanical, thermal and electrical properties of PTFE. Glass fibres improve mechanical properties, whereas bronze powder and graphite fillers are intended to improve bearing properties, for which PTFE is ideal because of its very low coefficient of friction. Bronze fillers improve thermal conductivity and are used in composites for high speed bearings. Graphite fillers give a very low coefficient of friction, good wear resistance and resistance to deformation.

Filled polyethylene-co-tetrafluoroethylene (PETFE)

This copolymer combines the good mechanical and thermal properties of PTFE with the capacity to be injection moulded. With glass fibre reinforcement this material may be used continuously to 200°C. It is used in pumps and gears as a bearing material.

Filled acetal copolymer (POM)

Glass fibre reinforcement in polyacetal (polyoxymethylene) and acetal copolymer marginally improves the impact strength but causes vast improvements to other properties, particularly heat deflection temperature. This is to be expected with a semi-crystalline material. The major applications are automotive, in aerosol containers and in domestic appliances. Over 80% of the applications are involved in the replacement of zinc, steel, aluminium and brass.

Mineral fillers and Teflon (PTFE) fibres are added to the matrix, the latter being used to promote improved lubrication for bearings.

TABLE 2 – PROPERTY VALUES FOR SHORT GLASS FIBRE REINFORCED, THERMOPLASTICS, EXCEPT WHERE STATED

Property	Unit	Standard				Fluoropolymers				Polyacetals		
		ASTM	DIN	ISO	Acrylonitrile-Butadiene-Styrene (ABS) 20-40%	Fluorinated Ethylene Propylene (FEP) 25%	Polytetrafluoroethylene (PTFE) 25%	Polytetrafluoroethylene 60% Bronze	Polyethylene-Co-Tetrafluoroethylene (PETFE) 25%	Acetal Copolymer 10-40%	Acetal Copolymer Mineral	Acetal Copolymer Teflon Fibre
Mechanical Properties												
Density	g/cm³	D-792	53479A	1183A	1.23-1.36	1.86	2.23	3.38	1.86	1.47-1.72	1.41	1.54-1.47
Elongation to break	%	D-638			—	8	280	80	9.0	14.0-2.0	40	15.0-10.0
Flexural strength	MPa	D-790	53452	178	6.0-9.0	45	—	—	6.5	87-141	53.1	55-61
Flexural modulus	GPa	D-790	53457	178	53-128	6.6	1.63	1.35	373	3.0-11.0	2.59	2.4-2.9
Notched Izod Impact (3.2 mm)	J/m	D-256	—	180/4A		47	117	—		35-40	53.4	37
Thermal Properties												
Maximum continuous service temperature	°C	UL	746		93	—	—	—	200	95-105	90	118-136
Heat deflection temperature	°C	D-648	53461	75	104	—	288	288	—	108-155	110	—
Flammability	—	(UL94)	—	—	—	—	—	—	—	HB	HB	HB
Coefficient of thermal expansion ($\times 10^{-5}$)	/°C	D-696	53752		—	—	8.0-4.0	—	1.6-3.2	11.0-1.0	8.5	12.1
Electrical												
Dielectric strength (3.2 mm)	kV/mm	D-149	—	IEC 250		2.26	2.6-2.9				19.7	15.8
Dielectric constant at 60 Hz		D-150	VDE0303/4	IEC 250		2-3			3.0		3.7	
" constant at 1 kHz									3.4	3.9		
" constant at 1 MHz		D-150					2.9			3.8		3.1-3.7
Dissipation factor at 60 Hz							0.0028-0.01		0.012			
" factor at 1 kHz										0.005		
" factor at 1 MHz												0.009-0.006

(continued)

* MD, Machine direction; TD, transverse direction.

TABLE 2 – PROPERTY VALUES FOR SHORT GLASS FIBRE REINFORCED THERMOPLASTICS, EXCEPT WHERE STATED – contd.

Property	Unit	Standard			Polyamides				Polyaryletherketones			Polycarbonate (PC) 10-40%
		ASTM	DIN	ISO	Polyamide 66 (DRY) 10-50%	Polyamide 66 (DRY) 30-50% Long Glass Fibre	Polyamide 66 (DRY) Mineral	Polyarylate 30% Arylon (DuPont)	Polyetheretherketone (PEEK) 20-30% Victrex (ICI)	Polyetherketone (PEK) 20-30% Victrex (ICI)	Polyketone (PK) 30-40% Kadel (Amoco)	
Mechanical Properties												
Density	g/cm³	D-792	53479A	1183A	1.21-1.57	1.37-1.57	1.35-1.5	1.44	1.43-1.49	1.44-1.53	1.47-1.55	1.25-1.52
Elongation to break	%	D-638	—	—	30.0	4.0	18.0-3.0	—	2.5-2.2	4.0	—	10.0-2.0
Flexural strength	MPa	D-790	53452	178	160-320	320-400	—	207	192-233	—	235-276	110-190
Flexural modulus	GPa	D-790	53457	178	4.2-12.0	10.0-15.8	4.4-5.9	8.4	6.66-10.31	7.5-9.0	1.00-12.4	3.7-10.0
Notched Izod Impact (3.2 mm)	J/m	D-256	—	180/4A	—	18-27	87-25	96	86-96	—	96-107	300-100
Thermal Properties												
Maximum continuous service temperature	°C	UL	746	—	80	80	80	171	250	260	326	≈130
Heat deflection temperature	°C	D-648	53461	—	232-250	255-261	95-215	—	285-315	352-358	—	140-150
Flammability	—	(UL94)	—	—	HB	HB	—	V-0	V-0	—	V-0	V-0,V-1
Coefficient of thermal expansion ($\times 10^{-5}$)	/°C	D-696	53752	—	4.0-3.0	4.0-3.0	—	2.7	2.4-2.2	3.7-1.7	—	4.0-2.0
Electrical												
Dielectric strength (3.2 mm)	kV/mm	D-149	VDE0303/4	IEC 250	—	—	—	15.9	—	—	14.8-16.8	18.20
Dielectric constant at 60 Hz		D-150	VDE0303/4		—	—	—	3.52	—	—	—	3.1-3.4
" constant at 1 kHz					—	4.0-4.1	4.0	3.52	—	3.5-3.8	3.74-4.01	—
" constant at 1 MHz					—	3.9-4.0	—	3.43	—	3.4-3.8	—	3.0-3.4
Dissipation factor at 60 Hz		D-150	VDE0303/4	IEC 250	—	0.007-0.006	—	0.003	—	—	—	—
" factor at 1 kHz					—	0.01	0.01	0.005	—	0.002-0.003	—	—
" factor at 1 MHz					—	—	—	0.018	—	0.005-0.004	—	—

(continued)

* MD, Machine direction; TD, transverse direction.

TABLE 2 – PROPERTY VALUES FOR SHORT GLASS FIBRE REINFORCED THERMOPLASTICS, EXCEPT WHERE STATED – contd.

Property	Unit	Standard			Polyesters (Thermoplastic)				Polyimides (Thermoplastic)				
		ASTM	DIN	ISO	Polybutyleneterephthal-ate (PBT) 10–50%	30% Long Glass Fibre PBT	Polyethyleneterephthal-ate (PET) 30–55%	Pet Mineral	Polyimide (PI) Envex (ERTA)	Polyimide (PI) Graphite	Polyamide (PAI) 30% Torlon (Amoco)	PAI Graphite Torlon (Amoco)	Polyetherimide (PEI) 10–40% Ultem (GEP)
Mechanical Properties													
Density	g/cm^3	D-792	53479A	1183A	1.52–1.71	1.69	1.56–1.80	—	1.8–1.9	1.40	1.61	1.46–1.50	1.34–1.61
Elongation to break	%	D-638		178	4.0–2.0	2.0–3.0	3.0–1.0	2.0	—	8.2	—	—	6.0–2.5
Flexural strength	MPa	D-790	53452	178	140–270	210–215	230–310	148	—	118.6	338	189–219	200–250
Flexural modulus	GPa	D-790	53457	178	—	—	6.0–17.9	9.6	13.7–16.5	4.0	11.7	6.3–7.3	4.5–11.7
Notched Izod Impact (3.2 mm)	J/m	D-256		180/4A	—	—	96	64	906	—	79	64–85	60–105
Thermal Properties													
Maximum continuous service temperature	°C	UL	746		140	145	—	—	260	288	282	279–280	≈170–180
Heat deflection temperature	°C	D-648	53461		200–222	217–220	220–229	215	348	—	—	—	207–213
Flammability		(UL94)			V-0, HB	HB,V-0	HB	HB	—	—	V-0	V-0	V-0
Coefficient of thermal expansion (×10^{-5})	/°C	D-696	53752		5.0–2.0	3.0	3.2–1.4	—	—	6.3	1.6	—	3.2–1.4
Electrical													
Dielectric strength (3.2 mm)	kV/mm	D-149	VDE0303/4	IEC 250	—	—	—	—	—	—	32.6	—	≈30
Dielectric constant at 60 Hz		D-150	VDE0303/4	IEC 250	—	—	—	—	—	—	—	—	—
" constant at 1 kHz					3.6–4.1	4.4	3.6–4.0	3.8	—	—	4.4	6.0–7.3	3.6–3.7
" constant at 1 MHz					3.4–3.9	3.8	3.5–3.9	3.7	—	—	—	—	—
Dissipation factor at 60 Hz		D-150	VDE0303/4	IEC 250	—	—	—	—	—	—	—	—	—
" factor at 1 kHz					0.003–0.0038	0.0024	0.005	0.008	—	—	0.022	0.037–0.059	0.0014
" factor at 1 MHz					0.017–0.013	0.0018	0.012–0.011	0.01	—	—	—	—	0.002

(continued)

* MD, Machine direction; TD, transverse direction.

TABLE 2 – PROPERTY VALUES FOR SHORT GLASS FIBRE REINFORCED THERMOPLASTICS, EXCEPT WHERE STATED – contd.

Property	Unit	Standard ASTM	Standard DIN	Standard ISO	Polyphenyleneoxide (Modified) (MPPO) Noryl (GEP) 10–30%	Polyphenylenesulphides — Polyarylenesulphide (PAS) PAS (Phillips)	Polyphenylenesulphides — Polyphenylenesulphide (SPS) Ryton (Phillips)	Polyphenylenesulphides — PPS Mineral Ryton (Phillips)	Polypropylene (PP) 30%	Polypropylene 30–60%	Polypropylene 25% Talc	Polystyrene (PS) 30%
Mechanical Properties												
Density	g/cm³	D-792	53479A	1183A	1.16–1.29	—	1.57–1.72	1.9–2.0	1.13	1.12–1.47	1.09	1.28–1.30
Elongation to break	%	D-638	—	178	6.0–3.0	—	1.0–0.8	0.7–0.5	—	1.0	—	0.75
Flexural strength	MPa	D-790	53452	178	120–160	179	134–193	—	—	117–138	—	136
Flexural modulus	GPa	D-790	53457	180/4A	4.0–9.0	9.6	9.6–14.0	14.0–17.0	—	5.5–11.0	—	7.6–8.3
Notched Izod Impact (3.2 mm)	J/m	D-256	—	—	100–80	—	53–69	31–53	200	373–479	50	160–200
Thermal Properties												
Maximum continuous service temperature	°C	UL	746	—	≃90	—	200–220	200–240	130	—	—	—
Heat deflection temperature	°C	D-648	53461	—	132–135	200	260	—	—	146–154	76	95
Flammability	—	(UL94)	—	—	V-1	—	V-0	—	—	—	—	—
Coefficient of thermal expansion (×10⁵)	/°C	D-696	53752	—	5.0–3.0	—	3.1–2.7	—	—	—	—	1.8
Electrical												
Dielectric strength (3.2 mm)	kV/mm	D-149	VDE0303/4	IEC 250	20–21	—	17	—	16–19	—	—	7–22
Dielectric constant at 60 Hz		D-150	VDE0303/4	IEC 250	3.0–3.2	—	—	—	2.2	—	—	—
" constant at 1 kHz					3.0	—	3.9–4.0	—	2.4	—	—	—
" constant at 1 MHz					—	—	3.8–3.9	—	—	—	—	—
Dissipation factor at 60 Hz		D-150			0.016–0.002	—	—	—	0.001	—	—	—
" factor at 1 kHz					—	—	0.002–0.003	—	0.003	—	—	—
" factor at 1 MHz					0.017–0.0021	—	0.0013–0.0014	—	—	—	—	—

(continued)

* MD, Machine direction; TD, transverse direction.

TABLE 2 – PROPERTY VALUES FOR SHORT GLASS FIBRE REINFORCED THERMOPLASTICS, EXCEPT WHERE STATED – contd.

Property	Unit	Standard ASTM	DIN	ISO	Polysulphone (PSO) 10-30% UDEL (Amoco)	Polyethersulphone (PES) 20-30% Victrex (ICI)	Polyethersulphone PES-HTA 20-30% Victrex (ICI)	Polyarylsulphone 10-30% Radel (Amoco)	Liquid crystal polymers Ekonol-E (Sumitomo) 40%	60% Long Glass Fibre Ekonol-E (sumitomo)
Mechanical Properties										
Density	g/cm³	D-792	53479A	1183A	1.33–1.49	1.51–1.6	1.50–1.58	1.45–1.58	1.7	1.92
Elongation to break	%	D-638			4.0–2.0	4.0–3.0	≈4.0	5.8–1.9	4.8	3.2
Flexural strength	MPa	D-790	53452	178	128–155		≈204	145–180	114	118
Flexural modulus	GPa	D-790	53457	178	3.79–7.59	5.8–8.4	6.4–8.5	4.06–8.07	11.3	16.0
Notched Izod Impact (3.2 mm)	J/m	D-256		180/4A	64–75	80		48–75	60	29
Thermal Properties										
Maximum continuous service temperature	°C	UL	746			180	200			
Heat deflection temperature	°C	D-648	53461		179–181	210–216	250–252	209–213	275	280
Flammability		(UL94)			V–0	V–0	V–0	V–0	V–0	V–0
Coefficient of thermal expansion (×10⁻⁵)	/°C	D-696	53752		3.6–1.9	3.6–2.3	2.6–2.3	3.6–3.1	1.3 MD 5.6 TD	1.9 MD 6.8 TD
Electrical										
Dielectric strength (3.2 mm)	kV/mm	D-149	VDE0303/4	IEC 250	19	16		17		
Dielectric constant at 60 Hz		D-150	VDE0303/4	IEC 250	3.3–3.5			3.68–4.11		
„ constant at 1 kHz					3.4–3.7	3.6		3.70–4.13	4.4	4.9
„ constant at 1 MHz								3.72–4.17	3.9	4.5
Dissipation factor at 60 Hz		D-150	VDE0303/4	IEC 250	0.001			0.015–0.0019		
„ factor at 1 kHz					0.005	0.002	0.001	0.00182	0.022	0.17
„ factor at 1 MHz								0.0072–0.0094	0.032	0.024

*MD, Machine direction; TD, transverse direction.

(continued)

TABLE 2 – PROPERTY VALUES FOR SHORT GLASS FIBRE REINFORCED THERMOPLASTICS, EXCEPT WHERE STATED – contd.

| Property | Unit | Standard | | | | Liquid crystal polymers | | | | |
		ASTM	DIN	ISO	Ultrason (BASF) 10–30%	Vectra (Hoechst-Celanese) A-130 to A-150 30–50%	E-130 30% Vectra (Hoechst-Celanese)	A-625 25% Graphite Vectra (Hoechst-Celanese)	Xydar (Amoco) 30–45%	Xydar (Amoco) Mineral
Mechanical Properties										
Density	g/cm³	D-792	53479A	1183A	1.31–1.45	1.6–1.8	1.6	1.5	1.62–1.76	1.84
Elongation to break	%	D-638			3.6–1.6	2.2–1.3	2.0	6.6	2.6–2.8	2.6
Flexural strength	MPa	D-790	53452	178		254–248	172	146	150–156	106
Flexural modulus	GPa	D-790	53457	178		15.0–20.0	13.8	10.3	11.6–13.1	10.35
Notched Izod Impact (3.2 mm)	J/m	D-256	—	180/4A	59–84	150–90	107	95	176–123	85
Thermal Properties										
Maximum continuous service temperature	°C	UL	746		160	200–220	≈250	200–220	≈217	≈217
Heat deflection temperature	°C	D-648	53461		≈179	230–232	270	180	252–254	244
Flammability	—	(UL94)			V-0	V-0	V-0	V-0	V-0	V-0
Coefficient of thermal expansion (×10⁻⁵)	/°C	D-696	53752		3.0–2.4	0.5 MD / 6.5 TD		0.6–4.5	0.47–0.9	1.15
Electrical										
Dielectric strength (3.2 mm)	kV/mm	D-149		IEC 250	>63	43–34	42	—	35.4	35.4
Dielectric constant at 60 Hz		D-150	VDE0303/4		3.2			—	4.2–4.1	3.8
„ constant at 1 kHz					3.2	4.1–4.3		—		
„ constant at 1 MHz					3.2	3.7–3.9	≈3.7	—	3.6–3.4	3.8
Dissipation factor at 60 Hz			D-150	IEC 250	0.0007			—	0.013–0.008	0.012
„ factor at 1 kHz					0.0008			—		
„ factor at 1 MHz					0.0014	≈0.006	≈0.006	—	0.033–0.031	0.034

* MD, Machine direction; TD, transverse direction.

Filled polyamides (PA)

There are numerous types and the values given are restricted to PA66, a popular engineering thermoplastic. This material may be reinforced with short or long glass fibres, or may be mineral filled. Glass fibres reinforcement reduces water absorption and hence gives better dimensional stability. Manufacturers' data should be sought to design for environments in which water vapour is present. Fibre reinforcement may be up to 50% (w/w). The main applications are automotive.

Filled polyarylates (PAR)

Like polyamides, polyarylates form a group of related materials, aromatic thermoplastic polyesters. They are UV and weather resistant but susceptible to environmental stress cracking. Applications include automotive and electrical hardware.

Filled polyaryletherketones

This group includes polyetheretherketone (PEEK), polyetherketone (PEK) and polyketone (PK), a fairly recent addition from AMOCO. These are expensive materials with excellent mechanical properties are high temperatures (~250°C). They are often used with glass or carbon short fibre reinforcement and PEEK is used as a matrix in continuous carbon fibre composites. Applications are in the automotive, aeronautical and aerospace fields, where good mechanical properties are required at high temperatures in harsh chemical environments.

Filled polycarbonate (PC)

The addition of 10–40% short glass fibres to PC gives a composite of improved mechanical and thermal properties, but reduced impact resistance, although it is still better than most untoughened polymers. Glass filled grades should be used when metal inserts are present in PC products.

Filled thermoplastic polyesters

This group includes polyethyleneterephthalate (PET) and polybutyleneterephthalate (PBT). The latter may be filled with 10–50% (w/w) short glass fibres or 30% long glass fibres; the former may be mineral filled or with 30–55% short glass fibres. The chemical resistance is good and PBT and in particular is competing with polyamides.

Filled thermoplastic polyimides

Included in this group is thermoplastic polyimide (PI), polyamideimide (PAI) and polyetherimide (PEI).

PI is difficult to process like the other thermoplastics and is often bought in as components. It has a very high continuous service temperature (260°C) and excellent chemical resistance. Graphite filled grades are intended for bearing applications and where low friction coefficients are required.

PAI is an injection mouldable polyimide with properties almost as good as those of PI. It is a good replacement for metals. Again graphite filled grades are available. PEI is a lower temperature matrix for glass fibre reinforcement and mineral reinforcement.

These materials are intended for high temperature use in harsh chemical environments. The most robust is PI and the least, but also the least expensive, is PEI.

Filled polyphenylenesulphides

Polyphenylenesulphide (PPS) is usually sold with a filler, mineral or glass fibre. It is highly crystalline and possesses excellent chemical resistance. Applications include corrosive fluid transport systems, pumps and impellors, but the greatest use is in electric and electronic components.

A newer related material, polyarylsulphide (PAS) has become available.

Filled polypropylene (PP)

Filling employed with PP includes short and long glass fibres and talc. These fillers improve properties and the latter gives rise to good flow characteristics and low mould shrinkage. Long glass fibre reinforcement improves properties further. If high temperatures are not encountered in service filled polypropylene grades may provide satisfactory and inexpensive products.

Filled polystyrene (PS)

Short glass fibre loadings of 30% give improvements to all mechanical properties and reduce mould shrinkage.

Filled polysulphones

This group of high temperature engineering thermoplastics includes polysulphone (PSO), polyethersulphone (PES) and polyarylsulphone. They are used in electrical and electronic applications. PES is used in circuit boards. The upper continuous service temperatures range from around 160 to 200°C.

Liquid crystal polymers

These materials, based on thermoplastic polyester copolymers, possess special properties due to the willingness of the molecules to align and remain in alignment after processing. This gives rise to a highly directional modulus and strength and low melt viscosity. These materials are often used with high percentages of glass fibre loading, made possible by the resins' low viscosity. The melts can form thin wall sections and provide high strength and stiffness in them. In general, these materials are intended for high temperature usage in harsh chemical environments. They have high flexural moduli and strengths. Copolymerisation of different polyester monomers gives a great array of products.

Carbon fibre reinforced plastics (CFRP)

AS A result of the high stiffness of carbon fibre, it is an excellent but expensive stiffening fibre when used with thermosets such as epoxy resins, or with the speciality thermoplastics. The resulting composite is much more competitive with structural metals. CFRP offers high stiffness with low density and strength comparability to or better than GRP.

There are three types of carbon fibre produced in the UK, usually in 7 μm diameter filaments.

High modulus   (Type 1 or HM-S)
High tensile   (Type 2 or HT-S)
Intermediate   (Type 3 or A-S)

Typical properties are shown in Table 1 for comparison with E-glass.

Carbon fibre yarns are difficult to handle directly for moulding and are prone to be damaged by abrasion. For practical use they are usually rendered in the form by prepreg yarns, sheets, tapes, etc. pre-impregnated with the resin system and partially cured. The resin system normally used in epoxide.

Typical forms of carbon filament prepregs are:

Unidirectional sheet – parallel carbon tows pre-impregnated with partially cured resin, yielding a carbon content of about 60% after moulding. Sheet sizes of up to 1220 mm × 920 mm with nominal moulded thicknesses from 0.05 mm.

TABLE 1

Fibre	Specific Gravity	Tensile Strength (GPa)	Tensile Modulus (GPa)	Specific Modulus (GPa)
Type 1	1.95	2.06	379	193
Type 2	1.75	2.55	241	138
Type 3	1.80	2.42	193	107
E-glass	2.55	2.76	68.9	27

Mat – parallel aligned short staple mat, with typical filament lengths of 3 mm pre-impregnated with partially cured resin. Sheet width is 290 mm with thicknesses from 0.25 mm to 0.5 mm. Moulded strength is about 80% of that of moulded unidirectional sheet.

Tows – single tows impregnated with resin for filament winding. Also available as sized tows with a protective coating of resin.

Unidirectional tape – continuous lengths pre-impregnated with partially cured resin in lengths of up to 304 m and widths from 11 mm to 510 mm.

TABLE 2 – TYPICAL PROPERTIES OF CFRP

Property	Composite	Value

Fatigue life
 	
Cantilever bend stress at 5000 rev/min
	Unidirectionally reinforced epoxy novolac	Type 1: Fatigue endurance limit 0.276 GPa
		
		Type 2: (i) cycling stress: 0.62 GPa
		    – 1.8×10^6 cycles to failure
		(ii) cycling stress 0.31 GPa
		    – 70×10^6 cycles without failure

Creep 3 point bend loading	Unidirectional Type 1 reinforced epoxy novolac	No measurable creep after 12 months at 60% of ultimate stress
Impact strength ASTM D256–56 (IZOD) narrow test specimens	Unidirectional Type 2 reinforced high strength cycloaliphatic epoxy	(i) notched 67.8 J/cm (ii) unnotched 81.3 J/cm
Coefficient of thermal expansion	Unidirectional reinforced epoxy novolac	(i) in fibre direction – 0.6×10^{-6} (ii) across fibre direction + 29×10^{-6}
Thermal conductivity	Unidirectional reinforced epoxy novolac	(i) in fibre direction 0.04 cal/cm/s/°C (ii) across fibre direction 0.002 cal/cm/s/°C
Wear rate Sliding against steel	PTFE reinforced with 30% chopped carbon fibre Unreinforced PTFE	1.6×10^{-10} cm^3/cm/kg 460×10^{-10} cm^3/cm/kg
Dielectric constant at 10^6 cycles	Unidirectional Type 1 reinforced epoxy	50–70

Source: Fothergill & Harvey Ltd.

Typical properties of CFRP mouldings are summarised in Table 2, whilst comparative tensile properties are given in Table 3. Properties of specific interest are:-

Impact strength – high and largely insensitive to notches or stress raisers.
Coefficient of thermal expansion – this is negative in the axial direction.
Fatigue properties – generally superior to GRP and relatively insensitive to surface condition or stress raisers.
Wear resistance – generally excellent, although single filaments or yarns are readily susceptible to damage by abrasion.
Frictional coefficient – generally of the same order as that of other thermoset laminates (typically 0.25).

**TABLE 3 – TENSILE AND SPECIFIC TENSILE PROPERTIES OF CFRP
AND METALS**

Property	Units	Type 1 High modulus carbon fibre reinforced*	Type 2 High strength carbon fibre reinforced*	E glass fibre reinforced*	Steel (Type S97)	Aluminium alloy (Type L65)	Titanium (DTD 5173)
Specific gravity		1.6	1.5	1.9	7.8	2.8	4.5
Tensile strength	GPa	0.813	1.102	1.310	0.999	0.462	0.930
Specific tensile strength	GPa	0.508	0.735	0.689	0.128	0.165	0.207
Tensile modulus	GPa	207	117	41	207	72	110
Specific tensile modulus	GPa	129	78	21.6	26.5	25.7	24.4

Source: Fothergill & Harvey Ltd.
* Unidirectional reinforcement, approximately 60% by volume fibre content.

Damping – generally excellent vibration damping characteristics, superior to aluminium and steel.

Chemical activity – generally inert, the chemical resistance of CFRP being the same as that of the resin system employed.

Bond compatibility – exceptionally good with special surface treatment of the fibres.

Fabrication methods

Pressure moulding is normally recommended for CFRP constructions, although filament winding can be used in the case of cylindrical or conical and similar shapes. Pressure moulding is normally done in matched metal dies or by autoclave moulding.

With *matched metal die moulding* pre-impregnated sheets are laid up in the appropriate shape and thickness and placed in a heated metal mould, typically at 170°C (340°F) for an epoxy resin prepreg. Pressure is then applied, usually about 700 kPa for 1 h. The moulding is then extracted from the mould and normally post-cured for several hours in an oven. This method is generally used for small or thick components, *eg* fan blades.

With *autoclave moulding* pre-impregnated sheets are placed in the mould and enclosed in a suitable, flexible bag material, from which air is evacuated. The assembly is placed in an autoclave, and pressure and heat are applied to mould and cure the component.

If required, post-curing can be carried out in a separate oven. This fabrication technique is normally used for large flat thin structures, *eg* aircraft wing skins.

Filament would components are produced by winding continuous pre-impregnated tow in a helical manner onto a suitably shaped mandrel, which may be hot or cold. Heat is then applied, usually by placing the wound mandrel in an oven, to cure the resin. The mandrel can then be removed using a variety of methods, to leave a carbon fibre reinforced filament wound component.

This fabrication method is used for tubes, pressure vessels and similar applications.

Hybrid composites

Hybrid composites employ two different types of reinforcement with the object of combining the potential advantages of each type. An attractive hybrid is a mixed glass fibre/carbon fibre composite aimed at combining the compliance of GRP with the stiffness of CFRP. Such a hybrid will modify the fracture characteristics of the material.

BFRP and TFRP

Other high modulus or high strength fibres may be used as reinforcement for thermosetting resins, Such composites are, at present, mainly of laboratory interest or used for highly specialised applications only. The commercial development of boron fibre reinforced plastics (BFRP) is largely inhibited by high cost, although its appearance preceded CFRP; the latter being developed as a less expensive alternative. Tungsten filament reinforced plastics (TFRP) and others remain in a similar category.

Carbon fibre reinforced thermoplastics

The main difficulty with carbon/epoxy composites is that the thermosetting resin cannot be reprocessed, making repair to products difficult. As a result continuous carbon reinforcement is used in some thermoplastics matrices at loadings of up to 68%.

The viscosity of the thermoplastics is much higher than the thermosets and impregnation of the carbon fibres is therefore more difficult. Sheets of carbon fibres in the thermoplastic matrix are obtained and thicker sheets are obtained by laminating sheets of different fibre or orientations, like an expensive plywood. The lay-ups are defined as follows. For instance 0_3 represents three plies joined together such that in each case the orientation of the fibres is parallel to a central axis. By contrast 90_3 represents three plies joined together with a common fibre orientation at right angles to the control axis. A set of brackets, ( ), represents a repeated sequence, with a suffix n to express the number of repeats. A symmetrical lay-up about the centre is denoted by s; *eg* $(0,90)_{3s}$ means $(0,90,0,90,0,90,90,0,90,0,90,0)$.

Symmetry about a control axis is essential to discourage warpage under thermal stresses. In the case of an odd number of plies, the central ply that is not repeated is denoted by a bar. Thus $(0,90,\overline{0})$ denotes $(0,90,0,90,0)$.

A popular form of lay-up called the quasi-isotropic laminate has roughly uniform properties in all directions in the plane of the sheet. This is often an advantage over the anisotropic properties provided by short and long fibre reinforced plastics. Common forms of the quasi-isotropic lay up are $(+45,90,-45,0)_{ns}$ and $(0,90,+45,-45)_{ns}$.

Table 4 shows property values for ICI's APC-2 – a continuous carbon fibre reinforced polyetheretherketone (68% w/w) with values for the carbon fibre and the matrix polymer.

TABLE 4 – THE MECHANICAL PROPERTIES OF APC-2 AS A FUNCTION OF LAY-UP, WITH VALUES FOR THE CARBON FIBRES USED AND PEEK

Material APC-2	Tensile Modulus (GNm^{-2})	Tensile Strength (MNm^{-2})
0°	134	2310
45°	19.2	300
90°	8.9	80
Carbon fibre	170–200	500–1000
Polyetheretherketone	3.6	62

Another matrix used in continuous carbon fibre reinforced composites is polyphenylsulphide. This is a highly crystalline material. Future advances in this type of composite will involve amorphous thermoplastics, particularly those such as polyethermide, which has excellent solvent resistance.

In many cases the continuous fibre composites do not have greatly elevated moduli or tensile strengths over the long fibre composites, the greatest advantage being in the dramatic improvement in Izod impact strength. As such, short and long carbon fibre reinforcement in thermoplastics is still the most prevalent means of improving the mechanical properties of these plastic matrices. Carbon fibre reinforcement is much more expensive than glass fibre reinforcement and is generally used in the more expensive thermoplastics. Table 5 shows property values for short carbon fibre reinforcement of thermoplastics matrices. All the composites are electrically conductive. For design purposes the manufacturers' data should be consulted.

Polyacetal and acetal copolymer (POM)

These materials are highly crystalline ($\sim$90%) and as such benefit greatly from fibre reinforcement. Short carbon fibres improve all mechanical properties, particularly heat deflection temperature.

Polyamides

These matrices are semi-crystalline and benefit from fibre reinforcement. Reinforced grades do not absorb moisture as readily as unreinforced grades, giving better dimensional stability. Values are given in Table 5 for PA66.

TABLE 5 – PROPERTY VALUES FOR SHORT CARBON FIBRE REINFORCEMENT

Property	Unit	ASTM	DIN	ISO	Acetal Copolymer 30%	Polyamide 66 (DRY) 30%	Polyetheretherketone 30% Victrex (ICI)	Polyetherketone 30% Victrex (ICI)	Polybutylene Terephthalate 30%	Polyamideimide 30% Torlon (Amoco)	Polyphenylenesulphide Ryton (Phillips)	Polysulphone UDEL (Amoco)	Polyethersulphone Victrex (ICI)	Liquid Crystal Polymer Vectra (Hoechst–Celanese)
Mechanical														
Density	g/cm^3	D-792	53479A	1183A	1.46	1.28	1.44	1.42	1.41	1.48	1.46	1.37	1.47	1.5
Elongation-to-break	%	D-638	53452	178	3.0	3.5	1.3	3.0	2.0	6.0	1.4	2.5	1.8	1.1
Flexural strength	MPa	D-790	53452	178			318			355				255
Flexural modulus	GPa	D-790	53457	178	9.3	21.0	13.0	17.9	16.0	19.9	17.2	14.1	15.2	24.0
Notched Izod Impact (3.2 mm)	J/m	D-256		180/4A	50	10	85	—	60	47	60	60	53	64
Thermal														
Maximum continuous service temperature	°C	UL	746	75			250	260				150	180	200–220
Heat deflection temperature	°C	D-648	53461		165	257	315	360	221	282	260	185	212	221
Flammability		(UL94)		—			V-0	V-0		V-0		V-0	V-0	V-0
Coefficient of thermal expansion ($\times 10^{-5}$)	/°C	D-696	53752	—			1.5	—		0.9		1.08	1.4	−0.2* MD 6.5 TD

*MD, machine direction; TD, transverse direction.

Three number rings in carbon fibre reinforced nylon for use in business machinery, calculators, cash registers.

Gear in 30% carbon fibre reinforced nylon 6.6.

Polyaryletherketones

These semi-crystalline materials benefit greatly from fibre reinforcement, and, as they are expensive polymers with excellent mechanical performance and resistance to chemicals, they are suitable matrices for carbon fibre reinforcement. The resulting composites are intended for aerospace applications, where high temperatures and harsh environments are involved. Data are given for PEEK and PEK in Table 5.

Thermoplastics polyesters

Carbon fibre reinforcement is used in polybutyleneterephthalate (PBT) matrices, where it improves mechanical and thermal properties. HDT is increased because the matrix is semi-crystalline.

Thermoplastic polyimides

This group includes carbon fibre reinforced polydamideimide (PAI). PAI is an injection mouldable imide, with properties slightly inferior to PI. It is good as a replacement for metals.

Polyphenylenesulphide (PPS)

Polyphenylenesulphide is highly crystalline with excellent chemical resistance. As such it is an ideal candidate for carbon fibre reinforcement and it may be used to high service temperatures (220–240°C).

Polysulphones

This group includes polysulphone (PSO) and polyethersulphone, which may be used at temperatures up to 160°C (PSO) and 180°C (PES). They are both amorphous polymers and carbon fibre reinforcement does not increase their HDTs much.

Liquid crystal polymers

These materials are based on polyester copolymers and a range of products is available with varying properties. Liquid crystal polymer molecules like to align and stay in alignment when processing ceases. This gives highly directional mechanical properties and low melt viscosity. The latter enables high percentages of fillers to be used without making injection moulding difficult. Carbon fibre reinforcement greatly enhances mechanical properties and the relatively low melt viscosities of the composites permits a reinforced thermoplastic to be injected into narrow wall sections. The moduli and strengths of these materials are very high.

SECTION 7

Materials for Specific Applications

ABRASIVES
BEARING METALS
DIAPHRAGM MATERIALS
HIGH TEMPERATURE STRUCTURAL MATERIALS
HONEYCOMB MATERIALS
INSULATING MATERIALS (THERMAL)
INSULATING MATERIALS (ACOUSTIC)
LOW DENSITY STRUCTURAL LAMINATES
LUBRICANTS
METALLIC WOOL
PERFORATED METALS
POWDERED METALS
PRINTING METALS
PYROMETRIC MATERIALS
REFRACTORY MATERIALS
SPRING MATERIALS
SURFACE FINISHING OF METALS
WOVEN WIRE

Abrasives

THE TWO chief types of abrasive used in the manufacture of grinding wheels are aluminium oxide and silicon carbide.

Aluminium oxide (Al_2O_3) is an extremely tough material with a sharp grain. It is particularly suitable for grinding and cutting high tensile strength materials such as alloy steels, high speed steels, high bronze, annealed malleable iron, etc. It can be rendered in various crystalline forms or different grades – *eg* crystals which fracture more readily and thus continually offer new and sharper cutting edges to the work.

Silicon carbide (SiC) is, in general, harder and sharper than aluminium oxide, but the crystals are not as tough (*ie* are more brittle). It is particularly effective for grinding materials which have a close, dense grain structure and for all materials with low tensile strength – *eg* aluminium, bronze, copper, cast iron, etc, and non-metallic materials. Again it can be produced in various grades, including special grades for grinding cemented carbide tools.

All types of abrasive are produced in a variety of grain sizes ranging from about 10 up to 600. Grain size is defined by the number of openings per lineal inch in the screen used to size the grains. Grain sizes can be related to the general description of abrasives as coarse, medium, fine, *eg*:

Coarse	–	10 to 24
Medium	–	30 to 60
Fine	–	70 to 180
Very fine	–	220 to 600

The most commonly used grain sizes are 40 to 80. Grain sizes 36 to 100 are used for commercial roughing and finishing. Grain sizes 120 to 600 are used for high and ultra-fine finishing.

The structure of a grinding wheel is defined as the relationship between the proportions of abrasive grain and bonding material, and the spaces or voids which may be separating them. The degree and extent of these voids is sometimes referred to as the porosity of the grinding wheel. All grinding wheels are

TABLE 1 – GRINDING WHEEL FAULTS RELATIVE TO CHOICE OF ABRASIVE

Fault	Cause	Action
Chatter marks	Wheel too hard	Use 1. Softer grade 2. Coarser grit 3. More open bond
Narrow and deep scratch marks	Wheel too coarse	Use finer grit
Varying depth shallow scratch marks	Wheel too soft	Use harder abrasive
Irregular marks	Loose grit	Re-dress or clean wheel
Grain marks	1. Wheel too soft 2. Wheel too coarse	Use harder grade wheel Select finer grit
Glazing of wheel	Wheel too hard	1. Increase speed or pressure 2. Re-dress wheel
Discolouration of work	Wheel too hard	Use softer wheel

porous to some degree, with a structure which can range from 'open' (*ie* a high percentage of voids) to dense or extra dense. The more open the structure, in general, the cooler the cutting qualities of a grinding wheel.

The strength of a grinding wheel is determined both by the type of bond material and the amount of bonding material present. For any given material the greater the proportion of bond the stronger the wheel and the greater its hardness. This characteristic is normally expressed as a *hardness grade* which may range from 'very soft' to 'very hard'.

The three principal types of bond used are:

(i)   *Vitrified* – the most used for general purpose grinding wheels. This type of bond can be expected to provide high stock removal and is not affected by water, oils, acids, etc, at normal temperatures.

(ii) *Resin bonds* – based on synthetic resins, are used when greater strength is required, *eg* for high speed wheels or thin wheels used for cutting-off. Resin bonded wheels cut coolly and can remove stock rapidly. Modern resin bonds embrace the organic *shellac* bonds originally used for light duty grinding when a high finish was required.

(iii) *Rubber bonds* – offer advantages when a good finish is required (*eg* on centreless grinding), and also for very thin wheels. Rubber bonds are strong and tough with a degree of resilience. Rubber bond wheels are normally used for wet cutting, using water or water with a small proportion of soluble oil as a coolant.

Bearing metals

THE ORIGINAL babbitt metal was a high-tin alloy containing 83 to 90% tin, patented in 1839. The name has subsequently become synonymous with white-metal linings for bearings although babbitt metals now embrace both tin-based and lead-based bearing alloys. These, in the main, have similar characteristics and are interchangeable to a large degree, provided specific attention is given to bearing design.

Antimony added to tin has a hardening effect, largely by introducing a proportion of hard grains embedded in the matrix. This harder constituent provides the resistance to wear and possibly the low coefficient of friction associated with anti-friction metals (due to the fact that the harder crystals stand out somewhat in relief). The softer matrix allows for some displacement of loading, even wear and low frictional heat.

A tin-antimony alloy is, however, brittle. The addition of copper serves to restore ductility and also toughens the alloy. Additional hardening elements may also be employed, in relatively small proportions, further to improve ductility or high temperature properties, or to act as grain refining agents, etc. These include cadmium, nickel, etc.

The addition of antimony to lead or tin in small proportions produces first a solid solution of homogeneous alloy. Once a certain proportion is reached the balance of the antimony will form subic crystals of antimony-tin (or antimony-lead), distributed throughout the solid solution matrix. Thus the micro-structure is quite distinct and readily recognised by microscopic examination. Both types of alloy, with innumerable variations, are manufactured.

The anti-friction properties of lead-based alloys are usually somewhat superior to those of tin-based alloys. Lead-based alloys, however, have a greater tendency to abrasive wear or segregation with increasing speeds and loads – hence their primary use is for low speed work and low to moderately loaded bearings. Tin-based alloys normally shrink less, run in more cleanly and have a more polished appearance. Tin-based alloys are therefore generally the preferred materials, although they are more costly than lead-based alloys.

TABLE 1 – TIN-BASED BEARING ALLOYS

Alloy No	\multicolumn{7}{c}{Composition (%)}							Brinell hardness		Tensile strength (ton/in^2)		Yield point (ton/in^2)	
	Sn	Cu	Sb	Pb	As	Fe	Ni	20 °C	100 °C	20 °C	100 °C	20 °C	100 °C
1	92.90	3.40	3.60	—	—	0.05	0.05	13.0	6.0	4.25	2.38	—	—
2	87.20	5.70	6.85	0.2	0.02	0.03	—	22.0	6.0	5.20	—	—	—
3	80.40	7.40	11.40	0.7	0.02	0.04	—	27.0	10.0	8.20	—	—	—
4	91.00	4.50	4.50	—	—	—	—	17.0	8.0	5.70		1.9	1.1
5	89.00	3.50	7.50	—	—	—	—	24.5	12.0	6.60	—	2.7	1.3
6	83.33	8.33	8.33	—	—	—	—	27.0	14.5	7.80	—	2.9	1.4
7	75.00	3.00	12.00	10.0	—	—	—	24.5	12.0	—	—	2.5	0.9
8	65.00	2.00	15.00	18.0	—	—	—	22.5	10.0	—	—	2.4	0.9

TABLE 2 – LEAD-TIN ANTIMONY BEARING ALLOYS

Alloy	Pb	Sn	Sb	Cu	Brinell hardness‡
1	90.0	5.00	5.00	—	15.2
2	89.4	0.11	9.90	0.12	14.5
3	85.0	10.00	5.00		15.1
4	84.7	0.09	14.80	0.19	15.0
5	84.6	5.00	9.90	0.06	19.0
6	82.0	2.05	15.70	0.12	17.5
7	80.0	5.00	15.00	—	16.7
8	79.4	5.20	14.90	0.14	20.0
9	75.0	10.90	14.50	0.11	22.5
10	63.7	19.80	14.60	1.50	21.0
11	65.0	30.00	5.00	—	15.1
12	85.0	5.00	10.00	—	23.2
13	80.0	10.00	10.00	—	25.4
14	75.0	15.00	10.00	—	26.4
15	70.0	20.00	10.00	—	23.2
16	65.0	25.00	10.00	—	24.0
17	75.0	10.00	15.00	—	31.0
18	70.0	15.00	15.00	—	32.0
19	65.0	20.00	15.00	—	27.6
20	75.0	5.00	20.00	—	26.7
21	70.0	10.00	20.00	—	37.0
22	65.0	15.00	20.00	—	35.6
23	70.0	5.00	25.00	—	27.8
24	65.0	10.00	25.00	—	33.6
25	65.0	5.00	30.00	—	28.8

‡ At room temperatures

TABLE 3 – LEAD-BASED BEARING ALLOYS

Alloy	Pb	Sn	Sb	Cu	As	Bi	
1	80.00	5.00	15.00	—	—	—	
2	90.00	—	10.00	—	—	—	
3	70.00–80.00	10.00–12.00	8.00–20.00	—	—	—	
4	77.25	6.00	16.00	—	—	0.25	
5	63.50	20.00	15.00	1.50	0.15	—	ASTM specification B, 23 to 26, Part 1, 1933
6	75.00	10.00	15.00	0.50 (max)	0.20	—	ASTM specification B, 23 to 26, Part 1, 1933
7	85.00	5.00	10.00	0.50 (max)	0.20	—	ASTM specification B, 23 to 26, Part 1, 1933
8	83.00	2.00	15.00	0.50 (max)	0.20	—	ASTM specification B, 23 to 26, Part 1, 1933
9	85.00		15.00	0.50 (max)	0.20	—	ASTM specification B, 23 to 26, Part 1, 1933
10	81.30–77.30	10.00–14.00	8.00	0.50 (max)	0.20 aluminium	—	AAR specification for journal bearing linings
11	98.63	0.69 calcium	0.62 sodium	0.04 lithium	0.02 ium	—	

Copper-based alloys

Copper-based alloys are stronger and harder than tin- or lead-based alloys and have greater resistance to wear. Their resistance to scoring is, however, generally inferior. Alloys of this type fall into four categories:

(i)   Copper-lead alloys
(ii) Lead-bronzes
(iii) Tin-bronzes
(iv) Aluminium bronzes

Copper-lead bearing alloys are based on a 20 to 50% lead content, together with up to 5% tin, iron and nickel. They are generally suitable for high speeds and high loads, particularly if plated with lead-tin or lead-indium. Tin, if present, does not usually exceed 2%, whilst a small amount of phosphorus may be employed as a de-oxidiser. The addition of tin to any large extent would tend to unduly harden and also to embrittle the mixture. The copper and the lead do not alloy beyond fractional percentages, and the resultant mixture is purely a mechanical one.

Strictly speaking, the description 'copper-lead' is more correctly applied to the softer alloys in the main group of lead-bronzes.

The best type of copper-lead mixture for use in bearings should possess (i) a high lead content, within certain limits; (ii) such lead to be finely distributed, and without visible segregation or formation of lead pools; (iii) the material should show a Brinell figure comparable with the average for a good quality tin-based babbitt.

Where copper-lead bearings are fitted, the observance of correct clearance and maintenance of first-rate lubrication are of paramount importance. Overheating does not bring about fusing of the liner (as with white-metal) thus preventing a seizure; on the contrary, if the bearings are allowed to become too hot, there is little to prevent them seizing. Should a copper-lead bearing become overheated the lead fuses and comes out from the bronze matrix and if the relief thus afforded is not sufficient to correct the distortion or tightness causing the excessive heating, the denuded copper matrix will seize on the shaft. The copper-lead material will not align or adjust itself to a journal as will white-metal.

The production of copper-lead bearings calls for specialised skill, different from that required in the manufacture of plain bearings in white-metal. Common defects in copper-lead linings are:

(i)   Poor adhesion between the liner and the steel shell.
(ii) Porosity in the copper-lead itself.
(iii) Segregation of the lead component, resulting in lead pools.

Proprietary copper-lead bearing alloys are designed to eliminate such faults.

Shafts with a minimum hardness of 300 Brinell are necessary with most lead-bronzes; in the case of some alloys, however, a minimum hardness of 200 to 220 Brinell is sufficient.

The following shows the effect of temperature on hardness in the case of copper-lead, compared with a typical tin-based white-metal:

	Brinell hardness number		
	Ambient temperature	at 100°C	at 130°C
Hoyt copper-lead mixture	31.0	26.0	23.8
Tin based white-metal	31.0–32.0	19.0–20.0	8.0–9.0

Copper-lead bearings cannot be re-bronzed in the same manner as they could be re-babbited; not only does bronzing call for the use of special furnaces and appliances beyond the capacity of the small workshop, but the heat necessary for the job would render the shells useless. This will be better understood when it is borne in mind that in manufacturing these bearings the shells are machined after bronzing and any scaling etc, due to heat, is removed.

When it is desired to replace the lead-bronze with white-metal the bearings should be bored out so as to allow for a finished thickness of white-metal of 0.05 mm (0.002 in). Clearance should be 0.01875 mm (0.00075 in) per inch diameter of shaft. Due allowance must, of course, be made for any re-grinding of the journals which has taken place.

Lead-bronze bearing alloys are another alternative to white-metal for lined bearings, with excellent casting and machining properties and high load capacity. They may range from soft alloys with 70/30 copper/lead, through low-lead bronze, to phosphor bronze. Tin may also be included, together with small proportions of zinc (see Table 4).

TABLE 4 – COPPER-LEAD BEARING ALLOYS

	Nominal composition (%)				Brinell hardness	Tensile strength (lb/in^2)	Maximum operating temp (°C)	Maximum load (lb/in^2)
	Cu	Sn	Pb	Zn				
Copper-lead	65	—	35	—	25	8000	175	2000
High-leaded tin-bronze	70	5	25	—	48	25 000	200+	3000+
Semi-plastic bronze	78	6	16	—	55	30 000	230	3000+
Leaded red brass	85	5	5	5	60	35 000	230	3500
Lead-bronze	83	7	7	3	60	35 000	230+	4000
Phosphor-bronze	80	10	10	—	63	35 000	230+	4000

Tin-bronzes are basically alloys of copper containing 5 to 20% tin, with a small phosphorus content. These alloys are generally described as phosphor-bronze (see *Copper and Copper Alloys* in Section 2). Phosphor-bronzes for bearing applications are of the high tin type, and may also contain lead. High tin phosphor-bronzes have good strength and hardness and excellent resistance to wear, combined with a low frictional coefficient. They are normally, however, used for lower speed applications where high loads are likely to be experienced. Low tin alloys, also containing zinc, are called *gun-metals*. These again are used for low speed bearing applications involving high loads (see also Table 5).

Aluminium-bronzes are essentially high copper alloys usually containing 5 to 10% aluminium, 4% iron (maximum) and up to 10% each of nickel and manganese – see also chapter on *Aluminium and Aluminium Alloys* (Section 2). They have excellent corrosion and wear resistance and high strength which is maintained at temperatures up to about 300°C. Their bearing properties are, however, rather more limited than those of the other accepted bearing metals, compatibility, conformability and embedding properties being poor. They are mainly employed for low speed, high load applications, particularly where corrosion resistance is important. Bearings of this type require particularly good lubrication.

Aluminium-tin alloys have more recently come to the fore (*ie* true aluminium-bronze, the so-called aluminium-bronzes (above) really being aluminium-brasses). Low tin alloys (5 to 7%) are relatively soft, although increased in strength and hardness by the addition of small proportions of copper. The main limitations of these alloys lie in their relatively high hardness, which means that shafts against which they run have to be specially hardened. The alloys are also sensitive to dirt particles in the lubricating oil and there is some difficulty in ob-

TABLE 5 – TYPICAL TIN-BRONZE BEARING ALLOYS

	Composition (%)				Brinell hardness	Tensile strength (lb/in^2)	Maximum operating temp (°C)	Maximum load (lb/in^2)
	Cu	Sn	Pb	Zn				
Gun-metal	88	10	–	2	65	45 000	260+	4000
Leaded gun-metal	88	10	2	–	70	40 000	260	4000+

taining a suitable direct bond in steel-backed bearings. Moreover, over the years, there have been occasional reports of seizures, leading to costly engine damage. Some of these may have arisen from the problems of maintaining adequate interference fit in the housings.

The introduction of aluminium-20% tin alloy bearings has been one of the most significant developments in bearing technology since the Second World War. The initial impetus for the development of this alloy came from the need

TABLE 6 – TYPICAL PROPERTIES OF ALUMINIUM-TIN BEARING ALLOY

Condition: Rolled and slightly annealed
Composition:
    Sn   17.5–22.5%
    Si   0.7% maximum
    Cu   0.7–1.3%
    Fe   0.7% maximum
    Mn   0.7% maximum
    Al   balance
(Elastic limit: 4–6 kg/mm^2 (5600–8500 lb/in^2))
Tensile strength: 11–13 kg/mm^2 (16 000–18 000 lb/in^2)
Elongation: 28–32%
Brinell hardness (HB 10/2.5): 28–35 kg/mm^2 (40 000–50 000 lb/in^2)
Cyclic bending fatigue strength: about 5.0 kg/mm^2 (7000 lb/in^2)
Modulus of elasticity: 6300 kg/mm^2 (9 $\times$ 10^6 lb/in^2)
Specific gravity: 3.12
Linear expansion coefficient, 20–200°C: 24 $\times$ 10^{-6}/°C

TABLE 7 – COMPATIBILITY OF BEARING ALLOYS
(STOP-START BUSHING TESTS)*

	Alloy			
	Tin based babbitt (SAE 12)	Lead based babbitt (SAE 15)	Al–20% Sn–1% Cu	Unplated Cu–Pb
Hardness HV 5 (typical)	27	17	35	40
Number of tests	8	2	31	14
Number of bushings surviving 144 cycles at following loads:				
(lb/in^2) (kg/mm^2)				
400 0.3	8	2	31	14
800 0.6	8	2	30	12
1200 0.8	8	2	27	12
1600 1.1	8	2	24	8
2000 1.4	8	2	16	2
Number of bushings surviving a further 1300 cycles at 2000 lb/in^2	8	2	14	1

* Tin Research Institute

TABLE 8 – POROUS METAL BEARINGS

	Surface speed (V ft/min)	Permissible load (P) (lb/in²)	Design PV limit	Maximum load capacity lb/in 2		Working temperature range (°C)
				Static (lb/in²)	Sliding (lb/in²)	
Porous bronze	Very low	4200	—	—	—	—
	25–50	2250	—	—	—	
	50–100	625	—	8500	4500	−15 – +65
	100–150	410	—	—	—	—
	150–200	300	—	—	—	—
	over 200	—	50 000	—	—	—
Porous iron	Very low	8000	—	—	—	—
	25–50	3000	—	—	—	—
	50–100	700	—	20 000	8000	−15 – +65
	100–150	400	—	—	—	—
	150–200	300	—	—	—	—
	over 200	—	50 000	—	—	—

to find a suitable material for use in the increasingly heavy duty encountered in main and connecting-rod bearings of small high performance petrol engines and high speed truck diesel and diesel-type engines. Aluminium-tin bearings have proved acceptable in service in such engines. They are usually employed in conjunction with unhardened steel or nodular iron crankshafts.

Aluminium-30% tin alloy bearings have also been produced for some years, and aluminium-40% tin alloys are currently under development in the UK and Japan.

For the particular applications of aluminium-tin bearings, virtually no other bearing material offers as good a compromise between fatigue strength, heat-

TABLE 9 – PHYSICAL PROPERTIES OF BEARING METALS

	Density (gm/cm³)	Brinell hardness	Tensile strength (lb/in²)	Modulus of elasticity (lb/in² × 10⁶)	Thermal conductivity (Btu/h/ft²/°F per ft thickness)	Coefficient of expansion (× 10⁶/°F)
Tin babbitt	7.4	25	11 000	7.6	32	13.0
Lead babbitt	10.1	21	10 000	4.2	14	14.0
Copper–lead	—	25+	8000	—	170	—
Leaded bronze	8.9	60	34 000	14.0	27	9.9
Tin–bronze	8.8	70	45 000	16.0	29	10.0
Aluminium alloy	2.9	35–45	22 000	10.3	119	13.5
Cadmium	8.6	35	—	8.0	53	16.6
Silver	10.5	25	23 000	11.0	238	10.9
Cast iron	7.2	180	35 000	23.0	30	5.7
Sintered iron	6.1	50	25 000	—	16	6.7
Sintered bronze	6.4	40	18 000	—	17	10.5

TABLE 10 – SPECIFIC CHARACTERISTICS OF BEARING METALS

Shaft	Minimum shaft hardness (Brinell)	Compat-ibility	Conform-ability and embedding properties	Corrosion resistance	Fatigue strength
Tin-based babbitt	150 or less	E	E	E	P
Lead-based babbitt	150 or less	E	E	F	P
Three-component bearings, babbitt surfaced	230 or less	E	G	G	G
Copper–lead	300	F	F	P	F
Copper–lead (overplated)	200	E	G	P	F
Lead–bronze	300	F	F	F	G
Tin–bronze	300–400	P	P	G	E
Aluminium alloy (6–7% tin)	200–300	F	F	E	E
Aluminium alloy (20% tin)	200	G	F	E	E
Cadmium-based	200–250	E	G	P	F
Silver (overplated)	300–400	G	F	E	E

and corrosion-resistance, embedding capability, conformability, hardness and surface properties (see Table 10).

Although aluminium-20% tin bearings have adequate strength to enable them to be used in the solid form, current automobile practice has been built up on bearing shells with a strong backing. The advantage of steel-backed bearings is that a thin bearing layer can be used on a strong steel support, giving higher fatigue strength and considerably reducing material costs. Differential thermal expansion problems between bearing and housing are also minimised.

Direct casting of the alloy onto the steel surface gives an unsatisfactory bond, due to the formation of a brittle inter-metallic aluminium-iron compound at the interface.

Roll-bonding has also proved unsatisfactory, due to the nature of the tin constituent, which meant that bonding had to be conducted at temperatures below the melting point of tin. However, under these conditions a strong bond on steel could not be obtained. The problem has been overcome by using an intermediate layer between the aluminium-tin alloy and the steel. A number of coating systems have been used, such as electroplated silver or platinum. However, a pure aluminium intermediate layer is usually adopted in the commercial process.

Cast iron

Cast iron is a long established material for plain bearings and bushings, the modern choice normally being a grey cast iron or Meehanite. Compatibility is very poor, calling for generous clearances and good alignment. Adequate lubrication with freedom from dust is also essential. Normally cast iron would be used only with steel shafts or sliding against steel, with a limiting pressure load of about 500 lb/in^2 and maximum rubbing speed of 130 ft/min.

The traditional use of cast iron for small bearings and bearing sleeves has largely been replaced by oil impregnated sintered metal bearings.

Cadmium

Cadmium bearing alloys have good mechanical properties, good high temperature characteristics and high fatigue resistance. They may be used on their own for lining steel bearings, or plated with indium to improve resistance to corrosion.

Cadmium-nickel alloys may contain 1.0 to 3.0% nickel, strength and hardness increasing with increasing nickel content. Hardness range is 33 to 48 Brinell. Alternative alloys are cadmium-silver and cadmium-copper-silver.

Cadmium lined bearings are particularly good with steel shafts, and are also suitable for heavy duty sliding bearings. Their relatively high cost generally limits their application to special designs.

Silver

Silver is an attractive material for bearings by virtue of its high resistance to fatigue. Speed and load ratings are quite high – 2000 ft/min and 4000 lb/in^2 respectively. It is, however, inferior to white-metal as regards embedding properties and also has a tendency to spot weld to the shaft in the event of localised failure of the lubricating film. To reduce the risk of seizure silver bearings may be overlaid with lead-indium or lead-tin.

Silver is also used as an alloying element in cadmium bearing alloys.

Porous metal bearings

Sintered metals are widely used to produce porous bearings which may be impregnated with lubricant and/or graphite or molybdenum disulphide. Metal powders most used are iron, bronze or leaded bronzes. Porosity may range from 10 to 35%, this volume being available to retain lubricant. Porous metal bearings may also be filled with a low friction thermoplastic for 'dry' running (*eg* PTFE).

Hard surface bearings

For particular applications hardened shafts rubbing on hardened bearing surfaces may show advantages – *eg* chrome coated (chrome diffused) shafts rubbing on hard chrome plated bearing surfaces. Such combinations of hard faced materials may show exceptional load carrying capacity with low frictional coefficients, when suitably lubricated. (See also *Non-metallic bearings* and *Jewel bearings*).

Diaphragm materials

MATERIALS FOR diaphragms include elastomeric sheet materials, rubber coated fabrics, fabric reinforced elastomeric sheet and thin metal sheet. The latter is largely confined to specialised applications (*ie* high pressures and/or temperatures) and is excluded from this chapter.

Of the fabric materials, cotton or rayon may be used for light duties where strength and heat resistance are not important factors. Nylon is the logical choice for greater strength with lightness, and also has greater displacement sensitivity. Woven glass fibre may be used where greater resistance to heat is required combined with great strength, or asbestos fibre for maximum heat resistance. Other synthetic fibres such as polyester and polyacrylonitrile may also be considered in this category.

Comparative properties of typical base fabrics are summarised in Table 1. Chemical resistance is included since although this is apparently not significant with a coated or impregnated fabric, in practice the base material may be exposed at the edges when wet, or at the periphery of clearance holes for fixing bolts.

Chemical and thermal properties of elastomeric materials are essentially the same whether used in sheet form, or as coatings or impregnants. The descriptions 'coating' or 'impregnant' are virtually synonymous in practice, since such a construction almost invariably yields the equivalent of a fabric-reinforced elastomer sheet material. The mechanical properties of the material are modified by the type of reinforcement used. The principal elastomeric materials are described under separate headings with more specific properties summarised in Table 2.

Natural rubbers

Natural rubbers retain their elasticity at lower temperatures than nearly all synthetic elastomers and may therefore be chosen for low temperature duties. They are also resistant to water and vegetable oils (*eg* castor), but are attacked by mineral oils and petroleum products, and many solvents (except alcohols and ketones). Maximum service temperature for natural rubbers is about 60°C.

TABLE 1 – BASE FABRICS FOR DIAPHRAGMS

Property	Cotton	Rayon	Polyester	Nylon	Glass
Adhesion to coatings	Excellent	Very good	Fair to very good	Good	Fair to Very good
Tensile strength	Medium	High	High	High	Very high
Elongation	Low	Medium	Medium	Medium	Very low
Elastic recovery	Good	Medium	Excellent	Good	Very good
Resistance to:					
Ageing resistance	Good	Very good	Good	Good	Excellent
Moisture	Fair	Poor	Excellent	Excellent	Excellent
Heat	Very good	Very good	Very good	Very good	Excellent
Mildew	Only fair	Good	Good	Good	Excellent
Strong acids	Poor	Poor	Fair to good	Poor	Poor
Weak acids	Poor	Poor	Good	Fair	Fair
Strong alkali	Excellent	Good	Fair	Good	Poor
Weak alkali	Excellent	Good	Fair to good	Good	Fair
Organic solvents	Excellent	Excellent	Good	Good	Excellent
Flammability	Poor	Poor	Fair	Fair	Excellent

‡Dupont Registered Trademark

TABLE 2 – COMPARATIVE PROPERTIES OF ELASTOMERS FOR DIAPHRAGMS

Property	Natural rubber	SBR	Butyl	Nitrile	Chloroprene	Fluorocarbon (Viton)	PTFE	Silicone	CSM (Hypalon)	Polysulphide	Polyurethane
Tensile strength	H	L	M	L	H	MH	MH	L	M	L	VH
Tear strength	VG	G	M	F	G	G	M	P	G	P	VH
Abrasion resistance	VH	VG	G	G	VG	VG	P	P	VG	P	VH
Compression set	VG	VG	F	M	M	G	P	F	F	P	G
Permeability	F	F	VL	L	L	L	VL	F	L	L	VL
Low temperature	VG	G	G	G	G	F	G	E	F	F	F
High temperature	F	F	G	G	F/G	E	E	E	G	F	G
Oxidation resistance	F	F	E	G	E	E	E	E	E	G	G
Ozone resistance	P	P	G	F	VG	E	E	E	E	E	E
Solvent resistance											
Water	G	G	VG	G	G	E	E	G	VG	F	G
Aliphatics	P	P	P	VG	G	VG	E	P	G	E	G
Aromatics	P	P	P	G	F	G	E	P	F	VG	F
Ketones	G	G	G	P	P	P	E	F	G	F	
Lacquer solvents	P	P	P	F	P	P	E	P	P	G	F
Petrol	P	P	P	VG	G	VG	E	P	F	E	F
Animal and vegetable oils	P to G	P to G	E	E	G	E	E	F	G	E	G
Concentrated acids	F to G	F to G	VG	F	F	VG	E	F	G	F	P
Dilute acids	F to G	F to G	VG	G	G	E	E	G	E	F	F
Alkalis	F	F	G	F	G	E	E	G	E	G	P
Price	L	L	L	M	M	VH	H	H	M	M	H

Key:
E – Excellent; F – Fair; G – Good; H – High; L – Low; M – Medium; MH – Medium high; P – Poor;
VG – Very good; VH – Very high.

SBR Buna S or GRS

SBR is a direct alternative to natural rubbers with a similar sphere of application but inferior low temperature characteristics.

Nitrile rubbers

Nitrile rubbers are the common choice when an oil resistant or solvent resistant rubber is required. Chemical resistance increases with increasing nitrile content, but the low temperature performance is degraded. Maximum service temperature ranges up to 100°C.

Chloroprene (Neoprene)

Chloroprene is noted for its excellent ozone, weathering and ageing resistance and is also oil resistant. Its chemical resistance is inferior to nitrile rubbers – eg it is swollen by aromatic solvents. Its maximum service temperature is about 80 to 90°C.

PTFE is a natural choice when extreme chemical resistance is required. Maximum service temperature can be as high as 250°C.

Silicone

Silicone rubber is a natural choice for high temperature working, but more specifically where good flexibility is required over a wide range of working temperatures – eg from –80°C up to about 200°C. Its chemical resistance is not outstanding. An unfavourable factor is its high cost.

Fluorocarbons (Viton)

Fluorocarbon rubbers have excellent resistance to ozone, weathering, ageing and most chemicals. These properties are well maintained up to a temperature of 200 to 250°C, hence fluorocarbons are favoured for higher temperature applications. Again cost is high.

Polyethylene copolymer

Chlorosulphated polyethylene (Hypalon) has properties which may be regarded as intermediate between chloroprene and the fluorinated rubbers (Viton). It has good resistance to acids and heat.

Acrylic rubbers

Polyacrylic rubbers (ACM) have excellent resistance to oils and greases up to a temperature of about 175°C, with excellent resistance to ageing and stress cracking. Resistance to water is only moderate and low temperature properties are not outstanding.

Polysulphide rubbers

Polysulphide rubbers have excellent resistance to ageing, ozone, fuels and solvents, so may be chosen for specialised chemical duties. In general their mechanical properties and heat resistance are only moderate.

Polyurethane rubbers

Polyurethane rubbers are particularly noted for their high strength and excellent tear and abrasion resistance. They also retain good flexibility at low temperatures. Their chemical resistance, however, may be limited. See also chapter on *Elastomers*.

High temperature structural materials

HIGH TEMPERATURE structural materials can be classified in four main groups:

(i)   Superalloys which have been developed to achieve service temperatures of about 1000 to 1200°C and are largely based on nickel (see Sections 1 and 2).
(ii) Ceramics and cermets which are basically brittle materials but have extremely high melting points (see Section 3).
(iii) Refractory metals and alloys.
(iv) Carbon, graphite and graphitic materials.

The main refractory metals comprise niobium (columbium), tantalum and tungsten. Rhenium also comes into this category but its rareness means that it is not a practical material except as an alloying element.

High strength *niobium alloys* tend to be more difficult to fabricate and weld than the lower strength alloys, and may also show reduced oxidation resistance – see also under *Niobium* in Section 2.

The high temperature properties of *molybdenum alloys* in general can be enhanced by precipitation hardening. This also improves weldability and reduces the likelihood of embrittlement following exposure to high temperatures – see also under *Molybdenum* in Section 2.

Tantalum alloys have been developed particularly for high temperature strength and have been produced with refined and stabilised grain structures which resist grain growth and embrittlement at temperatures of about 2000°C or higher – see also under *Tantalum* in Section 2.

Tungsten remains the strongest of the refractory metals, with the highest melting point and generally superior mechanical properties. Tungsten-rhenium alloys offer the same strength as unalloyed tungsten at very high temperatures, with improved room temperature properties.

Melting points of all the refractory metals are given in Table 1. For structural applications these materials can be considered to have little useful strength at temperatures in excess of 80 to 85% of their melting points. Nevertheless this extends their potentially usable range well above 2000°C.

TABLE 1 – MELTING POINTS OF REFRACTORY METALS

Metal	Melting point °C	Maximum service temperature* °C
Tungsten	3410	2900
Tantalum	*circa* 3000	*circa* 2550
Molybdenum	2610	2220
Niobium (columbium)	2470	1980

*To retain useful practical strength.

Tantalum has a favourably high maximum service temperature and is relatively easy to fabricate, but has generally inferior strength and strength/weight characteristics. For good strength at elevated temperatures it is only suitable for use in alloyed form. Niobium and molybdenum are favourable on a strength/weight basis at temperatures up to about 1700°C. Niobium and tantalum are the most easily contaminated and embrittled. Tungsten is overall the strongest of the refractory metals and the one with the highest service temperature, but also has the least favourable strength/weight ratio, as well as being the most difficult to work (see Figure 1).

A common characteristic of the refractory metals is an inability to withstand oxidation conditions, even at relatively moderate temperatures. Oxidation resistance may be improved by alloying, but this is usually achieved at the expense of high temperature strength. Coatings, based on aluminides, beryllides, silicides or glazes, have been the most successful method of overcoming this limitation.

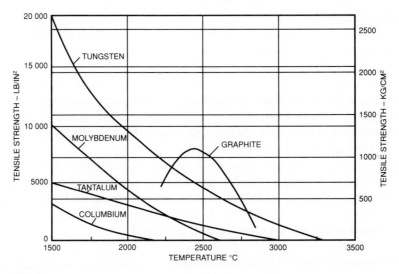

Fig. 1

Without coatings, molybdenum and tungsten in particular are subject to marked surface oxidation at high temperatures.

Graphite materials can prove superior to the refractory metals at the more elevated temperatures – *eg* good grades of commercial graphite can show higher strength than any of those metals and their alloys over the temperature range of about 2400 to 2800°C. Even better performance can be achieved with pyrolytic graphite whose strength may actually go on increasing with increasing temperature in the favourable (oriented) direction. The material is also considerably more resistant to erosion than plain graphite grades, although it is even more brittle and its strength in the non-oriented direction is relatively weak.

Carbon has the highest potential service temperature of any refractory material other than tantalum carbide and hafnium carbide, but the maximum service temperature of commercial grades of graphite is not necessarily superior to that of tungsten. Carbon sublimes at about 3700°C. At temperatures above 3870°C all materials are either molten or vaporised.

Some considerations affecting the use of materials in various temperature ranges are:

Range 230 to 480°C

Alloy steels and corrosion resistant stainless steels are the primary choice as structural materials. Platings which may be used in this temperature range include nickel, alloyed nickel-cadmium and nickel-zinc. Titanium alloys are attractive on account of their high strength/weight ratio.

Range 480 to 650°C

Suitable structural materials in this range largely embrace the heat treatable iron based alloys incorporating chromium and nickel, and also titanium alloys. Silver plating is a possibility on bearing surfaces.

Range 650 to 870°C

Nickel-based and cobalt-based alloys are the primary choice in this range, together with special alloy steels.

Range 870 to 1200°C

Nickel-based alloys are reaching the limit of their practical application as structural materials, with tensile strength decreasing quite rapidly in this range. The refractory metals such as columbium, tantalum and molybdenum, and molybdenum-titanium alloys, can offer favourable strength figures.

Range 1200 to 1600°C

All of the refractory metals, with the exception of tungsten, exhibit a drastic loss of tensile strength towards the upper temperature value, although still well within their acceptable working range. Ceramics may be a suitable alternative, particularly beryllides, which have good oxidation resistance and are less brittle than other ceramic materials.

Above 1600°C

Graphitic materials, carbides and nitrides come into their own in this range, although tungsten and some of the refractory metal alloys are still usable.

Above 2200°C

Borides are potentially competitive with graphites and carbides. Nitrides are less suitable, as their resistance to oxidation tends to be inferior – see also chapter on *Ceramics* (Section 3). Carbides have the highest melting points and therefore potentially the highest temperature applications, although their brittle behaviour under stress can limit their practical application.

Honeycomb materials

HONEYCOMB CORE structures are fabricated from sheet material. Paper honeycombs are employed for low cost laminates. They can be impregnated for increased wet strength, or filled for enhanced insulating properties. Such laminates would be used only for secondary load carrying in assemblies, or purely as insulating panels. Resin impregnated cotton cloth yields a honeycomb with a greater compressive strength and is a better secondary structural material, capable of being used for bulkheads, partitions, etc.

Somewhat greater stability is provided by metal honeycombs in either aluminium foil or stainless steel. The former is widely used, since the density of the honeycomb can be varied considerably to meet service requirements. The normal size of foil sheet material used is 0.002 in thickness with a cell size of ¼ in. The strength of the core is directly proportional to the density of the honeycomb, regardless of foil thickness or cell size. Stainless steel honeycombs are normally employed where service conditions are severe, for example in high temperature applications when bonded to heat-resisting face panels.

TABLE 1 – STRENGTH/WEIGHT OF HONEYCOMB SANDWICH COMPARED WITH OTHER MATERIALS*
(Weight of material for some deflection on 24 in span)

Material	Deflection (in)	Weight (lb)
Honeycomb sandwich	0.058	7.79
Nested beams	0.058	10.86
Steel angles	0.058	25.90
Magnesium plate	0.058	26.00
Aluminium plate	0.058	34.20
Steel plate	0.058	68.60
Glass reinforced plastics laminate	0.058	83.40

*E. C.Vicars

Honeycomb cores are also made from glass fibre possessing, in particular, excellent electrical qualities (using the electrical grade of glass fibre). The strength/weight ratio of such honeycombs is roughly twice that of foamed or cellular cores of comparable weight.

The latter are foamed in place between pre-formed faces and again, in general, have excellent dielectric properties. A particular advantage is that the expanded synthetic or cellular core eliminates cell joints and produces a material which is non-absorbent and consistent in density. Polystyrene is a favoured synthetic material, although hard cellular rubber and mineral foams are also employed.

Honeycombs may incorporate a wide variety of core configurations sandwiched between suitable skin materials. For high strength honeycombs to aircraft standards of requirement the core is made from thin metal foil or other high strength material in sheet form. This is fabricated to form a honeycomb structure, in which hexagonal shaped cells are arranged in a regular geometrical

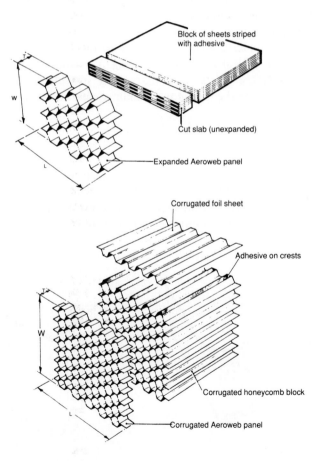

Fig. 1 (Ciba-Geigy)

pattern by bonding with specially developed adhesives. Its principal use is as a shear carrying core in lightweight sandwich structures, but its unusual physical and mechanical qualities make it suitable for many other industrial uses.

Two methods can be used to manufacture the honeycomb; the principles of both are illustrated in Figure 1. The expansion method is used for making most of the honeycombs used in sandwich structures. Very large panel sizes can be produced by this method, although the maximum panel thickness is limited. The principal advantage of the corrugation method is that much greater thickness can be produced, and it is particularly suitable for making high density honeycomb.

A variety of aluminium foils or non-metallic sheet materials is used, producing a wide range of honeycombs to meet the needs of the advanced technology industries. Honeycombs have excellent mechanical properties, which are used to greatest advantage in sandwich structures made by bonding thin facing skins onto the opposite faces of pieces of honeycomb, as illustrated in Figure 2.

When this type of structure is loaded, bending loads are resisted by the facing skins and shear loads are distributed between the skins and the honeycomb core. The honeycomb thus acts in the same way as the shear web of an I-beam (Figure 3). The sandwich beam is superior to the I-beam however because the core extends over the full width of the facing skins and stabilises them to inhibit wrinkling or buckling. Tests show that under suitable loading conditions the facing skins can be worked, without buckling, well above the 0.2% proof stress for the material.

An optimised sandwich design distributes the weight of material used in a manner which gives a close approximation to the theoretical ideals, yet is simple and economical to produce.

Honeycomb cored structures of this type are used wherever a limited weight of material must be used to maximum effect. Outstanding examples are the vehicles and other hardware used in space exploration, where structure weight must be reduced to a minimum without compromising strength and integrity. Exam-

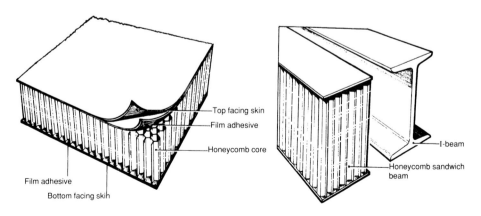

Fig. 2 (Ciba-Geigy)   *Fig. 3 (Ciba-Geigy)*

ples of the application of these principles can also be found in many other forms of transport, where cost conscious designers must balance the cost of structural refinements against resultant savings in operation. Examples to be found in aerospace work include aircraft fuselage shells, wing skin panels, flying control surfaces, engine structures and cowlings, tail structures, doors, hatches, floors, interior trim and furnishings. Many of these applications also take advantage of the extremely high resistance to sonic frequency fatigue that is characteristic of bonded honeycomb sandwich structures.

As with many techniques pioneered by the aerospace industry, bonded sandwich construction is being adopted by other industries. Honeycombs are making their contribution to the strength and reliability of hovercraft, road and rail vehicles, ships, radar antennae, electronic equipment housings of all kinds and many other engineering structures.

Metallic honeycomb has a remarkable ability to absorb mechanical energy by plastic deformation with constant resistance, thereby providing an excellent means of absorbing shock at constant deceleration and without rebound. It is also an efficient medium for fluid flow straightening, and for use in light collimating and heat exchanging devices. Its aesthetic qualities have attracted the attention of architects for use in light diffusing ceilings and other decorative schemes.

Honeycombs can also be made from plastic sheet and film, glass reinforced honeycombs offering comparable strength to metal honeycombs. Ceramic honeycombs have also been used successfully in heat exchangers for gas-turbine engines and other high temperature duties. Many of the composite (fibre or whisker reinforced) materials lend themselves to foaming or conversion into cellular form, offering a further choice of possible core materials for honeycomb construction.

Insulating materials (thermal)

SPECIFIC PROPERTIES of thermal insulation materials are given in Table 1. Materials are primarily chosen for their thermal conductivity value but due consideration must also be given to their working temperature range, resistance to specific ambiences (where applicable), density, durability, ease of manufacture and cost. It may also be necessary to consider their water absorption properties,

TABLE 1 – PROPERTIES OF HEAT INSULATING MATERIALS

Material	Specific gravity	Density lb/ft³	Working temperature range[1]		Thermal conductivity (k)[2]	
			Min °C	Max °C	Low temp.[3] °F	High temp.[4] °F
Asbestos fibre	0.24–0.29	15–18	−75	540	1.00 (0)	1.30 (400)
Balsa wood	0.10–0.25	6–16	−170	150	0.33 (50)	—
Cork	—	7–14	−150	120	0.28 (50)	—
Cotton	—	—	—	95	0.40 (50)	—
Calcium silicate	0.18–0.21	11–13	−20	1000	0.35 (100)	0.60 (700)
Diatomaceous earth	0.25–0.39	16–24		1000	0.36 (50)	0.67 (300)
						0.82 (1000)
Expanded ebonite	0.05–0.06	3– 4	—	150	0.25 (50)	—
Expanded polystyrene	0.02–0.09	1– 5	−150	70	0.25 (50)	—
Expanded polyurethane	0.03–0.05	2– 3	—	70	0.17 (50)	—
Expanded perlite	0.06–0.16	4–10	—	750	0.33 (100)	0.60 (400)
Felt	0.16–0.32	10–20	−75	150	0.40 (100)	—
Glass fibre	0.06–0.12	4– 8	−200	650	0.21 (50)	0.44 (300)
Glass foam	0.11–0.22	9–18	−150	600	0.38 (50)	0.55 (300)
85% magnesia	0.18–0.22	11–14	−20	540	0.40 (100)	0.45 (400)
Rock wool	0.22–0.39	14–16	−200	1200	0.31 (50)	0.90 (1200)
Slag wool	0.22–0.39	14–16	−200	1000	0.29 (100)	0.22 (700)
Sawdust	—	—	−175	93	0.40 (50)	—

Footnotes:
[1] This column is a good guide to the suitability of insulants for different working temperatures, and particularly for high temperature insulation where choice is more restricted.
[2] Typical k values are given in Btu/in/ft²/h/°F. For conversion, 1 Btu/in/ft²/h/°F = 1.44 W/cm/± °C.
[3] Value at a typical lower temperature as given in brackets.
[4] Value at a typical higher temperature as given in brackets.

**TABLE 2 – EXAMPLES OF VARIATIONS OF THERMAL CONDUCTIVITY
WITH TEMPERATURE**

Material	Temperature °F (°C)						
	−100 (−73)	0 (−18)	50 (10)	100 (43)	200 (93)	300 (150)	400 (200)
Calcium silicate	—	—	—	0.35	0.37	—	
Diatomaceous earth	—	—	0.36	0.37	0.40	0.44	
Expanded polystyrene	—	0.21	0.25	0.28	—	—	
Expanded perlite	—	—	—	0.33	0.36	0.48	0.60
Cork (granulated)	0.20	0.22	0.28	0.30	0.33	—	
Glass fibre	—	—	0.21	0.23	0.25	—	
Glass foam	—	—	0.38	0.42	0.46	0.55	
85% magnesia	—	—	—	0.40	0.43	0.46	
Rock wool	—	0.30	0.31	0.33	0.36	—	
Balsa:							
6 lb density	0.20	0.26	0.28	0.31	—	—	
12 lb density	0.28	0.36	0.41	0.45	—	—	

which could modify the thermal conductivity and may cause deterioration of the material. It should be remembered that the thermal conductivity of any material is variable with temperature (generally increasing with increasing temperature).

Where necessary, virtually any insulating material can be rendered waterproof by the application of a suitable surface barrier such as plastic film, metal foil, treated paper, etc. In such cases it is important that all joints in the insulant be sealed (usually with adhesive tape). Alternative treatment is to contain the insulant within a suitable impervious enclosure so that a vapour barrier is formed on the outside of the insulant.

For general purpose insulation glass fibre and mineral wools are the most widely used materials, although other materials may be more suitable for specific purposes. The general properties of specific insulants are described under separate headings, arranged in alphabetical order.

Asbestos

Three types of abestos may be employed – white (chrysolite) and blue (crocidolite and amosite) – although legislation has now largely ruled out blue asbestos. These are all natural mineral fibres which are refined and carded, usually with the addition of binders. Asbestos insulation is used in the form of loose wool, felts or made-up mattresses, etc.

The use of asbestos, particularly blue asbestos, has been drastically reduced since the introduction of legislation governing the handling and use of this material.

Balsa wood

This is a natural wood of very light weight, but with structural properties and good compressive strength. Its insulating properties are not as good as those of many other insulants but it finds application for specialised requirements such as the insulation of liquid gas containers, refrigerated containers, etc. Lowest k values are achieved with lower density balsa and in the direction of the grain.

Balsa insulation can be used down to –184°C, the k factor decreasing linearly with temperature (see Figure 1).

Calcium silicate

Calcium silicate is a combination of lime and silica to which may be added asbestos fibre to improve mechanical strength. It is particularly suitable for high temperature applications and is also fully resistant to water and moisture.

Ceramic fibres

These are particularly suitable for high temperature insulation (see chapter on *Ceramic Fibres* (Section 3)).

Cork

Cork is one of the traditional insulation materials but its maximum service temperature is strictly limited. The normal form is granulated cork which can be used as a filler material or formed with a bonding agent into sheets, slabs or special sections (*eg* pipe wrappings).

Diatomaceous earth

Diatomaceous earth is almost pure silica comprised of fossil diatoms of tiny size. Some 93% of the volume is void space. It is quarried as a natural material.

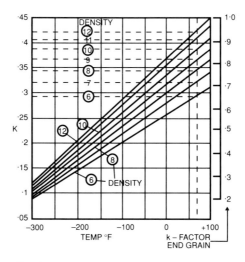

Fig. 1
Thermal conductivity of balsa.

Expanded ebonite

Expanded ebonite can be rendered in very low densities to give a strong, rigid, cellular material with a closed cell structure. It does not absorb water and has a low k value. Expanded ebonite is easy to cut and machine. It requires no special protection and is rot and vermin proof.

Expanded polystyrene

A generally excellent insulating material with low cost and ready availability, a particular advantage being ease of moulding in simple steam heated moulds to any shape required. It is also produced in sheet and block form, but it is difficult to re-work due to the brittle nature of the material. A particular advantage in some applications is that the material can be foamed *in situ* (*eg* between cavity walls).

Expanded polyurethane

Another readily mouldable material which is produced in slabs and shaped forms and is also suitable for foaming *in situ*. In the latter case it can be sprayed on and expanded by treatment with a fluorinated hydrocarbon. Somewhat tougher than expanded polystyrene, the material is more expensive, with superior insulating properties.

Expanded perlite

Perlite is a siliceous volcanic rock. The expanded form is used as a thermal insulation material and comprises mainly voids (air cells); density may range down to as low as 4 lb/ft^3.

Felt

Natural felt is composed of compressed animal or vegetable hairs or fibres, although they may also be mixed with other fibres. Its performance as a thermal insulation material is only fair and the material is prone to rot or liable to be attacked by vermin. Its main attraction is that it is a relatively inexpensive material, but it readily absorbs water and may need protection when used as a low temperature insulant.

Synthetic felts have improved performance and immunity to rot or attack but are much more expensive, and are therefore little used for thermal insulation.

Glass fibre

Glass fibre is particularly well known as an insulating material and is produced in a variety of mat and woven cloth forms. Its properties are particularly favourable for thermal insulation although its cost is relatively high. Foamed glass is a rigid insulant with structural possibilities.

Magnesia

The form used for thermal insulation is 85% light basic magnesia mixed with 15% asbestos fibre. It has the advantage that it can be applied as a plaster which dries out to form a solid coating, making it attractive for irregular shapes which are difficult to cover with other materials. The solid form is also pre-fabricated as slabs and sections. A major limitation is that magnesia is brittle and fragile and may need protecting with an outer cover. High temperature grades are suitable for contact with surface temperatures up to 675°C.

Mineral wools

Mineral wools include rock and slag wool. Rock wool is produced from siliceous rock and is normally combined with inorganic binders or bonding clay and may be mixed with asbestos fibre. Slag wool is produced from molten blast furnace slag by steam blasting and is characterised as a non-resilient material.

Phenolic foam

Closed cell phenolic foam is now coming into widespread use as a lightweight insulating material for cores of composite panels for building work. The material has a low k value which is consistent over a temperature range of –190 to +130°C. U values as low as 0.5 $W/m^2/°C$ can be realized with composite panels faced with aluminium, plywood, plasterboard or similar materials.

Specific advantages of phenolic foam are its infusibility, high fire resistance and negligible emission of smoke or fumes.

TABLE 3 – PHENOLIC FOAMS

Physical properties (typical values)	Units	Phenexpan 30	Phenexpan 40
Nominal density	kg/m^3	30/35	40/45
Closed cells (ASTM 1940–62T)	%	90+	90+
Compression strength			
Perpendicular to face	bars	1.0 to 1.4	1.6 to 2.0
Parallel to face	bars	1.6 to 2.0	2.0 to 3.0
Cross breaking strength	bars	2.8 to 3.5	3.5 to 4.5
Shear strength	bars	1.0	1.2
Water absorption at 20°C in one week	% increase v/v	5.5 to 7.7	4.5 to 6.8
Water vapour transmission at 88%			
R.H. 38°C	$gm/m^2/h/m$	0.019	0.026
Thermal properties			
Thermal conductivity at 20°C			
k value (guarded hot plate method)	W/m/°C	0.030 to 0.032	0.030 to 0.032
Thermal stability at 130°C	% shrinkage after 7 days	1.3	1.0
Maximum service temperature			
Continuous	°C	130	130
Intermittent	°C	180	180
Low temperatures			
Coefficient of linear contraction between +20°C and –190°C	$\times 10^{-6}$	30 to 35	30 to 35

Ack: APA Foam Products Ltd

Insulating materials (acoustic)

A CLEAR distinction is drawn between *sound insulation* materials which *reflect* incident sound energy and *sound absorption* materials which *absorb* incident sound energy (see Figure 1).

The basic requirement of a good sound insulating material is that it should be capable of reflecting a high proportion of the incident sound energy, *ie* the material acting as an insulating panel should experience low absorbed energy and low flanking energy. The ratio of the acoustic energy transmitted through a panel to the total energy incident on it is known as the transmission coefficient (T). Actual insulating performance is normally referred to in terms of the sound reduction index, when:

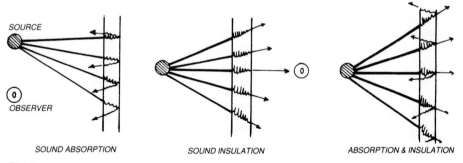

SOURCE

OBSERVER

SOUND ABSORPTION   SOUND INSULATION   ABSORPTION & INSULATION

Fig. 1

$$\text{Sound reduction index (R)} = 10 \log \frac{1}{10} \text{ dB}$$

In the case of single leaf partitions the sound reduction index is closely related to the mass and frequency by the following equation:

$$R = 20 \log Mf - 43 \text{ dB}$$

where:

M = superficial weight of the partition in kg/cm^2
f = frequency in Hz.

It follows that doubling the mass, or frequency, gives an increase in the sound reduction index of 6 dB.

With random incidence the sound reduction index will be decreased, as:

$$RRF = 20 \log Mf - 52 \text{ dB (approx)}.$$

This corresponds to reverberant field conditions, when sound energy may be incident at angles between 0 and 90g.

In a semi-reverberant field the corresponding relationship is:

$$RRS = 20 \log Mf - 49 \text{ dB (approx)}.$$

These relationships are shown graphically in Figure 2. It is a basic theoretical relationship which assumes that the partition has very low stiffness. In practice the performance of the panel will be frequency dependent. At low frequencies of incident sound energy, inertia forces will be small and the sound reduction index will be controlled by the stiffness of the material. Over a range of higher frequencies the panel will be subject to resonance effects. At still higher frequencies the panel will follow strongly the 'mass law' up to a critical frequency. At this point the bending waves produced in the panel will be travelling at the velocity of sound in air, or coincidence will occur. At higher frequencies there will be a substantial lowering of the sound reduction index of the panel.

Sound insulation materials are normally structural, *eg* panels of sheet material or sandwich construction, masonry walls, partitions, etc. As a general rule two materials of the same mass per unit area will provide the same sound insulation, but this generalisation does not take into consideration the phenomenon of coincidence, and for practical purposes the ability of a barrier to reduce noise depends also upon its stiffness or limpness, damping capacity and air impermeability.

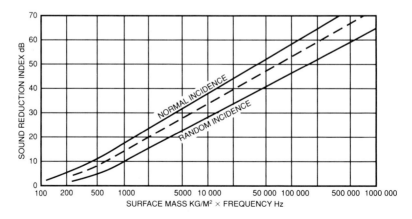

Fig. 2

When a pressure wave in air acts upon the surface of a panel, it bends the shear waves within the panel. While the speed of sound in air is constant, the velocity of the bending wave is proportional to the square root of the frequency. Therefore, as the frequency of the sound increases, so a condition will arise in which the bending wave in the panel coincides with the air borne wave, giving rise to a region of poor performance. In effect the sound wave in the air and the bending wave in the panel reinforce each other, giving a dip in the noise reduction curve.

For many of the materials used as inter-office partitions, *eg* plasterboard, chipboard and plywood, the start of this coincidence dip lies in the 200 to 1000 Hz frequency range at between 17 and 30 dB. Increasing the thickness of the panel will raise its insulation value, but the maximum increase will be only 5 to 6 dB for each doubling of the weight, and due to the change in bending stiffness with the increase in thickness, the coincidence dip will be moved to a lower frequency. This often brings the new dip in the curve within the critical frequency area, with the result that the overall improvement in noise reduction is small.

A particularly useful material for sound insulation is sheet lead. This does not exhibit coincidence within the frequency range applicable to noise in offices, and therefore each increase in weight gives the maximum improvement in sound insulation.

Comparative figures for different thicknesses of common panelling materials all of about $2 lb/ft^2$ weight are:

	Average sound reduction in dB for the range 100–4000 Hz
19 mm (¾ inch) plywood	9.8 kg/cm^2 (2.0 lb/ft^2) = 26
12.5 mm (½ inch) plasterboard	10.7 kg/cm^2 (2.2 lb/ft^2) = 30
0.8 mm ($^1/_{32}$ inch) sheet lead	9.8 kg/cm^2 (2.0 lb/ft^2) = 34.

If, therefore, the weight of a panel is increased by the addition of thin lead sheet rather than by doubling its thickness, the rigidity of the panel is not increased and the full effect of the increase in weight is achieved.

Figure 3 shows the effects of adding a sheet of 0.4 mm (1 lb/ft^2) to a sheet of 12 mm plasterboard. By increasing the weight without increasing the rigidity, the plateaux in the curve caused by coincidence have been removed and the sound reduction index increased from 24 to 28 dB.

Although double leaf barriers are superior to single skin panels, their sound insulation value can be reduced by the effects of coincidence as well as other factors.

Cavity resonance and reverberation

If there is no direct mechanical connection between the two leaves and the distance between the two panels is very large, the overall sound insulation will be about twice that of the single panels. In office partitions, however, where high value floor space is at a premium, the leaves will be set close together; low frequency cavity resonance and mid to high frequency reverberation will affect the

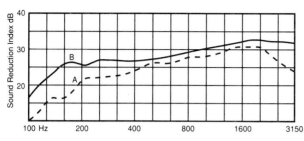

A. 12 mm plasterboard on one side of 50 mm studding weight 2 lb/ft² average TL 24 dB.

B. As 'A' with 0.4 mm lead sheet (lb/ft²) glued to inside face of plasterboard. Weight 3 lb/ft² average TL 28 dB.

Fig. 3

result, giving a sound reduction less than the total of the two separate leaves. In extreme cases these factors can reduce the sound insulation of the total partition to below that of a single panel of equal weight, and it is therefore important to select the correct materials to obtain the highest possible performance.

Cavity resonance is a function of the cavity width and the weight of each panel. For lightweight partitions it is usually based on the equation:

$$Fo = \frac{240}{\sqrt{Md}}$$

where:

Fo = resonance frequency
M = average panel weight (in lb/ft²) exclusive of support framing
d = width of cavity in inches.

For any given cavity width the most effective way of increasing the weight of the panel without adding substantially to overall partition width is by the addition of thin sheet lead.

Reverberation within the cavity itself can be visualised by considering each panel leaf independently. Assume a situation with an incident noise of 80 dB and a first leaf which gives insulation of 15 dB. Reverberation between the two leaves of the empty cavity can cause the actual noise level to build up from 65 dB to perhaps 70 dB, reducing the apparent insulation of the first leaf. The second leaf also provides an insulation of 15 dB, but the level within the cavity is now 70 dB and the perceived result is 55 dB, 5 dB worse than the anticipated insulation. The most effective method of reducing this reverberation is to incorporate a mineral wool or glass fibre quilt into the cavity space, fixed between the supporting studs.

Plenum drapes and overlays

No matter how high the insulation provided by the partition itself, unless it extends up to the structural floor slab there will be leakage of sound over the top of the partition. It is common practice in commercial buildings to erect partitions only up to a suspended ceiling, using the area above the ceiling for ser-

TABLE 1 – GENERAL CHARACTERISTICS OF SOUND ABSORBING MATERIALS

Material Type	Typical Absorption Coefficient*	Remarks
Porous panels	0.20 to 0.80; generally tend to increase with increasing frequency	Performance may be appreciably modified by surface treatment. Typical surface finishes — muslin or perforated covers; or membranes.
Porous materials (eg mineral, glass, wool and felts)	0.25 to 0.90; generally increasing with increasing frequency	Turnover frequency determined by depth of air volume.
Porous blocks	0.20 to 0.80 depending on cellular form and particularly thickness	May exhibit optimum performance at mid-frequencies.
Plain panels	Typically 0.40 maximum decreasing with increasing frequency	Low frequency performance improved by addition of absorbent backing. Absorption mainly by vibration causing energy damping.
Perforated panels and tiles	0.20 to 0.80 depending on thickness and size and disposition of perforations. Poor performance at low frequencies.	Normally exhibit a maximum value at mod-frequencies. Usually comprise porous material with hard face.
Perforated panels with absorbent backing	0.20 to 0.80; generally increasing with increasing frequency	Thin preformed face panels with absorbent backing behave largely like porous panels. Rigid constructions may act as resonant absorbers.
Carpets	0.05 to 0.65; poor at lower frequencies	Performance generally improved with underfelt or underlay.
Drapes	0.10 to 0.65	Performance varies considerabilty with thickness and weight or material, also texture and form.

*125 Hz to 4 000 Hz

vices, air condition ducting, etc. The sound generated in one area travels through the suspended ceiling via the plenum void to other areas, causing severe loss of privacy, unless these sources of sound leakage are dealt with.

Sound absorption materials

Sound absorption materials for buildings and similar applications can be broadly grouped as:

(i)   *Acoustic tiles* in various forms, sizes, shapes and thicknesses.
(ii) *Acoustic boards,* similar in construction to acoustic tiles, but larger in area.
(iii) *Plastic materials* – for trowel or spray application, as in the case of acoustic plasters.
(iv) *Blanket materials* – normally used in composite constructions.

TABLE 2 – PROPRIETARY ACOUSTIC MATERIALS

Material or Form	Types	Absorption Coefficient*	Remarks
Acoustic tiles cellulose	(i) regular perforations	½" thick 0.8—0.64 ¾" thick 0.12—0.65 1" thick 0.15—0.66	Widely used acoustic absorption material
	(ii) random perforations	½"thick 0.14—0.68 ¾" thick 0.18—0.62 1" thick 0.28—0.61	available in a large range of types. Smooth finished tiles
	(iii) textured	½" thick 0.11—0.52 ¾" thick 0.22—0.77 1¼" thick 0.10—0.46	may be membrane faced. In general increase of thickness increases sound
	(iv) slotted	¾" thick 0.18—0.71	absorption over a fairly
Acoustic tiles mineral fibre	(i) perforated	½" thick 0.08—0.58 ¾" thick 0.07—0.63 1" thick 0.11—0.67	narrow frequency range.
	(ii) slotted	¾" thick 0.10—0.080	
Acoustic boards	Generally as for tiles		Performance similar to tiles
Acoustic plasters	(i) various proprietary types	½" thick 0.05—0.60 ¾" thick 0.10—0.70	Typical performance but varies with backing
	(ii) vermiculite	⁷⁄₁₆" thick 0.50—0.75 1½" thick 0.25—0.95	Generally optimum values at mid-frequencies
Blanket materials	(i) rock wool	1" thick 0.10—0.90 2" thick 0.20—0.95 4" thick 0.50—0.95	Typical for laid in battens
	(ii) glass wool	1" thick 0.10—0.70 2" thick 0.20—0.80 4" thick 0.40—0.85	Typical for laid on battens
	(iii) wood wool	1" thick 0.10—0.75 2" thick 0.14—0.80	Typical for laid on battens

*125 Hz to 4 000 Hz

(v) *Acoustic roof decks* – capable of being used as an integral part of the structure.

(vi) *Acoustic assemblies* – pre-fabricated sections, etc usually incorporating (iv).

(vii) *Suspended absorbers* – independent assemblies mounted by suspension, including baffles and unit absorbers.

(viii) *Absorbent linings* – normally applicable to sound barriers or enclosures but may comprise materials classified above.

Sound produced by a continuous source within a room will be reflected off all surfaces on which it is incident, producing echoes which will tend to increase the overall noise level in the room. Sound absorbing materials are those which reduce the level of generally reflected sound and multiple sound reflections which persist with time. They may be used either for sound reflection within a room, reverberation control, or both. Each doubling of the total sound absorption in a room is equivalent to a reduction in overall sound pressure level of 3 dB and a halving of the reverberation time. It should be noted, however, that sound absorption has no effect on direct sound, nor on the sound insulation of the room (although some sound absorbent materials may be incorporated in composite sound insulating panels).

Hard surfaces such as glass, plaster and concrete are highly reflective for sound, absorbing appreciably less than 5% of the incident sound energy. A perfect acoustic absorption material would have an absorption of 100%, so the performance of a sound absorbing material is defined by the proportion of incident sound absorbed, normally expressed as a decimal rather than a percentage. This may also be called the noise reduction coefficient (NRC).

As with sound insulating materials, the absorption coefficient depends on the frequency of the sound, tending generally (but not universally) to increase with increasing frequency. It also varies with the angle of incidence of the sound. For practical purposes absorption coefficients are normally determined empirically in a reverberation chamber at specific frequencies in a series and are correctly referred to as random incidence absorption coefficients. For specific purposes normal absorption coefficients may be determined where the sound is incident at 90° to the test piece. There is no direct relationship between the two coefficients, although as a rough rule random incidence coefficients are approximately twice as great as normal coefficients for the same material at lower frequencies, about 4/3 times as high in the middle frequencies, and approximately the same at higher frequencies.

Empirical determination of sound reduction coefficients or acoustic absorption is commonly confined to six frequencies in the range 125 to 4000 Hz, although measurements may be extended to higher frequencies. In general, however, values for frequencies above 4000 Hz are similar to those for 4000 Hz. The noise reduction coefficient is commonly quoted as a single number average value, based on the sound reduction coefficients at 250, 500, 1000 and 2000 Hz. Values, in all cases, are rounded off to the nearest 0.05.

In the case of porous materials with inter-connecting pores, friction is the predominant factor in absorbing the energy of sound waves by progressive damping. In terms of electrical analogy, the material offers direct resistance and the damping effect is largely independent of frequency. However, there is an optimum value for the resistance. If too high, sound waves will be rejected or reflected rather than penetrating in depth. If too low, there will be insufficient friction to provide enough damping to make the material effective as a sound absorber.

With perforated materials opening into a body of porous material, solid materials and membranes, damping is provided by reaction rather than pure resistance. In consequence, the performance can be markedly dependent on frequency. This in turn is further dependent on the proportion of open area in the case of a perforated surface, or the mass in the case of an impervious membrane.

The total depth of the air volume between the face of the material and the rigid backing can also modify the frequency dependent characteristics; this air volume includes open pore volumes in the case of porous materials. Basically, this introduces the concept of a 'turnover frequency' at which the low frequency absorption characteristics will deteriorate rapidly. This turnover frequency (f_t) is given approximately by:

$$f_t = \frac{c}{2d}$$

where:

c =   the velocity of sound in air
d =   the total depth of air volume.

For d in inches this reduces to:

$$f_t = \frac{500}{d} \text{ Hz (approximately)}$$

This emphasises the importance of air volume in the absorption of lower frequencies, *ie* to produce suitable low values of turnover frequency.
The basic requirements of a sound absorbing material are that:

(i)   It should be sufficiently porous to allow sound waves to enter the material, and
(ii) The nature of the of the material should be such that the maximum proportion of sound energy will be transformed into heat energy by friction, providing dissipation of the sound energy.

Both can be related to the flow resistance, or the ratio of the pressure drop to the velocity of the air passing through the material. The pressure drop must be low enough to provide adequate transparency to sound waves, and high enough to provide sufficient friction.
Actual performance of an insulation panel is subject to modification in a number of ways:

(i)   The transmission coefficient, and therefore the sound reduction index, is frequency dependent; low frequencies are much more readily transmitted than higher frequencies.

   The performance of a sound insulating material is, therefore, only fully described by knowledge of its transmission coefficient at different frequencies. Specific information of this nature may be essential for dealing with critical insulation problems – *eg* when there is a predominance of a particular frequency or frequencies in the incident sound. For normal working average values are commonly used.

   Average values are based on the determination of the transmission coefficient, or sound reduction index, at a number of specific frequencies.

(ii) The transmission coefficient will depend on the angle of incidence of the sound waves received from the source. Since the great majority of cases of practical application involve random incidence, the sound reduction index is properly defined by sound waves of random incidence equivalent to hemispherical distribution from the source. There may be cases in which this does not apply and hence the transmission coefficient of the insulating panel may be modified.

(iii) At specific frequencies of incident sound coinciding with the natural frequencies of the panel, resonant vibration will occur, with a marked reduction in sound insulation. This is because the panel itself will now be acting as a generator of sound energy. Loss of performance as a sound insulating material will be most marked at the lowest natural frequency.

(iv) At certain frequencies the phase of incident sound will tend to coincide with the phase vibrations of the panel. This can introduce flexural vibration in the panel, again substantially reducing the sound insulation. This is known as *coincidence effect,* but can only occur at frequencies greater than a critical frequency (fe), defined as the frequency at which the flexural wave velocity equals the velocity of sound in air.

For any given homogeneous panel, the critical frequency is usually proportional to the thickness of the material. The problem of coincidence effect, therefore, is most likely to occur in the case of thin panels of homogeneous material with low damping. Increasing thickness or damping can render coincidence effects negligible.

(v) Sound transmission through the insulation panel may be supplemented by secondary transmissions through adjacent structures. These are known as *flanking transmissions,* and can increase the sound level on the receiving side of the panel.

Where the average insulation is about 35 dB or less, flanking transmission is usually negligible. This does not, however, preclude the possibility of sound being transmitted through vibration of the building structure (which may be considered a special case of flanking transmission) or, more particularly, through any gaps or openings which may be present in the receiving room.

Homogeneous insulating panels

In the case of homogeneous panels, the sound insulation provided (*ie* sound reduction index) is directly proportional to the superficial weight of the panel. The apparent relationship can be expressed by the formula:

$$R_{mean} = 20 + 14.5 \log_{10} w \text{ dB}$$

where:

w is the superficial weight in lb/ft^2.

In practice, the performance of actual materials may show appreciable departure from this so-called mass law because:

(i) The relationship is relative to a particular frequency or a mean frequency.
(ii) The panel may be subject to conditions outside the mass controlled region.

Effect of stiffness

Specifically, the transmission coefficient (and thus the sound reduction index) is dependent on frequency, the value of R increasing with increasing frequency (see Figure 4). Individual materials and different constructions may show departures from a steady increase. For practical purposes, therefore, it is necessary to determine values of R at specific test frequencies, which may then be used to determine the sound reduction achieved at specific frequencies, or to arrive at a mean or average value of R, which will give an average performance for the partition from a single calculation. Calculation of R on the mass law basis may be applied to homogeneous materials in the absence of empirical data.

The region in which a panel acts as a mass controlled attenuator is bounded by the resonant frequencies at the lower end and the coincident frequency at the upper end (see Figure 5). The first resonant frequency is usually well below 100 Hz in practical panel sizes and constructions, although this may not always be the case with small thin panels. As far as behaviour to incident frequencies below the resonant frequency is concerned, the insulation provided is largely stiffness controlled, with R increasing with decreasing frequency. The actual value of the resonant frequencies will depend on the size, mass, stiffness and method of edge fixing of the panel.

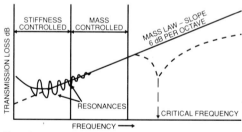

Fig. 4

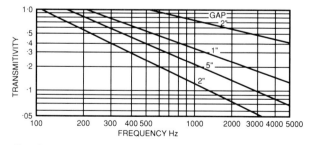

Fig. 5

Many practical panels exhibit coincident effects over part of the normal frequency range, and this largely accounts for the departure from the mass law behaviour.

The extent of this departure again depends primarily on the amount of damping present. The damping provided by studding and discontinuities in building partitions often reduces coincidence effects to negligible proportions. The significance of coincidence effects in practice also tends to be less than simple theory would predict because, with a random field, only a small proportion of the sound energy is incident at the correct angle to produce coincidence.

The insulation characteristics of partitions vary considerably with the type of construction, so individual characteristics can only be determined empirically. Double panel construction normally provides appreciably higher values of R at higher frequencies and similar values at lower frequencies (although for the same mean value of R, double panels will usually provide a lower value of R at lower frequencies). The type of rendering, or surface finish, can be extremely significant in the case of porous and semi-porous panels, porous materials generally being poor sound insulators, unless sealed.

Low density structural laminates

THE DESIGN of a structural laminate is analagous to that of an I-beam with high fatigue strength due to bonded assembly rather than riveting or welding. Rigid laminates may consist of wood, paper, synthetic, foamed or honeycombed cores faced with thin metal sheets, the whole being bonded permanently together in the form of a light, strong and rigid panel. One of the earliest commercial examples was the aluminium-balsa-aluminium laminate used for decks and catwalks in the American airships Akron and Macon, rigidity and durability being primary requirements alongside light weight. This was some 55 years ago. Today similar laminates are used extensively in modern aircraft, in all types of surface transport vehicles and for architectural purposes.

Similarity to an I-beam is given by the metal facings forming the flanges and the core of the web. Under bending loads the facings take the flexural stresses

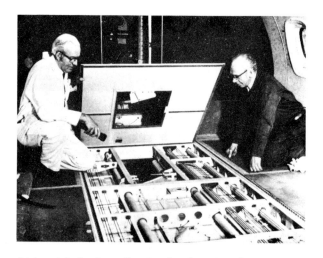

Lightweight laminate flooring fitted to aircraft.

whilst the lightweight core effectively increases the depth of the section to limit deflection and inhibit buckling. The latter is a particularly attractive feature from the point of view of architectural engineering because thin plate stressed skin structures are particularly prone to local wrinkling or buckling. A rigid laminate on the other hand, will withstand similar loading whilst maintaining a perfectly smooth face without imposing any significant weight penalty.

An ultra-lightweight rigid sandwich panel for carrying some of the solar cells used on the Ariel 3 satellite to provide power in orbit. (Ciba-Geigy).

Aeroweb-cored sandwich panels in aircraft galley unit. (Ciba-Geigy).

In general maximum flexural stiffness is achieved in lightweight panels when the core weight is approximately two-thirds of the total panel weight; and maximum bending strength is obtained when the core weight is roughly one-half of the total panel weight. The respective specific gravities of Alclad and 8 lb density balsa are 2.80 and 0.14, hence the ratio of core thickness to (total) metal surface thickness for maximum strength will be about 20:1.

Balsa, however, is but one of the many core materials used in modern laminates, and the facing material itself need not be metallic. Other natural woods, bonded plywoods, processed boards, cellular rubbers, glass fibre, metal and synthetic honeycombs, mineral and synthetic foams, etc are commonly used for core materials. The range of facing materials includes aluminium and aluminium alloys, magnesium, resin impregnated papers, glass fibre, steel, stainless steel, zinc-coated steel, enamelled steel and other metals.

Modern aircraft laminates are widely used for floor panels and interiors. Such panels are made from carefully matched combinations of lightweight core materials (balsa wood, rigid PVC foam or honeycomb) bonded to light alloy or fibre reinforced facing skins (glass and carbon fibre). They are designed to give the optimum combination of low weight and cost with a guaranteed service life appropriate to the structural specification for the aircraft to which they are to be fitted.

Balsa is a generally excellent core material for laminates because of its light weight, good strength and excellent bonding characteristics. Its main disadvantages are that it is a porous material which soaks up water readily and is subject to damage from moisture or fungus, although this can be overcome by proper edge sealing. Balsa wood also varies enormously in density as cut, ranging from as low as 4 lb/ft^3 up to 16 lb or higher. Stock for core material is generally selected from 8 to 9 lb timber, although again it is a characteristic of the material that the density may vary from end to end of a panel or sheet and so complete uniformity of the core is virtually impossible. Arrangement of the core material with end grain at the faces is generally the most consistent and also produces the most rigid laminate.

Facing material can be pure aluminium of any gauge thickness required, or Alclad for somewhat greater strength and durability. The latter material consists of hard aluminium alloy faced with pure aluminium to form a protective coating, the alloy being susceptible to corrosion. Adhesives commonly employed for bonding are phenol-formaldehyde and resorcinal resins. The strength and durability of the laminate will largely depend on the proper choice of adhesive for a specific duty. Thus the type of bond required for maximum tensile strength is rather more brittle than a bond which would give maximum fatigue strength or performance under cyclic loading.

The chief objection to the use of natural woods for core materials is their moisture absorption unless the edges are specially treated and sealed. Nevertheless a wide variety of laminates is made from such core stock including laminates of medium rather than light density for primary structural purposes.

Typical laminates employing three-ply wooden cores and steel faces have a stiffness comparable to the same thickness of aluminium plate for about two-thirds the weight, or roughly twice the thickness of solid oak panelling for the same weight and two-thirds the thickness of steel plate for one-third the weight. Typical density figures are summarised in the Tables.

TABLE 1 – MEDIUM-DENSITY LAMINATE WEIGHTS (lb/in²)

Construction			Overall Panel Thickness (inches)						
Face 1	Core	Face 2	¼	⅜	½	⅝	¾	⅞	1
Steel, 0.029 in	Plywood	None	1.80	2.20	2.50	2.90	3.20	3.60	3.90
Stainless steel, 0.019 in	Plywood	None	1.40	1.80	2.10	2.50	2.80	3.20	3.50
Stainless steel, 0.025 in	Plywood	None	1.70	2.10	2.40	2.80	3.10	3.50	3.80
Aluminium, 0.016 in	Plywood	None	0.86	1.22	1.57	1.92	2.27	2.63	2.98
Aluminium, 0.20 in	Plywood	None	0.93	1.29	1.64	1.99	2.34	2.70	3.05
Steel, 0.029 in	Plywood	Steel, 0.029 in	3.00	3.30	3.70	4.00	4.40	4.70	5.10
Stainless steel, 0.019 in	Plywood	Steel, 0.029 in	2.50	2.60	3.00	3.30	3.70	4.00	4.40
Stainless steel, 0.025 in	Plywood	Steel, 0.029 in	2.90	3.20	3.60	3.90	4.30	4.60	5.00
Stainless steel, 0.019 in	Plywood	Stainless steel, 0.025 in	2.70	3.10	3.50	3.80	4.20	4.50	4.90
Stainless steel, 0.019 in	Plywood	Aluminium, 0.016 in	1.60	2.00	2.30	2.70	3.00	3.40	3.80
Stainless steel, 0.025 in	Plywood	Aluminium, 0.020 in	2.00	2.30	2.70	3.00	3.40	3.80	4.1
Aluminium, 0.016 in	Plywood	Steel, 0.029 in	1.70	2.10	2.40	2.80	3.10	3.50	3.80
Aluminium, 0.020 in	Plywood	Steel, 0.029 in	2.10	2.50	2.80	3.20	3.50	3.90	4.20
Aluminium, 0.016 in	Plywood	Aluminium, 0.016 in	1.09	1.45	1.80	2.15	2.50	2.86	3.21
Aluminium, 0.020 in	Plywood	Aluminium, 0.070 in	1.21	1.57	1.92	2.27	2.62	2.98	3.33

TABLE 2 – CORE MATERIALS

Material	SG	Weight lb/in³
Balsa	0.14	0.005
Mahogany*	0.63–0.85	0.023–0.031
Plywood (AV) ¼ in	—	0.98 †
Plywood (AV) ⅜ in	—	1.56 †
Plywood (AV) ½ in	—	2.00 †
Plywood (AV) ⅝ in	—	2.50 †
Plywood (AV) ¾ in	—	3.00 †
		lb/sq ft per inch thickness
Aluminium foil		
0.002 — ¼ in cell	—	0.36
0.004 — ¼ in cell	—	0.74
0.006 — ¼ in cell	—	1.20
Fibreglass ⅜ in cell	—	0.95
Fibreglass ¼ in cell	—	0.70
Fibreglass ³⁄₁₆ in cell	—	0.88

* Used for edge banding only
† Weight in lb/sq ft

TABLE 3 – FACE MATERIALS

Material	SG	Weight lb/in³
Alclad	2.80	0.1015
Aluminium	2.70	0.0975
Brass	8.50	0.3070
Copper	8.82	0.3180
Magnesium (alloy)	1.80	0.0650
Monel	8.86	0.3200
Steel	7.85	0.2830
Stainless	7.93	0.2900
Micarta	1.38	0.0500
Formica	1.38	0.0500
Fibreglass	1.68	0.0600
Bakelised fabric	1.27	0.0460
Hard rubber	1.66	0.0600
Polystyrene (solid)	1.05–1.14	0.0380–0.0410

Choice of face material depends both on strength requirements and service conditions. Aluminium facings are the most common for general purpose use, or clad alloy with a minimum thickness of 36 swg. Magnesium alloy has the advantage of a lower density and greater stiffness, whilst stainless steel, titanium and glass fibre facings are coming to the fore for high temperature and corrosion resistant applications.

See also chapter on *Honeycomb materials.*

Lubricants

MINERAL LUBRICATING oils are derived from crude petroleum residues. Various base grades, ranging from thin spindle oils to heavy cylinder oils, are extracted by distillation under vacuum. Lubricating oils are reduced by further refining (normally solvent refining).

The original crudes from which the lubricating oils are obtained, are broadly classified as paraffinic, naphthenic or mixed bases. Modern solvent refining of lubricating oils has now so reduced the differences between the original crudes as to make them of little importance.

The relatively few grades of oil produced by the refineries can be increased by blending two or more of these bases together in order to produce the various viscosities and types of lubricants required by industry.

In addition to the now wide range of straight mineral oils further lubricants are prepared by additional blending with fixed oils, (*ie* animal or vegetable oils) and by the addition of chemical compounds which will confer some special properties on the final product. A limited number of lubricants for special applications are manufactured from synthetic materials.

All lubricating oils have physical properties which can be measured in the laboratory and which are included in the specifications for lubricants; they can also be subjected to chemical and mechanical tests which provide a criterion of how the oil will react under specific conditions.

Specific gravity

The specific gravity is usually measured by a hydrometer or, with very viscous oils, by weighing in a specific gravity bottle. The specific gravity of lubricating oils varies between 0.850 and 0.950. With modern solvent refining techniques gravity is of little significance where quality is concerned, whereas formerly it was accepted that a lower gravity indicated a better quality oil.

Viscosity

The viscosity of an oil is specific to its temperature – *ie* viscosity decreases with increasing temperature and *vice versa*. The viscosity of engine crankcase oils and gear oils is commonly specified in terms of SAE numbers, representing grading by viscosity range (see Table 1). Lubricating oils for general purpose applications may also be referred to by SAE number. Specific viscosity figures are quoted for oils used for special purposes (*eg* hydraulic oils).

TABLE 1 – SAE NUMBER CRANKCASE OILS

SAE Number	Viscosity	Viscosity Range			
		at 0°F*		at 210°F	
		Minimum	Maximum	Minimum	Maximum
5W	Centipoises	—	less than 1 200	—	—
	Centistokes	—	less than 1 300	—	—
	SUS	—	less than 6 000	—	—
10W	Centipoises	1 200	less than 2 400	—	—
	Centistokes	1 300	less than 2 600	—	—
	SUS	6 000	less than 12 000	—	—
20W	Centipoises	2 400	less than 9 600	—	—
	Centistokes	2 600	less than 10 500	—	—
	SUS	12 000	less than 48 000	—	—
20	Centipoises	—	—	5.7	less than 9.6
	Centistokes	—	—	45.0	less than 58.0
	SUS				
30	Centistokes	—	—	9.6	less than 12.9
	SUS	—	—	58.0	less than 70.0
40	Centistokes	—	—	12.9	less than 16.8
	SUS	—	—	70.0	less than 85.0
50	Centistokes	—	—	16.8	less than 22.7
	SUS	—	—	85.0	less than 110.0

* Prior to 1968, viscosity at 100°F was specified instead of 0°F.

Viscosity index

The rate of change of viscosity with temperature can be expressed in terms of a *viscosity index,* which is similar to the shape of the viscosity-temperature curve plotted on a log log-log scale (ASTM chart).

The viscosity index scale was originally established on the basis that a typical paraffinic oil showed a minimum change in viscosity with temperature, and was given a VI rating of 100; whilst a typical naphthenic oil showed a maximum change of viscosity with temperature and was given a VI rating of 0. The viscosity index of any other fluid was then determined on the basis of an equivalent mixture (*ie* the proportion of paraffinic-naphthenic mixture which would have the same viscosity-temperature characteristics as the fluid).

More specifically:

$$\text{Viscosity index} = \frac{L - U}{L + H} \times 100$$

where:

L = Viscosity at 100°F of an oil of VI 0 (naphthenic) having the same viscosity at 210°F as the fluid sample.
U = Viscosity of the oil sample at 100°F.
H = Viscosity at 100°F of an oil of VI 100 (paraffinic) having the same viscosity at 210°F as the fluid sample.

This formula enables the viscosity index of a fluid to be determined if its viscosity is known at two temperatures – 100°F and 210°F. Values of L and U are read from standard tables, consistent with the fluid viscosity at 210°F.

This method is valid for determining VI values up to 100. Certain fluids (*eg* water) and oils with viscosity index improvers, have VI values considerably in excess of 100. Extrapolation of the standard method to cover VI values above 100 yields inconsistent results and so a modern extension of the system has been devised to calculate them. These are correctly designated VI_E. A VI_E value quoted for high viscosity indices can therefore be taken as more truly indicative of the viscosity-temperature characteristics of a very high viscosity index fluid than a VI value.

Viscosity index extension (VI_E)

A new formula for calculating viscosity indices above 100 is based on the recommendations of the ASTM in Method D 2270-64. This goes a long way towards removing the anomalies in the significance of VI for high viscosity index oils. The effect of the new method is shown in Figure 1, where curve B shows the re-

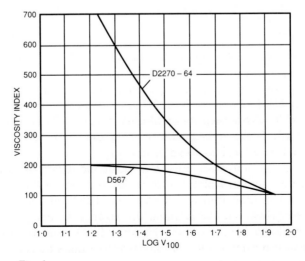

Fig. 1

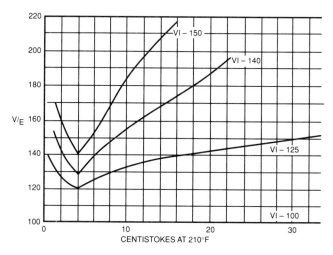

Fig. 2

TABLE 2 – SAE NUMBER TRANSMISSION AND AXLE LUBRICANTS

SAE Number	Viscosity Range			
	at 0°F		at 210°F	
	Minimum	Maximum	Minimum	Maximum
75	—	15 000	—	—
80	15 000	100 000	—	—
90	—	—	75	120
140	—	—	120	200
250	—	—	200	—

All viscosities in SUS

sults obtained by plotting the log of the viscosity at 100°F against VI. This curve is an asymptote, which for an oil having a viscosity of 10 centistokes at 210°F never exceeds approximately 208 VI, even if the viscosity at 100°F becomes equal to that at 210°F. Curve A shows VI_E plotted against the log of viscosity at 100°F for an oil having a viscosity of 10 centistokes at 210°F. This curve will be seen as asymptotic to the vertical axis, *ie* in the case of an ideal oil with similar viscosities at 100°F and 210°F would yield a VI_E of infinity.

The method is still not exact. Figure 2 shows the region in which anomalies occur, this being a series of plots for four different oils having the same VI by the original method, plotted against viscosity at 210°F on one axis, and against VI on the other. However, the relationship between VI and VI_E is reasonably consistent for oils having a viscosity above 4 centistokes at 210°F, and such anomalies are largely confined to the low viscosity ranges outside the normal application of hydraulic fluids.

Flash point

The flash point is the temperature at which a mixture of oil vapour and air will ignite when a flame is applied. The test is carried out in a Pensky Marten apparatus where the oil is heated in a closed metal cup into which a small flame is dipped at regular intervals during the test. When the flame ignites the vapour, the temperature is recorded as the closed flash point. The open flash point is determined by essentially the same test if required, except that an open cup is used and the oil is exposed to the flame for the duration of the test. The open flash point is usually 25 to 30°F higher than the closed. The flash point is an indication of the fire risk of the oil and to some extent indicates the volatility and initial vaporisation of the base oil.

Pour point

The pour point of a fluid is determined empirically, using standing apparatus, as 2.8°C (5°F) above that temperature at which the fluid ceases to flow. It represents the highest practical viscosity value where the substance is still fluid, or in practical terms the lowest temperature at which the substance can still be considered to be a fluid.

Pour points are specified in multiples of 5°F and are not strictly related to service or storage conditions where the actual pour point may be significant.

Spontaneous ignition temperature

The spontaneous ignition temperature (SIT) is a higher temperature still, at which a mixture of vapour and air will ignite without the application of an external spark or flame. It is totally unrelated to the flash point and is also dependent on the vapour and air mixture being within the requisite explosive limits. It is of obvious significance in the case of oils in systems subjected to adiabatic compression, where a 'diesel' explosion could be initiated when the SIT of the fluid corresponds with the peak temperature realized.

Wear tests

A variety of empirical test methods have been developed for measurement of the lubricating properties of oils, probably the best known being the 'ball' tests. A typical example is the 4-ball test. Briefly, this consists of spinning a ½ in diameter steel ball clamped in a chuck on a constant speed motor in the cavity formed by three other ½ in diameter balls squeezed together in a clamping ring. The assembly is maintained in position in a bowl filled with the oil to be tested. The bowl itself is free to rotate on a thrust bearing and a vertical load is applied to the top ball by means of pressure exerted by the underlying balls, which are forced upwards by weights hung on a counter-balanced lever. During rotation this load causes a turning moment to be applied to the bowl and this can be measured by means of a spring and recording drum actuated by a calibrating arm attached to the bowl.

The performance of the lubricant can thus be examined under several pre-determined loads, whilst an indication of its extreme pressure properties can be determined from the load at which seizure becomes apparent, as shown by violent movement of the torque indicator. The balls can also be examined for wear after a run of specified duration.

Such tests are mainly of value for assessing the qualities of conventional lubricants. A generally more satisfactory method for the evaluation of the lubricating qualities of hydraulic fluid is the Vickers test for Maximum Severity Oils. This employs a Vickers vane pump, type 105C, which is run at a speed of 1200 r/min, a fluid pressure of 140 kg/cm^2 with a flow rate of 5 gal/min. The fluid is maintained at a constant temperature of 65°C, simulating conditions of severe overload. The quality of the fluid can then be assessed on the wear (measured loss in weight) of the inner and outer port plates, cam ring and vanes of the pump, after a specified test duration. Test duration may be specified in hours requirement, or cease when excessive noise develops in the pump, or there is obvious loss of suction, due to excessive wear. Results are comparative (see Figure 3) an individual test normally being discontinued when the rate of wear becomes very rapid or noise becomes excessive (indicating excessive wear, which is subsequently measured). In addition to actual wear measurements, the condition of the pump after the test may also be given.

Carbon residue

If an oil is subjected to continuous heating so that it is completely volatilised, an amount of carbon or coke is left. The amount of carbon is expressed as a percentage by weight of the original sample.

The test therefore gives an indication of the tendency of oil to form carbonaceous deposits, but with the introduction of engine testing of crankcase oils the carbon residue test is not now regarded as significant.

Colour

The colour of unused lubricating oil is tested by comparison with glass colour standards. The principally used method is the Union colour scale, which ranges from No 1 lily white, No 5 light red to No 8 extra dark red. The colour of an oil

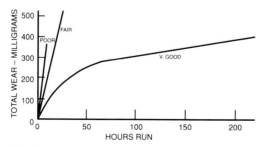

Fig. 3

is related to the method of refining and treatment and is useful for recognition purposes.

Additives

The main purposes of lubrication are the reduction of the coefficient of friction (by interposing a layer of oil between the two contact faces), prevention of wear, and as a cooling medium. Oil is also used in such applications as power transmission (as in the case of hydraulic systems), to radiate heat to form soluble and non-soluble cutting compounds and emulsions for engineering machine shops, to act as an insulator, and to provide protection against corrosion. For these and many other uses the oil is specially selected, blended and chemically compounded.

To give the required performance, selected oils are blended to give the correct viscosity or degree of fluidity at operating temperatures, and when required, are additive treated to impart certain characteristics (see Table 3).

TABLE 3 – ADDITIVES FOR LUBRICATING OILS

Additive	Purpose and Remarks
Oxidation inhibitors	To improve chemical stability and useful life of lubricants.
Rust inhibitors	To reduce or eliminate potential corrosive effect of water present in system.
Anti-wear additives	Boundary lubricant to reduce friction, stabilise lubricating film and reduce wear.
Detergent	Dispersion of contaminants.
Anti-foam agent	Elimation of foaming caused by presence of dissolved or entrained air.
Demulsifying agents	To separate water form oil; can introduce corrosion or need for rust inhibitors.
Viscosity index improver	Improving viscosity/temperature characteristics of oil.
Extreme pressure lubricant	Top maintain film lubrication under extreme local pressure – *eg* gear oils and hypoid gear lubricants.
Detergent dispersant	Detergent dispersion.
Pour point depressant	Improving fluidity at low temperatures.
Tackiness agent	Improving adhesion to metal surfaces.
Dyes	—
Colour stabilizer	—

Other additives may be used to promote:

Detergency
anti-oxidation
extreme pressure, or a property of film strength
demulsibility
non-glazing or prevention of lacquering
maximum oiliness
high dispersion rate
non-foaming
stabilisation
lower pour point
high flash point
water repellence, etc.

Some of these qualities are inherent in the base oils which are suitable for particular purposes. Other types have to be specially treated to attain the required characteristics, such as by the addition of silicon fluid to impart water repellent and anti-foam properties.

See also chapter on *Silicones* (Section 3).

Metallic wool

METALLIC WOOLS are widely used in the following industrial fields:

Cleaning.
Exhaust systems (acoustic packings).
Filtration units (filter media for air, gases and fluids).
Dust extractors.
Reinforcing materials.
Foundry uses.
Gold refining.

The main demand is for steel wool. Stainless steel wool is also coming into increasing use as an alternative to wools made from non-ferrous metals, where corrosion resistance and/or non-staining properties are required. Lead wool is particularly applicable as a caulking/sealing/jointing medium where metal-to-metal contact is important.

Applications in this field include:

Jointing cast iron and steel pipes.
Sealing off joints in interlocking steel piling.
Lead wool is also used for expansion joints in concrete, closing off cracks
  in concrete and anti-radiation screens in nuclear submarines.

Perforated metals

PERFORATED METALS are produced with an almost infinite variety of hole forms, sizes and spacings, although not necessarily as standard productions. The metals most commonly available in perforated form are aluminium, brass, copper, Monel, nickel, phosphor bronze, steel, stainless steel, titanium and titanium alloys, tin plate and zinc.

Standard forms of perforation include the following:

(i)   Round hole.
(ii)   Square hole.
(iii)   Slot hole – round end,   diagonal.
        Slot hole – square end,   herringbone.
(iv)   Diamond patterns.
(v)   Oval patterns.
(vi)   Hexagonal patterns.
(vii)   Embossed patterns – round slot.
(viii) Lipped perforations.
(ix)   Ornamental perforations.

They are not necessarily available in all metals, and in the case of specific metals the choice of perforation geometry may be controlled or limited by metal thickness.

In addition, perforation techniques may be used to produce sieves, sieve plates, screens, cable trays, etc, in suitable metals. Very fine mesh forms, comparable to woven wire mesh in geometry, can be produced by electroforming. These have the advantage over woven wire mesh that the mesh is integral and the edges do not unravel or distort when cut.

Perforated plastics

Thermoplastic sheet materials can readily be perforated by simple production techniques but are not widely produced as commercial materials, except to special requirements. The main exceptions are perforated PVC in sheets up to

36 in × 72 in (nominally 1 m × 2 m) for the chemical industry, interior decoration, display and other purposes, and perforated toughened polystyrene.

Perforated hardboards

Perforated hardboards, normally known as pegboards, are readily available as natural or surfaced hardboards with circular holes of various diameters and spacings. The most common is 5 mm ($^{3}/_{16}$ in) diameter at 12.56 mm (½ in), 19 mm (¾ in) and 25 mm (1 in) spacings.

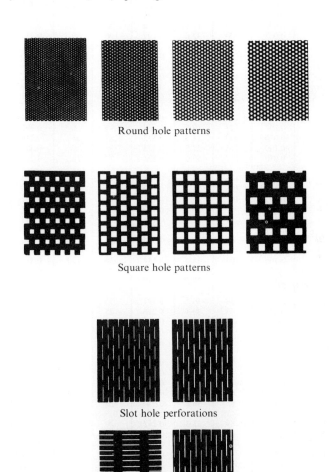

Round hole patterns

Square hole patterns

Slot hole perforations

Examples of perforated metal patterns.
(N. Greening (Warrington) Ltd).

TABLE 1 – PRODUCTION SIZES OF ROUND HOLE PATTERN PERFORATED METAL*

Hole Size in	Normal Maximum Thickness in Mild Steel BG	Close Pitch			Open Pitch 'Over and Over'			Special Staggered Pitch		
		Centres in	Holes per in²	Open Area %	Centre in	Holes per in²	Open Area %	Centres in	Holes per in²	Open Area %
0.026	26	0.052	427	22.6	0.090	142	7.5	0.156	47	2.5
0.029	24	0.059	333	21.9	0.102	111	7.3	0.177	37	2.4
0.032	22	0.070	236	18.9	0.121	79	6.3	0.210	26	2.1
0.032	22	0.090	143	11.5	0.156	48	3.8	0.270	16	1.3
0.037	26	0.062	300	32.6	0.107	100	10.7	0.186	33	3.6
0.037	22	0.080	180	19.3	0.138	60	6.4	0.240	20	2.1
0.038	24	0.063	291	33.0	0.109	97	11.0	0.189	32	3.7
0.039	22	0.084	164	19.5	0.145	55	6.5	0.252	18	2.2
0.039	20	0.090	143	17.0	0.155	48	5.7	0.270	16	1.9
0.041	26	0.068	250	32.9	0.118	83	11.0	0.204	28	3.7
0.041	20	0.088	149	19.5	0.152	50	6.5	0.264	17	2.2
0.045	22	0.097	123	19.5	0.168	41	6.5	0.291	14	2.2
0.047	26	0.076	200	34.5	0.132	67	11.5	0.228	22	3.8
0.050	18	0.100	115	22.5	0.173	38	7.5	0.300	13	2.5
0.055	24	0.070	217	50.0	0.126	72	16.7	0.219	24	5.6
0.055	24	0.089	146	34.7	0.154	49	11.6	0.267	16	3.9
0.055	20	0.105	105	25.0	0.182	35	8.3	0.315	12	2.8
0.058	24	0.093	103	35.2	0.161	44	11.8	0.297	15	3.9
0.058	16	0.109	97	25.5	0.189	32	8.5	0.327	11	2.8
0.063	16	0.115	87	27.5	0.199	29	9.2	0.345	10	3.1
0.063	16	0.130	67	21.3	0.225	22	7.1	0.390	7	2.4
0.066	22	0.105	105	36.0	0.182	35	12.0	0.315	12	4.0
0.069	22	0.106	103	38.4	0.184	34	12.8	0.318	11	4.3
0.069	18	0.121	79	29.5	0.209	26	9.8	0.363	9	3.3
0.069	16	0.126	73	27.2	0.218	24	9.1	0.378	8	3.0
0.072	22	0.114	89	36.2	0.197	30	12.1	0.342	10	4.0
0.072	18	0.125	75	30.5	0.216	25	10.2	0.375	8	3.4
0.075	22	0.118	83	36.5	0.204	28	12.2	0.354	9	4.1
0.075	20	0.128	70	31.0	0.221	23	10.3	0.384	8	3.4
0.077	20	0.121	79	36.7	0.210	26	12.2	0.363	9	4.1
0.079	18	0.124	75	36.9	0.215	25	12.3	0.372	8	4.1
0.079	14	0.158	46	22.7	0.274	15	7.6	0.474	5	2.5
0.081	16	0.136	63	32.5	0.236	21	10.8	0.408	7	3.6
0.083	20	0.129	69	37.5	0.223	23	12.5	0.387	8	4.2
0.085	18	0.133	65	36.8	0.230	22	12.3	0.399	7	4.1
0.085	18	0.140	59	33.3	0.242	20	11.1	0.420	7	3.7
0.088	20	0.137	62	37.5	0.237	21	12.5	0.411	7	4.2
0.088	16	0.144	56	33.8	0.249	19	11.3	0.432	6	3.8
0.090	18	0.140	59	37.5	0.242	20	12.5	0.420	7	4.2
0.092	18	0.143	56	37.5	0.247	19	12.5	0.429	6	4.2
0.095	18	0.143	56	40.0	0.247	19	13.3	0.429	6	4.4
0.095	14	0.165	42	30.0	0.286	14	10.0	0.495	5	3.3
0.095	13	0.188	33	22.5	0.325	11	7.5	0.564	4	2.5
0.097	18	0.149	52	38.4	0.258	14	12.8	0.347	5	4.2
0.099	14	0.154	49	37.5	0.267	16	12.5	0.462	5	4.2
0.099	14	0.190	32	24.5	0.328	11	8.2	0.570	4	2.7

* N. Greening (Warrington) Ltd.

(continued)

TABLE 1 – PRODUCTION SIZES OF ROUND HOLE PATTERN PERFORATED METAL* – *contd.*

Hole Size in	Normal Maximum Thickness in Mild Steel BG	Close Pitch			Open Pitch 'Over and Over'			Special Staggered Pitch		
		Centres in	Holes per in²	Open Area %	Centre in	Holes per in²	Open Area %	Centres in	Holes per in²	Open Area %
0.101	18	0.155	48	38.5	0.268	16	12.8	0.465	5	4.3
0.101	16	0.203	27	22.0	0.352	9	7.3	0.609	3	2.4
0.103	24	0.158	46	39.0	0.274	15	13.0	0.474	5	4.3
0.106	18	0.162	44	38.8	0.281	15	12.9	0.486	5	4.3
0.108	18	0.165	42	38.8	0.286	14	12.9	0.492	5	4.3
0.110	18	0.167	41	39.4	0.289	14	13.1	0.501	5	4.4
0.110	12	0.220	24	22.7	0.381	8	7.6	0.660	3	2.5
0.112	18	0.170	40	39.0	0.294	13	13.0	0.510	4	4.3
0.112	18	0.151	50	51.5	0.282	17	17.2	0.453	6	5.7
0.115	18	0.175	38	39.3	0.303	13	13.1	0.525	4	4.4
0.117	16	0.177	37	39.5	0.307	12	13.2	0.531	4	4.4
0.117	12	0.187	33	35.5	0.323	11	11.8	0.561	4	3.9
0.117	14	0.203	27½	32.5	0.352	9	10.8	0.609	3	3.6
0.117	12	0.237	24	26.0	0.350	8	8.7	0.711	3	2.9
0.120	18	0.180	36	40.3	0.313	12	13.4	0.540	4	4.5
0.120	12	0.240	20	22.7	0.416	7	7.6	0.720	2	2.5
0.122	18	0.184	34	40.0	0.319	11	13.3	0.552	4	4.4
0.125	16	0.188	32½	40.2	0.325	11	13.4	0.564	4	4.4
0.125	14	0.208	27	32.6	0.360	9	10.9	0.624	3	3.6
0.125	12	0.219	24	29.7	0.379	8	9.9	0.657	3	3.3
0.127	18	0.190	32	40.5	0.328	11	13.5	0.570	4	4.5
0.129	18	0.193	31	38.0	0.334	10	12.7	0.579	3	4.2
0.134	18	0.201	29	40.5	0.348	10	13.5	0.603	3	4.5
0.134	14	0.250	18½	26.2	0.433	6	8.7	0.750	2	2.9
0.136	18	0.203	27	38.0	0.352	9	12.7	0.609	3	4.2
0.139	18	0.193	31	47.5	0.334	10	15.8	0.579	3	5.3
0.139	16	0.207	27	40.9	0.359	9	13.6	0.621	3	4.5
0.139	10	0.238	21	30.8	0.412	7	10.3	0.714	2	3.4
0.141	16	0.210	26	39.8	0.364	9	13.3	0.630	3	4.4
0.143	18	0.213	25	40.8	0.369	8	13.6	0.539	3	4.5
0.146	16	0.217	24½	40.9	0.375	8	13.6	0.651	3	4.5
0.148	18	0.220	24	41.0	0.381	8	13.7	0.660	3	4.6
0.151	18	0.224	23	41.3	0.388	8	13.8	0.672	3	4.6
0.153	18	0.226	22½	41.5	0.391	8	13.8	0.678	3	4.6
0.155	18	0.232	21½	40.5	0.402	7	13.5	0.696	2	4.5
0.157	16	0.232	21½	41.5	0.402	7	13.8	0.696	2	4.6
0.157	10	0.257	17½	33.8	0.445	6	11.1	0.771	2	3.8
0.159	16	0.235	21	41.5	0.407	7	13.8	0.705	2	4.6
0.161	16	0.238	20½	41.5	0.412	7	13.8	0.714	2	4.6
0.161	10	0.380	8	16.0	0.658	3	5.3	1.140	–	2.8
0.164	16	0.243	19½	41.3	0.421	6	13.8	0.729	2	4.6
0.168	16	0.248	19	41.5	0.430	6	13.8	0.744	2	4.6
0.172	16	0.258	17½	40.5	0.447	6	13.5	0.774	2	4.5
0.175	16	0.257	17½	42.1	0.445	6	14.0	0.771	2	4.7
0.178	18	0.261	16½	42.0	0.452	5	14.0	0.783	2	4.7

* N. Greening (Warrington) Ltd.

(continued)

TABLE 1 – PRODUCTION SIZES OF ROUND HOLE PATTERN PERFORATED METAL* – *contd.*

Hole Size in	Normal Maximum Thickness in Mild Steel BG	Close Pitch			Open Pitch 'Over and Over'			Special Staggered Pitch		
		Centres in	Holes per in²	Open Area %	Centre in	Holes per in²	Open Area %	Centres in	Holes per in²	Open Area %
0.178	10	0.279	15	37.0	0.483	5	12.3	0.837	2	4.1
0.180	18	0.264	16½	42.0	0.457	5	14.0	0.792	2	4.7
0.182	18	0.267	16	42.0	0.462	5	14.0	0.801	2	4.7
0.185	18	0.272	15½	42.0	0.471	5	14.0	0.816	2	4.7
0.188	16	0.276	15	42.0	0.478	5	14.0	0.828	2	4.7
0.188	16	0.281	14½	40.5	0.486	5	13.5	0.843	2	4.5
0.188	3/16 in	0.437	6¼	16.5	0.757	2	5.5	1.311	—	1.8
0.191	18	0.279	15	42.0	0.483	5	14.0	0.837	2	4.7
0.194	16	0.283	14	42.0	0.490	5	14.0	0.849	2	4.7
0.197	16	0.287	14	42.5	0.497	5	14.2	0.861	2	4.7
0.203	16	0.297	13	42.5	0.513	4	14.2	0.891	1	4.7
0.207	16	0.303	12½	42.5	0.525	4	14.2	0.909	1	4.7
0.211	16	0.308	12½	43.0	0.533	4	14.3	0.924	1	4.8
0.219	16	0.317	11½	43.3	0.549	4	14.4	0.951	1	4.8
0.219	14	0.328	11	40.5	0.568	4	13.5	0.984	1	4.5
0.219	14	0.348	9½	35.8	0.603	3	12.0	1.044	1	4.0
0.227	16	0.329	10½	43.0	0.570	3	14.3	0.987	1	4.5
0.235	10	0.315	12½	45.0	0.546	4	15.0	0.945	1	5.0
0.235	10	0.369	8½	36.7	0.639	3	12.2	1.107	1	4.1
0.242	16	0.348	9½	44.0	0.603	3	14.6	1.044	1	4.9
0.250	18	0.343	9¾	48.2	0.594	3	16.1	1.029	1	5.4
0.250	10	0.359	9	43.0	0.622	3	14.3	1.077	1	4.8
0.255	14	0.367	8½	43.8	0.636	3	14.6	1.101	1	4.9
0.255	16	0.390	7½	40.0	0.675	2	13.3	1.170	—	4.4
0.262	14	0.377	8	43.8	0.653	2	14.6	1.131	—	4.9
0.265	14	0.381	8	44.0	0.660	2	14.7	1.143	—	4.9
0.269	14	0.394	7¼	45.0†	—	—	—	—	—	—
0.276	18	0.369	8½	50.0	0.639	3	16.7	1.107	1	5.6
0.276	10	0.419	6½	40.0	0.716	2	13.3	1.257	—	4.4
0.276	10	0.579	3½	19.0†	—	—	—	—	—	—
0.281	16	0.390	7½	47.0	0.675	2	15.6	1.170	—	5.2
0.285	12	0.407	7	44.0	0.705	2	14.6	1.221	—	4.9
0.293	12	0.419	6½	44.5	0.716	2	14.8	1.257	—	4.9
0.295	14	0.439	6	41.0	0.760	2	13.7	1.317	—	4.6
0.302	16	0.422	6½	46.0	0.731	2	15.3	1.266	—	5.1
0.312	3/16 in	0.444	5¾	44.7	0.769	2	14.9	1.332	—	5.0
0.316	10	0.449	5¾	45.5	0.778	2	15.2	1.347	—	5.1
0.322	14	0.422	6½	52.0	0.731	2	17.3	1.266	—	5.8
0.333	14	0.473	5	45.0	0.819	2	15.0	1.419	—	5.0
0.344	10	0.488	4¾	45.0	0.845	2	15.0	1.464	—	5.0
0.348	10	0.494	4¾	45.0	0.856	2	15.0	1.482	—	5.0
0.355	14	0.500	4½	45.0	0.866	1	15.0	1.500	—	5.0
0.365	14	0.515	4¼	45.5	0.890	1	15.2	1.545	—	5.1
0.375	14	0.500	4½	51.0	0.866	1	17.0	1.500	—	5.7
0.391	10	0.555	3¾	45.0	0.961	1	15.0	1.665	—	5.0
0.398	10	0.565	3½	45.0	0.979	1	15.0	1.695	—	5.0
0.406	10	0.564	3½	47.0	0.977	1	15.7	1.692	—	5.2

* N. Greening (Warrington) Ltd.

(continued)

TABLE 1 – PRODUCTION SIZES OF ROUND HOLE PATTERN PERFORATED METAL* – *contd.*

Hole Size in	Normal Maximum Thickness in Mild Steel BG	Close Pitch			Open Pitch 'Over and Over'			Special Staggered Pitch		
		Centres in	Holes per in²	Open Area %	Centre in	Holes per in²	Open Area %	Centres in	Holes per in²	Open Area %
0.430	14	0.594	3¼	48.0	1.029	1	16.0	1.782	—	5.3
0.437	14	0.581	3¼	48.0	1.006	1	16.0	1.743	—	5.3
0.500	16	0.625	3	58.0	1.082	1	19.3	1.875	—	6.4
0.500	16	0.662	2¾	52.0	1.146	—	17.3	1.986	—	5.8
0.500	10	0.750	2	40.3	1.299	—	13.4	2.250	—	4.5
0.516	14	0.696	2¼	44.0	1.204	—	14.7	2.088	—	4.9
0.531	14	0.649	2¾	57.0	1.124	—	19.0	1.947	—	6.3
0.531	10	0.796	1¾	40.3	1.38	—	13.4	2.388	—	4.5
0.531	10	1.000	—	27.0†	—	—	—	—	—	—
0.547	10	0.738	2	48.0	1.277	—	16.0	2.214	—	5.3
0.561	10	0.709	2¼	55.0	1.228	—	18.3	2.127	—	6.1
0.625	14	0.824	1¾	52.1	1.427	—	17.4	2.472	—	5.8
0.625	16	0.875	1½	46.3†	—	—	—	—	—	—
0.669	10	0.854	1¾	45.0	1.479	—	15.0	2.562	—	5.0
0.709	10	0.894	1½	57.0	1.548	—	19.0	2.682	—	6.3
0.750	18	0.921	1¼	60.1†	—	—	—	—	—	—
0.750	14	1.000	1	50.5†	—	—	—	—	—	—
0.787	10	1.037	—	50.0	1.796	—	16.7	3.111	—	5.6
0.875	10	1.128	—	54.8	1.954	—	18.3	3.384	—	6.1
1.000	18	1.188	—	64.3†	—	—	—	—	—	—
1.125	10	1.407	—	58.0	2.437	—	19.3	4.221	—	6.4
1.25	10	1.562	—	58.0	2.705	—	19.3	4.686	—	6.4
1.625	16	2.625	—	34.1	4.545	—	1.4	7.875	—	3.8

* N. Greening (Warrington) Ltd.

Powdered metals

METAL POWDERS are discrete particles of metallic elements or alloys. They are usually of irregular or near-spherical shape, although for some applications they may be fine flakes, fibres or of dendritic form (like snow flakes). Many combinations of particle composition and form are possible and the industries involved in producing and consuming metal powders serve many sectors of the economy. The powders may be produced chemically or mechanically, depending on the composition and form of power required.

Powders can be prepared chemically by reduction of oxides or other chemical compounds, by precipitation from solution, or by thermal decomposition from a gaseous phase. Of the powders used for making components by pressing and sintering, iron powder comprises the largest tonnage. For bulk production of cheap iron powder, iron oxide is reduced by reaction with carbon. Other metals, such as copper, molybdenum and tungsten can also be produced by reduction of the oxide, often using a gaseous reductant to prevent contamination of the powder with carbon or carbides. Metallic reducing agents are needed to make some elements; for example sodium is used to reduce $TiCl_4$. The high reaction temperatures in these processes often result in partial sintering, so that a spongy mass is formed which is then milled to a fine powder.

Production of powders from solution is also commonly practised. Fine particulate copper can be precipitated electro-chemically. There is also a process developed by Sherritt Gordon Mines for the refining of cobalt, nickel and copper, in which there elements are precipitated as powders by reaction with hydrogen under pressure.

Precipitation from a vapour phase may also produce particulate materials, probably the best known example being the production of pure nickel by thermal decomposition of nickel carbonyl ($Ni(CO)_4$).

Mechanical methods of powder production include crushing, grinding and milling, but more commonly atomising is used, fine particles being formed by disintegration of a stream of molten metal; as the metal is melted, alloying can be carried out during processing. The molten metal stream is broken up by jets of high pressure water or gas. Typically, water atomised powder is irregular and in the size range 50 to 100 μm. Gas atomised powder cools more slowly and surface tension may be sufficient to cause the droplets to form spheres before solidification. Although more expensive, inert gases can be used to minimise oxidation and other contamination of the product. The irregular water atomised particles are very suitable for die pressing for high volume, low cost production of components. The gas atomised powders are used for applications such as aerospace components, including jet engine components, for which high quality and integrity are vital. The rotating bar process, by arc melting the end of a rapidly rotating bar produces high quality powder. Droplets thrown off by the centrifugal forces solidify in flight to spherical particulates, and as the molten metal does not come into contact with any container, the product is free from contamination. This type of powder has been used to make titanium alloy compressor parts and turbine discs in nickel base superalloys.

Characteristics

The main features of powders are their chemical composition, mean particle size and size distribution, particle shape, flow rate, tap density and compressibility. The range of compositions available includes all the common metals and many alloys. By blending master alloys and elemental powders it is possible to produce alloys not easily made by conventional melting and casting methods.

Partial size (diameter) is dependent on the production process and can vary from less than 0.1 μm to more than 1 mm. However for production of conventional powder metallurgy components, 1 to 100 μm mean particle size is typical.

Particle shape is critical for many applications; for example fine flake is required for metallic paints, and most solid components are pressed form irregular shaped powders which have good green strength when consolidated in a die.

Commercially the flow characteristics of powders are very important. In die filling, powders are usually gravity fed and the powder must flow quickly and evenly into the die to ensure rapid continuous production. Similarly, uniform feeding behaviour is important in metal spraying, if uniform coatings are to be laid down.

Tap density (a measure of the weight of powder need to fill a given volume) and consistency are essential if all the components being pressed are required to come out the same. Compressibility is the extent to which the metal can be densified by applying a pressure and relates to shrinkage during sintering and final density, which is very important if the finished part is to be of the correct shape and size.

Powder metallurgy

Powder metallurgy is the production of solid objects from metal powders. Some special alloys can only be processed by the powder route, but these are few compared to the large numbers of components made in iron and copper alloys. Powder metallurgy is used to produce these components because it is a cheap, cost effective way of making large numbers of complex shaped components of good dimensional accuracy.

Consolidation processes

Production of components from powders is normally by die pressing and sintering. The appropriate powders are blended and pressed in a die to a workable 'green' compact. This is then heated (usually in a protective atmosphere), to strengthen the bond between the particles. Shrinkage will occur during sintering and careful control is needed to achieve specified dimensional tolerances.

Many parts are not sintered to their full theoretical density; typically a steel pressed at 450–500 MPa and sintered at 1120°C will shrink 1–2%. Residual porosity may be as high as 15%.

By pouring carefully sized nearly spherical particles into a mould and sintering without pressing it is possible to make porous filters and bearings with interconnected pores of a specific size. Another process becoming more readily available is metal powder injection moulding, in which the metal powder is suspended in a plastic and injected into a die. Parts made this way can be much more complex than simple die pressings.

Higher density and greater precision of dimensions can be achieved by post-sintering operations, such as re-pressing, either cold or hot. This also increases the strength and ductility of the component, although at increased cost.

Inclusion of low melting point alloys in the mixture, to give liquid phases during sintering, will increase the rate of densification and reduce the final porosity. Steel parts can also be infiltrated with molten copper to eliminate the porosity.

It is essential to define the properties required in the finished component, since the extra processing involved in achieving near zero porosity and thereby maximum strength and ductility, add to the cost. For many applications the properties of simple, cheap pressed and sintered parts are adequate, even though the ductility and impact resistance of porous materials are poor and the strength is lower than can be achieved in a similar alloy in wrought form.

As with components produced by other processes, the surface properties can be adjusted to improve corrosion or wear resistance by electro-plating, carburising or carbo-nitriding.

Applications

The most important use of metal powders is for the production of components. These range from small parts for business machines and magnets for the electronics industry, to sophisticated jet engine components, and surgical implants. The major user of powder metallurgy parts is the automotive industry, which requires bearings, filters, gears, clutch parts, cams and many other small, complex precision parts in large numbers, at low cost.

Large quantities of metal powder are used in applications other than the production of small components. They are a convenient starting point for the production of many metal salts and metal containing organic chemicals. They are added to plastics to render them conducting or for their decorative effect, and lead-loaded plastic is used for sound proofing. Aluminium powder is a rocket propellant and is used as a reagent in the thermit reaction. Powders are used in welding rods, for metal spraying, as a constituent in metallic paints and in the toner for copying machines.

Metal powders come in many compositions, shapes and sizes and further applications are developed for them as their many useful properties are recognised and exploited.

Printing metals

PRINTING METALS are lead based alloys containing 3 to 20% tin and 11 to 30% antimony, their composition varying with the application for which they are designed. The principal requirements for hot metal printing are:

(i)   Low melting point with clean melt.
(ii) Superior casting characteristics for reproduction of fine detail.
(iii) Sufficient hardness and resistance to abrasion for process printing.

Lead-tin-antimony alloys are noted for their clean melting and good casting characteristics, producing comparatively little dross, and can be used repeatedly. The melting point is dependent on the alloy composition, the minimum melting temperature being achieved with a 17% antimony content. Alloys of this type are normally used for slug casting but are softer than other alloys.

Since only eutectic alloys have a definite melting point, the actual freezing or solidification temperature of the alloy is also significant, as this determines the amount of cooling necessary before the casting can be ejected from the mould. This is normally specified as the solidification range or the difference between the actual melting point and the temperature at which the alloy has become fully solid. Thus in the cast of founder's type alloy (28% antimony) the solidification range is 74°C, compared with 4.4°C in the case of a Monotype alloy (15% antimony).

TABLE 1 – TYPICAL SHRINKAGE OF PRINTING METALS

Metal or Alloy	Shrinkage %
Lead	3.4
Tin	2.8
Antimony	0.9 expansion
Electro backing metal	2.6
Slug casting metal	2.3
Monotype metal	2.0

The eutectic alloy (4% tin, 12% antimony, 84% lead) has a melting point and freezing point of 240°C (*ie* has a zero solidification range). However it is generally too soft for printing purposes.

Virtually all the commonly used printing metals contract on solidification, although the contraction is very low (see Table 1). In general, the higher the antimony content the lower the shrinkage. The higher the antimony content, the greater the hardness of the metal and the higher the melting point. Alloys are therefore chosen according to the purpose for which they are required (see Table 2).

Impurities

It is important that impurities in printing metals should be minimal as these can produce bad castings, excessive dross and choked mouthpieces on the casting machines. When molten and freshly skimmed the printing metals should have a mirror-like surface. A clouded surface generally indicates the presence of impurities (and especially zinc).

TABLE 2 – TYPICAL COMPOSITIONS OF PRINTING METALS

Purpose	Antimony %	Tin %	Lead %
Slug casting	11–12	3–4	Balance
Monotype composition	15–16	6–10	Balance
Case type	up to 27	up to 14	Balance
Stereotyping	14–15	6–10	Balance
Founder's type	28	18	Balance

TABLE 3 – SLUG CASTING METALS*

Composition†		Liquidus		Solidus		Solidification Range		Hardness Brinell	Remarks
tin %	antimony %	°F	°C	°F	°C	°F	°C		
2	11	478	248.0			15	8.5	18.5	
3	11	477	247.0			14	7.5	19.5	Most used
3	12	470	243.0			7	3.5	20.0	
3	13	492	255.5			29	16.0	21.0	
				463	239.5				
4	10	485	251.5			27	12.0	20.5	
4	11	477	247.0			14	7.5	21.5	Most used
4	12	463	239.5			—	—	22.0	
5	11	475	246.0			12	6.5	22.0	

* Fry's Metal Ltd
† Balance lead

TABLE 4 – MONOTYPE METALS*

Composition†		Liquidus		Solidus		Solidification Range		Hardness Brinell	Remarks
tin %	antimony %	°F	°C	°F	°C	°F	°C		
6	15	502	261			39	21.5	23	
6	16	527	275			64	35.5	23	
7	15	503	261.5			40	22.0	24	For general composition
8	15	505	263	463	239.5	42	23.5	25	For general composition
8	17	520	271			57	31.5	27	
9	19	546	255.5			53	46.0	28.5	
10	15	518	270			55	30.5	26	Fine printing and long runs
10	16	525	274			62	34.5	27	Fine printing and long runs
13	17	542	283			79	43.5	29.5	High quality printing and superior face

* Fry's Metals Ltd † Balance lead

TABLE 5 – CASE TYPE METALS*

Composition†		Liquidus		Solidus		Solidification Range		Hardness Brinell	Remarks
tin %	antimony %	°F	°C	°F	°C	°F	°C		
9	19	546	255.5			83	46.0	28.5	
10	16	525	274			62	34.5	27	Dual-purpose metal
12	24	608	320	463	239.5	145	80.5	33	
13	17	542	283			79	43.5	29.5	
15	25	634	332			167	92.5	36	
18	27	696	341			183	101.5	38	

* Fry's Metal Ltd † Balance lead

TABLE 6 – TYPE-CASTING METALS*

Composition†		Liquidus		Solidus		Solidification Range		Hardness Brinell	Remarks
tin %	antimony %	°F	°C	°F	°C	°F	°C		
13	25	616	324.5			153	85.0		
14	27	635	335			172	95.5		
18	28	629	332	463	239.5	166	92.5		
22	27	626	330			163	90.5		
20	20	573	310.5			110	61		Script casting‡

* Fry's Metal Ltd † Balance lead ‡ with 0.8% copper.

TABLE 7 – STEREOTYPE METALS*

Composition†		Liquidus		Solidus		Solidification Range		Hardness Brinell	Remarks
tin %	antimony %	°F	°C	°F	°C	°F	°C		
5	15	509	265			46	25.5	23.0	
6	15	502	261	463	239.5	39	21.5	23.5	13.5% antimony content may also be used
7	15	503	261.5			40	22.0	24.0	
8	15	505	263			42	23.5	25.0	
9	15	509	265			46	25.5	25.5	
10	18	518	270			55	30.5	26.0	

* Fry's Metal Ltd   † Balance lead

TABLE 8 – ELECTRO BACKING METALS*

Composition†		Liquidus		Solidus		Solidification Range		Hardness Brinell	Remarks
tin %	antimony %	°F	°C	°F	°C	°F	°C		
2	2	555	307			122	67.5	10.0	Maximum antimony content for curved plates
2	3	572	300	463	239.5	109	60.5	11.6	
3	3	563	295			100	55.5	12.0	
4	4	545	255			82	45.5	13.5	

* Fry's Metal Ltd   † Balance lead

Impurities which may be present, and their effects, are listed under separate headings:

Zinc – can cause the most trouble and may be introduced into the melt from zincos or brass rules. Zinc will markedly increase the amount of dross, produce sluggish flow and seriously affect the quality of the castings.

Aluminium – has a similar effect to zinc, although it is not readily dissolved in the melt.

Iron – will probably be present in small quantities in even the virgin alloy and may increase slightly by absorption from iron melting pots. The effect is not usually significant, provided that only traces of iron are present.

Copper – is tolerable in small quantities, but if allowed to build up can cause choking of the throat or mouthpiece of the casting machine. Likely sources of copper contamination are cuttings from half tones and brass rules accidentally introduced into the melt.

Nickel – has a similar effect to copper, but can cause choking at lower concentration levels. However, nickel does not dissolve readily and may be skimmed off if the melt is not overheated. The most likely source of nickel contamination is when plated stereos are re-melted.

TABLE 9 – LIQUIDUS TEMPERATURES OF TIN-ANTIMONY-LEAD ALLOYS*

Antimony %		Tin %					
		2	**4**	**6**	**8**	**10**	**12**
2	°F	585	568	561	556	550	543
	°C	307	298	294	291	288	284
4	°F	559	545	540	536	529	522
	°C	293	285	282	280	276	272
6	°F	536	527	518	514	511	504
	°C	280	275	270	268	266	262
8	°F	513	507	498	495	491	482
	°C	267	264	259	257	255	250
10	°F	491	485	478	477	475	478
	°C	255	252	248	247	246	248
12	°F	482	463	477	485	489	493
	°C	250	239	247	252	254	256
14	°F	516	498	495	500	507	511
	°C	269	259	257	260	264	266
16	°F	550	531	527	514	525	531
	°C	288	277	275	268	274	277
18	°F	577	565	559	536	540	547
	°C	303	296	293	280	282	286
20	°F	604	601	590	565	552	558
	°C	318	316	310	296	289	292
22	°F	637	626	615	597	585	579
	°C	336	330	324	314	307	304
24	°F	662	646	635	626	612	608
	°C	350	341	335	330	322	320

*Journal of the institute of metals

Oxides – increase the amount of dross and can also cause choking. The best safeguard is to eliminate the possibility of oxides being carried forward to the casting pots, by careful cleaning and ingotting of the metal.

As a general rule, to minimise contamination of the printing metal, slugs or type should be re-melted as ingots rather than being put straight back into the machine. Brass rules, zincos and cuttings from copper plate should never be allowed to enter the melt. To maintain consistency, metals of different grades should be kept separate and never mixed either in the melt (*eg* a different grade added to make up the quantity), or in the ingot stock fed to the pots.

Temperature control is also important. Heating temperature must be high enough to ensure complete solution of tin-antimony crystals, but not so high that there is an excessive formation of dross. Casting temperature is a little below the heating temperature, typical values for the latter being:

Slug casting metals – 315°C.

Monotype composition metals – 371°C.

Melting should be rapid, with sufficient heating capacity to ensure that the correct temperature will be quickly recovered if the pot is chilled for any reason.

Pyrometric materials

PYROMETRIC CONES – commonly called Seger cones after their inventor Dr
Seger of the Royal Porcelain Factory in Charlottenburgh – are well known time-
and-temperature indicating devices. Pyrometric cones are also known under
other names, such as Staffordshire cones and Orton cones, their specific use
being to provide heat soak temperature information in kilns.

They consist of heat sensitive bars of triangular section made from fusible
ceramic materials. They are mounted on a fireclay base at an angle of 15°, when
they will soften and sag after soaking up a specific amount of heat. Different
formulations are used to produce a range of cones which sag at different heat
soak temperatures.

It should be appreciated that a pyrometric cone is not a *temperature* indicating device. A cone squats after it has soaked up a specific amount of *heat* and
thus really indicates heat work or time and temperature. The formulation of
cones is normally based on a typical and steady heat rise in a kiln of about 4°C
per minute and the cone will squat when reaching its specified temperature rating after being heated up at this rate. If the heat rise is *lower* the cone will soak
up more heat work and the cone will squat at a *lower* temperature. If the heat
rise is *greater,* the cone will squat at a *higher* temperature. This does not affect
the usefulness of cones since once the correct cone number has been established
for a particular kiln and firing temperature required, this cone number will always give identical readings. Cones cannot, however, replace a pryometer (high
temperature thermometer) as a means of indicating or measuring *actual* kiln
temperatures.

Pyrometric cones are made in up to sixty different numbers, each number having a different squatting point, in steps of 10, 20, 30 or 50°C between each number. The temperature range covered is from 600 to 1500°C, but most suppliers
offer a more restricted range. Cone numbering is not necessarily consistent between manufacturers and is sometimes hidden in a code number. A range of
numbers and their equivalent approximate squatting temperatures are given in
Table 1.

TABLE 1 – PYROMETRIC CONES – TEMPERATURE RATINGS

Cone no	Approximate squatting temperature °C	Cone no	Approximate squatting temperature °C
022	600	4	1160
0225 (or 020A)	625	4.5 (or 4A)	1170
021	650	5	1180
020	670	5.5 (or 5A)	1190
019	690	6	1200
018	710	6.5 (or 6A)	1215
017	750	7	1230
016	760	7.5 (or 7A)	1240
0155 (or 15A)	760	8	1250
015	790	8.5 (or 8A)	1260
0145 (or 014A)	800	8.7 (or 8B)	1270
014	815	9	1280
013	835	9.5 (or 9A)	1290
012	855	10	1300
011	880	10.5 (or 10A)	1310
010	900	11	1320
*		12	1350
1	1100	13	1380
1.5 (or 1A)	1110	14	1410
2	1120	15	1435
2.5 (or 2A)	1130	16	1460
3	1140	17	1480
3.5 (or 3A)	1150	18	1500

*Some manufacturers produce a range of cones for indicating temperatures between 900°C (No 010 cone) and 1100°C (No 1 cone).

TABLE 2 – PYROMETRIC BARS – TEMPERATURE RATINGS

Bar number	Approximate bending temperature °C	Bar number	Approximate bending temperature °C
010	600	20	1040
020	650	21	1060
030	670	22	1080
040	700	23	1100
050	730	24	1120
060	760	25	1140
070	790	25.5	1170
075	810	26	1200
080	840	26.5	1230
090	860	27	1250
10	875	27.5	1270
11	890	28	1280
12	905	29	1300
13	920	30	1325
14	935	31	1350
15	950	32	1380
16	960	33	1430
17	970	34	1460
18	985	35	1475
19	1000	36	1490

Cones are also made in two different sizes – standard (about 65 mm long) and miniature (about 30 mm long). Performance is identical for the same cone number. Standard size cones are normally used, but miniature cones may be a better choice for small kilns.

Pyrometric bars

Pyrometric bars are an alternative form to cones, work on the same principle and are constructed from similar materials. Pyrometric bars are mounted horizontally on a channel shaped fireclay stand and sag at specific temperatures, depending on the ceramic composition. Bar numbers are different from those for cones (see Table 2).

Heat fuses

Heat fuses are metallic elements with a precisely known melting point. They have a variety of applications for temperature determination and for thermal sensitive safety devices.

Heat fuses in the form of small rods may be inserted into drilled holes in components subject to high working temperatures. After the equipment has been operated for a specific period the temperature reached by the actual component can be gauged by observation of which plugs have melted. They provide a simple and effective method of temperature determination without the need for bulky equipment or external connections, particularly suitable for use on internal components, rotating components, etc.

Used as safety devices, heat fuses may take the form of plugs or links which fuse at a pre-determined temperature to operate a trip device, fire warning system, etc.

TABLE 3 – HEAT FUSES OF SPECIFIC MELTING TEMPERATURES IN THE RANGE 960 TO 1550°C

Centinel no	Nominal melting point °C	Constituent elements
960	960	Ag
1000	1000	Ag : Au
1063	1063	Au
1100	1100	Au : Pd
1150	1150	Au : Pd
1200	1200	Au : Pd
1237	1237	Pd : Ni
1250	1250	Au : Pd
1300	1300	Au : Pd
1350	1350	Au : Pd
1400	1400	Au : Pd
1500	1500	Au : Pd
1550	1550	Pd

Johnson Matthey Metals Ltd.

TABLE 4 – HEAT FUSES OF SPECIFIC MELTING TEMPERATURES IN THE RANGE 46.7 TO 900°C*

Centinel no	Nominal melting point °C	Constituent elements
47	46.7	Bi : Pb : Sn : Cd : In
58	58.0	Bi : Pb : Sn : In
70	70.0	Bi : Pb : Sn : Cd
73	73.0	Bi : Pb : Sn : Cd
79	78.8	Bi : Sn : In
92	91.5	Bi : Pb : Cd
95	95.0	Bi : Pb : Sn
103	102.5	Bi : Sn : Cd
117	117.0	Sn : In
124	124.0	Bi : Pb
130	130.0	Bi : Sn : Zn
139	138.5	Bi : Sn
143	143.0	Pb : Sn : Cd
155	155.0	In
176	176.0	Sn : Cd
183	183.0	Pb : Sn
199	199.0	Sn : Zn
215	215.0	Pb : Au
221	221.0	Sn : Ag
232	232.0	Sn
236	236.0	Pb : Cd : Sb
247	247.0	Pb : Sb
265	265.0	Cd : Zn
271	271.0	Bi
280	280.0	Sn : Au
290	290.0	Pb : Pt
296	296.0	Ag : Sn : Pb
304	304.0	Pb : Ag
321	321.0	Cd
327	327.0	Pb
360	360.0	Sb : Au
382	382.0	Zn : Al
420	420.0	Zn
434	434.0	Zn : Ag
451	451.0	In : Au
482	482.0	Ag : Sb
528	528.0	Sb : Cu
548	548.0	Al : Cu
577	577.0	Al : Si
620	620.0	Ag : Cu : Zn : Cd
660	660.0	Al
710	710.0	Zn : Ag : Cu
779	779.0	Ag : Cu
800	800.0	Ag : Cu
900	900.0	Ag : Au : Cu

*Johnson Matthey Metals Ltd.

Materials used for heat fuses may be pure metals or alloys. Pure metals and eutectic alloys have an exact melting point. In the case of non-eutectic alloys, fusing temperatures quoted are nominal and the exact melting point of any batch of material has to be determined empirically (normally specified by the manufacturer). Examples of heat fuse materials are given in Tables 3 and 4.

Refractory materials

THE METALLIC elements commonly defined as refractory are to be found in Groups VA and VIA of the periodic system and are distinguished by their very high melting and boiling points but not by oxidation resistance. The other high melting point metals are in Groups VIIA and VIII and are designated precious metals having resistance to oxidation as well as fairly high melting points.

The basic properties of the four prime refractory metals tungsten, molybdenum, tantalum and niobium are given in Table 1.

TABLE 1

Property		Tungsten	Molybdenum	Tantalum	Niobium
Symbol		**W**	**Mo**	**Ta**	**Nb**
Atomic number		74	42	73	41
Atomic weight		183.86	95.95	180.95	92.91
Melting point °C		3410	2610	2996	2468
Boiling point °C		5930	5660	5425	4927
Density gm/cm^3		19.3	10.22	16.6	8.6
Coefficient of thermal	25-500°C	4.52	5.19	6.6	7.1
expansion $\times$ 10^{-6}/°C	25-2500°C	5.86	8.00	8.7	circa 10.0
					(25–1500°C)
Specific heat	20°C	0.032	0.066	0.034	0.065
cal/gm°C	2500°C	0.043	0.128	0.053	0.080
					(1500°C)
Thermal conductivity	20°C	0.31	0.34	0.130	0.125
cal/sec cm°C	1500°C	0.253	0.22	0.188	—
Resistivity	20°C	5.65	5.5	12.4	15.1
microhm-cm	2000°C	65	60	87	—
Electrical conductivity Cu = 100%					
IACS%		31	34	13.9	13.3
Thermal neutron capture cross-section					
barns/atom		19.2	2.7	21	1.1

Although all four metals have considerable refractoriness their applications are often significantly different and will have to be treated as separate subjects; although the higher the melting point the greater the extent to which they are processed as metallic powders. All are used in the manufacture of carbides for the hard metal Industry and tungsten carbide is the majority constituent.

Tungsten was first discovered in 1755 and although it ranks 18th in relative abundance in the earth's crust, its use is limited by the cost of converting the metal into a solid workable mass. The Coolidge process was developed in 1909 to produce ductile tungsten wire for incandescent lamp filaments from tungsten powder by pressing, sintering and mechanically hot working. Subsequently it was discovered that the ductility and hence lamp filament application was derived from the accidental presence of trace impurities, which allowed the ultimate development of a non-sag structure. Research to improve filament quality is still continuing by controlling the known impurities and fixing their ratios, which has resulted in the introduction of 'halogen' lamps and 'long life' bulbs.

Pure tungsten rod and sheet of 99.96% plus purity is now produced by sintering fine tungsten powder without additions into blocks of up to about 20 kilograms weight at about 96% of the metal's boiling point. The consolidated metal bars are hot rolled and then hot rotary swaged into rods. These have a rough swaged finish and need to be centreless ground for most applications, which are now manifold. A large but declining application as is tungsten 'distribution points' for the automobile industry; these are abrasive wheel cut from rod, as tungsten is too tough for sawing. The resultant discs are rumbled, polished and cleaned before copper or silver alloy brazing to rivets, arms and brackets. Pure tungsten is also used for electrodes in starter and igniter tubes because of its coefficient of expansion similarity to glass and quartz.

Tungsten rods for argon or TIG non-consumable welding electrodes are made by the same swaging and centreless grinding route. Pure tungsten electrodes are now a minority usage and have been superseded by a form of MMC, in which refractory metal oxides are added to the tungsten powder prior to the sintering, and end up as elongated particles in the finished rods. These additions of 0.25% to 4% of oxides lower the work function of the electrodes, producing a more stable arc. The favoured oxides are of thorium, zirconium, lanthanum, cerium and yttrium, in decreasing order of use. Each refractory oxide has a preferential application according to materials being welded and the electric current form and configuration.

An affiliated production to non-sag tungsten wire is the manufacture of vacuum metallising filaments. Techniques have been developed for the non-sag tungsten structures to be formed at 2 to 0.5 mm diameter as well as the fine wire filament sizes. The wire is hot formed into coils, hairpins and baskets and used to retain metals such as aluminium, silver, gold and electrical alloys for evaporation in vacuum chambers for decorative finishes and electronic and electrical applications. Tungsten sheet is formed into trays to contain fluorides and other salts for coating lenses and reflective glass.

Tungsten sheet is only available as pure tungsten down to a few thousands of an inch thick and is made by cross rolling sintered slabs.

Pure tungsten wire and rods can be used as heating elements for furnaces operating at over 2000°C but it is essential to use a reducing atmosphere or high vacuum. The maximum temperature limitation is normally the refractory brickwork and not the tungsten. Crucibles, pots and dishes for high temperature applications can be made by a variety of techniques, including spot-welding and riveting, to supplement the routes already mentioned as well as chemical vapour deposition. A few minor alloys such as tungsten-molybdenum can be mechanically worked but in these tungsten is a minority constituent. Other tungsten rich metals are generally classified as composites and are made by powder metallurgy routes. The tungsten and silver, copper and additives are dealt with in the *Electrical Contacts* section. The generic group of Heavy Metals also defined as High Density Alloys are made by the powder metallurgy route and contain nickel, copper, iron and molybdenum as additions.

Molybdenum is the most commonly used of the refractory metals but is not so abundant in the earth's crust as tungsten. It was first isolated in 1790 by Hjelm but developed commercially about 1910 by the Coolidge process as used for tungsten. Current practice includes melting of molybdenum alloys as well as the powder metallurgy route. Molybdenum assets are an electrical conductivity of about one-third of that of copper, a good mechanical strength and a low vapour pressure of the same order as carbon.

Excluding use as ferro-molybdenum in steel making, the majority of applications are as pure molybdenum and the metal is much easier to work than tungsten; therefore sheet molybdenum is extensively used and fabricated. But the relative ease with which it is oxidised for a high melting point metal somewhat limits its use. Molybdenum wire and ribbon are used in the incandescent lamp industry because of the metal's matched low coefficient of expansion, and finds outlets in quartz and iodine lamps, as well as eyelets, hooks, etc, for supporting filaments. Sheet punched and also sintered molybdenum discs are used as mounting bases for silicon discs. The wire and ribbon, being more ductile than tungsten, is an excellent material for high temperature furnace windings, provided that a vacuum or protective atmosphere is available. Wire is metal sprayed to give protective and low friction coefficient coatings to other metals, particularly steels.

The sheet can be formed and bent to make crucibles, reflectors, furnace boats and missile components. The ease with which molybdenum sheet can be spun or flow turned has increased its application as nozzle and rocket inserts, and hot forged parts also find applications in this field. Molybdenum vacuum evaporation boats also have specific functions. There is a non-sag molybdenum wire available supplementing the use of tungsten non-sag wire.

Molybdenum alloys containing 10 and 30% tungsten are available for more critical applications and can be made by arc casting as well as powder metallurgy. TZM is used containing 0.5% titanium and 0.07% zirconium, which has a higher hot strength than the pure metal and is used for die casting cores and hot work tools.

Among the manifold uses of molybdenum rods are projection welding electrodes, glass making electrodes and high-rigidity boring bars. Most super alloys contain a significant molybdenum content to enhance their high temperature and wear resistance.

Tantalum as a refractory metal is usually thought of in terms of its great corrosion resistance and high ductility. Once again the main production route is by powder metallurgy, either sodium reduction or fused salt electrolysis. It is available also as fabricated sheet, wire and seamless tubing and as some specialised alloys, but again the greatest use is as pure metal. The corrosion resistance of tantalum has been compared to glass, being inert to all acids except hydrofluoric and suffering attack by hot strong alkalis. The metal is used to fabricate reactors, vapour condensers, bayonet heaters and multi-tube heat exchangers. For large installations, because of the high cost of tantalum, the parts are made of steel and coated with thin tantalum sheets of foil, which is feasible because of its high ductility.

Tantalum capacitors are made from powder and are favoured because of their small volume importance in high tech applications, but there is a high price penalty. It is the ability to form a stable thick electrolytic oxide film whose dielectric constant is even higher than that of aluminium which has advanced the use for capacitors. This oxide feature also relates to the use of tantalum and some alloys as surgical sutures and reinforcements, coupled with the resistance to body fluids and tolerance by animal tissues.

Niobium is often coupled with tantalum in technical literature and is still designated as Columbium in the USA, where they refuse to comply with current nomenclature. Although niobium has useful properties similar to tantalum and also akin to molybdenum, its use is restricted because of its high price as a result of extraction costs. It has been used as a fuel element canning materials in nuclear reactors, but all its main uses are as important minority additions to other materials. Niobium carbide is present in many hard metal blends. It is added to austenitic steels and nickel alloys as a weld decay inhibitor and to alloy steels to enhance their properties.

Spring materials

HARD DRAWN *MB spring steel* is a good quality, low cost spring steel used for many general purposes where high grade expensive steels are not required. It is satisfactory for low stresses and where commercial tolerances are permitted. It should not be used when a long fatigue life is necessary or above 120°C.

Music wire spring steel is tougher and more suitable for the manufacture of small springs subject to high stresses and frequent deflections. Again, it should not be used at temperatures above 120°C.

Oil tempered MB spring steel is a general purpose spring steel, commonly used for all types of coil spring where the stress range is not too high and where the springs are not subject to shock or impact loading. This is the most popular spring steel for general use, in diameters from 0.125 to 0.500 in.

The remainder of the spring steels are chrome-vanadium, silicon-chrome or silicon-manganese alloy steels.

Chrome-vanadium alloy steel is the most popular for withstanding shock or impact loading and is available in both oil tempered and annealed conditions. It

TABLE 1 – MODULI OF TYPICAL SPRING MATERIALS

Material	Modulus of rigidity (G) lb/cm$^2 \times 10^6$	Modulus of elasticity (E) lb/in$^2 \times 10^6$
Beryllium copper	6–7*	16–18.5*
Brass (70/30)	5.50	15
Carbon steel	11.50	29
18/8 stainless steel	9.15	28
Chrome vanadium	11.50	30
Inconel	11.00	31
Monel	9.50	26
Nickel silver	5.50	16
Z nickel	11.00	30

*Depending on heat treatment.

TABLE 2 – PROPERTIES OF NICKEL AND NICKEL ALLOY SPRING MATERIALS (SPRING WIRE AND STRIP)

	Monel	K Monel	Nickel	Inconel	Ni-Span C
Tensile strength lb/in² × 10³	strip 100–140 wire 140–170	strip 145–200* wire 145–200*	strip 90–130 wire 135–165	strip 145–170 wire 165–185	wire 90–200
Limit of proportionality lb/in² × 10³	wire 60–75	wire 60–85*	wire 50–65	wire 75–85	wire 15–110
Modulus of elasticity tension lb/in² tension lb/in²	26 × 10⁶ 9.5 × 10⁶	26 × 10⁶ 9.5 × 10⁶	30 × 10⁶ 11 × 10⁶	31 × 10⁶ 11 × 10⁶	24–27.5 × 10⁶ 10 × 10⁶

*age hardened

is particularly suitable for i/c engine valve springs and can be used at temperatures up to 220°C.

Silicon-manganese alloy steel is widely used for flat leaf springs, etc, but *silicon-chromium alloy steel* is generally to be preferred where optimum properties are required. It can be heat treated to yield higher hardness than any other spring steel and can be used at temperatures up to 220°C.

Three types of *stainless steel* (302, 304 and 316) may be interchanged for most spring applications. They are cold drawn to obtain their hardness and spring properties and cannot be hardened by heat treatment. They are slightly magnetised by the cold drawing operation. They may be used in elevated temperatures up to 290°C and also at sub-zero temperatures. They are usually used in diameters up to about 0.1875 in. Type 302 has the highest tensile strength and is most popular. Type 304 has lower tensile strength by about 5%, but has better bending properties. Type 316 has slightly better corrosion resistance.

TABLE 3 – EFFECT OF HEAT ON SPRING STIFFNESS*

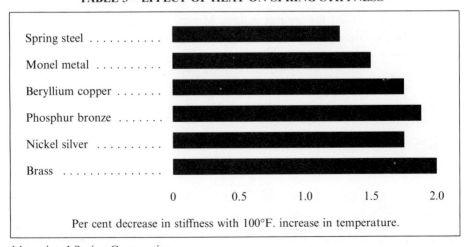

Per cent decrease in stiffness with 100°F. increase in temperature.

*Associated Spring Corporation.

Stainless steels types 414 and 420 can also be interchanged in many applications. They are generally used in the annealed state and then hardened and tempered after cooling. They are magnetic and are mostly used in diameters above 0.1875 in, but also in small diameters for springs with small spring indexes. They can be used up to 290°C, but not for sub-zero temperatures.

Stainless steel type 17-7 has high tensile strength and can be precipitation hardened, yielding particularly favourable spring characteristics. It is not as widely available as the other types of stainless spring steels. See also the chapter on *Spring steel.*

Of the non-ferrous materials, the following are the main types used for springs:

Spring brass (70-30) has the lowest spring qualities of the copper based alloys used for springs, but it can be easily formed, has high electrical conductivity and is the least expensive.

Phosphor-bronze (5% tin) is the most popular of the copper based group. It has fairly good tensile and electrical properties, good corrosion resistance, high resiliency, good fatigue resistance and is readily available from 0.005 to 0.250 in. It is used extensively in switches and other electrical apparatus.

Beryllium-copper (2¼%) is a precipitation hardened, copper based alloy that can be severely formed in the soft state and then hardened to quite high tensile strengths. Its principal use is to carry electrical current for brush springs and contact fingers in switches.

Inconel is a nickel based alloy which is recommended principally for high temperature service up to 371°C and for its excellent corrosion resistance. It is available mostly in sizes up to 0.250 in.

High nickel alloys are more expensive materials and may be chosen for high temperatures, sub-zero temperatures, damp or corrosive conditions and electrical applications.

Other materials used for springs include Monel, K-Monel, Z-nickel, stainless 431, Inconel X, Ni-Span, Elgiloy, Cobenium, nickel-silver, silicon-bronze, manganese-bronze, and manganese 720.

Surface finishing of metals

SURFACE FINISHING of metals covers a very wide range of processing, from the treatment of the base metal through to application of additional layers of metals, plastics on ceramics to the prepared surfaces of original metals, and including the conversion of that surface into chemical compounds. The reasons for the treatment are several and are sometimes multiple. They include:

Cleaning, which includes removal of scale, corrosion products and sand from castings.

Levelling, to offer a more presentable appearance or provide a better surface for subsequent coatings.

For decorous or decorative appeal on items varying from household goods to jewellery.

For protection, particularly in the case of plating against atmospheric or chemical environments.

Resistance, which is really an extension of protection and refers to hard coating such as carburising or nitriding of steels, hard chrome plating ion implantation of tools.

Functional, where the surface has a particular and sometimes localised use, as in electro-polishing of reflectors and the application of certain metals as sacrificial anodes to protect the prime equipment areas. Ablation coatings of silver and ceramic surfaces for space-craft are some of the latest examples.

Mechanical finishing

Embraces such methods as fettling and grinding, rumbling, shot-blasting, sand-blasting, wire-brushing, de-burring, burnishing, and shot-peening and involves many specialised or proprietary machines.

Surface treatment

These processes are considered here as those which involve preparing the metal surface and treating it with various chemicals to produce the salt or oxide of the

Sputter coating: physical vapour deposition of an almost unlimited range of coatings

Plasma spray

metal concerned. Many of the processes involve the use of heated aqueous solutions and one of the most important, for aluminium and its alloys is anodising which will be dealt with under a separate heading. Steels are 'blued' by air-heating or immersion in molten nitrates, usually followed by oil immersion. These processes give a pleasing appearance rather than much protection; solution or sulphides produce a variety of dark colours on steels. Bonderising and Walterising are protective coatings from acidic phosphate solutions containing manganese and other proprietary ingredients. Solutions of this type (often sprayed) are now the basis of coatings applied to automobile bodies prior to acrylic paint electrostatic spraying.

Anodising

Anodic oxidation of aluminium and its alloys (anodising) consists of producing a thin, invisible film on the surface by the passage of an electric current, and subsequent sealing of the film in boiling water and salt solutions. Prior to sealing the coating can be dyed, even in bright colours, by absorbing the dyestuff from an aqueous solution to form a 'lake' to fix the colour. The most high reflective, clear colours are achieved on pure aluminium, because the metallic additions made to improve the mechanical properties of the metal lead to a darkening of the surface and deterioration in the continuity of the anodised coating. Higher strength aluminium alloys are therefore often roll-clad with pure aluminum during fabrication, to produce the most effective anodised surfaces.

Anodising can be produced in a wide variety of solutions and changes can be made in the characteristics of the deposit by the solution temperatures, the magnitude of the current, and the process time. The solutions include 10 to 15% sulphuric acid, 10% chromic acid, Bengough-Stuart, 50 g/l chromic acid and sulphuric/oxalic acid solutions. Specifications are covered by British Standards 1615, 3987, AU89 and Defence Specification DEF-151. The type of anodising required is often defined as Decorative, Architectural, Hard or Corrosion Resistant.

There are non-electrical anodising processes such as the MBV method, but they produce inferior oxide coatings. The Alzak and Brytal proprietary processes are used to produce bright, reflective anodised surfaces on pure aluminium.

Chromating

This covers a range of surface treatments applied mainly to magnesium alloys, zinc castings and plating, and cadmium plating; as the name implies, a chromate of the metal is left on the surface. There are three main solutions used for magnesium. Acid chromate removes 0.005 in from the surfaces and leaves a coloured deposit, chrome–manganese solution hardly affects the dimensions overall and leaves a dark brown to black finish, whilst the RAE solution gives the best protection with a dark deposit and little dimensional change. If even greater protection is required, a fluoride anodising pre-treatment, the MEL process, is covered by British Patent.

A wide range of passivation processes is employed for zinc and cadmium, not all of which are based on chromates, but all have the same objective of protecting the corrosion-prone metal surface. The freshly deposited films are soft and easily damaged and up to 48 hours should be allowed to elapse before they are handled or painted after treatment. Cadmium usage has decreased because of its toxic salts, but it still finds limited use because of its protection against marine environments.

Electroplating

This process is extensively used on metals and in some cases on non-metals, as a finish. The coating can be applied for corrosion or tarnish protection, for wear resistance, or for its decorative value and it therefore covers a very wide range of metals; almost all those that are amenable to plating are used, with the exception of the alkali metals.

Most metals are electro-deposited from aqueous solutions, although molten salt baths are sometimes used. A range of metallic salts is employed and in the case of more extensively applied deposits there is a choice of salts, selected according to the finish, thickness and speed of plating required. Cyanides are probably the most widely used, in spite of their poison hazard, followed by sulphates, chlorides, fluoroborates, and then special solutions such as sulphumates. A range of other agents can be added to the solutions to confer particular properties on the electro deposits; these include surfactants, brighteners and other metals to give grain refinement and greater hardness.

Some metals are co-deposited for functional purposes, such as copper and zinc to form brass and tin and lead to form solders. Brittle metals such as chromium are under-plated with copper, to give a key and reduce porosity on decorative domestic articles. Indium is over-plated on lead and diffused, to reduce corrosion in thin wall bearings. Highly specialised. Hard Chrome is produced as a stressed deposit for wear resistant coatings and to build up worn parts.

Various components plated in electroless nickel and nickel–chrome

Galvanising

Galvanising is a single function coating applied to mild and low alloy steels from a molten zinc bath. It is generally considered to be a sacrificial coating, since it will be attacked in preference to the underlying steel; it is also cheap to apply to a low cost steel, such as for corrugated iron sheeting. The coating is often applied to nuts, bolts, washers, wire, fabricated tanks, hollow ware and other assemblies that can be drained. The process is simple and consists of degreasing and cleaning, pickling in hydrochloric acid, and fluxing prior to immersion in the galvanising bath at 445–465°C. An iron-zinc compound layer is formed which acts as a key for the pure zinc surface coating.

There is a range of British Standard specifications for the galvanising of various articles – 443, 729, 2989 and 3083. Sheradising is a declining process sometimes used for coating steel items with zinc, but is based on the use of zinc powder instead of the molten metal used in galvanising. The grit-blasted parts are rumbled in a heated container at 419°C and an iron-zinc alloy coating is formed on the parts. Sheradising is normally applied to small components and produces a thinner coating than galvanising; it is covered by British Standard 4921.

Tinning

Similar to galvanising in that it produces a nearly pure metal coating achieved by immersion in a molten bath, tinning is probably the oldest of the liquid metal processes and is still extensively used, particularly because of its compatibility with other materials, which is covered by British Standard 3788. Tin melts at 232°C and is therefore applied as a coating at a lower temperature than zinc, although the pre-treatment is somewhat more involved in order to form an adequate alloy layer on the steel. Vast quantities of tin plate are produced for the packaging industry, which in the case of food products has very high standards. The old hot tinning continuous production lines are rapidly being replaced by electro-tinning lines. In these, continuous steel strip is degreased, cleaned, electro-plated, surface melted and re-lubricated ready for tin container production, although for some critical applications a further lacquer coating has to be applied. British Standard 2920 lays down a requirement for cold-reduced tin plate.

Tinning is also a suitable process for the coating of copper vessels, sheet and wire.

Paint coatings

These probably account for more surface area than any other surface finishing process for metals. In general terms painting can be considered, as a process in which a fluid is applied to the prepared metal surface and is allowed to harden or cure to a solid form. This concept has to be broadened slightly when extended into the fields of plastics and some resins, but they will all be treated here as painting techniques.

Most paints consist of a pigment which is a metal, a metal oxide or salt or a stable organic compound in fine powder form, and this is blended with a vehicle which can be any oxidising oil or setting liquid. These blends are diluted with a thinner or solvent to the required viscosity, according to the method of application of the paint. The pigments give body colour and corrosion protection to the paint. The vehicle can be set by natural oxidation at room temperature, be hastened by heat or can be solidified by a chemical reaction, often above room temperature at about 100–250°C, known as stoving. Some curing, particularly of polyurethane resins, can be accelerated by using a two pack system, where the two ingredients are blended just before use and the chemical reaction occurs in a matter of hours at ambient temperature.

The traditional method of painting is by brushing; curtain painting is an extended version of brushing and dip painting is self-explanatory. Spray painting is from a gun fed from an inverted cup, or on a large scale from a hopper. Electrostatic coating is a development of spray painting, extensively used in the automobile industry, where the body or part is set at earth potential and the gun at up to 30 000 volts and negligible current. A low air pressure is used and thin coats of less than 0.005 in can be applied. The process is economical with paint, as it is attracted to the surface and is therefore good for irregular and rounded contours.

Stove enamelling has already been referred to and produces hard, resilient coatings, sometimes including deposits containing powdered glass, which are fritted at 400–1000°C.

Lacquers can be deposited by most techniques and the term tends to refer to varnishes and organic finishes (often clear).

The range of plastic and elastomer coating is very wide and is continually being expanded. Acrylic finishes are those largely used for automobile bodies and for this purpose metal powders are often added. Alkyd coatings are characteristic finishes for the cheaper metal components, including hardware. Cellulose is the generic term for nitrate, acetate butyrate and ethyl cellulose, from which the thinner is evaporated. Polyimide resins are wear resistant, have a low coefficient of friction and good strength and are widely used for metal objects.

Polyamides, such as nylon, are more often flame-sprayed or coated by fluidised bed dipping followed by fritting. The epoxy range of resins, which are catalytically hardened, are often mixed with other fillers and find application in electrical insulation. PVC (polyvinyl chloride) coatings, which are flexible, are also used for electrical insulation, extensively as wire coatings.

PTFE (polytetrafluoroethylene) bonded to metal finds use because of its non-stick characteristic and its low coefficient of friction.

Polystyrenes and silicones are also used as metal coatings with special applications, whilst polyesters and polyethylene can be applied as coatings, but mainly for glass-fibre and other non-metallic composites.

Fast drying and setting of paint coatings normally involves the use of long heated tunnels or lengthy time cycles, but the advent of radiation curing has greatly accelerated this for some applications.

Metal spraying

Two basic techniques are employed for metal spraying, one of which is by a wire of the metal required being fed through a concentrated flame, where it melts and is directed under pressure from an inert gas onto the surface being coated.

Most metals that are available as suitable wire and which will adhere to the base metal can be sprayed and a range if thickness can be applied. Copper, silver, aluminium and others, even to molybdenum (with a melting point of 2610°C) can be used.

The second basic method is by spraying metal powder, suspended in an inert gas, through a flame jet onto the surface. This method is available for metals or materials which are either not available or not suitable for wire-drawing.

There are also more advanced techniques for applying hard facing alloys. Union Carbide MCrAly alloys to surface when plasma torches are used to propel melted fine powders in argon onto the prepared surface. Some ceramic coatings can also be applied by this technique.

Case hardening

Case hardening covers all the furnace pack treatments and salt bath immersion processes for steel components. Packing of parts into metal boxes with treatment powders is one of the oldest finishing processes. White and black heart castings were made from white cast iron, by packing them in granulated iron oxide and heating the boxes for three or four days; when cooled the parts were sandblasted and ready for use. The outer layers of the castings were iron, with a stronger core of steel and in the case of black heart castings, contained free carbon. Pack carburising of mild steel is the reverse of the malleable iron process; generally machined parts are packed with energised charcoal and are heated for several hours at 900–950°C. The cooled parts then have a high carbon skin (up to 1 mm thick) and are given two quenching heat treatments to develop a very hard skin and a tough core.

When thin hard skins (0.2 mm or less) are required, mild steel components are immersed in molten sodium cyanide bearing salts for minutes and are then directly quenched in water or oil. The hardness of this surface is due to the presence of nitrides as well as carbides. The other surface hardening processes for steel are based on the use of carbon and nitrogen bearing gas atmospheres followed by quenching, to produce specified hard surface. If special steels are used together with decomposed ammonia atmospheres at 500°C, remarkably hard thin nitride coatings are formed.

The latest continuation of these surface hardening techniques, with availability for very localised hardening and other controls, is based on ion implantation and other high-tech systems. One of such methods is PAPVD (plasma assisted physical vapour deposition). This is carried out in a vacuum chamber using evaporated vapour sources and bent beam electron guns. Depositions range through argon and nitrogen to a range of hard oxides, nitrides and carbides.

Electro-polishing and vacuum metallising

These different processes, but both produce bright, highly reflective surfaces which have a range of applications in the jewellery and optical industries. Electro-polishing can be used on a range of metals, each of which requires a different chemical solution according to the application. The component is made the cathode in an electrical circuit with fluctuating current, which selectively dissolves the surface. The metals on which this is most frequently used are aluminium, stainless-steel, copper, brass, nickel-silver and tungsten.

Vacuum metallising is the deposition of an evaporated metal in a vacuum chamber onto rotating work pieces. The process is extensively used for plastic materials and glass, but also has application for metals, particularly for depositing thin coatings of silver, gold, aluminium and electrical alloys, often onto electro-polished surfaces.

Woven wire

WOVEN WIRE can be rendered in the form of cloth or gauze. The percentage of open area is governed by the diameter of the wire and the type of mesh. The latter can be specified by geometry (*eg* square mesh), number of apertures per unit area, number of apertures per unit length, clear space between the wires forming the mesh and distance between the centres of the wires forming the mesh. It also follows that:

(i)   For a given or required aperture size, a specific mesh (number per unit length or area) must be allied to a specific wire diameter size.

(ii) For a given percentage of open space or screening area the governing factors will be aperture size and wire diameter for any given weave.

Apart from its significance in filter applications, the percentage open area is also a rough indication of the strength of the wire cloth, viz:

Strength grade	Open area
Extra heavy	27% or less
Heavy	28 to 36%
Medium	37 to 45%
Light	46 to 60%
Extra light	61% or more

The following Tables summarise typical data for square mesh woven wire cloths.

TABLE 1 – SQUARE MESH CLOTH BY MESH AND GAUGE*

MESH no.	MESH Aperture		SWG	WIRE Diameter		Approx. percentage of space or screening area	MESH no.	MESH Aperture		SWG	WIRE Diameter		Approx. percentage of space or screening area
	Inch	m/m		Inch	m/m			Inch	m/m		Inch	m/m	
2	.34	8.63	8	.16	4.064	46	**4**	.122	3.099	10	.128	3.251	24
	.356	9.04	9	.144	3.658	51		.134	3.404	11	.116	2.946	29
	.372	9.44	10	.128	3.251	55	16	.146	3.708	12	.404	2.642	34
4	.384	9.75	11	.116	2.946	59	holes	.158	4.01	13	.092	2.337	40
holes	.396	10.05	12	.104	2.642	63	per	.17	4.32	14	.08	2.032	46
per	.408	10.36	13	.092	2.337	67	sq.	.178	4.52	15	.072	1.829	51
sq.	**.42**	**10.66**	**14**	**.08**	**2.032**	**71**	inch	**.186**	**4.72**	**16**	**.064**	**1.626**	**55**
inch	.428	10.87	15	.072	1.829	73		.194	4.93	17	.056	1.422	60
	.436	**11.07**	**16**	**.064**	**1.626**	**76**		**.202**	**5.13**	**18**	**.048**	**1.219**	**65**
	.444	11.27	17	.056	1.422	79		.21	5.33	19	.04	1.016	71
	.452	11.48	18	.048	1.219	82		**.214**	**5.45**	**20**	**.036**	**.914**	**73**
2½	.256	6.5	9	.144	3.658	41	**4½**	.1062	2.697	11	.116	2.946	23
	.272	6.9	10	.128	3.251	46		.1182	3.002	12	.104	2.642	28
	.284	7.21	11	.116	2.946	50		.1302	3.307	13	.092	2.337	34
6¼	.296	7.51	12	1.04	2.642	55	20¼	.1422	3.61	14	.08	2.032	41
holes	.308	7.82	13	.092	2.337	59	holes	.1502	3.81	15	.072	1.829	46
per	.32	8.12	14	.08	2.032	64	per	.1582	4.02	16	.064	1.626	51
sq.	.328	8.33	15	.072	1.829	67	sq.	.1662	4.22	17	.056	1.422	56
inch	.336	8.53	16	.064	1.626	71	inch	.1742	4.42	18	.048	1.219	61
	.344	8.73	17	.056	1.422	74		.1822	4.62	19	.04	1.016	67
	.352	8.94	18	.048	1.219	77		.1862	4.73	20	.036	.914	70
	.36	9.14	19	.04	1.016	81	**5**	.096	2.44	12	.104	2.642	23
3	.1893	4.80	9	.144	3.658	32		.108	2.74	13	.092	2.337	29
	.2053	5.21	10	.128	3.251	38	25	.12	3.05	14	.08	2.032	36
9	.2173	5.52	11	.116	2.946	43	holes	.128	3.25	15	.072	1.829	41
holes	.2293	5.82	12	.104	2.642	47	per	.136	3.45	16	.064	1.626	46
per	.2413	6.13	13	.092	2.337	52	sq.	**.144**	**3.65**	**17**	**.056**	**1.422**	**52**
sq.	.2533	6.43	14	.08	2.032	58	inch	.152	3.86	18	.048	1.219	58
inch	.2613	6.64	15	.072	1.829	61		**.16**	**4.06**	**19**	**.04**	**1.016**	**64**
	.2693	**6.84**	**16**	**.064**	**1.626**	**65**		.164	4.16	20	.036	.914	67
	.2773	**7.04**	**17**	**.056**	**1.422**	**69**		.168	4.26	21	.032	.813	71
	.2853	7.25	18	.048	1.219	73		.172	4.36	22	.028	.711	74
	.2933	7.45	19	.04	1.016	77		.176	4.46	23	.024	.61	77
3½	.1577	4.006	10	.128	3.251	30	**6**	.0866	2.201	14	.08	2.032	27
	.1697	4.31	11	.116	2.946	35		.0946	2.404	15	.072	1.829	32
	.1817	4.61	12	.104	2.642	40		.1026	2.607	16	.064	1.626	38
12¼	.1937	4.92	13	.092	2.337	46	36	.1106	2.81	17	.056	1.422	44
holes	.2057	5.22	14	.08	2.032	52	holes	**.1186**	**3.014**	**18**	**.048**	**1.219**	**51**
per	.2137	5.43	15	.072	1.829	56	per	.1266	3.217	19	.04	1.016	58
sq.	.2217	5.63	16	.064	1.626	60	sq.	**.1306**	**3.31**	**20**	**.036**	**.914**	**61**
inch	.2297	5.83	17	.056	1.422	65	inch	.1346	3.42	21	.032	.813	65
	.2377	6.04	18	.048	1.219	69		**.1386**	**3.52**	**22**	**.028**	**.711**	**69**
	.2457	6.24	19	.04	1.016	74		.1426	3.62	23	.024	.61	73
	.2497	6.34	20	.036	.914	76		.1446	3.67	24	.022	.559	75

* N. Greening (Warrington) Ltd.

(continued)

TABLE 1 – SQUARE MESH CLOTH BY MESH AND GAUGE* – *contd.*

MESH no.	Aperture Inch	Aperture m/m	SWG	Diameter Inch	Diameter m/m	Approx. percentage of space or screening area
7	.0948	2.409	18	.048	1.219	44
	.1028	2.612	19	.04	1.016	52
49	.1068	2.714	20	.036	.914	56
holes	.1108	2.815	21	.032	.813	60
per	.1148	2.91	22	.028	.711	65
sq.	.1188	3.018	23	.024	.61	69
inch	.1208	3.06	24	.022	.559	72
	.1228	3.12	25	.02	.508	74
8	.061	1.55	16	.064	1.626	24
	.069	1.753	17	.056	1.422	30
	.077	1.95	18	.048	1.219	38
64	.085	2.15	19	.04	1.016	46
holes	**.089**	**2.26**	**20**	**.036**	**.914**	**51**
per	.093	2.36	21	.032	.813	55
sq.	.097	2.46	22	.028	.711	60
inch	.101	2.56	23	.024	.61	65
	.103	2.616	24	.022	.559	68
	.105	2.66	25	.02	.508	71
9	.0631	1.603	18	.048	1.219	32
	.0711	1.806	19	.04	1.016	41
81	.0751	1.908	20	.036	.914	46
holes	.0791	2.009	21	.032	.813	51
per	.0831	2.11	22	.028	.711	56
sq.	.0871	2.212	23	.024	.61	61
inch	.0891	2.263	24	.022	.559	64
	.0911	2.314	25	.02	.508	67
	.0931	2.365	26	.018	.457	70
10	.052	1.32	18	.048	1.219	27
	.06	1.523	19	.04	1.016	36
100	.064	1.625	20	.036	.914	41
holes	**.068**	**1.726**	**21**	**.032**	**.813**	**46**
per	.072	1.828	22	.028	.711	52
sq.	.076	1.929	23	.024	.61	58
inch	**.078**	**1.98**	**24**	**.022**	**.559**	**61**
	.08	2.03	25	.02	.508	64
	.082	2.08	26	.018	.457	67
	.0836	2.123	27	.0164	.4166	70
	.0852	2.164	28	.0148	.3759	73
	.0864	2.194	29	.0136	.3454	75
	.0876	2.224	30	.0124	.313	77
	.0884	2.245	31	.0116	.2946	78
	.0892	2.265	32	.0108	.2743	80
12	.0473	1.202	20	.036	.914	32
	.0513	1.303	21	.032	.813	38
	.0553	**1.405**	**22**	**.028**	**.711**	**44**
144	.0593	1.506	23	.024	.61	51
holes	**.0613**	**1.557**	**24**	**.022**	**.559**	**54**
per	**.0633**	**1.608**	**25**	**.02**	**.508**	**58**
sq.	.0653	1.66	26	.018	.457	61
inch	.0669	1.7	27	.0164	.4166	64
	.0685	1.74	28	.0148	.3759	68
	.0697	1.77	29	.0136	.3454	70
	.0709	1.8	30	.0124	.315	72
	.0717	1.822	31	.0116	.2946	74
	.0725	1.84	32	.0108	.2743	76
14	.0394	1.001	21	.032	.813	30
	.0434	1.103	22	.028	.711	37
	.0474	1.204	23	.024	.61	44
196	.0494	1.255	24	.022	.559	48
holes	.0514	1.306	25	.02	.508	52
per	.0534	1.357	26	.018	.457	56
sq.	.055	1.397	27	.0164	.4166	59
inch	.0566	1.43	28	.0148	.3759	63
	.0578	1.46	29	.0136	.3454	65
	.059	1.5	30	.0124	.315	68
	.0598	1.52	31	.0116	.2946	70
	.0606	1.54	32	.0108	.2743	72
16	.0345	.876	22	.028	.711	30
	.0385	.977	23	.024	.61	38
	.0405	1.028	24	.022	.559	42
256	.0425	1.079	25	.02	.508	46
holes	**.0445**	**1.131**	**26**	**.018**	**.457**	**51**
per	**.0461**	**1.17**	**27**	**.0164**	**.4166**	**54**
sq.	**.0477**	**1.211**	**28**	**.0148**	**.3759**	**58**
inch	.0489	1.24	29	.0136	.3454	61
	.0501	1.27	30	.0124	.315	64
	.0509	1.29	31	.0116	.2946	66
	.0517	1.31	32	.0108	.2743	68
	.0525	1.33	33	.01	.254	71
	.0533	1.35	34	.0092	.2337	73
18	.0335	.852	24	.022	.559	36
	.0355	.9	25	.02	.508	41
	.0375	.954	26	.018	.457	46
324	.0391	.994	27	.0164	.4166	50
holes	.0407	1.035	28	.0148	.3759	54
per	.0419	1.065	29	.0136	.3454	57
sq.	.0431	1.096	30	.0124	.3150	60
inch	.0439	1.116	31	.0116	.2946	62
	.0447	1.136	32	.0108	.2743	65
	.0455	1.157	33	.01	.254	67
	.0463	1.177	34	.0092	.2337	69

* N. Greening (Warrington) Ltd.

(continued)

TABLE 1 – SQUARE MESH CLOTH BY MESH AND GAUGE* – *contd.*

MESH no.	MESH		SWG	WIRE		Approx. percentage of space or screening area	MESH no.	MESH		SWG	WIRE		Approx. percentage of space or screening area
	Aperture			Diameter				Aperture			Diameter		
	Inch	m/m		Inch	m/m			Inch	m/m		Inch	m/m	
20	.028	.711	24	.022	.559	31	**26**	**.0204**	**.5199**	**26**	**.018**	**.457**	**28**
	.03	.762	25	.02	.508	36		.022	.56	27	.0164	.4166	33
	.032	**.813**	**26**	**.018**	**.457**	**41**		.0236	.6	28	.0148	.3759	38
400	.0336	.853	27	.0164	.4166	45	676	.0248	.63	29	.0136	.3454	42
holes	**.0352**	**.894**	**28**	**.0148**	**.3759**	**50**	holes	.026	.66	30	.0124	.315	46
per	.0364	.924	29	.0136	.3454	53	per	.0268	.68	31	.0116	.2946	49
sq.	.0376	.954	30	.0124	.315	57	sq.	.0276	.702	32	.0108	.2743	52
inch	.0384	.975	31	.0116	.2946	59	inch	.0284	.72	33	.01	.254	55
	.0392	.995	32	.0108	.2743	61		.0292	.741	34	.0092	.2337	58
	.04	1.015	33	.01	.254	64		.03	.762	35	.0084	.2134	61
	.0408	1.036	34	.0092	.2337	67		.0308	.782	36	.0076	.193	64
	.0416	1.056	35	.0084	.2134	69							
22	.0255	.646	25	.02	.508	31	**28**	.0193	.492	27	.0164	.4166	29
	.0274	.7	26	.018	.457	36		**.0209**	**.532**	**28**	**.0148**	**.3759**	**34**
	.029	.737	27	.164	.4166	41		.0221	.561	29	.0136	.3454	38
484	.0306	.778	28	.1048	.3759	45	784	.0233	.592	30	.0124	.315	43
holes	.0318	.809	29	.0136	.3454	49	holes	.0241	.613	31	.0116	.2946	46
per	.033	.839	30	.0124	.315	53	per	.0249	.633	32	.0108	.2743	49
sq.	.0338	.859	31	.0116	.2946	55	sq	.0257	.653	33	.01	.254	52
inch	.0346	.88	32	.0108	.2743	58	inch	.0265	.674	34	.0092	.2337	55
	.0355	.9	33	.01	.254	61		.0273	.694	35	.0084	.2134	58
	.0363	.92	34	.0092	.2337	64		.0281	.714	36	.0076	.193	62
	.0371	.94	35	.0084	.2134	67		.0289	.734	37	.0068	.1727	65
	.0379	.96	36	.0076	.193	70		.0297	.754	38	.006	.1524	69
24	**.0236**	**.6**	**26**	**.018**	**.457**	**32**	**30**	.0185	.471	28	.0148	.3759	31
	.0252	.64	27	.0164	.4166	37		.0197	.501	29	.0136	.3454	35
	.0268	.68	28	.0148	.3759	41		.0209	.532	30	.1024	.315	39
576	**.028**	**.711**	**29**	**.0136**	**.3454**	**45**	900	**.0217**	**.553**	**31**	**.0116**	**.2946**	**42**
holes	.0292	.741	30	.0124	.315	49	holes	**.0225**	**.572**	**32**	**.0108**	**.2743**	**46**
per	**.03**	**.762**	**31**	**.0116**	**.2946**	**52**	per	.0233	.592	33	.01	.254	49
sq.	.0308	.782	32	.0108	.2743	55	sq.	.0241	.613	34	.0092	.2337	52
inch	.0316	.803	33	.01	.254	58	inch	.0249	.633	35	.0084	.2134	56
	.0324	.823	34	.0092	.2337	60		.0257	.653	36	.0076	.193	59
	.0332	.844	35	.0084	.2134	63		.0265	.674	37	.0068	.1727	63
	.034	.865	36	.0076	.193	67		.0273	.694	38	.006	.1524	67
25	.022	.56	26	.018	.457	30							
	.0236	.599	27	.164	.4166	35	**32**	.0177	.45	29	.0136	.3454	32
	.0252	.64	28	.0148	.3759	40		.0188	.478	30	.0124	.315	36
625	.0264	.67	29	.0136	.3454	44		.0196	.499	31	.0116	.2946	39
holes	.0274	.7	30	.0124	.315	47	1024	.0204	.519	32	.0108	.2743	43
per	.0284	.72	31	.0116	.2946	50	holes	.0212	.539	33	.01	.254	46
sq.	.0292	.741	32	.0108	.2743	53	per	.022	.56	34	.0092	.2337	50
inch	.03	.762	33	.01	.254	56	sq.	.0228	.58	35	.0084	.2134	53
	.0308	.782	34	.0092	.2337	59	inch	.0236	.6	36	.0076	.193	57
	.0316	.803	35	.0084	.2134	62		.0244	.62	37	.0068	.1727	61
	.0324	.823	36	.0076	.193	66		.0252	.64	38	.006	.1524	65

* N. Greening (Warrington) Ltd.

(continued)

TABLE 1 – SQUARE MESH CLOTH BY MESH AND GAUGE* – contd.

MESH no.	MESH Aperture Inch	m/m	SWG	WIRE Diameter Inch	m/m	Approx. percentage of space or screening area	MESH no.	MESH Aperture Inch	m/m	SWG	WIRE Diameter Inch	m/m	Approx. percentage of space or screening area
34	.017	.432	30	.0124	.315	33	**46**	.0117	.298	33	.01	.254	29
	.0178	.452	31	.0116	.294	37	2116	.0125	.318	34	.0092	.2337	33
1156	.0186	.472	32	.0108	.274	40	holes	.0133	.338	35	.0084	.2134	37
holes	.0194	.493	33	.01	.254	44	per	.0141	.359	36	.0076	.193	42
per	.0202	.513	34	.0092	.233	47	sq.	.0149	.38	37	.0068	.1727	47
sq.	.021	.533	35	.0084	.213	51	inch	.0157	.399	38	.006	.1524	52
inch	.0218	.554	36	.0076	.193	55							
	.0226	.574	37	.0068	.172	59	**48**	.0108	.275	33	.01	.254	27
	.0234	.594	38	.006	.152	63		.0116	.295	34	.0092	.2337	31
							2304	.0124	.315	35	.0084	.2134	35
36	.0153	.391	30	.0124	.315	30	holes	.0132	.336	36	.0076	.193	40
	.0161	.4109	31	.0116	.2946	34	per	.014	.356	37	.0068	.1727	45
1296	.0169	.431	32	.0108	.2743	37	sq.	.0148	.376	38	.006	.1524	50
holes	**.0177**	**.451**	**33**	**.01**	**.254**	**41**	inch	.0156	.397	39	.0052	.3121	56
per	.0185	.471	34	.0092	.2337	44		.016	.407	40	.0048	.1219	59
sq.	.0193	.492	35	.0084	.2134	48							
inch	.0201	.512	36	.0076	.193	52	**50**	.01	.254	33	.01	.254	25
	.0209	.532	37	.0068	.1727	57		.0108	.275	34	.0092	.2337	29
	.0217	.553	38	.006	.1524	61	2500	**.0116**	**.295**	**35**	**.0084**	**.2134**	**34**
							holes	**.0124**	**.315**	**36**	**.0076**	**.193**	**38**
38	.0147	.373	31	.0116	.2946	31	per	.0132	.336	37	.0068	.1727	44
	.0155	.394	32	.0108	.2743	35	sq.	.014	.356	38	.006	.1524	49
1444	.0163	.414	33	.01	.254	38	inch	.0148	.376	39	.0052	.1321	55
holes	.0171	.434	34	.0092	.2337	42		.0152	.386	40	.0048	.1219	58
per	.0179	.455	35	.0084	.2134	46							
sq.	.0187	.475	36	.0076	.193	51	**52**	.01	.254	34	.0092	.2337	27
inch	.0195	.495	37	.0068	.1727	55		.0108	.275	35	.0084	.2134	32
	.0203	.516	38	.006	.1524	60	2704	.0116	.295	36	.0076	.193	36
							holes	.0124	.315	37	.0068	.1727	42
40	.0134	.34	31	.0116	.2946	29	per	.0132	.336	38	.006	.1524	47
	.0142	.361	32	.0108	.2743	32	sq.	.014	.356	39	.0052	.1321	53
1600	.015	.381	33	.01	.254	36	inch	.0144	.366	40	.0048	.1219	56
holes	**.0158**	**.401**	**34**	**.0092**	**.2337**	**40**							
per	.0166	.421	35	.0084	.2134	44	**54**	.0093	.237	34	.0092	.2337	25
sq.	**.0174**	**.441**	**36**	**.0076**	**.193**	**48**		.0101	.257	35	.0084	.2134	30
inch	.0182	.462	37	.0068	.1727	53	2916	.0109	.277	36	.0076	.193	35
	.019	.482	38	.006	.1524	58	holes	.0117	.298	37	.0068	.1727	40
	.0198	.502	39	.0052	.1321	63	per	.0125	.318	38	.006	.1524	46
							sq.	.0133	.338	39	.0052	.1321	52
42	.013	.33	32	.0108	.2743	30	inch	.0137	.348	40	.0048	.1219	55
	.0138	.35	33	.01	.254	34							
1764	.0146	.371	34	.0092	.2337	38	**56**	.0094	.241	35	.0084	.2134	28
holes	.0154	.391	35	.0084	.2134	42	3130	.0102	.261	36	.0076	.193	33
per	.0162	.412	36	.0076	.193	46	holes	.011	.28	37	.0068	.1727	38
sq.	.017	.432	37	.0068	.1727	51	per	.0118	.301	38	.006	.1524	44
inch	.0178	.452	38	.006	.1524	56	sq.	.0126	.32	39	.0052	.1321	50
							inch	.013	.33	40	.0048	.1219	53
44	.0119	.303	32	.0108	.2743	27							
	.0127	.323	33	.01	.254	31	**58**	.0088	.224	35	.0084	.2134	26
1936	.0135	.343	34	.0092	.2337	35	3364	.0096	.254	36	.0076	.193	31
holes	.0143	.364	35	.0084	.2134	40	holes	.0104	.265	37	.0068	.1727	36
per	.0151	.384	36	.0076	.193	44	per	.0112	.285	38	.006	.1524	42
sq.	.0159	.404	37	.0068	.1727	49	sq.	.012	.306	39	.0052	.1321	48
inch	.0167	.424	38	.006	.1524	54	inch	.0124	.315	40	.0048	.1219	52

* N. Greening (Warrington) Ltd.

(continued)

TABLE 1 – SQUARE MESH CLOTH BY MESH AND GAUGE* – *contd.*

MESH no.	Aperture Inch	Aperture m/m	SWG	Diameter Inch	Diameter m/m	Approx. percentage of space or screening area	MESH no.	Aperture Inch	Aperture m/m	SWG	Diameter Inch	Diameter m/m	Approx. percentage of space or screening area
60	**.099**	**.23**	**36**	**.0076**	**.193**	**29**	**130**	.00409	.104	43	.0036	.0914	28
3600	**.0098**	**.251**	**37**	**.0068**	**.1727**	**35**	16900	.00449	.114	44	.0032	.0813	34
holes	.0106	.271	38	.006	.1524	40	holes	.00489	.124	45	.0028	.0711	41
per	.0114	.291	39	.0052	.1321	47	per						
sq.	.0018	.301	40	.0048	.1219	50	sq.in.						
inch	.0122	.311	41	.0044	.1188	54							
	.0126	.322	42	.004	.1016	57	**140**	.0035	.09	43	.0036	.0914	24
							19600	.0039	.1	44	.0032	.0813	30
							holes	.0043	.11	45	.0028	.0711	36
							per						
64	.0088	.224	37	.0068	.1727	32	sq.in.						
4096	.0096	.25	38	.006	.1524	38							
holes	.0104	.265	39	.0052	.1321	44	**150**	.00346	.088	44	.0032	.0813	27
per	.0108	.275	40	.0048	.1219	48	22500	.00386	.098	45	.0028	.0711	34
sq.	.0112	.285	41	.0044	.1118	51	holes	.00426	.108	46	.0024	.061	41
inch	.0116	.295	42	.004	.1016	55	per						
							sq.in.						
70	.0074	.19	37	.0068	.1727	27	**160**	.003	.077	44	.0032	.0813	23
4900	.0082	.21	38	.006	.1524	33	24600	.0034	.087	45	.0028	.0711	30
holes	.009	.237	39	.0052	.1321	40	holes	.0038	.097	46	.0024	.061	37
per	.0094	.241	40	.0048	.1219	43	per						
sq.	.0098	.251	41	.0044	.1118	47	sq.in.						
inch	.0102	.261	42	.004	.1016	51							
							180	.0027	.07	45	.0028	.0711	24
							35400	.0031	.08	46	.0024	.961	31
							holes	.0035	.09	47	.002	.0508	40
							per						
80	.0065	.166	38	.006	.1524	27	sq.in.						
6400	**.0073**	**.186**	**39**	**.0052**	**.1321**	**34**							
holes	**.0077**	**.197**	**40**	**.0048**	**.1219**	**38**	**200**	.0026	.066	46	.0024	.061	27
per	.0081	.206	41	.0044	.1118	42	40000	.003	.077	47	.002	.0508	36
sq.in.	.0085	.216	42	.004	.1016	46	holes						
							per						
							sq.in.						
85	.0065	.166	39	.0052	.1321	31							
7225	.0069	.177	40	.0048	.1219	34	**240**	.00256	.0650	48	.00i6	.0406	37
holes	.0073	.186	41	.0044	.1118	39	57600						
per	.0077	.197	42	.004	.1016	43	holes						
sq.in.							per						
							sq.in.						
90	.0059	.15	39	.0052	.1321	28	**250**	.0024	.061	48	.0016	.0406	36
8100	.0063	.16	40	.0048	.1219	32	62500						
holes	.0067	.17	41	.0044	.1118	36	holes						
per	.0071	.18	42	.004	.1016	41	per						
sq.in.							sq.in.						
100	**.0056**	**.142**	**41**	**.0044**	**.1118**	**31**	**300**	.00213	.054	49	.0012	.0304	41
10000	**.006**	**.152**	**42**	**.004**	**.1016**	**36**	90000						
holes	.0064	.162	43	.0036	.0914	41	holes						
per	.0068	.172	44	.0032	.0813	46	per						
sq.in.							sq.in.						
110	.00469	.119	41	.0044	.1118	27	**350**	.00175	.043	49½	.0011	.028	35
12100	.00509	.129	42	.004	.1016	31	122500						
holes	.00549	.1395	43	.0036	.0914	37	holes						
per	.00589	.150	44	.0032	.0813	42	per						
sq.in.							sq.in.						
120	.0043	.11	42	.004	.1016	27	**400**	.0015	.038	50	.001	.025	36
14400	.0047	.12	43	.0036	.0914	32	160000						
holes	.0051	.13	44	.0032	.0813	37	holes						
per	.0055	.14	45	.0028	.0711	44	per						
sq.in.							sq.in.						

* N. Greening (Warrington) Ltd.

TABLE 2 – SQUARE MESH WIRE CLOTH BY APERTURE SIZE

Aperture		Mesh no.	Wire S W G	Aperture		Mesh no.	Wire S W G
Decimals of inch	m/m			Decimals of inch	m/m		
.0024	.061	250	48	.0085	.216	80	42
.00256	.065	240	48	.0088	.224	58	35
.0026	**.066**	**200**	**46**	.0088	.224	64	37
.0027	.07	180	45	**.009**	**.23**	**60**	**36**
.003	.077	160	44	.009	.23	70	39
.003	.077	200	47	.0092	.233	48	31
.0031	**.08**	**180**	**46**	.0092	.233	50	32
.0034	**.087**	**160**	**45**	.0092	.233	52	33
.00346	.088	150	44	.0093	.237	46	30
.0035	.09	140	43	.0093	.237	54	34
.0035	.09	180	47	.0094	.241	56	35
.0038	.097	160	46	.0094	.241	70	40
.00386	**.098**	**150**	**45**	.0096	.245	58	36
				.0096	.245	64	38
				.0098	**.251**	**60**	**37**
.0039	**.1**	**140**	**44**	.0098	.251	70	41
.00409	.104	130	43	.01	.254	48	32
.00426	.108	150	46	.01	.254	52	34
.0043	.11	120	42	.01	.254	50	33
.0043	.11	140	45	.0101	.257	46	31
.00449	.114	130	44	.0101	.257	54	35
.00469	.119	110	41	.0102	.26	40	28
.0047	**.12**	**120**	**43**	.0102	.26	42	29
.00489	.124	130	45	.0102	.261	56	36
.00509	**.129**	**110**	**42**	.0102	.26	70	42
.0051	.13	90	38	.0103	.262	44	30
.0051	.13	120	44	.0104	.265	58	37
.00549	.139	110	43	.0104	.265	64	39
.0055	.14	120	45	.0106	.271	60	38
.0056	**.142**	**100**	**41**	.0108	.275	48	33
.0057	.144	80	37	.0108	.275	50	34
.00589	.15	110	44	.0108	.275	52	35
.0059	.15	90	39	.0108	.275	64	40
.006	**.152**	**100**	**42**	.0109	.277	46	32
.0063	.16	90	40	.0109	.277	54	36
.0064	.162	100	43	.011	.28	56	37
.0065	.166	80	38	.0111	.282	44	31
.0065	.166	85	39	.0112	.285	58	38
.0067	.17	70	36	.0112	.285	64	41
.0067	.17	90	41	.0114	.291	60	39
.0068	.172	100	44	.0114	.291	36	27
.0069	.177	85	40	.0114	.291	40	29
.0071	.18	90	42	.0114	.291	42	30
.0073	**.186**	**80**	**39**	.0115	.293	38	28
.0073	.186	85	41	.0116	.295	48	34
.0074	.19	70	37	**.0116**	**.295**	**50**	**35**
.0077	**.197**	**80**	**40**	.0116	.295	52	36
.0077	.197	85	42	.0116	.295	64	42
				.0117	.298	46	33
				.0117	.298	54	37
				.0118	.301	56	38
.008	.203	64	36	.0118	.301	60	40
.0081	.206	80	41	.0119	.303	44	32
.0082	.21	70	38	.012	.306	58	39

TABLE 2 – SQUARE MESH WIRE CLOTH BY APERTURE SIZE – *contd.*

Aperture		Mesh no.	Wire S W G	Aperture		Mesh no.	Wire S W G
Decimals of inch	m/m			Decimals of inch	m/m		
.0122	.31	42	31	.0158	.401	34	29
.0122	.311	60	41	**.0158**	**.401**	**40**	**34**
.0124	.315	48	35	.0159	.404	44	37
.0124	**.315**	**50**	**36**	.016	.407	48	40
.0124	.315	52	37	.016	.407	26	24
.0124	.315	58	40	.016	.407	28	25
.0125	.318	46	34	.0161	.411	36	31
.0125	.318	54	38	.0162	.412	42	36
.0126	.32	40	30	.0163	.414	38	33
.0126	.32	56	39	.0165	.419	32	28
.0126	.322	60	42	.0166	.421	40	35
.0127	.323	38	29	.0167	.424	44	38
.0127	.323	44	33	.0169	.43	30	27
.013	.33	42	32	.0169	.431	36	32
.013	.33	56	40	.017	.432	22	22
.013	.33	30	25	.017	.432	34	30
.013	.33	32	26	.017	.432	42	37
.013	.33	34	27	.0171	.434	38	34
.013	.33	36	28	**.0174**	**.441**	**40**	**36**
.0132	.336	50	37	.0177	.450	32	29
.0132	.336	48	36	**.0177**	**.450**	**36**	**33**
.0132	.336	52	38	.0178	.452	42	38
.0133	.338	46	35	.0178	.452	34	31
.0133	.338	54	39	.0179	.455	38	35
.0134	.34	40	31	.018	.457	20	21
.0135	.343	44	34	.018	.457	24	23
.0137	.348	54	40	.018	.457	26	25
.0138	.35	42	33	.018	.457	28	26
.0139	.353	38	30	.0182	.462	40	37
.014	.356	48	37	.0185	.471	30	28
.014	.356	50	38	.0185	.471	36	34
.014	.356	52	39	.0186	.472	34	32
.0141	.359	36	29	.0187	.475	38	36
.0141	.359	46	36	.0188	.478	32	30
.0142	.361	40	32	.019	.482	40	38
.0143	.364	44	35	.0193	.492	28	27
.0144	.366	52	40	.0193	.492	36	35
.0146	.371	34	28	.0194	.493	34	33
.0146	.371	42	34	.0195	.495	38	37
.0147	.373	38	31	.0196	.499	32	31
.0148	.376	32	27	.0197	.501	30	29
.0148	.376	50	39	.0198	.502	40	39
.0148	.376	48	38	.02	.508	24	24
.1049	.38	46	37	.02	.508	25	25
.015	.381	30	26	.0201	.512	36	36
.015	.381	40	33	.0202	.513	34	34
.0151	.384	44	36	.0203	.516	38	38
.0152	.386	50	40	**.0204**	**.519**	**26**	**26**
.0153	.391	36	30	.0204	.519	32	32
.0154	.391	42	35	**.0209**	**.532**	**28**	**28**
.0155	.394	38	32	.0209	.532	30	30
				.0209	.532	36	37
.0156	.397	48	39	.021	.533	22	23
.0157	.399	46	38				

(continued)

TABLE 2 – SQUARE MESH WIRE CLOTH BY APERTURE SIZE – *contd.*

Aperture		Mesh no.	Wire S W G	Strength	Aperture		Mesh no.	Wire S W G
Decimals of inch	m/m				Decimals of inch	m/m		
.021	.533	34	35	D	.0289	.734	28	37
.0213	.539	32	33	D	.029	.737	22	27
.0216	.548	24	25	A	.0292	.741	26	34
.0217	**.553**	**30**	**31**	**C**	.0292	.741	25	32
.0217	.553	36	38	E	.0292	.741	24	30
.0218	.554	34	36	D	.0297	.754	28	38
.022	.56	20	22	A	.03	.762	26	35
.022	.56	25	26	B	.03	.762	25	33
.022	.56	26	27	B	.03	.762	20	25
.022	.56	32	34	D	.03	.762	24	31
.0221	.561	28	29	C	.0306	.778	22	28
.0225	**.572**	**30**	**32**	**D**	.0308	.782	26	36
.0226	.574	34	37	D	.0308	.782	25	34
.0228	.58	32	35	D	.0308	.782	24	32
.0233	.592	28	30	C				
.0233	.592	30	33	D	.0316	.803	25	35
					.0316	.803	24	33
.0234	.594	34	38	E	.0318	.809	22	29
.0235	.597	22	24	A	.032	.813	18	23
.0236	**.6**	**24**	**26**	**B**	**.032**	**.813**	**20**	**26**
.0236	.6	25	27	B	.0324	.823	25	36
.0236	.6	26	28	C	.0324	.823	24	34
.0236	.6	32	36	D	.033	.839	22	30
.0241	.613	28	31	D	.0332	.844	24	35
.0241	.613	30	34	D	.0335	.852	18	24
.0244	.62	32	37	E	.0336	.853	20	27
.0248	.63	26	29	C	.0338	.859	22	31
.0249	.633	28	32	D	.034	.865	24	36
.0249	.633	30	35	D	.0345	.876	16	22
.0252	.64	24	27	C	.0346	.88	22	32
.0252	.64	25	28	C	.035	.888	12	18
.0252	.64	32	38	E	**.0352**	**.894**	**20**	**28**
.0255	.646	22	25	B	.0355	.9	22	33
.0257	.653	28	33	D	.0355	.9	18	25
.0257	.653	30	36	D	.036	.914	10	16
.026	.66	20	23	A	.0363	.92	22	34
.026	.66	26	30	D	.0364	.924	20	29
.0264	.67	25	29	C	.0371	.94	22	35
.0265	.674	28	34	D	.0375	.954	18	26
.0265	.674	30	37	E	.0376	.954	20	30
.0268	.68	26	31	D	.0379	.96	22	36
.0268	.68	24	28	C	.0384	.975	20	31
.027	.685	12	17	A	.0385	.977	16	23
.0273	.694	28	35	D				
.0273	.694	30	38	E	.0391	.994	18	27
.0274	.7	22	26	B	.0392	.995	20	32
.0274	.7	25	30	D	.0394	1.001	14	21
.0276	.702	26	32	D	.04	1.015	20	33
.028	.711	18	22	A	.0405	1.028	16	24
.028	.711	20	24	B	.0407	1.035	18	28
.028	**.711**	**24**	**29**	**C**	.0408	1.036	20	34
.0281	.714	28	36	E	.0416	1.056	20	35
.0284	.72	25	31	D	.0419	1.065	18	29
.0284	.72	26	33	D	.0425	1.079	16	25

(continued)

TABLE 2 – SQUARE MESH WIRE CLOTH BY APERTURE SIZE – *contd.*

Aperture Decimals of inch	Aperture m/m	Mesh no.	Wire S W G
.0431	1.096	18	30
.0434	1.103	14	22
.0439	1.116	18	31
.044	1.117	10	17
.044	1.117	12	19
.0445	**1.13**	**16**	**26**
.0477	1.136	18	32
.0455	1.157	18	33
.0461	**1.17**	**16**	**27**
.0463	1.177	18	34
.0473	1.202	12	20
.0474	1.204	14	23
.0477	**1.211**	**16**	**28**
.0489	1.24	16	29
.0494	1.255	14	24
.0501	1.27	16	30
.0509	1.29	16	31
.0513	1.303	12	21
.0514	1.306	14	25
.0517	1.31	16	32
.052	1.32	10	18
.0525	1.33	16	33
.0533	1.35	16	34
.0534	1.357	14	26
.055	1.397	14	27
.0553	**1.405**	**12**	**22**
.0566	1.43	14	28
.0578	1.46	14	29
.059	1.5	14	30
.0593	1.506	12	23
.0598	1.52	14	31
.06	1.523	10	19
.0606	1.54	14	32
.061	1.55	8	16
.0613	**1.557**	**12**	**24**
.0631	1.603	9	18
.0633	**1.608**	**12**	**25**
.064	1.625	10	20
.0653	1.66	12	26
.0669	1.7	12	27
.068	**1.726**	**10**	**21**
.0685	1.74	12	28
.069	1.753	8	17
.0697	1.77	12	29
.0709	1.8	12	30
.0711	1.806	9	19
.0717	1.822	12	31
.072	1.828	10	22
.0725	1.84	12	32
.0751	1.908	9	20
.076	1.929	10	23
.077	1.95	8	18
.078	**1.98**	**10**	**24**
.0791	2.009	9	21
.08	2.03	10	25
.082	2.08	10	26
.0831	2.11	9	22
.0836	2.123	10	27
.085	2.15	8	19
.0852	2.164	10	28
.0864	2.194	10	29
.0866	2.201	6	14
.0871	2.212	9	23
.0876	2.224	10	30
.0884	2.245	10	31
.089	**2.26**	**8**	**20**
.0891	2.263	9	24
.0892	2.265	10	32
.0911	2.314	9	25
.093	2.36	8	21
.0931	2.365	9	26
.0946	2.404	6	15
.0948	2.409	7	18
.096	2.44	5	12
.097	**2.46**	**8**	**22**
.101	2.56	8	23
.1015			
.1026	2.607	6	16
.1028	2.612	7	19
.103	2.616	8	24
.105	2.66	8	25
.1062	2.697	4½	11
.1068	2.714	7	20
.108	2.74	5	13
.1106	2.81	6	17
.1108	2.815	7	21
.1148	2.91	7	22
.1182	3.002	4½	12
.1186	**3.014**	**6**	**18**
.1188	3.018	7	23
.12	3.05	5	14

(continued)

TABLE 2 – SQUARE MESH WIRE CLOTH BY APERTURE SIZE – *contd.*

Aperture		Mesh no.	Wire S W G	Aperture		Mesh no.	Wire S W G
Decimals of inch	m/m			Decimals of inch	m/m		
.1208	3.06	7	24				
.122	3.099	4	10	.2217	5.63	3½	16
.1228	3.12	7	25	.2293	5.82	3	12
				.2297	5.83	3½	17
.1266	3.217	6	19				
.128	3.25	5	15	.2377	6.04	3½	18
.1302	3.307	4½	13	.2413	6.13	3	13
.1306	**3.31**	**6**	**20**	.2457	6.24	3½	19
.134	3.404	4	11	.2497	6.34	3½	20
.1346	3.42	6	21				
.136	3.45	5	16	.2533	6.43	3	14
.1386	**3.52**	**6**	**22**	.256	6.5	2½	9
				.2613	6.64	3	15
.1422	3.61	4½	14				
.1426	3.62	6	23	**.2693**	**6.84**	**3**	**16**
.144	**3.65**	**5**	**17**	.272	6.9	2½	10
.1446	3.67	6	24	.2773	7.04	3	17
.146	3.708	4	12				
.1502	3.81	4½	15	.284	7.21	2½	11
.152	3.86	5	18	.2853	7.25	3	18
				.2933	7.45	3	19
.1577	4.006	3½	10	.296	7.51	2½	12
.158	4.01	4	13				
.1582	4.02	4½	16	.308	7.82	2½	13
.16	**4.06**	**5**	**19**	.3125			
.164	4.16	5	20	.32	8.12	2½	14
.1662	4.22	4½	17	.328	8.33	2½	15
.168	4.26	5	21	.3281			
.1697	4.31	3½	11	.336	8.53	2½	16
.17	4.32	4	14	.34	8.63	2	8
.172	4.36	5	22	.344	8.73	2½	17
.1742	4.42	4½	18	.352	8.94	2½	18
.176	4.46	5	23	.356	9.04	2	9
.178	4.52	4	15				
.1817	4.61	3½	12	.36	9.14	2½	19
.1822	4.62	4½	19	.372	9.44	2	10
.186	**4.72**	**4**	**16**				
.1862	4.73	4½	20	.384	9.75	2	11
				.396	10.05	2	12
.1893	4.8	3	9				
.1937	5.92	3½	13	.408	10.36	2	13
.194	4.93	4	17	.42	10.66	2	14
.202	**5.13**	**4**	**18**	.428	10.87	2	15
				.436	**11.07**	**2**	**16**
.2053	5.21	3	10				
.2057	5.22	3½	14	.444	11.27	2	17
.21	5.33	4	19	.452	11.48	2	18
.2137	5.43	3½	15				
.214	**5.435**	**4**	**20**				
.2173	5.52	3	11				

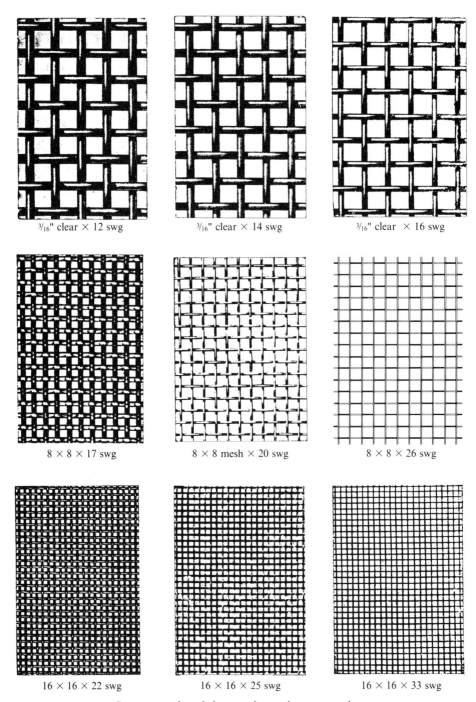

³/₁₆" clear × 12 swg	³/₁₆" clear × 14 swg	³/₁₆" clear × 16 swg
8 × 8 × 17 swg	8 × 8 mesh × 20 swg	8 × 8 × 26 swg
16 × 16 × 22 swg	16 × 16 × 25 swg	16 × 16 × 33 swg

Some examples of clear meshes and square meshes.

SECTION 8

Materials for Electrical Applications

ELECTRICAL CONTACT MATERIAL
ELECTRICAL TAPES
FUSIBLE ELEMENTS (ELECTRICAL)
INSULATING MATERIALS (ELECTRICAL)
MAGNETIC MATERIALS
PERMANENT MAGNET MATERIALS
SOFT MAGNETIC MATERIALS
MAGNETOSTRICTION – THE JOULE EFFECT
PIEZOELECTRIC MATERIALS

Electrical contact material

SILVER IS the most widely used contact material and also has the highest conductivity of all metals. It is suitable for both make and break contacts and certain types of sliding contact, as well as being the cheapest of true contact materials. It is limited by the fact that it forms a tarnish of sulphide which requires a minimum circuit voltage of about 20 and a contact pressure of at least 15 grams to break down and ensure satisfactory performance. With this reservation it is suitable for light duty contacts, but will generally perform quite adequately under medium duty conditions.

Other pure metal contact materials are gold, platinum, palladium and molybdenum, as well as various alloys.

Silver alloys

1715N alloy* – An alloy of silver with small amounts of nickel and magnesium, this material is capable of being hardened, after forming, by a simple irreversible oxidation process. Hardened, it combines the contact properties of silver with moderate tensile strength.

1715P alloy* – This alloy has all the desirable contact properties of hardened 1715N alloy, plus the advantage of not needing to be oxidation hardened. However, 1715P is not as easily formed as 1715N alloy.

7½%, 10%, 20% and 50% copper-silver – The addition of increasing amounts of copper to silver progressively increases its hardness and, in turn, its resistance to mechanical wear. As the copper content is increased the alloy becomes more prone to tarnish.

10% gold-silver – This alloy has slightly greater resistance to tarnishing than fine silver.

5%, 10% and 20% palladium-silver – The addition of palladium to silver gives progressively greater resistance to tarnishing and to mechanical wear, although conductivity is reduced.

*Johnson Matthey Metals Ltd

Gold and gold alloys

Gold – Gold remains tarnish free and has a high conductivity but, when unalloyed and in wrought form, it is soft and has few contact applications. Electro-deposited gold attains a useful hardness and is widely used for plug and socket contacts.

5% nickel-gold – This alloy provides the excellent properties of fine gold and is much harder and more resistant to wear and to welding.

*625 alloy** – Complete resistance to tarnish at ordinary temperatures, with spring qualities comparable with those of phosphor-bronze, make this copper-silver-gold alloy ideal for light duty sliding contacts operating against rhodium, palladium or palladium alloys.

*625R alloy** – Similar in contact behaviour to 625 alloy, this material is recommended for headed rivets and inlaid bi-metal.

30% silver-gold-platinum-silver-gold (PGS alloy) – These materials are occasionally used as substitutes for platinum, to which they are inferior in resistance to electrical wear. 30% silver-gold is sometimes used as an alternative to 625 alloy for headed contacts of 'difficult' dimension ratios.

Platinum and palladium and their alloys

Platinum – Platinum in both fabricated and electro-deposited forms remains completely free from tarnish and is ideal where voltage and mechanical pressure are low.

Palladium – Palladium can sometimes be used as a cheaper alternative to platinum. Its disadvantages are that it is softer and begins to tarnish at a temperature of about 400°C. Electro-deposited palladium is harder than plated gold, whilst thickness for thickness, it is of similar cost.

10%, 20% and 25% iridium-platinum – The addition of increasing amounts of iridium to platinum brings about a progressive increase in its hardness without affecting its complete resistance to tarnishing.

40% copper-palladium – Generally similar to 40% silver-palladium, this alloy is more resistant to material transfer.

*77 alloy** – This alloy is ideal for wiping contacts against noble metal potentiometer wires and slip rings. It can be formed into complex shapes in a solution-treated condition and then age hardened to develop optimum spring properties similar to those of phosphor-bronze.

*Johnson Matthey Metals Ltd

Rhodium

Rhodium (electro-deposited) – Electro-deposited rhodium combines complete freedom from tarnishing with great hardness, although its use for make and break contacts is restricted to applications involving only very light electrical and mechanical duty. Rhodium is also very successful in a variety of light duty sliding contact applications.

Elkonites‡

Silver-cadmium oxide D54, D54X and D55X – These combine much of the conductivity and low contact resistance of fine silver with excellent anti-welding properties. The 54 group of materials are all similar in composition. The 55 group are more resistant to welding.

Silver-nickel D56, D520 and D510 – These compare with silver in conductivity although they are considerably more resistant to arc erosion.

Silver-graphite D58-2% and D58-1% – The small graphite contents of these materials make them ideal for light duty sliding contacts operating against silver.

Recommended materials for light duty contacts are detailed in Table 1, with copper included for comparison. Typical properties of a wider range of contact materials for medium duties are given in Table 2.

‡Mallory Metallurgical Products Ltd.

Forms of contacts

Contact materials are generally available in the following forms:

Solid headed rivet contacts – These are manufactured by cold forging, the most economical method of mass producing solid contact rivets.

Solid turned rivet contacts – Such contacts are manufactured when the quantity ordered is insufficient to justify setting up a heading machine (usually 5000), when the sign of the contacts is unsuitable for heading, or when the contact material is not ductile enough for cold forging.

Composite turned rivet contacts – These are similar to solid turned rivet contacts, except that the contact material is brazed or welded to a base metal backing either for economy or because the contact material is unsuitable for riveting.

Tubular rivet contacts – A special form of mass produced headed composite contact is made in which a facing of silver is bonded to a copper backing with a hollow shank.

'Optecon' silver faced rivet contacts – This is another type of mass produced silver on copper composite contact. For quantities of more than 100 000 they are the cheapest form of silver faced rivet available.

Composite button contacts – These coined contacts consist of a facing of silver on a nickel plated mild steel or copper nickel disc. The backing has a projection for easy resistance welding to the contact support.

Solid button contacts – These are silver buttons which have a projection on one face to facilitate resistance welding onto copper based backings.

Inlaid contact bi-metal – This strip material consists of base metal inlaid or faced on one or both sides with a contact material. It is used for economical mass production of certain types of contact part.

Onlaid contact bi-metal – This form is similar to inlaid contact bi-metal, but the contact material stands proud of the base metal to which it is bonded.

Rod and wire – These forms are available in any required diameter for the manufacture of contacts.

TABLE 1 – LIGHT DUTY CONTACT MATERIALS

Material	Specific gravity g/cm³	Hardness (annealed), HV	Specific resistance, μ Ω cm	Electrical conductivity, % IACS	Thermal conductivity, cgs units	Remarks
Silver	10.5	26	1.6	106.0	1.000	Minimum 20 volts and 15 grams contact pressure
40% silver-palladium	11.0	95	43.0	4.0	0.075	Lowest cost tarnish-free material. Good wear resistance.
40% copper-palladium	10.4	145	35.0	4.9	0.170	Harder and more resistant to wear
Platinum	21.3	65	11.6	15.0	0.170	Exceptional resistance to tarnishing.
Palladium	11.9	40	10.7	16.0	0.170	Less tarnish resistant than platinum, but cheaper.
10% iridium-platinum.	21.6	120	24.5	7.0	0.074	Tarnish free and highly resistant to wear.
20% iridium-platinum	21.7	200	30.0	5.7	0.042	Tarnish free and extremely resistant to wear, but more difficult to fabricate.
5% nickel-gold	18.3	100	14.1	12.0		Harder and more resistant to wear and welding than gold.
Copper*	8.9	35	1.7	100	0.94	Not a suitable contact material*.

*Properties shown for comparison only

TABLE 2 – TYPICAL PROPERTIES AND AVAILABLE FORMS OF CONTACT MATERIALS*

Material	Specific gravity g/cm³	Hardness (annealed), HV	Melting point (solidus), °C	Specific resistance, μ Ω cm	Electrical conductivity, % IACS	Thermal conductivity, cgs units
Tarnish-free materials						
Platinum*	21.3	65 400**	1769	11.6	15	0.17
10% iridium-platinum	21.6	120	1780	24.5	7.0	0.074
20% iridium-platinum	21.7	200	1815	30.0	5.7	0.042
25% iridium-platinum	21.7	240	1845	32.0	5.4	0.039
Iridium-ruthenium-platinum (Irru)	20.8	310	1890	39.0	4.4	†
Rhodium	12.4	800**	1960	4.9	35	0.21
Palladium	12.0	40 350**	1552	10.7	16	0.17
40% silver-palladium	11.0	95	1290	43.0	4.0	0.075
40% copper-palladium	10.4	145	1200	35.0	4.9	†
JMM 77 alloy	12.0	210††	1085	37.5††	4.6††	†
Fine gold	19.3	60**	1063	2.4	68	0.7
JMM hard gold	19.1	115**	1057	2.4	68	0.7
5% nickel-gold	18.3	100	998	14.1	12	†
Copper-silver-gold, JMM 625 alloy	13.7	160–190***	861	14.0	12	†
JMM 625R alloy	14.4	95	1014	12.5	14	†
30% silver-gold	16.6	32	1025	10.4	16	0.16
Platinum-silver-gold (PGS alloy)	17.1	60	1100	16.8	11	†
Silver-based Materials						
Silver	10.5	26 80–110**	961	1.6	106	1.00
JMM 1715 alloy†††	10.4	145	961	2.8	60	0.66
7½% copper-silver (Standard silver)	10.3	56	778	1.9	90	0.83
10% copper-silver*	10.3	62	778	2.0	86	0.82
20% copper-silver	10.2	85	778	2.1	82	0.80
50% copper-silver	9.7	95	778	2.1	82	0.75
10% gold-silver	11.0	20	965	3.6	48	0.47
5% palladium-silver	10.5	33	965	3.8	45	0.53
10% palladium-silver	10.6	40	1000	5.8	30	0.34
20% palladium-silver	10.7	55	1070	10.1	17	0.22
Silver-cadmium oxide, Elkonite® D54	9.8	58	—	2.1	82	0.88
Elkonite® D54X	10.0	58	—	2.1	82	0.88
Elkonite® D55X	9.8	60	—	2.3	75	0.85
Silver-nickel, Elkonite® D56	9.9	68	—	2.4	72	0.74
Elkonite® D520	10.1	48	—	2.1	82	0.85
Elkonite® D510	10.3	40	—	2.0	87	0.87
Silver-graphite, Elkonite® D58-2%	9.7	10	—	2.0	86	0.90
Elkonite® D58-1%	9.9	40	—	1.8	96	0.92

*   Properties given here relate to platinum to which 0.5% of another noble metal has been added to give slightly greater hardness. This grade is invariably used for electrical contacts.

**   These figures apply to electrodeposited forms of the metals.

***   Can also be supplied hard, 200-240 HV or 250 HV minimum.

†   Where thermal conductivity figures are not available, an approximate value can be obtained from the electrical conductivity figures, which vary in direct proportion.

††   Figure for solution-treated material: hardness can be increased to 320—370 HV minimum by age hardening.

†††   Oxidation hardened.

*Johnson Matthey Metals Ltd.

Strip – Rolled strip is supplied in a wide range of thicknesses for contact facings and special contact parts.

D-section rod – D-shaped rod in silver or copper-silver alloy is supplied in random lengths or cut into short pieces ready for brazing onto contact backings.

Electro-deposits – Electro-deposits of certain metals are used when it is uneconomical or impossible to produce the contact surface in any other way.

Facings and inserts – These include shaped pieces, wrought or sintered, ready for brazing to backings.

Bi-metal clad wires – The wire has a thin clad surface of the contact material on a copper or copper based alloy core.

Choice of contact material

The ideal characteristics of low cost, high electrical and thermal conductivity, high resistance to mechanical and electrical wear and complete freedom from the formation of surface films are not realised by any single contact material. For example, silver, which is attractive from the point of view of cost and electrical and thermal conductivity, is prone to tarnish and has only moderate resistance to wear. On the other hand, 5% nickel-gold, which remains completely free from tarnish (thus giving very low contact resistance) and has superior resistance to wear, is many times more costly than silver and has relatively low conductivity. A comparatively inexpensive contact material having excellent resistance to electrical wear is tungsten, but it forms a hard film of oxide that requires appreciable mechanical force and wipe, coupled with adequate circuit voltage, to break it down, and its use is therefore limited.

When selecting a material, it is necessary to assess the relative importance of these four characteristics in conjunction with the physical and electrical conditions under which the contacts are to be used.

TARNISH

Inorganic surface films

The presence of a tarnish film on the surface of a contact may cause high and unstable contact resistance. A tarnish film is particularly troublesome if the electrical and mechanical duties are light, since under these conditions it can prevent electrical contact from being made. In heavier duty applications, where the voltage across the contacts is sufficiently high to penetrate the film and where the higher mechanical pressure may be assisted by wiping action as the contacts close, it is less likely to cause open circuit conditions. Nevertheless, it may cause undue heating at the contact interface.

As would be expected from the characteristics of the parent metals, silver-copper alloys are subject to tarnish. Superior, though incomplete, tarnish resistance is shown by certain of the silver alloys containing a small proportion of a more noble metal. An example is 10% gold-silver alloy. Complete tarnish resistance (at normal temperatures) can be obtained from platinum, palladium and the majority of the alloys containing a preponderance of platinum, palladium or gold.

The readiness of a material to tarnish is obviously affected by the ambient conditions; for example, a sulphur laden atmosphere will tarnish silver more rapidly and more heavily than a clean atmosphere.

Organic surface films

In the presence of a hydrocarbon vapour it is known that platinum and platinum group metals and alloys can form, under certain conditions, insulating layers of an organic polymer on their contact surfaces. Gold and alloys having a high gold content, or silver, are passive in this respect. Since silver is not generally a suitable alternative, gold alloys such as 5% nickel-gold, 25% silver-gold and 625 alloy are recommended substitutes when these conditions apply.

Electrical and thermal conductivity

Apart from the resistance at the point of contact, the current carrying capacity of a pair of contacts of given size is largely dependent on the conductivity and melting temperature of the material from which they are made. High thermal conductivity ensures that the heat generated in the contacts due to construction resistance and contact resistance is kept to a minimum. It also helps in the dissipation of heat caused by arcing.

For light duty contacts electrical and thermal conductivity are of little consequence though they are of first importance when a material is chosen for heavier duty contacts. The conductivity of platinum and palladium is about one-seventh that of silver, while the alloys of platinum have even lower conductivity.

Resistance to wear

The resistance of a material to electrical wear must be considered in terms or its resistance to material transfer or its resistance to arc erosion. For a given voltage there exists for every material a value of current, known as the limiting current, below which a stable arc cannot be maintained, even with a small contact gap. In d.c. circuits, even though the contacts may be interrupting currents below the limiting values, some slight transfer of material from one contact to the other takes place, this transfer setting a limit to the useful life of the contacts. Above the limiting value, arcing will occur in both a.c. and d.c. circuits and the problem of electrical wear becomes more significant. Normally the harder and denser materials and those having the highest melting temperatures are the most resistant to arc erosion. Tungsten and sintered combinations of tungsten or molybdenum with silver are generally employed when arcing is particularly severe.

In the case of make and break contacts, resistance to mechanical wear is usually related to the hardness of the material. However, most of the contact materials available are satisfactory from this point of view, unless the operating frequency of the contacts is so high that it becomes essential to employ one of the harder alloys, such as one of the iridium-platinum series.

Cost

The importance of cost is obvious. Silver and the silver based alloys have the lowest intrinsic values of the precious metal contact materials. Of the materials that remain completely free from tarnish films at room temperature, 40% silver-palladium is the least costly; next to this in order of increasing intrinsic value are 625 alloy, 25% silver-gold, platinum-silver-gold, palladium, 5% nickel-gold, platinum and the iridium-platinum alloys.

When comparing the initial cost of finished contacts in various materials, however, the relative ease of fabrication of each in the particular form require must also be taken into account.

Electrical tapes

MODERN ELECTRICAL tapes and their typical applications are summarised in Table 1. Self-adhesive tapes may be coated with various rubber based, thermoplastic resin or thermosetting resin adhesives, including adhesives which are electrically conducting, solvent resisting, flame retardant, etc. The choice is wide. In the case of thermosetting adhesives, the normal curing time is about 1 hour at 150°C, 2 hours at 135°C or 3 hours at 120°C.

The cure times will vary according to metal mass, oven temperature control and similar variables. To do the job for which it has been developed, a thermosetting tape should have a complete cure. For example, partial cure will give improved solvent resistance; a full cure will yield greater solvent resistance. Thermosetting silicon adhesive systems have normal cure cycles of three hours at 260°C. Curing at this temperature for 24 hours provides maximum solvent resistance.

Curing is generally accomplished when the unit is pre-baked to drive off moisture prior to impregnation. The adhesive mass is not rigid when completely thermoset. Having flexibility allows the tape to absorb the shock of metal expansion, contraction or vibration.

Storage recommendations

Electrical tapes should be stored below 23°C, avoiding extremes of humidity. They should be stored away from dust and moisture, and protected from direct sunlight and corrosive or solvent fumes. Storage in the cartons in which the tapes are supplied is recommended.

Under reasonable storage conditions, electrical tapes with thermoplastic or silicon adhesives may be kept for up to 12 months, except for certain grades which may have a lower shelf life specified (eg six months). Polyester electrical tapes with thermosetting adhesives may also be stored for up to 12 months. Other tapes with thermosetting adhesives should not be kept more than six months.

Electrolytic corrosion factor

The electrolytic corrosion factor of a tape is an expression of both the chemical purity of an insulation and its stability at an elevated temperature, high humidity and electrical stress. The higher the factor the less likely is the insulation to be the cause of corrosive damage to windings or components, even under adverse environmental conditions.

TABLE 1 – TYPES AND APPLICATIONS AND ELECTRICAL TAPES

Type	Form	BS 3924 classification	Typical applications
Paper	Smooth, crepe backed	20/105 Ts	Coil covers, insulating and holding windings, mechanical protection
	Smooth, flat backed	21/105Ts	Coils, solenoids, transformers, edge binding and slot insulation on rotors.
	Resin/fibre backed	20/105 Ts	Coil insulation, end trim and interplace insulation on rotors.
	Double sided	21/105 Ts	Stick-wound coils, holding capacitors, electrical grade adhesive
PTFE	Adhesive film	36/180 Si	Harness wrapping, insulation for coils, solenoids, relays, transformers, motors, etc for service temperatures up to 310°C (short term) or 180°C (continuous).
Aluminium	Foil	—	Screening and shielding, electrical connections and antennas.
	Embossed foil	—	Cable shielding; contractors.
Copper	Foil	—	As for aluminium
	Embossed foil	—	As for aluminium
Vinyl Glass cloth	PVC Woven tape	31/90 Tp 13/180 Ts, Sl, Tp	General purpose insulating tape Insulating applications up to 300°C (short term) or 130°C continuous (Class B temperatures). Also Class H & F temperatures.
Acetate	Film	—	Low cost insulation for coils, transformers, relays, solenoids, etc.
	Cloth	12/105 Ts	Lead holding, interwinding insulation, out wrap insulation.
	Film and glass filament	12/105 Ts	High strength tape for coil holding, etc.
Polyester	Film	35/130 Ts	Coils, transformers, rotors, solenoids, etc, requiring high resistance to oils and solvents.
	Film, double sided	35/130 Ts	Layer winding, holding wires, coils capacitors, etc.
	Resin backed	35/130 Ts	Flame retardant tapes
Polyimide	Film with silicon adhesive	—	Insulation with superior heat resistance — Class H (180°C) insulation requiring minimum tape shrinkage.

TABLE 2 – TYPES AND PROPERTIES OF ELECTRICAL TAPES (3M)

TYPE	PAPER				P.T.F.E.		ALUMINIUM			COPPER		VINYL	GLASS CLOTH		
BS 3924 Classification	21/105 Ts	20/105 Ts	20/105 Ts	21/105 Ts	36/180 Si	36/180 Si	—	—	—	—	—	31/90 Tp	13/180 Ts	13/180 Si	13/180 Tp
"Scotch" number	3	9	38	43	60	61	49	1170	1267	1181	1245	33	27	69	79
Backing	Buff Flat Back Paper	Brown Micro-creped Paper	Brown Creped Paper	Double Coated Paper	P.T.F.E. Film	P.T.F.E. Film	Foil	Foil	Embossed Foil	Foil	Embossed Foil	Black P.V.C.	White woven Glass Cloth	White woven Glass Cloth	White woven Glass Cloth
Adhesive	RT	RT	RT	RT	ST	ST	AN	AN	AN	AN	AN	RN	RT	ST	AN
Thickness (mm)	0.13	0.43	0.19	0.15	0.09	0.17	0.13	0.10	0.11	0.09	0.10	0.18	0.20	0.19	0.19
Tensile Strength (kg/10mm)	8.0	9.0	3.2	5.4	3.6	8.0	5.4	2.8	3.2	3.6	4.5	3.6	26.8	25.0	25.0
Elongation (% at break)	4	15	12	6	150	200	10	10	10	10	10	200	6	6	6
Adhesion to metal (g/10mm)	800	350	400	800	300	300	600	400	400	400	420	300	400	400	250
Breakdown Voltage (volts)	2000	3000	1600	1500	9000	12000	—	—	—	—	—	9000	2500	2500	2000
Dielectric Strength (Kv/mm)	15.4	7.0	8.5	10.0	100.0	70.0	—	—	—	—	—	50.0	12.5	13.1	10.5
Insulation Resistance (megohms)	100	50	150	100	>1×10⁶	>1×10⁶	—	—	—	—	—	>1×10⁶	1000	1000	500
Electrolytic Corrosion Factor	.85	.90	.90	.90	1.0	1.0	—	—	—	—	—	1.0	.97	.97	.97
Temperature Class (°C)	105A	105A	105A	105A	180H	180H	105A	130B	130B	130B	130B	90Y	130B	180H	155F

(continued)

TABLE 2 – TYPES AND PROPERTIES OF ELECTRICAL TAPES (3M) – *contd.*

TYPE	ACETATE					POLYESTER						POLY-IMIDE	FLAME RETARDANT		DIFF. PLATING TAPE
BS 3924 Classification	—	12/105 Ts	12/105 Ts	—	—	35/130 Ts	35/130 Ts	35/130 Ts	35/130 Ts	35/130 Ts	35/130 Ts	—	—	35/130 Ts	35/130 Tp
"Scotch" number	7	11	28	30	46	5	56	57	58	74	75	92	1266	1284	1280
Backing	Transparent Acetate Film	Black Acetate Cloth	White Acetate Cloth	Acetate Film Glass Filament		Transparent Film	Yellow Film			Yellow Film	Yellow Double Coated Film	Poly-imide	White Resin Fibre	Red Polyester Film	Red Polyester Film
Adhesive	RT	RT	RT	OT	RX	AT	RT	RT	RT	RT	RT	ST	RT	RT	RN
Thickness (mm)	0.09	0.20	0.20	0.20	0.18	0.05	0.05	0.09	0.09	0.02	0.09	0.05	0.18	0.06	0.10
Tensile Strength (kg/10mm)	3.6	8.0	8.0	35.5	32.0	4.0	3.8	8.0	8.0	2.1	3.7	6.2	3.2	4.0	6.0
Elongation (% at break)	15	15	15	5	5	70	70	100	100	70	70	100	20	70	100
Adhesion to metal (g/10mm)	600	400	400	470	550	330	460	460	460	200	500	230	400	450	400
Breakdown Voltage (volts)	5000	2500	2500	5500	5000	5000	5000	6500	6500	2500	5500	5000	5000	5500	—
Dielectric Strength (Kv/mm)	55.6	12.5	12.5	27.5	27.8	100.0	100.0	72.2	72.2	125.0	61.0	100.0	33.3	91.5	—
Insulation Resistance (megohms)	$>1\times10^6$	1×10^5	1×10^5	1000	10 000	$>1\times10^6$	$>1\times10^6$	$>1\times10^6$	$>1\times10^6$	$>1\times10^6$	$>1\times10^6$	$>1\times10^6$	$>1\times10^5$	$>1\times10^6$	—
Electrolytic Corrosion Factor	1.0	1.0	1.0	1.0	1.0	1.0	1.0	1.0	1.0	1.0	1.0	1.0	1.0	1.0	1.0
Temperature Class (°C)	105A	105A	105A	105A	105A	130B	130B	130B	130B	130B	130B	180H	130B	130B	—

TABLE 3 – ADHESIVE RATINGS FOR SOLVENT AND OIL RESISTANCE

Adhesive	Type	Solvent		Oil	
		Before	After	Before	After
R = Rubber resin	RN	0	0	0	0
A = Acrylic	RT	0	6	0	0
S = Silicon	RX	3	6	0	0
O = Oil resistant	AN	4	6	6	6
T = Thermosetting	AT	4	8	4	7
N = Non-thermosetting	ST	0	5	0	0
X = 4X	OT	3	9	5	10

Rating 10 = excellent, 5 = fair, 0 = poor

Figure 1 shows clearly the superior quality of electrical grade tapes over non-electrical cellophane and paper masking tapes. It gives, therefore, an indication of both the performance and the reliability of these insulations, particularly when used in fine wire applications. A guide to the upper temperature limits of electrical tapes is given in Figure 2.

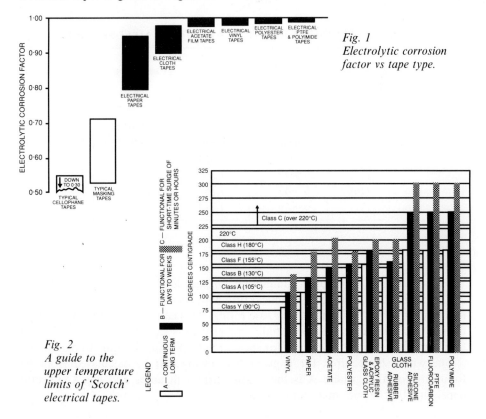

Fig. 1
Electrolytic corrosion factor vs tape type.

Fig. 2
A guide to the upper temperature limits of 'Scotch' electrical tapes.

Fusible elements (electrical)

THE STANDARD tin-lead alloy for fusible elements up to about 10 to 13 amps capacity comprises 63% tin and 37% lead. Equivalent sizes are shown in Table 1 with the current rating based on semi-enclosure of the fuse wire. A heat insulating tube around the fuse wire reduces the rating to about 90% in the case of a single wire.

TABLE 1 – FUSING CURRENT OF 63/37 TIN/LEAD FUSE WIRE

Wire	Diameter size		Current rating
swg	in	mm	amps
27	0.0164	0.417	1.8(2)
23	0.0240	0.610	3.0
21	0.0320	0.813	4.0
(18)	0.0510	1.300	10.0
(16)	0.0613	1.560	13.0
(15)	0.0674	1.710	15.0

Fusing currents for copper, aluminium, tin and lead wires are given in Table 2. Copper is not considered a practical fuse wire because of the high temperature it will reach when used with a low fuse rating. This can result in overheating, reducing the fusing current substantially. Silver wire is suitable only for ratings up to 5 amps, but has the advantage for such circuits of being immune to oxidation.

The wire size required for a given fuse rating can be calculated from the following formula, where A is the fusing current in amps and f is a constant, depending on the wire material (see Table 3):

TABLE 2 – FUSING CURRENT OF WIRES

Wire size swg	Diameter		Fusing current			
	in	mm	Copper	Aluminium	Tin	Lead
40	0.0048	0.122	3.41	2.520	0.546	0.46
39	0.0052	0.132	3.84	2.840	0.616	0.52
38	0.0060	0.152	4.76	3.525	0.760	0.64
37	0.0068	0.172	5.74	4.250	0.920	0.77
36	0.0076	0.193	6.79	5.025	1.090	0.92
35	0.0084	0.213	7.88	5.840	1.250	1.06
34	0.0092	0.233	9.04	6.690	1.440	1.21
33	0.0100	0.259	10.20	7.585	1.640	1.38
32	0.0108	0.274	11.50	8.510	1.840	1.55
31	0.0116	0.294	12.80	9.480	2.060	1.72
30	0.0124	0.315	14.15	10.470	2.270	1.90
29	0.0136	0.345	15.50	11.400	2.520	2.19
28	0.0148	0.376	18.44	13.600	2.960	2.48
27	0.0164	0.416	21.54	15.900	3.450	2.89
26	0.0180	0.457	24.74	18.300	3.960	3.33
25	0.0200	0.508	29.00	21.400	4.650	3.90
24	0.0220	0.559	33.43	24.700	5.360	4.50
23	0.0240	0.610	38.10	28.200	6.100	5.13
22	0.0280	0.711	48.00	35.530	7.690	6.46
21	0.0320	0.813	58.60	43.400	9.400	7.88
20	0.0360	0.914	69.97	51.700	11.220	9.41
19	0.0400	1.020	81.50	60.300	13.070	10.95
18	0.0480	1.220	107.70	79.700	17.270	14.50
17	0.0560	1.420	132.50	98.000	21.250	17.84
16	0.0640	1.630	165.80	122.000	26.580	22.32
15	0.0720	1.830	198.00	146.000	31.750	26.60
14	0.0800	2.030	232.00	171.000	37.150	31.20

$$\text{Diameter (inches)} \quad = \quad \sqrt[3]{\left(\frac{A}{f}\right)^2}$$

$$\text{Diameter (mm)} \quad = 25.4 \quad \sqrt[3]{\left(\frac{A}{f}\right)^2}$$

TABLE 3 – 'f' VALUES FOR FORMULA CALCULATION

Metal	f
63/37 Tin/lead	857
Aluminium	7585
Copper	10 244
Lead	1379
Tin	1642

Insulating materials (electrical)

AN (ELECTRICAL) insulating material is defined as one which offers a relatively high resistance to the passage of electrical current. The electrical properties of significance are insulation resistance or resistivity, dielectric strength, permittivity and power factor.

Resistivity embraces volume resistivity or specific resistance, and surface resistivity. *Volume resistivity* (specific resistance) is a measure of the resistance to leakage currents through the bulk of the material, expressed in the unit ohm/cm^3. *Surface resistivity* is a measure of the resistance to leakage currents developing across the face of the material, between two opposite sides of a square. The unit in this case is ohms per unit square (*eg* ohm/cm^2) – but since the quantitative value is independent of the actual unit, it can equally well be expressed in ohms. Surface resistivity, in particular, is very dependent on the amount of water present on the surface of the material.

Dielectric strength is a measure of the ability of the material to withstand electrical stress without breakdown and hence is measured in terms of potential difference per unit dimension to cause breakdown. The standard unit is kilovolts per millimetre (kV/mm). Dielectric strength is temperature dependent, standard measuremeng being at 20°C.

Permittivity is defined as the ratio of the electric flux density produced in the material to that produced in the same free space by the same electric force. Qualitatively this is the ratio of the capacitance of a capacitor where the material is the dielectric to the capacitance of the same capacitor with an air dielectric. This is a property specific to the material and not its dimensions, and so has no units. Permittivity may also be quoted as *specific induction capacity* or *dielectric constant.*

Power factor is a measure of the departure from the behaviour of a perfect dielectric when the material is subjected to alternating electrical stresses – *ie* causing the capacitor circuit to lead the voltage by a phase angle of less than 90°, due to *dielectric hysteresis.* By definition, power factor is the cosine of the phase angle, but for small loss angles may be taken as the tangent of the *loss angle.*

TABLE 1 – ELECTRIC INSULATORS* (VOLUME RESISTIVITY > 10^9 ohm cm)

Material	Density g/cm^3 at 20°C	Flexural strength kg/cm^2 at 20°C	Impact strength cmkg/cm^2 at 20°C
ABS polymers	1.04–1.08	650–950	>80
Terluran	1.06	700–1000	>80
Cellulose acetate and acetobutyrate injection moulding compounds	1.18–1.32	380–600	>55 to no fracture
Ceramics			
Hard porcelain	2.30–2.60	400–1000	2.0–2.5
Kondensa (containing rutile)	3.50–3.90	900–1500	2.5–3.5
Steatite	2.60–2.80	200–600	3.0–5.0
Ebonite	1.10–1.20	900–1000	20.0–30.0
Epoxy resins without extenders	1.10–1.40	950–1500	15.0–30.0
with ground quartz	1.50–2.00	550–1000	10.0–20.0
with about 60% wt glass	1.60–1.80	4000–6000	150.0–200.0
Glass	2.50–3.80	1000–2000	slight
Guttapercha	0.96–1.00	—	—
Hard paper	1.30–1.40	1000–1500	20.0–40.0
Insulating paper, insulating linen	1.00–1.20	—	—
Insulating silk	1.40	—	—
Melamine resin compression-moulding compounds	1.50–2.00	400–800	2.5–15.0
Melamine (containing mica)	2.00–2.60	—	—
Mica	2.50–3.20	—	—
Mycalex (lead borate with mica)	3.00–3.30	1000–1200	4.0–6.0
Nylon (dry to moist): 6 and 6 : 6	1.12–1.15	300–1100	no fracture
10	1.07–1.09	350–750	no fracture
Paraffin wax	0.90	—	—
Phenolic compression-moulding compounds	1.40–2.00	400–1200	2.0–25.0

*Derived from various sources, *eg* leaflets and booklets from various companies.

TABLE 1 – ELECTRIC INSULATORS (VOLUME RESISTIVITY > 10^9 ohm cm) – *contd.*

Marten's value melting point °C	Resistance to oil	Dielectric strength[1] KV/mm at 20°C	Dielectric constant at 20°C	Dissipation factor[2] at 20°C	
				50 c/s	10^6 c/s
65–80	limited	50	3–4	0.005–0.040	0.0020–0.0700
70–80	limited	100	3.5	0.005–0.010	0.0020–0.0030
40–55	limited	24–40	3.3–6.0	0.006–0.010	0.0180–0.0660
very high	yes	35	5.0–6.5	0.020–0.030	0.0070–0.0100
very high	yes	10–20	40.0–80.0	0.001	0.0003–0.0006
very high	yes	20–45	5.5–6.5	0.001–0.003	0.0003–0.0005
50–70	yes	2–3	3.0–4.0	0.003–0.005	0.0050–0.0090
50–180	yes	40	3.5–5.0	0.002–0.060	0.0020–0.0600
70–180	yes	40	3.2–4.5	0.008–0.030	0.0080–0.0300
>200	yes	30	3.2–4.5	0.008–0.030	0.0080–0.0300
very high	yes	10–40	3.0–15.0	0.004–0.030	0.0010–0.0100
softens 53–60	no	1.5–2.0	2.6–3.2	0.002–0.006	—
120–150	yes	10–20	3.5–5.0	0.050–0.100	0.0200–0.0800
—	yes	5–10	3.0–6.0	<0.010	—
—	yes	30–40	—	0.040	—
120–150	yes	10–15	6.0–10.0	800 c/s: 0.1–0.3	—
—	no	20–30	4.5–5.5	0.010–0.100	—
<500	no	30–70	5.0–9.0	0.002	0.0005
400	slight	ca 1.5	8.0	0.002	0.0020
melting point A~250	yes	30–50	3.5–6.0	0.010	0.0200–0.0300
melting point B~215					
melting point S~210	yes	50	3.0–4.0	0.020	0.0300
melting point~55		ca 30	2.0–2.2	0.005	—
100–160	yes	5–20	4.0–10.0	800c/s: 0.03–0.2	—

[1] Dependent on thickness and temperature
[2] Dependent on frequency and temperature

(continued)

TABLE 1 – ELECTRIC INSULATORS* (VOLUME RESISTIVITY > 10⁹ ohm cm) – *contd.*

Material	Density g/cm³ at 20°C	Flexural strength kg/cm² at 20°C	Impact strength cmkg/cm² at 20°C
Polyethylene typical	0.92	80–100	no fracture
crosslinked	0.92	70–90	no fracture
high temperature	0.96	350–400	no fracture
Polyester resins without filler	1.20–1.30	700–1500	5.0–20.0
Polyester resin dough moulding compounds	1.70–2.10	500–1700	5.0–60.0
Polyester resins with 30% wt glass mat	1.40–1.50	1800–2000	150.0–200.0
Polyisobutylene	0.92	—	—
Polymethylmethacrylate	1.18	1000	15.0–35.0
Polypropylene	0.906	460–470	no fracture
Polystyrene general-purpose types	1.05	1000–1100	16.0–25.0
impact tyres	1.06	500–700	55.0–85.0
Polytetrafluorethylene	2.20	110	—
Presspahn	1.20–1.40	700	—
PVC rigid	1.38	750–1000	no fracture
plasticised	1.20–1.35	—	—
Quartz	2.00–2.20	700	slight
Quartzitic glass	2.10–2.50	700	slight
Shellac	1.10	—	—
Silicon rubber	1.20–2.30	—	—
Soft rubber without filler based on			
butadiene-styrene rubber	0.90–1.00	—	—
butadiene-acrylonitrile rubber	0.90–1.00	—	—
butyl rubber	0.90–1.00	—	—
chlorobutadiene rubber	1.20–1.30	—	—
natural rubber	0.90–1.00	—	—

*Derived from various sources, *eg* leaflets and booklets from various companies.

INSULATING MATERIALS (ELECTRICAL)   701

TABLE 1 – ELECTRIC INSULATORS (VOLUME RESISTIVITY > 10⁹ ohm cm) – *contd.*

Marten's value melting point °C	Resistance to oil	Dielectric strength[1] KV/mm at 20°C	Dielectric constant at 20°C	Dissipation factor[2] at 20°C	
				50 c/s	**10⁶ c/s**
melting point ~112	limited	100	2.3	0.0001	0.0001
–	limited	70–100	2.3	0.0002–0.0010	0.0002–0.0010
melting point ~132	limited	100	2.3	0.0001	0.0001
50–100	yes	30–55	2.7–3.6	0.0030–0.0100	0.0080–0.0200
125–160	yes	10–15	3.0–6.0	800 c/s: 0.01–0.010	—
80–<200	yes	30	3.0–4.5	0.0080–0.0300	0.0080–0.0300
—	no	23	2.3	0.0004	0.0004
65–90	yes	40	3.4	0.0500	0.0300
melting point ~ 160	limited	50	2.4	0.0004	0.0004
63–75	limited	120	2.6	0.0002	0.0002
65–73	limited	100	2.6–2.9	—	0.0005–0.0010
120	yes	>35	2.0	0.0002	0.0002
~90	absorbs oil	10–13	2.5	0.0600	—
65–70	yes	50	3.3–4.0	0.0150–0.0200	0.0200–0.0300
—	limited	40	3.5–7.5	80 c/s: 0.1	0.0600–0.1200
very high	yes	25–40	3.5–4.5	0.0010	0.0005
very high	yes	10–15	3.5–4.0	0.0005	0.0002
softens 80	no	10–15	3.0–4.0	0.0100–0.0400	—
—		20–30	3.0–9.0	—	0.0010–0.1000
—	no	20–25	2.7	100 c/s : 0.005	—
—	yes	15–20	10.0	100 c/s : 0.100	—
—	no	15–20	2.3	100 c/s : 0.002	—
—	yes		6.7	100 c/s : 0.020	—
—	no	20–25	2.5	100 c/s : 0.005	—

[1] Dependent on thickness and temperature
[2] Dependent on frequency and temperature

The *dielectric loss* or energy loss under alternating electrical stress is dependent on the power factor and the permittivity and may be quoted as $K \times \tan d$ where K is the permittivity and d the loss angle. The power factor is frequency dependent.

Significant physical properties

The following physical properties may be of significance in influencing the choice of a suitable insulation material:

Specific gravity – mostly important in the case of liquid dielectric as far as the influence of this factor on electrical properties is concerned.

Moisture absorption – is a very important property since the amount of water present or absorbed can markedly affect electrical properties. The physical effects of moisture absorption can also limit the choice of insulation materials for particular ambiences or service conditions.

Thermal properties – the softening or melting point of a material will determine its maximum service temperature, but consideration may also have to be given to the effect of heat ageing or other permanent loss of properties at specific temperature levels, specific heat, thermal conductivity and thermal expansion. In the case of elastomers, the loss of flexibility at lower (normally sub-zero) temperatures may be significant.

Significant mechanical properties

Mechanical properties will have varying significance, depending on the particular application of the insulating material (*eg* whether stressed or not, and/or in which modes the material is stressed). The formability or machinability of the material may also have to be considered.

Mechanical properties which may or may not be significant are:

Tensile strength
Flexural strength
Compressive strength and deformation characteristics under load
Cross-breaking strength
Tear strength
Impact strength
Elasticity, elongation and resilience.

Mechanical properties are seldom quoted in detail in data tables, etc, but may be found elsewhere by reference to individual chapters covering the materials concerned.

Significant chemical properties

The chemical properties of main interest in the case of insulating materials are:

(i)   Resistance to oxidation and weathering under normal ambient or service conditions.

(ii) Ageing characteristics under conditions of use (*eg* ageing effects may be accelerated by excessive UV light, high ambient temperatures, etc).

TABLE 2 – ELECTRICAL PROPERTIES OF PLASTICS*

	Volume resistivity ohm cm ASTM D257	Dielectric strength volts/mil ASTM D149	Dielectric constant ASTM D150		Power factor ASTM D150	
			10^3 c/s	10^6 c/s	10^3 c/s	10^6 c/s
Polythene (low density)	$>10^{17}$	460–690	2.25–2.35	2.25–2.35	<0.0005	<0.0005
Polythene (high density)	$>10^{17}$	>800	2.25–2.35	2.25–2.35	<0.0005	<0.0005
Polypropylene	$>10^{17}$	>800	2.00–2.10	2.00–2.50	<0.0010	<0.0005
Polyvinyl chloride PVC (rigid)	$>10^{15}$	400–500	3.00–3.20	3.00–3.20	<0.0200	<0.0100
Vinyl chloride/vinyl acetate copolymer (rigid)	$>10^{15}$	>1000		3.00–3.50		<0.0100
Plasticized PVC (low plasticiser content)	10^{14}–10^{15}	>400	4.00–5.00		<0.1000	
Plasticized PVC (high plasticiser content)	10^{12}–10^{13}	>300	5.00–6.00		<0.1000	
Polyvinylidene chloride	10^{14}–10^{16}	400–600		3.00–400		0.0500–0.0800
Polymethyl methacrylate	$>10^{14}$	450–550	3.00–4.00	2.50–3.50	0.0400	0.0200–0.0300
Polytetrafluoroethylene	$>10^{17}$	400–600	2.00–2.10	2.00–2.10	<0.0003	<0.0003
Polystyrene	$>10^{13}$	500–700		2.40–3.00		<0.0010
Polystyrene (toughened)	$>10^{13}$	300–600		2.40–3.50		<0.0020
Acrylonitrile/butadiene/styrene terpolymer ABS	$>10^{15}$	310–410		2.50–3.50	0.0060–0.0200	0.0070–0.0200
Styrene/acrylonitrile copolymer SAN	$>10^{13}$	400–500		2.75–3.40		0.0070–0.0100
Cellulose acetate	10^{10}–10^{13}	260–365		3.20–7.00		0.0100–0.1000
Nylon 6	10^{12}–10^{15}	440–510	4.00–6.00	3.00–7.00		0.0200–0.1300
Nylon 6.6	10^{13}–10^{15}	250–450	3.9	3.40–6.00	0.0100–0.0300	0.0200–0.0600
Nylon 6.6 (glass filled)	10^{14}	500		3.40	0.0200	0.0200
Nylon 6.10	10^{14}–10^{16}	>250	3.60–6.00	3.00–4.00	0.0200–0.0400	0.0200–0.0300
Nylon 11	$>10^{13}$		3.30		0.0300	0.0300
Polycarbonates	$>10^{16}$	400	2.80–2.90	2.60–2.70	0.0010–0.0020	0.0100–0.0150
Polyacetals	$>10^{14}$	500	3.80	3.70	0.0040	0.0040
Phenol formaldehyde (unfilled)	10^{11}–10^{12}	300–400		4.50–5.00		0.0150–0.0300
Phenol formaldehyde (wood flour filled)	10^{9}–10^{13}	200–425		4.00–7.00		0.0300–0.0700
Phenol formaldehyde (asbestos filled)	10^{9}–10^{12}	300–400		4.00–6.00		0.0300–0.0700
Phenol formaldehyde (glass filled)	10^{13}–10^{15}	150–350		4.50–6.00		0.0100–0.0300
Urea formaldehyde (cellulose filled)	10^{12}–10^{14}	300–400	8.00–10.0	6.00–8.00	0.0500–0.1000	0.2500–0.3500
Melamine formaldehyde (cellulose filled)	10^{12}–10^{14}	300–400		7.20–8.30		0.0270–0.0450
Polyesters (rigid and unfilled)	10^{12}–10^{14}	325–450	2.80–5.20	2.80–5.00	0.0100–0.0300	0.0250–0.0350
Epoxides (rigid and unfilled)	$>10^{15}$	350–450		3.00–4.00		0.0100–0.0200
Epoxides (flexible and unfilled)	$>10^{15}$	300–400		3.00–4.00		0.0100–0.0200
Equivalent or comparable British Standard Test	BS 2782 Pt2/202	BS 2782 Pt2/201		Pt2/205	BS 2782 Pt2/206	Pt2/207

*Design and Components in Engineering.

TABLE 3 – INSULATION PROPERTIES OF MISCELLANEOUS MATERIALS

Material	Specific gravity	Resistivity ohm cm	Dielectric strength volts/mil	Relative permittivity
Asbestos	3.0	16×10^2	100–150	—
Bakelite products	1.3–1.9	10^3–10^5	200–800	4.0–8.0
Bitumen, vulcanised	1.2	2×10^7	330	4.0
Cambric, varnished	—	4×10^6	500–1 500	4.0–6.0
Celluloid	1.4	7×10^2	300–5009	1.2–2.6
Cotton, varnished	—	4×10^6	270–350	4.0–6.0
Ebonite	1.2	$(2–1000) \times 10^7$	500–1 000	2.5–3.5
Fibre, red	1.3	1–50	150–300	—
Glass (according to kind)	2.5–3.5	$(1–10) \times 10^5$	200–250	5.0–9.0
Gutta-percha	0.98	up to 4×10^6	200–500	3.0–4.5
Marble	2.5–2.8	10–10^3	165	8.3
Mica	—	10^8–10^9	500–1 500	5.0–8.0
Micanite	—	$(1–7) \times 10^7$	400–1 000	—
Paper, dry	0.7–1.0	up to 10^3	150–300	1.5–2.5
Paper, impregnated	—	10^6–10^7	500–1 200	2.5–4.0
Paraffin wax	0.9	$(3–5) \times 10^{10}$	300	2.0–2.3
Pitch	1.1	—	50	1.8
Porcelain	2.2–2.7	10^4–10^7	150–300	4.4–6.8
Presspahn	—	10^2	200–250	—
Quartz	2.5–2.8	2×10^6	—	3.5–4.5
Resin	1.1	5×10^8	280	2.6
Rubber, pure	0.9–1.0	$(10–15) \times 10^7$	500–700	2.2–2.4
Rubber, vulcanised	1.2–1.8	$(1–5) \times 10^7$	400–650	3.5
Shellac	—	$(2–10) \times 10^7$	40	3.0–3.5
Slate	2.5–3.0	up to 75×10^6	30	6.5–7.4
Sulphur	1.9	$(4–100) \times 10^7$	—	2.5–3.5
Wood (according to kind)	0.4–1.3	$(1–50) \times 10^4$	10–20	2.5–5.0

(iii) Resistance to chemical attack where the material has to be used in an aggressive ambience.

(iv) Compatibility with other materials.

The increasing use of synthetic resins as insulating materials has appreciably reduced ageing, chemical resistance and compatibility problems in many fields. The chemical properties of these materials can be determined by reference to the individual chapters in which they are described.

Magnetic materials

THE BEST known soft magnetic metal is Mumetal – a 77% nickel alloy having high permeability at low field strengths. The hysteresis loss is small and the total electrical losses are very low. Mumetal is particularly suitable for transformers operating at low energy levels, where full advantage may be taken of the high effective permeability to produce high impedance value. Fabricated parts must be heat treated to develop the required magnetic properties.

TABLE 1 – TYPICAL MAGNETIC PROPERTIES*

	Initial permeability (dcμ_s)	Maximum permeability	Saturation ferric induction (Tesla)	Remanence,B_{rem}, from saturation (Tesla)	Coercivity H_c (A/m)	Hysteresis loss at B_{sat} (J/m^3/Hz)	Curie point (°C)
Mumetal	55 000	240 000	0.77	0.37	1.00	3.2	350
Mumetal Plus	69 000	300 000	0.77	0.37	0.80	1.3	350
Supermumetal	127 000	350 000	0.77	0.40	0.55	0.9	350
Orthomumetal			0.80	0.70	2.40	7.5	350
Satmumetal	65 000	240 000	1.50	0.70	2.00	12.0	550
Radiometal 50	6000	30 000	1.60	1.00	8.00	40.0	525
Super Radiometal	11 000	100 000	1.60	1.10	3.20	20.0	525
Radio 36	3000	20 000	1.20	0.50	16.00	76.0	275
Hyrho Radiometal	3500	60 000	1.40	1.00	8.00	45.0	525
Hyrem Radiometal	1000	10 000	1.50	1.35	8.00	50.0	525
HCR Alloy	500	100 000	1.54	1.50	10.00	65.0	525
Permendur	1000	7 000	2.35	1.50	135.00	1270.0	975
Supermendur		70 000	2.35	2.05	19.00	170.0	975
Permendur 24	250	2 000	2.35	1.65	950.00		925
Vicalloy			1.50	1.00	20 000.00	12 × 10^4	
R2799			0.32 at 0°C				60
			0.22 at 20°C				
R2800			0.76 at 0°C				140
			0.68 at 20°C				

*Telcon Metals Ltd

TABLE 2 – TYPICAL MECHANICAL PROPERTIES*

	Hardness HV		UTS					
	Hard rolled	Annealed	Hard rolled			Annealed		
			MN/m²	kg/mm²	Ton/in²	MN/m²	kg/mm²	Ton/in²
Mumetal	290	110	910	93	59	540	55	35
Mumetal Plus	290	110	910	93	59	540	55	35
Supermumetal	290	110	910	93	59	540	55	35
Orthomumetal	300	115	930	94	60	540	55	35
Satmumetal	250	105	850	87	55	430	43	28
Radiometal 50	250	105	700	71	45	430	43	28
Super Radiometal	250	105	700	71	45	430	43	28
Radiometal 36	250	115	970	99	63	530	54	34
Hyrho Radiometal	250	105	700	71	45	430	43	28
Hyrem Radiometal	250	105	700	71	45	430	43	28
HCR Alloy	260	105	1310	134	85	430	43	28
Permendur	400	240	1230	126	80	540	55	35
Supermendur	400	220	1230	126	80	490	50	32
Permendur 24	350	170	1080	110	70	540	55	35
Vicalloy	400	900 (aged)	1500	153	97	2200 (aged)	224	142
R2799	190	110	660	68	43	430	43	28
R2800	170	110	660	68	43	430	43	28

*Telcon Metals Ltd

TABLE 3 – TYPICAL PHYSICAL PROPERTIES*

	Coefficient of linear expansion °C	Resistivity μ ohm m	Specific gravity	Thermal conductivity W/m/°C	Specific heat J/g/°C
Mumetal	13×10^{-6}	0.62	8.80	33	0.44
Mumetal Plus	13×10^{-6}	0.60	8.80	33	0.44
Supermumetal	13×10^{-6}	0.60	8.80	33	0.44
Orthomumetal	13×10^{-6}	0.58	8.75	33	0.44
Satmumetal	11×10^{-6}	0.45	8.30	16	0.46
Radiometal 50	10×10^{-6}	0.45	8.30	13	0.48
Super Radiometal	10×10^{-6}	0.40	8.30	13	0.48
Radiometal 36	2×10^{-6}	0.80	8.10	8	0.47
Hyrho Radiometal	10×10^{-6}	0.75	8.30	13	0.48
Hyrem Radiometal	10×10^{-6}	0.40	8.25	16	0.48
HCR Alloy	10×10^{-6}	0.40	8.25	16	0.48
Permendur	9×10^{-6}	0.47	8.05		
Supermendur	9×10^{-6}	0.40	8.15		
Permendur 24	10×10^{-6}	0.20	7.95		
Vicalloy		1.00	8.00		
R2799	10×10^{-6}	0.85	8.00	13	0.50
R2800	5×10^{-6}	0.83	8.00	13	0.50

*Telcon Metals Ltd

Various other proprietary alloys have been developed with compositions similar to Mumetal but yielding substantially higher permeabilities and lower losses – *eg* Mumetal Plus, Supermumetal, Orthomumetal, Satmumetal, Sanbolt, etc. All nickel-iron alloys with a nickel content of about 12%, exhibit interesting magnetic characteristics. With a nickel content from 12–30% the alloys are virtually non-magnetic. Alloys with 35% or greater nickel, suitably heat treated, exhibit very high permeability. Alloys with about 50% nickel show high initial permeability, high saturation induction and low electrical losses, permitting operation at higher flux densities than Mumetal (and are used where a higher permeability than that given by silicon iron is desirable). Peak permeability is normally realised with a nickel content of about 75%. The addition of silicon up to about 4% reduces eddy current losses.

Examples of soft magnetic alloys and their properties are given in Tables 1, 2 and 3. These are normally available as strip materials. International standard thicknesses of cold rolled strip are 0.050 mm (0.002 in), 0.10 mm (0.004 in), 0.15 mm (0.006 in), 0.35 mm (0.0138 in), 0.5 m (0.20 in). 0.20 mm (0.008 in) is the preferred intermediate size between 0.15 mm (0.006 in) and 0.35mm (0.138 in). The strip is supplied either in coils weighing up to 150 kg (330 lb) or in cut lengths up to 2.5 m.

Soft magnetic materials may also be available in the form of sheet, bar, rod, section and wire. See also chapter on *Nickel and Nickel Alloys.*

Permanent magnet materials

THE MAGNET industry and the availability of permanent magnet materials have undergone enormous change over the past fifteen years. A new family of materials has been added to the existing alnicos and ferrites known collectively as rare earth cobalt or rare earth iron. In simple terms, the new rare earth magnets are between 6 and 8 times stronger than ferrite.

Permanent magnets are measured by the amount of flux they can develop, combined with their resistance to demagnetising influences such as alternating electric (magnetic) currents.

Three parameters are identified; Br (remanence), bHc (coercive force) and BHmax (energy product) and are used to compare and contrast materials.

Alnico or Alcomax magnets are complex alloys of Al, Ni, Co, Cu and Fe with small additions of Zr, Ti, Nb and S where necessary. The Fe constitutes about 50% by weight of the alloy.

Magnets from these elements are produced in two ways:

(i)   Melting and casting to shape.
(ii)   Powder pressing and sintering.

The as-cast or as-sintered component is subjected to a controlled heat treatment. By cooling from above the Curie temperature in a magnetic field, anisotropy or preferred direction of magnetic orientation can be induced. Anisotropic magnets have enhanced magnetic characteristics in the direction of orientation compared with the non-aligned isotropic ones. Alcomax III is anisotropic; Alnico is isotropic. With the casting process additional magnetic gains can be made by preferential crystalline orientation – Columax. However this technique is shape-dependent. Magnetic properties can be varied by adjustment to composition, heat treatment and crystallisation (see Table 1). Properties of sintered magnets tend to be slightly lower than the equivalent cast materials.

Alnico magnets have densities in the range 6.8 to 7.3 g/cm^3, are electrical conductors, have a high Curie temperature, 860°C and can be used as sources of magnetic energy up to 550°C. Flux losses do occur at temperatures above 20°C;

TABLE 1

Material	Magnetic properties			Form	Other names		
	Br (Tesla)	BHmax (Kj/m³)	bHc (KA/m)		USA	Germany	France
Alnico	0.80	13.5	40	Isotropic	Alnico 2	Alnico 160	Nialco 1
Alcomax III	1.27	42	52	Anisotropic	Alnico 5	Alnico 500	Alnico 600
Hycomax III	0.90	45	125	Anisotropic	Alnico 8B	Alnico 450	Alnico 1500
Columax	1.35	60	59	Anisotropic crystalline oriented.	Alnico 5—7	Alnico 700	Super Ugimax

TABLE 2

Material	Magnetic properties			Form	Other names		
	Br (Tesla)	BHmax (Kj/m³)	bHc (KA/m)		USA	Germany	France
Feroba 1 (D1)	0.22	8.0	136	Isotropic	Ferrite 1	Oxit 100	
Feroba 2	0.39	28.8	176	Anisotropic	Ferrite 5	Oxit 300R	Spinalor 3B
Feroba 3	0.37	26.4	240	Anisotropic	Ferrite 7	Oxit 330	Spinalor 4H

typically 2% per 100°C rise. The materials are easily magnetised after assembly, using field strengths of 250 to 500 kA/m (3000 to 6000 Oersted).

Alnico magnets are physically hard (Rockwell C45–C60), brittle and most easily machined by alumina grit grinding. Manufacturing techniques impose limits on shapes/sizes and tolerances, as outlined.

	Cast	**Sintered**
Shapes	Slugs, blocks, thick walled rings, rods, horse shoes.	Dictated by complexity of press tools.
Sizes	Typically 50 g–10 kg.	1 g–50 g+
Tolerances	As cast ±0.8 mm.	As sintered ±0.15 mm.
	Ground ±0.05 mm.	Ground ±0.5 mm.
Production		
Quantities	100–1000+.	10000–100000+.

Applications include moving coil instruments, magnetrons, klystrons, TWTs, Watt-hour meters, sensors, actuators, and d.c. professional motors.

Ferrite or ceramic magnets provide the bulk of magnet production in the Western World. Based on inexpensive plentiful raw materials, processing is via a powder metallurgy route. Iron oxide FE_2O_3 in fine powder form is blended with either Ba/Sr carbonate and fired or calcined to form a hexaferrite $BaO.6Fe_2O_3$.

The calcine is ground down/milled to approximately single domain size (lu) by both dry crushing and milling in water, in which the material is inert. The slurry may be fed directly to a press, or spray dried and the dry powder pressed. The dry powder is often coated with wax.

Once pressed, components are fired again at about 1250°C in an oxidising atmosphere. During the firing cycle the powder particles fuse together and the pressing may shrink by up to 20%. The fired components are physically hard, extremely brittle and must be handled with care to prevent cracking or chipping. Although the raw materials are inexpensive, process costs are heavily dependent on energy usage, and efficient plant is essential.

With wet pressing, tool costs for each shape can reach £20K (1989) leading to some degree of standardisation with suppliers, and use in high volume applications.

The magnetic characteristics of ferrite differ greatly from Alnico, providing a lower flux density but higher resistance to demagnetisation. The two types of magnet material are therefore not directly interchangeable and need to be designed into the application for optimum performance.

Ferrite properties

Sintered ferrites have a density of 4.8 to 5.0 g/cm³, are electrical insulators, have a Curie temperature of 450°C and an upper operating temperature of 250°C.

Flux losses of 20% per 100°C temperature rise from ambient can occur. Most of the losses are recovered when returning to ambient.

Due to shrinkage, good dimensional control is difficult, as the fired dimension can vary by ±2%. However product can be ground with diamond to ±0.05 mm especially on pole faces.

TABLE 3

Material	Magnetic properties			Form	Other names
	Br (Tesla)	BHmax (Kj/m³)	bHc (KA/m)		
Supermagloy S2	0.90	160	680	Anisotropic (1:5)	Res, Recoma, Seco, Crucore Tascore, Vacomax
Supermagloy S3	1.08	220	800	Anisotropic (2:17)	

TABLE 4

Material	Magnetic properties			Form	Other names
	Br (Tesla)	BHmax (Kj/m³)	bHc (KA/m)		
Neomagloy N28	1.10	224	740	Anisotropic	Res, Refema, Crumax, Vacodym
Neomagloy N35	1.20	275	800	Anisotropic	

NB This list is not comprehensive, product standardisation is evolving.

TABLE 5

Material	Magnetic properties			Form	Other names
	Br (Tesla)	BHmax (Kj/m^3)	bHc (KA/m)		
Neomagloy 6B	0.56	48	344	Isotropic Compression/ injection moulded.	Neobond, Magnequench, Dynamag Neofer.
Neomagloy 8B	0.62	64	432	Isotropic Comp mould.	
Neomagloy 6F	0.53	48	36B	Isotropic Flexible.	

TABLE 6 – RELATIVE COST/COMPARISON OF ENERGY PRODUCT

Parameter/Material	Columax	Feroba III	(2:17)	Sintered NdFeB	Bonded NdFeB
Energy product (Kj/m^3)	60	28	220	280	64
Relative cost (/kg)	6	1	45	27	20
Relative cost/unit energy (/kg)	0.1	0.035	0.2	0.09	0.3

Magnet sizes and shapes have been influenced by the application industry needing them:

For loudspeaker designs, rings are often used, such as:

o.d. (mm)	i.d. (mm)	L. (mm)	
32	16	6	In car speakers.
225	122	23	In hi-fi.

(For magnetic separators, large blocks are produced up to $150 \times 100 \times 25$ mm, which can be stacked together to develop high magnetic field volumes. These building blocks can also be cut or carved with diamond saws to smaller pieces for actuators and relays.

A series of arc segments have been developed for use in commercial d.c. motors for the automotive, consumer electronics and durable goods industries.

Ferrite magnets are normally supplied un-magnetised to facilitate handling, packing, storing and assembly into the magnetic device.

Rare earth cobalt magnets are available based upon two compositions:

(i)   Samarium – 33%, balance cobalt, by weight, identified as $SmCo_5$, or just 1:5.

(ii) Samarium, Zr, Fe, Cu, balance Co, referred to as Sm_2Co_{17}, or 2:17.

The manufacturer of either type of material follows the powder metallurgical-sinter route.

The base alloys are extremely expensive and readily oxidisable, especially as a fine powder. Finished magnets, however, are chemically stable in normal environments. Components can be plated with Zn or Ni if additional protection is required.

The materials are anisotropic and shape dependent upon the die cavity.

Magnet density exceeds 8.1 g/cm^3 and the materials are brittle and hard (Rockwell C50). Both will chip easily and can spark if struck with a hard metal (similar to lighter flints).

1:5 magnets have a Curie temperature of 725°C but would not be recommended for use above 250°C. Flux losses of 4% per 100°C temperature rise occur.

2:17 magnets have better temperature stability and higher operating temperature to 350°C with 3% flux loss per 100°C. However, 2:17 is a more difficult material to magnetise and is normally supplied fully saturated.

Due to their excellent properties, rare earth cobalt magnets are used for miniaturisation in aerospace d.c. motors, voice coil motors for disc drives, microwave devices, magnetic bearings, watch stepping motors and tiny switching and sensing devices.

Simple shapes, blocks, rings and discs can be produced, up to 100 mm in diameter or 100×100 mm square section. Due to high material cost, finished magnets often weigh as little as 0.1 g. For precise sizes magnets can be ground with diamond, spark eroded, and sliced/diced.

TABLE 7

Material	Magnetic properties			Form	Other names
	Br (Tesla)	BHmax (Kj/m³)	bHc (KA/m)		
1% Carbon Steel	0.90	1.6	4	Wrought	
5% Chrome Steel	0.94	2.4	5.2	Wrought	
35% Cobalt Steel	0.90	7.6	20	Wrought	
Cunife	0.57	11	40	Wrought	
Cunico	0.53	7.2	36	Wrought	
Crofeco	1.27	41	49	Cast	Chromindur
Vicalloy	0.90	8	24	Forged & Drawn	
P.B. Alnico	0.45	10.8	81	Plastic bonded	Prac, Tromalit
P.B. Ferrite	0.15	4.8	96	Plastic bonded	Sprox, C1
I.M. Ferrite	0.28	15.2	176	Injection moulded	Dynamag, Sprox
R.E. Ferrite	0.15	4.8	104	Rubber extruded	Ferriflex, Plastiform
P.B. Sm-Co	0.59	60	400	Plastic bonded	Supermagloy B2/B3

NB This list is not comprehensive, other specialist materials in limited manufacture can be found.

Rare earth iron (sintered)

The most recent addition to the magnet manufacturer's portfolio is the production of neodymium-iron-boron magnets. Nd is another element in the rare earth family but is more plentiful than Sm and less expensive. The terniary alloy is sometimes doped with Dy, Co, or Al.

Processed by a powder metallurgical route these materials provide the strongest known permanent magnets to date. Compared to Sm-Co magnets, NdFeB is more ductile, less prone to chipping, and easier to machine. There are, however, some serious disadvantages. NdFeB will rust if exposed to moist air, and is readily attacked by salts/chlorides. In addition the alloy has a low Curie temperature (312°C) and poor performance at elevated temperature.

Much effort has been focused on these problems and the development continues.

Techniques have been introduced to coat finished magnets before final magnetising. These include aluminium vapour deposition. Epoxy coating and stove enamelling. They are only successful as long as the coated surface remains free from damage or scratches.

The new material has a density of 7.5 g/cm^3, loses flux at a rate of 13% per 100°C temperature increase, and has an upper temperature limit of 140°C. This limit is constantly being extended with further research.

Rare earth iron (bonded)

At the same time that sintered NdFeB was being developed, General Motors (USA), perfected a method of melt-spinning a similar alloy.

Molten metal is rapidly solidified by allowing it to impinge onto a cooled spinning copper wheel. Cooling rates of 10°C per second are achieved. The resultant ribbon is semi-crystalline or amorphous.

Crushed ribbon (300) can be coated with a resin (epoxy) pressed to form a solid shape and subsequently magnetised. Magnets produced in this matter are the strongest isotropic type available. Bonded magnets are easy to press to tight tolerances ±0.05 mm, and into complex shapes unachievable by the sintering route. The magnets, however, still suffer from rusting and need coating to prevent degradation.

Table 6 compares and contrasts the above materials in terms of their magnetic energy and relative cost.

The materials described above form the bulk of magnet production and use. However, for completeness, Table 7 identifies other materials available.

Soft magnetic materials

SOFT MAGNETIC materials carry and often concentrate flux generated by external magnetic fields. However they do not generate flux themselves (as opposed to permanent hard magnetic materials). When the external magnetic field is removed, the flux in the material should reduce to zero. The three major groups which exhibit soft ferro/ferri magnetic properties are:

(i)   Iron alloys.
(ii) Iron oxides – ferrites.
(iii) Amorphous alloys – glasses.

Soft magnetic materials can operate under d.c. and a.c. conditions as a means of flux transfer, energy conversion, suppression, and shielding of unwanted magnetic fields. For d.c, softness can be quantitatively described by reference to permeability (m), saturation (Bsat), and coercive force (Hc). For a.c. conditions those parameters plus eddy current losses (We), hysteresis losses (Wh), loss factor, and Curie temperature (Tc) are used with reference to the operating frequency range.

Iron based alloys, Fe, FeSi, FeSiAl, FeCo, FeNi

These materials can be processed by vacuum melting, casting, rolling to form strip or sheet, and annealing or reducing to powder (dust) before pressing to shape, with a suitable binder, and curing.

High purity Fe alloys are used predominantly to transfer flux as pole-pieces in magnetic circuits, as laminations in high power transformers, contacts in biased relays, and pole shoes in motors and generators, where frequency of field change is nominally between 0 and 100 hZ. For pole-pieces, materials need to have high Bsat, high μ, good homogeneity and temperature stability. Permendur has the best Bsat. Additionally, for power transformers, motors, and generators, alloys need a small area enclosed by the hysteresis loop (since this represents energy lost during cyclic magnetisation), and a form which maximises internal resistance to eddy currents, whilst enabling component shapes to be fabricated. As a consequence, sheet, foil and strip are produced by hot and cold rolling down to 30 μm. Many ferro al-

TABLE 1 – ALLOYS BASED ON FE WITH TYPICAL SOFT MAGNETIC PROPERTIES

Material	Permeability $\mu(i)$	$\mu(m)$ x $\mu(o)$	Coercivity A/m	Bsat T	Energy Loss W/kg*	Tc C	Product
Pure Fe	—	88 400	3.2	2.16	—	770	—
Comm Fe	200	20 000	100	2.16	—	770	—
FeSi	800	30 000	8–40	2.0	.5/.9	750	Hyperm Trafoperm
FeAl	6000	70 000	2–8	1.2	—	350	Alperm Vacodur
FeCo	300	60 000	16–200	2.4	.8/2.0	950	Permendur
FeNi (36–48)%	$2\text{–}5\times 10^{-3}$	$7\text{–}20\times 10^{-3}$	12–60	1.4	.5/1.0	250–400	Radiometal
FeNi (50)%Ni	$.4\text{–}10\times 10^{-3}$	$35\text{–}200\times 10^{-3}$	4–16	1.45	.25/.4	450–550	Permalloy
FeNi (60–80)%	80×10^{-3}	$70\text{–}250\times 10^{-3}$	.25–5	0.9	–	350–520	μmetal

Permeability measurements are relative to μ_o(air)
Energy loss at 50 Hz in a 1 Tesla field*

loys are brittle and difficult to work. Rolling operations are expensive when compared to the alloys. Strip may be resin coated or interleaved with insulation before being wound to form toroids. Laminations are stamped out of sheet and stacked together to form cores. Total energy losses, Wh + We, should be minimised by correct choice of soft magnetic material: these losses are dissipated as heat. For relays, weak control currents convert into strong flux densities via soft magnetic materials; the magnetic elements attract and close an independent electrical circuit. Alloys need a high μ, plus a low Hc to ensure that the relay will release when the control current is removed. They also need mechanical ductility and resistance to work hardening to survive many million operations.

Densities range from 7.8 gm/cm³ for pure Fe to 8.8 gm/cm³ for some FeNi. As metal alloys, specific resistivity is in the range 9 to 145 μohm/cm.

Soft magnetic dust components are produced from fine powders of Fe and its alloys compacted together to form the required shape. The powders 2 to 10 μm in size can be made by the carbonyl process, by mechanical reduction from the cast ingot, or by electrolytic means. In the carbonyl process, Fe is reacted with CO at 200° to form iron penta-carbonyl. The reaction produces a very fine dust which can be returned to the metal by radiation heating. Manufacturers generally

TABLE 2 – TYPICAL PROPERTIES OF THREE KINDS OF DUST

Material	Initial $\mu(i)$	Coercivity A/m	Bsat T	Referred loss v freq factor		
Carbonyl Fe	50	—	1.4	7×10^{-4}	—	1kHz
				7×10^{-4}	—	10kHz
				4×10^{-4}	—	100kHz
Mo perm-alloy	125	0.4	1.03	9×10^{-6}	—	1kHz
				4×10^{-4}	—	10kHz
				4×10^{-3}	—	100kHz
FeAlSi-Sendust	80	1.76	1.05	9×10^{-6}	—	1kHz
				4×10^{-4}	—	10kHz
				4×10^{-3}	—	100kHz

offer varieties of dust cores to meet frequency of operation needs. However the μ depends not only on the properties of the dust, but also on the packing density achievable. Dusts of Fe, FeNi, Mo permalloy, FeAlSi (Sendust) are available.

Prior to pressing, the dust is insulated with a resin and/or phosphated to produce a surface oxide, to increase resistivity and so reduce eddy current losses.

Soft magnetic properties of dust cores can be adjusted to the application by varying the resin content, the pressing tonnage and hence the density, and the size of the dust particle.

Dust cores have mostly been superseded by soft ferrites, which offer higher permeability and lower loss factors.

Soft ferrites

This term is used collectively to describe materials based on iron oxide blended with either Mn, Zn, Ni, oxides. Soft ferrites are tailor-made to satisfy the desired characteristics of electronic circuits and components. Detailed control of chemistry and crystallography is essential.

With the enormous growth of the communications industry it became necessary to establish a range of small, inexpensive, soft magnetic components, to operate at low power range (10^{+2} to 10^{-10}W), but high operating frequencies (from 10^{+2} to 10^{+10} Hz).

Soft ferrites were developed to fulfil this need. This development continues as users extend frequency ranges and increase power levels, whilst attempting to reduce physical size. Typical applications for these materials include use as inductor cores, transformer cores, high frequency switching, and both RFI and EMI shielding. The components are used in most electronic industries, such as radio, TV, telecommunications, and the computer industry.

Manufacture is complex and capital intensive, a standard process route being powder processing, blending, pellitising, calcining or firing, followed by milling, spray drying, pressing/moulding, before sintering to shape and size and final grinding to achieve tolerance control.

Two basic compositions are manufactured Mn Zn ferrite, and Nβi Zn ferrite. MnZn is used predominantly for telecommunications, entertainment, and industrial applications up to 10^{+6} Hz frequency.

NiZn is used for entertainment/industrial applications between 10^{+6} and 10^{+8} Hz.

The magnetic performance of ferrites is dependent upon frequency of use, temperature, and physical stress which may distort the crystalline structure. As with alloy materials, losses occur due to hysteresis and eddy currents, the loss factor being dependent upon the initial permeability of the material, frequency, inductance and resistivity. Most ferrite makers specify loss factor *versus* frequency for each grade of material.

The physical needs for inductors or transformer cores for communications are high permeability, low power loss combined with high electrical resistivity, consistency, and simplicity of shape.

High permeability implies fewer turns on a coil for a given inductance.

High constancy and good stability can be achieved by partial surrender of permeability with the introduction of air gaps in the magnetic circuit; gapping also reduces eddy currents.

Toroidal cores are less frequently used for inductors/transformers, being replaced by pot cores. Pot cores are easier to wind, can be accurately gapped, and provide shielding from external magnetic fields. With the addition of a screw core, adjustable inductance-tuning can be made available.

Other soft ferrite shapes include RM cores (rectangular mode), U cores, E cores, beads, bobbins, screw cores, rods and tubes.

The following line drawings indicate the shapes available (Figure 1):

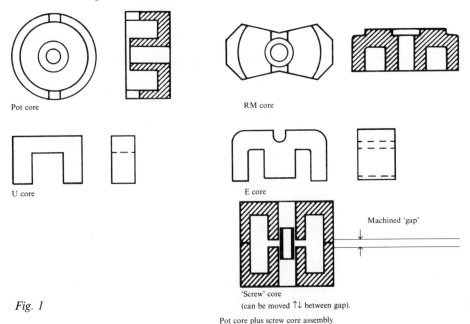

Pot core

RM core

U core

E core

Machined 'gap'

'Screw' core
(can be moved ↑↓ between gap).

Fig. 1

Pot core plus screw core assembly

Table 3 identifies some of the range of available soft ferrites with typical Trade names. This should not be considered a comprehensive list, but only a sample of products and manufacturers.

Amorphous alloys and glasses

These materials were developed in the 1950s by a Rapid Solidification Method. They have metallic properties, but without the crystal lattice of conventional metallic material. They have no atomic order (*ie* they are amorphous like glass), and are frequently referred to as metallic glasses.

Manufacturing requires high frequency melting of the alloy, and forcing the molten liquid through a small orifice onto a rotating wheel made of a good thermal conductor. Heat is extracted from the alloy so quickly that crystallisation is unable to occur (10^{+6} °C/s) and the alloy solidifies in the form of a tape. The thickness and width of the tape can be controlled by the speed of the wheel and the feed of the metal. Thicknesses of between 0.025 and 0.04 mm are available in widths up to 25 mm. Excellent tolerance control is achievable (±0.005 mm). Alloys

TABLE 3 – POT CORE PLUS SCREW CORE ASSEMBLY

Material	Initial μ	Bsat T	Loss factor v Freq	Tc °C	p Ωcms	Trade names
NiZn	650	.34	50 — 250kHz 65 — 500kHz 130 — 1MHz	180	3.10^{-4}	Neosid F13 Allen. BradH Stackpole 7D
NiZn	125	.34	60 — 1MHz 65 — 5MHz 100 — 10MHz	270	10^{-5}	Allen.BradQ1 Ferroxcube4C4 NeosidF16
NiZn	50	—	50 — 1MHz 75 — 10MHz 300 — 40MHz	450	10^{-5}	Allen. BradQ3 Stackpole12 NeosidF25
NiZn	12	—	100 — 10MHz 200 — 100MHz 1000 — 200MHz	500	10^{-5}	SiemensU17 Stackpole14 NeosidF29
MnZn	2500	.5	10 — 100kHz	230	10^{-2}	T.D.K. H7C1 NeosidF4 SiemensN67
MnZn	4400	.38	20 — 100kHz	140	50	T.D.K.H5A SiemensN41 NeosidF9
MnZn	2000	—	6 — 10kHz 15 — 100kHz	150	10^{-2}	NeosidP10 Stackpole24B SiemensN27 Allen. Brad05

TABLE 4 – SOME OF THE SOFT MAGNETIC AMORPHOUS
MATERIALS AVAILABLE AND A RANGE OF TYPICAL PROPERTIES

Basic material	Permeability $\mu(max)$†	Coercivity	Bsat T	Energy loss W/kg*	Tc °C	Names
Fe	100 000	<4A/m	1.5	10	420	Vitrovac 7505 ACO–AM
FeNi	250 000	<1A/m	0.8	6	260	Vitrovac 4040 ACO–3H ACO–3M
Co	600 000	<4A/m	0.55	4	250	Vitrovac 6025 ACO–4M

† 50Hz
* Core loss at 20kHz, and 0.2 Tesla field.

containing Fe, Co, Ni, at approximately 75 atomic % are produced, combined with B, Si, C, to assist in the formation of the amorphous state.

In addition to the excellent magnetic properties of these tapes, due to their lack of structure, their mechanical properties are improved. For example, silicon iron, normally very brittle, can be made ductile and flexible, easily wound into toroids. To obtain optimum magnetic properties, amorphous alloys must be heat treated, sometimes in an external magnetic field.

Applications for these soft magnetic tapes include strip wound cores for miniature transformers up to 400 Hz, flexible magnetic shielding of cables, electro-magnetic detection devices, and RFI chokes.

Armophous tapes have higher specific resistivity than their crystalline counterparts, which combined with their lower thickness, enable core losses in the kHz range to be less than those of metal alloys. In addition the shape of the hysteresis loop can be altered by suitable heat treatment.

Table 4 illustrates some of the soft magnetic amorphous materials available and a range of typical properties.

Magnetostriction – the Joule effect

THIS IS the term given to changes in shape/size of a ferromagnetic body upon magnetisation. Changes are reversible as the applied field is cycled. In lamination steel for transformers, magnetostriction can cause noise on the a.c. output combined with physical movement weakening the structure.

Until quite recently the phenomenon was a physical curiosity as the greatest changes which could be generated were small. However with the development of rare earth alloying and production technology some extremely advanced magnetostrictive alloys are now available.

In addition to the construction of magnetostrictive transducers for underwater echo sounding, the new alloys are being developed for use as magnetic sensors, precision actuators, and rigid supports.

At room temperature, magnetostriction coefficients greater than 1000 microstrain can exhibit highly efficient conversion of magnetic to mechanical energy and *vice versa*. Three other less well known effects are also associated with magnetostriction:

> *The Guillemin effect* – The tendency of an elastically bent rod to straighten when placed in a longitudinal magnetic field.
>
> *The Weidemann effect* – The twisting of a rod carrying an electric current when it is subject to a magnetic field.
>
> *The Villari effect* – The change in magnetic induction when a magnetised rod is subjected to longitudinal stress.

For Joule magnetostriction, the symbol λ is used to denote incremental changes in length.

Magnetostriction changes as a function of the applied magnetic field and ultimately reaching a level of saturation λ_s. λ_s, exhibits anisotropy; having different values in different crystallographic directions. Absolute values are therefore made on single crystals. λ_s for polycrystalline materials is the average of the values determined for the three principal directions in single crystals. Values for polycrystalline materials are therefore variable, dependent upon composition, grain size, surface condition etc.

Typical values for the three ferromagnetic element single crystals are shown below:

Element	λ_{100}	λ_{111}	at saturation and 20°C
Fe	20	−20	
Co	−46	−24	
Ni	−51	−28	
	Values $\times$ 10^{-6}		

100, 111 are the crystalline directions.
Put more simply, a rod of Ni as a single crystal, 100 mm long, will contract along its length by 0.005 mm when subjected to a saturation magnetic field.

For useful devices the level of saturation magnetic field must be small and easily attained. In single crystals there are directions of easy and difficult magnetisation. This characteristic is known as magnetocrystalline anisotropy. For Fe, a body centered cubic, the easy axis is parallel to the edge of the cube, the difficult axis along the diagonals. The magnetising force required to saturate Fe is approximately:

8 kA/m (100oe) parallel to edge.
40 kA/m (500oe) along the diagonal.

The excess energy needed when comparing the easy/dificult axes is referred to as the Anisotropy Energy (K). The anisotropy constant can be positive or negative dependent upon the crystalline directions chosen.

For practical applications, high strains at low fields are desirable, combined with low anisotropy.

Recently a new family of materials based on the binary $REFe_2$ compounds have been produced. These compounds possess the largest known magnetostriction coefficients at 20°C, of about 2000 microstrain (c.f. 20 for Fe).

Magnetostriction and magnetic anisotropy polarity are shown for some of the rare earth elements at room temperature:

RE element	Polarity magnetostriction	Polarity anisotropy
Samarium Sm	– negative	– negative
Terbium Tb	+ positive	– negative
Dysprosium Dy	+ positive	+ positive
Holmium Ho	+ positive	+ positive

The most significant material from this new family has the composition Tb.$_{27}$Dy.$_{73}$Fe$_2$ and is known as Terfenol D. The material exhibits a large positive magnetostriction in the <111> direction combined with easy magnetisation. To fully exploit this property the ideal material would consist of a single crystal with a <111> direction aligned to the direction of magnetisation.

Terfenol-D can be produced as a single crystal by vertical RF zone solidification. However the rod so produced has the <112> direction parallel to its physical length. With the applied magnetic field along the length of the rod the magnetostrictive strain is slightly reduced by the angular difference in direction, <111> c.f. <112>.

Investigations of the Villari Effect for Terfenol (magneto-mechanical behaviour), have shown that dramatic changes to Young's Modulus can be created in moderate applied magnetic fields. By pre-stressing a rod of Terfenol, and applying a biasing magnetic field from a permament magnet, large changes in strain can be created by small fluctuations to that field. The reverse of this process enables extremely sensitive strain gauges to be developed.

Another use for magnetostrictive materials is in the construction of low frequency transducers. These are used to radiate acoustic energy in water (sonar).

Terfenol is brittle in nature, and should therefore be mechanically pre-stressed to protect it against tensional forces generated under dynamic conditions. A sketch of a form of transducer is shown in Figure 1.

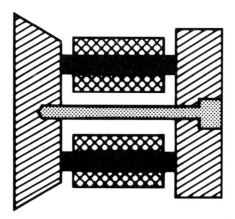

Fig. 1

A d.c. current through the solenoid provides a static magnetic bias, upon which is superimposed an alternating current (magnetic field) to generate the vibration. At frequencies below 2 kHz this type of transducer is better than piezo-electric systems in terms of power output per unit weight of active material.

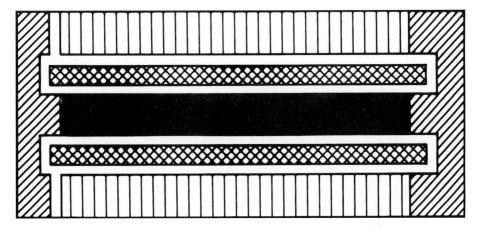

Fig. 2

Another potential use for Terfenol type materials is as fast acting actuators, relays and switches, where high contact pressure is required. A possible arrangement is shown in Figure 2.

The newer materials described above are relatively expensive, due to both their raw material content and their cost of manufacture, however they do offer significant advances in magnetostrictive properties. It is the role of applications engineers to exploit these characteristics.

Piezoelectric materials

THE PIEZOELECTRIC effect is an inter-relationship between mechanical distortion and electrical effects peculiar to certain crystalline materials. If such materials are distorted by mechanical loading, they generate electricity; conversely, if charged with electricity they undergo a dimensional change. The piezoelectric crystal, in other words, is a simple form of transducer for converting one form of energy into another.

Piezoelectric effects in any crystalline substance are dependent on the direction of the applied force or electric field relative to the crystallographic axis of the material. The number of materials which exhibit marked piezoelectric effects is limited to certain natural crystals and their synthesised counterparts, together with certain polycrystalline ceramic materials (see Table 1).

The application of piezoelectric materials is divided broadly into:

(i) Devices operating at mechanical resonant frequency, utilising the motor effect of the phenomenon.

(ii) Devices operating under non-resonant conditions, utilising the generator effect.

In certain applications both motor and generator effects may be employed, such as one crystal operating under resonant conditions to 'drive' a second crystal working as a generator.

Bimorphs

Bimorphs are flexing type piezoelectric elements which have a higher capacity for handling larger motions and smaller forces than single piezoelectric plates. The element consists of two face-shear plates or two transverse-expander plates cemented together face-to-face in such a manner that a voltage applied to the electrodes causes the plates to deform in opposite directions, resulting in either a twisting or bending action. Conversely, depending on its construction, mechanical bending or twisting of the element will cause it to develop a corresponding

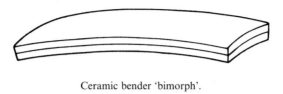

Ceramic bender 'bimorph'.

Crystal bender 'bimorph'.

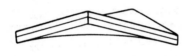

Crystal twister 'bimorph'.

Fig. 1
Curvature of bimorphs.

voltage between electrode terminals. The curvatures developed in bimorphs are shown greatly exaggerated in Figure 1.

Twister bimorphs, which are made up of two shear plates, and bender bimorphs, which are made up of two expander plates, are of either series or parallel construction, depending on the electrical relationship between the two plates of the assembly. Parallel bimorphs are generally used in electrically driven applications because of the higher sensitivity resulting from the application of full voltage to each plate. Since the parallel relationship of the two plates provides maximum capacitance or minimum electrical impedance, this type of element is also preferred to a series bimorph for those mechanically driven applications where circuit loading is sufficient to cause loss of output. Series bimorphs are used in mechanically driven applications to take advantage of the addition effect of generated voltages in the two plates, but only in those cases where circuit loading is not severe; for example, in hearing aid microphones, which are usually installed in the hearing aid immediately adjacent to the input amplifier.

TABLE 1 – PROPERTIES OF PIEZOELECTRIC MATERIAL

Material	Cut	Basic plate action	Free dielectric constant	Density in g/cm³	Volume resistivity ohm metres	Acoustic impedance in kilograms/metre³ – metre/seconds	Freq constant free-free in cycle-metre open circuit	Coefficient of coupling %	d Constant, strain out field in metres/metre volt/metre
Multiply figure in Table by:			1	1	10^9	10^6		1	10^{-12}
	0°X	FS	350*	1 770	>10	—	—	76	550
	45°X	LE	350*	1 770	>10	5.6	2 100	73	275
Rochelle salt @ 30°C	0°Y	FS	9.4	1 770	>10	—	—	32	54
	45°Y	LE	9.4	1 770	>10	4.3	1 200	29	27
	L	TE	125	1 770	>10	7.6	—	24	150
Ammonium dihydrogen phosphate	0°Z	FS	15.3	1 880	>51	—	—	32	48
	45°Z	LE	15.3	1 800	>1	6.1	1 680	28	24
ADP	L	TE	42	1 800	>1	8.5	—	3.5	8.7
Lithium sulphate	0°Y	TE	10.3	2 060	>10	11.2	2 730	38	16
LH	0°Y	VE	10.3	2 060	>10	—	—	—	13.5
Dipotassium	45°Z	LE	6.49	1 990	10	7.2	1 800	23	11
tartrate	0°Z	TS	6.49	1 990	10	—	1 030	13	11
DKT	0°Z	FS	6.49	1 990	10	—	1 110	28	22
Potassium dihydrogen phosphate KDP	45°Z	LE	21.3	2 340	>1	7.0	1 500	11	10.5
	0°Z	FS	21.3	2 340	>1	—	810	12	21
Quartz	0°X	TE	4.5	2 650	>1 000	15.2	2 870	11	2.3
	0°X	LE	4.5	2 650	>1 000	14.3	2 700	10	2.3
Tourmaline	0°Z	TE	6.6	3 100	>100	22.3	3 600	10	1.9
	0°Z	VE	6.6	3 100	>100	—	—	—	2.4
Barium titanate	—	TE	1200–1700	5 500–5 700	>100	30	2 600	50	190
Nominal values @ 25°C	—	LE	1200–1700	5 500–5 700	>100	25	2 200	21	78
	—	VE	1200–1700	5 500–5 700	>100	—	—	—	34

FS : Face shear plate   * Low field
LE : Length expander plate

TABLE 2 – PROPERTIES OF PIEZOELECTRIC MATERIAL

g Constant field out stress in $\frac{\text{volts/metre}}{\text{Newton/metre}^2}$	Coefficient of expansion parallel to thickness in strain/degC	Coefficient of expansion parallel to width in strain/degC	Coefficient of expansion parallel to length in strain/degC	Maximum safe stress N/m²	Maximum safe operating temperature °C	Minimum safe relative humidity unprotected crystal %	Maximum safe relative humidity unprotected crystal %	Typical applications
10^{-3}	10^{-6}	10^{-6}	10^{-6}	10^6	1	1	1	
180	28	20	17	14.7	45	40	70	Twister bimorphs
90	28	18.5	18.5	14.7	45	40	70	Bender bimorphs, underwater sound transducers
664	17	28	20	14.7	45	40	70	
332	28	24	24	14.7	25	40	70	Underwater sound transducers
100	21	22	17	14.7	45	40	70	
354	1.7	35	35	20.6	12.5	0	94	Twister bimorphs, material also used for electro-optic properties
177	1.7	35	35	20.6	125	0	94	Underwater sound transducers. Bender bimorphs
77.5	12.6	35	24	20.6	125	0	94	
175	28	−5	16	—	75	0	95	
148	28	−5	16	—	75	0	95	Hydrophones
192	32	29	29	—	100	0	70	Frequency control, filters and bender bimorphs (tentative)
192	32	45	13	—	100	0	70	Frequency control and filters (tentative)
384	32	45	13	—	100	0	70	Frequency control, filters twister bimorphs (tentative)
55.8	45	27	27	—	150	0	95	
102	45	27	27	—	150	0	95	Used for electro-optical properties
58	14.8	9	14.8	98	550	0	100	Ultrasonic transducers, underwater sound transducers
58	14.8	9	14.8	98	550	0	100	
—	—	—	—	—	—	0	100	
—	—	—	—	—	—	0	100	Pressure gauge
12.6	19	19	19	80	100	0	100	Ultrasonic transducers, underwater sound transducers
5.2	19	19	19	80	100	0	100	Underwater sound transducers, Bender bimorphs
2.2	19	19	19	80	100	0	100	

SECTION 9

Joining Materials

ADHESIVES
LOW MELTING POINT ALLOYS (FUSIBLE ALLOYS)
SEALANTS
SOLDERS
BRAZING MATERIALS

Adhesives

THE TRADITIONAL types of adhesive – animal and vegetable glues, etc – are mechanical in function, their strength relying on mechanical adhesion by interlocking between surfaces porous enough to absorb the glue. The introduction of casein glues marked the appearance of adhesives exhibiting a molecular attraction to the substrates similar to the adhesive force in a single substance – or specific adhesion as it has come to be called. The later introduction of synthetic adhesives (thermosetting resins) provided further marked improvements in specific adhesion. There has been parallel development of natural and synthetic rubber adhesives with high specific adhesion for bonding smooth, non-porous surfaces, mainly for lightly stressed joints not subject to severe weathering, grease, oil or solvent except where a special synthetic elastomer is used. The strongest rubber-based types are thermosetting or curing and can be used for bonding friction linings and other hard, non-porous surfaces to metal.

In general, polar adhesives are required for bonding polar materials, and non-polar adhesives for non-polar materials so that the adhesive is capable of wetting the surface, which is a primary requirement for a good bond. Some materials can be simultaneously polar and non-polar and a few can be changed by chemical treatment from one state to the other. An adhesive should be relatively free from residual stresses when set, or if they are unavoidable, the adhesive must not disrupt or craze under their action. Characteristics established on setting are also influenced by time (ageing) and temperature. Service conditions include the strength required in a specific temperature range, the degree of exposure, type of loading, bond life and resistance to weathering and solvents. Production requirements include whether or not force drying and/or pressure are needed to close the joint during setting and working times.

Types of industrial adhesive now in common use are described under the separate headings which follow.

Acrylics

Acrylic resins have applications as specialised adhesives, *eg* non-structural metal-to-metal bonding. Acrylic emulsions are widely used for bonding plasticised PVC to concrete, cement, timber, etc (*eg* floor coverings), where they exhibit excellent dimensional stability. Acrylic emulsions are also used for bonding scrim to balsa to make flexible panels as core material for GRP mouldings, a particular advantage being that the adhesive is soluble in polyester resin. Acrylic copolymers are produced as lower cost adhesives for PVC tiles and floor coverings. Polyester-acrylic adhesives are used for bonding metals and hard glazed surfaces, and as thread locking adhesives.

Casein

Casein – the original resin-type woodworking glue – is currently used mainly for bonding asbestos. Latex/casein adhesives are used for foil/paper lamination.

Ceramics

Ceramic adhesives have been developed for high temperature applications requiring high strength, particularly metal-to-metal bonding. They are based on borosilicate or other glasses compounded with alkaline earths and oxides of alkaline metals. Ceramic adhesives are set by firing at temperatures between 700 and 1200°C and in this respect resemble ceramic glazes.

Cyanoacrylate

This is an anaerobic synthetic resin where polymerisation is brought about in a matter of seconds by pressure on a thin film of the adhesive between two surfaces. Cyanoacrylate will adhere with a high bond strength to most materials and surfaces. The bond is resistant to oils and many solvents. It is only moderately resistant to water and is broken down by steam. It is also a characteristic of cyanoacrylate adhesives that they set rapidly in the presence of water. Maximum service temperature rating is 80°C for continuous or 100°C for limited exposure.

Epoxy

Epoxy resin adhesives are produced either as two-part mixtures (resin and hardener) for self-curing at room temperatures or one-part resins for heat-curing. Two-part epoxy resins may have long setting times (up to 24 hours) or be formulated for rapid setting (*eg* setting time may be reduced to as little as five minutes). Originally developed primarily as metal-to-metal adhesives, epoxy adhesives are suitable for bonding most materials including glass, ceramics, wood, many rubbers and some plastics. They are also noted for their excellent resistance to oils and good resistance to water and most solvents.

Pure epoxy resin is solid at room temperature and becomes plastic at 40 to 50°C. Above this the resin undergoes 'curing', eventually becoming a solid mass. At the same time it fulfils the major requirements of a good adhesive. When the temperature is raised to the flow point the resin readily 'wets' metals and other non-porous surfaces, whilst the film strength when cured is high and the film properties stable. Curing can be achieved simply by holding the assembly at the required temperature for the specified time, no pressure being needed. The curing temperature is largely influenced by the materials; below about 180°C, however, the curing time is long.

Cold-setting adhesives employ an accelerator and/or hardener to promote curing at room temperature. The more viscous cold-setting resin-hardener mixtures are sometimes thinned with a solvent for spray application; up to one hour may be allowed for evaporation of the solvent before the joint faces are brought together and clamped. In general, the thicker the resin used for cold setting (ie the more viscous the resin) the stronger the bond produced. Heat curing tends to promote a higher bond strength. For maximum bond strength the full heat curing type is preferred. Further improvements can be had with two-part adhesives, one of the best known being a mixture of polyvinyl formal powder and a phenol formaldehyde liquid resin (Redux). This gives a metal-to-metal joint which is normally stronger than riveting or spot welding and appreciably better than that achieved with the epoxy resins. The joint strength is, however, temperature dependent and begins to show a marked decrease above approximately 80°C. For maximum strength at elevated temperatures the formulation may be modified to provide a maximum service temperature of up to 250°C for stressed applications. The normal process with Redux bonding is to coat both joint surfaces with the liquid resin, after which powdered forvar is sprinkled or dusted on or applied by dip. No reaction takes place until the two surfaces are brought together under heat and pressure. The amount of pressure needed is not critical, with 100 lb/in^2 a typical figure, although up to twice this may be employed with distinct advantages in specific applications. The time cycle for curing varies with temperature but is normally from 4 minutes at 180°C to 20 minutes at 145°C.

Modified epoxy resins, notably epoxy-phenolic and epoxy-polyamide, all have enhanced properties. The addition of nylon improves both the shear and peel strengths, and they can be cold cured under contact pressure, although far better and more rapid curing is produced by heating at 150 to 175°C, the curing time then being about one hour. Adhesives of this type are in the form of non-woven mat, plain film or film reinforced with glass fibre or nylon cloth. This two-part adhesive is sensitive to humidity and is also costly. It has been used successfully for bonding stainless steel, aluminium and titanium in aircraft and in aerospace and cryogenic structures.

The addition of phenolic resin materially improves the service temperature range of a straight epoxy resin, and epoxy-phenolic adhesives retain up to 50% of their room temperature strength at up to 290°C or more. They are also suitable for short term exposure to temperatures as high as 565°C. Besides maintaining relatively low bond strength, creep is exceptionally low at high temperatures. The bond is brittle and the peel strength low. Curing is under heat and pressure, 160 to 175°C at 7 kg/cm^2 for 45 minutes to one hour. The mixture is active at room temperature, so the shelf life is limited to a few months, although it may be improved by refrigerated storage. The addition of polyamides to epoxy resins greatly improves flexibility of the bond and gives higher peel strength, although mechanical strength and creep resistance are both reduced.

The adhesive is also sensitive to humidity and tends to become brittle at high temperatures. Maximum service temperature is only 66°C. The adhesive is cheap and simple to use. It can be a two-part solution or paste, or a cast or calendered film, and will set at contact pressure. At room temperature, setting time is three to five days, but only three to five minutes at 200°C. Such adhesives are finding increasing industrial application, notably in the automobile industry.

Epoxy-polysulphide was one of the original modified forms of epoxy with enhanced elasticity and peel strength. Typical shear strength is 210 to 280 kg/cm^2. It continues to be used extensively in America for bonding metal plated end beams to concrete. Epoxy-silicon resins offer greater heat resistance than other epoxy-type adhesives, with a maximum service temperature rating of up to 300°C continuous or 500 to 550°C for intermittent exposure. Shear strength is only moderate, however, at about 100 to 140 kg/cm^2.

Furane

Limited use is made of furane copolymers as adhesives for thermoset plastics and laminated plastics.

Natural rubber

Adhesives based on natural rubber are either of solvent or heat curing type. The latter yields a vulcanised bond of improved strength and higher service temperature rating (90°C as opposed to 60°C is typical of a solvent-set natural rubber adhesive).

Solvent-set rubber adhesives and solutions may also incorporate resin for improved bond strength and are useful general purpose adhesives, strictly limited for engineering and structural applications because of generally low bond strength and the thermo-plastic nature of the base material. They have good resistance to water and mould growth but are attacked by oils, solvents and many chemicals. The choice and performance of a natural rubber adhesive will be dependent on the type and formulation.

Substances for which natural rubber adhesives are generally suitable are:

Natural rubber (not synthetic rubbers).
Some plastics (*eg* acrylic, GRP, PTFE and ABS).
Expanded rubbers – natural rubber and natural latex foam.
Expanded elastomers – polyurethane and polystyrene.
Metals – aluminium, aluminium alloys and iron and steel.
Fabrics – especially cotton, wool, glass fibre cloth, felt rayon and asbestos cloth, cardboard and paper.
Wood, chipboard and hardboard.
Leather.
Glass.
Ceramic and glazed surfaces.
Concrete.
Brickwork.

TABLE 1 – SERVICE TEMPERATURES FOR ADHESIVES

Adhesive base and type	Minimum temperature °C	Maximum service temperature	
		Continuous °C	Intermittent or short term °C
Cyanoacrylate	—	80	100
Epoxy	—	90	—
Epoxy-phenolic	—	up to 200	up tp 260
Epoxy-polyamide	—	100	—
Epoxy-polysulphide	—	90	—
Epoxy-silicon	—	300	up to 550
Natural rubber	–40	65	—
Natural rubber (vulcanised)	–30	90	—
Neoprene	–50	90	—
Nitrile	–50	up to 150	—
Polyurethane	–200	95 to 150	—

Neoprene

Neoprene based adhesives have good resistance to water, solvents, chemicals and oils (with the exception of strong oxidising agents and aromatic hydrocarbons). They can be formulated for solvent setting or vulcanising (either by heat curing or by the addition of a catalyst), with or without resin modifiers.

Polyimides

Polyimide adhesives are of the thermosetting type, curing at temperatures of 260 to 371°C. Post-curing is also required to develop maximum bond strength. They have high bond strength with service temperatures of about 290 to 400°C. They are a recently introduced group of structural adhesives which will bond most materials, but cost is high.

Polybenzimiazoles

These are thermally stable adhesives with a maximum service temperature of about 400 to 565°C for continuous rating (although some degradation or ageing effects may be apparent above 450°C). They require high temperature curing and post-curing, so their application is relatively limited.

Polyurethane

Polyurethane adhesives produce flexible bonds with good pull strength and good resistance to shock and vibration. Bond strength is high at normal room temperatures but decreases rapidly with increasing temperature. Polyurethane has excellent resistance to oils, acids, alkalis and many solvents, but resistance to moisture and water tends to be poor.

TABLE 2 – TYPICAL STRENGTH FIGURES FOR ADHESIVES

Adhesive	Representative strength figures			Remarks
	Shear kg/cm^2	Shear lb/in^2	Peel	
Epoxy (unmodified)	350	5000	—	
Filled epoxy	140–210	2000–3000	—	Aluminium oxide filler
Epoxy-polyamide	245	3500	28 piw	On aluminium
Epoxy-nylon	420	6000	20–100 piw	Requires clamping pressure
Epoxy-polysulphide	200–280	3000–4000	—	
Epoxy-silicon	105–140	1500–2000	—	
Neoprene	20	1000	—	
Nitrile	70	1000	—	
Phenolic-neoprene	140–200	2000–3000	High	Creeps at high loads
Phenolic-nitrile	up to 280	up to 4000	—	
Phenolic-polyamide	350	5000	—	
Phenolic-vinyl	350	5000	—	
PVA	up to 200	up to 3000	—	Metal-to-metal bond.
Polyamide	140–175	2000–2500	—	Stainless steel to stainless steel
Polyurethane	35–105	300–15000	20 to 40 piw	
Silicon (unmodified)	up to 140	up to 2000		

Fluorocarbon (Viton)

Adhesives based on fluorocarbon polymers are suitable for higher temperature duties, but are normally used only for special applications. Fluorocarbon adhesives are particularly suitable for the following substrates:

> Plastics – fluorocarbons.
> Metals – most.
> Fabrics – felt and asbestos cloth.
> Glass and glazed surfaces.
> Concrete.
> Brickwork.

Polyester

Polyester resins have limited application as adhesives. They are not, for example, good adhesives to use with cured GRP (epoxy resins preferred in this case), but specific substrates for which polyester adhesives are suitable include:

> Plastics – PVC (plasticised and unplasticised), polyester films, polystyrene.†
> Metals – copper and copper alloys.
> Fabrics – most (except acrylic, asbestos and rayon).

† Depends on formulation; some polyester adhesives may be unsuitable.

Polysulphide

Polysulphide mixtures have a particular application as sealing compounds. They are also formulated as adhesives for the following substrates:

Plastics – GRP mouldings and polycarbonate.‡
Metals – aluminium and aluminium alloys, copper and copper alloys, iron and steel.
Fabrics – suitable for most (except acrylic, polyester and polyamide).
Wood, chipboard and hardboard.
Leather.
Glass and glazed surfaces.
Concrete.
Brickwork.

‡ Depends on formulation; some polysulphide adhesives may be unsuitable.

Polyamides

Whilst polyamides in adhesives are mainly used as modifying agents in epoxy resins and phenolic resins, they have a limited use for bonding metals and many plastics. They are produced as solutions, films and hot melts. Heat sealing polyamides set immediately on cooling. Chemical resistance is similar to that of nylons.

Phenolic resins

Thermosetting phenolic resins (PF resins) are major adhesives used in the woodworking industry, particularly in the manufacture of plywood. They provide excellent resistance to water, solvents and oils, etc. Structural or engineering adhesives are also based on mixtures of phenolic and other synthetic resins, *eg*:

Phenolic-neoprene – heat-curing adhesives for metal-to-metal or metal-to-wood joints, etc. Shear strength of the order of 175 to 210 kg/cm^2 can be achieved.
Phenolic-nitrile – heat-curing adhesives with shear strengths up to 280 kg/cm^2 and improved maximum service temperature (up to 175°C for continuous or up to 250°C for intermittent exposure). These can be regarded as special adhesives for metals and for bonding non-metallic components to metals (*eg* brake linings to brake shoes).
Phenolic-polyamide – generally used in the form of a (thermoplastic) polyamide film in conjunction with a (thermosetting) phenolic resin solution. Bond strengths of up to 350 kg/cm^2 are attainable, and good strength is maintained up to quite high temperatures – *eg* 14 kg/cm^2 at 150°C.
Phenolic-vinyl – generally used in the form of a (thermoplastic) vinyl powder dusted on to a liquid phenolic resin film, although one-part mixtures are also available. Shear strengths of up to 350 kg/cm^2 are obtainable, but strength degrades rapidly above 100°C. This type of adhesive is widely used for bonding honeycomb sandwich constructions, metal-to-plastic bonding and bonding cyclised rubbers to metal.

Polyvinyl acetate (PVA)

Thermoplastic resin or 'white glue' is widely used in the modern woodworking industry, but is also suitable for use with metals, many plastics, glass, ceramics, leather, etc. It has a high bond strength, is non-staining and has good resistance to oils and mould growth. It has very poor resistance to heat, however, and strictly limited resistance to water and moisture. PVA emulsions are widely used as ceramic tile adhesives.

Until recently all PVA adhesives were slow setting, requiring joints to be clamped for up to 24 hours, but modern formulations are fast setting.

Resorcinol formladehyde

Resorcinol formaldehyde is another thermosetting resin used mainly for wood-working joints, with a superior strength, water resistance and maximum service temperature to UF resins. It can also be used for bonding acrylics, nylons, phenolics and urea plastics.

Silicone

Basic silicone adhesives have excellent resistance to high temperatures but rela-tively low strength. They are normally formulated with stronger adhesives to provide high temperature stability with good mechanical strength. Epoxy silicone adhesives may be rated for continuous service temperatures up to 340°C and up to 510°C for intermittent exposure.

Sodium silicate

Sodium silicate based adhesives are used for bonding asbestos cloth lagging in high temperature insulation.

Urea formaldehyde

UF is a thermosetting resin widely used in woodworking. It is available as a one-part mixture activated by mixing with water, or as a two-part mixture of resin (or resin powder for mixing with water) and catalyst. It has good resistance to oils, solvents and water (but not boiling water).

In addition to woodworking joints it can also be used as an adhesive for phenolic, melamine and urea thermoset plastics.

HANDLING INDUSTRIAL ADHESIVES

The following notes have been prepared by the British Adhesive Manufacturers' Association as a guide to precautionary measures which should be observed to ensure the safe handling of industrial adhesives.

Hygiene

The ingestion of adhesives should be avoided, and the consumption or storage of food or drink should be prohibited in areas where adhesives are handled or used.

Certain powder adhesives and those giving off toxic vapours present a hazard when inhaled. Suitable dust masks, respirators, and/or ventilation should be provided.

Skin contact should be minimised, and the suppliers will advise on barrier and cleansing creams appropriate to their products. Suppliers will also advise on suitable protective clothing and eye protection.

Allergies

There is always the possibility that an individual may be allergic to one of the substances used in a particular adhesive. There is generally no solution to situations of this sort, except transfer to work which does not require exposure to that adhesive.

Spillage and waste disposal

Spillages of any type should be attended to immediately. Aqueous products can be washed away with water before they dry, provided that the appropriate authority permits the discharge of this type of effluent to the drains and waterways concerned. An alternative method is to soak up the spillage with an inert material which can then be placed in a suitable container for disposal in accordance with the requirements of the local authority. This latter technique is particularly appropriate for solvent-based adhesives. Sand is often a suitable absorbent. In addition, for solvent-based adhesive spillage, the use of an emulsifier is often beneficial prior to absorption.

Care should be taken in the disposal of empty containers for solvent based adhesives in order not to present a latent explosion and/or fire hazard. The empty containers should, with due precaution, be either punctured or have the lids left off to ensure that no solvent vapour is trapped under pressure. (This advice does not apply to aerosol packages; these are pressurised and must not be punctured.)

Empty and partially empty drums constitute as great a hazard as full drums and should be handled with similar precautions.

Storage

The storage of adhesives should be restricted to NO SMOKING areas, since even for non-flammable adhesives there is sometimes a risk that vapours can be given off which can be converted to toxic pyrolysis products by a burning cigarette.

The majority of dry constituents of adhesives are combustible and present a possible fire hazard, but in normal applications this is small compared with the risk from substrates (*ie* paper, wood, plastic film, etc).

All adhesives should be stored at reasonable temperatures, *ie* preferably between 10 and 25°C, and away from damp or wet.

Stock rotation

Adhesive manufacturers take care to deliver products in good condition, and their handling instructions should be followed. Some adhesives have a limited shelf life, and practising strict rotation of stock will ensure not only that the material reaches the production stage in good condition, obviating many problems, but also that possible safety hazards are avoided.

Conditions of sale

Most adhesive manufacturers stipulate conditions for the supply and use of their products.

Adhesive applications cover a very wide field and it will be appreciated that once an adhesive has been delivered it may be subjected to conditions of storage, application and use beyond the control of the supplier and for which he cannot be considered responsible. Users should therefore ensure that they are fully aware of the conditions under which particular manufacturers supply to them.

Solvent-based and solvent-containing adhesives

Adhesives in this class represent the most obvious hazards to users.

The use and application of adhesives containing flammable solvents is controlled by legislation such as the High Flammable Liquids and Liquefied Petroleum Gases Regulations 1972 and in some cases the Petroleum (Consolidation) Act (which necessitates a licence to store). Additionally, some local authorities specify the conditions to be followed. All persons should be aware of the statutory regulations involved. All containers for flammable adhesive must only be used and stored under strict supervision in restricted NO SMOKING flameproof areas. Containers above 5 litres in volume are identified by a red diamond to indicate the fire and explosion hazard.

It is a wise precaution to adopt a similar procedure for adhesives containing halogenated solvents. Although not flammable, dangerous by-products can be evolved under certain conditions.

All solvent-based adhesives, including those containing non-flammable materials, act as narcotics and/or anaesthetics and must only be used in adequately ventilated areas.

The toxicity of solvents varies considerably even within chemical groups. The Threshold Limit Values (TLV) of a wide variety of chemicals is the subject of a biannual booklet issued by the Department of Employment. The lower the quoted TLV factor, the higher the toxicity, and consequently, more effective ventilation is required to maintain the solvent vapour concentration in the working area below the limit. The only satisfactory method of establishing that the safe limit is not being exceeded is measurement. Methods such as Draeger Tubes or other gas analysers are usually used, and the adhesive user should ensure that solvent concentrations in the working area are monitored and information such as date, time, location, type of solvent and level is recorded.

The prime consideration should be to control the emissions of solvent from the adhesive close to the source. The most effective method is to enclose the adhesive in a suitable dispenser, but this is not always practicable. If 'dilution ventilation', such as deliberately opened windows and doors, is inadequate to reduce solvent vapour to the safe level, or where adhesives are being sprayed, supplementary local exhaust ventilation will be necessary.

The emphasis should always be on protecting the operative, and portable extraction should be used in confined, badly ventilated spaces. If this is completely impracticable, suitable breathing apparatus should be worn.

In controlling the emission of solvents, it should be noted that the Health and Safety at Work, Act 1974 places a duty on every manufacturer to carry out his operations in such a way as not to cause risk to the public.

Contact of solvent-based adhesive with the skin must be avoided. Apart from a general de-fatting effect and irritation of the skin and mucous membranes, certain solvents and other ingredients present may be absorbed through the skin and act as systemic poisons. Care should obviously be taken to prevent contact, and this is best accomplished by the application of a suitable barrier cream and/or the use of protective gloves. The use of a barrier cream will facilitate cleaning if the skin should subsequently become contaminated. A suitable antiseptic cleaner should be used to remove the adhesive. Do not use solvents.

When handling low viscosity adhesives, suitable approved goggles should be worn to protect the eyes from splashes.

Water-based adhesives; emulsions, latices and solutions

Being water-based, this class of product is not normally flammable. The dry adhesive film, formed when water is removed, may be capable of burning but does not usually sustain combustion. Some synthetic polymers depolymerise under heat, liberating volatile, toxic, and/or flammable vapours.

Solvents are incorporated in many emulsion adhesives for special applications and this may affect toxicity and flammability. Synthetic latices or emulsions contain free monomer which, although normally present at low levels, can be a potential hazard to health. Other volatile ingredients such as ammonia, formaldehyde, etc, can also be troublesome. All such products should therefore be used with adequate ventilation.

Contact with the skin should be avoided but if this does occur, aqueous adhesives should be washed off with cold water before they can dry. This is to avoid discomfort rather than injury from adherent polymer. However, repeated contact may cause dermatitis in sensitive individuals and the use of barrier cream and/or protective gloves is advisable. If adhesives dry on unprotected skin, some will be found to pull off without inconvenience. Others, particularly pressure sensitive films, are less easy to remove and may require a special skin cleanser. Do not use solvents.

Splashes into eyes, mouth or nose should be washed without delay with copious quantities of water, and medical advice should be obtained immediately.

Hot melt adhesives

The greatest hazards associated with the use of hot melt adhesives occur when they are molten. Severe burns can result if skin contact occurs, and adequate protective clothing should therefore be worn. Suitable approved eye protection should also be used, if molten adhesive is being transferred or if there is a danger of splashing.

If burns do occur, the recommended procedure is as follows:

(a) Immerse the affected area in cold, clean water immediately.
(b) Do not attempt to remove the cold adhesive from the skin.
(c) Cover the affected area with a wet compress and obtain medical advice immediately.

Hot melts may fume during operation. The vapours given off are not normally considered to be toxic but they should not be inhaled, and suitable, preferably forced draught ventilation should be provided.

Hot melt adhesives should be used at their recommended operating temperatures. If overheating occurs, there could be a fire risk, as the vapours evolved might be ignitable by a spark. If a fire does start, a dry powder extinguisher should be used and under no circumstances should water be allowed to come into contact with the molten adhesive.

Resin-based adhesives

Liquid resin adhesives, such as epoxides and formaldehyde condensation products, present no particular fire or explosion hazards unless they contain flammable solvents. Under such circumstances, the precautions already outlined for solvent based adhesives should be observed. The handling technique should ensure that un-cured resin or hardener does not come in contact with the skin. Operators should be provided with suitable gloves, the insides of which must be kept scrupulously clean, and care should be taken to prevent cuffs becoming contaminated. Damaged gloves must be replaced. Barrier creams applied to the skin before work begins offer additional protection. If, despite all precautions, the skin does become contaminated, the affected area should be washed immediately with a suitable antiseptic hand cleanser and disposable towels used for drying. Do not use solvents.

The mixing of adhesive formulations should only be carried out in suitably ventilated areas. As a precaution, the wearing of approved dust respirators is advised when handling powder fillers and/or hardeners, in order to avoid inhalation. (HM Inspector of Factories produces a list of suitable respirators.)

Cleanliness and tidiness in the working area are of the utmost importance. Benches should be covered with replaceable paper which should be removed and destroyed when contaminated. Containers should be kept as far away as practicable in a clearly marked off area of the work space. Spillage and contamination of tools and equipment or of the outsides of containers are naturally to be avoided. If these occur, the affected area must be cleaned up immediately.

Powder adhesives

This covers a very wide range of products from powdered starch and animal glues, which are relatively innocuous, to the more hazardous synthetic resin powders.

Under certain circumstances, dust can present an explosion hazard. This risk can be avoided by paying careful attention to good housekeeping and maintaining low dust levels. For further details, reference should be made to HM Stationery Office Publication *Dust Explosions in Factories*, New Series No 22.

Inhalation of dust is another potential hazard and exposure should be minimised with suitable respirators being worn where necessary.

Some powdered glues are acknowledged to be dermatitic, and in such cases manufacturers' recommendations must be strictly followed. As a general precaution in all cases, strict attention should be paid to personal hygiene, and direct handling should be avoided. The use of barrier creams and/or protective gloves is recommended.

Low melting point alloys (fusible alloys)

THE DESCRIPTION *fusible alloy* is given to alloys of bismuth, cadmium, iridium, lead and tin which melt at relatively low temperatures. Melting points of these pure metals are:

	°C	°F
bismuth	271.3	520.3
cadmium	320.9	609.6
iridium	156.4	313.5
lead	327.4	621.3
tin	231.9	449.9

Alloys consisting of two or more of these pure metals melt at much lower temperatures, as low as 70°C (158°F) in the case of bismuth–lead–tin–cadmium or Wood's metal. Even lower melting points can be achieved with gallium and iridium for highly specialised applications, the tin-gallium eutectic, for example, having a melting point of only 20°C (68°F).

Eutectic alloys have a specific melting point. Non-eutectic alloys melt over a range of temperatures or have a plastic or 'pasty' temperature range (see chapter on *Solders*).

It is also a general characteristic of many of these fusible alloys that they are non-shrinking or expand on solidification, making them excellent for casting work, foundry patterns, etc. Many of the fusible alloys also make excellent soft solders. They can be applied in the same way as the conventional tin–lead alloys, but offer a greater margin of safety when the assemblies being soldered must not be overheated. Typical applications are in soldering close to joints already made with tin–lead solder or to materials easily damaged by heat, such as plastics, fabrics, lacquer, glass; soldering hardened steels and metals of low melting point such as pewter; soldering containers holding inflammable liquids.

Specific-purpose fusible alloys include tube bending alloy (melting point 70°C), matrix alloy (melting point 102°C), and others. *Tube bending alloy* is used for filling tubes to prevent kinking or cracking during bending. The alloy melts

TABLE 1 – EUTECTIC FUSIBLE ALLOYS

Alloy	Melting point	Weight lb/in²	Tensile strength tons/in²	Elongation %	Brinell hardness	Joint strength on brass tons/in²	Soldering qualities	Special features
Bismuth-lead-tin-cadmium	70°C 158°F	0.34	1.6	200	7.2	1.3	Good with all types of flux.	Wood's metal. Melts in warm water. Expands on solidification.
Bismuth-cadmium-lead	91°C 196°F	0.37	2.2	100	7.5	0.9	Does not tin readily; requires an active flux.	
Bismuth-lead-tin	95°C 203°F	0.35	2.6	130	9.6	0.5	Fair; active flux recommended.	Expands on solidification; just melts in boiling water.
Bismuth-cadmium-tin	103°C 217°F	0.32	3.9	160	16.0	1.1	Fair; active flux recommended.	
Bismuth-lead	124°C 256°F	0.38	2.6	70	9.6	1.0	Does not tin readily; requires an active flux.	Non-shrinking alloy for foundry patterns.
Bismuth-tin	138°C 281°F	0.31	4.3	0.2	9.6	0.7	Good with all types of flux.	Expands on solidification; gives accurate reproduction of the mould.
Tin-lead-cadmium	142°C 288°F	0.29	3.4	78	13.2	1.8	Good with all types of flux.	Excellent alloy for low temperature soldering.
Bismuth-cadmium	144°C 291°F	0.34	3.3	0.5	14.2	1.2	Good with all types of flux.	Expands on solidification.
Tin-cadmium	177°C 351°F	0.28	4.2	250	14.0	2.1	Good with all types of flux.	Lead-free solder.
Tin-lead	183°C 361°F	0.30	4.6	20	13.8	2.6	Good with all types of flux.	Lowest melting point tin-lead solder.

readily in hot water; only simple equipment is required and the temperature employed does not affect the temper of the tube metal. The alloy is poured into the tube and solidified rapidly. It expands on cooling and thus provides firm internal support for the tube walls. After bending, the alloy is melted out by immersion in hot water and can be used over and over again.

Matrix alloy provides a rapid and inexpensive method of locating punches in dies. The die recesses are rough machined oversize and the punches are placed in position using a template; matrix alloy is poured into the clearance space and, on setting, prevents lateral movement of the punch. The alloy expands on cooling, thus gripping the die securely.

Matrix alloy can be poured at 170°C (338°F) and thus does not draw the temper of the punch, but if possible, a higher pouring temperature of 270°C (518°F) is desirable, especially where the casting has thin sections. Hardening operations can be carried out before mounting. Anchorage should be provided on the punch, for example, by a keyway or ground flat, to prevent rotation. The punch should be bedded down firmly on the die so that the alloy does not penetrate beneath the punch.

TABLE 2 – COMPOSITIONS AND MELTING TEMPERATURE OF SOME FUSIBLE ALLOYS AND EUTECTICS

Type of Alloy or name	Composition %					Melting range			
						Solidus		Liquidus	
	Sn	Bi	Pb	Cd	Others	°C	°F	°C	°F
Binary eutectic	99.25				0.75 Cu	227	441	227	441
Binary eutectic	96.50				3.50 Ag	221	430	221	430
Binary eutectic				17.00	83.00 Tl	203	397	203	397
Binary eutectic	92.00				8.00 Zn	199	390	199	390
Binary eutectic		47.50			52.50 Tl	188	370	188	370
Binary eutectic	62.00		38.00			183	361	183	361
Binary eutectic	67.00			33.00		176	349	176	349
Binary eutectic	56.50				43.50 Tl	170	338	170	338
	40.00		42.00	18.00		145	293	160	320
Ternary eutectic	51.20		30.60	18.20		145	293	145	293
Binary eutectic		60.00		40.00		144	291	144	291
	48.80	10.20	41.00			142	288	166	331
Binary eutectic	43.00	57.00				138	281	138	281
	41.60	57.40	1.00			134	273	135	275
Ternary eutectic	40.00	56.00			4.00 Zn	130	266	130	266
Ternary eutectic	46.00			17.00	37.00 Tl	128	262	128	262
Binary eutectic		55.50	44.50			124	255	124	255
Binary eutectic	48.00				52.00 In	117	243	117	243
	50.00				50.00 In	117	243	127	260
	1.00	55.00	44.00			117	243	120	248
	14.50	48.00	28.50		9.00 Sb	103	217	227	440
	34.50	44.50		21.00		103	217	120	248
	25.00	50.00		25.00		103	217	113	235
Ternary eutectic	25.90	53.90		20.20		103	217	103	217
	33.00	34.00	33.00			96	205	143	289
Malotte's	34.20	46.10	19.70			96	205	123	253
Rose's	22.00	50.00	28.00			96	205	110	230
D'Arcet's	25.00	50.00	25.00			96	205	98	208
Newton's	18.80	50.00	31.20			96	205	97	207
Onion's or Lichtenberg's	20.00	50.00	30.00			96	205	100	212
Ternary eutectic	15.50	52.50	32.00			96	205	96	205
Ternary eutectic		51.70	40.20	8.10		92	198	92	198
	15.40	38.40	30.80	15.40		70	158	97	207
	11.30	42.50	37.70	8.50		70	158	90	194
	24.50	45.30	17.90	12.30		70	158	88	190
	13.00	40.00	37.00	10.00		70	158	85	185
	13.00	42.00	35.00	10.00		70	158	80	176
Lipowitz's or Cerrobend	13.30	50.00	26.70	10.00		70	158	73	163
Wood's	12.50	50.00	25.00	12.50		70	158	72	162
Quaternary eutectic	13.10	49.50	27.30	10.10		70	158	70	158
	13.20	49.30	26.30	9.80	1.40 Ga	65	149	66	151
Cerrolow 147	12.77	48.00	25.63	9.60	4.00 In	61	142	65	149
Cerrolow 136	12.00	49.00	18.00		21.00 In	57	136	57	136
Cerrolow 117	8.30	44.70	22.60	5.30	19.10 In	47	117	47	117
Binary eutectic	8.00				92.00 Ga	20	68	20	68

See also chapter on *Printing metals*.

The good strength and hardness of matrix alloy, combined with its low casting temperature and expansion on solidification, make it suitable for many other purposes, for example production of experimental machine parts, foundry patterns, keying of shafts in magnet castings, filling of castings.

Further data on fusible alloys are summarised in the Tables.

Sealants

SEALANTS FOR joints are best classified by chemical type. Each type may be produced in a variety of modified compositions both as general and specific purpose sealants, and also in different types according to the intended method of application – *eg* strips, knife grades and gun grades. The main chemical types are as follows:

Acrylic sealants

These are a comparatively recent introduction based on mixtures of acrylic polymers. They are plastic in nature and their long-term performance has yet to be established.

Butyl sealants

These are based on butyl rubbers or degraded butyl rubbers used alone or in combination with solvents, oils, fillers, pigments and extenders. They are produced in a very wide variety of compositions ranging from very soft sticky compounds to hard rubber strips. They can also be made skin-forming, non-drying or self-vulcanising (curing) by solvent evaporation.

Oleo-resin sealants

This group is based on drying and non-drying oils, with resin fillers and extenders, and can be skin- or non-skin forming. They usually require primers to give adhesion to porous surfaces but can be tamped into butt joints.

Polysulphide sealants

Polysulphide sealants may be one-part or two-part compositions based on polysulphide polymers. Like butyl sealants their composition can range from soft sticky compounds to hard (thermoset) rubbers. They can also be pigmented in a wide range of colours. Primers are usually required for satisfactory adhesion to most porous or friable surfaces, but they are also widely used as a basis for caulking compounds.

TABLE 1 – TYPICAL TOLERANCES OF SEALANTS

Type	Maximum movement in tension %	Tolerance in shear %
Rubber-bitumen	10	
Oleo-resin	15	40
Butyl	20	50
Silicone	25	60
Polysulphide	25–33	70–80

TABLE 2 – EXAMPLES OF PROPRIETARY SEALANTS

Sealant	Description	Application method	Service† temp limits Low °C	High °C	Resistance Oil	Weathering	Water	Solvent	Colour	Solvent
Industrial sealant 750°C	Tough, flexible airtight sealant for ducting and metal fabrication. Can be used as weld-thru' sealer.	Flow, Caulking-Gun	−55	95	E	E	G	E	Red	Ketone Blend
Industrial sealant 802	Brushable version of Scotchseal 750°C.	Brush	−55	95	E	E	G	E	Red-Brown	Ketone Blend
Ribbonseal gun grade sealant	Flexible, weatherproof sealant for building-joints with limited movement.	Flow, Caulking-Gun	−30	70		E	E		Grey (2185) Black (7836)	Xylol
Preformed sealant strip 5313	High tack, non-shrinking strip which performs both as a gasket type seal and as an adhesive.	Preformed Strip	−40	135	F	E	E		Black	None
Ribbonseal preformed sealant strip 79	Permanently-tacky, non-shrinking strip for use as a gasket seal.	Preformed Strip	−20	95		E	E		Grey	None

† Service temperature limits may vary according to the application.
E – Excellent
G – Good
F – Fair
'Blank' – Not recommended

Polyurethane sealants

These are more recent types of two-part sealants based on polyurethane polymers curing to a fully elastic rubber. Primers are required on most porous surfaces for satisfactory adhesion and they may exhibit adhesive failure under prolonged stress, even at low level stress levels. Their long-term properties are not yet established.

Rubber–bitumen

The 'traditional' group of general purpose sealants is based on natural or synthetic rubber compounded with pitch or bitumen, resulting in a characteristic black colour. Such mixtures can be cold-setting by solvent action or hot poured. The amount of movement they can tolerate is smaller than that of most of the other synthetic sealants – see Table 1.

Silicone sealants

Silicone sealants are one-part chemically curing silicon elastomers, with or without fillers and pigments. They are available in a wide range of colours (including 'clear') and remain fully elastic when set. They have excellent chemical and thermal resistance but primers are required for satisfactory adhesion to most porous and friable surfaces. Like polyurethane sealants they can suffer from adhesive failure under prolonged low level stress.

Solders

SOLDERING IS the method of joining metals whereby the parent metal is heated up to or beyond the melting point of a filler metal of different composition, causing the filler metal to bond to the parent metal. The same process is involved in brazing, the main difference being that in soldering the filler metal is an alloy with a relatively low melting point, known as solder. The process is generally described as *soft soldering*. *Silver soldering* and *hard soldering* imply the use of higher melting point filler metals, the process in this case being described as low temperature brazing – see also chapter on *Brazing materials*.

Common soft solders are alloys of tin and lead, the melting point depending on the tin:lead proportions. The eutectic alloy is 63:37 tin:lead, which has a melting point of 183°C. The melting points for other alloy proportions are summarised in Table 1. It will be noted that higher tin contents than 63:37 result in an increase in melting point and so offer no practical advantages. Higher lead content offers higher melting point solders – for higher temperature duties – at some expense in physical properties (see Table 3). Another significant point is that solders with a lower tin content that the eutectic (63% tin) have a plastic

TABLE 1 – LEAD/TIN ALLOYS – THERMAL DATA

Tin	Lead	Melting point		Plastic range	
		°F	°C	degF	degC
0	100	621	327	260	144
20	80	527	275	125	92
30	70	491	255	130	72
40	60	453	234	92	51
45	55	435	224	74	41
50	50	414	212	53	29
60	40	370	188	9	5
63	37	361	183	—	—
100	0	452	232	91	49

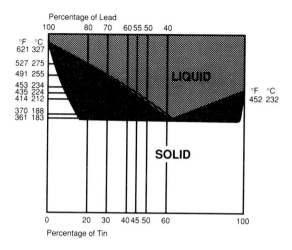

Fig. 1

TABLE 2 – SOFT SOLDER ALLOYS

Tin %	Lead %	Antimony %	Copper %	Cadmium %	Silver %	Specification
100.0	—	—	—	—	—	BS 3252
60.0	Balance	—	—	—	—	BS 219 and DIN 1707
50.0	Balance	—	—	—	—	BS 219 and DIN 1707
45.0	Balance	—	—	—	—	BS 219
40.0	Balance	—	—	—	—	BS 219
35.0	Balance	—	—	—	—	BS 219
30.0	Balance	—	—	—	—	BS 219
20.0	Balance	—	—	—	—	BS 219
15.0	Balance	—	—	—	—	BS 219
63.0	Balance	0.25	—	—	—	QQ-S-571 E
60.0	Balance	0.25	—	—	—	QQ-S-571 E and DIN 1707
50.0	Balance	0.25	—	—	—	QQ-S-571 E and DIN 1707
40.0	Balance	0.25	—	—	—	QQ-S-571 E and DIN 1707
95.0	—	Balance	—	—	—	BS 219 and DIN 1707
50.0	Balance	—	1.50	—	—	DTD900/4535 and DIN 1707
50.0	Balance	—	—	17.0	—	DIN 1707 and BS 219, Grade T
5.0	Balance	—	—	—	1.5	BS 219 Grade 5S
62.0	Balance	—	—	—	2.0	DIN 1707 and BS 219, Grade 62S
62.0	Balance	0.25	—	—	2.0	QQ-S-571E
96.3	—	—	—	—	3.7	BS 219 and QQ-S-571 E

range – see Figure 1 and Table 1. For much work a plastic range is an advantage although for certain work (*eg* tag jointing on electrical equipment) the smaller the plastic range the less the chance of the joint being fractured by slight vibration whilst the solder is setting. Thus 60/40 alloy is widely used for high quality electrical work (the small plastic range and small difference in melting point compared with the eutectic being negligible).

Principal types of soft solders in use are summarised in Table 2, together with typical application data in Table 3. Physical and electrical properties of tin/lead solders are given in Table 4. All solders used for electrical work should be non-antimonial, or any traces of antimony should be within the limits specified by BS 219.

TABLE 3 – SOFT SOLDERS – APPLICATION DATA

Alloy	Melting temperature		Recommended minimum bit temperature °C	Uses
	Solidus °C	Liquidus °C		
60/40	183	188	248	High quality work requiring low melting point alloy.
*Savbit 1	183	215	275	Hand soldering radio,
50/50	183	212	272	telephone and electrical
45/55	183	224	284	equipment; batteries.
40/60	183	234	294	
30/70	183	255	315	Fuses, motors, dynamos.
20/80	183	275	335	Lamps, motors, dynamos.
15/85	227	288	348	Lamps.
*TLC tin/lead/cadmium	145	145	205	Specially low melting point solder. Soldering on gold.
*LMP with 2% silver content and SN62 alloy	179	179	239	Particularly useful when soldering ceramics or other silver coated surfaces.
PT pure tin	232	232	292	Used when a lead-free solder is required.
*High melting point (HMP)	296	301	356	Special high melting point solder to BS 219, Grade 5S.
95A	236	243	303	High melting lead-free alloy.
*96S	221	221	281	Bright strong non-toxic solder.

*Proprietary alloys of Multicore Solders Ltd.

TABLE 4 – PHYSICAL AND ELECTRICAL PROPERTIES OF TIN/LEAD SOLDERS

Alloy Tin-Lead	Tensile tons/in²	Shear tons/in²	Specific gravity	Electrical conductivity (% of conductivity copper)
60/40	3.8	2.8	8.5	11.5
50/50	3.0	2.6	8.9	10.9
45/55	2.8	2.5	9.1	10.5
40/60	2.7	2.4	9.3	10.1
30/70	2.5	2.3	9.7	9.3
20/80	2.4	2.1	10.0	8.7

Silver-bearing soft solders

Silver-bearing soft solders have two important advantages over ordinary tin–lead solders:

(a) Because of their higher melting temperatures they have higher safe service temperatures.
(b) Some of them have special properties of corrosion resistance, particularly in contact with supply waters.

The principal properties of some alloys in general use are given in Table 5.

Intending users should not assume that the strength properties of silver-bearing soft solders are akin to those of silver brazing alloys. Except in the case of

TABLE 5 – PROPERTIES OF TYPICAL SOFT SILVER SOLDERS*

Alloy	Composition	Melting range °C	Tensile strength tons/in²	Elongation %	Electrical conductivity % IACS
Comsol	Silver-tin-lead	296	2.5	40	8.0
A 25	Silver-lead	304	2.5	35	7.2
A 5	Silver-lead	304–370	2.5	35	7.2
Plumbsol	Silver-tin	221–225	1.7	60	13.3
P 5	Silver-tin	221–235	4.0	30	14.3
LM 10A	Silver-tin-copper	214–275	4.8	16	13.0
LM15	Silver-cadmium-zinc	280–320	11.5	5	15.0
LM 5	Silver-cadmium	338–390	8.4	25	22.4
63/37 tin lead (For comparison only)		183	3.4	30	11.9

The physical properties given above relate to the alloys in the cast condition.
The strength of butt joints is dependent upon a number of conditions and may appreciably exceed these figures.

*Johnson Matthey Metals Ltd.

TABLE 6 – PLUMBERS' SOLDER* (For lead pipe and cable jointing)

Grade	Solid at	Liquid at	Weight lb/in³	Uses
Belfry	185°C 365°F	252°C 486°F	0.352	For general plumbing.
Abbey	185°C 365°F	248°C 478°F	0.348	High grade plumbing and cable jointing.
Prior	185°C 365°F	245°C 473°F	0.343	Rich tin quality for mains and cable work.
Grade J	183°C 361°F	255°C 491°F	0.350	Low antimonial solder for dip soldering.
Grade H	183°C 361°F	244°C 471°F	0.343	For specialised cable jointing.

*Fry's Metals Ltd.

LM 15 the use of these solders is not economically justified by the increase in room temperature strength of the joints they produce compared with tin–lead solders. Their employment is justified when the user's need is either for greater strength at elevated service temperatures or for increased resistance to corrosion.

None of these solders is subject to what is known as 'tin-pest'. This is the state induced by exposure to low temperatures resulting in a change in tin from its normal, white, lustrous, malleable condition to a grey, friable powder.

Workshop practice

Silver-bearing soft solders are applied by the same methods as are customarily employed when soldering with tin–lead solders. It is important to bear in mind that the higher flow temperatures of the silver-bearing alloys make necessary the use of higher application temperatures, whether the work is heated by a soldering bit, by high frequency induction, by oven or by a torch flame.

Fluxes for soft soldering applications are available from manufacturers specialising in this field. The most effective type of flux is that which chemically attacks the parent metal. Such fluxes are mostly based on zinc chloride although organic halogen compounds are coming increasingly into use. For use up to a maximum temperature of about 300°C, resin based fluxes may be used, with or without activating agents that increase their fluxing power.

The recommended range of joint clearances between clean, bare metal parts is from 0.003 in. up to about 0.008 in., depending on the type of flux used. The fluidity of these solders at a given temperature interval above the liquidus is approximately in inverse ratio to the liquidus temperatures; joint clearances should therefore be appropriately regulated within the range recommended.

TABLE 7 – SOLDERS FOR HOT DIPPING*

Grade	Melting range		Weight lb/in³	Recommended bath temperature	
	°C	°F		°C	°F
TR	185–204	365–399	0.317	270–320	518–608
1	185–227	365–441	0.332	280–330	536–626
2	185–248	365–478	0.348	310–360	590–680
3	185–275	365–527	0.370	340–400	644–752
TM5	240–282	464–540	0.390	360–400	680–752
Flowsolder	183–185	361–365	0.302	260–300	500–570
LS4	294–305	561–581	0.400	380–420	720–790

*Fry's Metals Ltd.

Joints have maximum soundness and maximum strength when the work is heated evenly to a temperature that exceeds the liquidus of the solder employed by 60–120°C. If temperature gradients are inescapable, the lowest temperature should be where the solder is applied and the highest at the point to which the solder is required to penetrate.

Because soldering temperatures appreciably in excess of 300°C are sometimes required with these materials, organic fluxes based on resin cannot always be used, particularly if the work is heated by a torch flame. Reduced risk of attack resulting from incomplete removal of corrosive flux residues is obtained if the parts are first flushed on the joint areas with the solder aided by an active zinc chloride flux whose residues can be completely removed. The preflushed parts should then be assembled with a relatively 'safe' flux or, at worst, a much smaller quantity of the active flux. Obviously, the best course is to use joints that are completely accessible for flux residue removal after the soldering operation has been completed. Joint clearances between preflushed components need, for the best results, to be only of the order of 0.001–0.002 in.

Aluminium solders

Soft soldering offers the possibility of joining aluminium or its alloys with less distortion (due to thermal expansion or local stress relief) than may occur with hard soldering or welding, since the temperatures required are not so high. For the same reason, strength changes in the aluminium joint members may also be less than with hard soldering or welding. Hitherto, the problems of soldering aluminium and its alloys – slow wetting, poor fluidity of the solder, and low corrosion resistance of the joint – have limited the applications of aluminium soldering to a few special cases involving pure aluminium because of its zinc content. The tin–zinc solder commonly used formed a strong oxide skin when melted, and this skin, together with that on the aluminium itself, required the use of very corrosive fluxes to allow the solder to flow.

TABLE 8 – SOLDERABILITY OF VARIOUS METALS AND ALLOYS USING ALU-SOL 'D' FLUX

Wrought aluminium alloys: 'Pure' aluminium (up to 1% impurities) aluminium-manganese	Excellent
Aluminium + up to 3% magnesium Aluminium + up to 1.5% each magnesium and silicon Aluminium-copper	Good
Aluminium + over 3% magnesium	Poor
Cast aluminium alloys: 'Pure' aluminium (up to 0.5% impurities) aluminium-copper	Good if rough cast surface is first machined off
Aluminium-silicon	Unsolderable
Other aluminium finishes: Anodised surfaces	Not solderable without first removing the anodising*
Aluminium-silicon coatings	Unsolderable
Other metals and alloys: Tin plate Copper Brass Nickel and nickel silver	Excellent
Steel, stainless steel and zinc alloy die-castings	Good
Chromium Titanium	Unsolderable

*Anodising cannot be applied over solder

Modern silver-bearing solders have made it possible to solder pure or lightly alloyed aluminium almost as easily as copper, with a joint strength at least equal to tin–zinc. Particular attention has also been paid to improved capillary penetration and corrosion resistance, and the development of special fluxes to give satisfactory wetting of the more difficult aluminium alloys.

Alu-Sol 45D (Multicore Solders Ltd.) is one of a complete patented range of tin–lead–silver alloys of a special grade. The alloys vary somewhat in price, the more expensive alloys being of lower melting temperature and therefore easier to work. They can be classed as general purpose aluminium solders offering good working properties at moderate cost.

Brazing materials

BRAZING IS a method of joining metals using filler metals penetrating the capillary gaps between the components being joined. *Low temperature* brazing is brazing in the temperature range 600 to 850°C using filler metals mostly based on copper and silver. Alternative descriptions are hard soldering, silver soldering, silver brazing, etc. *Braze welding*, another type of low temperature brazing, employs a copper-zinc filler metal where the strength of the joint depends primarily on the building up of a fillet of deposited metal.

The majority of brazing is done with the aid of a flux to dissolve the metallic oxides as they are formed by heating, allowing the filler metal to wet the surfaces of the components being brazed. Fluxes used are normally solid at room temperatures (mostly applied in the form of a water-mixed paste) and liquefy on heating. They are normally based on alkali metal fluoborates.

The following is a general guide to the brazeability of engineering metals:

Parent metal	Brazeability	Specific notes
Aluminium Aluminium alloys	Generally impracticable with silver, copper or noble metal alloys.	Soldering of aluminium and certain aluminium alloys can be performed using zinc based or cadmium based alloys, which may contain silver, and by employing a suitably active flux. Aluminium and some of its alloys can also be brazed with aluminium-silicon alloys.
Beryllium	No technique in general use.	The brazing of beryllium to itself and to other metals such as nickel-iron alloys is practised on a small scale, using silver-copper eutectic or pure silver as the brazing material. They should preferably be heated in an atmosphere of dry hydrogen but sometimes it is done in air, use being made of an active flux.
Chromium	May demand special precautions	Although the need to braze elemental, massive chromium seldom arises, it can be accomplished in an atmosphere of relatively dry hydrogen or in air with the aid of a special flux. Brazing of

Parent metal	Brazeability	Specific notes
		chromium plated surfaces usually results in the destruction or discolouration of the plating.
Cobalt	Straightforward.	
Copper	Straightforward.	Copper can easily be brazed by orthodox methods using a flux and can also be brazed in air, without a flux, using phosphorus-bearing brazing alloys.
Copper-aluminium alloys	May demand special precautions.	Copper alloys containing more than about 2% aluminium and up to about 10% aluminium can readily be silver brazed, but the use of a special flux is desirable or necessary. Complications resulting in relatively brittle joints can arise when such alloys are brazed to mild or carbon steel. The latter comment applies also to copper alloys containing beryllium.
Copper-beryllium	Straightforward.	Copper alloys containing up to about 2% beryllium are easily silver brazed but consideration should be given to the possible effect on the ability of the beryllium-copper to be age-hardened.
Copper-cadmium	Straightforward.	
Copper-chromium	Straightforward.	The effect of brazing on precipitation-hardening characteristics has to be considered when brazing copper-chromium.
Copper-nickel	May demand special precautions.	Nickel and some nickel-rich alloys are subject to intergranular penetration by certain silver brazing alloys when brazed in a state of stress.
Copper-palladium	Straightforward.	
Copper-silver	Straightforward.	
Copper-tin (bronze)	Straightforward.	
Copper-tungsten	May demand special precautions.	Copper-tungsten sintered powder-metallurgy compacts sometimes need special cleaning prior to brazing; a special flux may be used to obviate this.
Copper-zinc (brass)	Straightforward.	
Copper-zinc-nickel	Straightforward.	
Gold/gold alloys	Straightforward.	
Iridium	Straightforward.	
Carbon steel	Straightforward.	
Cast iron	May demand special precautions.	Engineering grades of cast iron, or malleable cast iron, are readily brazed on surfaces freed from cast skin by shot-blasting, filing, grinding, or machining. Grey cast iron will not be wetted by silver alloys unless the graphite inclusions are first oxidised and the oxidised surface subsequently reduced. An alternative

Parent metal	Brazeability	Specific notes
		method involves the electrolytic removal of graphite and siliceous inclusions by treatment in a molten salt bath.
Low alloy steels	Straightforward.	
Mild steels	Straightforward.	
Iron, wrought	Straightforward.	
Iron-chromium	May demand special precautions.	Austenitic stainless steel alloys sometimes require a special-purpose flux and certain precautions, dependent on the size of the parts and on the design of the joint. The same comments apply to stainless iron alloys substantially free of nickel, as well as to nickel-chromium alloys. It is advisable in all cases where silver brazing is to be used to make the components from a stabilised welding grade of material. Crevice corrosion may occur if the work is subsequently exposed to wet or moist conditions.
Magnesium/ magnesium alloys	Generally impracticable with silver, copper or noble-metal alloys.	
Molybdenum	May demand special precautions.	Few applications have so far arisen in which silver brazing is the desired method of brazing molybdenum to itself or to other metals. Success has been achieved in a number of cases, but sometimes special brazing alloys and/or fluxes are necessary, and it is recommended that advice should be taken on each application.
Nickel	May demand special precautions.	Nickel and nickel-rich alloys are subject to intergranular penetration by certain silver brazing alloys when brazed in a state of stress.
Nickel-beryllium	Straightforward.	Nickel alloys containing up to about 2% beryllium are easily silver brazed but consideration should be given to the possible effect on the ability of the beryllium-nickel to be age-hardened.
Nickel-chromium	May require special precautions.	See remarks for Iron-chromium.
Nickel-chromium-aluminium	May require special precautions.	Nickel-chromium alloys containing aluminium and/or titanium are used mainly for applications requiring good strength at high temperatures. Because of the high stability of the natural oxide films of these metals, special fluxes are required in order to achieve maximum joint strength.

Parent metal	Brazeability	Specific notes
Nickel-chromium-iron	May require special precautions.	See remarks for iron-chromium.
Nickel-copper	May require special precautions.	Nickel-copper and some nickel-rich alloys are subject to intergranular penetration by certain silver brazing alloys when brazed in a state of stress.
Nickel-iron	May require special precautions.	See remarks for Nickel-copper.
Nickel-manganese	May require special precautions.	See remarks for Nickel-copper.
Nickel-silver (German silver)	Straightforward.	
Palladium/palladium alloys	Straightforward.	
Platinum/platinum alloys	Straightforward.	
Rhenium	May require special precautions.	Due to its comparative rarity, rhenium has not often been silver brazed and special advice should be sought on proposed applications.
Rhodium	Straightforward.	
Ruthenium	Straightforward.	
Silver	Straightforward.	
Silver alloys	Straightforward.	
Tantalum	No technique in general use.	Tantalum readily absorbs most gases at elevated temperatures. Consequently brazing in air or in a protective atmosphere is not practised although, with the aid of a special flux, silver brazed joints have been made in air where the size of the assembly has limited the time of heating to a few seconds. Tantalum can be brazed successfully in a vacuum.
Titanium/titanium based alloys	No technique in general use.	Silver alloy brazing of titanium in air is not practised regularly due to the propensity of the the metal to form brittle intermetallic compounds with orthodox brazing alloys. Suitably active fluxes for the purpose are not pleasant to use. Titanium can be brazed in a vacuum or in an atmosphere of argon or helium using pure silver, silver-copper or silver-manganese as the joining material. (Similar comments apply to zirconium).
Tungsten	May require special precautions.	See remarks for Titanium.
Zirconium	No technique in general use.	See remarks for Titanium.

TABLE 1 – EXAMPLES OF PROPRIETARY SILVER BRAZING ALLOYS

Description	Silver percentage and composition	Melting range, (°C)	BS and DIN specification	Remarks
East-flo	50-Ag Cu Zn Cd	620 to 630	BS 1845 AG1	These two alloys have similar low melting temperatures and quick flowing characteristics. Easy-flo No.2 is the popular choice. Easy-flo is used where maximum ductility and smooth joint fillets are required. Both alloys need well-fitted joints with small gaps for best performance.
Easy-flo No.2	42-Ag Cu Zn Cd	608 to 617	BS 1845 AG2	
DIN Argo-flo	40-Ag Cu Zn Cd	595 to 630	DIN 8513 L-Ag 40 Cd	This series of five alloys of silver-copper-zinc-cadmium gives a range of fillet-forming materials designed for use where wide joint gaps may arise or where pronounced fillets may be required. They are not suitable for applications where slow heating may produce liquidation.
Argo-flo	38-Ag Cu Zn Cd	605 to 651	BS 1845 AG3	
Mattibraze 34	34-Ag Cu Zn Cd	612 to 668	DIN 8513 L-Ag 30 Cd	
Argo-swift	30-Ag Cu Zn Cd	600 to 690		
Argo-bond	23-Ag Cu Zn Cd	616 to 735		
Sil-fos	15-Ag Cu P	640 to 700‡	BS 1845 CP1	Designed primarily for brazing copper without flux, these alloys can be used with flux on copper alloys, but should not be used on ferrous or nickel-based metals. Sil-fos is a relatively ductile alloy; Silbralloy is inexpensive.
Silbralloy	2-Ag Cu P	640 to 740‡	BS 1845 CP2	
Easy-flo No.3	50-Ag Cu Zn Cd Ni	634 to 656		Easy-flo No.3 is a special-purpose alloy for brazing tungsten carbide, and for some types of stainless steel possessing good resistance to corrosion by sea water. Argo-braze 50 and Argo-braze 56 have been developed for brazing stainless steel low in nickel.
Argo-braze 50	50-Ag Cu Cd Zn Ni Mn	639 to 668		
Argo-braze 56	56-Ag Cu In Ni	600 to 711		In addition to the Silver-flo range of materials, this series of silver-copper-zinc alloys includes the well-known BS 1845 types AG4 and AG5, whose chief recommendation is their long history of successful use in miscellaneous engineering applications. These ternary alloys find uses where their melting temperatures or their colours make them particularly suitable. G.6 is substantially white, having been developed for use by silversmiths. Silver-flo 16 is an excellent colour match for brass. Silver-flo 12 can be used for the first step in a three stage brazing operation when Silver-flo 24 and Easy-flo No.2 are used as the second and third stages respectively.
G.6*	67-Ag Cu Zn	705 to 723		
1845 AG4*	61-Ag Cu Zn	690 to 737	BS 1845 AG4	
1845 AG5*	43-Ag Cu Zn	698 to 788	BS 1845 AG5	
Silver-flo 55*	55-Ag Cu Zn Sn	630 to 660		
Silver-flo 45*	45-Ag Cu Zn	670 to 700		
Silver-flo 44*	44-Ag Cu Zn Sn	675 to 735	DIN 8513 L-Ag 44	
Silver-flo 40*	40-Ag Cu Zn Sn	640 to 700		
Silver-flo 33* (formerly D.3 alloy)	33-Ag Cu Zn	700 to 740		
Silver-flo 30*	30-Ag Cu Zn	695 to 770	DIN 8513 L-Ag 25	
Silver-flo 25*	25-Ag Cu Zn	700 to 800		
Silver-flo 24* (formerly C.4 alloy)	24-Ag Cu Zn	740 to 780		
Silver-flo 16* (formerly B.6 alloy)	16-Ag Cu Zn	790 to 830		
Silver-flo 12*	12-Ag Cu Zn	810 to 835	BS 1845 AG7 and BS 1845 AG7 (V)	
Silver-copper eutectic	72-Ag Cu	778		This eutectic alloy of silver and copper is widely used for making flux-free furnace-brazed joints in vacuum tube assemblies.

‡small percentage of higher melting point phase remains at this temperature.
**These alloys can contain a maximum of 0.05% cadmium.

Ack. Johnson Matthey Metals Ltd

TABLE 2 – PALLADIUM-CONTAINING ALLOYS‡

Alloy (Pallabraze)	Composition with % palladium content	Melting range (°C)	Usual joint gaps (mm)	Usual joint gaps (in)	Conforms to BS 1845
810	5 Pd-Ag-Cu	807 to 810	0.025–0.1	0.001–0.004	PD 1 V
840	10 Pd-Ag-Cu	830 to 840	0.025–0.1	0.001–0.004	PD 3 V
850	10 Pd-Ag-Cu	824 to 850	0.075–0.2	0.003–0.008	PD 2 V
880	15 Pd-Ag-Cu	856 to 880	0.075–0.2	0.003–0.008	PD 4 V
900	20 Pd-Ag-Cu	876 to 900	0.075–0.2	0.003–0.008	PD 5 V
950	25 Pd-Ag-Cu	901 to 950	0.075–0.2	0.003–0.008	PD 6 V
1010	5 Pd-Ag	970 to 1010	0.075–0.2	0.003–0.008	PD 7 V
1090	18 Pd-Cu	1080 to 1090	0.075–0.2	0.003–0.008	PD 8 V
1225	30 Pd-Ag	1150 to 1225	0.075–0.2	0.003–0.008	—
1237	60 Pd-Ni	1237	0.025–0.1	0.001–0.004	PD 14 V

‡ Johnson Matthey Metals Ltd

TABLE 3 – SILVER-COPPER EUTECTIC AND PURE METALS*

Alloy or metal	Nominal melting point (°C)	Usual joint gaps (mm)	Usual joint gaps (in)	Conforms to BS 1845
Silver-copper eutectic	778	0.025–0.1	0.001–0.004	AG 7 V
Titanium-cored Ag Cu eutectic	960‡	0.025–0.1	0.001–0.004	—
Silver	962	0.025–0.1	0.001–0.004	AG 8
Gold	1064	0.025–0.1	0.001–0.004	—
Palladium	1554	0.025–0.1	0.001–0.004	—
Platinum	1772	0.025–0.1	0.001–0.004	—

*.Johnson Matthey Metals Ltd

TABLE 4 – GOLD-CONTAINING ALLOYS*

Alloy (Orobraze)	Composition with % gold content	Melting range (°C)	Usual joint gaps (mm)	Usual joint gaps (in)	Conforms to BS 1845
910	80 Au-Cu-Fe	908–910	0.025–0.1	0.001–0.004	AU 1 V
940	62.5 Au-Cu	930–940	0.025–0.1	0.001–0.004	AU 2 V
998	37.5 Au-Cu	980–998	0.025–0.1	0.001–0.004	AU 3 V
1018	30 Au-Cu	996–1018	0.025–0.1	0.001–0.004	AU 4 V
1040	70 Au-Ag	1030–1040	0.025–0.1	0.001–0.004	—
950	82.5 Au-Ni	950	0.025–0.1	0.001–0.004	AU 5 V
980	68 Au-Cu-Ni-Cr-B	950–980	0.025–0.1	0.001–0.004	—
990	75 Au-Ni	950–990	0.075–0.2	0.003–0.008	AU 6 V

*Johnson Matthey Metals Ltd

Noble-metal brazing materials

High purity noble-metal brazing materials may be used where freedom from surface contamination is required when brazing in protective atmospheres or in a vacuum, or where high melting points are required with good mechanical properties at elevated temperatures. Brazing materials in this category include silver-copper eutectic, pure noble-metals and gold-containing and palladium-containing alloys. These are mostly used for fluxless brazing in a vacuum or in a protective atmosphere.

SECTION 10

Engineering Data

METALS DATA
USEFUL CORROSION INFORMATION
COMPARISON OF TEST FIGURES FOR PLASTICS
SUMMARY OF ASTM TEST METHODS

Metals data

EQUIVALENTS OF STANDARD GAUGE SIZES IN MILLIMETRES

Gauge No.	I.S.W.G. Imperial Standard Wire gauge	B.W.G. Birmingham Wire gauge	B.S. American Brown & Sharp Wire gauge	A.S.W.G. American Steel Wire gauge	P.G. Paris gauge	B.G. Birmingham Hoop & Sheet gauge	U.S.S.G. United States Sheet gauge	N.W.S.G. Nettlefolds Wood Screw gauge	Gauge No.
10 0	—	—	—	—	—	20.11	—	—	10 0
9 0	—	—	—	—	—	19.05	—	—	9 0
8 0	—	—	—	—	—	17.99	—	—	8 0
7 0	12.70	—	—	12.45	—	16.93	12.45	—	7 0
6 0	11.79	—	14.73	11.72	—	15.88	11.67	—	6 0
5 0	10.97	12.70	13.12	10.93	—	14.94	10.89	—	5 0
4 0	10.16	11.53	11.68	10.00	—	13.76	10.12	1.37	4 0
3 0	9.45	10.80	10.40	9.21	—	12.70	9.34	1.45	3 0
2 0	8.84	9.65	9.27	8.41	—	11.31	8.56	1.52	2 0
0	8.23	8.64	8.25	7.79	—	10.07	7.78	1.60	0
1	7.62	7.62	7.35	7.19	.60	8.97	7.00	1.68	1
2	7.01	7.21	6.54	6.67	.70	7.99	6.61	2.03	2
3	6.40	6.58	5.83	6.19	.80	7.12	6.23	2.39	3
4	5.89	6.05	5.19	5.72	.90	6.35	5.84	2.74	4
5	5.38	5.59	4.62	5.26	1.00	5.65	5.45	3.10	5
6	4.88	5.16	4.11	4.88	1.10	5.03	5.06	3.45	6
7	4.47	4.57	3.67	4.50	1.20	4.48	4.67	3.81	7
8	4.06	4.19	3.26	4.11	1.30	3.99	4.28	4.17	8
9	3.66	3.46	2.91	3.77	1.40	3.55	3.89	4.52	9
10	3.25	3.70	2.59	3.43	1.50	3.18	3.50	4.88	10
11	2.95	3.05	2.30	3.06	1.60	2.83	3.11	5.23	11
12	2.64	2.77	2.05	2.68	1.80	2.52	2.72	5.59	12
13	2.34	2.41	1.83	2.32	2.00	2.24	2.33	5.94	13
14	2.03	2.11	1.63	2.03	2.20	1.99	1.95	6.30	14
15	1.83	1.83	1.45	1.83	2.40	1.78	1.75	6.65	15
16	1.63	1.65	1.29	1.59	2.70	1.59	1.56	7.01	16
17	1.42	1.47	1.15	1.37	3.00	1.41	1.40	7.37	17
18	1.22	1.24	1.02	1.21	3.40	1.26	1.24	7.72	18
19	1.02	1.07	.912	1.04	3.90	1.12	1.09	8.08	19
20	.914	.889	.813	.884	4.40	.996	.935	8.43	20
21	.813	.813	.724	.805	4.90	.886	.856	8.79	21
22	.711	.711	.643	.726	5.40	.794	.777	9.14	22
23	.610	.635	.574	.655	5.90	.707	.701	9.50	23
24	.559	.559	.511	.584	6.40	.629	.622	9.86	24
25	.508	.508	.455	.518	7.00	.560	.544	10.21	25
26	.457	.457	.404	.460	7.60	.498	.467	10.57	26
27	.417	.406	.361	.439	8.20	.443	.429	10.92	27
28	.376	.356	.320	.411	8.80	.397	.389	11.28	28
29	.345	.330	.287	.381	9.40	.353	.351	11.63	29
30	.315	.305	.254	.356	10.00	.312	.312	11.99	30
31	.295	.254	.226	.335	—	.279	.272	12.34	31
32	.274	.229	.203	.325	—	.249	.254	12.70	32
33	.254	.203	.180	.300	—	.221	.234	—	33
34	.234	.178	.160	.264	—	.196	.213	13.41	34
35	.213	.127	.142	.241	—	.175	.196	—	35
36	.193	.102	.127	.229	—	.155	.175	14.12	36
37	.173	—	.114	.216	—	.137	.165	—	37
38	.152	—	.102	.203	—	.122	.155	14.83	38
39	.132	—	.089	.191	—	.109	.145	—	39
40	.122	—	.079	.178	—	.098	.137	15.54	40
41	.112	—	.071	.168	—	.087	.132	—	41
42	.102	—	.064	.157	—	.078	.127	—	42
43	.091	—	.052	.152	—	.069	.122	—	43
44	.081	—	.050	.147	—	.061	.117	—	44
45	.071	—	.045	.140	—	.055	—	—	45
46	.061	—	.040	.132	—	.049	—	—	46
47	.051	—	.036	.127	—	.043	—	—	47
48	.041	—	.031	.122	—	.039	—	—	48
49	.030	—	.028	.117	—	.034	—	—	49
50	.025	—	.025	.112	—	.030	—	19.10	50
51	—	—	—	—	—	.027	—	—	51
52	—	—	—	—	—	.024	—	—	52

HARDNESS CONVERSION TABLE: FERROUS METALS

This Table shows the approximate relationship between hardness values determined by different test apparatus. It is applicable only to steels of uniform chemical composition and uniform heat treatment. It is not suitable for applying to case hardened steels, nor to non-ferrous metals. Figures in italic are an approximation only, and should only be used with reservation. This applies particularly to equivalent tensile strength figures given in the last column, which is an approximate guide only to the equivalent tensile strength of SAE carbon and alloy constructional steels.

Brinell		Vickers or Firth Diamond Hardness No.	Rockwell		Shore Scleroscope No.	Approx. Tensile Strength 1000 lb/in^2
Diameter in mm 3000 kg 10 mm Ball	Hardness No.		C 150 kg load 120° Diamond Cone	B 100 kg load $\frac{1}{16}$ in Diamond Ball		
2.05	898	—	—	—	—	440
2.10	857	—	—	—	—	420
2.15	817	—	—	—	—	401
2.20	780	1 150	70	—	106	384
2.25	745	1 050	68	—	100	368
2.30	712	960	66	—	95	352
2.35	682	885	64	—	91	337
2.40	653	820	62	—	87	324
2.45	627	765	60	—	84	311
2.50	601	717	58	—	81	298
2.55	578	675	57	—	78	287
2.60	555	633	55	120	75	276
2.65	534	598	53	119	72	266
2.70	514	567	52	119	70	256
2.75	495	540	50	117	67	247
2.80	477	515	49	117	65	238
2.85	461	494	47	116	63	229
2.90	444	472	46	115	61	220
2.95	429	454	45	115	59	212
3.00	415	437	44	114	57	204
3.05	401	420	42	113	55	196
3.10	388	404	41	112	54	189
3.15	375	389	40	112	52	182
3.20	363	375	38	110	51	176
3.25	352	363	37	110	49	170
3.30	341	350	36	109	48	165
3.35	331	339	35	109	46	160
3.40	321	327	34	108	45	155
3.45	311	316	33	108	44	150
3.50	302	305	32	107	43	146
3.55	293	296	31	106	42	142
3.60	285	287	30	105	40	138
3.65	277	279	29	104	39	134
3.70	269	270	28	104	38	131
3.75	262	263	26	103	37	128

(continued)

HARDNESS CONVERSION TABLE: FERROUS METALS – *contd.*

Brinell		Vickers or Firth Diamond Hardness No.	Rockwell		Shore Scleroscope No.	Approx. Tensile Strength 1000 lb/in^2
Diameter in mm 3000 kg 10 mm Ball	Hardness No.		C 150 kg load 120° Diamond Cone	B 100 kg load ¹⁄₁₆ in Diamond Ball		
3.80	255	256	25	102	37	125
3.85	248	248	24	102	36	122
3.90	241	241	23	100	35	119
3.95	235	235	22	99	34	116
4.00	229	229	21	98	33	113
4.05	223	223	20	97	32	110
4.10	217	217	18	96	31	107
4.15	212	212	17	96	31	104
4.20	207	207	16	95	30	101
4.25	202	202	15	94	30	99
4.30	197	197	13	93	29	97
4.35	192	192	12	92	28	95
4.40	187	187	10	91	28	93
4.45	183	183	9	90	27	91
4.50	179	179	8	89	27	89
4.55	174	174	7	88	26	87
4.60	170	170	6	87	26	85
4.65	166	166	4	86	25	83
4.70	163	163	3	85	25	82
4.75	159	159	2	84	24	80
4.80	156	156	1	83	24	78
4.85	153	153	—	82	23	76
4.90	149	149	—	81	23	75
4.95	146	146	—	80	22	74
5.00	143	143	—	79	22	72
5.05	140	140	—	78	21	71
5.10	137	137	—	77	21	70
5.15	134	134	—	76	21	68
5.20	131	131	—	74	20	66
5.25	128	128	—	73	20	65
5.30	126	126	—	72	—	64
5.35	124	124	—	71	—	63
5.40	121	121	—	70	—	62
5.45	118	118	—	69	—	61
5.50	116	116	—	68	—	60
5.55	114	114	—	67	—	59
5.60	112	112	—	66	—	58
5.65	109	109	—	65	—	56
5.70	107	107	—	64	—	55
5.75	105	105	—	62	—	54
5.80	103	103	—	61	—	53
5.85	101	101	—	60	—	52
5.90	99	99	—	59	—	51
5.95	97	97	—	57	—	50
6.00	95	95	—	56	—	49

HARDNESS CONVERSION TABLE: NON-FERROUS METALS

This table can be used, with reservations, to convert the hardness quoted for a non-ferrous metal from one scale to another. It should be most accurate in the case of brass (70:30 alloy)

Brinell Hardness Number 500 kg load 10 mm Ball	Diamond Pyramid Hardness Number	Rockwell hardness number		Rockwell superficial hardness number		
		B Scale 100 kg load 1/16 in Ball	F Scale 60 kg load 1/16 in Ball	15-T Scale 15 kg load 1/16 in Ball	30-T Scale 30 kg load 1/16 in Ball	45-T Scale 45 kg load 1/16 in Ball
42	45	—	40.0	—	—	—
43	46	—	43.0	—	—	—
44	47	—	45.0	—	—	—
45	48	—	47.0	53.5	—	—
46	49	—	49.0	54.5	—	—
47	50	—	50.5	55.5	—	—
48	52	—	53.5	57.0	—	—
50	54	—	56.5	58.5	12.0	—
52	56	—	58.8	60.0	15.0	—
53	58	—	61.0	61.0	18.0	—
55	60	10.0	63.0	62.5	20.5	—
57	62	12.5	65.0	63.5	23.0	—
59	64	15.5	66.8	65.0	25.5	—
61	66	18.5	68.5	66.0	28.0	—
62	68	21.5	70.0	67.0	30.0	—
63	70	24.5	71.8	68.0	32.0	—
64	72	27.5	73.2	69.0	34.0	—
66	74	30.0	74.8	70.0	36.0	1.0
68	76	32.5	76.0	70.5	38.0	4.5
70	78	35.0	77.4	71.5	39.5	7.5
72	80	37.5	78.6	72.0	41.0	10.0
74	82	40.0	80.0	73.0	43.0	12.5
76	84	42.0	81.2	73.5	44.0	14.5
77	86	44.0	82.3	74.5	45.5	17.0
79	88	46.0	83.5	75.0	47.0	19.0
80	90	47.5	84.4	75.5	48.0	21.0
82	92	49.5	85.4	76.5	49.0	23.0
83	94	51.0	86.3	77.0	50.5	24.5
85	96	53.0	87.2	77.5	51.5	26.5
86	98	54.0	88.0	78.0	52.5	28.0
88	100	56.0	89.0	78.5	53.5	29.5
90	102	57.0	89.8	79.0	54.5	30.5
92	104	58.0	90.5	79.5	55.0	32.0
94	106	59.5	91.2	80.0	56.0	33.0
95	108	61.0	92.0	—	57.0	34.5
97	110	62.0	92.6	80.5	58.0	35.5
99	112	63.0	93.0	81.0	58.5	37.0
101	114	64.0	94.0	81.5	59.5	38.0
103	116	65.0	94.5	82.0	60.0	39.0
105	118	66.0	95.0	82.5	60.5	40.0

(continued)

HARDNESS CONVERSION TABLE: NON-FERROUS METALS – *contd.*

Brinell Hardness Number 500 kg load 10 mm Ball	Diamond Pyramid Hardness Number	Rockwell hardness number		Rockwell superficial hardness number		
		B Scale 100 kg load $\frac{1}{16}$ in Ball	F Scale 60 kg load $\frac{1}{16}$ in Ball	15-T Scale 15 kg load $\frac{1}{16}$ in Ball	30-T Scale 30 kg load $\frac{1}{16}$ in Ball	45-T Scale 45 kg load $\frac{1}{16}$ in Ball
106	120	67.0	95.5	—	61.0	41.0
108	122	68.0	96.0	83.0	62.0	42.0
110	124	69.0	96.5	—	62.5	43.0
112	126	70.0	97.0	83.5	63.0	44.0
113	128	71.0	97.5	—	63.5	45.0
114	130	72.0	98.0	84.0	64.5	45.5
116	132	73.0	98.5	84.5	65.0	46.5
118	134	73.5	99.0	—	65.5	47.5
120	136	74.5	99.5	85.0	66.0	48.0
121	138	75.0	100.0	—	66.5	49.0
122	140	76.0	100.5	85.5	67.0	50.0
124	142	77.0	101.0	—	67.5	51.0
126	144	77.5	101.5	86.0	68.0	51.5
128	146	78.0	102.0	—	68.5	52.5
129	148	79.0	102.5	—	69.0	53.0
131	150	80.0	—	86.5	69.5	53.5
133	152	80.5	103.0	—	—	54.0
135	154	81.5	103.5	—	70.0	54.5
136	156	82.0	104.0	87.0	70.5	55.5
138	158	83.0	104.5	—	71.0	56.0
139	160	83.5	—	—	71.5	56.5
141	162	84.0	105.0	87.5	—	57.5
142	164	85.0	105.5	—	72.0	58.0
144	166	85.5	—	—	72.5	58.5
146	168	86.0	106.0	88.0	73.0	59.0
147	170	87.0	—	—	—	59.5
149	172	87.5	106.5	—	73.5	60.0
150	174	88.0	—	88.5	74.0	60.5
152	176	88.5	107.0	—	—	61.0
154	178	89.0	—	—	74.5	61.5
156	180	90.0	107.5	—	75.0	62.0
157	182	90.5	108.0	89.0	—	62.5
159	184	91.0	—	89.5	75.5	63.0
161	186	91.5	108.5	—	76.0	63.5
162	188	92.0	—	—	—	64.0
164	190	92.5	109.0	—	76.5	64.5
166	192	93.0	—	—	77.0	65.0
167	194	—	109.5	—	—	65.5
169	196	93.5	110.0	90.0	77.5	66.0

WEIGHTS OF METALS

Metal	g/cm^3	Pounds/in^3	Approx. pounds ft^3	in^3 per pound
Alclad	2.80	0.1015	175	9.86
Aluminium	2.70	0.0975	166	10.26
Aluminium bronze	8.13	0.2930	506	3.41
Aluminium manganese	2.70	0.0975	167	10.26
Barronia	8.80	0.3170	548	3.16
Birmabright	2.69	0.0970	167	10.31
Bismuth	9.85	0.3560	613	2.81
Brass (average)	8.50	0.3070	530	3.26
Bronze (typical)	8.54	0.3080	534	3.25
Copper	8.82	0.3180	550	3.15
Copper, cast or rolled	8.80–9.00	0.3170–0.3250	530–560	3.16–3.08
Duralumin	2.80	0.1015	175	9.86
Gold (typical)	19.20	0.6970	1200	1.44
Gunmetal, 79%	8.69	0.3140	542	3.19
Gunmetal, 83%	8.48	0.3060	530	3.27
Hiduminium	2.80	0.1015	175	9.86
Immadium bronze	7.80	0.2800	486	3.57
Inconel	8.81	0.3170	550	3.16
Iridium	21.80	0.7980	1360	1.25
Iron, cast	7.20	0.2600	450	3.85
Iron, wrought	7.85	0.2830	490	3.53
Lead	11.37	0.4100	708	2.44
Magnesium	1.74	0.0630	108	15.87
Magnesium alloys	1.80	0.0650	112	15.38
MG5, MG7 (light alloy)	2.63	0.0950	164	10.53
Mercury (at 15.6°C)	13.60	0.4910	849	2.04
Monel	8.86	0.3200	553	3.13
Nickel	8.81	0.3180	550	3.15
Niobium	8.56	0.3090	534	3.24
Phosphor bronze	8.85	0.3190	552	3.14
Platinum	21.50	0.7790	1340	1.28
Silver (typical)	10.40	0.3790	650	2.64
Solder (typical)	9.40	0.3400	585	2.94
Steel	7.85	0.2830	488	3.53
Stainless steel, 304	7.89	0.2850	—	3.50
310	7.87	0.2850	—	3.51
316	7.84	0.2840	—	3.52
321	7.89	0.2860	—	3.50
347	7.89	0.2860	—	3.50
446	7.53	0.2730	—	3.66
Tantalum	16.60	0.5970	1035	1.68
Tin	7.29	0.2630	454	3.80
Titanium	4.51	0.1630	281	6.14
Tungsten	19.20	0.6940	1200	1.44
Tungum	8.43	0.3040	525	3.29
Vanadium	6.11	0.2200	381	4.55
Zinc	7.00	0.2530	436	3.95
Zirconium	6.52	0.2350	407	4.26

Useful corrosion information

TYPES OF CORROSION

Corrosion can occur in many forms for which there are different terms. The following is a simplified and brief explanation of the more common types.

General corrosion

This is the general wastage of a metal by reaction with its environment to produce a chemical compound of the metal. General corrosion proceeds evenly over the surface of a metal and the rate of reduction in metal thickness is often predicatable.

Galvanic corrosion

If dissimilar metals are connected in the presence of a conducting solution an electrolytic cell can be produced. The most active metal will then become an anode and be corroded away at a greater rate than if not coupled to the dissimilar metal. Conversely the less active metal will be protected. Table 1 shows a typical 'Galvanic Series'; the further apart the connected metals are in this series, the greater the corrosion of the more active metal. This is a simplified explanation as factors such as the relative surface areas of the two metals in contact with the electrolyte also influence the amount of corrosion observed.

Demetallification

The corrosion action on an alloy can be highly selective, and effectively remove only one metal, for instance zinc, from a copper–zinc alloy. The metal remaining after this form of attack generally has similar dimensions to the original component, but is porous and lacking in mechanical strength.

Intergranular corrosion

This is a highly selective form of attack, in which the material between the grains of metal is either corroded away or weakened to the extent that the individual metal crystals can separate.

TABLE 1 – TYPICAL GALVANIC SERIES OF METALS AND ALLOYS

	Protected end (cathodic or most noble)
↑ Increasing reactivity	Platinum
	Gold
	Graphite
	Silver
	Hastelloy C
	18-8 Mo Stainless steel
	18-8 Stainless steel[a]
	Chromium irons
	Inconel
	Nickel
	Silver solder
	Monel
	Copper–nickel alloys
	Bronzes
	Copper
	Brasses
	Stainless steels[a]
	Tin
	Lead
	Lead-tin solders
	Nickel cast irons
	Cast iron
	Carbon steel
	Aluminium alloys (2017, 2024)
	Cadmium
	Aluminium alloy 1100
	Zinc
	Magnesium alloys
	Magnesium
	Corroded end (anodic or least noble)

[a] Stainless steels are shown as occupying two positions because they exhibit erratic potential depending upon the incidence of pitting. The two positions are intended to represent possible extremes of behaviour.

Pitting corrosion

This term is used to describe all those forms of corrosion where the attack is localised into small but possibly deep holes. The total amount of metal removed may be small but a metal section thickness may be perforated. The pits can be either randomly distributed or associated with changes in section, chemical composition or microstructure of the alloy.

Crevice corrosion

The differences in chemical composition between the environment inside a crevice and that outside a crevice can stimulate pitting from the crevice. The crevice may for instance, be a sharp corner where debris can accumulate or a gap at the edge of a gasket or packing material.

Deposit attack

This is a form of pitting where the different chemical environment under a scale or deposit gives rise to pits.

Waterline attack

Pitting or grooving can occur at the waterline of a container due to the higher oxygen concentration in the surface layers of solution.

Impingement attack

Where a flowing stream of fluid strikes a plate or pipe bend a combination of the greater supply of reactive species, mechanical abrasion by suspended particles and the moving fluid can give localised attack.

Cavitation attack

Sudden changes in pressure in a liquid can produce vapour-filled cavities, which subsequently collapse. The collapsing cavities release energy, producing mechanical damage. Propellers and pumps suffer from this form of attack which is mainly mechanical, but otherwise has similarities to impingement.

Stress corrosion

Metals supporting stresses well within their normal capabilities can crack in some specific environments. Where the environments that produce stress corrosion cracking in a particular alloy are known they can be avoided. Environments which are likely to cause stress corrosion can sometimes be predicted from existing information.

Corrosion fatigue

Metals subjected to cyclic loading are prone to failure by the gradual growth of cracks. This process of metal fatigue is frequently accelerated in corrosive environments and the term 'corrosion fatigue' is then used.

Liquid metal embrittlement

Some metals undergo cracking in the presence of liquid metals and care is therefore required in handling liquid metals.

Comparison of test figures for plastics

THE FOLLOWING table (originated by BASF AG) will be useful in correlating quoted properties in plastics determined by DIN and ASTM test methods. British Standard methods of testing plastics are detailed in BS 2782.

TABLE 1 – COMPARISON OF TEST FIGURES FOR PLASTICS

Property	Test method	Symbol	Unit	Correlation
Dichte Rohdichte Density	DIN 1306 DIN 53479 ASTM D 792-66	ρ d_R D^{23c}	 g/cm^3 at 20°C g/cm^3 at 23°C	Comparable at the same temperature
Schmelzindex Melt index (for PE) Flow rate (for other thermoplastics)	DIN 53735 (Etnwurf) (~ISO / R292) ASTM D 1238-65 T	MFI —	g/10 min g/10 min	Comparable under identical conditions
Mechanical properties Elastizitätsmodul Elastic modulus (Young's modulus)	 DIN 53 457 ASTM D 638-68 (tensile) ASTM D 695-68 T (compression) ASTM D 790-66 (flexural) ASTM D 882-67 (thin sheeting)	 E — — E_B —	 kp/cm^2 bar or lb/in^2 bar or lb/in^2 bar or lb/in^2 bar or lb/in^2	Comparable only under identical conditions: same shape and condition of specimen, length measured, and rate of stressing
Schubmodul Shear modulus	DIN 53 445 (ISO/R 537) ASTM D 2236-67 T	G G	dyn/cm^2 or kp/cm^2 $dyne/cm^2$ lb/in^2	Comparable
Torsionssteifheit Stiffness properties (Apparent modulus of rigidity)	DIN 53 447 (ISO/R 458) ASTM D 1043-69	T G	kp/cm^2 bar or lb/in^2	Comparable

(continued)

TABLE 1 – COMPARISON OF TEST FIGURES FOR PLASTICS – *contd.*

Property	Test method	Symbol	Unit	Correlation
Zugversuch **tensile properties**				
Zugestigkeit	DIN 53 455 (ISO/R 527)	σ_B	kp/cm^2	
Tensile strength, average (at maximum load)	ASTM D 882-67		bar or lb/in^2	
Dehnung bei Höchstkraft	DIN 53 455	ϵ_B	%	Comparable only under
Percentage elongation	ASTM D 638-68	% El	%	identical conditions:
	ASTM D 882-67	—	%	same shape and
Reissfestigkeit	DIN 53 455	σ_R	kp/cm^2	conditions of specimen,
Tensile strength (at break)	ASTM D 638-68	σ_U	bar or lb/in^2	length measured, and
	ASTM D 882-67	—		rate of stressing
Reissdehnung	DIN 53 455	ϵ_R	%	
Percentage elongation				
(at break)	ASTM D 638-68	—	%	
	ASTM D 882-67	—	%	
Streckspannung (Streckgrenze)	DIN 53 455	σ_S	kp/cm^2	
Tensile strength at yield	ASTM D 638-68	—	bar or lb/in^2	Comparable, but only
Yield strength	ASTM D 882-67	—	bar or lb/in^2	under identical
Dehnung bei				conditions: same shape
Streckspannung	DIN 53 455	ϵ_S	%	and condition of
Percentage elongation				specimen, length
(at yield)	ASTM D 638-68	—	%	measured, and rate of
	ASTM D 882-67	—	%	stressing
Biegeversuch **flexural properties**				
Biegefestigkeit	DIN 53 452	σ_{bB}	kp/cm^2	
Flexural strength	ASTM D 790-66	—	bar or lb/in^2	Limited comparability
Grenzbiegespannung	DIN 53 452	σ_{bG}	kp/cm^2	
Flexural stress	ASTM D 790-66	—	bar or lb/in^2	
(at given strain)				
Schlagbiegeversuch **impact resistance**				
Schlagzähigkeit (Charpy)	DIN 53 453 (ISO/R 179)	a_n	cm kp/cm^2	Not comparable
Impact strength (Charpy)	ASTM D 256-56	—	ft lb/in	
Kerbschlagzähigkeit (Charpy)	DIN 53 453 (ISO/R 179)	a_k	cm kp/cm^2	Not comparable
Impact strength (notched)	ASTM D 256-56		ft lb/in	
(Izod)	(ISO/R 180)			
Shore Härte	DIN 53 505	—	Shoreskala A,	
Indentation hardness	ASTM D 2240-68		C, D units on	Comparable
(Durometer)			A and D scales	
Kugeldruckhärte	DIN 53 456	H	kp/mm^2 or kp/cm^2	
α Rockwell hardness	ASTM D 785-65	—	Rockwell R scale	Not comparable
(= Procedure B)				

(continued)

TABLE 1 – COMPARISON OF TEST FIGURES FOR PLASTICS – *contd.*

Property	Test method	Symbol	Unit	Correlation
Thermal properties				
Vicat-Erweichungstemp (VST) (formerly ~punkt) (VSP)	DIN 53 460 (ISO/R 306)	VST	°C	Comparable if measurement made under the same load
Vicat softening temperature (formerly ~point)	ASTM D 1525-65 T	—	°C	and at the same rate of heating in a liquid bath
Formbeständigkeit in der Wärme nach Martens —	DIN 53 458 —	— —	°C —	
Formbestädnigkeit in der Wärme nach ISO/R 75	DIN 53 461 (ISO/R 75)	F_{ISO}	°C	Comparable if the load, shape and condition of
Deflection temperature (formerly HDT)	ASTM D 648-56 —	—	°C, °F —	specimen are identical
Wasseraufnahme	DIN 53 472 DIN 53 475 (ISO/R 62)	— —	Conditions 22°C, 4d, mg 23°C, 24h, mg	Comparable when converted if pretreatment and post-treatment,
Water absorption	ASTM D 570-63	—	23°C, 2h, 24h, %	the specimens, and the times concerned are identical
Electrical properties				
Spezifischer Durchgangswiderstand	DIN 53 482 (VDE 0303 Teil 3)	ρ_D —	Ω cm	Comparable
Volume resistivity	ASTM D 257-66	ρ_V	Ω cm	
Oberflächenwiderstand	DIN 53 482 (VDE 0303 Teil 3)	R_O	Ω	Comparable
Surface resistivity	ASTM D 257-66	ρ_s	Ω	
Dielektrizitätszahl (relative Dieletrizitätskonstante)	DIN 53 483 (VDE 0303 Teil 4)	ϵ_r	—	Comparable
Dielectric constant	ASTM D 150-68	χ	—	
Dielektrischer Verlustfaktor (tan δ)	DIN 53 483 (VDE 0303 Teil 4)	—	—	Comparable
Dissipation factor (tan δ)	ASTM D 150-68	D	—	
Dielektrischer Schweissfaktor	—	—	—	
Dielectric loss index (= χ tan δ)	ASTM D 150-68	—	—	
Durchschlagfestigkeit	DIN 53 481 (VDE 0303 Teil 2)	E_d	kV/mm or kV/cm	Not comparable
Dielectric strength	ASTM D 149-64	—		
Lichtbogenfestigkeit	DIN 53 484 (VDE 0303 Teil 5)	—	—	Not comparable
Arc resistance	ASTM D 495-61	—	sec	
Kreichstromfestigkeit	DIN 53 480 (VDE 0303 Teil 1)	—	—	Not comparable
Tracking resistance	ASTM D 2132-67 T	—	—	

Summary of ASTM test methods

ASTM test methods are widely used for the determination of the properties of plastics in the United States and the UK. The other main methods are DIN (Continental Europe) and British Standard BS 2782: 1970 (UK). The following is a summary of some of the standard tests and their purpose, together with a summary of appropriate ASTM test methods and possible test variables and their significance (extracted from a bulletin published by Borg Warner).

Deflection temperature

Purpose of test. To provide an indication of the temperature range in which a plastic material might be useful.

Test method. A test specimen, usually a rectangular bar, is supported at two points in an oil bath. The test apparatus includes a device to measure the deflection of the specimen.

A constant load, calculated to give 66 lb/in^2 or 264 lb/in^2 outer fibre stress in the specimen, is applied at mid-span and the bath temperature is raised at a steady rate of 2°C per minute.

The deflection of the specimen is observed during test. When this deflection reaches a specified amount (0.010 in) the oil bath temperature is noted and is then reported as the deflection temperature.

Test variables and their effect

A *Fibre stress loading.* The higher the fibre stress, the lower the deflection temperature.

B *Specimen thickness.* Thickness may vary between ⅛ in and ½ in. The thicker the specimen, the higher deflection temperature. This is because a thick specimen requires a longer time to heat through.

C *Specimen preparation.* Compression moulded specimens usually have higher deflection temperatures than do injection moulded specimens. This is because any warping caused by 'frozen in' stress in injection moulded and extruded bars cannot be distinguished from the distortion due to heat softening, resulting in an earlier or lower reading.

D   *Annealing.* Conditioning at elevated temperature prior to test (annealing) will relieve much of the 'frozen in' stress, thereby reducing the effect of specimen preparation on test results. Tests made on annealed bars yield higher deflection temperature values compared with unannealed specimens. The annealed deflection temperature is the preferred comparison, since it is affected least by moulding conditions, and therefore approximates to the inherent property of the plastic.

Deflection temperature values should be compared only when the previous parameters are defined and are the same for the materials being compared.

Tensile properties

Purpose of test. To obtain an indication of how a material will respond to a slowly applied tensile load.

Test method. A specimen is clamped securely in two grips. One grip is fastened to the stationary head of a test machine and the other grip is fastened to the movable head. The machine heads are pulled apart, elongating the specimen. The rate of head travel is usually constant, and for rigid plastics is of the order of 0.2 in/min. Some test machines have facilities whereby the load on the specimen and the elongation are measured simultaneously. A graphic plot of the applied load versus elongation of the specimen is termed a stress-strain curve.

Test results. The slope of the initial portion of the stress-strain curve is used to compute the tensile modulus, which is the applied stress divided by the resultant elongation. Two types of tensile strength may be reported – at yield, or at failure. The yield strength is computed from the load at which the specimen continues to elongate *without additional load.* The failure strength is the load when the specimen fractures.

The graphs in Figure 1 illustrate these points. Figure 2 represents the stress-strain performance of ABS.

Some plastics, however, particularly crystalline polymers such as nylon, get stronger as they are deformed or stretched (work hardening). For these plastics, the ultimate tensile strength is greater than the yield strength.

Therefore, it is important when comparing tensile data to know the shape of the stress-strain curve. For most applications it is the yield strength which is most important, since straining the material beyond this point results in permanent deformation of the part.

Test variables and their effect

A   *Rate of straining.* Strain rate is determined by crosshead movement and specimen geometry. As the strain rate is increased, the tensile strength in-

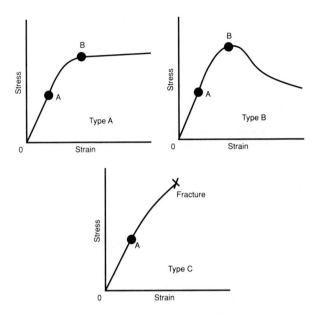

Fig. 1
Typical stress vs strain curves for gradual yielding (Type A),
abrupt yielding (Type B), and fracture at low strains before
yielding (Type C).

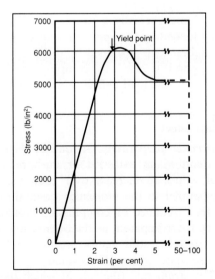

Fig. 2
Strain vs strain curve for ABS injection
moulding resin.

creases and elongation decreases. The normal crosshead speed is (5.1 mm/min to 6.3 mm/min) 0.2 in/min to 0.25 in/min.

B   *Temperature.* Thermoplastics, like most other materials, are effected by temperature variations. Tensile strength and modulus are decreased and elongation is increased as temperature increases.

C   *Specimen preparation.* Molecular orientation can cause injection moulded and extruded specimens to vary in their properties. A load applied parallel to the molecular orientation may yield higher values than if applied perpendicular to the orientation. The opposite is true for elongation.

   Machining may lower tensile and elongation values because of small irregularities introduced into the bar.

D   *Specimen size.* It is necessary to know the size of the specimen, since, even though the cross-head speed may be identical, the percentage rate of strain will vary if different specimen sizes are tested. This may give different numbers for the tensile strength and elongation values reported.

E   *Other variables.* There are several factors which affect the tensile strength and elongation of an injection moulded specimen, such as gate and runner size and position, temperature of the material and mould, condition of the mould, and injection pressure used. Since these factors are seldom assessed in the production of injection moulded test specimens, differences in test results are apt to occur from run to run.

Flexural properties

Purpose of test. To measure the performance characteristics of a material in bending (rigidity and strength).

Test method. A specimen, usually a rectangular bar, is supported at points near its ends and loaded by the application of weight at its midpoint.

Test results. The highest load sustained by the bar without failing or slipping through the supports is used to compute the apparent flexural yield strength. The slope of the load deflection curve is used to compute the flexural modulus.

Test variables and their effect

A   *Method of specimen preparation.* A load applied perpendicularly to the orientation found in an extruded or injection moulded specimen will cause flexural values to increase over compression moulded values. A load applied parallel to the molecular orientation will result in slightly lower values than obtained on compression moulded specimens. The differences become more apparent as the orientation is increased.

B   *Temperature.* Higher testing temperatures can be expected to produce lower flexural strength and modulus values.

C   *Strain rate.* The strain rate (the rate at which the specimen is stretched) depends on three factors: the machine test speed, the specimen thickness and the span. The strain rate must be held constant if comparable values are to be obtained. ASTM test methods attempt to specify the test speeds and spans for each thickness range that will result in the same strain rate for all specimens but some thicknesses (near the upper limit of a range) yield higher strain rates regardless of test speeds and spans.

Impact resistance

Purpose of test. To measure the relative susceptibility of materials to failure under suddenly applied loads.

Test method. There are two related methods of pendulum impact tests using two types of test machines – Charpy and Izod. In each instance a free swinging arm with a hammer at its end is allowed to strike a specimen. The energy lost in breaking the specimen is measured and defined as the impact strength of the specimen.

The results obtained by these different methods cannot be directly compared.

Test results. The impact energy required to fracture the specimen is reported in ft/lbs/in of specimen thickness, at the fracture point.

Test variables and their effect

A   *Specimen thickness.* Since a thick specimen is relatively restricted in its ability to twist and bend and therefore less able to dissipate energy in this manner, a thick specimen usually yields lower impact data than a thin one.

B   *Notch.* The reason for notching impact specimens is to provide an area of high but carefully controlled stress concentration. The notch should be correctly and cleanly cut, since the radius of the notch and any imperfections of cut may affect the results.

C   *Rate of straining.* Strain rate is determined by the speed of the hammer, and specimen geometry. In general, it may be said that the higher the strain rate, the lower the impact value.

D   *Temperature.* Impact resistance tends to increase with increasing temperature.

E   *Specimen preparation.* Molecular orientation causes injection moulded and extruded specimens to vary in their properties.

Specimens tested across flow lines will absorb more impact energy than those tested parallel to flow lines.

F   *Other variables.* Machine differences such as distance of striking point from the notch, bar position, and the alignment of machine components are all factors which often go undefined but may influence impact results.

Flow orientation

Flow orientation has been a variable since the introduction of injection moulding. In many cases it seriously handicaps a material because of the effects it has on physical properties.

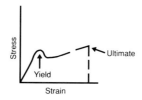

Fig. 3
Stress-strain curve — nylon.

Flow orientation is the partial alignment of 'stringing out' of the molecules during flow of the melt. Prior to flow, the molecules form a somewhat tangled mass, very similar to a bowl of spaghetti. As the melt is forced through a restriction such as a gate, this molecular arrangement is changed. The material at the wall of the cavity moves relatively slowly, while the material in the centre moves more rapidly. This differential flow results in a shearing action which tends to strain and 'iron out' the molecules in the direction of flow. If flow is stopped prior to cooling, the strained molecules will tend to relax back to a random orientation. If the material is solidified before most of the relaxation can take place, however, the molecules will remain strained. This most generally occurs in the portion of the moulded part closest to the mould surface.

Although this orientation strain can be somewhat controlled by melt temperature during flow, gate location, injection pressure, mould temperature, injection speed, etc, many materials are more sensitive to flow orientation than others.

To investigate further the effect of flow orientation and strain on physical properties, a special moulding and testing procedure for tensile strength, impact strength and dimensional stability was established. It is described in some detail here, and results are indicated. The results are reflected in the testing information that preceded this section.

PREPARATION OF SPECIMENS

By keeping the moulding temperature and over-all cycle time constant, a number of bars are moulded at the lowest possible cylinder temperature using full ram pressure. The cylinder temperatures are then increased in steps of –39°C (25°F) and the ram pressure is reduced accordingly to obtain good specimens. In this manner, test specimens are moulded at temperatures over a range from –17°C to 378°C (0°F to 100°F) above the minimum moulding temperature.

Dimensional stability

In order to determine the dimensional stability of a moulded part, an oven shrinkage test is conducted. This test is designed to discover strains and also to measure the relative heat resistance of the plastic. Tensile bars moulded over the entire test range are placed in an oven preheated to 100°C or other suitable temperature. They are removed after 24 hours and the shrinkage measured. By plot-

ting the results, a curve showing the effect of moulding temperature on shrinkage may be drawn.

Tensile strength

Tensile properties decrease as injection moulding temperatures increase, indicating less orientation at the higher moulding temperatures. The amount of decrease is
a measure of the sensitivity of a material to flow orientation. The values may be
plotted using the highest value as unity, with the other values appearing as percentages. By comparing the slope of the curves, relative comparisons between
different materials can be made. Materials that have a nearly horizontal slope
possess little flow orientation over the temperature range studied.

Impact strength

The decrease in impact strength with increasing mould temperature does not
mean that at a lower temperature the best overall impact strength is obtained. As a
higher moulding temperature is approached flow orientation in the specimen is
decreased – thus giving a lower apparent Izod impact strength across the direction of flow. If the specimens are broken parallel to the flow, however, a steady
increase in impact strength is found as the temperature is increased. The over-all
part strength is therefore higher. Some typical results are shown in Table 1.

TABLE 1

Moulding temperature		Impact strength	
		Broken across flow orientation ft lb/in	Broken with flow orientation ft lb/in
°C	°F		
205	400	5.0	2.3
220	425	4.7	3.0
230	450	4.3	3.4
245	475	3.8	3.8

Impact data are plotted using the same procedure as for tensile data. Materials that have a nearly horizontal slope possess little orientation. From an inspec-
tion of all the curves, often overlooked comparisons may be made. Parts
moulded from orientation-sensitive materials, unless moulding conditions are
carefully controlled, will fluctuate in property measurements. Sometimes these
fluctuations will seriously affect the mechanical suitability of the parts. A material with less flow sensitivity may be moulded with less rigid control of moulded
conditions and still produce parts essentially constant in properties.

Editorial Index

A